Geometric Optimal Control

Interdisciplinary Applied Mathematics

Editors
S.S. Antman   P. Holmes
K. Sreenivasan

Series Advisors
L. Glass   P. S. Krishnaprasad
R.V. Kohn   J.D. Murray
S.S. Sastry

Problems in engineering, computational science, and the physical and biological sciences are using increasingly sophisticated mathematical techniques. Thus, the bridge between the mathematical sciences and other disciplines is heavily traveled. The correspondingly increased dialog between the disciplines has led to the establishment of the series: *Interdisciplinary Applied Mathematics*.

The purpose of this series is to meet the current and future needs for the interaction between various science and technology areas on the one hand and mathematics on the other. This is done, firstly, by encouraging the ways that mathematics may be applied in traditional areas, as well as point towards new and innovative areas of applications; and, secondly, by encouraging other scientific disciplines to engage in a dialog with mathematicians outlining their problems to both access new methods and suggest innovative developments within mathematics itself.

The series will consist of monographs and high-level texts from researchers working on the interplay between mathematics and other fields of science and technology.

For further volumes:
http://www.springer.com/series/1390

Heinz Schättler • Urszula Ledzewicz

Geometric Optimal Control

Theory, Methods and Examples

 Springer

Heinz Schättler
Washington University
St. Louis, MO, 63130
USA

Urszula Ledzewicz
Southern Illinois University
Edwardsville IL, 62026
USA

Series Editors:
S.S. Antman
Department of Mathematics
and
Institute for Physical Science
  and Technology
University of Maryland
College Park, MD, 20742
USA
ssa@math.umd.edu

P. Holmes
Department of Mechanical and Aerospace
  Engineering
Princeton University
215 Fine Hall
Princeton, NJ, 08544
USA
pholmes@math.princeton.edu

K. Sreenivasan
Department of Physics
New York University
70 Washington Square South
New York City, NY, 10012
USA
katepalli.sreenivasan@nyu.edu

ISSN 0939-6047
ISBN 978-1-4899-8680-1   ISBN 978-1-4614-3834-2 (eBook)
DOI 10.1007/978-1-4614-3834-2
Springer New York Heidelberg Dordrecht London

Mathematics Subject Classification (2010): 49K15, 49L20, 34H05, 93C15

Printed on acid-free paper

Springer is part of Springer Science+Business Media (www.springer.com)

To our parents,

   Elfriede and Oswald, Teresa and Jerzy,
and our sons,

   Filip and Max

Preface

Optimal control theory is a discipline that has its historical origin in the calculus of variations, dating back to the formulation of Johann Bernoulli's brachistochrone problem more than 300 years ago. It developed into a field of its own in the 1960s in connection with the development of space exploration. In fact, it was the engineering problem of launching a satellite into a sustained orbit—*Sputnik*—that generated the activities in the Soviet Union that led to the early developments of the theory. This theory provides techniques to analyze far-reaching problems in various fields of science, engineering, economics, and more recently also biomedicine. Generally, in these problems the underlying task is to transfer the state of a dynamical system from a given initial position into a desired terminal condition, e.g., deploy a satellite into a prescribed orbit; guide a spacecraft to some remote planet, possibly even making a soft landing (Mars rover); perform tasks using robotic manipulators; achieve positions of wealth in economic endeavors through investment decisions; eliminate, if possible, cancer cells from our body or infected cells from a biological host; and so on, with the number of realistic examples limitless. All these problems have in common that the dynamics of the system can be influenced ("controlled") by means of some external variable, e.g., the fuel burned inside a rocket to generate thrust, the allocation of economic resources between consumption and investment, the amounts of therapeutic agents given to treat cancer. Naturally, there always exist practical constraints that are imposed by a particular situation—the amount of fuel that a rocket can carry is limited, drugs cannot be given without careful consideration of their side effects, and so on. Still, and within the physical and other limits imposed by a particular situation, generally there exists tremendous freedom in the choice of the controls over time to achieve a desired objective. This leads to optimization problems. Sometimes, problems are naturally associated with an objective function to be minimized or maximized; in other instances, there is no such choice, and imposing a criterion may simply be a means to generate procedures (i.e., through necessary conditions for optimality) that allow one to come up with a reasonable solution to the underlying problem. The problem of transferring the state of a dynamical system from a given initial condition into a set of desired terminal conditions, while at the same time minimizing some objective associated with the motion, and possibly a penalty on the terminal state, thus is a most natural one. These belong to the general type of problems that are analyzed with the tools and techniques of optimal control theory.

More precisely, the kinds of problems that will be considered in this text are *finite-dimensional deterministic optimal control problems*. These are characterized by the fact that the time evolution of the underlying dynamical system is described by solutions to ordinary differential equations ("finite-dimensional") as opposed to partial differential equations ("infinite-dimensional"), and stochastic effects related to noise and other modeling perturbations associated with random effects are not included in the modeling ("deterministic"). Clearly, these are important aspects as well. However, methods that deal with these structures are of a very different type and are well represented in the literature. The problems we are analyzing here are among the most classical ones in mathematics and physics and have their origin in the calculus of variations. In fact, a calculus of variation problem simply is a special optimal control problem with a trivial dynamics (given by the derivative of the curve) and no constraints imposed on the control. While such constraints are important in practical problems, these do not, from our point of view, constitute the main difference between problems in the calculus of variations and optimal control problems. Rather, it is the presence of a nontrivial and typically nonlinear *dynamics* that connects the controls and states. For this reason, optimal control problems become much more difficult than mere extensions of optimization problems from a finite- to an infinite-dimensional setting. While there has been tremendous progress in numerical methods in optimal control over the past fifteen years that has led to the solutions of some specific and very difficult problems—the design of optimal controls by NASA for the positioning of the international space station using gyros with pseudospectral techniques or the experimental design of highly complicated pulse sequences in nuclear magnetic resonance (NMR) spectroscopy, to mention just two of the outstanding achievements—there still do not exist reliable numerical procedures that could simply be applied to any optimal control problem and give the solution. Specific methods, such as pseudospectral techniques, shooting methods, and arc parameterization techniques, have their strengths and shortcomings, simply because there exists far too great a variety in the dynamics. Nonlinear systems defy simple classifications, and from the practical side, problems often have to be solved on a case-by-case basis.

Yet, there does exist a common framework that can be used to tackle these problems, and it is this framework that we describe in our text. We give a comprehensive treatment of the fundamental necessary and sufficient conditions for optimality and illustrate how these can be used to solve optimal control problems. Our emphasis is on the *geometric aspects of the theory*, and in this context, we also provide tools and techniques that go well beyond standard conditions (including a comprehensive treatment of envelopes and singularities in the flow of extremals as well as a Lie-algebra-based framework for explicit computations in canonical coordinates) and can be used to obtain a full understanding of the *global* structure of solutions for the underlying problem, not just an isolated numerical computation for specific parameter values. We include a palette of examples that are worked out in detail and range from classical to novel and from elementary to the highly nontrivial. All these examples, in one way or another, illustrate the power of geometric techniques and methods.

The text is quite versatile and contains material on different levels ranging from the introductory and elementary to the quite advanced. In this sense, some parts of our text can be viewed as a comprehensive textbook for both advanced undergraduate and all levels of graduate courses on optimal control in both mathematics and engineering departments. In fact, this text grew out of lecture notes of the authors for courses taught at the Departments of Systems Science and Mathematics and Electrical and Systems Engineering in the School of Engineering and Applied Science at Washington University in St. Louis and the Department of Mathematics and Statistics and various engineering departments at Southern Illinois University Edwardsville. The variety of fully solved examples that illustrate the theory, rarely present to this extent in more advanced textbooks and monographs in this field, makes this text a strong educational asset. The text moves smoothly from the more introductory topics to those parts that are in a monograph style where more advanced topics are presented. While this presentation is mathematically rigorous, it is carried out in a tutorial style that makes the text accessible to a wide audience of researchers and students from various fields, not just the mathematical sciences and engineering. In a sequel, in which applications of geometric optimal control to biomedical problems will be analyzed, the tools and techniques developed in this text will be used to solve various optimal control problems that arise in cancer treatments that range from classical procedures such as chemo- and radiotherapy to novel approaches that include antiangiogenic agents and immunotherapy.

We are greatly indebted to not only our teachers who have influenced our views on the subject, especially to our doctoral advisors, Hector Sussmann and Stanislaw Walczak, but also to H.W. Knobloch, who introduced the first author to the fields of differential equations and optimal control. This book would not have come into existence without the guiding influence and passion for the subject instilled in us by our mentor and good friend Hector Sussmann, who introduced us to the beauty of the geometric approach to optimal control. Many professional colleagues have been instrumental in our academic careers, and we would like to take the opportunity to thank some of them, in particular E. Roxin, A. Nowakowski, A. Krener, V. Lakshmikantham, and T.J. Tarn. Especially we would like to acknowledge the late J. Zaborszky, a true engineer who appreciated mathematics, but always insisted on a practical connection. Thanks are also due to all our students who at one stage or another have contributed to the writing of this text. We also would like to thank our universities, Washington University in St. Louis and Southern Illinois University Edwardsville, and the National Science Foundation, which has supported our research at various stages for by now over 20 years. Finally, we would like to thank David Kramer, who so carefully read our text, and all the editors at Springer, especially Achi Dosanjh and Donna Chernyk, who have been very helpful throughout the entire production process.

Edwardsville, Illinois, USA
Heinz Schättler
Urszula Ledzewicz

Outline of the Chapters of the Text

Below we give a brief outline of the chapters that can serve as a road map for the scientific journey through our text.

Chapter 1 introduces the fundamental results of the calculus of variations organized around complete solutions of two cornerstone classical examples: the brachistochrone problem and the problem of surfaces of revolution of minimum area. The ideas and concepts presented in this chapter serve both as an introduction to and as a motivation for the corresponding notions in optimal control theory to be discussed in subsequent chapters.

The *Pontryagin maximum principle*, which gives the fundamental necessary conditions for optimality in optimal control problems, will be introduced in Chap. 2 with the focus on illustrating how this result can be used to solve problems. To this end, we introduce important Lie-derivative-based techniques that form the basis for geometric optimal control and use them to give a detailed derivation of H. Sussmann's results on the structure of time-optimal controls for nonlinear control-affine systems in the plane. These results serve as a first illustration of the power of geometric methods that go well beyond the conditions of the maximum principle and lead to deep results about the structure of optimal solutions.

While the emphasis of our text is on methods for nonlinear systems, in Chaps. 2 and 3 we also give some of the classical results about linear time-invariant systems. They include a proof of the convexity of the reachable sets and two formulations of the celebrated bang-bang theorem.

In Chap. 4 we then prove the Pontryagin maximum principle. Necessary conditions for optimality follow from separation results about convex cones that approximate the reachable set and the set of points where the objective decreases, respectively. These constructions equally apply to the classical needle variations used by Pontryagin et al. and to high-order variations. Specific variations will be made to prove the Legendre–Clebsch condition, the Kelley condition, and the Goh condition for optimality of singular controls. For this, an adequate computational framework is needed that is provided by exponential representations of solutions to differential equations and the associated Lie-algebraic formalism related to the Baker–Campbell–Hausdorff formula.

Chapters 5 and 6 then deal with sufficient conditions for optimality, both local and global. In Chap. 5 we introduce *parameterized families of extremals*, i.e., collections of controlled trajectories that satisfy the conditions of the maximum principle. Throughout the text, we emphasize the role they play in the construction of solutions to the *Hamilton–Jacobi–Bellman equation*, a first-order partial differential equation coupled with the solution of a minimization problem for the controls that describe the minimum value of the optimal control problem as a function of the initial data. We adapt the method of characteristics, a classical solution procedure for first-order partial differential equations, to the optimal control setting and use it to construct the *value function* associated with a parameterized family of extremals. For example, in this way we give an elementary proof of the optimality of the synthesis for the Fuller problem for which optimal solutions consist of chattering arcs whose controls switch infinitely often on finite intervals. These geometric constructions provide the generalization of the concept of a field of extremals from the calculus of variations to optimal control theory and clearly bring out the relationships between the necessary conditions of the maximum principle and the sufficient conditions of the dynamic programming principle.

While the results in Chap. 5 have a mostly local character and are all developed in the context of continuous controls, in Chap. 6 we extend the constructions to broken extremals that are finite concatenations of bang and singular controls. Geometric transversality and matching conditions will be developed that allow us to investigate the optimality of the flow of extremals as various patches are combined. The main result of this chapter is a *verification theorem* due to H. Sussmann that implies the optimality of a synthesis of controlled trajectories if the associated value function satisfies some weak continuity properties and is a continuously differentiable solution of the Hamilton–Jacobi–Bellman equation away from a locally finite union of embedded submanifolds of positive codimension. The results that will be developed in Chaps. 5 and 6 precisely lead to these piecewise differentiability properties. It is not required that the value function be continuous.

Chapter 7 concludes our text with illustrating how these techniques can also be used in low dimensions to determine small-time reachable sets exactly. This also provides an alternative geometric viewpoint to the results on time-optimal control for nonlinear systems in the plane that were derived in Chap. 2. The material in this chapter has never been presented before in book form. Bits and pieces are available in the research literature, and here these approaches are unified, and for the first time an accessible account of this subject is given.

Throughout our presentation, the text is as much self-contained as possible, and we do include more technical and difficult computations if they are required in the proofs or to give complete solutions for some of the examples. At various stages, we revisit the same topic from different angles, and below is a short road map to some of these topics:

- *Linear-Quadratic Regulator and Perturbation Feedback Control*: Sects. 2.1, 2.4, and 5.3
- *Time-Optimal Control for Linear Systems*: Sects. 2.5 and 2.6 and Chap. 3

- *Conjugate Points and Envelopes*: Sects. 1.4–1.5, 5.3–5.4, and 6.1
- *Singular Controls*, Legendre–Clebsch, Kelley, and Goh conditions: Sects. 2.8–2.9, 4.6, and 6.2
- *Chattering Controls and the Fuller Problem*: Sects. 2.11, 5.1 and 5.2.3
- *Time-Optimal Control for Single-Input, Control-Affine Nonlinear Systems* and Small-Time Reachable Sets in Low Dimensions: Sects. 2.9–2.10 and 4.5, and Chap. 7.

Contents

Chapter 1
The Calculus of Variations: A Historical Perspective

We begin with an introduction to the historical origin of optimal control theory, the calculus of variations. But it is not our intention to give a comprehensive treatment of this topic. Rather, we introduce the fundamental necessary and sufficient conditions for optimality by fully analyzing two of the cornerstone problems of the theory, the *brachistochrone problem* and the problem of determining surfaces of revolution with minimum surface area, so-called *minimal surfaces*. Our emphasis is on illustrating the methods and techniques required for getting *complete* solutions for these problems. More generally, we use the so-called fixed-endpoint problem, the problem of minimizing a functional over all differentiable curves that satisfy given boundary conditions, as a vehicle to introduce the classical results of the theory: (a) the Euler–Lagrange equation as the fundamental first-order necessary condition for optimality, (b) the Legendre and Jacobi conditions, both in the form of necessary and sufficient second-order conditions for local optimality, (c) the Weierstrass condition as additional necessary condition for optimality for so-called strong minima, and (d) its connection with field theory, the fundamental idea in any sufficiency theory. Throughout our presentation, we emphasize geometric constructions and a geometric interpretation of the conditions. For example, we present the connections between envelopes and conjugate points of a fold type and use these arguments to give a full solution for the minimum surfaces of revolution.

The classical ideas and concepts presented here will serve us both as an introduction to and motivation for the corresponding notions in optimal control theory to be discussed in subsequent chapters. Since geometric content is most easily visualized in the plane—and since the classical problems we are going to analyze are of this type—we restrict our introductory treatment here to one-dimensional problems. This mostly simplifies the notation, and only to a small extent the mathematics. We include a brief treatment of the multidimensional case in Sect. 2.3 as a corollary to the Pontryagin maximum principle from optimal control theory.

Chapter 1 is organized as follows: In Sect. 1.1 we introduce Johann Bernoulli's brachistochrone problem, the very first problem in the calculus of variations posed as a challenge to the mathematical community of its time in 1696. The fundamental first-order necessary condition for optimality, the Euler–Lagrange equation, will be

H. Schättler and U. Ledzewicz, *Geometric Optimal Control: Theory, Methods and Examples*, Interdisciplinary Applied Mathematics 38, DOI 10.1007/978-1-4614-3834-2_1, © Springer Science+Business Media, LLC 2012

developed in Sect. 1.2 and used to compute the extremals for the brachistochrone problem, the cycloids. In general, extremals are curves that satisfy the Euler–Lagrange equation. In Sect. 1.3, we formulate the problem of finding surfaces of minimum area of revolution for positive differentiable functions and show that the catenaries are the only extremals. We also include a detailed analysis of the mapping properties of the family of catenaries and their envelope, a rather technical mathematical argument, which, however, is essential to the understanding of the full solutions to this problem that will be given in Sect. 1.7. This requires that we develop second-order necessary conditions for optimality and the notion of conjugate points, which will be done in Sects. 1.4 and 1.5, which naturally leads to results about the local optimality of trajectories. A global solution of problems in the calculus of variations requires the notion of a field of extremals, which will be developed in Sect. 1.6, and leads to the Weierstrass condition. In Sects. 1.7 and 1.8, we then return to the problem of minimum surfaces of revolution and the brachistochrone problem, respectively, and give complete global solutions. This requires a further and nontrivial analysis of the geometric properties of the flows of extremals for these problems, which will be carried out in detail in these sections. In fact, both of these classical problems cannot be analyzed directly with standard textbook results of the theory. For the problem of minimum surfaces of revolution, the reason is that a problem formulation within the class of positive continuously differentiable functions is not wellposed, and a second class of candidates for optimality, the so-called Goldschmidt extremals, which are only piecewise differentiable, rectifiable curves, needs to be taken into account. For the brachistochrone problem, the extremals, the cycloids, have singularities at their initial point that require special attention. We include complete and mathematically rigorous solutions for these two benchmark problems of the calculus of variations. All these arguments foreshadow similar constructions that will be carried out more generally in the solutions of optimal control problems in Chaps. 2 and 5. We close this introductory chapter with a brief discussion of the Hamilton–Jacobi equation in Sect. 1.9 and provide in Sect. 1.10 an outlook on how the conditions of the calculus of variations developed here as a whole foreshadowed the Pontryagin maximum principle of optimal control, the fundamental necessary conditions for optimality for an optimal control problem.

1.1 The Brachistochrone Problem

The origins of the subject, and in some sense for much of the further development of calculus as a whole, lie with the statement of the following problem posed in 1696 by Johann Bernoulli:

[Brachistochrone] "Given two points A and B in the vertical plane, for a moving particle m, find the path AmB descending along which by its own gravity and beginning to be urged from the point A, it may in the shortest time reach B."

Fig. 1.1 The brachistochrone problem

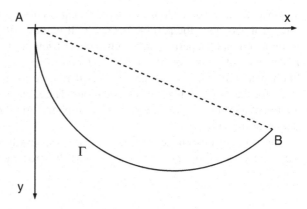

This problem, which sounds much better in Latin than in its English translation, is the so-called brachistochrone problem, named by combining the two Greek words for "shortest" ($\beta\rho\alpha\chi\iota\sigma\tau\sigma\varsigma$, brachistos) and "time" ($\chi\rho\sigma\nu\sigma\varsigma$, chronos). Johann Bernoulli, who already had solved this problem with a rather ingenious ad hoc argument, challenged the scientific community of his time, or as he called it "the sharpest mathematical minds of the globe," to give solutions to this problem. In 1697, several solutions, including Johann's own and solutions by his older brother Jakob, Newton, and others also including a note by Leibniz, whose solution was similar to Johann's, were published in the *Acta Eruditorum*. The solution is a *cycloid*, the locus that a point fixed on the circumference of a circle traverses as the circle rolls on a line, not a line itself, a rather implausible candidate Johann Bernoulli had warned against in his posting, nor a circle, the solution suggested by Galileo more than a hundred years earlier. In hindsight, it was the newly developed ideas of calculus, of which Galileo was deprived, and their relations with physics problems that made it possible to give these solutions. A common thread in all of them was the use of arguments—mathematics and physics still being the same discipline at that time—based on Fermat's principle or the laws of refraction by Snell and Huygens. It was up to Euler, and even more so to his student Lagrange, to consider the problem as the minimum-time problem it is and give a solution based on variational ideas that eventually grew into the calculus of variations. We refer the reader to the article [245] by H. Sussmann and J. Willems for an excellent and detailed exposition of the historical context and the developments that led from the posing of this problem to the formulation of the maximum principle of optimal control, one of the main topics of this book.

In order to give a mathematical formulation, let A be the origin of a two-dimensional coordinate system and denote the horizontal coordinate by x and the vertical coordinate by y, but orient it downward (see Fig. 1.1). It is clear from conservation of energy that the terminal point B cannot lie higher than the initial point A and thus it needs to have coordinates $B = (x_0, y_0)$ with y_0 nonnegative. Without loss of generality, we also assume that $x_0 > 0$, ignoring the trivial case of free fall ($x_0 = 0$). The objective is to minimize the time it takes for the particle to move from A to B along a curve Γ that connects A with B when the only force that acts upon the particle is gravity. It is implicitly assumed in the problem formulation that the particle descends without friction. It seems obvious that curves Γ that are not graphs of functions $y : [0,x_0] \to \mathbb{R}_+$, $x \mapsto y(x)$, cannot be optimal. Mathematically, this can be proven, but it requires a different and more general setup for the problem than the classical one. Here, we wish to follow the classical argument and thus, a priori, restrict the problem formulation to curves Γ that are graphs of functions y.

The time of descent along such a curve can easily be computed with some elementary facts from physics. The speed of the particle is the change of distance traveled in time,

$$v = \frac{ds}{dt},$$

and the total time can formally be computed as

$$T = \int dt = \int \frac{ds}{v}.$$

In our case, s represents the arc length of the graph of a function $y : [0,x_0] \to \mathbb{R}_+$, $x \mapsto y(x)$, and thus

$$s(z) = \int_a^z \sqrt{1 + y'(x)^2}\, dx, \qquad 0 \le z \le x_0,$$

which gives

$$T = \int_0^{x_0} \frac{\sqrt{1 + y'(x)^2}}{v(x)}\, dx.$$

We then need to express the velocity v as a function of x. In the absence of friction, the decrease in potential energy is accompanied by an equal increase in kinetic energy, i.e.,

$$mgy = \frac{1}{2} mv^2,$$

and thus the velocity at the point $(x, y(x))$ is given by

$$v(x) = \sqrt{2gy(x)}.$$

Summarizing, mathematically, the brachistochrone problem therefore can be formulated as the following minimization problem:

[Brachistochrone] Find a "function"

$$y : [0, x_0] \to \mathbb{R}_+, \qquad y(0) = 0, \qquad y(x_0) = y_0 > 0,$$

that minimizes the integral

$$I(y) = \int\limits_0^{x_0} \sqrt{\frac{1 + y'(x)^2}{2gy(x)}} \, dx.$$

Note that we are dealing with a minimization problem over a set of functions; that is, the functions themselves are the variables of the problem. This raises some immediate questions, and at the very least, we should specify exactly what this class of "functions," the domain of the minimization problem, actually is. It turns out that this is not always an obvious choice, and it certainly is not in the case of the brachistochrone problem. For instance, we have the boundary condition $y(0) = 0$, and thus this is an improper integral. So we probably might want to require that the integral converge. On the other hand, if the integral diverges to ∞, obviously this is not going to be an "optimal" choice, and if we allow for this, then why not keep these functions, since no harm will be done. Another obvious requirement is that y be differentiable, at least on the open interval $(0, x_0)$. But do we need differentiability at the endpoints from the right at 0 and from the left at x_0? This is not quite clear. In fact, all we need is that the integral remain finite. Indeed, the solutions to the problem, the cycloids, are functions whose derivative $y'(x)$ diverges to ∞ as x decreases to 0. As this example shows, the choice of the class of functions over which to minimize a given functional is not always a simple issue.

On the other hand, the choice of functions to minimize over is intimately connected with the important question of the existence of a solution. The theory of existence of solutions is well-established, and there exist numerous classical sources on this topic (for example, see [70, 118, 260], to mention just a couple of textbooks). The techniques used in existence proofs are very different from the methods pursued in this book, and therefore we will not address the issue of existence of solutions. Instead, we simply proceed, assuming that solutions to the problems we consider exist within a reasonably nice class of functions (as will be the case for the problems we shall consider) and try to single out candidates for a minimum. Our interest in this text is to characterize such a minimizer through necessary conditions for optimality and provide sufficient conditions that allow us to conclude the local or even global optimality of a candidate found through application of these necessary conditions. For the brachistochrone problem, and, more generally, for problems in the calculus of variations, a natural approach is, as in classical calculus, to ask whether the objective I might be "differentiable" and then develop first- and second-order derivative tests. This is the approach of Lagrange, who formalized and generalized the main necessary condition for optimality derived earlier for the brachistochrone problem with geometric means by his teacher, Leonhard Euler, himself a student of Johann Bernoulli in Basel.

1.2   The Euler–Lagrange Equation

The brachistochrone problem is a special case of what commonly is called the *simplest problem in the calculus of variations*, and we use this problem to develop the fundamental results of the theory. As is customary, we denote the space of all continuous real-valued functions $x : [a,b] \to \mathbb{R}$, $t \mapsto x(t)$, defined on the compact interval $[a,b]$ by $C([a,b])$. We use the notation $C^r([a,b])$ for functions that are r-times continuously differentiable on the open interval (a,b) and have derivatives that extend continuously to the compact interval $[a,b]$. Furthermore, when describing intervals, we consistently use brackets to denote that the boundary point is included and parentheses to indicate that the boundary point is not part of the set. For example, $(a,b] = \{t \in \mathbb{R} : a < t \le b\}$, etc.

Definition 1.2.1. We denote by $\mathscr{X}$ the Banach space $C([a,b])$ equipped with the supremum norm

$$\|x\|_{\mathbf{C}} = \|x\|_\infty = \max_{a \le t \le b} |x(t)|.$$

Convergence in the supremum norm is uniform convergence (see Appendix A). It is well-known that $C([a,b])$ with the supremum norm is a Banach space, i.e., if $\{x_n\}_{n \in \mathbb{N}}$ is a sequence of continuous functions that is Cauchy in the supremum norm, then there exists a continuous function x such that x_n converges to x uniformly on $[a,b]$ (see Proposition A.2.2).

Definition 1.2.2. We denote by $\mathscr{Y}$ the Banach space that is obtained by equipping $C^1([a,b])$ with the norm

$$\|x\|_D = \|x\|_\infty + \|\dot{x}\|_\infty.$$

It easily follows that convergence in the norm $\|\cdot\|_D$ is equivalent to uniform convergence of both the curves and their derivatives on the compact interval $[a,b]$. Figure 1.2 gives an example of a low-amplitude, high-frequency oscillation that lies in a small neighborhood of $x \equiv 0$ in $\mathscr{X}$, but not in $\mathscr{Y}$.

We can now formulate the "simplest problem in the calculus of variations."

[CV] Let $L : \mathbb{R} \times \mathbb{R} \times \mathbb{R} \to \mathbb{R}$, $(t,x,y) \mapsto L(t,x,y)$, be an r-times continuously differentiable function, $r \ge 2$. Among all functions $x \in C^1([a,b])$ that, for two given points A and B in $\mathbb{R}$, satisfy the boundary conditions $x(a) = A$ and $x(b) = B$, find one that minimizes the functional

$$I[x] = \int\limits_a^b L(t,x(t),\dot{x}(t))dt.$$

The integrand L is called the *Lagrangian* of the problem. Note that the derivative $\dot{x}$ of the function x takes the place of the argument y when the objective is evaluated, and it is therefore customary (although a bit confusing in the beginning) to simply denote this variable by $\dot{x}$ instead of y.

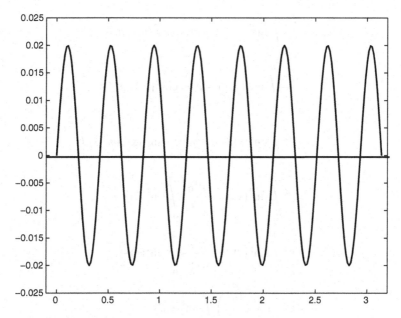

Fig. 1.2 A low-amplitude, high-frequency oscillation not close to $x \equiv 0$ in $\mathscr{Y}$

Definition 1.2.3 (Weak and strong minima). A function $x_* \in C^1([a,b])$ is called a *weak* local minimum if it minimizes the functional I over some neighborhood of x_* in $\mathscr{Y}$; it is said to provide a *strong* local minimum if it provides a minimum of I over some neighborhood of x_* in $\mathscr{X}$.

Note that strong local minimizers x_* minimize the functional I over all functions x that have the property that they are close to the reference at every time $t \in [a,b]$, whereas weak local minimizers are optimal only relative to those functions that in addition have their derivatives $\dot{x}$ close to the derivative of the reference $\dot{x}_*$ and thus minimize over a smaller collection of functions.

Example. The function $x \equiv 0$ is a weak, but not a strong, local minimum for the functional

$$I[x] = \int_0^\pi x(t)^2 (1 - \dot{x}(t)^2) dt, \quad x(0) = 0, \ x(\pi) = 0,$$

defined on $C^1([0,\pi])$. By inspection, the functional $I[x]$ is nonnegative on the open unit ball $B_1(0) = \{x \in \mathscr{Y} : \|x\|_D < 1\}$, and thus $x_* \equiv 0$ is a weak local minimum. But clearly, if $\|\dot{x}\|_\infty$ becomes too large, the functional can be made negative. For $\varepsilon > 0$ and $n \in \mathbb{N}$ simply consider the low-amplitude, high-frequency oscillations $x_n \in C^1([0,\pi])$ given by $x_n(t) = \varepsilon \sin(nt)$. Then we have

$$I[x_n] = \varepsilon^2 \int_0^{\pi} \sin^2(nt)\left(1 - \varepsilon^2 n^2 \cos^2(nt)\right) dt$$

$$= \frac{\varepsilon^2}{n} \int_0^{n\pi} \sin^2(s)\left(1 - \varepsilon^2 n^2 \cos^2(s)\right) ds$$

$$= \varepsilon^2 \int_0^{\pi} \sin^2(s) ds - \varepsilon^4 n^2 \int_0^{\pi} \sin^2(s) \cos^2(s) ds$$

$$= \varepsilon^2 \int_0^{\pi} \sin^2(s) ds - \frac{\varepsilon^4 n^2}{4} \int_0^{\pi} \sin^2(2s) ds$$

$$= \varepsilon^2 \int_0^{\pi} \sin^2(s) ds - \frac{\varepsilon^4 n^2}{4} \int_0^{2\pi} \sin^2(r) \frac{dr}{2}$$

$$= \varepsilon^2 \left(1 - \frac{\varepsilon^2 n^2}{4}\right) \int_0^{\pi} \sin^2(s) ds,$$

and thus $I[x_n] < 0$ for $n > \frac{2}{\varepsilon}$. In fact, $I[x_n] \to -\infty$ as $n \to \infty$, and this functional does not have a minimum. $\square$

We now develop necessary conditions for a function x_* to be a weak local minimizer. The fundamental necessary conditions for optimality follow from the well-known conditions for minimizing a function in calculus by perturbing the reference x_* with a function $h \in C^1([a,b])$, $h(a) = h(b) = 0$. Let $C_0^1([a,b])$ denote this class of functions. Clearly, if x_* is a local minimizer, then for $|\varepsilon|$ small enough, the function $x_* + \varepsilon h$ is also admissible for the minimization problem and lies in the neighborhood over which x_* is minimal. Thus, given any $h \in C_0^1([a,b])$, for ε in some small neighborhood of 0, we have that

$$I[x_*] \le I[x_* + \varepsilon h],$$

and it thus is a necessary condition for minimality of x_* that for all $h \in C_0^1([a,b])$, the first derivative of the function $\varphi(\varepsilon; h) = I(x_* + \varepsilon h)$ at $\varepsilon = 0$ vanish and that the second derivative be nonnegative:

$$\varphi'(0; h) = \frac{d}{d\varepsilon}_{|\varepsilon=0} I[x_* + \varepsilon h] = 0 \tag{1.1}$$

and

$$\varphi''(0; h) = \frac{d^2}{d\varepsilon^2}_{|\varepsilon=0} I[x_* + \varepsilon h] \ge 0. \tag{1.2}$$

The quantities in Eqs. (1.1) and (1.2) are called the *first and second variation* of the problem, respectively, and are customarily denoted by $\delta I(x_*)[h]$ and $\delta^2 I(x_*)[h]$. Differentiating under the integral gives that

$$\delta I[x_*](h) = \int_a^b L_x(t,x_*,\dot{x}_*)h + L_{\dot{x}}(t,x_*,\dot{x}_*)\dot{h}\,dt \qquad (1.3)$$

with the curve x_*, its derivative $\dot{x}_*$, and the variation h all evaluated at t, and similarly,

$$\delta^2 I[x_*](h) = \int_a^b (h,\dot{h}) \begin{pmatrix} L_{xx}(t,x_*,\dot{x}_*) & L_{x\dot{x}}(t,x_*,\dot{x}_*) \\ L_{\dot{x}x}(t,x_*,\dot{x}_*) & L_{\dot{x}\dot{x}}(t,x_*,\dot{x}_*) \end{pmatrix} \begin{pmatrix} h \\ \dot{h} \end{pmatrix} dt. \qquad (1.4)$$

The following lemma is fundamental in the calculus of variations and makes it possible to eliminate the variational directions h from the necessary conditions.

Lemma 1.2.1. *Suppose α and β are continuous functions defined on a compact interval $[a,b]$, $\alpha, \beta \in C([a,b])$, with the property that*

$$\int_a^b \alpha(t)h(t) + \beta(t)\dot{h}(t)dt = 0 \qquad (1.5)$$

for all $h \in C_0^1([a,b])$. Then β is continuously differentiable, $\beta \in C^1([a,b])$, and

$$\dot{\beta}(t) = \alpha(t) \quad \text{for all } t \in (a,b).$$

Proof. If we define $A(t) = \int_a^t \alpha(s)ds$, then, it follows from integration by parts that for all $h \in C^1([a,b])$ and any constant $c \in \mathbb{R}$, we have that

$$0 = \int_a^b (-A(t) + \beta(t) - c)\dot{h}(t)\,dt.$$

Choosing

$$c = \frac{1}{b-a}\int_a^b (\beta(t) - A(t))\,dt$$

and taking

$$h(t) = \int_a^t (\beta(s) - A(s) - c)\,ds,$$

it follows that $h \in C_0^1([a,b])$ and

$$\int_a^b \dot{h}(t)^2 dt = 0.$$

Since $\dot{h}$ is continuous, h is constant, and $h(a) = 0$ implies $h \equiv 0$. Hence $\beta(t) = A(t) + c$, and therefore β is differentiable with derivative $\dot{\beta}(t) = \alpha(t)$.   □

Applying this to Eq. (1.3), we obtain the Euler–Lagrange equation, the fundamental first-order necessary condition for optimality for problems in the calculus of variations.

Corollary 1.2.1 (Euler–Lagrange equation). *If x_* is a weak minimum for the functional I, then the partial derivative $\frac{\partial L}{\partial \dot{x}}$ is differentiable along the curve $t \rightarrow (t,x_*(t), \dot{x}_*(t))$ with derivative given by*

$$\frac{d}{dt}\left(\frac{\partial L}{\partial \dot{x}}(t,x_*(t),\dot{x}_*(t))\right) = \frac{\partial L}{\partial x}(t,x_*(t),\dot{x}_*(t)). \tag{1.6}$$

Admissible functions $x : t \mapsto x(t)$ that satisfy this differential equation are called *extremals*. Note that it is not required that L be twice differentiable for the Euler–Lagrange equation to hold. It is a part of the statement of the lemma that the composition $t \rightarrow L_{\dot{x}}(t,x_*(t),\dot{x}_*(t))$ will always be differentiable in t (regardless of whether the second partial derivatives of L exist) and that its derivative is given by $L_x(t,x_*(t),\dot{x}_*(t))$.

If the Lagrangian L is twice differentiable, and if also the extremal x_* under consideration lies in C^2, then we can differentiate Eq. (1.6) to obtain

$$L_{t\dot{x}} + L_{\dot{x}x}\dot{x}_* + L_{\dot{x}\dot{x}}\ddot{x}_* = L_x.$$

In this case, the Euler–Lagrange equation becomes a possibly highly nonlinear second-order ordinary differential equation. This equation, however, will be singular if the second derivative $L_{\dot{x}\dot{x}}(t,x_*(t),\dot{x}_*(t))$ vanishes at some times $t \in (a,b)$. We shall see in Sect. 1.4 that it is a second-order necessary condition for optimality, the so-called Legendre condition, that this term be nonnegative, but generally, it need not be positive. If $L_{\dot{x}\dot{x}}(t,x_*(t),\dot{x}_*(t))$ is positive for all $t \in [a,b]$, then the Euler–Lagrange equation is a regular second-order ordinary differential equation, and if we specify as initial conditions $x(a) = A$ and $\dot{x}(a) = p$ for some parameter p, then a local solution $x(t;p)$ exists near $t = a$. However, this solution need not necessarily exist on the full interval $[a,b]$. In addition, we are interested in the solution to the two-point boundary value problem with boundary condition $x(b) = B$. Thus there may be no solutions, the solution may be unique, or there may exist several, even infinitely many, solutions. The analysis of the Euler–Lagrange equation therefore generally is a challenging and nontrivial problem.

The following result, known as the Hilbert differentiability theorem, often is useful for establishing a priori differentiability properties of extremals.

Proposition 1.2.1 (Hilbert differentiability theorem). *Let $x_* : [a,b] \to \mathbb{R}$ be an extremal with the property that*

$$L_{\dot{x}\dot{x}}(t, x_*(t), \dot{x}_*(t)) \neq 0 \qquad \text{for all} \quad t \in [a,b].$$

Then x_* has the same smoothness properties as the Lagrangian L for problem [CV]. That is, if $L \in C^r$, then also $x_* \in C^r$.

Proof. For some constant c, the extremal x_* is a solution to the Euler–Lagrange equation in integrated form,

$$\frac{\partial L}{\partial \dot{x}}(t, x_*(t), \dot{x}_*(t)) - \int_a^t \frac{\partial L}{\partial x}(s, x_*(s), \dot{x}_*(s)) ds - c = 0.$$

If we define a function $G(t,u)$ as

$$G(t,u) = \frac{\partial L}{\partial \dot{x}}(t, x_*(t), u) - \int_a^t \frac{\partial L}{\partial x}(s, x_*(s), \dot{x}_*(s)) ds - c,$$

then the equation $G(t,u) = 0$ has the solution $u = \dot{x}_*(t)$. By the implicit function theorem (see Theorem A.3.2), this solution is locally unique and k times continuously differentiable if the function $G(t, \cdot)$ is k times continuously differentiable and if the partial derivative

$$\frac{\partial G}{\partial u}(t, \dot{x}_*(t)) = \frac{\partial^2 L}{\partial \dot{x}^2}(t, x_*(t), \dot{x}_*(t))$$

does not vanish. Hence $x_* \in C^r$ if L is r times continuously differentiable. $\qquad \square$

For integrands L of the form $L(x, \dot{x}) = x^\alpha \sqrt{1 + \dot{x}^2}$ with $\alpha \in \mathbb{R}$, we have that

$$L_{\dot{x}\dot{x}}(x, \dot{x}) = \frac{x^\alpha}{(1 + \dot{x}^2)^{\frac{3}{2}}} > 0,$$

and thus all extremals that lie in $x > 0$ are twice continuously differentiable. This applies, for example, to the brachistochrone problem with $\alpha = -\frac{1}{2}$.

Corollary 1.2.2 (First integral). *If x_* is a weak minimum for the functional I that is twice continuously differentiable and if the function L is time-invariant, then it follows that the function*

$$t \mapsto L(t, x_*(t), \dot{x}_*(t)) - \dot{x}_*(t) \frac{\partial L}{\partial \dot{x}}(t, x_*(t), \dot{x}_*(t)) \tag{1.7}$$

is constant over the interval $[a,b]$, *i.e., the Euler–Lagrange equation has a "first integral" given by*

$$L - \dot{x}L_{\dot{x}} = const.$$

Proof. Since L does not explicitly depend on t, we have, omitting the arguments, that

$$\frac{d}{dt}(L - \dot{x}L_{\dot{x}}) = L_x\dot{x} + L_{\dot{x}}\ddot{x} - \ddot{x}L_{\dot{x}} - \dot{x}\frac{d}{dt}(L_{\dot{x}})$$

$$= \dot{x}\left(L_x - \frac{d}{dt}(L_{\dot{x}})\right) \equiv 0.$$

$\square$

Thus, for a time-invariant system, all twice continuously differentiable extremals need to lie in the level curves of the function $L - \dot{x}L_{\dot{x}}$. This often provides a quick path to finding extremals for a given problem, and we will use formula (1.7) to find extremals for the brachistochrone problem. Deleting the constant, in this case L is given by

$$L(x,\dot{x}) = \sqrt{\frac{1+\dot{x}^2}{x}},$$

and thus the first integral takes the form

$$\sqrt{\frac{1+\dot{x}^2}{x}} - \dot{x}\left(\frac{1}{2}\sqrt{\frac{1}{x(1+\dot{x}^2)}}2\dot{x}\right) = c,$$

where c is some constant. Simplifying this expression yields

$$(1+\dot{x}^2 - \dot{x}^2)\sqrt{\frac{1}{x(1+\dot{x}^2)}} = c,$$

or equivalently, and renaming the constant,

$$x(1+\dot{x}^2) = 2\xi > 0. \tag{1.8}$$

While it is possible to solve the resulting ordinary differential equation with standard methods (albeit a bit tedious), it is more elegant and quicker to introduce a reparameterization $t = t(\tau)$ of the time scale so that the derivative $\dot{x}$ becomes

$$\frac{dx}{dt} = -\tan\left(\frac{\tau}{2}\right). \tag{1.9}$$

For in this case, we then have

$$1 + \dot{x}^2 = \frac{1}{\cos^2\left(\frac{\tau}{2}\right)}$$

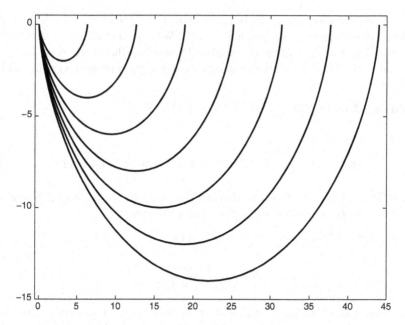

Fig. 1.3 The field of cycloids for the brachistochrone problem

and thus

$$x = \frac{2\xi}{1+\dot{x}^2} = 2\xi \cos^2\left(\frac{\tau}{2}\right) = \xi(1+\cos\tau). \tag{1.10}$$

Furthermore,

$$\frac{dt}{d\tau} = \frac{dt}{dx}\frac{dx}{d\tau} = -\frac{1}{\tan\left(\frac{\tau}{2}\right)}\xi(-\sin\tau) = \xi\frac{2\sin\left(\frac{\tau}{2}\right)\cos\left(\frac{\tau}{2}\right)}{\tan\left(\frac{\tau}{2}\right)}$$

$$= 2\xi\cos^2\left(\frac{\tau}{2}\right) = \xi(1+\cos\tau),$$

which gives

$$t(\tau) = \xi(\tau + \sin\tau) + c.$$

Since extremals start at the origin, $x(a) = 0$, we choose $a = -\pi$, and then the constant is $c = \xi\pi$. Thus, overall, the required time reparameterization is given by

$$t(\tau) = \xi(\pi + \tau + \sin\tau). \tag{1.11}$$

Equations (1.10) and (1.11) represent, as already mentioned, a family of curves called *cycloids*. Geometrically, a cycloid is the locus that a point fixed on the circumference of a circle of radius ξ traverses as the circle rolls on the lower side of the line $x = 0$. Examples of such curves are drawn in Fig. 1.3 below. Note that although we represent the cycloids as curves in $\mathbb{R}^2_+ = \{(t,x) : t > 0, x > 0\}$, each cycloid is the graph of some function $t \mapsto x(t)$. We conclude this discussion of the extremals for the brachistochrone problem by showing that there exists one and only one cycloid in this family that passes through a given point (t,x) with t and x positive.

Theorem 1.2.1. *The family $\mathscr{C}$ of curves defined by Ξ,*

$$\Xi : (-\pi, \pi) \times (0, \infty) \to \mathbb{R}^2_+,$$

$$(\tau, \xi) \mapsto (t(\tau; \xi), x(\tau; \xi)) = (\xi(\pi + \tau + \sin \tau), \xi(1 + \cos \tau)), \qquad (1.12)$$

covers $\mathbb{R}^2_+ = \{(t,x) : t > 0, x > 0\}$ diffeomorphically; that is, the map Ξ is one-to-one, onto, and has a continuously differentiable inverse.

Proof. (Outline) Define $f : (-\pi, \pi) \to (0, \infty)$ by

$$f(\tau) = \frac{1 + \cos \tau}{\pi + \tau + \sin \tau}.$$

It is a matter of elementary calculus to verify that f is $1 : 1$ and onto. (Establish that f has a simple pole at $\tau = -\pi$, is strictly monotonically decreasing over the interval $(-\pi, \pi)$, and is continuous at π with value 0.) Thus, given $(\alpha, \beta) \in \mathbb{R}^2_+$, there exists a unique $\tau \in (-\pi, \pi)$ such that $f(\tau) = \frac{\beta}{\alpha}$; the parameter ξ is then given by solving either Eqs. (1.10) or (1.11) for ξ. Furthermore, an explicit calculation verifies that the Jacobian determinant of the transformation is everywhere positive, and thus by the inverse function theorem (see Theorem A.3.1), Ξ has a continuously differentiable inverse as well.   □

This verifies that the family of cycloids forms what later on will be called a *central field* on the region $\mathbb{R}^2_+$. We shall prove in Sect. 1.8 that these curves are indeed the *optimal solutions*. Note that in the original parameterization $x = x(t)$ of the curves as functions of t, the derivative $\dot{x}$ converges to ∞ as t converges to 0 from the right (cf. Eq. (1.9)).

1.3 Surfaces of Revolution of Minimum Area

A second classical example in the calculus of variations that dates back to Euler in the mid-eighteenth century is to find surfaces of revolution that have a minimum area, so-called *minimal surfaces*: Given $t_1 > 0$ and x_0 and x_1 positive numbers, among all curves that join two given points $(0, x_0)$ and (t_1, x_1) and have positive values, find the one that generates a surface that has the smallest surface area of revolution when rotated around the t-axis (see Fig. 1.4).

Fig. 1.4 Surfaces of
revolution

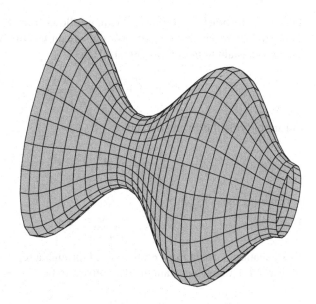

As in the brachistochrone problem, for starters, we restrict the curves to be graphs
of differentiable functions, and we do not allow the functions to touch 0. There is no
difficulty in finding the solutions experimentally: dip two circular loops into soap
water and remove them, holding them close to each other; the soap film that forms
between the circles is the desired solution. Unfortunately, this is a significantly
harder problem mathematically. But then again, its solution offers important insights
into calculus of variations problems as a whole, and in many ways, in the words
of Gilbert Bliss, this example is *"the most satisfactory illustration which we have
of the principles of the general theory of the calculus of variations"* [39, p. 85].
Mathematically, the problem can be formulated as follows:

[Minimal Surfaces] Given $t_1 > 0$, let x_0 and x_1 be positive reals. Minimize the
integral

$$I[x] = 2\pi \int\limits_{0}^{t_1} x\sqrt{1+\dot{x}^2}dt$$

over all functions $x \in C^1([0,x_1])$ that have positive values, $x : [0,t_1] \rightarrow (0,\infty)$, and
satisfy the boundary conditions

$$x(0) = x_0 \qquad \text{and} \qquad x(t_1) = x_1.$$

The integrand L of the general problem formulation is given by

$$L(x,\dot{x}) = 2\pi x\sqrt{1+\dot{x}^2},$$

and, as for the brachistochrone problem, it follows from the Hilbert differentiability theorem that extremals lie in C^2. We thus again use the first integral $L - \dot{x}L_{\dot{x}} = c$ with c a constant to find the extremals. Here

$$x\sqrt{1+\dot{x}^2} - \dot{x}x\frac{\dot{x}}{\sqrt{1+\dot{x}^2}} = c,$$

and simplifying this expression gives

$$\frac{x}{\sqrt{1+\dot{x}^2}}(1+\dot{x}^2 - \dot{x}^2) = c,$$

or renaming the constant,

$$\frac{x}{\sqrt{1+\dot{x}^2}} = \beta. \tag{1.13}$$

This equation is again most easily solved through a reparameterization of the time scale, $t = t(\tau)$. Here we want the derivative $\dot{x}$ to be

$$\frac{dx}{dt} = \sinh \tau, \tag{1.14}$$

so that

$$1+\dot{x}^2 = \cosh^2 \tau$$

and thus

$$x(\tau) = \beta \cosh \tau.$$

This also implies

$$\frac{dt}{d\tau} = \frac{dt}{dx}\frac{dx}{d\tau} = \frac{1}{\sinh \tau}\beta \sinh \tau = \beta,$$

and thus we simply have

$$t(\tau) = \alpha + \beta\tau$$

for some constant α. Hence, the general solutions to the Euler–Lagrange equation are given by the following family of catenaries:

$$x(t) = \beta \cosh\left(\frac{t-\alpha}{\beta}\right), \qquad \alpha \in \mathbb{R}, \qquad \beta > 0. \tag{1.15}$$

However, as illustrated in Fig. 1.5, the mapping properties of this family of extremals are drastically different from those of the cycloids for the brachistochrone problem. Now, given a fixed initial condition $(0, x_0)$, it is no longer true that an extremal necessarily passes through any point $(t_1, x_1) \in \mathbb{R}_+^2$, while there exist two extremals through other points (t_1, x_1). The theorem below summarizes the mapping properties of the family of catenaries. We call an open and connected subset of $\mathbb{R}^n$ a *region*.

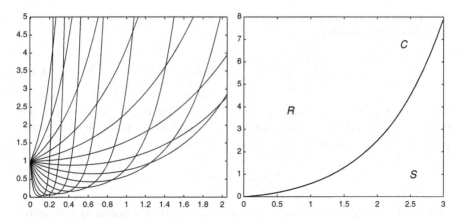

Fig. 1.5 The family of catenaries and its envelope

Theorem 1.3.1. *There exists a differentiable, strictly monotonically increasing, and strictly convex function* $\gamma: [0,\infty) \to [0,\infty)$, $t \mapsto \gamma(t)$, *that satisfies*

$$\gamma(0) = 0, \ \dot{\gamma}(0+) = 0, \text{ and } \lim_{t \to +\infty} \gamma(t) = +\infty, \ \lim_{t \to +\infty} \dot{\gamma}(t) = +\infty,$$

with the property that if C denotes the graph of γ, *then there exist exactly two catenaries that pass through any point* (t_1, x_1) *in the region R above the graph of* γ, *and there exists no catenary that passes through a point* (t_1, x_1) *in the region S below the graph of* γ. *Through any point* (t_1, x_1) *on the graph C itself there exists a unique catenary that passes through the point.*

This material is classical, and we follow the presentation of Bliss [39, pp. 92–98] in our proof below. Since we shall return to this example numerous times for illustrative purposes, we also include some intricate details of the calculations to give the complete picture. The reader who is mostly interested in the theoretical developments may elect to skip this technical proof without loss of continuity, but some of the properties established here will be used later on.

Proof. The initial condition x_0 imposes the relation

$$x_0 = \beta \cosh\left(-\frac{\alpha}{\beta}\right)$$

and reduces the collection of catenaries defined by Eq. (1.15) to a 1-parameter family. Introduce a new parameter p, $p = -\frac{\alpha}{\beta}$, so that we have

$$\beta = \frac{x_0}{\cosh p} \quad \text{and} \quad \alpha = -p\beta = -\frac{px_0}{\cosh p}.$$

This gives the following parameterization of all extremals through the point $(0, x_0)$ in terms of the parameter p:

$$x(t; p) = \frac{x_0}{\cosh p} \cosh \left(p + \frac{t}{x_0} \cosh p \right), \quad p \in \mathbb{R}, \quad t > 0. \tag{1.16}$$

We fix the time t and compute the range of $x(t; p)$ for $p \in \mathbb{R}$. As $p \to \pm\infty$, the term $p + \frac{t}{x_0} \cosh p$ diverges much faster to $+\infty$ than p does, and it follows from the properties of the hyperbolic cosine that

$$\lim_{p \to \pm\infty} x(t; p) = +\infty.$$

This also is readily verified using L'Hospital's rule. Hence the real analytic function $x(t; \cdot)$ has a global minimum at some point $p_* = p_*(t)$ and

$$\frac{\partial x}{\partial p}(t; p_*) = 0.$$

The key technical computations are given in the lemma below, where, as is customary, we denote partial derivatives with respect to t by a dot.

Lemma 1.3.1. *If p_* is a stationary point, $\frac{\partial x}{\partial p}(t; p_*) = 0$, then*

$$\dot{x}(t; p_*) = \frac{\partial x}{\partial t}(t; p_*) > 0 \quad \text{and} \quad \frac{\partial^2 x}{\partial p^2}(t; p_*) > 0. \tag{1.17}$$

Proof. In these computations, it is important to combine terms that arise properly to recognize sign relations that are not evident a priori. We have that

$$\dot{x}(t; p) = \sinh \left(p + \frac{t}{x_0} \cosh p \right),$$

and for later reference note that

$$\frac{x(t; p)}{\dot{x}(t; p)} = \frac{\cosh \left(p + \frac{t}{x_0} \cosh p \right)}{\sinh \left(p + \frac{t}{x_0} \cosh p \right)} \frac{x_0}{\cosh p}. \tag{1.18}$$

The partial derivatives of x with respect to p are best expressed in terms of this quantity at times t and 0. We have that

$$\frac{\partial x}{\partial p}(t;p) = \frac{x_0}{\cosh^2 p}\left[\sinh\left(p+\frac{t}{x_0}\cosh p\right)\left(1+\frac{t}{x_0}\sinh p\right)\cosh p\right.$$

$$\left. - \cosh\left(p+\frac{t}{x_0}\cosh p\right)\sinh p\right]$$

$$= \frac{\sinh\left(p+\frac{t}{x_0}\cosh p\right)\sinh p}{\cosh p}$$

$$\times \left[\frac{x_0+t\sinh p}{\sinh p} - \frac{\cosh\left(p+\frac{t}{x_0}\cosh p\right)}{\sinh\left(p+\frac{t}{x_0}\cosh p\right)}\frac{x_0}{\cosh p}\right]$$

$$= \frac{\dot{x}(t;p)\dot{x}(0;p)}{\cosh p}\left[t+\frac{x(0;p)}{\dot{x}(0;p)} - \frac{x(t;p)}{\dot{x}(t;p)}\right]. \tag{1.19}$$

This quantity has an interesting and useful geometric interpretation due to Lindelöf: For the moment, fix p and consider the catenary $x(s,p)$ as a function of s and let ℓ_0 and ℓ_t denote the tangent lines to the graph at the initial point $(0,x_0)$ and at the point $(t,x(t;p))$, respectively, i.e.,

$$\ell_0(s) = x_0 + s\dot{x}(0;p) \quad\text{and}\quad \ell_t(s) = x(t;p)+(s-t)\dot{x}(t;p).$$

These tangent lines intersect for

$$s = \frac{x(t;p)-x_0-t\dot{x}(t;p)}{\dot{x}(0;p)-\dot{x}(t;p)},$$

and the value at the intersection is given by

$$\varkappa(t;p) = x_0 + \frac{x(t;p)-x_0-t\dot{x}(t;p)}{\dot{x}(0;p)-\dot{x}(t;p)}\dot{x}(0;p)$$

$$= \frac{\dot{x}(t;p)\dot{x}(0;p)}{\dot{x}(t;p)-\dot{x}(0;p)}\left[t-\frac{x(t;p)}{\dot{x}(t;p)}+x_0\left(\frac{\dot{x}(t;p)-\dot{x}(0;p)}{\dot{x}(t;p)\dot{x}(0;p)}+\frac{1}{\dot{x}(t;p)}\right)\right]$$

$$= \frac{\dot{x}(t;p)\dot{x}(0;p)}{\dot{x}(t;p)-\dot{x}(0;p)}\left[t-\frac{x(t;p)}{\dot{x}(t;p)}+\frac{x_0}{\dot{x}(0;p)}\right].$$

We can therefore express the partial derivatives of x with respect to p as

$$\frac{\partial x}{\partial p}(t;p) = \varkappa(t;p)\frac{\dot{x}(t;p)-\dot{x}(0;p)}{\cosh p}. \tag{1.20}$$

Since the function $t \mapsto x(t;p)$ is strictly convex, it follows that $\dot{x}(t;p) > \dot{x}(0;p)$, and thus p_* is a stationary point of the function $p \mapsto x(t;p)$, now with t fixed, if and only if $\varkappa(t;p_*) = 0$, that is, if and only if the tangent line ℓ_0 to the catenary at the initial

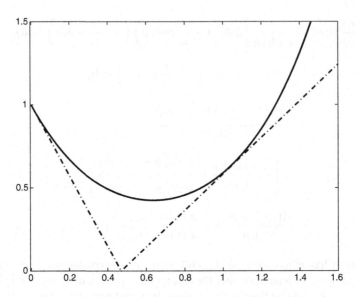

Fig. 1.6 Tangent lines to the catenary $x(\cdot\,;p_*)$ for a stationary p_*

point $(0,x_0)$ and the tangent line ℓ_t through the point $(t,x(t;p))$ intersect on the horizontal, or time, axis (see Fig. 1.6). In particular, this is possible only if $\dot{x}(t;p_*)$ is positive. Geometrically, the point $x(t,p_*)$ always lies on the increasing portion of the catenary $s \mapsto x(s,p_*)$. Furthermore, at any stationary point p_*, we have that

$$\frac{x(t;p_*)}{\dot{x}(t;p_*)} - \frac{x(0;p_*)}{\dot{x}(0;p_*)} = t > 0. \tag{1.21}$$

We now compute the second partial derivative with respect to p, but evaluate it only at a stationary point p_*. Since the term in brackets in Eq. (1.19) vanishes, we get that

$$\frac{\partial^2 x}{\partial p^2}(t;p_*) = \frac{\dot{x}(t;p_*)\dot{x}(0;p_*)}{\cosh p_*}\frac{\partial}{\partial p}\left(\frac{x(0;p)}{\dot{x}(0;p)} - \frac{x(t;p)}{\dot{x}(t;p)}\right)_{|p_*}.$$

Differentiating Eq. (1.18) gives

$$\frac{\partial}{\partial p}\left(\frac{x(t;p)}{\dot{x}(t;p)}\right) = -\frac{x_0 + t\sinh p}{\sinh^2\left(p + \frac{t}{x_0}\cosh p\right)\cosh p} - x_0\frac{\cosh\left(p + \frac{t}{x_0}\cosh p\right)}{\sinh\left(p + \frac{t}{x_0}\cosh p\right)}\frac{\sinh p}{\cosh^2 p}$$

$$= -\frac{(x_0 + t\sinh p)\cosh p + x_0\cosh\left(p + \frac{t}{x_0}\cosh p\right)\sinh\left(p + \frac{t}{x_0}\cosh p\right)\sinh p}{\sinh^2\left(p + \frac{t}{x_0}\cosh p\right)\cosh^2 p}.$$

At a stationary point p_*, we have that (cf. the derivation of Eq. (1.19))

$$x_0 + t \sinh p_* = x_0 \frac{\cosh\left(p_* + \frac{t}{x_0}\cosh p_*\right)\sinh p_*}{\sinh\left(p_* + \frac{t}{x_0}\cosh p_*\right)\cosh p_*}, \tag{1.22}$$

and thus

$$\frac{\partial}{\partial p}\left(\frac{x(t;p)}{\dot{x}(t;p)}\right)_{|p_*} = -\frac{x_0 \cosh\left(p_* + \frac{t}{x_0}\cosh p_*\right)\sinh p_*}{\sinh^3\left(p_* + \frac{t}{x_0}\cosh p_*\right)\cosh^2 p_*}$$

$$-\frac{x_0 \cosh\left(p_* + \frac{t}{x_0}\cosh p_*\right)\sinh p_*}{\sinh\left(p_* + \frac{t}{x_0}\cosh p_*\right)\cosh^2 p_*}$$

$$= -x_0 \frac{\cosh\left(p_* + \frac{t}{x_0}\cosh p_*\right)\left[1 + \sinh^2\left(p_* + \frac{t}{x_0}\cosh p_*\right)\right]\sinh p_*}{\sinh^3\left(p_* + \frac{t}{x_0}\cosh p_*\right)\cosh^2 p_*}$$

$$= -x_0 \frac{\cosh^3\left(p_* + \frac{t}{x_0}\cosh p_*\right)\sinh p_*}{\sinh^3\left(p_* + \frac{t}{x_0}\cosh p_*\right)\cosh^2 p_*}$$

$$= -\frac{1}{x_0^2}\left(\frac{x(t;p_*)}{\dot{x}(t;p_*)}\right)^3 \cosh p_* \sinh p_*.$$

Hence, overall,

$$\frac{\partial^2 x}{\partial p^2}(t;p_*) = \frac{\dot{x}(t;p_*)\sinh p_*}{\cosh p_*}\frac{1}{x_0^2}\left[\left(\frac{x(t;p_*)}{\dot{x}(t;p_*)}\right)^3 - \left(\frac{x(0;p_*)}{\dot{x}(0;p_*)}\right)^3\right]\cosh p_* \sinh p_*$$

$$= \frac{\dot{x}(t;p_*)}{x_0^2}\left[\left(\frac{x(t;p_*)}{\dot{x}(t;p_*)}\right)^3 - \left(\frac{x(0;p_*)}{\dot{x}(0;p_*)}\right)^3\right]\sinh^2 p_*. \tag{1.23}$$

We already have shown above that $\dot{x}(t;p_*) > 0$, and it follows from Eq. (1.21) that

$$\frac{x(t;p_*)}{\dot{x}(t;p_*)} > \frac{x(0;p_*)}{\dot{x}(0;p_*)}.$$

Clearly, $p_* \neq 0$, and thus

$$\frac{\partial^2 x}{\partial p^2}(t;p_*) > 0.$$

$\square$

This lemma implies that every stationary point of the function $p \mapsto x(t;p)$ is a local minimum. But then this function $x(t;\cdot)$ has a unique stationary point $p_* = p_*(t)$ corresponding to its global minimum over $\mathbb{R}$, and $p_* : [0,\infty) \to [0,\infty)$, $t \mapsto p_*(t)$, is well-defined, i.e., is a single-valued function. Furthermore, p_* is given by the solution of the equation

$$\frac{\partial x}{\partial p}(t;p) = 0.$$

Since $\frac{\partial^2 x}{\partial p^2}(t;p_*(t)) > 0$, it follows from the implicit function theorem that $p_*(t)$ is continuously differentiable with derivative given by

$$\dot{p}_*(t) = -\frac{\frac{\partial^2 x}{\partial t \partial p}(t;p_*(t))}{\frac{\partial^2 x}{\partial p^2}(t;p_*(t))}.$$

It is clear that for t fixed, $x(t;p)$ is strictly decreasing in p for $p < p_*(t)$ and strictly increasing for $p > p_*(t)$ with limit ∞ as $p \to \pm\infty$. Hence, for any point (t_1,x_1) with $t_1 > 0$ and $x_1 > 0$, the equation $x(t_1;p) = x_1$ has exactly two solutions $p_1 < p_*(t_1) < p_2$ if $x_1 > x(t_1;p_*(t_1))$, the unique solution $p_*(t_1)$ if $x_1 = x(t_1;p_*(t_1))$, and no solutions for $x_1 < x(t_1;p_*(t_1))$. The minimum value $x(t_1;p_*(t_1))$ is positive, and since

$$\frac{\partial x}{\partial p}(t;0) = x_0 \sinh\left(\frac{t}{x_0}\right) > 0,$$

$p_*(t)$ is always negative.

Consequently, the curve

$$\gamma : [0,\infty) \to [0,\infty), \quad t \mapsto \gamma(t),$$

is defined by

$$\gamma(t) = x(t;p_*(t)).$$

The derivatives of γ are then given by

$$\dot{\gamma}(t) = \frac{d\gamma}{dt}(t) = \frac{\partial x}{\partial t}(t;p_*(t)) + \frac{\partial x}{\partial p}(t;p_*(t))\dot{p}_*(t) = \frac{\partial x}{\partial t}(t;p_*(t)) > 0,$$

so that γ is strictly increasing, and

$$\ddot{\gamma}(t) = \frac{d^2\gamma}{dt^2}(t) = \frac{\partial^2 x}{\partial t^2}(t;p_*(t)) + \frac{\partial^2 x}{\partial p \partial t}(t;p_*(t))\dot{p}_*(t)$$

$$= \frac{\partial^2 x}{\partial t^2}(t;p_*(t)) - \frac{\left(\frac{\partial^2 x}{\partial t \partial p}(t;p_*(t))\right)^2}{\frac{\partial^2 x}{\partial p^2}(t;p_*(t))}.$$

In general, we have that

$$\frac{\partial^2 x}{\partial t^2}(t;p) = \cosh\left(p + \frac{t}{x_0}\cosh p\right)\frac{\cosh p}{x_0} = x(t;p)\frac{\cosh^2 p}{x_0^2}$$

and

$$\frac{\partial^2 x}{\partial p \partial t}(t;p) = \cosh\left(p + \frac{t}{x_0}\cosh p\right)\left(1 + \frac{t}{x_0}\sinh p\right).$$

Using Eq. (1.22) to eliminate $x_0 + t\sinh p_*$ at the critical point p_*, we obtain

$$\frac{\partial^2 x}{\partial p \partial t}(t;p_*) = \frac{\cosh^2\left(p_* + \frac{t}{x_0}\cosh p_*\right)}{\sinh\left(p_* + \frac{t}{x_0}\cosh p_*\right)}\frac{\sinh p_*}{\cosh p_*}$$

$$= \frac{1}{x_0^2}\frac{x(t;p_*)^2}{\dot{x}(t;p_*)}\cosh p_* \sinh p_*.$$

Putting all this together, and using Eq. (1.23), we therefore have that

$$\frac{d^2\gamma}{dt^2}(t) = x(t;p_*)\frac{\cosh^2 p_*}{x_0^2} - \frac{\left(\frac{1}{x_0^2}\frac{x(t;p_*)^2}{\dot{x}(t;p_*)}\cosh p_* \sinh p_*\right)^2}{\frac{\dot{x}(t;p_*)}{x_0^2}\left[\left(\frac{x(t;p_*)}{\dot{x}(t;p_*)}\right)^3 - \left(\frac{x(0;p_*)}{\dot{x}(0;p_*)}\right)^3\right]\sinh^2 p_*}$$

$$= x(t;p_*)\frac{\cosh^2 p_*}{x_0^2}\left[1 - \frac{\left(\frac{x(t;p_*)}{\dot{x}(t;p_*)}\right)^3}{\left(\frac{x(t;p_*)}{\dot{x}(t;p_*)}\right)^3 - \left(\frac{x(0;p_*)}{\dot{x}(0;p_*)}\right)^3}\right]$$

$$= -\frac{x(t;p_*)}{x_0^2}\frac{\left(\frac{x(0;p_*)}{\dot{x}(0;p_*)}\right)^3}{\left(\frac{x(t;p_*)}{\dot{x}(t;p_*)}\right)^3 - \left(\frac{x(0;p_*)}{\dot{x}(0;p_*)}\right)^3}\cosh^2 p_*.$$

By Eq. (1.21), the denominator is positive. Since $p_* < 0$, it follows that $\dot{x}(0;p_*) = \sinh p_* < 0$, and thus $\ddot{\gamma}(t)$ is positive. Hence the function γ is strictly convex over $(0,\infty)$.

In particular, these properties imply that $\lim_{t\to+\infty}\gamma(t) = +\infty$. But also $\dot{\gamma}(t) \to +\infty$ as $t\to+\infty$: The function $\kappa(p) = \frac{p}{\cosh p}$ is bounded over $\mathbb{R}$, and therefore,

$$p_*(t) + \frac{t}{x_0}\cosh p_*(t) = \left(\frac{p_*(t)}{\cosh p_*(t)} + \frac{t}{x_0}\right)\cosh p_*(t)$$

$$\geq \frac{p_*(t)}{\cosh p_*(t)} + \frac{t}{x_0} \to \infty \quad \text{as } t\to\infty.$$

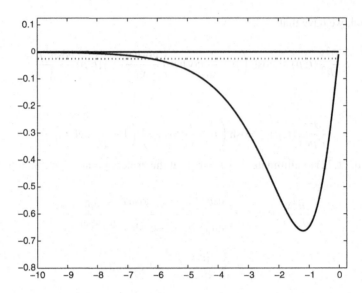

Fig. 1.7 Asymptotic properties of $\gamma(t)$

Thus, also

$$\dot{\gamma}(t) = \frac{\partial x}{\partial t}(t; p_*(t)) = \sinh\left(p_*(t) + \frac{t}{x_0}\cosh p_*(t)\right) \to \infty \quad \text{as } t \to \infty.$$

We conclude the proof with an analysis of the asymptotic properties of $\gamma(t)$ as $t \to 0+$. An arbitrary catenary $x(s; p)$ has its minimum when $p + \frac{s}{x_0}\cosh p = 0$, and the minimum value is given by $\frac{x_0}{\cosh p}$. For $t > 0$ small enough, there exists a solution $\tilde{p} = \tilde{p}(t)$ to the equation

$$\frac{p}{\cosh p} = -\frac{t}{x_0}$$

with the property that $\tilde{p}(t) \to -\infty$ as $t \to 0+$ (see Fig. 1.7).

For any time t, we thus have that

$$\gamma(t) = \min_{p \in \mathbb{R}} x(t; p) \le x(t; \tilde{p}(t)) = \frac{x_0}{\cosh \tilde{p}(t)},$$

and consequently $\lim_{t \to 0+} \gamma(t) = 0$. Furthermore, since the graph of γ lies below the curve ς of minima,

$$\varsigma : t \mapsto \left(t, \frac{x_0}{\cosh \tilde{p}(t)}\right) = \left(-\frac{x_0\tilde{p}(t)}{\cosh \tilde{p}(t)}, \frac{x_0}{\cosh \tilde{p}(t)}\right),$$

it also follows that there exists a time $\tau \in (0, t)$ at which the slope $\dot{\gamma}(\tau)$ must be smaller than the slope of the line connecting the origin with this minimum point on the catenary, i.e.,

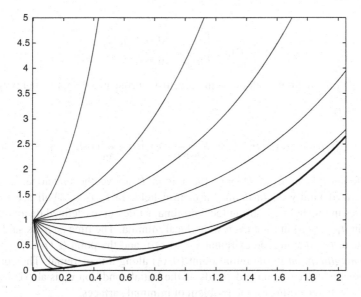

Fig. 1.8 The restricted family of catenaries with envelope C

$$\dot{\gamma}(\tau) \le \frac{1}{t} \frac{x_0}{\cosh \tilde{p}(t)} = -\frac{\cosh \tilde{p}(t)}{x_0 \tilde{p}(t)} \frac{x_0}{\cosh \tilde{p}(t)} = -\frac{1}{\tilde{p}(t)} \to 0 \quad \text{as } t \to 0+.$$

Since $\dot{\gamma}(t)$ is strictly increasing, we therefore also have that $\lim_{t \to 0+} \dot{\gamma}(t) = 0$. This concludes the proof.   □

For endpoints (t_1, x_1) in the region R above the curve C, there exist two solutions, and thus these geometric properties naturally raise the question as to which of the catenaries is optimal. The curve $C : [0, \infty) \to [0, \infty)$, $t \mapsto (t, \gamma(t))$, is regular (i.e., the tangent vector $\dot{\gamma}$ is everywhere nonzero) and is everywhere tangent to exactly one member of the parameterized family of extremals, $x : [0, \infty) \to [0, \infty)$, $t \mapsto x(t, p)$ (with $p < 0$), without itself being a member of the family. In geometry such a curve is called an *envelope* (see Fig. 1.8). The catenary $x(t; 0) = x_0 \cosh \left(\frac{t}{x_0} \right)$ for $p = 0$ is asymptotic to C at infinity, and for $p > 0$ the catenaries do not intersect the curve C. In fact, for each $p < 0$, there exists a unique time $t = t_c(p) > 0$ when the corresponding catenary (that is, the graph of the curve) is tangent to the envelope C, and if we were to allow curves that lie in $x_1 < 0$ as well, then for $p > 0$ the catenaries touch the curve that is obtained from C by reflection along the x-axis at time $t_c(p) = -t_c(-p) < 0$. The structure of the solutions for $x_1 < 0$ is symmetric to the one for $x_1 > 0$, and thus we consider only this scenario. By restricting the times along the catenaries appropriately, we can eliminate the overlaps. If we define

$$D = \{(t, p) : p \in \mathbb{R}, 0 < t < \tau(p)\},$$

where

$$\tau(p) = \begin{cases} \infty & \text{if } p \geq 0, \\ t_c(p) & \text{if } p < 0, \end{cases}$$

then, as with the field of cycloids in the brachistochrone problem, the mapping

$$\Xi : D \to R,$$

$$(t, p) \mapsto (t, x(t; p)) = \left(t, \frac{x_0}{\cosh p} \cosh \left(p + \frac{t}{x_0} \cosh p \right) \right), \qquad (1.24)$$

is a diffeomorphism from D onto the region R above the envelope, and this parameterized family again defines a central field. In fact, we shall see that all trajectories in this field are strong local minima, while curves defined on an interval $[0, t_1]$ with $t_1 \geq t_c(p)$ are not even weak local minima. The time $t_c(p)$ is said to be conjugate to $t = 0$ along the extremal $x = x(\cdot; p)$, and the point $(t_c(p), x(t_c(p); p))$ is the *conjugate point* to the initial point $(0, x_0)$ along this extremal. The curve C, whose precise shape of course depends on the initial condition x_0, is also called the curve of conjugate points for the problem of minimal surfaces.

In the next sections, we develop the tools that allow us to prove these statements about optimality and, more generally, explain the significance of conjugate points for the local optimality of curves in the calculus of variations.

1.4   The Legendre and Jacobi Conditions

If x_* is a weak local minimum for problem [CV], then besides being an extremal, by Eq. (1.2), we also have for all functions

$$h \in C_0^1([a, b]) = \{ h \in C^1([a, b]) : h(a) = h(b) = 0 \}$$

that the second variation is nonnegative,

$$\delta^2 I[x_*](h) = \int_a^b L_{xx}(t, x_*, \dot{x}_*) h^2 + 2 L_{\dot{x}x}(t, x_*, \dot{x}_*) h\dot{h} + L_{\dot{x}\dot{x}}(t, x_*, \dot{x}_*) \dot{h}^2 dt \geq 0.$$

This implies the following second-order necessary condition for optimality.

Theorem 1.4.1 (Legendre condition). *If x_* is a weak local minimum for problem [CV] (see Fig. 1.9), then*

$$L_{\dot{x}\dot{x}}(t, x_*(t), \dot{x}_*(t)) \geq 0 \qquad \text{for all} \quad t \in [a, b].$$

Proof. This condition should be clear intuitively. Choosing functions $h \in C_0^1([a, b])$ that keep the norm $\|h\|_\infty$ small while having large derivatives $\|\dot{h}\|_\infty$, it follows that

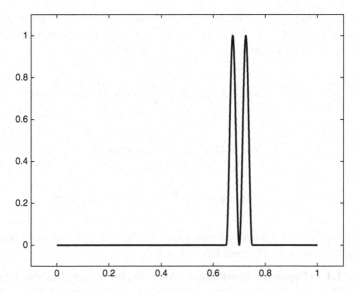

Fig. 1.9 The variation for the proof of the Legendre condition

the term multiplying $\dot{h}^2$ will become dominant and thus needs to be nonnegative. We prove this by contradiction. Suppose there exists a time $\tau \in (a,b)$ at which $L_{\dot{x}\dot{x}}(\tau, x_*(\tau), \dot{x}_*(\tau)) = -2\beta < 0$. Choose $\varepsilon > 0$ such that $L_{\dot{x}\dot{x}}(t, x_*(t), \dot{x}_*(t)) < -\beta$ for $t \in [\tau - \varepsilon, \tau + \varepsilon] \subset [a,b]$ and pick the function

$$h(t) = \begin{cases} \sin^2\left(\frac{\pi}{\varepsilon}(t - \tau)\right) & \text{for } |t - \tau| \le \varepsilon, \\ 0 & \text{otherwise.} \end{cases}$$

Then, since

$$\dot{h}(t) = 2\sin\left(\frac{\pi}{\varepsilon}(t - \tau)\right)\cos\left(\frac{\pi}{\varepsilon}(t - \tau)\right)\frac{\pi}{\varepsilon} = \frac{\pi}{\varepsilon}\sin\left(\frac{2\pi}{\varepsilon}(t - \tau)\right),$$

we have $h \in C_0^1([a,b])$ and

$$\delta^2 I[x_*](h) = \int_{\tau-\varepsilon}^{\tau+\varepsilon} L_{xx}(t, x_*(t), \dot{x}_*(t))\sin^4\left(\frac{\pi}{\varepsilon}(t - \tau)\right)$$

$$+ \int_{\tau-\varepsilon}^{\tau+\varepsilon} 2L_{\dot{x}x}(t, x_*(t), \dot{x}_*(t))\frac{\pi}{\varepsilon}\sin^2\left(\frac{\pi}{\varepsilon}(t - \tau)\right)\sin\left(\frac{2\pi}{\varepsilon}(t - \tau)\right)$$

$$+ \int_{\tau-\varepsilon}^{\tau+\varepsilon} L_{\dot{x}\dot{x}}(t, x_*(t), \dot{x}_*(t))\frac{\pi^2}{\varepsilon^2}\sin^2\left(\frac{2\pi}{\varepsilon}(t - \tau)\right) dt.$$

If we now let M and N, respectively, be upper bounds on the absolute values of the continuous functions $L_{xx}(t,x_*(t),\dot{x}_*(t))$ and $L_{\dot{x}x}(t,x_*(t),\dot{x}_*(t))$ on the interval $[a,b]$, then we have the contradiction that

$$\delta^2 I[x_*](h) \le 2\varepsilon M + 4N\pi - \beta \int\limits_{\tau-\varepsilon}^{\tau+\varepsilon} \frac{\pi^2}{\varepsilon^2} \sin^2\left(\frac{2\pi}{\varepsilon}(t-\tau)\right) dt$$

$$= 2\varepsilon M + 4N\pi - \beta \frac{\pi}{2\varepsilon} \int\limits_{-2\pi}^{2\pi} \sin^2(s)\,ds$$

$$= 2\varepsilon M + 4N\pi - \beta \frac{\pi^2}{\varepsilon} \to -\infty \qquad \text{as } \varepsilon \to 0+ .$$

Hence $L_{\dot{x}\dot{x}}(t,x_*(t),\dot{x}_*(t))$ must be nonnegative on the open interval (a,b) and thus, by continuity, also on the compact interval $[a,b]$.   □

Definition 1.4.1 (Legendre conditions). An extremal x_* satisfies the Legendre condition along $[a,b]$ if the function $t \mapsto L_{\dot{x}\dot{x}}(t,x_*(t),\dot{x}_*(t))$ is nonnegative; it satisfies the strengthened Legendre condition if $L_{\dot{x}\dot{x}}(t,x_*(t),\dot{x}_*(t))$ is positive on $[a,b]$.

Legendre mistakenly believed that the strengthened version of this condition also would be sufficient for a local minimum. However, considering the geometric properties of the extremals for the minimum surfaces of revolution, this clearly cannot be true, since all extremals for this problem satisfy the strengthened Legendre condition. It was up to Jacobi to rectify Legendre's argumentation and come up with the correct formulations, now known as the Jacobi conditions, which we develop next. For convenience, throughout the rest of this and the next section, we make the *simplifying assumption* that the function L is three times continuously differentiable, $L \in C^3$.

Let x_* be an extremal that satisfies the strengthened Legendre condition everywhere on $[a,b]$. Then it follows from Hilbert's differentiability theorem, Proposition 1.2.1, that $x_* \in C^2$, and we can integrate the mixed term in the second variation by parts to get

$$\int\limits_a^b 2L_{\dot{x}x}(t,x_*(t),\dot{x}_*(t))\dot{h}(t)h(t)dt = -\int\limits_a^b \left(\frac{d}{dt}[L_{\dot{x}x}(t,x_*(t),\dot{x}_*(t))]\right) h(t)^2 dt.$$

This allows us to rewrite the second variation in the simpler form

$$\delta^2 I[x_*](h) = \int\limits_a^b Q(t)h(t)^2 + R(t)\dot{h}(t)^2 dt,$$

where

$$Q(t) = L_{xx}(t, x_*(t), \dot{x}_*(t)) - \frac{d}{dt}\left(L_{\dot{x}x}(t, x_*(t), \dot{x}_*(t)) \right)$$

and

$$R(t) = L_{\dot{x}\dot{x}}(t, x_*(t), \dot{x}_*(t)).$$

Under the assumption that $L \in C^3$, it follows that Q is continuous and R is continuously differentiable, $Q \in C([a, b])$ and $R \in C^1([a, b])$.

Definition 1.4.2 (Quadratic form). Given a symmetric bilinear form $\mathscr{B}$ defined on a real vector space X, $\mathscr{B} : X \times X \to \mathbb{R}$, $(x, y) \mapsto \mathscr{B}(x, y)$, the mapping $\mathscr{Q} : X \to \mathbb{R}$, $h \mapsto \mathscr{Q}(h) = \mathscr{B}(h, h)$, is called a quadratic form. The quadratic form $\mathscr{Q}$ is said to be positive semidefinite if $\mathscr{Q}(h) \geq 0$ for all $h \in X$; it is positive definite if it is positive semidefinite and $\mathscr{Q}(h) = 0$ holds only for the function $h \equiv 0$. The kernel of a quadratic form, $\ker \mathscr{Q}$, consists of all $h \in X$ for which $\mathscr{Q}(h) = 0$. On a normed space X, a quadratic form is said to be strictly positive definite if there exists a positive constant c such that for all $h \in X$,

$$\mathscr{Q}(h) \geq c \|h\|^2.$$

We formulate necessary and sufficient conditions for a quadratic form $\mathscr{Q} : C_0^1([a, b]) \to \mathbb{R}$ given by

$$\mathscr{Q}(h) = \int_a^b Q(t)h(t)^2 + R(t)\dot{h}(t)^2 dt \tag{1.25}$$

with $Q \in C([a, b])$ and $R \in C^1([a, b])$ to be positive semidefinite, respectively positive definite, on the space $C_0^1([a, b])$. These are classical results that can be found in most textbooks on the subject. Our presentation here follows the one in Gelfand and Fomin [105]. These positivity conditions then translate into second-order necessary and sufficient conditions for a weak local minimum for problem [CV]. Henceforth $\mathscr{Q}$ will always denote this quadratic form defined by Eq. (1.25). Theorem 1.4.1 immediately implies the following condition:

Corollary 1.4.1. *If the quadratic form $\mathscr{Q}(h)$ is positive semidefinite over $C_0^1([a, b])$, then $R(t)$ is nonnegative on $[a, b]$.* □

Assuming that $R(t)$ is positive on $[a, b]$, Legendre's idea was to complete the square in the quadratic form $\mathscr{Q}$ and write $\mathscr{Q}$ as a sum of positive terms. For any differentiable function $w \in C^1([a, b])$, we have that

$$0 = \int_a^b \frac{d}{dt}(wh^2)dt = \int_a^b \dot{w}h^2 + 2whh\,dt,$$

and thus adding this term to the quadratic form gives

$$\mathcal{Q}(h) = \int_a^b (R\dot{h}^2 + 2wh\dot{h} + (Q+\dot{w})h^2)dt$$

$$= \int_a^b R\left(\dot{h} + \frac{w}{R}h\right)^2 + \left(\dot{w} + Q - \frac{w^2}{R}\right)h^2 dt.$$

If we now choose w as a solution to the differential equation

$$\dot{w} = \frac{w^2}{R} - Q, \tag{1.26}$$

and if this solution were to exist over the full interval $[a,b]$, then we would have $\mathcal{Q}(h) \geq 0$ for all $h \in C_0^1([a,b])$ and $\mathcal{Q}(h) = 0$ if and only if $\dot{h} + \frac{w}{R}h = 0$. But since $h(a) = 0$, this is possible only if $h \equiv 0$. Thus, in this case the quadratic functional $\mathcal{Q}(h)$ would be positive definite. However, (1.26) is a Riccati differential equation, and while solutions exist locally, in general there is no guarantee that solutions exist over the full interval $[a,b]$. For example, for $R = \alpha$ and $Q = -\frac{1}{\alpha}$ we get $\dot{w} = \frac{1}{\alpha}\left(1 + w^2\right)$, with general solution $w(t) = \tan(\frac{t}{\alpha})$, which exists only on intervals of length at most $b - a < \alpha\pi$, which can be arbitrarily small. Riccati differential equations arise as differential equations satisfied by quotients of functions that themselves obey linear differential equations. As a consequence of this fact, solutions to a Riccati differential equation can be related to the solutions of a second-order linear differential equation by means of a classical change of variables. In the result below, this connection is used to give a characterization of the escape times of solutions to Riccati equations in terms of the existence of nonvanishing solutions to a corresponding second-order linear differential equation.

Proposition 1.4.1. *Suppose the function R is positive on the interval $[a,b]$. Then the Riccati differential equation*

$$R(Q + \dot{w}) = w^2$$

has a solution over the full interval $[a,b]$ if and only if there exists a solution to the second-order linear differential equation

$$\frac{d}{dt}(R\dot{y}) = Qy \tag{1.27}$$

that does not vanish over $[a,b]$.

Proof. If there exists a solution y to (1.27) that does not vanish on the interval $[a,b]$, then the function $w = -\frac{\dot{y}}{y}R$ is well-defined on $[a,b]$, and it is simply a matter of verification to show that w solves the Riccati equation:

$$R(Q+\dot{w}) = R\left(Q + \frac{-\frac{d}{dt}(\dot{y}R)y + \dot{y}R\dot{y}}{y^2}\right)$$

$$= R\left(\frac{Qy^2 - Q\dot{y}^2 + R\dot{y}^2}{y^2}\right) = \left(\frac{R\dot{y}}{y}\right)^2 = w^2.$$

Conversely, if w is a solution to the Riccati equation (1.27) that exists on the full interval $[a,b]$, let y be a nontrivial solution to the linear differential equation $R\dot{y} + wy = 0$. Then $R\dot{y}$ is continuously differentiable and we have that

$$\frac{d}{dt}(R\dot{y}) = \frac{d}{dt}(-wy) = -w\dot{y} - \dot{w}y = -w\left(-\frac{wy}{R}\right) - \left(\frac{w^2}{R} - Q\right)y = Qy.$$

Furthermore, as a nontrivial solution, y does not vanish.   □

Note that if y is a solution to Eq. (1.27) that vanishes at some time c, then so is $y_\alpha(t) = \alpha y(t)$ for any $\alpha \in \mathbb{R}$, $\alpha \neq 0$. In particular, to see whether there exist nonvanishing solutions, without loss of generality we may normalize the initial condition on the derivative so that $\dot{y}(a) = 1$. This leads to the following definition.

Definition 1.4.3 (Conjugate points and Jacobi equation). A time $c \in (a,b]$ is said to be conjugate to a if the solution y to the initial value problem

$$\frac{d}{dt}(R\dot{y}) = Qy, \qquad y(a) = 0, \qquad \dot{y}(a) = 1, \tag{1.28}$$

vanishes at c; in this case the point $(c, x_*(c))$ on the reference extremal is called a conjugate point to $(a, x_*(a))$. The equation

$$\frac{d}{dt}(R\dot{y}) = Qy \tag{1.29}$$

is called the *Jacobi equation*.

We now show that the absence of conjugate points on the interval $(a,b]$ is equivalent to the quadratic form $\mathcal{Q}(h)$ being positive definite.

Theorem 1.4.2. *Let x_* be an extremal for which the strengthened Legendre condition is satisfied on the interval $[a,b]$, i.e.,*

$$R(t) = L_{\dot{x}\dot{x}}(t, x_*(t), \dot{x}_*(t)) > 0 \qquad \text{for all} \quad t \in [a,b].$$

Then the quadratic functional

$$\mathcal{Q}(h) = \int_a^b R(t)\dot{h}(t)^2 + Q(t)h(t)^2 dt$$

is positive definite for $h \in C_0^1([a,b])$ if and only if the interval $(a,b]$ contains no time conjugate to a.

Proof. Essentially, the sufficiency of this condition has already been shown: If there exists no time conjugate to a in $(a,b]$, then, since the solution to the Jacobi equation depends continuously on the initial time, for $\varepsilon > 0$ sufficiently small, the solution to the initial value problem

$$\frac{d}{dt}(R\dot{y}) = Qy, \qquad y(a-\varepsilon) = 0, \qquad \dot{y}(a-\varepsilon) = 1,$$

still exists and does not vanish on the interval $(a-\varepsilon,b]$. Hence, by Proposition 1.4.1, there exists a solution to the Riccati equation $R(Q+\dot{w}) = w^2$ over the full interval $[a,b]$ and thus

$$\mathscr{Q}(h) = \int_a^b R\left(\dot{h}+\frac{w}{R}h\right)^2 \geq 0,$$

with equality if and only if $\dot{h}+\frac{w}{R}h = 0$, i.e., $h \equiv 0$.

The condition about the nonexistence of conjugate times is also necessary for $\mathscr{Q}(h)$ to be positive definite: Consider the convex combination of the quadratic form $\mathscr{Q}(h)$ and the quadratic form given by $\int_a^b \dot{h}(t)^2 dt$, i.e., set

$$\mathscr{Q}_s(h) = s\int_a^b \left(R(t)\dot{h}(t)^2 + Q(t)h(t)^2\right) dt + (1-s)\int_a^b \dot{h}(t)^2 dt. \qquad (1.30)$$

Note that the quadratic form $\mathscr{Q}_0(h) = \int_a^b \dot{h}(t)^2 dt$ is positive definite on $C_0^1([a,b])$. For if $\mathscr{Q}_0(h) = 0$, then it follows that h is constant, and thus $h \equiv 0$ by the boundary conditions. Since both quadratic forms $\mathscr{Q}_1(h)$ and $\mathscr{Q}_0(h)$ are positive definite on $C_0^1([a,b])$, it follows that the convex combination $\mathscr{Q}_s(h)$ is positive definite for all $s \in [0,1]$.

Let $y : [0,1] \times [a,b] \to \mathbb{R}$ denote the solutions to the corresponding Jacobi equation (1.29) given by

$$\frac{d}{dt}([sR+(1-s)]\dot{y}) = sQy, \qquad y(s,a) = 0, \qquad \dot{y}(s,a) = 1.$$

For $s = 0$, we have the trivial equation $\ddot{y} = 0$ and thus $y(0,t) = t - a$, i.e., there exists no time conjugate to a. We now show that this property is preserved along the convex combination, and thus also for $s = 1$ there does not exist a time conjugate to a in $(a,b]$.

Consider the zero set $\mathscr{Z}$ of y away from the trivial portion for $t = a$ (see Fig. 1.10),

$$\mathscr{Z} = \{(s,t) \in [0,1] \times (a,b] : y(s,t) = 0\}.$$

Since $\dot{y}(s,a) = 1$, there exists an $\varepsilon > 0$ such that we actually have

$$\mathscr{Z} \subset [0,1] \times [a+\varepsilon,b].$$

Fig. 1.10 The zero set $\mathscr{Z}$

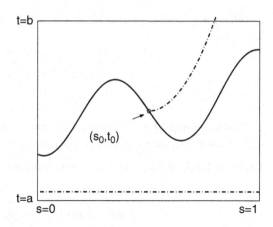

(For every $s \in [0,1]$ there exists a neighborhood U_s of (s,a) in $[0,1] \times [a,b]$ such that $y(s,t) > 0$ for $(s,t) \in U_s$ and $t > 0$. By compactness we can choose a uniform bound $\varepsilon > 0$.) If $(s_0,t_0) \in \mathscr{Z}$, then $\dot{y}(s_0,t_0)$ cannot vanish, since otherwise $y(s,\cdot)$ vanishes identically in t as a solution to a second-order linear differential equation that vanishes with its derivative at t_0. But this is not possible, since $\dot{y}(s,a) = 1$. Hence, by the implicit function theorem, the equation $y(s,t) = 0$ can locally be solved for t with a continuously differentiable function $t = t(s)$ near any point $(s_0,t_0) \in \mathscr{Z}$. Thus, if $(s_0,t_0) \in \mathscr{Z}$, then it follows that there exists a curve $C \subset \mathscr{Z}$ that passes through (s_0,t_0) and can be described as the graph of some function $\varsigma : I \to [a,b], s \mapsto \varsigma(s)$, defined on some maximal interval $I \subset [0,1]$. But this interval I must be all of $[0,1]$. For we always have $\varsigma(s) \geq a + \varepsilon$, and thus C cannot leave the set $[0,1] \times [a,b]$ through $t = a$. But it cannot escape through $t = b$ either. For if for some $\bar{s} \in [0,1]$ we have $y(\bar{s},b) = 0$, then the function h defined by $h(t) = y(\bar{s},t)$ lies in $C_0^1([a,b])$, and by integration by parts we have that

$$\int_a^b [sR(t) + (1-s)]\dot{h}(t)^2 + sQ(t)h(t)^2 dt$$

$$= \int_a^b \left(-\frac{d}{dt}\left[(sR(t) + (1-s))\dot{h}(t)\right] + sQ(t)h(t) \right) h(t)\, dt = 0.$$

Since $\dot{h}(a) = 1$, $h(\cdot) = y(\bar{s},\cdot)$ is not identically zero, contradicting the fact that $\mathscr{Q}_s(h)$ is positive definite. But then the curve C must extend over the full interval $I = [0,1]$ with both $\varsigma(0)$ and $\varsigma(1)$ taking values in the open interval (a,b). But for $s = 0$ we have $y(0,t) = t - a$, and thus $\mathscr{Z}$ cannot intersect the segment $\{0\} \times [a + \varepsilon, b]$. Contradiction. Thus the set $\mathscr{Z}$ is empty. $\qquad\square$

We separately still formulate the following fact, which was shown in the proof of Theorem 1.4.2.

Corollary 1.4.2. *If $h \in C_0^1([a,b])$ is a solution to the Jacobi equation, then $h \in$* $\ker \mathscr{Q}$, *i.e.,*

$$\mathscr{Q}(h) = \int\limits_a^b R(t)\dot{h}(t)^2 + Q(t)h(t)^2 dt = 0.$$

□

This fact, coupled with the preceding proof, also allows us to give the desired second-order necessary condition for optimality of extremals:

Theorem 1.4.3. *If $R(t) > 0$ and the quadratic functional*

$$\mathscr{Q}(h) = \int\limits_a^b R(t)\dot{h}(t)^2 + Q(t)h(t)^2 dt$$

is positive semidefinite over $C_0^1([a,b])$, then the open interval (a,b) contains no time conjugate to a.

Proof. In this case, the convex combination $\mathscr{Q}_s(h)$ defined in Eq. (1.30) is still positive definite for $s \in [0,1)$, and thus for these values there does not exist a time c conjugate to a in $(a,b]$. In particular, the set $\mathscr{Z}$ therefore does not intersect $[0,1) \times (a,b]$. If there were to exist a time $c \in (a,b)$ such that $(1,c) \in \mathscr{Z}$, then, since $\dot{y}(1,c) \neq 0$, there would exist a differentiable curve in $\mathscr{Z}$ that starts at the point $(1,c)$ and takes values in (a,b). Specifically, there would exist a differentiable function ς defined over some interval $[1-\varepsilon,1]$, $\varsigma : [1-\varepsilon,1] \to (a,b)$, $s \mapsto \varsigma(s)$, such that $(s,\varsigma(s)) \in \mathscr{Z}$. Contradiction. Note, however, that if $\mathscr{Q}(h)$ is only positive semidefinite, then it is possible that b is conjugate to a and this gives rise to a ˙conjugate point $(b,x_*(b))$. This happens if the Jacobi equation has a nontrivial solution that vanishes at $t = b$.   □

We summarize the necessary conditions for a weak minimum:

Corollary 1.4.3 (Necessary conditions for a weak local minimum). *Suppose x_* : $[a,b] \to \mathbb{R}$ is a weak local minimum for problem [CV]. Then*

1. *x_* is an extremal, i.e., satisfies the Euler–Lagrange equation*

$$\frac{d}{dt}\left(\frac{\partial L}{\partial \dot{x}}(t,x_*(t),\dot{x}_*(t)) \right) = \frac{\partial L}{\partial x}(t,x_*(t),\dot{x}_*(t));$$

2. *x_* satisfies the Legendre condition, i.e.,*

$$\frac{\partial^2 L}{\partial \dot{x}^2}(t,x_*(t),\dot{x}_*(t)) \geq 0 \qquad \text{for all} \quad t \in [a,b];$$

3. *if x_* satisfies the strengthened Legendre condition, i.e.,*

$$\frac{\partial^2 L}{\partial \dot{x}^2}(t, x_*(t), \dot{x}_*(t)) > 0 \qquad \text{for all} \quad t \in [a,b],$$

then the open interval (a,b) contains no times conjugate to a.

Definition 1.4.4 (Jacobi condition). An extremal x_* that satisfies the strengthened Legendre condition over the interval $[a,b]$ satisfies the Jacobi condition if the open interval (a,b) contains no time conjugate to a; if the half-open interval $(a,b]$ contains no time conjugate to a, the strengthened Jacobi condition is satisfied.

Theorem 1.4.4 (Sufficient conditions for a weak local minimum). *An extremal $x_* : [a,b] \to \mathbb{R}$ that satisfies the strengthened Legendre and Jacobi conditions is a weak local minimum for problem [CV].*

Proof. Let $\mu = \min_{[a,b]} R(t) > 0$ and for $\alpha \in [0,\mu)$ consider the quadratic form

$$\mathcal{Q}_\alpha(h) = \int_a^b R(t)\dot{h}(t)^2 + Q(t)h(t)^2 dt - \alpha \int_a^b h(t)^2 dt$$

with corresponding Jacobi equation

$$\frac{d}{dt}[(R(t) - \alpha)\dot{y}(t)] = Q(t)y(t), \qquad y(a) = 0, \qquad \dot{y}(a) = 1.$$

The solutions $y_\alpha(t)$ to this initial value problem depend continuously on the parameter α, and by assumption no conjugate time c to a exists in the interval $(a,b]$ for $\alpha = 0$. Hence it follows that the solution $y_\alpha(t)$ does not vanish in the interval $(a,b]$ for sufficiently small $\alpha > 0$, and thus, by Theorem 1.4.2, the quadratic form $\mathcal{Q}_\alpha(h)$ is positive definite on $C_0^1([a,b])$, i.e., for all $h \in C_0^1([a,b])$, $h \neq 0$, we have that

$$\int_a^b R(t)\dot{h}(t)^2 + Q(t)h(t)^2 dt > \alpha \int_a^b h(t)^2 dt.$$

This relation allows us to conclude that x_* is a weak local minimum: Let $h \in C_0^1([a,b])$ with small norm $\|h\|_D = \|h\|_\infty + \|\dot{h}\|_\infty$. Using Taylor's theorem, the value $I[x_* + h]$ of the objective can be expressed in the form

$$I[x_* + h] = I[x_*] + \delta I[x_*](h) + \frac{1}{2}\delta^2 I[x_*](h) + r(x_*; h)$$

with $r(x_*; h)$ denoting the remainder. Since we are assuming that L is three times continuously differentiable, the remainder can be expressed as a sum of bounded terms multiplying a cubic expression in h and $\dot{h}$. This implies that $r(x_*; h)$ can be written in the form

$$r(x_*;h) = \int\limits_a^b \left(\xi(t)h(t)^2 + \rho(t)\dot{h}(t)^2 \right) dt$$

and the terms ξ and ρ are of order $O(\|h\|_D)$, i.e., can be bounded by $C\|h\|_D$ for some positive constant C. It follows from Hölder's inequality (see Proposition D.3.1) that

$$h^2(t) = \left(\int\limits_a^t h(s)ds \right)^2 \le (t-a) \int\limits_a^t \dot{h}(s)^2 ds \le (t-a) \int\limits_a^b \dot{h}^2(s)ds$$

and thus

$$\int\limits_a^b h^2(t)dt \le \frac{1}{2}(b-a)^2 \int\limits_a^b \dot{h}^2(s)ds.$$

Hence

$$|r(x_*,h)| \le C \left(1 + \frac{1}{2}(b-a)^2 \right) \left(\int\limits_a^b \dot{h}(t)^2 dt \right) \|h\|_D.$$

Since x_* is an extremal, we have $\delta I[x_*](h) = 0$, and from the calculation above, the second variation can be bounded below as

$$\delta^2 I[x_*](h) > \alpha \int\limits_a^b \dot{h}(t)^2 dt.$$

Thus, overall, we have that

$$I[x_* + h] > I[x_*] + \frac{1}{2}\alpha \int\limits_a^b \dot{h}(t)^2 dt - C \left(1 + \frac{1}{2}(b-a)^2 \right) \left(\int\limits_a^b \dot{h}(t)^2 dt \right) \|h\|_D$$

$$= I[x_*] + \left(\frac{1}{2}\alpha - C(1 + \frac{1}{2}(b-a)^2) \|h\|_D \right) \left(\int\limits_a^b \dot{h}(t)^2 dt \right)$$

$$> I[x_*]$$

for sufficiently small $\|h\|_D$. Hence x_* is a weak local minimum. $\square$

1.5 The Geometry of Conjugate Points and Envelopes

The Jacobi equation (1.29) provides a simple and efficient means to calculate conjugate points numerically. As an example, consider the problem of minimum surfaces of revolution. In this case,

$$L(x,\dot{x}) = x\sqrt{1+\dot{x}^2},$$

and thus

$$\frac{\partial L}{\partial \dot{x}}(x,\dot{x}) = x\frac{\dot{x}}{\sqrt{1+\dot{x}^2}}, \qquad \frac{\partial^2 L}{\partial x \partial \dot{x}}(x,\dot{x}) = \frac{\dot{x}}{\sqrt{1+\dot{x}^2}}, \qquad \frac{\partial^2 L}{\partial \dot{x}^2}(x,\dot{x}) = \frac{x}{(1+\dot{x}^2)^{\frac{3}{2}}}.$$

Along the extremal $x_*(t) = \beta \cosh\left(\frac{t-\alpha}{\beta}\right)$, $t \geq 0$, we have that

$$R(t) = \frac{\partial^2 L}{\partial \dot{x}^2}(x_*(t),\dot{x}_*(t)) = \frac{x_*(t)}{(1+\dot{x}_*(t)^2)^{\frac{3}{2}}} = \frac{\beta}{\cosh^2\left(\frac{t-\alpha}{\beta}\right)}$$

and

$$Q(t) = \frac{\partial^2 L}{\partial x^2}(x_*(t),\dot{x}_*(t)) - \frac{d}{dt}\left(\frac{\partial^2 L}{\partial x \partial \dot{x}}(x_*(t),\dot{x}_*(t))\right)$$

$$= -\frac{\partial^3 L}{\partial x \partial \dot{x}^2}(x_*(t),\dot{x}_*(t)) \cdot \ddot{x}_*(t) = -\frac{\ddot{x}_*(t)}{(1+\dot{x}_*(t)^2)^{\frac{3}{2}}}$$

$$= -\frac{\frac{1}{\beta}\cosh\left(\frac{t-\alpha}{\beta}\right)}{\cosh^3\left(\frac{t-\alpha}{\beta}\right)} = -\frac{1}{\beta}\frac{1}{\cosh^2\left(\frac{t-\alpha}{\beta}\right)}.$$

Thus, in differentiated form, the Jacobi equation becomes

$$R(t)\ddot{u} + \dot{R}(t)\dot{u}(t) = Q(t)u$$

with

$$\dot{R}(t) = -2\frac{\sinh\left(\frac{t-\alpha}{\beta}\right)}{\cosh^3\left(\frac{t-\alpha}{\beta}\right)}.$$

Multiplying all terms by $\frac{\beta^2}{R(t)} = \beta\cosh^2\left(\frac{t-\alpha}{\beta}\right)$, we obtain the following equivalent formulation of the Jacobi equation:

$$\beta^2\ddot{y} - 2\beta\tanh\left(\frac{t-\alpha}{\beta}\right)\dot{y} + y = 0, \qquad y(0) = 0, \quad \dot{y}(0) = 1.$$

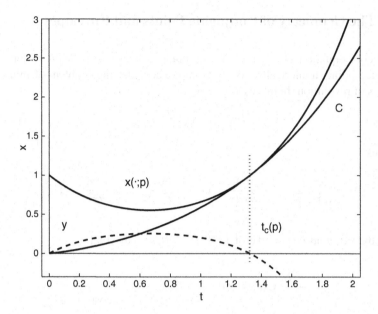

Fig. 1.11 Solution to the Jacobi equation

Since we are interested only in the zeros of nontrivial solutions to the Jacobi equation, we can arbitrarily normalize the initial condition for the derivative, and thus there is no need to multiply by $\frac{1}{R(t)}$. This equation is easily solved numerically.

Figure 1.11 shows the solution of the Jacobi equation for the extremal with values $\alpha = -p\beta$ and $\beta = \frac{x_0}{\cosh p}$ for $x_0 = 1$ and $p = -1.2$ as a dashed curve and the extremal as a solid curve. Note that the zero of the Jacobi equation exactly identifies the point of tangency between the extremal and the envelope of the family of catenaries as the conjugate point. This geometric feature is a general property of certain conjugate points and is related to the fact that *the Jacobi equation is the variational equation of the Euler–Lagrange equation*. We now develop these geometric properties.

We henceforth *assume* that $L \in C^3$ and that the strengthened Legendre condition is satisfied along an extremal x_*. Under these conditions, the extremal x_* can be embedded into a 1-parameter family of extremals: it follows from Hilbert's differentiability theorem that $x_* \in C^2([a,b])$, and the Euler–Lagrange equation can be rewritten in the form

$$L_{\dot{x}t} + L_{\dot{x}x}\dot{x} + L_{\dot{x}\dot{x}}\ddot{x} = L_x.$$

This is a regular second-order ordinary differential equation with continuous coefficients which, in a neighborhood of the reference extremal $t \mapsto (t, x_*(t), \dot{x}_*(t))$, can be written in the form

$$\ddot{x} = F(t, x, \dot{x})$$

with a continuously differentiable function F. For p in some neighborhood $(-\varepsilon, \varepsilon)$ of 0, there exists a solution $x = x(t; p)$ to this equation for the initial conditions

$$x(a;p) = x_*(a) = A \quad \text{and} \quad \dot{x}(a;p) = \dot{x}_*(a) + p.$$

For $p = 0$, this is the reference extremal x_* defined over $[a,b]$. Again, without loss of generality we may assume that x_* is defined on some open interval containing $[a,b]$, and thus by continuous dependence of the solutions of an ordinary differential equation on initial conditions and parameters, for $\varepsilon > 0$ small enough all the solutions $x(\cdot;p)$ will exist on some open interval containing $[a,b]$. Under our assumptions, the solutions are also continuously differentiable functions of the initial conditions. Thus $x(t;p)$ is continuously differentiable in both t and p, and the partial derivative with respect to the parameter p can be computed as the solution to the corresponding variational equation. While this construction of a parameterized family that contains the reference extremal is canonical for the problem [CV], sometimes (like in the case of minimum surfaces of revolution) other parameterizations may be preferred or may be more natural. We thus, more generally, formalize the construction in the definition below.

Definition 1.5.1 (Parameterized family of extremals for problem [CV]). A C^1-parameterized family $\mathscr{E}$ of extremals for problem [CV] consists of a family $x = x(\cdot;p)$ of solutions to the Euler–Lagrange equation defined over some domain $D = \{(t,p) : p \in P, \, a \leq t < t_f(p)\}$ with P an open interval parameterizing the extremals and $t_f : P \to \mathbb{R}$, $p \mapsto t_f(p)$, a function determining the interval of definition of the extremal $x(\cdot;p)$, so that the following properties hold:

1. for all $t \in [a,b]$, $x(t;0) \equiv x_*(t)$;
2. all extremals in the family satisfy the initial conditions

$$x(a;p) = A \quad \text{and} \quad \frac{\partial \dot{x}}{\partial p}(a;p) \neq 0 \quad \text{for all} \quad p \in P; \qquad (1.31)$$

3. the partial derivatives of x with respect to the parameter p exist, are continuous as functions of both variables t and p, and satisfy

$$\frac{d}{dt}\left(\frac{\partial x}{\partial p}(t;p)\right) = \frac{\partial}{\partial p}\left(\frac{\partial x}{\partial t}(t;p)\right) = \frac{\partial \dot{x}}{\partial p}(t;p). \qquad (1.32)$$

Any member $x(\cdot;p)$ of this family is said to be embedded in a family of extremals.

The trivial (and useless) parameterization of the form $x(t,p) \equiv x_*(t)$ for some fixed extremal x_* is excluded by the requirement that $\frac{\partial \dot{x}}{\partial p}(a;p) \neq 0$ for all $p \in P$.

Note that the terminal condition is left unspecified or free in this definition. The function t_f could be in principle an arbitrary function that defines the limits of the parameterization, possibly $t_f(p) \equiv \infty$. It may define the full lengths of the intervals for which the curves are extremals or it may somewhat arbitrarily limit these intervals to make the parameterization injective due to some overlaps that otherwise would occur. Note that if the strengthened Legendre condition is satisfied, then solutions to the Euler–Lagrange equation are unique. But this is a second-order equation, and so uniqueness holds in $(x, \dot{x})$-space. But we are interested in the projections of the solutions into x-space, and here overlaps may occur as in the family of catenaries. Observe that the tangent vectors of the two catenaries that intersect are always different at the point of intersection. For the family of catenaries, we thus may take $t_f(p) \equiv \infty$ to parameterize the full family or simply limit the domains to $[0, t_c(p))$ to have an injective parameterization. But for the moment, injectivity of the parameterization is not required in this definition, and it is allowed that extremals in the family intersect. Summarizing the earlier construction, we have shown the following result:

Corollary 1.5.1. *If x_* is an extremal that satisfies the strengthened Legendre condition over $[a, b]$, then there exist an $\varepsilon > 0$ and a continuously differentiable function t_f that satisfies $t_f(p) > b$ for all $p \in (-\varepsilon, \varepsilon)$, so that x_* can be embedded into a C^1-parameterized family of extremals with domain $D = \{(t, p) : p \in (-\varepsilon, \varepsilon), \, a \leq t < t_f(p)\}$ and $x(t; 0) = x_*(t)$ for all $t \in [a, b]$.* ∎

These parameterizations are closely connected with the solutions to the Jacobi equation.

Theorem 1.5.1. *Let $x = x(t; p)$ be a C^1-parameterized family of extremals. Then for every $p \in P$, the function*

$$y(t; p) = \frac{\partial x}{\partial p}(t; p)$$

is a non-trivial solution to the Jacobi equation on the interval $[a, b]$ that satisfies $y(a; p) = 0$.

Proof. Clearly, since $x(a; p) = A$, we have $y(a; p) = 0$ and also

$$\dot{y}(a; p) = \frac{d}{dt}_{|t=a} \left(\frac{\partial x}{\partial p}(t; p) \right) = \frac{\partial \dot{x}}{\partial p}(a; p) \neq 0$$

for all $p \in P$. It follows that none of the functions $y(\cdot; p)$ vanishes identically, and thus it remains only to verify that y satisfies the Jacobi equation. We differentiate the Euler–Lagrange equation,

$$\frac{d}{dt} \left(\frac{\partial L}{\partial \dot{x}}(t, x(t; p), \dot{x}(t; p)) \right) = \frac{\partial L}{\partial x}(t, x(t; p), \dot{x}(t; p)),$$

with respect to p and interchange the derivatives with respect to p and t. This gives (and for simplicity, we drop the arguments $(t; p)$ in x and $\dot{x}$)

$$\frac{d}{dt} \left(\frac{\partial^2 L}{\partial \dot{x} \partial x}(t, x, \dot{x}) \frac{\partial x}{\partial p} + \frac{\partial^2 L}{\partial \dot{x}^2}(t, x, \dot{x}) \frac{\partial \dot{x}}{\partial p} \right) = \frac{\partial^2 L}{\partial x^2}(t, x, \dot{x}) \frac{\partial x}{\partial p} + \frac{\partial^2 L}{\partial x \partial \dot{x}}(t, x, \dot{x}) \frac{\partial \dot{x}}{\partial p}.$$

Computing the time derivative on the left and writing y and $\dot{y}$ for the partial derivatives in p yields

$$\frac{d}{dt}\left(\frac{\partial^2 L}{\partial \dot{x}\partial x}(t,x,\dot{x})\right)y + \left(\frac{\partial^2 L}{\partial \dot{x}\partial x}(t,x,\dot{x})\right)\dot{y} + \frac{d}{dt}\left(\frac{\partial^2 L}{\partial \dot{x}^2}(t,x,\dot{x})\right)\dot{y} + \left(\frac{\partial^2 L}{\partial \dot{x}^2}(t,x,\dot{x})\right)\ddot{y}$$

$$= \frac{\partial^2 L}{\partial x^2}(t,x,\dot{x})y + \frac{\partial^2 L}{\partial x\partial \dot{x}}(t,x,\dot{x})\dot{y},$$

and upon simplification, thus

$$\left(\frac{\partial^2 L}{\partial \dot{x}^2}(t,x,\dot{x})\right)\ddot{y} + \frac{d}{dt}\left(\frac{\partial^2 L}{\partial \dot{x}^2}(t,x,\dot{x})\right)\dot{y} = \left(\frac{\partial^2 L}{\partial x^2}(t,x,\dot{x}) - \frac{d}{dt}\left(\frac{\partial^2 L}{\partial \dot{x}\partial x}(t,x,\dot{x})\right)\right)y,$$

i.e.,

$$\frac{d}{dt}\left(R(t)\dot{y}(t)\right) = R(t)\ddot{y}(t) + \dot{R}(t)\dot{y}(t) = Q(t)y(t).$$

Hence, y is a solution to the Jacobi equation.   □

Corollary 1.5.2. *Given a C^1-parameterized family of extremals, $x = x(t;p)$, let $y(t;p)$ be the solution of the corresponding Jacobi equation with initial condition $y(a;p) = a$ and $\dot{y}(a;p) = \frac{\partial x}{\partial p}(a;p)$. Then*

$$x(t;p) = x_*(t) + y(t;p)p + \rho(t;p),$$

where the remainder ρ satisfies $\lim_{p \to 0} \frac{\rho(t;p)}{p} = 0$ uniformly over the compact interval $[a,b]$.

We furthermore have the following characterization of conjugate points:

Corollary 1.5.3. *Given a C^1-parameterized family of extremals, $x = x(t;p)$, defined over $D = \{(t,p) : p \in P, a \le t < t_f(p)\}$, a point $(t_c, x(t_c;p_c))$ with $(t_c,p_c) \in D$ is a conjugate point to (a,A) if and only if*

$$\frac{\partial x}{\partial p}(t_c;p_c) = 0. \tag{1.33}$$

For a C^1-parameterized family of extremals, conjugate points are thus characterized by one inequality condition, the transversality condition $\dot{y}(a;p) \ne 0$ that guarantees that the solution of the Jacobi equation is nontrivial, and one equality condition. Conjugate points can then be classified further by means of singularity theory according to the order of contact that the zero of $\frac{\partial x}{\partial p}(t_c;\cdot)$ has with zero at p_c. The least degenerate situation arises if the order of contact is 1, i.e., if the second partial derivative in p does not vanish at the conjugate point. This is the most typical scenario, and we shall now show that these so-called fold points exhibit the same geometry as occurs for the problem of minimum surfaces of revolution.

Definition 1.5.2 (Fold points). Let $\mathscr{E}$ be a C^1-parameterized family of extremals defined on the set $D = \{(t,p) : p \in P, \ a \leq t < t_f(p)\}$. A conjugate point $(t_c, x(t_c; p_c))$ with $(t_c, p_c) \in D$ is called a fold point if

$$\frac{\partial^2 x}{\partial p^2}(t_c; p_c) \neq 0.$$

If a fold occurs at the point (t_c, p_c), then the function $p \mapsto x(t_c; p)$ has a local minimum or maximum at p_c, and for points p close enough to p_c it resembles a parabola, whence the name for this singularity.

The conjugate points for the problem of minimum surfaces of revolution are all fold points: In this case, the parameterization of the smooth extremals given earlier in Eq. (1.16),

$$x(t;p) = \frac{x_0}{\cosh p} \cosh\left(p + \frac{t}{x_0}\cosh p\right), \quad t > 0, \ p \in \mathbb{R},$$

is real analytic, and we have $x(0; p) \equiv x_0$ and

$$\frac{d}{dt}_{|t=0}\left(\frac{\partial x}{\partial p}(t;p)\right) = \frac{\partial \dot{x}}{\partial p}(0;p) = \frac{\partial}{\partial p}(\sinh p) = \cosh p \geq 1.$$

The condition $\frac{\partial^2 x}{\partial p^2}(t_c; p_c) > 0$ was verified for any stationary point in the proof of Theorem 1.3.1, and it follows from this proof that every extremal $t \mapsto x(t; p)$ has a unique conjugate point at which its graph is tangent to the envelope C and the locus of all conjugate points is this envelope C.

The same geometric properties are generally valid near fold points $(t_c, x_c) = (t_c, x(t_c; p_c))$ of a C^1-parameterized family of extremals: since $\frac{\partial^2 x}{\partial p^2}(t_c; p_c) \neq 0$, by the implicit function theorem, there exist an $\varepsilon > 0$ and a continuously differentiable function p_* defined on the interval $(t_c - \varepsilon, t_c + \varepsilon)$ with values in P,

$$p_* : (t_c - \varepsilon, t_c + \varepsilon) \to P, \qquad t \mapsto p_*(t),$$

such that

$$\frac{\partial x}{\partial p}(t, p_*(t)) = 0 \quad \text{for all} \quad t \in (t_c - \varepsilon, t_c + \varepsilon). \tag{1.34}$$

Thus, the curve

$$C : (t_c - \varepsilon, t_c + \varepsilon) \to [a, \infty) \times \mathbb{R}, \qquad t \mapsto (t, x(t, p_*(t))),$$

that is, the graph of the function $t \mapsto x(t, p_*(t))$, is *an envelope to the parameterized family of extremals at the fold point* (t_c, x_c) *that entirely consists of conjugate points* *for the problem* [CV]. The envelope condition (1.34) expresses the geometric fact that the derivative of the function

$$\gamma : (t_c - \varepsilon, t_c + \varepsilon) \to \mathbb{R}, \qquad t \mapsto \gamma(t) = x(t, p_*(t)),$$

at time t is the same as the time derivative of the extremal corresponding to the parameter $p_*(t)$. In other words, the curve C and the graph of the extremal $x(\cdot; p_*(t))$ not only are tangent, but even have the same tangent vector at time t. This allows us to construct a 1-parameter family of *continuously differentiable functions* Γ_t for $t \in (t_c - \varepsilon, t_c]$ by concatenating the extremal $x(\cdot; p_*(t))$ corresponding to the parameter value $p_*(t)$ and restricted to the interval $[a, t]$ with the restriction of the function γ defining the envelope to the interval $[t, t_c]$, i.e.,

$$\Gamma_t(s) = \begin{cases} x(s; p_*(t)) & \text{if } a \leq s \leq t, \\ x(s; p_*(s)) & \text{if } t \leq s \leq t_c. \end{cases}$$

This curve Γ_t is continuously differentiable at time t because of the envelope condition (1.34). Thus, each of these curves is admissible for the problem [CV] with boundary conditions

$$x(a) = A \quad \text{and} \quad x(t_c) = x_c = x(t_c; p_c).$$

The significant observation in this context now is the fact that all these curves have the same value for the objective,

$$I[x] = \int_a^{t_c} L(s, x(s), \dot{x}(s)) ds.$$

Theorem 1.5.2 (Envelope theorem). *For all $t \in (t_c - \varepsilon, t_c]$ we have that*

$$I[\Gamma_t] = \text{const.} \tag{1.35}$$

This result typically allows us to exclude the optimality of the extremal Γ_{t_c}, i.e., of the extremal in the field corresponding to parameter value p_c defined over the interval $[a, t_c]$. For if this curve is optimal, then by the envelope theorem, so are all the other curves Γ_t for $t \in (t_c - \varepsilon, t_c)$. But this requires Γ_t to be an extremal. Thus, if the envelope is not a solution to the Euler–Lagrange equation (as, for example, is the case for the problem of minimum surfaces of revolution), this already provides a contradiction. A sufficient condition for this to be the case in general again is that the extremal $x(\cdot; p_c)$ satisfy the strengthened Legendre condition over the closed interval $[a, t_c]$. For in this case, the Euler–Lagrange equation with terminal conditions

$$x(t_c) = \gamma(t_c) = x(t_c; p_c) \quad \text{and} \quad \dot{x}(t_c) = \dot{x}(t_c; p_c) \tag{1.36}$$

has a unique solution that is given by the extremal $x(\cdot; p_*(t))$. But this extremal cannot be equal to the curve $t \mapsto (t, x(t; p_*(t)))$, since otherwise,

$$\frac{\partial x}{\partial p}(t; p_c) = \frac{\partial x}{\partial p}(t; p_*(t)) = 0$$

implies that $\frac{\partial x}{\partial p}(\cdot; p_c)$ vanishes identically (as solution to the Jacobi equation), and this contradicts the condition $\frac{\partial x}{\partial p}(a; p_c) \neq 0$ in the definition of a parameterized family of extremals. Hence, $x(\cdot; p_c)$ and Γ_t are different solutions to the Euler–Lagrange equation with terminal data (1.36). Contradiction. We thus have the following:

Corollary 1.5.4. *Let x_* be an extremal that satisfies the strengthened Legendre condition over the interval $[a, b]$ and suppose x_* is embedded into a C^1-parameterized family of extremals. If the conjugate point $(t_c, x_c) = (t_c, x(t_c; p_c))$ is a **fold** point of the parameterization, then the restriction of x_* to the interval $[a, t_c]$ is not a weak local minimum for the problem $[CV]$ with boundary data $x(a) = A$ and $x(t_c) = x_c$.* ∎

It remains to prove the envelope theorem: We verify that $I[\Gamma_t]$ is differentiable with derivative 0,

$$I[\Gamma_t] = \int_a^t L(s, x(s; p_*(t)), \dot{x}(s; p_*(t))) ds + \int_t^{t_c} L(s, x(s; p_*(s)), \dot{x}(s; p_*(s))) ds.$$

Recall that the curve $s \mapsto x(s; p_*(s))$ describes the envelope and that its derivative is simply given by $\dot{x}(s; p_*(s))$, since $\frac{\partial x}{\partial p}(s; p_*(s)) \equiv 0$. Thus, by the regularity assumptions made, this function is continuously differentiable in t and we have that

$$\dot{I}[\Gamma_t] = \frac{dI}{dt}[\Gamma_t] = L(t, x(t; p_*(t)), \dot{x}(t; p_*(t)))$$

$$+ \left[\int_a^t \left(\frac{\partial L}{\partial x}(s, x(s; p_*(t)), \dot{x}(s; p_*(t))) \right) \frac{\partial x}{\partial p}(s; p_*(t)) ds \right] \dot{p}_*(t)$$

$$+ \left[\int_a^t \left(\frac{\partial L}{\partial \dot{x}}(s, x(s; p_*(t)), \dot{x}(s; p_*(t))) \right) \frac{\partial \dot{x}}{\partial p}(s; p_*(t)) ds \right] \dot{p}_*(t)$$

$$- L(t, x(t; p_*(t)), \dot{x}(t; p_*(t))).$$

Integrate the term $\frac{\partial L}{\partial \dot{x}} \frac{\partial \dot{x}}{\partial p}$ by parts to get

$$\int\limits_a^t \frac{\partial L}{\partial \dot{x}}(s,x(s;p_*(t)),\dot{x}(s;p_*(t)))\frac{\partial \dot{x}}{\partial p}(s;p_*(t))ds$$

$$= \left(\frac{\partial L}{\partial \dot{x}}(s,x(s;p_*(t)),\dot{x}(s;p_*(t)))\frac{\partial x}{\partial p}(s;p_*(t))\right)\Bigg|_{s=a}^{s=t}$$

$$- \int\limits_a^t \frac{d}{dt}\left(\frac{\partial L}{\partial \dot{x}}(s,x(s;p_*(t)),\dot{x}(s;p_*(t)))\right)\frac{\partial x}{\partial p}(s;p_*(t))ds.$$

For any C^1-parameterized family of extremals and any parameter p, we have that

$$\frac{\partial x}{\partial p}(a;p) = 0$$

and since the point $(t,x(t;p_*(t))$ is a conjugate point to (a,A)—this is precisely the meaning of the envelope condition (1.34)—we also have that

$$\frac{\partial x}{\partial p}(t;p_*(t)) = 0.$$

Hence

$$\dot{I}[\Gamma_t] = \frac{dI}{dt}[\Gamma_t]$$

$$= \left(\int\limits_a^t \left[\left(\frac{\partial L}{\partial x} - \frac{d}{dt}\frac{\partial L}{\partial \dot{x}}\right)(s,x(s;p_*(t)),\dot{x}(s;p_*(t)))\right]\frac{\partial x}{\partial p}(s;p_*(t))ds\right)\dot{p}_*(t),$$

and this term vanishes, since $x(\cdot;p_*(t))$ is an extremal.   □

Thus, in a C^1-parameterized family of extremals, local optimality ceases at conjugate points $(t_c,x_c) = (t_c,x(t_c;p_c))$ that are fold points. If the conjugate point (t_c,x_c) is not a fold point, then it is possible that the restriction of an extremal x_* to the interval $[a,t_c]$ still provides a weak local minimum. For example, this can happen for the next degenerate type of conjugate points, so-called *simple cusp* points. These are conjugate points such that

$$\frac{\partial^2 x}{\partial p^2}(t_c;p_c) = 0, \qquad \text{but} \qquad \frac{\partial^3 x}{\partial p^3}(t_c;p_c) \neq 0.$$

We will describe the geometric properties of optimal solutions near this singularity more generally in Sect. 5.5, for the optimal control problem. In contrast to the geometry of fold points, this structure seems to have attracted little attention in the calculus of variations, but it is equally important, if not more so, in optimal syntheses for control problems.

We close this section with another classical geometric characterization of general conjugate points, which, at times, has been taken as a definition of conjugate points.

Proposition 1.5.1. *Let x_* be an extremal along which the strengthened Legendre condition is satisfied. Then $(c, (x_*(c))$ is a conjugate point to $(a, x_*(a))$ if and only if $(c, (x_*(c))$ is the limit point of points of intersection between the graph of x_* and graphs of neighboring extremals drawn from the same initial point $(a, x_*(a))$.*

Proof. For p in some neighborhood $(-\varepsilon, \varepsilon)$ of 0, let $x = x(t; p)$ be the embedding of the extremal x_* constructed in Proposition 1.5.1 and denote the difference with the reference extremal x_* by

$$\Delta(t; p) = x(t; p) - x(t; 0) = x(t; p) - x_*(t).$$

For $p = 0$, we have $\Delta(t; 0) \equiv 0$, and thus it is possible to factor p from this equation and rewrite Δ in the form $\Delta(t; p) = p\Omega(t; p)$. Let $y(t)$ be the solution to the Jacobi equation and note that

$$y(t) = \frac{\partial x}{\partial p}(t; 0) = \frac{\partial \Delta}{\partial p}(t; 0) = \Omega(t; 0).$$

A time c is conjugate to a if and only if $y(c) = 0$. Since y does not vanish identically, we have $\dot{y}(c) \neq 0$ and thus

$$\frac{\partial \Omega}{\partial t}(c; 0) = \dot{y}(c) \neq 0.$$

Hence, by the implicit function theorem, there exists a unique solution $t = t_c(p)$ of the equation $\Omega(t; p) = 0$ near the point $(c, 0)$. Consequently, for all sufficiently small $|p|$, $p \neq 0$, there exist zeros of $\Delta(\cdot; p)$. That is, there exist points of intersection of the graphs of x_* and $x(\cdot; p)$ that converge to $(c, (x(c))$ as $p \to 0$.

Conversely, if such a sequence $\{(t_n, p_n)\}_{n \in \mathbb{N}}$ with $(t_n, p_n) \to (c, 0)$ exists, then

$$y(c) = \Omega(c; 0) = \lim_{n \to \infty} \Omega(t_n; p_n) = 0,$$

and thus c is a conjugate time to a for x_*. $\square$

This geometric characterization is sometimes useful in recognizing conjugate points that are not easily characterized otherwise (see Fig. 1.12). The best example for this is given by the cycloids of the brachistochrone problem. In the field Ξ defined in Eq. (1.12),

$$\Xi : (-\pi, \pi) \times (0, \infty) \to \mathbb{R}^2_+ = \{(\alpha, \beta) : \alpha > 0, \beta > 0\},$$
$$(\tau, \xi) \mapsto (t(\tau), x(\tau)) = (\xi(\pi + \tau + \sin \tau), \xi(1 + \cos \tau)),$$

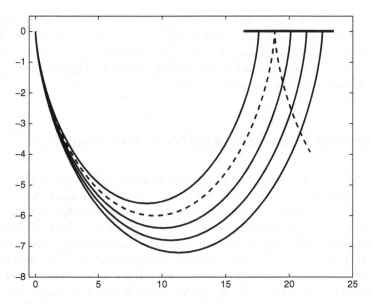

Fig. 1.12 "Conjugate" point for the family of cycloids

we somewhat justifiably, but nevertheless mathematically arbitrarily, restricted the parameterizations to traverse exactly one cycle. Obviously, otherwise the cycloids start to overlap, and we do not want that. Nevertheless, these curves are extremals, and the natural range for the parameterization would be $\tau \geq -\pi$. Also, the full range of terminal points for the problem is not $\mathbb{R}^2_+$, but $\{(b,B) : b \geq 0, B \geq 0\}$. Clearly, if the endpoint (b,B) lies on the x-axis, $b = 0$, the optimal solution is simply the straight line corresponding to free fall. If the endpoint lies on the t-axis, $B = 0$, naturally we expect the catenary that reaches this point after exactly one cycle to be the optimal one. Being mathematically rigorous, however, once we allow curves to lie in $\{(\alpha,\beta) : \alpha \geq 0, \beta \geq 0\}$, we see that for many points there exist multiple extremals that pass through (b,B) consisting of cycloids that hit the line $x = 0$ orthogonally and reflect. However, it is clear that these extremals intersect with nearby extremals for parameter values $\tau > \pi$. It follows from the argument in the proof of Proposition 1.5.1 that the times $\tau = \pi$ are all conjugate times to the initial time $-\pi$ chosen for the parameterization. (Note that this direction of the argument is generally valid.) Reasoning by means of analogy, we would expect that the cycloids are no longer optimal for values $\tau > \pi$, and this is indeed correct. However, this does not follow from the theory developed so far. For if we insist on a formulation of the curves as graphs of differentiable functions, the cycloids have singularities at both their initial and terminal points, and thus the theory developed above is not applicable. If we use parameterizations, there are no problems at the endpoints $\pm\pi$ of the parameter interval, but we need to consider a two-dimensional version of Jacobi's theorem, and this simply would lead us too far astray here given the introductory nature of this chapter. Thus, we simply present the very convincing picture and leave these statements unproven. But note that the cycloids are still optimal for $\tau = \pi$, in contrast to the situation for the minimum surfaces of revolution. We shall encounter the very same geometric properties in Sect. 6.1.3, when we consider *transversal folds*, the generalization of conjugate points to bang-bang extremals in the optimal control problem.

1.6   Fields of Extremals and the Weierstrass Condition

We once again return to the problem of minimum surfaces of revolution. By now, we know that for every extremal $x(\cdot;p)$, there exists a unique conjugate time $t_c = t_c(p)$ and that the extremal $x(\cdot;p)$ is no longer optimal on intervals $[a,b]$ if $b \geq t_c(p)$. For $b = t_c(p)$ this was just verified in Corollary 1.5.4, and for $b > t_c(p)$ this is clear by the Jacobi necessary condition. This leaves us with the central field $\mathscr{F}$ defined by Ξ in Eq. (1.24) with the domain restricted to the intervals $(0,t_c(p))$. This field covers the region R above the envelope C diffeomorphically. It is tempting to conjecture that all extremals in this field are optimal, but as will be seen in Sect. 1.7, this is *not* true. By the way, existence theory is not of any help here, since this problem does not have a solution for all terminal points in the class of positive (or more generally, nonnegative) differentiable functions. Clearly, also for points (t_1,x_1) in the region S below the curve C of conjugate points, differentiable curves exist that connect $(0,x_0)$ with (t_1,x_1), but there is no optimal one. For if there were one, it would need to be an extremal. But no catenaries pass through $(0,x_0)$ and points $(t_1,x_1) \in S$; and catenaries are the only extremals. It is correct, and this will be shown below, that the catenaries in the field $\mathscr{F}$ are better than any other differentiable curve whose graph lies in the region R; that is, the catenaries provide a strong local minimum. But this minimum is not global.

Having a C^1-parameterized family of extremals that diffeomorphically covers some region R in itself is not sufficient to have local minima, but an additional necessary condition for optimality needs to be satisfied, the so-called Weierstrass condition. The reason lies with the fact that so far, we have developed necessary conditions for optimality only for a weak local minimum, but optimality over a region R of endpoints is a strong minimum property, and the Weierstrass condition is necessary for strong minima, but not for weak ones. We now develop these results and again consider the scalar fixed endpoint problem [CV] as defined earlier, that is, the problem to minimize the functional

$$I[x] = \int_a^b L(t,x(t),\dot{x}(t))\,dt$$

over all continuously differentiable curves $x \in C^1([a,b])$ that satisfy the boundary conditions $x(a) = A$ and $x(b) = B$. However, now we not only aim at solving this problem for one specific point (b,B), but want to solve the problem for arbitrary terminal points. We keep the notation of (t,x) for the curves and change the notation for the terminal point to the more convenient (s,z), i.e., $x(s) = z$.

Definition 1.6.1 (Central field). A central field of extremals, $\mathscr{F}$, is a parameterized family of extremals for the problem [CV] with domain $D = \{(t,p) : p \in P,\ a \le t < t_f(p)\}$ for which the associated flow map restricted to the interior of D, $\mathring{D}$,

$$F : \mathring{D} = \{(t,p) : p \in P,\ a < t < t_f(p)\} \to R,\ (t,p) \mapsto F(t,p) = (t,x(t,p)),\ (1.37)$$

is a diffeomorphism onto the simply connected[1] region $R = F(\mathring{D})$. Recall, by the definition of a C^1-parameterized family of extremals, that we have $x(a;p) \equiv A$ for all $p \in P$.

Thus, we now assume that except for the initial point (a,A), for which all extremals coincide, there are no further overlaps in the family of extremals and away from (a,A), the graphs of the extremals in the field cover the region R bijectively. Thus, if (s,z) is a point in R, then there exists a unique extremal $t \mapsto x(t;p)$ of the field $\mathscr{F}$ that passes through the point z at time s, $z = x(s;p)$. Collectively, the tangent vectors $\dot{x}(s;p)$ at the points $z = x(s;p)$ define a *field of directions* on R,

$$\dot{z}(s) = \Psi(s,z), \tag{1.38}$$

which is referred to as a *central field*. Note that both the field of cycloids for the brachistochrone problem and the field of restricted catenaries defined in Eq. (1.24) are central fields in this sense. For the catenaries, this is immediate from the parameterization (1.16) given in Sect. 1.3; for the field of cycloids defined in Eq. (1.12) this follows upon a suitable reparameterization that inverts the map $\tau \mapsto t(\tau)$. In general, we have the following fundamental local embedding result:

Proposition 1.6.1. *Let x_* be an extremal that satisfies the strengthened Legendre condition over $[a,b]$ for which there does not exist a conjugate time $t_c \in (a,b]$. Then x_* can be embedded into a central field of extremals, $\mathscr{F}$, with domain $D = \{(t,p) : p \in (-\varepsilon,\varepsilon),\ a \le t < t_f(p)\}$, where $\varepsilon > 0$ and t_f is a continuously differentiable function that satisfies $t_f(p) > b$ for all $p \in (-\varepsilon,\varepsilon)$. By construction, $x(t;0) = x_*(t)$ for all $t \in [a,b]$.*

Proof. It was already shown in the proof of Proposition 1.5.1 that for p in some neighborhood $(-\varepsilon,\varepsilon)$ of 0, the solutions $x = x(t;p)$ to the Euler–Lagrange equation with initial conditions

$$x(a;p) = x_*(a) = A \quad \text{and} \quad \dot{x}(a;p) = \dot{x}_*(a) + p$$

---

[1] A precise definition of this term lies beyond the scope of this text, but the common calculus characterization as a connected set "without holes" suffices for our purposes.

define a C^1-parameterized family of extremals for the problem [CV] with domain $D = \{(t,p) : p \in (-\varepsilon,\varepsilon),\ a \le t < t_f(p)\}$ that satisfies $t_f(p) > b$ for all $p \in (-\varepsilon,\varepsilon)$. If there does not exist a conjugate time along the reference trajectory, then it also follows from Proposition 1.5.1 that for sufficiently small $\varepsilon > 0$ the graphs of these extremals do not intersect, or in other words, the flow map

$$F : \mathring{D} \to R,\ (t,p) \mapsto F(t,p) = (t,x(t,p))$$

is a diffeomorphism. This is a direct consequence of the relation

$$x(t;p) = x_*(t) + py(t) + o(t;p),$$

where $y(\cdot)$ is the solution of the Jacobi equation along the extremal x_*.   □

Definition 1.6.2 (Value function). Let $\mathscr{F}$ be a central field of extremals for the problem [CV] that covers a simply connected region R of $(a,\infty) \times \mathbb{R}$. The associated value function $V^{\mathscr{F}}$,

$$V^{\mathscr{F}} : R \to \mathbb{R},\quad (s,z) \mapsto V^{\mathscr{F}}(s,z),$$

is defined as

$$V^{\mathscr{F}}(s,z) = V^{\mathscr{F}}(s,x(s;p)) = \int_a^s L(t,x(t;p),\dot{x}(t;p))\,dt,$$

where $x(\cdot;p)$ is the unique extremal in the field that passes through z at time s, $z = x(s;p)$.

If we define the parameterized cost along the extremals of the field as

$$C(s;p) = \int_a^s L(t,x(t;p),\dot{x}(t;p))\,dt,$$

then it is clear that $V \circ F = C$. The function C is continuously differentiable ($\mathscr{F}$ is a C^1-parameterized family of extremals), and since F is a diffeomorphism, it follows that V is continuously differentiable on R.

Note that the mapping

$$\mathfrak{gr}\, V^{\mathscr{F}} : D \to [a,\infty) \times \mathbb{R} \times \mathbb{R},\quad (t,p) \mapsto (t,x(t,p),C(t,p)),\qquad (1.39)$$

gives a parameterization of the *graph of the value function* associated with the family, and more generally, even for a C^1-parameterized family of extremals with overlaps, this construction provides a simple and efficient way of generating the graph of this possibly multivalued function. Figure 1.13 shows a portion of this multivalued graph for the family of catenaries based on Eqs. (1.46) and (1.47) given

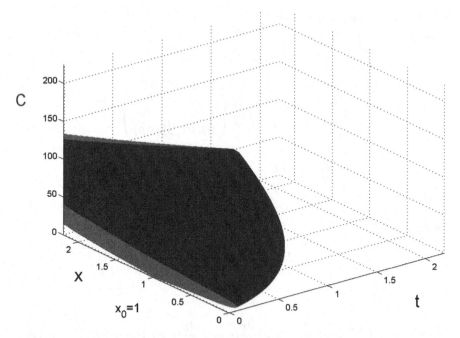

Fig. 1.13 The multivalued graph $gr\,V_{\mathscr{F}}$ of the value function $V_{\mathscr{F}}$ for the family of catenaries

later in Sect. 1.7. We shall also use this correspondence extensively in Sects. 5.4 and 5.5 when we study singularities for the value function of an optimal control problem.

Theorem 1.6.1. *The value function* $V^{\mathscr{F}}$ *is continuously differentiable on R and the differential of* $V^{\mathscr{F}}$ *is given by*

$$dV^{\mathscr{F}} = \left(L(s,z,\Psi(s,z)) - \frac{\partial L}{\partial \dot{x}}(s,z,\Psi(s,z))\Psi(s,z) \right) ds + \frac{\partial L}{\partial \dot{x}}(s,z,\Psi(s,z))dz.$$

$$(1.40)$$

Proof. To simplify the notation, we write V for $V^{\mathscr{F}}$. It follows from $V(s,x(s;p)) = C(s,p)$ that

$$\frac{\partial V}{\partial s}(s,x(s;p)) + \frac{\partial V}{\partial z}(s,x(s;p))\dot{x}(s;p) = \frac{\partial C}{\partial s}(s,p)$$

and

$$\frac{\partial V}{\partial z}(s,x(s;p))\frac{\partial x}{\partial p}(s;p) = \frac{\partial C}{\partial p}(s,p).$$

Solving for the gradient of V gives

$$\left(\frac{\partial V}{\partial s}(s, x(s; p)), \frac{\partial V}{\partial z}(s, x(s; p)) \right)$$

$$= \left(\frac{\partial C}{\partial s}(s, p), \frac{\partial C}{\partial p}(s, p) \right) \begin{pmatrix} 1 & 0 \\ \dot{x}(s; p) & \frac{\partial x}{\partial p}(s; p) \end{pmatrix}^{-1}$$

$$= \left(\frac{\partial C}{\partial s}(s, p), \frac{\partial C}{\partial p}(s, p) \right) \frac{1}{\frac{\partial x}{\partial p}(s; p)} \begin{pmatrix} \frac{\partial x}{\partial p}(s; p) & 0 \\ -\dot{x}(s; p) & 1 \end{pmatrix}.$$

We have that

$$\frac{\partial C}{\partial s}(s, p) = L(s, x(s; p), \dot{x}(s; p))$$

and

$$\frac{\partial C}{\partial p}(s, p) = \int_a^s \left(\frac{\partial L}{\partial x}(t, x(t; p), \dot{x}(t; p)) \frac{\partial x}{\partial p}(t; p) + \frac{\partial L}{\partial \dot{x}}(t, x(t; p), \dot{x}(t; p)) \frac{\partial \dot{x}}{\partial p}(t; p) \right) dt.$$

Integrating by parts, exactly as in the proof of the envelope theorem, we get for this term that

$$\frac{\partial C}{\partial p}(s, p) = \int_a^s \left[\frac{\partial L}{\partial x}(t, x(t; p), \dot{x}(t; p)) - \frac{d}{dt} \left(\frac{\partial L}{\partial \dot{x}}(t, x(t; p), \dot{x}(t; p)) \right) \right] \frac{\partial x}{\partial p}(t; p) dt$$

$$+ \frac{\partial L}{\partial \dot{x}}(s, x(s; p), \dot{x}(s; p)) \frac{\partial x}{\partial p}(s; p)$$

$$= \frac{\partial L}{\partial \dot{x}}(s, x(s; p), \dot{x}(s; p)) \frac{\partial x}{\partial p}(s; p),$$

since $x(\cdot; p)$ is an extremal. Thus, overall, we have that

$$\left(\frac{\partial V}{\partial s}(s, x(s; p)), \frac{\partial V}{\partial z}(s, x(s; p)) \right)$$

$$= \left(L(s, x(s; p), \dot{x}(s; p)), \frac{\partial L}{\partial \dot{x}}(s, x(s; p), \dot{x}(s; p)) \frac{\partial x}{\partial p}(s; p) \right) \frac{1}{\frac{\partial x}{\partial p}(s; p)} \begin{pmatrix} \frac{\partial x}{\partial p}(s; p) & 0 \\ -\dot{x}(s; p) & 1 \end{pmatrix}$$

$$= \left(L(s, x(s; p), \dot{x}(s; p)) - \frac{\partial L}{\partial \dot{x}}(s, x(s; p), \dot{x}(s; p)) \dot{x}(s; p), \frac{\partial L}{\partial \dot{x}}(s, x(s; p), \dot{x}(s; p)) \right).$$

If we now replace $x(s;p)$ by z and $\dot{x}(s;p)$ by the direction $\Psi(s,z)$ of the field, see Eq. (1.38), then we get that

$$\left(\frac{\partial V}{\partial s}(s,z), \frac{\partial V}{\partial z}(s,z) \right)$$

$$= \left(L(s,z,\Psi(s,z)) - \frac{\partial L}{\partial \dot{x}}(s,z,\Psi(s,z))\Psi(s,z), \frac{\partial L}{\partial \dot{x}}(s,z,\Psi(s,z)) \right)$$

and the result follows.   □

Deviating slightly from historical custom, for $(t,x) \in R$ and $\lambda \in \mathbb{R}$ we define the *Hamiltonian* function H as

$$H : R \times \mathbb{R} \to \mathbb{R}, \quad (t,x,\lambda) \mapsto H(t,x,\lambda) = L(t,x,\Psi(t,x)) + \lambda \Psi(t,x),$$

and also define a function

$$\lambda : R \to \mathbb{R}, \quad (t,x) \mapsto \lambda(t,x) = -\frac{\partial L}{\partial \dot{x}}(t,x,\Psi(t,x)). \tag{1.41}$$

The Hamiltonian defined in this way is simply the negative of the classical form in physics and the calculus of variations. It will become clear in Sect. 1.10 why we prefer this formulation. Using this notation, we then can employ the differential $dV^{\mathscr{F}}$ to compute $V^{\mathscr{F}}(s,z)$ as a line integral along any rectifiable curve Γ that lies in R (except for its initial point which needs to be given by (a,A)) and has terminal point (s,z) as

$$V^{\mathscr{F}}(s,z) = \int_{\Gamma} dV^{\mathscr{F}} = \int_{\Gamma} H(t,x,\lambda(t,x))dt - \lambda(t,x)dx$$

$$= \int_{\Gamma} L(t,x,\Psi(t,x))dt + \lambda(t,x)\left[\Psi(t,x)dt - dx\right]. \tag{1.42}$$

This integral is called the *Hilbert invariant integral* and is the key tool in proving strong local optimality of the extremals in the field relative to any other function whose graph lies in R. In fact, these functions need not be continuously differentiable, but can have corners. We say that a continuous function $\tilde{x} : [a,s] \to \mathbb{R}$, $t \mapsto \tilde{x}(t)$, is piecewise continuously differentiable if there exists a finite partition $a = t_0 < t_1 < \cdots < t_r < t_{r+1} = s$ of the interval $[a,s]$ such that the restrictions of $\tilde{x}$ to the intervals $[t_i,t_{i+1}]$ lie in $C^1([t_i,t_{i+1}])$, and we denote the space of all these functions by $C_{pc}^1([a,s])$. We call a function $\tilde{x} \in C_{pc}^1([a,s])$ admissible for the region R if $\tilde{x}(a) = A$ and $(t,\tilde{x}(t)) \in R$ for all $t \in (a,s]$. Clearly, we can perform the integration in (1.42) piecewise, and thus we can take as the curve Γ the graph of any admissible function $\tilde{x} \in C_{pc}^1([a,s])$.

Let $\tilde{x} \in C_{pc}^1([a,s])$ be any admissible function for the region R (in particular $\tilde{x}(s) = z$) and denote its graph by $\tilde{\Gamma}$. We then have that

Fig. 1.14 The Weierstrass
excess function

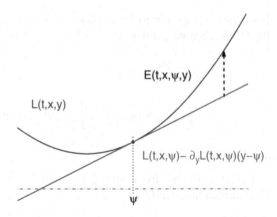

$$I[\tilde{x}] = \int_{\tilde{\Gamma}} L(t,\tilde{x},\dot{\tilde{x}})dt = \int_{a}^{s} L(t,\tilde{x}(t),\dot{\tilde{x}}(t))\,dt$$

and

$$V^{\mathscr{F}}(s,z) = \int_{\tilde{\Gamma}} L(t,\tilde{x},\Psi(t,\tilde{x}))dt + \lambda(t,\tilde{x})\left[\Psi(t,\tilde{x})dt - d\tilde{x}\right].$$

By construction, $V^{\mathscr{F}}(s,z)$ is the cost of the extremal x_* in the field that passes through z at time s, and thus the difference $I[\tilde{x}] - I[x_*]$ can be expressed as

$$I[\tilde{x}] - I[x_*] = \int_{\tilde{\Gamma}} \left[L(t,\tilde{x},\dot{\tilde{x}}) - L(t,\tilde{x},\Psi(t,\tilde{x}))\right]dt - \frac{\partial L}{\partial \dot{x}}(t,\tilde{x},\Psi(t,\tilde{x}))\left[d\tilde{x} - \Psi(t,\tilde{x})dt\right].$$

$$(1.43)$$

Definition 1.6.3 (Weierstrass excess function). The Weierstrass excess function

$$E : (a,\infty) \times \mathbb{R} \times \mathbb{R} \times \mathbb{R} \to \mathbb{R}, \quad (t,x,\psi,y) \mapsto E(t,x,\psi,y)$$

is defined as

$$E(t,x,\psi,y) = L(t,x,y) - L(t,x,\psi) - \frac{\partial L}{\partial \dot{x}}(t,x,\psi)(y - \psi).$$

Geometrically, if we regard $L(t,x,y)$ as a function of y for (t,x) fixed, then the excess function E is simply the vertical distance between the point $L(t,x,y)$ on the graph and the value at y of the tangent line through the point $L(t,x,\psi)$, i.e., $L(t,x,\psi) + \frac{\partial L}{\partial \dot{x}}(t,x,\psi)(y - \psi)$ (see Fig. 1.14).

Definition 1.6.4 (Weierstrass condition). Let $\mathscr{F}$ be a central field of extremals for the problem [CV] that covers a simply connected region R of $(a,\infty) \times \mathbb{R}$ with field of directions $\dot{x} = \Psi(t,x)$. The Weierstrass condition is satisfied on the region R if for all $(t,x) \in R$ and all $y \in \mathbb{R}$ we have that

$$E(t,x,\Psi(t,x),y) \geq 0.$$

Thus, the Weierstrass condition is a convexity statement that requires that for every $(t,x) \in R$, the graph of the function $y \mapsto L(t,x,y)$ lie above the tangent line to this graph through the point $\Psi(t,x)$. This will automatically be satisfied if the function $L(t,x,\dot{x})$ is convex in $\dot{x}$. For example, this applies to integrands of the form $L(x,\dot{x}) = f(x)\sqrt{1+\dot{x}^2}$ for some positive function f. Hence the Weierstrass condition is satisfied for both the brachistochrone problem and the problem of minimum surfaces of revolution.

Theorem 1.6.2 (Sufficient condition for a strong minimum). *Let $\Gamma \in C^1([a,b])$ be an extremal for the problem [CV] and suppose there exists a central field of extremals, $\mathscr{F}$, that covers a simply connected region R of $(a,\infty) \times \mathbb{R}$ with field of directions given by $\dot{x} = \Psi(t,x)$ such that Γ is a member of the field $\mathscr{F}$. If the Weierstrass condition is satisfied on R, then Γ is minimal when compared with any other function from $C^1_{pc}([a,b])$ that satisfies the same boundary conditions as Γ and whose graph (with the exception of the initial condition) lies in the region R. In particular, Γ is a strong local minimum.*

Proof. Let $\tilde{x} \in C^1_{pc}([a,s])$ be any admissible function for problem [CV] that satisfies the same boundary conditions as Γ and whose graph $\tilde{\Gamma}$ lies in R except for the initial condition. By Eq. (1.43), the difference in the objectives between the curve $\tilde{\Gamma}$ and the extremal Γ can be expressed as

$$I[\tilde{x}] - I[x_*] = \int\limits_a^s E\left(t, \tilde{x}(t), \Psi(t,\tilde{x}(t)), \dot{\tilde{x}}(t)\right)\, dt \geq 0, \qquad (1.44)$$

and thus the result follows.   □

We close this section by showing that the Weierstrass condition also is a necessary condition for strong optimality for extremals.

Theorem 1.6.3 (Weierstrass condition as necessary condition for a strong minimum). *Suppose $x_* \in C^1_{pc}([a,b])$ is a broken extremal that is a strong local minimum for problem [CV]. Then for all $y \in \mathbb{R}$ and all $t \in [a,b]$, we have that*

$$E(t, x_*(t), \dot{x}_*(t), y) \geq 0. \qquad (1.45)$$

Proof. It suffices to prove the inequality for all times t when x_* is continuously differentiable. By taking appropriate limits from the left and right, Eq. (1.45) then holds everywhere in $[a,b]$.

Let $t_0 \in (a,b)$ be a time such that x_* is continuously differentiable at t_0 and set $x_0 = x_*(t_0)$ and $\dot{x}_0 = \dot{x}_*(t_0)$. Let $y \in \mathbb{R}^n$ and pick $\delta > 0$ such that $t_0 + \delta < b$ and let $\varepsilon \in (0,1)$ (see Fig. 1.15). Define a 2-parameter family of arcs $x(\cdot; \varepsilon, \delta) \in C^1_{pc}([a,b])$ as

$$x(t; \varepsilon, \delta) = \begin{cases} x_*(t) & \text{for } a \leq t \leq t_0 \text{ and } t_0 + \delta \leq t \leq b, \\ x_*(t) + (t - t_0)(y - \dot{x}_0) & \text{for } t_0 \leq t \leq t_0 + \varepsilon\delta, \\ x_*(t) + \frac{\varepsilon}{1-\varepsilon}(t_0 + \delta - t)(y - \dot{x}_0) & \text{for } t_0 + \varepsilon\delta \leq t \leq t_0 + \delta. \end{cases}$$

Fig. 1.15 The variation
$x(t;\varepsilon,\delta)$

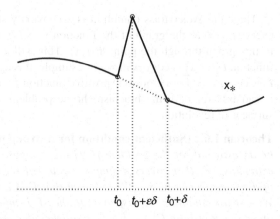

$$t_0 \quad t_0+\varepsilon\delta \quad t_0+\delta$$

Since x_* is a strong local minimum, there exists an open neighborhood R of the restriction of the graph of x_* to the half-open interval $(a,b]$ such that the value $I[x_*]$ is no smaller than the value $I[\tilde{x}]$ for any other admissible curve $\tilde{x} \in C^1_{pc}([a,b])$ whose graph lies in R. Since R is open, for sufficiently small ε and δ, $0 \le \varepsilon \le \varepsilon_0$, $0 \le \delta \le \delta_0$, the graph of the perturbed curve $x(\cdot;\varepsilon,\delta)$ still lies in R, and by construction it is an admissible curve. Therefore we have that

$$0 \le I[x(\cdot;\varepsilon,\delta)] - I[x_*] = A(\varepsilon,\delta) + B(\varepsilon,\delta)$$

where

$$A(\varepsilon,\delta) = \int_{t_0}^{t_0+\varepsilon\delta} L(t,x_*(t)+(t-t_0)(y-\dot{x}_0),\dot{x}_*(t)+(y-\dot{x}_0))$$
$$- L(t,x_*(t),\dot{x}_*(t))\, dt$$

and

$$B(\varepsilon,\delta) = \int_{t_0+\varepsilon\delta}^{t_0+\delta} L\left(t,x_*(t)+\frac{\varepsilon}{1-\varepsilon}(t_0+\delta-t)(y-\dot{x}_0),\dot{x}_*(t)-\frac{\varepsilon}{1-\varepsilon}(y-\dot{x}_0)\right)$$
$$- L(t,x_*(t),\dot{x}_*(t))\, dt.$$

Taking the limit $\delta \to 0$ in the first term, it follows that

$$\lim_{\delta \to 0} \frac{A(\varepsilon,\delta)}{\varepsilon\delta} = L(t_0,x_0,y) - L(t_0,x_0,\dot{x}_0),$$

and in the second term,

$$\lim_{\delta \to 0} \frac{B(\varepsilon, \delta)}{\varepsilon \delta} = \left(\frac{1}{\varepsilon} - 1 \right) \left[L \left(t_0, x_0, \dot{x}_0 - \frac{\varepsilon}{1 - \varepsilon} (y - \dot{x}_0) \right) - L(t_0, x_0, \dot{x}_0) \right].$$

Consequently, setting $\xi = \frac{\varepsilon}{1 - \varepsilon}$ yields

$$0 \leq L(t_0, x_0, y) - L(t_0, x_0, \dot{x}_0) + \frac{L(t_0, x_0, \dot{x}_0 - \xi (y - \dot{x}_0)) - L(t_0, x_0, \dot{x}_0)}{\xi}.$$

Taking the limit $\xi \longrightarrow 0$, it follows that the second term converges to

$$-\frac{\partial L}{\partial \dot{x}} (t_0, x_0, \dot{x}_0)(y - \dot{x}_0),$$

and thus overall, we have that

$$0 \leq L(t_0, x_0, y) - L(t_0, x_0, \dot{x}_0) - \frac{\partial L}{\partial \dot{x}} (t_0, x_0, \dot{x}_0)(y - \dot{x}_0) = E(t_0, x_0, \dot{x}_0, y).$$

$\square$

The variations made in this proof are fundamentally different from the classical variations made to calculate the first and second variations. Weierstrass's variations have corners, and in fact, are designed to have spikes. In essence, they already contain the main ideas of the needle variations made by Pontryagin et al. in their proof of the maximum principle for optimal control problems. These constructions will be developed in Sect. 4.2. In this chapter, we still show how the methods developed in this section can be used to (a) find the globally optimal solutions for the problem of minimal surfaces and (b) prove optimality of the cycloids for the brachistochrone problem.

1.7 Optimal Solutions for the Minimum Surfaces of Revolution

We give a full solution to the problem of minimum surfaces of revolution. Our exposition here follows Bliss [39], but condenses the reasoning and includes the calculations. Theorem 1.6.2 directly applies to the restricted field of catenaries and gives the following statement:

Corollary 1.7.1 (Strong local minimum property of the restricted catenaries). *Let $\mathscr{F}$ be the central field of catenaries defined by*

$$\Xi : D = \{(t, p) : p \in \mathbb{R}, 0 < t < \tau(p)\} \to R,$$

$$(t, p) \mapsto (t, x(t; p)) = \left(t, \frac{x_0}{\cosh p} \cosh \left(p + \frac{t}{x_0} \cosh p \right) \right).$$

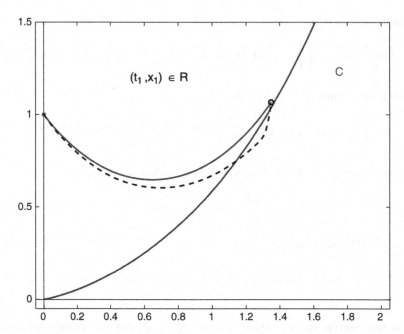

Fig. 1.16 A C^1-curve that does better than the catenary through a point (t_1, x_1) just above the envelope C

Then every extremal in the field $\mathscr{F}$ is a strong local minimum with respect to any other continuous and piecewise continuously differentiable function that has the same boundary values and whose graph lies in R.   ∎

Note, however, that we have only an optimality result relative to other curves that lie in R. So this result is local. We can make no statement about optimality of the catenaries in comparison to curves that would have portions that lie outside of R, and thus the question about the global optimality of these trajectories arises.

In fact, it should already be clear from the earlier results that not all of these extremals can be globally optimal. For $p < 0$ fixed, let $\gamma_s(p)$ denote the extremal $x(t; p)$ defined for the compact interval $0 \leq t \leq s$. Since the extremal $\gamma_{t_c(p)}$ defined up to and including the conjugate time is not a weak local minimum, given any neighborhood of this curve in $\mathscr{Y}$ there exists another continuously differentiable function $\tilde{x}$, close to $\gamma_{t_c(p)}$ and with the same boundary conditions $\tilde{x}(0) = x_0$ and $\tilde{x}(t_c(p)) = x(t_c(p); p)$, that gives a better value, $I[\tilde{x}] < I[\gamma_{t_c(p)}(p)]$. But the value $I[\gamma_s(p)]$ depends continuously on s and thus also for times s close to $t_c(p)$, $s < t_c(p)$, $I[\tilde{x}] < I[\gamma_s(p)]$. It is possible to deform the curve $\tilde{x}$ slightly to satisfy the new boundary conditions for the extremal $\gamma_s(p)$ and still provide a smaller value (see Fig. 1.16). Thus $\gamma_s(p)$ cannot be optimal. We shall now derive the globally optimal solution for the minimal surfaces.

However, in order to do so, we need to revise the classical problem formulation. Recall that we formulated this problem to minimize the integral

$$I[x] = \int\limits_0^{t_1} 2\pi x(t)\sqrt{1+\dot{x}(t)^2}\,dt$$

over all positive curves $x \in C^1([0,x_1])$, $x : [0,t_1] \to (0,\infty)$, that satisfy given boundary conditions $x(0) = x_0$ and $x(t_1) = x_1$. If there exists a solution for given boundary conditions (t_1,x_1) within this class of functions, it needs to be one of the catenaries in the family $\mathscr{E}$ of extremals,

$$x(t;p) = \frac{x_0}{\cosh p}\cosh\left(p + \frac{t}{x_0}\cosh p\right), \quad p \in \mathbb{R}. \tag{1.46}$$

But this family has an envelope C, the graph of the function $\gamma : [0,\infty) \to [0,\infty)$, $t \mapsto \gamma(t)$, defined in Theorem 1.3.1, and thus does not contain any point in the region S below the envelope C in its image. Clearly, there exist positive continuously differentiable functions x that satisfy $x(t_1) = x_1$, and thus the problem is one of existence of optimal solutions. Since any optimal solution in this class needs to be an extremal, it simply follows that *there does not exist an optimal solution in the class of positive continuously differentiable functions if* $(t_1,x_1) \in C \cup S$. In fact, as we shall show now, for these points, and even for some points above the envelope C, the optimal solution no longer is the graph of a function, but is only a piecewise continuously differentiable *curve*. Its structure is easily revealed in a simple experiment with soap films. The minimal surfaces problem is equivalent to determining the shape of a soap film stretched between two circles whose planes are parallel and whose centers are on a common third axis perpendicular to these circles [39, p. 119]. As the two disks are slowly moved apart, this surface stretches until at some point it suddenly ruptures, and all that is left are two disconnected circles of soap film. This breakage happens exactly as the conjugate point is reached, and the subsequent result is our desired optimal solution, the so-called *Goldschmidt extremal* (see Fig. 1.17). It corresponds to the following parameterized curve,

$$\mathfrak{z}_{(t_1,x_1)} : [0, x_0 + t_1 + x_1] \to [0,\infty) \times [0,\infty),$$

$$\tau \mapsto (t(\tau), x(\tau)) = \begin{cases} (0, x_0 - s) & \text{if } 0 \le s \le x_0, \\ (s - x_0, 0) & \text{if } x_0 < s \le x_0 + t_1, \\ (t_1, s - t_1 - x_0) & \text{if } x_0 + t_1 < s \le x_0 + t_1 + x_1. \end{cases}$$

Indeed, the "correct" domain for this minimization problem is the space $\mathscr{Z}$ of all rectifiable piecewise continuously differentiable curves

$$z : [\alpha, \beta] \to \mathbb{R} \times [0,\infty), \quad \tau \mapsto (t(\tau), x(\tau)),$$

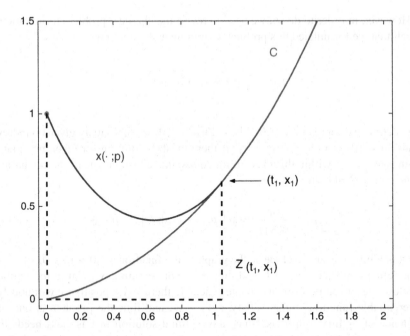

Fig. 1.17 The Goldschmidt extremal through a point (t_1, x_1) on the envelope C

defined on some compact interval $[\alpha, \beta] \subset \mathbb{R}$. Given any such curve, we denote by

$$s = s(\tau) = \int_{\alpha}^{\tau} \sqrt{t'(\xi)^2 + x'(\xi)^2} \, d\xi$$

its arc length. If z is the graph of some function x defined on an interval $[0, t_1]$, say $z(t) = (t, x(t))$, then

$$I[x] = \int_0^{t_1} 2\pi x(t) \sqrt{1 + \dot{x}(t)^2} \, dt = \int_0^{t_1} 2\pi x(\tau) \, ds(\tau) = \int_0^{t_1} 2\pi x \, ds.$$

We can thus reformulate the problem to find minimum surfaces of revolution more generally as follows:

[Min Surf] Given (t_1, x_1) with $x_1 \geq 0$, find a rectifiable piecewise continuously differentiable curve z,

$$z : [\alpha, \beta] \to \mathbb{R} \times [0, \infty), \quad \tau \mapsto (t(\tau), x(\tau)),$$

that satisfies the boundary conditions

$$t(\alpha) = 0, \quad x(\alpha) = x_0 \quad \text{and} \quad t(\beta) = t_1, \quad x(\beta) = x_1,$$

and minimizes the objective

$$I[z] = 2\pi \int\limits_{\alpha}^{\beta} x(t)ds(t).$$

Compared with the original formulation [Minimal Surfaces] in Sect. 1.3, now parameterized curves with a finite number of corners are allowed that do not have to be graphs of functions. Also, these curves can lie anywhere in the closed half-plane $H = \{(t,x) : x \geq 0\}$. In particular, the endpoint (t_1,x_1) can lie to the left or simply above or below the initial conditions as well. Then, as already mentioned in Sect. 1.3, the catenaries $t \mapsto x(t,p)$ for $p > 0$ have their conjugate point for $t_c(p) = -t_c(-p) < 0$. It is clear that if we consider the family Eq. (1.46) also for negative times, this family is invariant under the symmetry $(t,p) \to (-t,-p)$, and it follows that the associated value function has the symmetry

$$V^{\mathscr{F}}(t,x) = V^{\mathscr{F}}(-t,x) \qquad \text{for all} \quad (t,x) \in R.$$

Thus there is no need to consider the case of terminal conditions with $t_1 < 0$ separately. But clearly, the family of catenaries defined for all t has a singularity for $t = 0$ and there is a new situation for terminal conditions that satisfy $t_1 = 0$. Actually, the optimal solutions for this case are trivial, but it becomes necessary, and does indeed provide some insight, to add this solution to the family of catenaries.

Lemma 1.7.1. *If $t_1 = 0$, then the optimal solution to the problem [Min Surf] consists of the straight line segment connecting x_0 with x_1 on the line $t = 0$, and the value of the objective is given by*

$$I[z] = \pi \left| x_1^2 - x_0^2 \right|.$$

Proof. It is geometrically clear that the straight line segment generates an annulus of surface area $\pi \left(x_0^2 - x_1^2 \right)$ if x_1 lies below x_0, respectively, surface area $\pi \left(x_1^2 - x_0^2 \right)$ if x_1 lies above x_0. Formally, set

$$z : [0,1] \to \mathbb{R} \times [0,\infty), \quad \tau \mapsto (0, \tau x_1 + (1-\tau)x_0),$$

and thus

$$I[z] = \int\limits_0^1 2\pi \left(\tau x_1 + (1-\tau)x_0 \right) \sqrt{(x_1 - x_0)^2} d\tau$$

$$= \pi \left| x_1 - x_0 \right| \frac{\left(\tau x_1 + (1-\tau)x_0 \right)^2}{x_1 - x_0} \Bigg|_{\tau=0}^{\tau=1}$$

$$= \pi \left| x_1 - x_0 \right| \frac{x_1^2 - x_0^2}{x_1 - x_0} = \pi \left| x_1^2 - x_0^2 \right|.$$

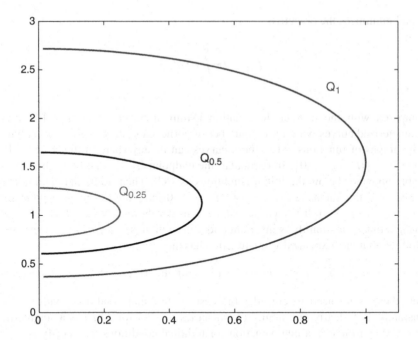

Fig. 1.18 Examples of the level sets Q_α for $x_0=1$

If $\tilde{z}$ is any other curve in $\mathscr{Z}$, then clearly, its arc length is strictly greater than the length of the straight line segment connecting the points x_0 and x_1. If we parameterize both curves in terms of their arc length s, then z is defined over the interval $[0, |x_1 - x_0|]$ and $\tilde{z}$ is defined on an interval $[0, \beta]$ with $\beta > |x_1 - x_0|$. For the sake of argument, let us consider the case $x_1 < x_0$. Then in this parameterization, the value of $\tilde{x}(s)$ can never be smaller than the value of $x(s)$, simply because the straight line segment provides the maximum possible decay for the variable x. Hence, we have that

$$I[z] = \int_0^{|x_1-x_0|} 2\pi x(s)ds \leq \int_0^{|x_1-x_0|} 2\pi \tilde{x}(s)ds < \int_0^{\beta} 2\pi \tilde{x}(s)ds = I[\tilde{z}].$$

Thus the straight line segment provides the global minimum for the problem [Min Surf] for these boundary data.   □

Adding the vertical line segments with appropriate arrows to the field of the catenaries gives the natural extension of this field onto the x-axis and provides useful closure properties. Clearly, this is the limiting behavior of the curves in the field as $t \to 0+$ (see Fig. 1.8). We now show that the value for the vertical line segments provides a continuous extension of the value function for the field of catenaries to the x-axis.

Proposition 1.7.1. *Let $V^{\mathscr{F}}$ be the value function corresponding to the field of catenaries defined in Eq. (1.46). Then V extends continuously to the x-axis with value*

$$V^{\mathscr{F}}(0,x_1) = \pi \left| x_1^2 - x_0^2 \right|.$$

Proof. We first describe the limiting properties of the parameterization in some detail. Clearly, as $t \to 0+$, for all extremals we have that $x(t;p) \to x_0$, and so we need to desingularize this behavior. Let $\alpha \in \mathbb{R}_+$ be a positive constant and consider the level sets $Q_\alpha = \{(t,p) : t\cosh p = \alpha x_0\}$ (see Fig. 1.18).

Rewriting the catenaries as

$$x(t;p) = \frac{x_0}{\cosh p}\cosh\left(p + \frac{t}{x_0}\cosh p\right) = x_0\left[\cosh\alpha + \tanh p \sinh\alpha\right],$$

it follows that for $(t,p) \in Q_\alpha$ we have that

$$\lim_{p\to\pm\infty} x(t;p) = x_0\left[\cosh\alpha \pm \sinh\alpha\right] = x_0 e^{\pm\alpha},$$

and thus we obtain the vertical line segment above x_0 in the limit $p \to \infty$ and the vertical line segment below x_0 in the limit $p \to -\infty$ along Q_α. In this sense, the entire x-axis arises as limiting behavior of the parameterization for $t \to 0+$ and $p \to \pm\infty$ with $(t,p) \in Q_\alpha$.

Now suppose $\{(t_n,x_n)\}_{n\in\mathbb{N}}$ is a sequence of points in R that converges to some point $(0,x_1)$ with $x_1 > 0$ and $x_1 \neq x_0$. We can easily compute the limit of the value function $V(t_n,x_n)$ as $n \to \infty$ using the parameterization. Since $\varXi$ is a diffeomorphism, there exists a unique sequence $\{p_n\}_{n\in\mathbb{N}}$ such that

$$V(t_n,x_n) = V(t_n,x_n(t_n,p_n)) = C(t_n,p_n).$$

But $C(t,p)$ is easily computed:

$$C(t,p) = \int_0^t 2\pi \frac{x_0}{\cosh p}\cosh^2\left(p + \frac{s}{x_0}\cosh p\right)ds$$

$$= 2\pi\left(\frac{x_0}{\cosh p}\right)^2 \int_p^{p+\frac{t}{x_0}\cosh p} \cosh^2 r\, dr$$

$$= \pi\left(\frac{x_0}{\cosh p}\right)^2 \left[\cosh r \sinh r + r\right]\Big|_p^{\left|p+\frac{t}{x_0}\cosh p\right|}$$

$$= \pi \left(\frac{x_0}{\cosh p} \right)^2 \left[\cosh \left(p + \frac{t}{x_0} \cosh p \right) \sinh \left(p + \frac{t}{x_0} \cosh p \right) \right.$$

$$\left. - \cosh p \sinh p + \frac{t}{x_0} \cosh p \right]$$

$$= \pi \left[x(t;p)^2 \tanh \left(p + \frac{t}{x_0} \cosh p \right) - x_0^2 \tanh p + \frac{t x_0}{\cosh p} \right]. \qquad (1.47)$$

Since $(t_n, x_n) \to (0, x_1)$, we have that $p_n \to \pm\infty$ with the plus sign valid for $x_1 > x_0$ and the minus sign valid for $x_1 < x_0$. Without loss of generality, we may assume that all x_n satisfy either $x_n > x_0$ or $x_n < x_0$. It then follows that the points x_n all lie on ascending portions of the catenaries for $x_1 > x_0$ and on descending portions for $x_1 < x_0$. Consequently, as $n \to \infty$, we accordingly have

$$p_n + \frac{t_n}{x_0} \cosh p_n \to \pm\infty.$$

This is clear if $p_n > 0$ and follows from the fact that x_n/x_0 converges to a positive limit for $p_n < 0$. Hence

$$\lim_{n \to \infty} \tanh \left(p_n + \frac{t_n}{x_0} \cosh p_n \right) = \pm 1.$$

Obviously, we also have

$$\lim_{n \to \infty} \frac{t_n x_0}{\cosh p_n} = 0 \qquad \text{and} \qquad \lim_{n \to \infty} \tanh p_n = \pm 1,$$

and thus altogether

$$\lim_{n \to \infty} V(t_n, x_n) = \lim_{n \to \infty} C(t_n, p_n) = \pi \left| x_1^2 - x_0^2 \right|.$$

$\square$

Proposition 1.7.2. *Let z be any rectifiable piecewise continuously differentiable curve connecting $(0, x_0)$ with a point (t_1, x_1), $t_1 > 0$, and suppose z has a point in common with the envelope C of the family of catenaries. Then z generates a larger surface area of revolution than the Goldschmidt solution $\varsigma_{(t_1, x_1)}$ through the point (t_1, x_1),*

$$I[z] > I[\mathfrak{z}_{(t_1, x_1)}]. \qquad (1.48)$$

Proof. Without loss of generality, we may assume that the curve z lies in the first quadrant. (If it doesn't, we replace the portion of the curve that lies in $t < 0$ by the corresponding vertical line segment and in this way generate another curve that has a smaller surface area of revolution and consider this as our original curve z.) Thus, suppose z has a parameterization $z : [\alpha, \beta] \to [0, \infty) \times [0, \infty)$, $\tau \mapsto (t(\tau), x(\tau))$, and let θ be the first time for which $(t(\theta), x(\theta)) \in C$ (see Fig. 1.19).

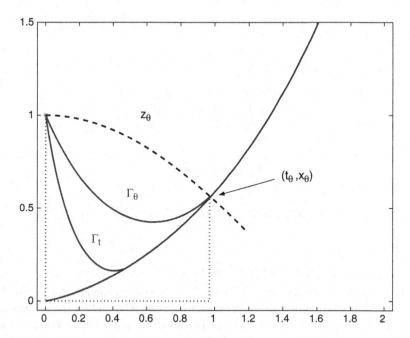

Fig. 1.19 The curves $\Gamma_t * C_{[t,\theta]}$

Denote by z_θ the restriction of z to the parameter interval $[\alpha,\theta]$ and let (t_θ,x_θ) denote the point on the envelope C. We first show that the corresponding Goldschmidt solution $\mathfrak{z}_{(t_\theta,x_\theta)}$ through the point (t_θ,x_θ) gives a better value for I than z_θ. Since except for the endpoint, the curve z_θ lies entirely in the region R covered by the field of catenaries, it follows from Corollary 1.7.1 that the value $I[z_\theta]$ is at least as large as the value for the unique extremal in the field that passes through the point (t_θ,x_θ). If we call the corresponding parameter p_θ and denote the catenary $x(\cdot;p_\theta)$ defined over the interval $[0,t_\theta]$ by Γ_θ, we thus have that

$$I[z_\theta] \geq I[\Gamma_\theta].$$

More generally, for $0 < t \leq \theta$, let Γ_t denote the catenary $x(\cdot;p_*(t))$ of the field restricted to the interval $[0,t]$ that passes through the point $(t,\gamma(t))$ of the envelope (see Theorem 1.3.1) and denote the concatenation of the corresponding graph with the portion of the envelope C over the interval $[t,\theta]$ by $\Gamma_t * C_{[t,\theta]}$. For $t = 0$, we use the same notation to denote the concatenation of the vertical line segment from $(0,x_0)$ to the origin with the envelope C. It then follows from the envelope theorem (Theorem 1.5.2), and taking the limit as $t \to 0+$, that all these curves have the same value, i.e.,

$$I[\Gamma_\theta] = I[\Gamma_t * C_{[t,\theta]}] = I[\Gamma_0 * C_{[0,\theta]}] \qquad \text{for all} \quad t \in [0,\theta].$$

But the latter value is easily estimated: the curve $t \mapsto x_*(t) = x(t; p_*(t))$ is positive and increasing, and therefore

$$I[\Gamma_0 * C_{[0,\theta]}] = I[\Gamma_0] + I[C_{[0,\theta]}]$$

$$= \pi x_0^2 + 2\pi \int_0^\theta x_*(r)\sqrt{1 + \dot{x}_*(r)^2}\, dr$$

$$> \pi x_0^2 + 2\pi \int_0^\theta x_*(r)\dot{x}_*(r)\, dr$$

$$= \pi x_0^2 + \pi x_*^2\big|_0^\theta = \pi \left[x_0^2 + x(\theta; p_*(\theta))^2\right] = I\left[\mathfrak{z}_{(t_\theta, x_\theta)}\right].$$

Hence

$$I[z_\theta] \geq I[\Gamma_\theta] = I[\Gamma_0 * C_{[0,\theta]}] > I[\mathfrak{z}_{(t_\theta, x_\theta)}].$$

But this implies that we also have $I[z] > I[\mathfrak{z}_{(t_1, x_1)}]$. For more generally, for $\tau \in [\alpha, \beta]$ let z_τ be the restriction of z to the parameter interval $[\alpha, \tau]$ and let (t_τ, x_τ) denote the corresponding point on the curve z. Then the difference in the value of the objective between the curve z_τ and the Goldschmidt extremal $\mathfrak{z}_{(t_\tau, x_\tau)}$ through (t_τ, x_τ) is given by

$$\Delta(\tau) = \int_\alpha^\tau 2\pi x(t)\, ds(t) - \pi \left[x_0^2 + x(\tau)^2\right],$$

and since

$$\frac{d\Delta}{d\tau}(\tau) = 2\pi x(\tau)\, ds(\tau) - 2\pi x(\tau)\frac{dx(\tau)}{d\tau}$$

$$= 2\pi x(\tau)\left(\sqrt{\left(\frac{dt}{d\tau}(\tau)\right)^2 + \left(\frac{dx}{d\tau}(\tau)\right)^2} - \frac{dx}{d\tau}(\tau)\right) \geq 0,$$

this difference is nondecreasing along the curve z. Thus we have

$$I[z] - I[\mathfrak{z}_{(t_1, x_1)}] = \Delta(\beta) \geq \Delta(\theta) > 0.$$

This proves the result.   □

Corollary 1.7.2. *If (t_1, x_1) is a point that lies on or below the envelope C, then the optimal solution for problem [Min Surf] is given by the Goldschmidt extremal $\mathfrak{z}_{(t_1, x_1)}$ with value $V(t_1, x_1) = \pi(x_0^2 + x_1^2)$.*   ■

This now leaves us with the question of optimal solutions for terminal points (t_1, x_1) in the region R above the envelope C. Clearly, there are only two candidates, the catenaries and the Goldschmidt extremals. For if z is any rectifiable piecewise continuously differentiable curve that connects $(0,x_0)$ with (t_1,x_1), then it either intersects the envelope C or it doesn't. If it does, it can do no better than the Goldschmidt extremal; if it doesn't, the curve lies in the region R, and thus the catenary provides a lower bound. Thus, quite simply, we have the following result:

Proposition 1.7.3. *If (t_1,x_1) is a point that lies in the region R above the envelope C, then the optimal solution for problem [Min Surf] is given by either the catenary of the field $\mathscr{F}$ that passes through (t_1,x_1) or the Goldschmidt extremal $\mathfrak{z}_{(t_1,x_1)}$, whichever provides the smaller value.* ∎

It remains to find the exact location of the *cut-locus* between the corresponding two value functions, i.e., the set of points for which the surface area of revolution for the catenary agrees with the value of the Goldschmidt extremal. For this, as above, consider how the difference in values between the catenaries and the Goldschmidt extremal changes along a fixed extremal $x(\cdot\,;p)$, $p \in \mathbb{R}$. Let $\Delta : [0,\infty) \to \mathbb{R}$, $t \mapsto \Delta(t)$, be given by

$$\Delta(t;p) = \pi \left(\int_0^t 2x(r;p)\sqrt{1+\dot{x}(r;p)^2}dr \right) - \pi \left(x_0^2 + x(t;p)^2 \right).$$

Clearly, we have $\Delta(0;p) = -2\pi x_0^2 < 0$, so that the catenary is better for small times t. Furthermore, for all times $t > 0$ we have that

$$\dot{\Delta}(t;p) = 2\pi x(t;p) \left(\sqrt{1+\dot{x}(t;p)^2} - \dot{x}(t;p) \right) > 0,$$

and thus Δ is strictly monotonically increasing. For $p < 0$, we already know that $\Delta(t_c) > 0$ for the conjugate time t_c, and thus it follows that along each catenary $x(\cdot\,;p)$ for $p < 0$ there exists a unique positive time $t = t_g(p) < t_c(p)$ such that $\Delta(t_g(p),p) = 0$. By the implicit function theorem, see Theorem A.3.2, t_g is a continuously differentiable function. The graph G of this function,

$$G : (-\infty,0) \to (0,\infty) \times (0,\infty), \quad p \mapsto (t_g(p),x(t_g(p),p))$$

is the desired *cut-locus*. Since the catenaries are the better solutions in the limit $(t_1,x_1) \to (0,x_1)$ for $x_1 > 0$ and since the Goldschmidt extremals are the better solutions in the limit $(t_1,x_1) \to (t_1,0)$ for $t_1 > 0$, it follows that the cut-locus has the origin in its boundary. Thus, for $p \to -\infty$, we must have that

$$\lim_{p \to -\infty} (t_g(p),x(t_g(p),p)) = (0,0).$$

Generally, the cut-locus G and the envelope C of conjugate points have a very similar shape (see Fig. 1.20). While there is no explicit formula for G, the parameterized costs for the two families are easily compared numerically. Recall that for the family

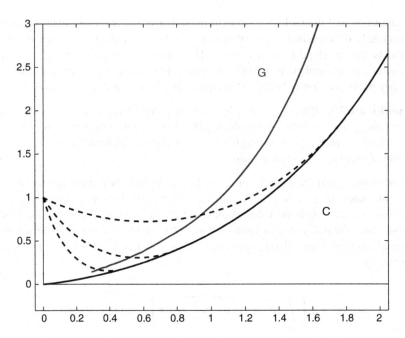

Fig. 1.20 The cut-locus G for $x_0 = 1$

of catenaries we have that (see the derivation of (1.47))

$$C(t;p) = \pi \left(\frac{x_0}{\cosh p} \right)^2 \left[\cosh \left(p + \frac{t}{x_0} \cosh p \right) \sinh \left(p + \frac{t}{x_0} \cosh p \right) \right.$$
$$\left. - \cosh p \sinh p + \frac{t}{x_0} \cosh p \right],$$

and for the Goldschmidt extremals the objective can be written as

$$G(t;p) = \pi \left[x(t;p)^2 + x_0^2 \right] = \pi \left(\frac{x_0}{\cosh p} \right)^2 \left[\cosh^2 \left(p + \frac{t}{x_0} \cosh p \right) + \cosh^2 p \right].$$

If we introduce the abbreviation

$$q = p + \frac{t}{x_0} \cosh p,$$

then in the parameter space (t,p), an equation that defines the cut-locus can be written in the following simple but unfortunately highly transcendental form

$$q + \sinh q \cosh q - \cosh^2 q = p + \sinh p \cosh p + \cosh^2 p.$$

Numerically, however, this zero set is easily computed, and we plot it in Fig. 1.20 for $x_0 = 1$.

While the Goldschmidt extremals may appear an oddity, and they are somewhat in the sense of the calculus of variations, these are the more natural of the two candidates from an optimal control point of view. By going to the formulation [Min Surf] with curves, in this more general formulation, the t-axis with $x = 0$ provides a "free pass," which is the best one can do for the objective. Thus, it is an obvious heuristic strategy to move toward the curve with as little cost as possible—and these are the vertical straight line segments that generate the circles—and then to follow the curve $\{x = 0\}$ to some optimal point where this curve is left to get to the prescribed terminal point, again with as little cost as possible, that is, by a vertical line segment. We shall encounter such a structure in the form of bang-singular-bang controls as a typical and in some sense common behavior of optimal controlled trajectories in optimal control problems (e.g., Sect. 2.9). For example, this also becomes the dominant feature in the design of optimal protocols for tumor antiangiogenic treatments for cancer considered in [160]. Generally, the only reason why such a strategy would not be optimal is that it simply is too "expensive" to connect with this special curve $\{x = 0\}$ and that a direct path exists that gives a better value. These are the catenaries. Naturally, the true optimal solution then strikes a balance between these two classes of candidates for optimality. This, of course, is done through a direct comparison of their corresponding values, a global and thus much more involved geometric argument. Cut-loci, the sets of points where the values corresponding to different extremal strategies agree, in general are a dominant structure for global solutions of calculus of variations and optimal control problems. We shall return to this topic in Sect. 5.5. Figure 1.21 shows the globally optimal synthesis for the minimum surfaces of revolution problem.

1.8 Optimality of the Cycloids for the Brachistochrone Problem

We next establish the global optimality of the field of cycloids for the brachistochrone problem. Also here the reasoning is not entirely straightforward. Note that Theorem 1.6.2 does not directly apply, since the field has a singularity at the origin. But the basic constructions of Sect. 1.6 can easily be modified to deal with this issue.

Theorem 1.8.1 (Global minimum property of the cycloids). *Let $\mathscr{F}$ be the central field of cycloids defined by*

$$\Xi : (-\pi, \pi) \times (0, \infty) \to \mathbb{R}_+^2 = \{(t_1, x_1) : t_1 > 0, x_1 > 0\},$$

$$(\tau, \xi) \mapsto (t(\tau; \xi), x(\tau; \xi)) = (\xi(\pi + \tau + \sin \tau), \xi(1 + \cos \tau)).$$

Then every cycloid in this family is the global minimum for the brachistochrone problem for the corresponding terminal condition.

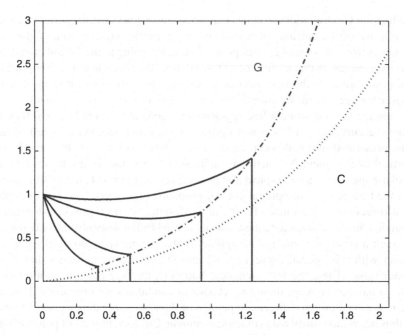

Fig. 1.21 The optimal synthesis for the minimum surfaces problem

Proof. Since extremals have a singularity as $t \to 0+$ in (t,x)-space, we need to carry out a limiting argument as the initial point converges to the origin. Recall that the Hilbert invariant integral (1.42) is given by

$$I^* = \int_\Gamma L(t,x,\Psi(t,x))dt + \frac{\partial L}{\partial \dot{x}}(t,x,\Psi(t,x))\,(dx - \Psi(t,x)dt),$$

where $\Psi(t,x)$ defines the directions for the field $\mathscr{F}$ of extremals. For the brachistochrone problem, we have that

$$I^* = \int_\Gamma \sqrt{\frac{1+\Psi(t,x)^2}{x}}dt + \Psi(t,x)\sqrt{\frac{1}{x(1+\Psi(t,x)^2)}}\,(dx - \Psi(t,x)dt)$$

$$= \int_\Gamma \sqrt{\frac{1}{x(1+\Psi(t,x)^2)}}\left[\left(1+\Psi(t,x)^2\right)dt + \Psi(t,x)\,(dx - \Psi(t,x)dt)\right]$$

$$= \int_\Gamma \sqrt{\frac{1}{x(1+\Psi(t,x)^2)}}\,[dt + \Psi(t,x)dx]. \tag{1.49}$$

This integral will take the value 0 if we choose Γ as an integral curve of the differential equation

$$\frac{dx}{dt} = -\frac{1}{\Psi(t,x)},$$
(1.50)

that is, along a curve that is everywhere orthogonal to the cycloids of the field.

Given a point $(t_1, x_1) \in \mathbb{R}_+^2$, let $x : [0, t_1] \to [0, \infty)$, $t \mapsto x(t)$, be any continuous function that satisfies the boundary conditions $x(0) = 0$ and $x(t_1) = x_1$ and is positive and continuously differentiable in the open interval $(0, t_1)$. For simplicity of notation, we denote the graph of the function x by C and write $C_{s,t}$ for the portion of the graph of x defined over the interval $[s,t]$. Also, we denote the portion of the cycloid that connects the origin with $(t, x(t))$ by Γ_t. We then define a 1-parameter family of curves $\Omega_\ell = \Gamma_\ell * C_{\ell, t_1}$, $0 \leq \ell \leq t_1$, by concatenating Γ_ℓ with C_{ℓ, t_1}, the remaining portion of the graph of x over the final interval $[\ell, t_1]$. Specifically, since the field $\mathscr{F}$ covers $\mathbb{R}_+^2$ diffeomorphically, there exist differentiable functions $\tau = \tau(\cdot)$ and $\xi = \xi(\cdot)$ defined on the half-open interval $(0, t_1]$, so that in terms of the parameterization of the field, we have that

$$p(\ell) = (\ell, x(\ell)) = (t(\tau(\ell); \xi(\ell)), x(\tau(\ell); \xi(\ell))) \qquad \text{for all } \ell \in (0, t_1].$$

The curve Ω_ℓ is then given in parameterized form by

$$\Omega_\ell(u) = \begin{cases} (t(u; \xi(\ell)), x(u; \xi(\ell))) & \text{for} \quad -\pi \leq u \leq \tau(\ell), \\ (u - \tau(\ell) + \ell, x(u - \tau(\ell) + \ell)) & \text{for} \quad \tau(\ell) \leq u \leq \tau(\ell) + t_1 - \ell. \end{cases}$$

Let φ be the value of the objective I along the continuous and piecewise continuously differentiable curve Ω_ℓ, i.e.,

$$\varphi : [0, t_1] \to \mathbb{R}, \quad \ell \mapsto \varphi(\ell) = I[\Omega_\ell] = I[\Gamma_\ell * C_{\ell, t_1}] = I[\Gamma_\ell] + I[C_{\ell, t_1}].$$

It suffices to show that φ is monotonically nonincreasing. For then we have that

$$I[x] = I[C_{0, t_1}] = \varphi(0) \geq \varphi(1) = I[\Gamma_{t_1}],$$

and thus the curve C is no better than the cycloid Γ_{t_1}.

In order to prove this monotonicity property, we first show that for times r and s close to each other, we have that

$$I[\Gamma_r] + I^*[C_{r,s}] = I[\Gamma_s].$$
(1.51)

If we simply ignored the fact that the field of extremals has a singularity at the origin and apply Hilbert's invariant integral anyway, this identity would follow. Rigorous reasoning, however, requires a limiting argument (see Fig. 1.22).

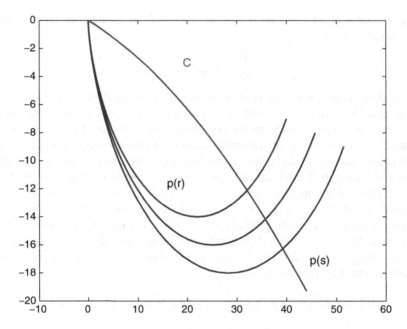

Fig. 1.22 The curves $\Gamma_s * C_{[s,\beta]}$

Fix $s \in (0,t_1)$ and suppose r is close to s, $r < s$. Choose ε close to $-\pi$ and let Θ be the integral curve to the differential equation (1.50) with initial condition

$$(t_\varepsilon, x_\varepsilon) = (t(\varepsilon; \xi(s)), \, x(\varepsilon; \xi(s))),$$

i.e., the initial condition is a point on the cycloid Γ_s (through $(s, x(s))$) close to the origin. In the limit $(t,x) \to (0+, 0+)$, the field $\Psi(t,x)$ converges to the direction $(0,1)$ and the perpendicular field has limiting direction $(-1,0)$, i.e., is horizontal at the origin. It follows that for an initial condition $(t_\varepsilon, x_\varepsilon)$ near the origin, there exists a $\delta = \delta(\varepsilon) > 0$ such that the solution Θ is defined over some interval $(-\delta, \delta)$ and is orthogonal to all the cycloids it intersects. Thus, for times r close enough to s, the cycloid Γ_r passing through the point $(r, x(r))$ also intersects this integral curve Θ. Hence we can construct a piecewise continuously differentiable closed curve Ω by concatenating the following arcs:

1. The portion $\Theta_{\varepsilon, \eta}$ of the integral curve Θ that goes from the point $(t_\varepsilon, x_\varepsilon)$ on the cycloid Γ_s to the point (t_η, x_η) of intersection of Θ with the cycloid Γ_r,
2. The portion $\tilde{\Gamma}_{\eta, r}$ of the cycloid Γ_r that connects the point (t_η, x_η) with the point $(r, x(r))$ on the curve C,
3. The portion $C_{r,s}$ of the graph C of x corresponding to the time interval $[r, s]$, and
4. The portion $\tilde{\Gamma}_{\varepsilon, s}$ of the cycloid Γ_s defined over the parameter interval $[\varepsilon, \tau(s)]$ that connects the point $(t_\varepsilon, x_\varepsilon)$ with the point $p(s) = (s, x(s))$, but run backward.

Thus, altogether we have that

$$\tilde{\Gamma}_{\varepsilon,s} = \Theta_{\varepsilon,\eta} * \tilde{\Gamma}_{\eta,r} * C_{r,s}$$

and the full curve lies in $\mathbb{R}^2_+$. Since the Hilbert invariant integral I^* is defined by an exact differential, it follows that

$$I^*[\tilde{\Gamma}_{\varepsilon,s}] = I^*[\Theta_{\varepsilon,\eta}] + I^*[\tilde{\Gamma}_{\eta,r}] + I^*[C_{r,s}].$$

By construction, $\Theta_{\varepsilon,\eta}$ is an integral curve of the field perpendicular to Ψ, and it therefore follows from (1.49) that $I^*[\Theta_{\varepsilon,\eta}] = 0$. Furthermore, since $\tilde{\Gamma}_s$ and $\tilde{\Gamma}_r$ are integral curves of the field $\mathscr{F}$ itself, we have that

$$I^*[\tilde{\Gamma}_{\varepsilon,s}] = I[\tilde{\Gamma}_{\varepsilon,s}] \qquad \text{and} \qquad I^*[\tilde{\Gamma}_{\eta,r}] = I[\tilde{\Gamma}_{\eta,r}].$$

If we now take the limit $\varepsilon \to -\pi+$, then it follows that $(t_\eta, x_\eta) \to (0+, 0+)$, and thus

$$\lim_{\varepsilon \to 0} I[\tilde{\Gamma}_{\varepsilon,s}] = I[\Gamma_s] \qquad \text{and} \qquad \lim_{\varepsilon \to 0} I[\tilde{\Gamma}_{\eta,r}] = I[\Gamma_r].$$

Here we use the fact that the integrals for the cycloids converge. Thus, as claimed, for r close enough to s we have that

$$I[\Gamma_s] = I[\Gamma_r] + I^*[C_{r,s}]. \tag{1.52}$$

But it follows from the Weierstrass condition that $I^*[C_{r,s}] \leq I[C_{r,s}]$. (This precisely is the significance of the Weierstrass condition.) For everywhere in the region $\mathbb{R}^2_+$, the Weierstrass condition is satisfied and thus for all $(t,x) \in \mathbb{R}^2_+$, $\psi = \Psi(t,x)$, and all $y \in \mathbb{R}$ we have that

$$E(t,x,\psi,y) = L(t,x,y) - L(t,x,\psi) - \frac{\partial L}{\partial \dot{x}}(t,x,\psi)(y - \psi) \geq 0.$$

Hence it follows along the arc $C_{r,s}$ that

$$I^*[C_{r,s}] - I[C_{r,s}] = \int_{C_{r,s}} [L(t,x,\Psi(t,x)) - L(t,x,\dot{x}(t))] \, dt$$

$$+ \frac{\partial L}{\partial \dot{x}}(t,x,\Psi(t,x)) [dx - \Psi(t,x)dt]$$

$$= -\int_{C_{r,s}} E(t,x,\Psi(t,x),\dot{x}) \, dt \leq 0.$$

Consequently,

$$I[\Gamma_s] \leq I[\Gamma_r] + I[C_{r,s}], \tag{1.53}$$

and overall

$$\varphi(r) = I[\Gamma_r] + I[C_{r,t_1}] = I[\Gamma_r] + I[C_{r,s}] + I[C_{s,t_1}] \geq I[\Gamma_s] + I[C_{s,t_1}] = \varphi(s).$$

This proves the result.   □

1.9   The Hamilton–Jacobi Equation

We briefly formulate an alternative approach to fields of extremals due to Hamilton. As in Sect. 1.6, let $\mathscr{F}$ be a central field of extremals for the general problem [CV] with field of directions given by $\dot{x} = \Psi(t,x)$ that covers a simply connected region R of $(a,\infty) \times \mathbb{R}$. It is more convenient, and customary in this context, to denote the points in the region R by (t,x), and these now stand for the endpoint of the extremals in the field. Recall that by Theorem 1.6.1, the differential of the value function $V^{\mathscr{F}}$ associated with the field $\mathscr{F}$ is given by

$$dV^{\mathscr{F}} = \left(L(t,x,\Psi(t,x)) - \frac{\partial L}{\partial \dot{x}}(t,x,\Psi(t,x))\Psi(t,x) \right) dt + \frac{\partial L}{\partial \dot{x}}(t,x,\Psi(t,x))dx.$$

In other words, the value function is a solution to the following first-order partial differential equation determined solely by the integrand L and the central field Ψ:

$$\frac{\partial V}{\partial t}(t,x) + \frac{\partial V}{\partial x}(t,x)\Psi(t,x) = L(t,x,\Psi(t,x)).$$

In terms of the Hamiltonian function H defined earlier,

$$H : [a,\infty) \times \mathbb{R} \times \mathbb{R} \to \mathbb{R}, \ (t,x,\lambda) \mapsto H(t,x,\lambda) = L(t,x,\Psi(t,x)) + \lambda\Psi(t,x), \ (1.54)$$

this partial differential equation can then be written in the form

$$\frac{\partial V}{\partial t}(t,x) = H\left(t,x,-\frac{\partial V}{\partial x}(t,x) \right). \tag{1.55}$$

Definition 1.9.1 (Hamiltonian, Hamilton–Jacobi equation). The function H defined in Eq. (1.54) is called the Hamiltonian, and Eq. (1.55) is the Hamilton–Jacobi equation.

This formalism, due to Hamilton and developed by him in connection with what now is called Hamiltonian dynamics in physics, allows for an important reformulation of the Euler–Lagrange equation as a pair of first-order ordinary differential equations, sometimes also called the *canonical form of the Euler–Lagrange equation*: we have that

$$\dot{x} = \Psi(t,x) = \frac{\partial H}{\partial \lambda}(t,x,\lambda),$$

and if we define $\lambda(t)$ by the derivative that the variable λ represents, i.e.,

$$\lambda(t) = -\frac{\partial L}{\partial \dot{x}}(t,x(t),\Psi(t,x(t))),$$

then by the Euler–Lagrange equation, we also have that

$$\dot{\lambda}(t) = \frac{d}{dt}\left(-\frac{\partial L}{\partial \dot{x}}(t,x(t),\Psi(t,x(t)))\right) = -\frac{\partial L}{\partial x}(t,x(t),\Psi(t,x(t)))$$

$$= -\frac{\partial L}{\partial x}(t,x(t),\Psi(t,x(t))) - \left[\frac{\partial L}{\partial \dot{x}}(t,x(t),\Psi(t,x(t))) + \lambda(t)\right]\frac{\partial \Psi}{\partial x}(t,x(t))$$

$$= -\frac{\partial H}{\partial x}(t,x(t),\lambda(t)).$$

Thus, if a function $t \mapsto x(t)$ is a solution to the Euler–Lagrange equation, then the pair $(x(t),\lambda(t))$ is a solution to the following system of Hamiltonian differential equations:

$$\dot{x} = \frac{\partial H}{\partial \lambda}(t,x,\lambda) \quad \text{and} \quad \dot{\lambda} = -\frac{\partial H}{\partial x}(t,x,\lambda). \tag{1.56}$$

We summarize the main relations between solutions to the Euler–Lagrange equation and the Hamilton–Jacobi equation in the propositions below.

Proposition 1.9.1. *Let $V = V(t,x;\alpha)$ be a solution of the Hamilton–Jacobi equation that depends on a parameter $\alpha \in \mathbb{R}$ and is continuously differentiable as a function of (t,x) and the parameter α. Then the partial derivative of V with respect to the parameter α,*

$$\frac{\partial V}{\partial \alpha}(t,x;\alpha),$$

is a first integral for the Hamiltonian system (1.56), i.e., along its solutions

$$\frac{d}{dt}\left(\frac{\partial V}{\partial \alpha}(t,x(t);\alpha)\right) = 0.$$

Proof. This is a direct computation: we have that

$$\frac{d}{dt}\left(\frac{\partial V}{\partial \alpha}(t,x(t);\alpha)\right) = \frac{\partial^2 V}{\partial \alpha \partial t}(t,x(t);\alpha) + \frac{\partial^2 V}{\partial \alpha \partial x}(t,x(t);\alpha)\dot{x}(t),$$

and differentiating Eq. (1.55) with respect to α gives

$$\frac{\partial^2 V}{\partial \alpha \partial t}(t,x(t);\alpha) = -\frac{\partial H}{\partial \lambda}\left(t,x(t),-\frac{\partial V}{\partial x}(t,x(t);\alpha)\right)\frac{\partial^2 V}{\partial \alpha \partial x}(t,x(t);\alpha)$$

$$= -\frac{\partial H}{\partial \lambda}(t,x(t),\lambda(t))\frac{\partial^2 V}{\partial \alpha \partial x}(t,x(t);\alpha)$$

where we use the trivial identity

$$\frac{\partial H}{\partial \lambda}\left(t,x(t),-\frac{\partial V}{\partial x}(t,x(t);\alpha)\right) = \Psi(t,x(t)) = \frac{\partial H}{\partial \lambda}(t,x(t),\lambda(t)).$$

Substituting into the previous formula yields

$$\frac{d}{dt}\left(\frac{\partial V}{\partial \alpha}(t,x(t);\alpha)\right) = \frac{\partial^2 V}{\partial \alpha \partial x}(t,x(t);\alpha)\left(\dot{x}(t) - \frac{\partial H}{\partial \lambda}(t,x(t),\lambda(t))\right) = 0$$

since by assumption,

$$\dot{x}(t) = \Psi(t,x(t)) = \frac{\partial H}{\partial \lambda}(t,x(t),\lambda(t)).$$

$\square$

These calculations can easily be reversed and lead to the following result of Jacobi:

Proposition 1.9.2. *Let $V = V(t,x;\alpha)$ be a solution to the Hamilton–Jacobi equation depending on a parameter $\alpha \in \mathbb{R}$ and suppose*

$$\frac{\partial^2 V}{\partial \alpha \partial x}(t,x;\alpha) \neq 0.$$

Then the general solution $x = x(t;\alpha,\beta)$ to the Euler–Lagrange equation can be obtained by solving for given constants β the equation

$$\frac{\partial V}{\partial \alpha}(t,x;\alpha) = \beta \tag{1.57}$$

for $x = x(t;\alpha,\beta)$ and then setting

$$\lambda(t;\alpha,\beta) = -\frac{\partial V}{\partial x}(t,x(t;\alpha,\beta);\alpha). \tag{1.58}$$

Proof. We only need to reverse the above calculations. Since $\frac{\partial^2 V}{\partial \alpha \partial x}(t,x;\alpha) \neq 0$, by the implicit function theorem we can locally solve the Eq. (1.57) for x with a continuously differentiable function $x = x(t; \alpha, \beta)$, and along this function, as above, we have the following identity shown in the proof of Proposition 1.9.1:

$$0 = \frac{d}{dt}\left(\frac{\partial V}{\partial \alpha}(t, x(t; \alpha, \beta), \alpha)\right)$$

$$= \frac{\partial^2 V}{\partial \alpha \partial x}(t, x(t; \alpha, \beta), \alpha)\left(-\frac{\partial H}{\partial \lambda}(t, x(t; \alpha, \beta), \lambda(t; \alpha, \beta)) + \frac{dx}{dt}(t; \alpha, \beta)\right).$$

Since $\frac{\partial^2 V}{\partial \alpha \partial x}(t, x; \alpha) \neq 0$, it follows that $x = x(t; \alpha, \beta)$ satisfies

$$\dot{x}(t; \alpha, \beta) = \frac{\partial H}{\partial \lambda}(t, x(t; \alpha, \beta), \lambda(t; \alpha, \beta)).$$

Furthermore,

$$\dot{\lambda}(t; \alpha, \beta) = \frac{d}{dt}\left(-\frac{\partial V}{\partial x}(t, x(t; \alpha, \beta); \alpha)\right)$$

$$= -\frac{\partial^2 V}{\partial t \partial x}(t, x(t; \alpha, \beta); \alpha) - \frac{\partial^2 V}{\partial x^2}(t, x(t; \alpha, \beta); \alpha)\dot{x}(t; \alpha, \beta)$$

and differentiating Eq. (1.55) with respect to x gives

$$\frac{\partial^2 V}{\partial t \partial x}(t, x(t; \alpha, \beta); \alpha)$$

$$= \frac{\partial H}{\partial x}\left(t, x(t; \alpha, \beta), -\frac{\partial V}{\partial x}(t, x(t; \alpha, \beta); \alpha)\right)$$

$$- \frac{\partial H}{\partial \lambda}\left(t, x(t; \alpha, \beta), -\frac{\partial V}{\partial x}(t, x(t; \alpha, \beta); \alpha)\right)\frac{\partial^2 V}{\partial x^2}(t, x(t; \alpha, \beta); \alpha).$$

As above, substitute this relation into the prior equation to get

$$\dot{\lambda}(t; \alpha, \beta) = -\frac{\partial H}{\partial x}\left(t, x(t; \alpha, \beta), -\frac{\partial V}{\partial x}(t, x(t; \alpha, \beta); \alpha)\right)$$

$$+ \left[\frac{\partial H}{\partial \lambda}\left(t, x(t; \alpha, \beta), -\frac{\partial V}{\partial x}(t, x(t; \alpha, \beta); \alpha)\right) - \dot{x}(t; \alpha, \beta)\right]$$

$$\times \frac{\partial^2 V}{\partial x^2}(t, x(t; \alpha, \beta); \alpha)$$

$$= -\frac{\partial H}{\partial x}(t, x(t; \alpha, \beta), \lambda(t; \alpha, \beta)) + [\Psi(t, x(t; \alpha, \beta)) - \dot{x}(t; \alpha, \beta)]$$

$$\times \frac{\partial^2 V}{\partial x^2}(t,x(t;\alpha,\beta);\alpha)$$

$$= -\frac{\partial H}{\partial x}(t,x(t;\alpha,\beta),\lambda(t;\alpha,\beta)).$$

This proves the result.   □

Example. Suppose $f \in C([a,b])$ is positive and consider the fixed-endpoint problem [CV] with objective

$$I[x] = \int\limits_a^b f(x)\sqrt{1+\dot{x}^2}dt.$$

If we define

$$\lambda = -\frac{\partial L}{\partial \dot{x}}(x,\dot{x}) = -f(x)\frac{\dot{x}}{\sqrt{1+\dot{x}^2}},$$

then, solving for $\dot{x}$, we have that

$$\dot{x} = -\frac{\lambda}{\sqrt{f(x)^2 - \lambda^2}},$$

and thus the Hamiltonian H becomes

$$H(x,\lambda) = L(x,\dot{x}) + \lambda\dot{x} = f(x)\sqrt{1+\dot{x}^2} + \lambda\dot{x}$$

$$= f(x)\frac{f(x)}{\sqrt{f(x)^2 - \lambda^2}} - \frac{\lambda^2}{\sqrt{f(x)^2 - \lambda^2}}$$

$$= \sqrt{f(x)^2 - \lambda^2}.$$

The Hamilton–Jacobi equation then simply has the form

$$\frac{\partial V}{\partial t} = \sqrt{f(x)^2 - \left(\frac{\partial V}{\partial x}\right)^2},$$

or in a more symmetric expression,

$$\left(\frac{\partial V}{\partial t}\right)^2 + \left(\frac{\partial V}{\partial x}\right)^2 = f(x)^2.$$

Typically, as in this case, the Hamilton–Jacobi equation is nonlinear and does not fall into any established class of partial differential equations. However, in this particular case, the equation is separable and can easily be solved. With an ansatz of the form $V(t,x) = A(t) + B(x)$, the PDE reduces to

$$\dot{A}(t)^2 = f(x)^2 - B'(x)^2,$$

and since one side depends only on t, the other only on x, this quantity must be a positive constant, say α^2. But then the solution simply is

$$V(t,x;\alpha) = V(t_0,x_0;\alpha) + \int_{t_0}^{t} \frac{\partial V}{\partial t}(s,x_0;\alpha)\,ds + \int_{x_0}^{x} \frac{\partial V}{\partial x}(t,\xi;\alpha)\,d\xi$$

$$= V(t_0,x_0;\alpha) + \alpha(t-t_0) + \int_{x_0}^{x} \sqrt{f(\xi)^2 - \alpha^2}\,d\xi.$$

By the above proposition, extremals are the level curves of the function

$$\frac{\partial V}{\partial \alpha}(t,x;\alpha) = t - \alpha \int_{x_0}^{x} \frac{d\xi}{\sqrt{f(\xi)^2 - \alpha^2}}.$$

1.10 From the Calculus of Variations to Optimal Control

In the previous section we defined the Hamiltonian as it was done historically (except for a change in sign that is immaterial). In optimal control theory, a different—and in the view of the optimal control problem, a more natural—formulation is used. It was argued by Sussmann and Willems in [245, p. 39] that these are the form of *"Hamilton's equations as he should have written them."* At first sight, the difference is inconspicuous. Rather than using a field of extremals to define a vector field of directions, $\Psi(t,x)$, we instead replace the derivative $\dot{x}$ by introducing a new variable u, $\dot{x} = u$, the control, and write the Hamiltonian (also changing the order between x and λ to make it consistent with the usage in optimal control) in the form

$$H : [a,\infty) \times \mathbb{R} \times \mathbb{R} \times \mathbb{R} \to \mathbb{R}, \quad (t,\lambda,x,u) \mapsto H(t,\lambda,x,u) = L(t,x,u) + \lambda u. \quad (1.59)$$

In order to distinguish this function from its classical form, in this section we call it the *control Hamiltonian*. This formulation allows to combine the fundamental necessary conditions for optimality in the calculus of variations, the Euler–Lagrange equation and the Weierstrass condition, into a coherent structure as it was never achieved in the classical approach. For if we define

$$\lambda(t) = -\frac{\partial L}{\partial u}(t,x(t),\dot{x}(t)),$$

then, by the Euler–Lagrange equation, we have the following three conditions:

$$\dot{x}(t) = u(t) = \frac{\partial H}{\partial \lambda}(t, \lambda(t), x(t), \dot{x}(t)),$$

$$\dot{\lambda}(t) = \frac{d}{dt}\left(-\frac{\partial L}{\partial u}(t, x(t), \dot{x}(t))\right) = -\frac{\partial L}{\partial x}(t, x(t), \dot{x}(t)) = -\frac{\partial H}{\partial x}(t, \lambda(t), x(t), \dot{x}(t)),$$

and automatically

$$0 = \frac{\partial H}{\partial u}(t, \lambda(t), x(t), \dot{x}(t)).$$

Furthermore, the Weierstrass condition states that for any $u \in \mathbb{R}$ we have that

$$L(t, x(t), \dot{x}(t)) - \frac{\partial L}{\partial u}(t, x(t), \dot{x}(t))\dot{x}(t) \leq L(t, x(t), u) - \frac{\partial L}{\partial u}(t, x(t), \dot{x}(t))u$$

and this can be rewritten as

$$H(t, \lambda(t), x(t), \dot{x}(t)) = \min_{u \in \mathbb{R}} H(t, \lambda(t), x(t), u).$$

Thus, *it is a necessary condition of optimality for the minimization problem* [CV] *that the control Hamiltonian be minimized in the variable u at the point $\dot{x}(t)$*. If we had retained the classical formulation of the Hamiltonian with its sign from physics, at this point a necessary condition for minimizing in the problem [CV] would be to maximize this function. To us, it thus seems more natural to have the signs as we have chosen them. An immediate corollary of this minimization property is the Legendre condition,

$$0 \leq \frac{\partial^2 H}{\partial u^2}(t, \lambda(t), x(t), \dot{x}(t)) = \frac{\partial^2 L}{\partial u^2}(t, x(t), \dot{x}(t)).$$

Thus, by introducing the formulation of the control Hamiltonian, the necessary conditions for optimality of the calculus of variations take the following canonical form:

[CV-OC]   If a function $x : [a, b] \to \mathbb{R}$, $t \mapsto x(t)$, is a minimum for the calculus of variations problem [CV], then there exists a function $\lambda : [a, b] \to \mathbb{R}$, $t \mapsto \lambda(t)$, such that with $u(t) = \dot{x}(t)$ we have that

$$\dot{x}(t) = \frac{\partial H}{\partial \lambda}(t, \lambda(t), x(t), u(t)), \qquad \dot{\lambda}(t) = -\frac{\partial H}{\partial x}(t, \lambda(t), x(t), u(t)),$$

and the function $v \mapsto H(t, \lambda(t), x(t), v)$ is minimized at the point $u(t)$,

$$H(t, \lambda(t), x(t), u(t)) = \min_{v \in \mathbb{R}} H(t, \lambda(t), x(t), v).$$

This formulation succinctly combines the Weierstrass condition, the Euler–Lagrange equation, and the Legendre condition and shows their interdependencies. But there is much more to this formulation: any reference to differentiability of the "control" u has disappeared, and in the formulation itself, the condition $\dot{x} = u$ is immaterial. More generally, if we replace the notion of differentiable curves x by one of "controlled trajectories," that is curves x that are solutions to a *dynamical equation* $\dot{x} = f(t,x,u)$ driven by some control, and define the Hamiltonian now as

$$H(t,\lambda,x,u) = \lambda_0 L(t,x,u) + \lambda f(t,x,u)$$

with an extra multiplier $\lambda_0 \in \{0,1\}$ at the objective, then basically the same necessary conditions are valid with the one exception that the new multiplier λ_0 actually can be zero. These so-called *abnormal* extremals do not occur in the simplest problem in the calculus of variations considered here, but they are also possible for other more complicated problem formulations in the calculus of variations [67]. Furthermore, it no longer needs to be allowed that the control can take values anywhere, but the values can be restricted to lie in an *arbitrary control set* $U \subset \mathbb{R}$. These are the fundamental necessary conditions for optimality for an optimal control problem to be developed in subsequent chapters.

1.11 Notes

The material that was presented in this chapter is classical, and its pieces can be found in one formulation or other in many textbooks on the topic, e.g., [39, 40, 67, 70, 105, 118, 136, 185, 260]. Our exposition is somewhat different with its emphasis on the geometric aspects of the theory, and we also have provided the more technical and, at times, difficult computational details. Our main source for the historical comments was the article [245] by H. Sussmann and J. Willems, and we also based our presentation of the connection between Hamilton–Jacobi theory and optimal control theory on this excellent article, which we highly recommend as additional reading. Our treatment of the problem of minimum surfaces of revolution is based on the nicely written, but unfortunately somewhat out-of-fashion, text by G. Bliss [39]. But we have added numerous computational details to make his treatment more accessible to the reader. We have chosen this particular problem as a vehicle to develop the theoretical results, since it beautifully illustrates the geometric properties that really stand behind the conditions of the calculus of variations. These were the concepts that we wanted to emphasize in this introductory chapter. We will further develop these geometric properties for the optimal control problem, and in this sense our presentation was guided by the material still to come. Our source for the material on quadratic forms in Sect. 1.4 was the textbook by Gelfand and Fomin [105].

Chapter 2
The Pontryagin Maximum Principle:
From Necessary Conditions to the Construction
of an Optimal Solution

We now proceed to the study of a finite-dimensional optimal control problem, i.e., a dynamic optimization problem in which the state of the system, $x = x(t)$, is linked in time to the application of a control function, $u = u(t)$, by means of the solution to an ordinary differential equation whose right-hand side is shaped by the control. We now consider multidimensional systems in which both the state and the control variables no longer need to be scalar. In particular, the results presented here also provide high-dimensional generalizations for the classical theorems of the calculus of variations developed in Chap. 1. So far, we have considered only the simplest problem in the calculus of variations in which the functional is minimized over all curves that satisfy prescribed boundary conditions. Much more than in the calculus of variations, an optimal control problem is determined by its *constraints*. Of these, the most important one is represented by the *dynamics*, which in this text will always be given by an ordinary differential equation,

$$\dot{x} = f(t, x, u(t)),$$

and the optimization is carried out over a subset of solutions to this differential equation, so-called *admissible controlled trajectories*, not just simply over all differentiable curves. In most optimal control problems, the controls are required to satisfy *control constraints* in the form

$$u \in U$$

requiring that the control function $u(t)$ take values in a prescribed set U at (almost) all times t. This set U is called the control set and in our formulations will always be taken as a subset of $\mathbb{R}^m$, but otherwise arbitrary. For example, the choice $U = \{u_1, \ldots, u_r\}$ would define a control system that is allowed to switch between r possible settings. We also consider *terminal constraints* of the form

$$(T, x(T)) \in N,$$

H. Schättler and U. Ledzewicz, *Geometric Optimal Control: Theory, Methods and Examples*, Interdisciplinary Applied Mathematics 38, DOI 10.1007/978-1-4614-3834-2_2, © Springer Science+Business Media, LLC 2012

where T denotes the final time on the trajectory and N is a subset of the combined time–state space $\mathbb{R} \times \mathbb{R}^n$. Restrictions on the final time T, for example a fixed terminal time, will be included in this constraint. We shall impose assumptions that make N a "nice" geometric object. Many more types of constraints are conceivable and occur in real systems. For example, state-space constraints restrict the state of the system from entering prohibited regions. Mixed control-state constraints are simultaneous requirements on the state and control in the sense that if the state of the system has a specific value, then only a limited choice of control actions is available. These clearly are realistic and important scenarios. However, the inclusion of constraints of this type leads to a more complex theory, and in this text we restrict our treatment to what is *a finite-dimensional optimal control problem with control and terminal constraints*. Given these constraints, we then consider an *objective* of the form

$$ \mathscr{J}(u) = \int_{t_0}^{T} L(s, x(s), u(s)) ds + \varphi(T, x(T)) $$

with the integral representing the running cost along the controlled trajectory and the function φ defined on N defining a penalty term on the final state. A precise problem formulation including all assumptions will be given in Sect. 2.2, which also contains a statement of the main necessary conditions for optimality, the Pontryagin maximum principle [193].

The rest of the chapter will then be devoted to illustrating the use of this result, with the proof deferred until Chap. 4. Among the illustrations we provide, we include a statement of the necessary conditions for optimality for the calculus of variations problem in $\mathbb{R}^n$ (Sect. 2.3), the classical linear-quadratic regulator (Sects. 2.1 and 2.4), several examples of optimal solutions for the time-optimal control problem to the origin in $\mathbb{R}^2$ for time-invariant linear systems (Sects. 2.5 and 2.6) and some classical examples of optimal control problems with a time-varying or nonlinear dynamics (Sect. 2.7). General properties of optimal solutions for the time-optimal control problem for nonlinear systems that are affine functions of the control(s) will be developed in Sect. 2.8, which provides an introduction to some of the Lie derivative-based techniques that form the basis for geometric methods in optimal control. This section also includes a discussion of singular controls and additional necessary conditions for optimality of the corresponding controlled trajectories, such as the Legendre–Clebsch condition. We then use the developed theory to analyze some generic cases for the time-optimal control problem in the plane (Sects. 2.9 and 2.10). These results, due to H. Sussmann [230, 236], serve as a first illustration of the power of geometric methods in the solution of optimal control problems. We close this chapter (Sect. 2.11) with a derivation of the optimal controls for the Fuller problem, a classical optimal control problem whose solutions are given by chattering arcs, i.e., the associated controls switch infinitely often on a finite interval and are no longer piecewise continuous.

In this chapter, the emphasis is on illustrating the use of the necessary conditions for optimality of the maximum principle. We simplify the presentation by making the mathematically unjustified, but in practical problems often satisfied, assumption that optimal controls are piecewise continuous. With only minor modifications, all the results presented in this chapter remain valid for the more general class of locally bounded Lebesgue measurable functions, and in subsequent chapters we then shall work with this, for our purpose, adequately general class of controls.

We close these introductory comments with establishing our notation. The equations of the maximum principle and many of the involved computations can be written in a concise and elegant form that avoids the use of matrix transpositions so common in the classical textbooks if a proper notation is established. In this chapter, our state space will always be $\mathbb{R}^n$ or some open subset of it, and we write the state x as a column vector. However, in our notation, we already here distinguish between what in the formulation on manifolds will be tangent vectors, which we write as column vectors, and cotangent vectors (or covectors for short), which we write as row vectors. For example, $\dot{x}$ is a tangent vector, and thus the right-hand side of the dynamics, $f(t,x,u)$ in the formulations above, is a column vector. On the other hand, geometrically, multipliers λ represent linear functionals and thus are covectors. We denote the space of n-dimensional covectors or row vectors by $(\mathbb{R}^n)^*$, but do not distinguish between $\mathbb{R}$ and $\mathbb{R}^*$. For a scalar continuously differentiable function $h : \mathbb{R}^n \to \mathbb{R}$, $x \mapsto h(x)$, we consistently write the gradient with respect to x as a row vector and denote it by $\nabla h(x)$ or $\frac{\partial h}{\partial x}(x)$, i.e.,

$$\nabla h(x) = \frac{\partial h}{\partial x}(x) = \left(\frac{\partial h}{\partial x_1}(x), \ldots, \frac{\partial h}{\partial x_n}(x) \right).$$

For a vector-valued continuously differentiable map H,

$$H : \mathbb{R}^k \to \mathbb{R}^\ell, \qquad x \mapsto H(x) = \begin{pmatrix} h_1(x) \\ \ldots \\ h_k(x) \end{pmatrix},$$

we denote the Jacobian matrix of the partial derivatives of the components $h_i(x)$ with respect to the variables x_j by

$$DH(x) = \frac{\partial H}{\partial x}(x) = \begin{pmatrix} \frac{\partial h_1}{\partial x_1}(x) & \cdots & \frac{\partial h_1}{\partial x_k}(x) \\ \vdots & \frac{\partial h_i}{\partial x_j}(x) & \vdots \\ \frac{\partial h_k}{\partial x_1}(x) & \cdots & \frac{\partial h_k}{\partial x_k}(x) \end{pmatrix}_{1 \le i,j \le k},$$

with i as row index and j as column index. Thus, the Jacobian matrix is the matrix whose ith row is given by the gradient of the component h_i. The Hessian matrix of a twice continuously differentiable function $h : \mathbb{R}^n \to \mathbb{R}$, $x \mapsto h(x)$, is the matrix of the second-order partial derivatives of h and will be denoted by $D^2 h(x) = \frac{\partial^2 h}{\partial x^2}(x)$. With the convention above, the Hessian of h is the Jacobian matrix of the transpose of the gradient of h,

$$D^2 h(x) = \frac{\partial^2 h}{\partial x^2}(x) = \frac{\partial (\nabla h)^T}{\partial x}(x).$$

If $\Lambda = (\lambda_1, \ldots, \lambda_n)$ is a row vector of continuously differentiable functions $\lambda_j : \mathbb{R}^n \to \mathbb{R}$, $x \mapsto \lambda_j(x)$, $j = 1, \ldots, n$, then, and consistent with the notation just introduced, we denote the matrix of the partial derivatives $\left(\frac{\partial \lambda_j}{\partial x_i}\right)_{1 \le i, j \le n}$ with row index i and column index j by $\frac{\partial \Lambda}{\partial x}$, that is,

$$\frac{\partial \Lambda}{\partial x}(x) = \left(\frac{\partial \Lambda^T}{\partial x}(x)\right)^T \quad \text{or} \quad D\Lambda(x) = \left(D\left(\Lambda^T(x)\right)\right)^T.$$

Not only does this formalism properly distinguish the different geometric meanings of the variables involved, but it also allows us to write almost all formulas without having to use transposes and simplifies the notation considerably.

Finally, we denote the space of all $k \times \ell$ matrices of real numbers by $\mathbb{R}^{k \times \ell}$. We assume that the reader is familiar with the basic concepts of matrix algebra and recall that a matrix $P \in \mathbb{R}^{n \times n}$ is *positive semidefinite* if it is symmetric and if $v^T P v \ge 0$ for any vector $v \in \mathbb{R}^n$; P is said to be *positive definite* if P is positive semidefinite and if in addition, $v^T P v = 0$ holds only for $v = 0$. It is well-known from linear algebra that a matrix P is positive definite/semidefinite if and only if all eigenvalues are positive/nonnegative. Note that as a symmetric matrix, P has a full set of n real eigenvalues.

2.1 Linear-Quadratic Optimal Control

Before formulating the general optimal control problem, we first fully solve by elementary means what, from an applications point of view, justifiably may be considered the single most important optimal control problem, the so-called *linear-quadratic regulator*. Mathematically, this is but a small extension of the simplest problem in the calculus of variations—neither control nor terminal constraints are imposed—in the sense that the trivial dynamics $\dot{x} = u$ is replaced by a linear differential equation $\dot{x} = Ax + Bu$ and the objective to be minimized is a positive definite quadratic form in x and u. Standard calculus of variations techniques suffice to solve this problem. In fact, Legendre's idea of "completing the square" presented in Sect. 1.4 works to perfection here and in this section we give an elementary and self-contained derivation of the optimal solution based on Legendre's argument.

The importance of the problem lies in its practical applications. Essentially, this is the problem to regulate a typically nonlinear system around some reference trajectory. In the mathematical formulation below, the reference trajectory and control are normalized to be $x \equiv 0$ and $u \equiv 0$. As such, but also due to the simplicity of its solution and the fact that this solution easily allows the inclusion of stochastic effects (e.g., noisy measurements and estimation of the states from an incomplete set of measurements by means of the Kalman filter), the linear-quadratic regulator is the theoretical basis for many practical control schemes whose aim is to regulate a system around some set point. Real systems based on this principle range from autopilots in commercial aircraft to advanced stability control systems in cars to standard chemical process control. Naturally, this problem, and its manifold extensions, are the subject of numerous textbooks on automatic control, one of the best still being the classical text by Kwakernaak and Sivan [144]. For this reason, this topic is not in the focus of our presentation in this text, and we shall limit ourselves to its connection with conjugate points and perturbation feedback control for nonlinear optimal control problems. These will be discussed in the context of sufficient conditions for a strong local minimum in Sect. 5.3.

Let $[0,T]$ be a finite and fixed time horizon and suppose

$$A : [0,T] \to \mathbb{R}^{n \times n}, \ t \mapsto A(t), \qquad B : [0,T] \to \mathbb{R}^{n \times m} \ t \mapsto B(t),$$

$$Q : [0,T] \to \mathbb{R}^{n \times n}, \ t \mapsto Q(t), \qquad R : [0,T] \to \mathbb{R}^{m \times m} \ t \mapsto R(t),$$

are continuous matrix-valued functions defined on $[0,T]$. We assume that the matrices $Q(t)$ and $R(t)$ are symmetric and in addition that $Q(t)$ is positive semidefinite and $R(t)$ is positive definite for all $t \in [0,T]$. Furthermore, let $S_T \in \mathbb{R}^{n \times n}$ be a constant, symmetric, and positive semidefinite matrix. The *linear-quadratic regulator* then is the following optimal control problem:

[LQ] Find a continuous function $u : [0,T] \to \mathbb{R}^m$, the control, that minimizes a quadratic objective of the form

$$J(u) = \frac{1}{2} \int_0^T \left[x^T(t)Q(t)x(t) + u^T(t)R(t)u(t) \right] dt + \frac{1}{2} x^T(T)S_T x(T) \qquad (2.1)$$

subject to the linear dynamics

$$\dot{x}(t) = A(t)x(t) + B(t)u(t), \qquad x(0) = x_0. \qquad (2.2)$$

It follows from well-known results about ordinary differential equations (see Appendix B) that the initial value problem for the homogeneous linear matrix differential equation

$$\dot{X}(t) = A(t)X(t) \qquad \text{and} \qquad X(s) = \mathrm{Id}$$

has a unique solution $\Phi(t,s)$, called its fundamental solution. For any initial time $s \in [0,T]$, this solution exists over the full interval $[0,T]$. The unique solution $x(t;x_0)$ to the homogeneous vector equation $\dot{x}(t) = A(t)x(t)$ with initial condition $x(0) = x_0$ is then given by $x(t;x_0) = \Phi(t,0)x_0$, and as is easily verified, the solution to the inhomogeneous equation (2.2) is obtained by variation of constants as

$$x(t;x_0) = \Phi(t,0) \left(x_0 + \int_0^t \Phi(0,s)B(s)u(s)ds \right).$$

This solution is called the *trajectory corresponding to the control u*. If the matrix A is time-invariant, then $\Phi(t,s)$ is simply given by the absolutely convergent matrix exponential,

$$\Phi(t,s) = \exp(A(t-s)) = \sum_{k=0}^{\infty} \frac{A^k}{k!}(t-s)^k.$$

In the time-varying case, for scalar problems, it is still possible to write down an explicit formula as

$$\Phi(t,s) = \exp\left(\int_s^t A(r)dr\right),$$

but in dimensions $n \geq 2$ this formula no longer is valid, since generally $A(t)$ and $\exp\left(\int_s^t A(r)dr\right)$ do not commute. Series expansions of the solution can still be given in higher dimensions and are related to Lie-algebraic formulas in connection with the Baker–Campbell–Hausdorff formula involving commutators [256] (see also Sect. 4.5), but will not be needed here. The important fact simply is that the fundamental matrix Φ exists and is unique.

Theorem 2.1.1. *The solution to the linear-quadratic optimal control problem* [LQ] *is given by the linear feedback control*

$$u_*(t,x) = -R(t)^{-1}B(t)^T S(t)x,$$

where S is the solution to the Riccati terminal value problem

$$\dot{S} + SA(t) + A^T(t)S - SB(t)R(t)^{-1}B^T(t)S + Q(t) \equiv 0, \qquad S(T) = S_T. \qquad (2.3)$$

This solution S exists on the full interval $[0,T]$ *and is positive semidefinite. The minimal value of the objective is given by* $\frac{1}{2}x_0^T S(0)x_0$.

Proof. This is Legendre's argument from the calculus of variations adjusted to this setting. Let $u : [0,T] \to \mathbb{R}^m$ be any continuous control and let $x : [0,T] \to \mathbb{R}^n$ denote the corresponding trajectory. Dropping the argument t from the notation, we have for any differentiable matrix function $S \in \mathbb{R}^{n \times n}$ that

$$\frac{d}{dt}\left(x^T Sx\right) = \dot{x}^T Sx + x^T \dot{S}x + x^T S\dot{x}$$

$$= (Ax+Bu)^T Sx + x^T \dot{S}x + x^T S(Ax+Bu),$$

and thus, by adjoining this quantity to the Lagrangian in the objective, we can express the cost equivalently as

$$J(u) = \frac{1}{2}\int_0^T \left[x^T(Q+A^TS+\dot{S}+SA)x + x^T SBu + u^T B^T S^T x + u^T Ru\right] dt$$

$$+ \frac{1}{2}x^T(T)[S_T - S(T)]x(T) + \frac{1}{2}x_0^T S(0)x_0.$$

Take S as a symmetric matrix and complete the square to get

$$J(u) = \frac{1}{2} \int_0^T \left[x^T (\dot{S} + SA + A^T S - SBR^{-1}B^T S + Q)x \right.$$
$$\left. + (u + R^{-1}B^T Sx)^T R(u + R^{-1}B^T Sx) \right] dt$$
$$+ \frac{1}{2} x^T (T)[S_T - S(T)]x(T) + \frac{1}{2} x_0^T S(0)x_0.$$

For the moment, let us assume that there exists a solution S to the matrix Riccati equation (2.3) over the full interval $[0,T]$. Then the objective simplifies to

$$J(u) = \frac{1}{2} \int_0^T (u + R^{-1}B^T Sx)^T R(u + R^{-1}B^T Sx)dt + \frac{1}{2} x_0^T S(0)x_0.$$

Since the matrix R is continuous and positive definite over $[0,T]$, the minimum is realized if and only if

$$u(t) = -R^{-1}(t)B^T (t)S(t)x(t),$$

and the minimum value is given by

$$\frac{1}{2} x_0^T S(0)x_0.$$

Thus the optimal solution to the linear-quadratic control problem is given as a linear *feedback* function, i.e., a function $u_* : [0,T] \times \mathbb{R}^n \to \mathbb{R}^m$ defined in the time–state space, given by

$$u_*(t,x) = -R^{-1}(t)B^T (t)S(t)x.$$

For this argument to be valid, it remains to argue that such a solution S to the initial value problem (2.3) indeed does exist on all of $[0,T]$. It follows from general results about the existence of solutions to ordinary differential equations that such a solution exists on some maximal interval $(\tau,T]$ and that as $t \searrow \tau$ (i.e., $t \to \tau$ and $t > \tau$), at least one of the components of the solution $S(t)$ needs to diverge to $+\infty$ or $-\infty$. For if this were not the case, then by the local existence theorem on ODEs, the solution could be extended further onto some small interval $(\tau - \varepsilon, \tau + \varepsilon)$, contradicting the maximality of the interval $(\tau,T]$. In general, however, this explosion time τ could be nonnegative, invalidating the argument above. That this is not the case for the linear-quadratic regulator problem is a consequence of the positivity assumptions on the objective, specifically, the definiteness assumptions on the matrices R, Q, and S_T.

In order to see this, suppose the explosion time τ of the solution to the Riccati equation is nonnegative, $\tau \geq 0$, and consider the linear-quadratic regulator problem for variable initial conditions $(t_0, x_0) \in [0,T] \times \mathbb{R}^n$. If $t_0 > \tau$, then the reasoning above is valid; thus the solution to the minimization problem [LQ] is given by the feedback control $u_*(t,x)$, and the minimal value is $J(u_*) = \frac{1}{2} x_0^T S(t_0)x_0$.

This holds for arbitrary initial conditions x_0. Since $J(u)$ is always nonnegative by our assumptions on the matrices in the objective, the matrix $S(t_0)$ must be positive semidefinite. But we can choose t_0 arbitrarily in the interval $(\tau, T]$, and thus it follows that the matrix $S(t)$ is positive semidefinite on this interval. Furthermore, since for any other control u defined on $[t_0, T]$ we have that

$$J(u) \geq \frac{1}{2} x_0^T S(t_0) x_0,$$

using the control $u \equiv 0$, we obtain an upper bound in the form

$$0 \leq \frac{1}{2} x_0^T S(t_0) x_0 \leq \frac{1}{2} x_0^T \left(\int_{t_0}^T \Phi(t,t_0)^T Q(t) \Phi(t,t_0) dt + \Phi(T,t_0)^T S_T \Phi(T,t_0) \right) x_0$$

(2.4)

for every $x_0 \in \mathbb{R}^n$. Choosing for the initial condition x_0 the ith coordinate vectors, $e_i = (0, \ldots, 0, 1, 0, \ldots, 0)^T$, with the 1 in the ith position, the lower estimate in Eq. (2.4) gives $S_{ii}(t_0) \geq 0$. The upper estimate is continuous in t_0 on the full interval $[0, T]$ and thus remains bounded over the full interval. Hence there exists a positive constant C such that

$$0 \leq S_{ii}(t_0) \leq C \quad \text{for all} \quad t_0 \in (\tau, T].$$

Choosing $x_0 = e_i \pm \theta e_j$, we furthermore obtain

$$0 \leq (e_i \pm \theta e_j)^T S(t_0)(e_i \pm \theta e_j) = S_{ii}(t_0) + 2\theta S_{ij}(t_0) + \theta^2 S_{jj}(t_0)$$

for all $\theta \in \mathbb{R}$, which is equivalent to

$$S_{ij}^2(t_0) \leq S_{ii}(t_0) S_{jj}(t_0).$$

But then all entries $S_{ij}(t_0)$ of the matrix $S(t_0)$ take values in the interval $[-C, C]$ for all times t_0 from the interval $(\tau, T]$. Hence there cannot be an explosion of the solution as $t_0 \searrow \tau$. This contradicts the fact that $(\tau, T]$ is the maximal interval of existence for the solution S of Eq. (2.3). Thus we must have $\tau < 0$, and the solution to the Riccati equation exists over the full interval $[0, T]$.   □

2.2 Optimal Control Problems

We now formulate the optimal control problem to be considered in this text and introduce the main necessary conditions for optimality, the Pontryagin maximum principle [193].

2.2.1 Control Systems

We think of a control system as a collection of time-dependent vector fields on a differentiable manifold parameterized by controls that by means of the solutions of the corresponding ordinary differential equations, give rise to a family of controlled trajectories. An optimal control problem then is the task to minimize some functional over these controlled trajectories subject to additional constraints. We shall postpone a precise definition along these lines until Chap. 4, where we actually prove the maximum principle. Here, in view of the still introductory character of this chapter, we retain the more elementary formulation of optimal control problems with state space $\mathbb{R}^n$. However, we already arrange the material according to this framework.

Definition 2.2.1. A **control system** is a 4-tuple $\Sigma = (M, U, f, \mathscr{U})$ consisting of a state space M, a control set U, a dynamics f, and a class $\mathscr{U}$ of admissible controls.

Throughout this chapter, we make the following assumptions about the data defining the control system:

1. The *state space* M is an open and connected subset of $\mathbb{R}^n$.
2. The *control set* U is a subset of $\mathbb{R}^m$. No further regularity conditions on the structure of U need to be imposed, although in many practical situations U is compact and convex.
3. The *dynamics* $\dot{x} = f(t, x, u)$ is defined by a family of time-varying vector fields f parameterized by the control values $u \in U$,

$$f : \mathbb{R} \times M \times U \to \mathbb{R}^n, \quad (t, x, u) \mapsto f(t, x, u),$$

   i.e., f assigns to every point $(t, x, u) \in \mathbb{R} \times M \times U$ a (tangent) vector $f(t, x, u) \in \mathbb{R}^n$. We assume that the time-varying vector fields are continuous in (t, x, u), differentiable in x for fixed $(t, u) \in \mathbb{R} \times U$, and that the partial derivatives $\frac{\partial f}{\partial x}(t, x, u)$ are continuous as a function of all variables; no differentiability assumptions in the control variable u are made.
4. The class $\mathscr{U}$ of *admissible controls* is taken to be piecewise continuous functions u defined on a compact interval $I \subset \mathbb{R}$ with values in the control set U. Without loss of generality, we assume that controls are continuous from the left.

These specifications are simplifications of the setting considered in Chap. 4. Here our aim is to formulate the fundamental necessary conditions for optimality and then to illustrate how these conditions can be put to work. For this, the simpler framework formulated above that requires only some knowledge of advanced calculus and ordinary differential equations is adequate, and it simplifies the technical aspects of the theory. In the more general framework considered in Chap. 4, the state space M will be a C^r-manifold, and the class $\mathscr{U}$ of admissible controls will consist of all locally bounded Lebesgue measurable functions u that take values in the control set U, i.e., given a compact interval $I \subset \mathbb{R}$, there exists a compact subset

V of U such that u takes values in V almost everywhere on I. In particular, if the control set U already is compact, then admissible controls are simply Lebesgue measurable functions that take values in U almost everywhere. The need for taking as admissible controls the class of Lebesgue measurable functions lies in the fact that the class of piecewise continuous controls simply is too small, and this will already be seen in Sect. 2.11 of this chapter, to guarantee the existence of optimal solutions. Greater generality is required for several important and fundamental results to be valid. Locally bounded Lebesgue measurable functions are pointwise limits of piecewise continuous functions and provide the required closure properties needed for many arguments. A brief exposition of Lebesgue measurable functions is given in Appendix D, but this will be needed only in Chaps. 3, 4, and some of 6. Similarly, many control systems, especially those connected with mechanical systems (e.g., robotic manipulators) have natural state-space descriptions that are manifolds. Clearly, the circle S^1 is a far superior model for the state space of a fixed-amplitude oscillation than $\mathbb{R}^2$. The sphere S^2 is the only reasonable model to calculate the shortest air route from Paris to Sydney. But these generalizations will be considered only in Chap. 4.

In the same spirit, we always impose conditions on the dynamics that for a given admissible control, guarantee not only the existence of solutions to the differential equation, but also its uniqueness. From an engineering perspective, this is as important[1] a condition as existence of solutions, and we will insist on it being satisfied. Using the practical class of piecewise continuous controls in this chapter suffices for our arguments and simplifies the reasoning. Given any piecewise continuous control $u \in \mathscr{U}$ defined over some open interval J, it follows from standard local existence and uniqueness results for ordinary differential equations (see Appendix B) that for any initial condition $x(t_0) = x_0$ with $t_0 \in J$, there exists a unique solution x to the initial value problem

$$\dot{x}(t) = f(t, x, u(t)), \qquad x(t_0) = x_0, \qquad (2.5)$$

defined over some maximal interval $(\tau_-, \tau_+) \subset J$ that contains t_0.

Definition 2.2.2 (Admissible controlled trajectory). Given an admissible control $u \in \mathscr{U}$ defined over an interval J, let x be the unique solution to the initial value problem (2.5) with maximal interval of definition $I = (\tau_-, \tau_+)$. We call this solution x the trajectory corresponding to the control u and call the pair (x, u) an admissible controlled trajectory over the interval I.

An optimal control problem then consists in finding, among all admissible controlled trajectories, one that minimizes an objective, possibly subject to additional constraints. In this text, in addition to the control constraints that are implicit in the definition of the control set, we consider only **terminal constraints** in the form of a target set into which the controls need to steer the system. However, we restrict the

---

[1]From our point of view, uniqueness may be the more important of the two conditions.

terminal set to have the regular geometric structure of a k-dimensional embedded submanifold N in $\mathbb{R} \times M$ (see Appendix C). More specifically, we assume that

$$N = \{(t,x) \in \mathbb{R} \times M : \Psi(t,x) = 0\},$$

where $\Psi : \mathbb{R} \times M \to \mathbb{R}^{n+1-k}$, $(t,x) \mapsto \Psi(t,x) = (\psi_0(t,x), \ldots, \psi_{n-k}(t,x))^T$, is a continuously differentiable mapping and the matrix $D\Psi$ of the partial derivatives with respect to (t,x) is of full rank $n + 1 - k$ everywhere on N, i.e., the gradients of the functions $\psi_0(t,x), \ldots, \psi_{n-k}(t,x)$ are linearly independent on N.

Finally, the **objective** is given in so-called Bolza form as the integral of a Lagrangian L plus a penalty term φ. For the Lagrangian we make the same regularity assumptions as on the dynamics f, i.e., the function

$$L : \mathbb{R} \times M \times U \to \mathbb{R}, \quad (t,x,u) \mapsto L(t,x,u),$$

is continuous in (t,x,u), differentiable in x for fixed $(t,u) \in \mathbb{R} \times U$, and the derivative $\frac{\partial L}{\partial x}(t,x,u)$ is continuous as a function of all variables. The penalty term φ is given by a continuously differentiable function

$$\varphi : \mathbb{R} \times M \to \mathbb{R}, \quad (t,x) \mapsto \varphi(t,x).$$

Clearly, this function needs to be defined only on N. Since we assume that N is an embedded submanifold of $\mathbb{R}^{n+1}$, if necessary, we can always extend φ to a differentiable function $\varphi : \mathbb{R} \times M \to \mathbb{R}$ locally, and thus for simplicity we assume that φ is defined in the ambient state space. The objective or cost functional is then given as

$$\mathscr{J}(u) = \int_{t_0}^{T} L(s,x(s),u(s))ds + \varphi(T,x(T)), \tag{2.6}$$

where x is the unique trajectory corresponding to the control u. The terminal time T can be fixed or free. A fixed terminal time simply will be modeled as the equation $\varphi_0(t,x) = t - T$ in the mapping Ψ defining the constraint in N. The initial time t_0 and initial condition x_0 are fixed, but arbitrary. Then the optimal control problem is the following one:

[OC] Minimize the objective $\mathscr{J}(u)$ over all admissible controlled trajectories (x,u) defined over an interval $[t_0,T]$ that satisfy the terminal constraint $(T,x(T)) \in N$.

2.2.2 The Pontryagin Maximum Principle

The maximum principle of optimal control gives the fundamental necessary conditions for a controlled trajectory (x,u) to be optimal. It was developed in the mid 1950s in the Soviet Union by a group of mathematicians under the leadership of L.S. Pontryagin, also including V.G. Boltyanskii, R.V. Gamkrelidze, and E.F. Mishchenko, and is known as the Pontryagin maximum principle [41, 193]. Below, and consistent with our choice of admissible controls, we give its formulation under the additional assumption that the optimal control is piecewise continuous. Recall that we write tangent vectors as column vectors and cotangent vectors (i.e., multipliers) as row vectors.

Definition 2.2.3 (Hamiltonian). The (control) *Hamiltonian* function H of the optimal control problem [OC] is defined as

$$H : \mathbb{R} \times [0, \infty) \times (\mathbb{R}^n)^* \times \mathbb{R}^n \times \mathbb{R}^m \to \mathbb{R}$$

with

$$H(t, \lambda_0, \lambda, x, u) = \lambda_0 L(t, x, u) + \lambda f(t, x, u). \tag{2.7}$$

Theorem 2.2.1 (Pontryagin maximum principle). [193] *Let (x_*, u_*) be a controlled trajectory defined over the interval $[t_0, T]$ with the control u_* piecewise continuous. If (x_*, u_*) is optimal, then there exist a constant $\lambda_0 \geq 0$ and a covector $\lambda : [t_0, T] \to (\mathbb{R}^n)^*$, the so-called adjoint variable, such that the following conditions are satisfied:*

1. Nontriviality *of the multipliers:* $(\lambda_0, \lambda(t)) \neq 0$ *for all $t \in [t_0, T]$.*
2. Adjoint equation*: the adjoint variable λ is a solution to the time-varying linear differential equation*

$$\dot{\lambda}(t) = -\lambda_0 L_x(t, x_*(t), u_*(t)) - \lambda(t) f_x(t, x_*(t), u_*(t)). \tag{2.8}$$

3. Minimum condition: *everywhere in $[t_0, T]$ we have that*

$$H(t, \lambda_0, \lambda(t), x_*(t), u_*(t)) = \min_{v \in U} H(t, \lambda_0, \lambda(t), x_*(t), v). \tag{2.9}$$

If the Lagrangian L and the dynamics f are continuously differentiable in t, then the function

$$h : t \mapsto H(t, \lambda_0, \lambda(t), x_*(t), u_*(t))$$

is continuously differentiable with derivative given by

$$\dot{h}(t) = \frac{dh}{dt}(t) = \frac{\partial H}{\partial t}(t, \lambda_0, \lambda(t), x_*(t), u_*(t)). \tag{2.10}$$

4. Transversality condition*: at the endpoint of the controlled trajectory, the covector*

$$(H + \lambda_0 \varphi_t, -\lambda + \lambda_0 \varphi_x)$$

is orthogonal to the terminal constraint N, i.e., there exists a multiplier $\nu \in (\mathbb{R}^{n+1-k})^*$ such that

$$H + \lambda_0 \varphi_t + v D_t \Psi = 0, \qquad \lambda = \lambda_0 \varphi_x + v D_x \Psi \qquad at \ (T, x_*(T)). \qquad (2.11)$$

The following statement is an immediate special case.

Corollary 2.2.1. *If the Lagrangian L and the dynamics f are time-invariant (do not depend on t), then the function $h : t \mapsto H(t, \lambda_0, \lambda(t), x_*(t), u_*(t))$ is constant. If φ and Ψ also do not depend on t (and in this case the terminal time T necessarily is free), then for any multiplier (λ_0, λ) that satisfies the conditions of the maximum principle, the Hamiltonian H vanishes identically along the optimal controlled trajectory (x_*, u_*):*

$$H(t, \lambda_0, \lambda(t), x_*(t), u_*(t)) \equiv 0. \qquad \square$$

We start our discussions of the maximum principle by introducing some useful terminology and give a brief and somewhat informal description of the significance of each condition.

Definition 2.2.4 (Extremals; normal and abnormal). We call controlled trajectories (x, u) for which there exist multipliers λ_0 and λ such that the conditions of the maximum principle are satisfied *extremals*, and the triples $(x, u, (\lambda_0, \lambda))$ including the multipliers are called *extremal lifts* (to the cotangent bundle in case of manifolds). If $\lambda_0 > 0$, then the extremal lift is called *normal* while it is called *abnormal* if $\lambda_0 = 0$.

1. **Normal and abnormal extremal lifts.** The maximum principle takes the form of a multiplier rule with multiplier $(\lambda_0, \lambda(t))$. The nontriviality condition precludes a trivial solution of these conditions with $(\lambda_0, \lambda(t)) = (0, 0)$. Since the conditions are linear in the multipliers (λ_0, λ), it is always possible to normalize this vector. For example, if $\lambda_0 > 0$, then the conditions do not change if we divide by λ_0 and instead consider as the new multiplier $(1, \tilde{\lambda}(t))$, where $\tilde{\lambda}(t) = \lambda(t)/\lambda_0$. Thus, without loss of generality, we may always assume that $\lambda_0 = 1$ if the extremal lift is normal. Note that it is a property of the extremal lift, not the controlled trajectory, to be normal or abnormal. It is possible that both normal and abnormal extremal lifts exist for a given controlled trajectory (x, u). For this reason, controlled trajectories for which only abnormal extremal lifts exist are sometimes called *strictly abnormal*. We shall see in Sect. 2.3 that all extremals for the simplest problem in the calculus of variations are normal, and this fact actually is the source of the terminology, which goes back to Carathéodory [67]. In spite of their name, abnormal extremals are by no means pathological situations, and if they exist, they often play an important role in determining the structure of optimal solutions. We shall see in Sect. 2.6 that the synthesis of optimal trajectories for the problem of steering points to the origin time-optimally for the harmonic oscillator with bounded controls, a simple and standard text book example, contains optimal, strictly abnormal extremals and that these play a crucial role in determining the overall structure of the solutions.

2. **Adjoint system.** First note that as a solution to a linear time-varying ordinary differential equation with piecewise continuous entries, the adjoint variable $\lambda(\cdot)$

exists over the full interval $[t_0, T]$. We shall see in the proof of the maximum principle in Sect. 4.2 that $(\lambda_0, \lambda(t))$ arises as a normal vector to a hyperplane in (t,x)-space (hence also the nontriviality condition) that evolves in time according to the adjoint equation. This equation arises as the adjoint in the sense of linear ordinary differential equations of the so-called *variational equation*

$$\dot{y} = f_x(t, x_*(t), u_*(t))y, \tag{2.12}$$

which transports tangent vectors (that will be generated by means of variations) along a reference controlled trajectory $t \mapsto (x_*(t), u_*(t))$. Solutions of the adjoint system provide the corresponding transport for covectors along this curve. In terms of the Hamiltonian H, the coupled system consisting of the dynamics and the adjoint equation can be written as

$$\dot{x}_*(t) = \frac{\partial H}{\partial \lambda}(t, \lambda_0, \lambda(t), x_*(t), u_*(t)) \quad \text{and} \quad \dot{\lambda}(t) = -\frac{\partial H}{\partial x}(t, \lambda_0, \lambda(t), x_*(t), u_*(t)) \tag{2.13}$$

and thus forms a *Hamiltonian system* that is coupled with the control u_* through the minimization condition (2.9).

3. **Minimum condition.** In the original formulation of the theorem by Pontryagin et al. [193], this condition was formulated as a maximum condition and gave the result its name. In fact, depending on the choice of the signs associated with the multipliers λ_0 and λ, the maximum principle can be stated in four equivalent versions. Here, since most of the problems we will be considering are cast as minimization problems, we prefer this more natural formulation, but retain the classical name. The minimum condition (2.9) states that in order to solve the minimization problem on the function space of controls, the control u_* needs to be chosen so that for some extremal lift, it minimizes the Hamiltonian H pointwise over the control set U, i.e., for every $t \in [t_0, T]$, the control $u_*(t)$ is a minimizer of the function $v \mapsto H(t, \lambda_0, \lambda(t), x_*(t), v)$ over the control set U. Note that it is not required just that the control satisfy the necessary conditions for minimality—and this is how a weak version of the maximum principle is formulated—but that the control $u_*(t)$ be a true minimizer over the control set U. This condition typically is the starting point for any analysis of an optimal control problem. Formally, we first try to "solve" the minimization condition (2.9) for the control u as a function of the other variables, $u = u(t, x_*; \lambda_0, \lambda)$, and then substitute the "result" into the differential equations for dynamics and adjoint variable to get

$$\dot{x} = f(t, x, u(t, x_*; \lambda_0, \lambda)), \qquad x(t_0) = x_0,$$

$$\dot{\lambda}(t) = -\lambda_0 L_x(t, x_*(t), u(t, x_*; \lambda_0, \lambda)) - \lambda(t)f_x(t, x_*(t), u(t, x_*; \lambda_0, \lambda)).$$

Since multiple solutions to the minimization problem can exist, this is not in general a unique specification of the control. Even if the minimization problem does have a unique solution, this solution depends on the multiplier, i.e., lives in the cotangent bundle, and thus need not give rise to unique controlled trajectories.

4. **Transversality conditions.** Equations (2.5) and (2.8) form a system in $2n+1$ variables (the state x, the multiplier λ, and the terminal time T) with the initial condition x_0 specified for the state at time t_0. Information about the remaining $n+1$ conditions is contained in the transversality conditions at the endpoint. The requirement that the terminal state lie on the manifold N, $(T,x(T)) \in N$, imposes $n+1-k$ conditions and thus leaves k degrees of freedom. The adjoint variable $\lambda(T) \in (\mathbb{R}^n)^*$ at the terminal time T is determined on the k-dimensional tangent space to N at $(T,x_*(T))$ by the relation

$$\lambda(T) = \lambda_0 \varphi_x(T,x_*(T)) + \nu D_x \Psi(T,x_*(T)$$

and the multiplier $\nu \in \left(\mathbb{R}^{n+1-k} \right)^*$ in this equation accounts for $n - (n+1-k) = k-1$ degrees of freedom, with the last degree of freedom taken up by the equation

$$H(T,\lambda_0,\lambda(T),x_*(T),u_*(T)) + \lambda_0 \varphi_t(T,x_*(T)) + \nu D_t \Psi(T,x_*(T)) = 0$$

which gives information about the terminal time T. Overall, there thus are $2n+1$ equations for the boundary values $x(T)$, $\lambda(T)$, and T. Hence, at least in nondegenerate situations, the transversality conditions provide the required information about the missing boundary conditions for both the adjoint variable and the terminal time T.

The geometric statement that the vector $(H + \lambda_0 \varphi_t, -\lambda + \lambda_0 \varphi_x)$ is orthogonal to the terminal constraint N at the endpoint of the controlled trajectory is valid for any embedded submanifold N. For since the condition is local, it is always possible to choose a collection of functions ψ_i, $i = 0,\ldots,n-k$, so that $N = \{(t,x) : \Psi(t,x) = 0\}$ and the gradients of the functions ψ_i are linearly independent at $(T,x_*(T))$. The gradients $\nabla \psi_i$ are all orthogonal to N, and since they are linearly independent, they span the space normal to N. Thus any covector normal to N at $(T,x_*(T))$ is a linear combination of these covectors. Since the gradients are the rows of the matrix $D\Psi(T,x_*(T))$, there exists a row vector $v = (v_0,\ldots,v_{n-k})$ such that

$$(H + \lambda_0 \varphi_t, -\lambda + \lambda_0 \varphi_x) = -v (D_t \Psi, D_x \Psi).$$

This is equivalent to the formulation given in the theorem.

Summarizing, in order to solve an optimal control problem, in principle, we need to *find all solutions to a boundary value problem on state and costate, coupled by a minimization condition, and then compare the costs that the projections of these solutions onto the controlled trajectories give.* Clearly, this is not an easy problem, and thus the rest of this chapter will be spent on illustrating how one may go about doing this for some classes of optimal control problems, namely (i) once more the simplest problem in the calculus of variations, but now in dimension n, (ii) the linear-quadratic regulator, but now deriving its solution using the maximum principle,

(iii) the time-optimal control problem for linear time-invariant systems, and (iv) time-optimal control for general single-input, nonlinear, control-affine systems in the plane.

2.3 The Simplest Problem in the Calculus of Variations in $\mathbb{R}^n$

We once more consider the simplest problem in the calculus of variations, but now in arbitrary dimension n. This is a special case of an optimal control problem, and we illustrate how far-reaching the conditions of the maximum principle are by briefly deriving the highdimensional versions of the necessary conditions for optimality developed in Chap. 1.

Let $L : [a,b] \times \mathbb{R}^n \times \mathbb{R}^n \to \mathbb{R}$, $(t,x,u) \mapsto L(t,x,u)$, be a continuous function that for fixed $t \in [a,b]$, is differentiable in (x,u) with the partial derivatives $\frac{\partial L}{\partial x}(t,x,u)$ and $\frac{\partial L}{\partial u}(t,x,u)$ continuous in all variables. Also, let A and B be two given points in $\mathbb{R}^n$. We then consider the following problem:

[CV] Find, among all continuously differentiable curves $x : [a,b] \mapsto \mathbb{R}^n$ that satisfy the boundary conditions $x(a) = A$ and $x(b) = B$, one that minimizes the functional
$$I(x) = \int_a^b L(t,x(t),\dot{x}(t))dt.$$

Calculus of variations problems are optimal control problems with a *trivial dynamics*, $\dot{x} = u$, and *no restrictions on the control set*: the state space is given by $M = \mathbb{R}^n$, the control set U is all of $\mathbb{R}^n$, and within our framework, the class $\mathscr{U}$ of admissible controls is given by all piecewise continuous functions; the terminal manifold N is zero-dimensional given by the point B. If $x_* : [a,b] \mapsto \mathbb{R}^n$ is an optimal solution, then with $u_*(t) = \dot{x}_*(t)$, the conditions of the maximum principle state that there exist a constant $\lambda_0 \geq 0$ and an adjoint variable $\lambda : [a,b] \to (\mathbb{R}^n)^*$ satisfying

$$\dot{\lambda}(t) = -\lambda_0 \frac{\partial L}{\partial x}(t,x_*(t),u_*(t))$$

such that $(\lambda_0,\lambda(t)) \neq 0$ for all $t \in [a,b]$ and

$$\lambda_0 L(t,x_*(t),u_*(t)) + \lambda(t)u_*(t) = \min_{v \in \mathbb{R}^n}[\lambda_0 L(t,x_*(t),v) + \lambda(t)v] = \text{const}. \quad (2.14)$$

Since the interval $[a,b]$ and the endpoint are fixed, no transversality conditions apply: the vector v can be any vector in $(\mathbb{R}^{n+1})^*$ leaving the terminal values of λ and $H(b,\lambda_0,\lambda(b),x_*(b),u_*(b))$ free. But *extremals for the simplest problem in the calculus of variations are always normal*: If $\lambda_0 = 0$, then the minimum condition (2.14) implies that $u_*(t)$ minimizes the linear function $v \mapsto \lambda(t)v$ over $\mathbb{R}^n$. But such a minimum exists only if $\lambda(t) = 0$, and this then contradicts the nontriviality of the multipliers. Thus λ_0 cannot vanish, and without loss of generality we may normalize it as $\lambda_0 = 1$.

The first-order necessary conditions for minimizing the function

$$v \mapsto L(t, x_*(t), v) + \lambda(t)v$$

over $\mathbb{R}^n$ then imply that

$$\frac{\partial L}{\partial u}(t, x_*(t), u_*(t)) + \lambda(t) = 0.$$

Combining this relation with the adjoint equation, while identifying $\dot{x}_*$ with the control u, gives the standard form of the *Euler–Lagrange equation*, now valid for the coordinates of the respective gradients of the Lagrangian

$$\frac{d}{dt}\left(\frac{\partial L}{\partial \dot{x}}(t, x_*(t), \dot{x}_*(t))\right) = \frac{\partial L}{\partial x}(t, x_*(t), \dot{x}_*(t)).$$

The actual minimum condition (2.14) of the maximum principle is the *Weierstrass condition* of the calculus of variations: recall that the Weierstrass excess function E was defined as

$$E(t, x, y, u) = L(t, x, u) - L(t, x, y) - \frac{\partial L}{\partial \dot{x}}(t, x, y)(u - y);$$

thus condition (2.14) states that

$$E(t, x_*(t), \dot{x}_*(t), u) \geq 0 \qquad \text{for all } u \in \mathbb{R}^n.$$

As shown in Sect. 1.6, this is a necessary condition for a *strong* local minimum of a very different character from that of the Euler–Lagrange equation. Recall that the piecewise continuous variations used in its proof allowed the derivatives to diverge, and thus this no longer is a necessary condition for a weak minimum. We shall see in Sect. 4.2 that Weierstrass's variations pointed the path to the variations used in the proof of the maximum principle.

If the Lagrangian L is twice continuously differentiable, additional regularity statements about extremals easily follow from the maximum principle. For example, the second-order necessary condition for the function $v \mapsto L(t, x_*(t), v) + \lambda(t)v$ to have a minimum over $\mathbb{R}^n$ at $\dot{x}_*(t)$ implies that the Hessian matrix

$$\frac{\partial^2 L}{\partial \dot{x}^2}(t, x_*(t), \dot{x}_*(t))$$

is positive semidefinite for $t \in [a, b]$. This is the multi-dimensional version of the *Legendre condition*. The *strengthened Legendre condition* holds over the interval $[a, b]$ if this matrix is positive definite for $t \in [a, b]$. In this case, as in the scalar case, the *Hilbert differentiability theorem* is valid, and the extremal x_* is twice continuously differentiable. The argument is the same as in the scalar case: for some constant c the extremal x_* is a solution to the Euler–Lagrange equation in integrated form,

$$\frac{\partial L}{\partial \dot{x}}(t, x_*(t), \dot{x}_*(t)) - \int_a^t \frac{\partial L}{\partial x}(t, x_*(s), \dot{x}_*(s)) ds - c = 0,$$

and defining a function $F(t, w)$ as

$$F(t, w) = \frac{\partial L}{\partial \dot{x}}(t, x_*(t), w) - \int_a^t \frac{\partial L}{\partial x}(t, x_*(s), \dot{x}_*(s)) ds - c,$$

the equation $F(t, w) = 0$ has the solution $w(t) = \dot{x}_*(t)$. By the implicit function theorem, this solution is continuously differentiable if the partial derivative

$$\frac{\partial F}{\partial w}(t, \dot{x}_*(t)) = \frac{\partial^2 L}{\partial \dot{x}^2}(t, x_*(t), \dot{x}_*(t))$$

is nonsingular. Hence x_* is twice continuously differentiable at all points where the strengthened Legendre condition holds.

The connections between optimal control and problems in the calculus of variations can be carried further including generalizations of the Jacobi condition and field theory. These aspects will be developed in Chap. 5.

2.4   The Linear-Quadratic Regulator Revisited

We briefly return to the linear-quadratic regulator and give a derivation of the optimal feedback control law from the conditions of the maximum principle. This argument is instructive and will be expanded further in Sect. 5.3 in connection with conjugate points for the optimal control problem. Also, in low dimensions, explicit solutions of the Riccati equation for the feedback gain S can be computed using these constructions, and we illustrate this with two scalar examples. As in the calculus of variations, there are no restrictions on the control set, i.e., $U = \mathbb{R}^m$, but now a dynamics (albeit a simple linear one) is involved. In fact, since the Hamiltonian H is strictly convex in the control u, for this case, variational arguments as they were developed in Chap. 1 would still be sufficient to characterize the minimum.

2.4.1   A Derivation of the Optimal Control from the Maximum Principle

Recall that the linear quadratic regulator [LQ] is the problem of minimizing a quadratic objective of the form

$$J(u) = \frac{1}{2} \int_0^T \left[x^T(t)Q(t)x(t) + u^T(t)R(t)u(t) \right] dt + \frac{1}{2} x^T(T)S_T x(T)$$

over all (piecewise) continuous functions $u : [0,T] \to \mathbb{R}^m$ defined over a fixed interval $[0,T]$ subject to a linear dynamics

$$\dot{x}(t) = A(t)x(t) + B(t)u(t), \qquad x(0) = x_0.$$

The entries of the matrices $A(\cdot)$, $B(\cdot)$, $R(\cdot)$, and $Q(\cdot)$ are continuous functions on the interval $[0,T]$, and the matrices $R(\cdot)$ and $Q(\cdot)$ are symmetric; $R(\cdot)$ is positive definite, and $Q(\cdot)$ positive semidefinite; S_T is a constant positive definite matrix.

As in the simplest problem of the calculus of variations, all *extremals are normal*: formally, since the problem is a minimization over a fixed interval $[0,T]$ without terminal constraints, the submanifold N is described by a single function $\Psi : [0,\infty) \times M \to \mathbb{R}^1$, $(t,x) \mapsto \Psi(t,x) = t - T$, defining the final time T, and the transversality condition (2.11) reduces to $\lambda(T) = \lambda_0 x^T(T)S_T$. Thus, the adjoint equation with terminal condition is given by

$$\dot{\lambda} = -\lambda_0 x^T Q(t) - \lambda A(t), \qquad \lambda(T) = \lambda_0 x^T(T)S_T.$$

If $\lambda_0 = 0$, then λ is a solution to a homogeneous linear equation with 0 boundary conditions, hence identically zero. But this contradicts the nontriviality statement of the maximum principle. Thus, without loss of generality, we set $\lambda_0 = 1$. The Hamiltonian function H then takes the form

$$H = \frac{1}{2} x^T Q(t)x + \frac{1}{2} u^T R(t)u + \lambda(A(t)x + B(t)u),$$

and since the matrix R is positive definite, is strictly convex with a unique minimum given by the stationary point of the gradient in u,

$$\frac{\partial H}{\partial u} = u^T R(t) + \lambda B(t) = 0,$$

i.e.,

$$u = -R^{-1}(t)B^T(t)\lambda^T. \tag{2.15}$$

For the subsequent calculation it is more convenient to write the equations in terms of λ^T, and we therefore define $\mu = \lambda^T$. Substituting Eq. (2.15) into the system and adjoint equation gives the following classical linear two-point boundary value problem for x and μ:

$$\begin{pmatrix} \dot{x} \\ \dot{\mu} \end{pmatrix} = \begin{pmatrix} A(t) & -B(t)R(t)^{-1}B(t)^T \\ -Q(t) & -A(t)^T \end{pmatrix} \begin{pmatrix} x \\ \mu \end{pmatrix}, \qquad \begin{pmatrix} x(0) \\ \mu(T) \end{pmatrix} = \begin{pmatrix} x_0 \\ S_T x(T) \end{pmatrix}.$$

This is the n-dimensional analogue of the linear Hamiltonian system considered in Sect. 1.4. Its solution is easily obtained from the solution of the associated matrix differential equation

$$\begin{pmatrix} \dot{X} \\ \dot{Y} \end{pmatrix} = \begin{pmatrix} A(t) & -B(t)R(t)^{-1}B(t)^T \\ -Q(t) & -A(t)^T \end{pmatrix} \begin{pmatrix} X \\ Y \end{pmatrix}, \qquad \begin{pmatrix} X(0) \\ Y(T) \end{pmatrix} = \begin{pmatrix} \text{Id} \\ S_T \end{pmatrix}.$$

The following classical result generalizes Proposition 1.4.1 to the multidimensional case and establishes the connections between solutions to Riccati equations and quotients of solutions to linear differential equations in general. In the engineering literature, e.g., [64], this technique and its generalizations are known as the *sweep method*.

Proposition 2.4.1. *Suppose $A(\cdot)$, $B(\cdot)$, $M(\cdot)$, and $N(\cdot)$ are continuous $n \times n$ matrices defined on $[0,T]$ and let $(X,Y)^T$ be the solution to the initial value problem*

$$\begin{pmatrix} \dot{X} \\ \dot{Y} \end{pmatrix} = \begin{pmatrix} A & -M \\ -N & -B \end{pmatrix} \begin{pmatrix} X \\ Y \end{pmatrix}, \qquad \begin{pmatrix} X(0) \\ Y(0) \end{pmatrix} = \begin{pmatrix} X_0 \\ Y_0 \end{pmatrix}. \tag{2.16}$$

Suppose X_0 is nonsingular. Then the solution $X(t)$ is nonsingular on the full interval $[0,T]$ if and only if the solution S to the Riccati equation

$$\dot{S} + SA(t) + B(t)S - SM(t)S + N(t) \equiv 0, \qquad S(0) = Y_0 X_0^{-1}, \tag{2.17}$$

exists on the full interval $[0,T]$, and in this case we have that

$$Y(t) = S(t)X(t). \tag{2.18}$$

The solution S to the Riccati equation (2.17) has a finite escape time at $t = \tau$ if and only if τ is the first time when the matrix $X(t)$ becomes singular.

Proof. [$\Longrightarrow$] Suppose $X(t)$ is nonsingular for all $t \in [0,T]$. Then $S(t) = Y(t)X(t)^{-1}$ is well-defined over $[0,T]$, and we need only verify that S satisfies the Riccati equation (2.17). This is shown with a direct calculation: omitting the variable t, we have that

$$\dot{S} = \frac{d}{dt}(YX^{-1}) = \dot{Y}X^{-1} + Y\frac{d}{dt}(X^{-1}).$$

Since $X(t)X(t)^{-1} = \text{Id}$, it follows that

$$0 = \frac{d}{dt}(XX^{-1}) = \dot{X}X^{-1} + X\frac{d}{dt}(X^{-1}),$$

or

$$\frac{d}{dt}(X^{-1}) = -X^{-1}\dot{X}X^{-1},$$

and thus

$$\dot{S} = \dot{Y}X^{-1} - YX^{-1}\dot{X}X^{-1}.$$

Substituting the differential equations for $\dot{X}$ and $\dot{Y}$ gives

$$\dot{S} = (-NX - BY)X^{-1} - S(AX - MY)X^{-1} = -SA - BS + SMS - N.$$

$[\Longleftarrow]$ Conversely, suppose a solution S to the Riccati equation exists on all of $[0, T]$. The linear equation

$$\dot{U} = (A(t) - M(t)S(t))U, \qquad U(t_0) = X_0,$$

has a solution $U = U(t)$ defined over the full interval $[0, T]$. Setting $V(t) = S(t)U(t)$, we have $V(0) = S(0)X_0 = Y_0$ and

$$\dot{V} = \dot{S}U + S\dot{U} = (-SA - BS + SMS - N)U + S(A - MS)U = -NU - BV. \quad (2.19)$$

Thus the pair $(U, V)^T$ is a solution to the initial value problem (2.16). But so is $(X, Y)^T$, and by the uniqueness of solutions we have $(X, Y) = (U, V)$, i.e., $Y(t) = S(t)X(t)$.

Suppose that there exists a time τ for which $X(\tau)$ is singular. Pick $x_0 \neq 0$ such that $X(\tau)x_0 = 0$ and let $x(t) = X(t)x_0$ and $y(t) = Y(t)x_0$. Then $x(\tau) = 0$ and $y(\tau) = Y(\tau)x_0 = S(\tau)x(\tau) = 0$, and thus since $(x, y)^T$ satisfies a homogeneous linear differential equation, both x and y vanish identically. But $x(0) = X(0)x_0 \neq 0$, since $X(0)$ is nonsingular. Contradiction. Thus $X(t)$ is nonsingular over all of $[0, T]$. $\quad \square$

For the linear-quadratic problem we already have seen in Theorem 2.1.1 that the associated Riccati equation has a solution over the full interval $[0, T]$, and thus we have $\mu(t) = S(t)x(t)$, or in the original notation, $\lambda^T(t) = S(t)x(t)$. Hence, as we already know, the optimal control is given by

$$u(t) = -R(t)^{-1}B(t)^T S(t)x(t).$$

This argument, however, is based only on necessary conditions and thus by itself does not prove the optimality of this control law. But of course, we already know that the control is optimal from Sect. 2.1.

2.4.2 Two Scalar Examples

We illustrate the solution procedure with two scalar examples in which the Riccati equation can be solved in analytic form.

Example 2.4.1. Let x and u be scalar and consider the problem to minimize the objective

$$J(u) = \frac{1}{2}\int_0^T \left(qx^2 + u^2\right) dt$$

subject to the dynamics

$$\dot{x} = ax + bu, \qquad x(0) = x_0, \qquad 0 \le t \le T.$$

For example, this is a simple model of regulating the pH value of some chemical component [139]. The variable x denotes the deviation of the pH value from a preset nominal value and the pH value is regulated through a controlling agent with the rate of change in pH proportional to a weighted sum of its current value and the strength of the controlling ingredient u, also measured by its deviation from the nominal pH value; a and b are known positive constants and x_0 is the known initial value.

This formulation fits the model exactly, and thus the optimal control is given in feedback form as

$$u_*(t) = -bS(t)x_*(t), \qquad 0 \le t \le T,$$

where $S(t)$ is the solution to the Riccati equation (2.3),

$$\dot{S} + 2aS - b^2S^2 + q = 0, \qquad S(T) = 0.$$

A scalar Riccati equation can always be reduced to a second-order homogeneous linear differential equation by making the substitution

$$\frac{\dot{\phi}}{\phi} = -b^2 S,$$

which gives

$$-\frac{1}{b^2}\left(\frac{\ddot{\phi}\phi - (\dot{\phi})^2}{\phi^2}\right) = \dot{S} = \frac{2a}{b^2}\left(\frac{\dot{\phi}}{\phi}\right) + \frac{1}{b^2}\left(\frac{\dot{\phi}}{\phi}\right)^2 - q.$$

Equivalently,

$$\frac{1}{b^2}\left(\frac{\ddot{\phi}}{\phi}\right) = -\frac{2a}{b^2}\left(\frac{\dot{\phi}}{\phi}\right) + q,$$

and thus we obtain the following second-order homogeneous equation with constant coefficients:

$$\ddot{\phi} + 2a\dot{\phi} - qb^2\phi = 0.$$

From the terminal condition on S we get $\dot{\phi}(T) = 0$, and since we are just looking for a nontrivial solution, we may take $\phi(T) = 1$. Setting $\kappa = \sqrt{a^2 + qb^2}$, the explicit solution is given as

$$\phi(t) = e^{-a(t-T)}\left[\cosh(\kappa(t-T)) + \frac{a}{\kappa}\sinh(\kappa(t-T))\right], \qquad 0 \le t \le T,$$

and thus

$$S(t) = -\frac{1}{b^2}\left(\frac{\dot{\phi}(t)}{\phi(t)}\right) = \frac{1}{b^2}\left(a - \kappa\frac{\kappa\sinh\left(\kappa\left(t-T\right)\right)+a\cosh\left(\kappa\left(t-T\right)\right)}{\kappa\cosh\left(\kappa\left(t-T\right)\right)+a\sinh\left(\kappa\left(t-T\right)\right)}\right),$$

with the optimal time-varying feedback gain given by $-bS(t)$.

Example 2.4.2 (Inventory control). [139] In most regulator problems, the variables are normalized as deviations from predetermined set points. In this example, a simple inventory control problem, the desired values are left as predetermined time-varying quantities, and we illustrate the changes that arise in the argument for such a model that involves a modified form of the Lagrangian in the objective. The reasoning given here easily extends to the general case (for example, see [64, 144]).

Consider a company that produces some good and has desired levels for the production and inventory over a planning horizon $[0,T]$ represented by $u_d(t)$ and $x_d(t)$, respectively. If the demand at time t is denoted by $d(t)$, then the rate of change of the inventory level $x(t)$ is given by

$$\dot{x}(t) = u(t) - d(t), \qquad x(0) = x_0.$$

If the firm's objective is to maintain the inventory and production levels, then it is reasonable to minimize a functional of the form

$$J(u) = \frac{1}{2}\int_0^T q\left[x(t) - x_d(t)\right]^2 + r\left[u(t) - u_d(t)\right]^2 dt,$$

where r and q are positive weights selected by the company. In this problem, we have the restrictions $x(t) \geq 0$ and $u(t) \geq 0$ that do not fit into the linear-quadratic regulator model, but making the natural assumption that x_d and u_d are positive continuous functions, for sufficiently high weights r and q we can assume that these conditions will not be violated. In other words, we solve the problem ignoring these constraints, but then need to verify that the optimal solution does not violate them. The other, less significant change to the model formulation analyzed so far is that the objective, when multiplied out, contains linear terms in x and u as well. These are easily incorporated into the sweep method described above. (This topic will still be picked up in greater generality in Sect. 5.3.)

The above change in the problem formulation does not alter the fact that extremals are normal, and the Hamiltonian for the problem is

$$H(t,\lambda,x,u) = \frac{q}{2}\left(x - x_d(t)\right)^2 + \frac{r}{2}\left(u - u_d(t)\right)^2 + \lambda\left(u - d(t)\right).$$

Minimization of the Hamiltonian over $u \in \mathbb{R}$ leads to $\lambda = -r(u_* - u_d)$ and hence

$$u_*(t) = -\frac{\lambda_*(t)}{r} + u_d(t), \qquad 0 \leq t \leq T.$$

Substituting this relation into the dynamics and combining with the adjoint equation gives the inhomogeneous linear system

$$\dot{x}_*(t) = -\frac{\lambda_*(t)}{r} + u_d(t) - d(t),$$

$$\dot{\lambda}_*(t) = -qx_*(t) + qx_d(t),$$

with boundary conditions $x_*(0) = x_0$ and $\lambda_*(T) = 0$. In this case, the solutions are related by

$$\lambda_*(t) = a(t) + b(t)x_*(t), \tag{2.20}$$

for some C^1-functions a and b that satisfy the terminal conditions $a(T) = 0$ and $b(T) = 0$. Differentiating Eq. (2.20), we get that

$$\dot{\lambda}_* = \dot{a} + \dot{b}x_* + b\dot{x}_*,$$

which, upon substituting for $\dot{\lambda}_*$ and $\dot{x}_*$, yields

$$\dot{a} + b(u_d(t) - d(t)) - qx_d(t) - \frac{ab}{r} + \left(\dot{b} - \frac{b^2}{r} + q\right)x_* = 0.$$

This equation will be satisfied if we choose a and b such that

$$\dot{a} + b(u_d(t) - d(t)) - qx_d(t) - \frac{ab}{r} = 0, \qquad a(T) = 0, \tag{2.21}$$

$$\dot{b} - \frac{b^2}{r} + q = 0, \qquad b(T) = 0. \tag{2.22}$$

Equation (2.22) is the Riccati equation for a related standard linear-quadratic optimal control problem [LQ] and has a solution over the full interval $[0,T]$ because of the positivity of q and r. Equation (2.21) then is a time-varying linear ODE defined over the full interval and thus also has a solution over the interval $[0,T]$. As for Example 2.4.1, the solution b to the Riccati equation can be calculated explicitly by making a substitution of the form

$$\frac{\dot{\phi}}{\phi} = -\frac{1}{r}b,$$

yielding the second-order equation

$$\ddot{\phi} = \frac{q}{r}\phi.$$

Setting $\kappa = \sqrt{\frac{q}{r}}$, the solution for terminal conditions $\phi(T) = 1$ and $\dot{\phi}(T) = 0$ is given by $\phi(t) = \cosh(\kappa(t - T))$ and thus

$$b(t) = -r\kappa \tanh(\kappa(t-T)) = r\kappa \tanh(\kappa(T-t)).$$

We still need to find $a(t)$. This solution depends on the demand $d(t)$ and the specified production levels $x_d(t)$. If these are constants, say the firm controlling the inventory desires to have the production rate equal to the demand rate, $u_d = d = \text{const}$, while at the same time maintaining a constant level of inventory, $x_d = \text{const}$, then Eq. (2.21) becomes

$$\dot{a} - \frac{b(t)}{r}a - qx_d = 0, \qquad a(T) = 0.$$

Solving this equation, it follows that

$$a(t) = \frac{qx_d}{\kappa} \tanh(\kappa(t-T)) = -\sqrt{qr}x_d \tanh(\kappa(T-t)).$$

Hence the optimal feedback control $u_*(t,x)$ is given by

$$u_*(t,x) = -\frac{\lambda_*(t)}{r} + d = -\frac{a(t) + b(t)x_*(t)}{r} + d$$
$$= \kappa \tanh(\kappa(T-t))(x_d - x_*(t)) + d, \qquad 0 \le t \le T.$$

Thus the optimal control equals the constant demand rate d plus a time-varying inventory correction factor proportional to the deviation from the set point.

2.5 Time-Optimal Control for Linear Time-Invariant Systems

The two classes of problems considered so far, the simplest problem in the calculus of variations and the linear-quadratic regulator, were both problems without constraints on the control set and as such, are examples that still could be fully analyzed with techniques from the calculus of variations. We now consider examples from another class of classical problems for which this no longer is the case: time-optimal control to a point for a time-invariant linear system with bounded controls.

[LTOC]   Given a time-invariant linear control system

$$\Sigma: \qquad \dot{x} = Ax + Bu, \qquad A \in \mathbb{R}^{n \times n}, \quad B \in \mathbb{R}^{n \times m},$$

find among all piecewise continuous controls u that take values in the hypercube

$$U = \left\{ u \in \mathbb{R}^m : \|u\|_\infty = \max_{i=1,\ldots,m} |u_i| \le 1 \right\},$$

one that steers a given (but arbitrary) initial point $x_0 \in \mathbb{R}^n$ into the origin 0 in minimum time.

In the formulation of Sect. 2.2, we have $M = \mathbb{R}^n$, $L(t,x,u) \equiv 1$, $f(t,x,u) = Ax + Bu$, $\varphi \equiv 0$, and Ψ is given by $\Psi : [0,\infty) \times M \to \mathbb{R}^n$, $(t,x) \mapsto \Psi(t,x) = x$, i.e., $N = \{x \in \mathbb{R}^n : x = 0\}$. Since both initial and terminal points on the state are specified, in this case the transversality conditions give no information about the multiplier λ. Formally, we have $\frac{\partial \Psi}{\partial x}(t,x) = \text{Id}$ and thus $\lambda(T) = \nu$, an arbitrary covector from $(\mathbb{R}^n)^*$. But the transversality condition on the final time T implies that $H(T,\lambda_0,\lambda,x,u) = 0$. Since

$$H = \lambda_0 + \lambda(Ax + Bu)$$

is time-invariant, it follows that along any extremal, we have

$$H(t,\lambda_0,\lambda(t),x_*(t),u_*(t)) \equiv 0.$$

In particular, $\lambda(t)$ can never vanish, since otherwise also $\lambda_0 = 0$. The adjoint equation is given by

$$\dot{\lambda} = -\lambda A,$$

and the minimum condition implies that for each $i = 1,\ldots,m$, the ith component $u_*^{(i)}(t)$ of an optimal control must satisfy

$$u_*^{(i)}(t) = \begin{cases} +1 & \text{if } \lambda(t)b_i < 0, \\ -1 & \text{if } \lambda(t)b_i > 0, \end{cases}$$

where b_i is the ith column of B. Summarizing, we thus have the following version of the Maximum Principle for the optimal control problem [LTOC]:

Theorem 2.5.1 (Maximum principle for problem [LTOC]). *Let (x_*,u_*) be a controlled trajectory defined over the interval $[t_0,T]$ that minimizes the time of transfer from $x_0 \in \mathbb{R}^n$ to the origin. Then there exists a nontrivial solution $\lambda : [t_0,T] \to (\mathbb{R}^n)^*$ to the adjoint equation $\dot{\lambda} = -\lambda A$ so that the control u_* satisfies*

$$u_*^{(i)}(t) = \begin{cases} +1 & \text{if } \lambda(t)b_i < 0, \\ -1 & \text{if } \lambda(t)b_i > 0, \end{cases} \tag{2.23}$$

and the Hamiltonian is identically zero on $[t_0,T]$, $H(t,\lambda_0,\lambda(t),x_*(t),u_*(t)) \equiv 0$.

The necessary conditions of this theorem are also sufficient for optimality under some easily verifiable controllability assumption on the system Σ. For the moment, consider a system Σ of the form

$$\Sigma : \quad \dot{x} = Ax + Bu, \quad x(0) = p, \quad u \in U, \tag{2.24}$$

with a general control set $U \subset \mathbb{R}^m$. Since the system is time-invariant, without loss of generality we normalize the initial time to $t_0 = 0$, and the solution $x(\cdot;p)$ to the initial value problem (2.24) is given by

$$x(t;p) = e^{At}p + \int_0^t e^{A(t-s)}Bu(s)ds. \tag{2.25}$$

Definition 2.5.1 (Reachable and controllable sets). The time-t-reachable set from p is the set of all points $q \in \mathbb{R}^n$ that can be reached from p by means of an admissible control defined on the interval $[0,t]$,

$$\text{Reach}_{\Sigma,t}(p) = \left\{ q \in \mathbb{R}^n : \exists\, u \in \mathcal{U} \text{ such that } q = e^{At}p + \int_0^t e^{A(t-s)}Bu(s)ds \right\}.$$

The *reachable set* from p is the union of all time-t-reachable sets for $t > 0$,

$$\text{Reach}_{\Sigma}(p) = \bigcup_{t>0} \text{Reach}_{\Sigma,t}(p).$$

The time-t-controllable set to q is the set of all points $p \in \mathbb{R}^n$ that can be steered into q by means of an admissible control defined on the interval $[0,t]$,

$$\text{Contr}_{\Sigma,t}(q) = \left\{ p \in \mathbb{R}^n : \exists\, u \in \mathcal{U} \text{ such that } q = e^{At}p + \int_0^t e^{A(t-s)}Bu(s)ds \right\}.$$

The *controllable set* to q is the union of all time-t-controllable sets for $t > 0$,

$$\text{Contr}_{\Sigma}(q) = \bigcup_{t>0} \text{Contr}_{\Sigma,t}(q).$$

Clearly, a point q is reachable from p in time t if and only if p is controllable to q in time t. Thus, generally, we restrict out attention to reachable sets. It is clear from Eq. (2.25) that
$$\text{Reach}_{\Sigma,t}(p) = e^{At}p + \text{Reach}_{\Sigma,t}(0)$$
and henceforth we consider only the case $p = 0$.

A special situation arises if there are no restrictions on the control set, i.e., if $U = \mathbb{R}^m$. That is, we are considering the problem of whether *in principle* it is possible to steer a point p into another point q. In this case, for every $t > 0$ the reachable set $\text{Reach}_{\Sigma,t}(0)$ is a linear subspace (and in fact, the same one regardless of the size of the interval), known as the *controllable subspace* $\mathscr{C}(A,B)$. This is the subspace spanned by the columns of the so-called Kalman matrix, i.e.,

$$\mathscr{C}(A,B) = \text{Im}\,(B, AB, A^2B, \dots, A^{n-1}B). \tag{2.26}$$

Theorem 2.5.2. *If $U = \mathbb{R}^m$, then for every $t > 0$*

$$\text{Reach}_{\Sigma,t}(0) = \mathscr{C}(A,B) = \text{Contr}_{\Sigma,t}(0).$$

Proof. We fix t and first show that the reachable set $\text{Reach}_{\Sigma,t}(0)$ is given by the image $\text{Im}\, W(t)$ of the matrix

$$W(t) = \int_0^t e^{A(t-s)} BB^T e^{A^T(t-s)} ds.$$

Choosing continuous time-varying controls of the form

$$u(s) = B^T e^{A^T(t-s)} p,$$

it follows that

$$x(t) = \int_0^t e^{A(t-s)} Bu(s)ds = W(t)p,$$

and thus $\text{Im}\, W(t) \subset \text{Reach}_{\Sigma,t}(0)$. But $W(t)$ is a symmetric matrix, and hence the full space $\mathbb{R}^n$ is the direct sum of the image and the kernel of $W(t)$ [113],

$$\mathbb{R}^n = \text{Im}\, W(t) \oplus \ker W(t).$$

Furthermore, the kernel is the orthogonal complement of the image, $\ker W(t) = \text{Im}\, W(t)^\perp$, and it therefore suffices to show that

$$\ker W(t) \subset \text{Reach}_{\Sigma,t}(0)^\perp.$$

Given any point $y \in \ker W(t)$, we have that

$$0 = \langle y, W(t)y \rangle = \int_0^t y^T e^{A(t-s)} BB^T e^{A^T(t-s)} yds = \int_0^t \left\| B^T e^{A^T(t-s)} y \right\|_2^2 ds,$$

and thus $y^T e^{A(t-s)} B \equiv 0$ on the interval $[0,t]$. Since any point q in the reachable set $\text{Reach}_{\Sigma,t}(0)$ is of the form

$$q = \int_0^t e^{A(t-s)} Bu(s)ds$$

for some control u, we thus have that

$$\langle y, q \rangle = \int_0^t y^T e^{A(t-s)} Bu(s)ds = 0$$

for all $q \in \text{Reach}_{\Sigma,t}(0)$. Hence $y \in \text{Reach}_{\Sigma,t}(0)^\perp$ as claimed. Overall, it therefore follows that

$$\text{Reach}_{\Sigma,t}(0) = \text{Im} W(t) \quad \text{for all} \quad t > 0.$$

It remains to compute this image, or equivalently, the kernel of $W(t)$. If $y \in \ker W(t)$, then as shown above, $y^T e^{A(t-s)} B \equiv 0$ on the interval $[0,t]$. Since this function is real-analytic, this is equivalent to the fact that all derivatives vanish at $t = 0$, i.e.,

$$y^T A^k B = 0 \quad \text{for all} \quad k \in \mathbb{N}.$$

By the Cayley–Hamilton theorem [113], A^n can be expressed as a linear combination of the powers A^i for $i = 0, 1, \ldots, n-1$, and thus this is equivalent to

$$y^T (B, AB, A^2 B, \ldots, A^{n-1} B) = 0.$$

Hence the columns of the Kalman matrix

$$K = (B, AB, A^2 B, \ldots, A^{n-1} B)$$

span the orthogonal complement to $\ker W(t)$; that is, they span $\text{Im} W(t)$. This proves the result.   □

Definition 2.5.2 (Completely controllable). The linear system Σ is said to be completely controllable if $\mathscr{C}(A, B) = \mathbb{R}^n$.

Thus, if the system Σ is completely controllable, then in principle, it is possible to go from any point $p_0 \in \mathbb{R}^n$ to any other point $p_1 \in \mathbb{R}^n$ in arbitrarily short time T. (Simply take the control that steers the point 0 into the point $p_1 - e^{AT} p_0$ in time T.) Obviously, the shorter the time-interval is, the larger the control values need to become, and if the controls are bounded, then complete controllability no longer ensures that such a transfer is possible. In fact, as we shall see in the examples in the next section, with a bound on the controls, it may no longer be possible to steer p_0 into p_1 at all. However, for the system Σ with control set given by the hypercube

$$U = \left\{ u \in \mathbb{R}^m : \|u\|_\infty = \max_{i=1,\ldots,m} |u_i| \leq 1 \right\}$$

(more generally, for any control set U that has 0 as interior point), this notion of complete controllability makes the conditions of the maximum principle also sufficient for optimality.

Theorem 2.5.3. *Consider the time-optimal control problem to the origin for the time-invariant linear system*

$$\Sigma : \qquad \dot{x} = Ax + Bu, \qquad A \in \mathbb{R}^{n \times n}, \quad B \in \mathbb{R}^{n \times m},$$

with control set

$$U = \left\{ u \in \mathbb{R}^m : \|u\|_\infty = \max_{i=1,\ldots,m} |u_i| \leq 1 \right\}.$$

If the system Σ is completely controllable, then a control $u : [0,T] \rightarrow U$ is time-optimal if and only if there exists a nontrivial solution $\lambda : [0,T] \rightarrow (\mathbb{R}^n)^*$ to the adjoint equation $\dot{\lambda} = -\lambda A$ such that

$$\lambda(t)Bu(t) = \min_{v \in U} \lambda(t)Bv. \tag{2.27}$$

This theorem will be proven in Sect. 3.5. Thus, for the time-optimal control problem for linear time-invariant systems, the conditions of the maximum principle are both necessary and sufficient for optimality under an easily verifiable algebraic condition. Note that the minimum condition (2.27) gives no information about $u_*^{(i)}(t)$ for times t when $\lambda(t)b_i = 0$. The function $\Phi_i(t) = \lambda(t)b_i$ is called the ith *switching function*, and its properties determine the structure of optimal controls. For instance, if $\Phi_i(t)$ has a simple zero at time τ, then the control switches between $+1$ and -1 at τ. Controls that oscillate only between the upper and lower values ± 1 are called *bang-bang controls*. For general nonlinear systems with locally bounded Lebesgue measurable functions as controls, the switching functions may have complicated zerosets (see Sect. 2.8). But for linear systems, as we shall show in Chap. 3, these phenomena play a minor role, and we therefore do not discuss these features here. Rather, we close this section with a useful criterion on the eigenvalues of the matrix A that ensures that optimal controls are bang-bang and gives a bound on the number of switching times.

Proposition 2.5.1. *If all eigenvalues of the matrix A are real, then optimal controls for the single-input linear control system*

$$\dot{x} = Ax + bu, \qquad A \in \mathbb{R}^{n \times n}, \quad b \in \mathbb{R}^n, \qquad |u| \le 1,$$

are bang-bang with at most $n-1$ switching times.

Proof. It follows from the adjoint equation, $\dot{\lambda} = -\lambda A$, that the derivatives of the switching function $\Phi(t) = \lambda(t)b$ are given by

$$\dot{\Phi}(t) = -\lambda(t)Ab, \quad \ddot{\Phi}(t) = \lambda(t)A^2b, \quad \ldots, \quad \Phi^{(r)}(t) = (-1)^r A^r b, \quad \ldots.$$

By the Cayley–Hamilton theorem, the matrix $-A$ is a root of its characteristic polynomial, $\chi_{-A}(-A) = 0$, say

$$\chi_{-A}(t) = \det(t \cdot \mathrm{Id} + A) = t^n + a_{n-1}t^{n-1} + \cdots + a_1 t + a_0.$$

Thus $(-A)^n$ can be written as a linear combination of lower powers of A,

$$(-A)^n = -a_{n-1}(-A)^{n-1} - \cdots - a_1(-A) - a_0 \mathrm{Id}.$$

The switching function therefore satisfies the nth-order linear differential equation

$$\Phi^{(n)}(t) + a_{n-1}\Phi^{(n-1)}(t) + \cdots + a_1\Phi(t) + a_0 \equiv 0$$

with constant coefficients, where the polynomial is the characteristic polynomial of the matrix $-A$. Since all eigenvalues of A, and thus also those of $-A$, are real, the general solution Φ to this differential equation is of the form

$$\Phi(t) = \sum_{i=1}^{k} p_i(t) e^{-\alpha_i t}, \qquad (2.28)$$

where $\alpha_1, \ldots, \alpha_k$ are the *distinct* eigenvalues of the matrix A and p_i are polynomials of degree at most d_i, where d_i is the algebraic multiplicity of the eigenvalue α_i, that is, the multiplicity of α_i as a zero of the characteristic polynomial of A. Expressions of this type are called exponential polynomials, and the result of the proposition follows from a general property of these functions. Define the degree Deg Φ of an exponential polynomial of the form (2.28) as

$$\text{Deg } \Phi = \sum_{i=1}^{k} (1 + \deg p_i),$$

where $\deg p_i$ denotes the usual degree of the polynomial p_i. Then the proposition follows from the following lemma:

Lemma 2.5.1. *A nontrivial exponential polynomial of the form*

$$\Phi(t) = \sum_{i=1}^{k} p_i(t) e^{-\alpha_i t}$$

of degree Deg $\Phi = r$ *has at most* $r - 1$ *zeros.*

Proof. The proof is by induction on the degree r. If Deg $\Phi = 1$, then Φ is of the form $\Phi(t) = ce^{-\alpha t}$ with $c \neq 0$ and hence Φ has no zeros. Thus, inductively, assume that the statement is correct for all exponential polynomials of degree at most r and assume that Φ is of degree $r + 1$. Then

$$\Psi(t) = \Phi(t) e^{\alpha_1 t} = p_1(t) + \sum_{i=2}^{k} p_i(t) e^{(\alpha_1 - \alpha_i)t}$$

also is an exponential polynomial of degree $r + 1$, and we have

$$\dot{\Psi}(t) = \dot{p}_1(t) + \sum_{i=2}^{k} (\dot{p}_i(t) + (\alpha_i - \alpha_1)p_i(t)) \, e^{(\alpha_1 - \alpha_i)t}.$$

Since differentiation of the polynomial lowers the degree, $\deg \dot{p}_1 = \deg p_1 - 1$, the derivative $\dot{\Psi}$ is an exponential polynomial of strictly smaller degree, Deg $\dot{\Psi} \leq r$. Hence, by the inductive assumption, $\dot{\Psi}$ has at most $r - 1$ zeros. By the mean value theorem, Ψ therefore has at most r zeros. Hence so does Φ. $\square$

2.6 Time-Optimal Control for Planar Linear Time-Invariant Systems: Examples

We give several examples that illustrate how the conditions of the maximum principle can be used to construct optimal solutions for linear time-optimal control problems. The examples are two-dimensional, but the procedures are generally applicable. We start with the classical model of time-optimal control to the origin for the double integrator.

2.6.1 The Double Integrator

The double integrator is a mathematical model of an object moving along a horizontal line without friction, and the goal is to bring it to rest at the origin in minimum time. Here $x(t)$ denotes the position of the object at time t, $\dot{x}(t)$ its velocity, and $u(t)$ the external force applied to the object. Mathematically, we take as variable $x = (x_1, x_2)^T = (x, \dot{x})^T$, and the dynamics can be written in the form

$$\dot{x} = \begin{pmatrix} 0 & 1 \\ 0 & 0 \end{pmatrix} x + \begin{pmatrix} 0 \\ 1 \end{pmatrix} u.$$

The Hamiltonian H is given by

$$H = \lambda_0 + \lambda \left[\begin{pmatrix} 0 & 1 \\ 0 & 0 \end{pmatrix} x + \begin{pmatrix} 0 \\ 1 \end{pmatrix} u \right] = \lambda_0 + \lambda_1 x_2 + \lambda_2 u,$$

and thus the minimum condition implies that

$$u(t) = \begin{cases} +1 & \text{if } \lambda_2(t) < 0, \\ -1 & \text{if } \lambda_2(t) > 0. \end{cases}$$

Obviously, the matrix A has the double eigenvalue 0, and thus by Proposition 2.5.1, optimal controls are bang-bang with at most one switching. Naturally, for this simple model this also is easily seen directly: The adjoint equation is given by

$$\dot{\lambda} = -\lambda \begin{pmatrix} 0 & 1 \\ 0 & 0 \end{pmatrix},$$

or

$$\dot{\lambda}_1 = 0, \qquad \dot{\lambda}_2 = -\lambda_1,$$

and thus

$$\ddot{\lambda}_2 \equiv \frac{d}{dt}(-\lambda_1) = 0.$$

Hence any solution of the adjoint equation is an affine function $\lambda_2(t) = \alpha t + \beta$ and has at most one zero. Therefore *optimal controls are bang-bang with at most one switching*.

Once this structure is known, it is straightforward to synthesize all possible extremals. We simply need to analyze the phase portraits of the two systems corresponding to the constant controls $u \equiv +1$ and $u \equiv -1$ and then consider all possible combinations that steer the system into the origin and have no more than one switching. Let X denote the vector field corresponding to control $u \equiv -1$, i.e., $\dot{x}_1 = x_2$ and $\dot{x}_2 = -1$. Forming $\frac{dx_1}{dx_2} = -x_2$, we see that the integral curves have the form $x_1 = -\frac{1}{2}x_2^2 + a$ with $a \in \mathbb{R}$ some constant. Analogously, if Y denotes the vector field corresponding to control $u \equiv +1$, then we have $\dot{x}_1 = x_2$ and $\dot{x}_2 = 1$, and now the integral curves are given by $x_1 = \frac{1}{2}x_2^2 + b$ with $b \in \mathbb{R}$ another constant. Thus, all integral curves are parabolas opening left for $u = -1$ and right for $u = +1$. Among all these curves, however, there are only two that steer the system into the origin directly, namely

$$\Gamma_+ : x_1 = \frac{1}{2}x_2^2 \quad \text{for} \quad x_2 < 0$$

and

$$\Gamma_- : x_1 = -\frac{1}{2}x_2^2 \quad \text{for} \quad x_2 > 0.$$

Only these two half-parabolas are integral curves that steer the system into the origin; the other two halves that were dropped steer the system away from the origin. Thus any optimal trajectory needs to arrive at the origin along either Γ_+ or Γ_-. Bang-bang trajectories that have exactly one switching are now constructed by integrating the vector field X backward from any point in Γ_+ and integrating Y backward from any point in Γ_-.

Denote the resulting family of extremal controlled trajectories by $\mathscr{F}$. It is clear that away from Γ_+ and Γ_-, this family $\mathscr{F}$ covers the entire state space injectively and for every initial condition $(x_1^0, x_2^0) \neq (0,0)$ there exists a unique extremal in $\mathscr{F}$ that is bang-bang with at most one switching and steers the system into the origin *forward* in time. This family is shown in Fig. 2.1. In general, such a family of controlled trajectories is called an *extremal synthesis* (and this will be the main topic of Chap. 6). Note that for each trajectory, the control at (x_1, x_2) depends only on the actual point (x_1, x_2), but not on the path along which this point was reached and thus we can describe the controls associated with this family as a discontinuous feedback control. If we define regions

$$G_+ = \left\{ (x_1, x_2) : x_1 < -\text{sgn}(x_2)\frac{1}{2}x_2^2 \right\}$$

and

$$G_- = \left\{ (x_1, x_2) : x_1 > -\text{sgn}(x_2)\frac{1}{2}x_2^2 \right\},$$

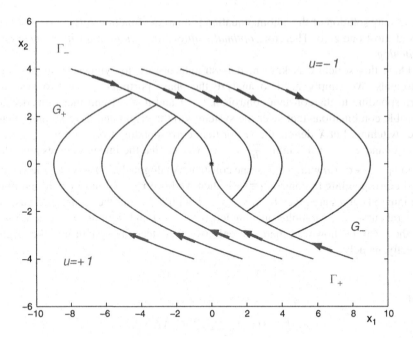

Fig. 2.1 Synthesis of optimal controlled trajectories for the time-optimal control problem to the origin for the double integrator

then the corresponding controls are given by

$$u_*(x) = \begin{cases} +1 & \text{for } x \in \Gamma_+ \cup G_+, \\ -1 & \text{for } x \in \Gamma_- \cup G_-. \end{cases}$$

It follows from Theorem 2.5.3 that the controlled trajectories in this family are optimal. For this simple example, this can also easily be verified directly [41]: Let $(\bar{x}_1, \bar{x}_2) \neq (0,0)$ be an arbitrary initial condition and let T denote the time it takes for the system to reach the origin along the controlled trajectory in the family $\mathcal{F}$. Suppose there exists another control $\bar{u}$ that steers $(\bar{x}_1, \bar{x}_2)$ into the origin in time $\bar{T} < T$. Without loss of generality, consider the case that the control in the family $\mathcal{F}$ is given by

$$u(t) = \begin{cases} -1 & \text{for } 0 < t \leq \alpha, \\ +1 & \text{for } \alpha < t \leq T. \end{cases}$$

Define functions

$$\Phi(t) = -x_1(t) + x_2(t)(t - \alpha)$$

and

$$\Psi(t) = -\bar{x}_1(t) + \bar{x}_2(t)(t - \alpha),$$

where $(x_1(\cdot), x_2(\cdot))$ is the solution from the family $\mathcal{F}$ and $(\bar{x}_1(\cdot), \bar{x}_2(\cdot))$ is the solution corresponding to the control $\bar{u}$. Then we have that

$$\dot{\Phi}(t) = -\dot{x}_1(t) + \dot{x}_2(t)(t - \alpha) + x_2(t) = u(t)(t - \alpha) = |t - \alpha|$$

and

$$\dot{\Psi}(t) = -\dot{\bar{x}}_1(t) + \dot{\bar{x}}_2(t)(t - \alpha) + \bar{x}_2(t) = \bar{u}(t)(t - \alpha).$$

Since $U = [-1, 1]$, it follows that $\dot{\Psi}(t) \leq |\dot{\Psi}(t)| \leq \dot{\Phi}(t)$ and thus

$$\int_0^{\bar{T}} \dot{\Phi}(t)dt \geq \int_0^{\bar{T}} \dot{\Psi}(t)dt.$$

Hence

$$\Phi(\bar{T}) - \Phi(0) \geq \Psi(\bar{T}) - \Psi(0).$$

But by construction,

$$\Phi(0) = -\bar{x}_1 - \alpha \bar{x}_2 = \Psi(0),$$

and since $\Psi(\bar{T}) = 0$ (the system is at the origin at time $\bar{T}$), we have $\Phi(\bar{T}) \geq 0$. But then

$$0 < \int_{\bar{T}}^T |t - \alpha|dt = \int_{\bar{T}}^T \dot{\Phi}(t)dt = \Phi(T) - \Phi(\bar{T}) = -\Phi(\bar{T}) \leq 0.$$

Contradiction. This proves that the family $\mathcal{F}$ is an *optimal synthesis of controlled trajectories*.

The general question of optimality of an extremal synthesis will be considered in the context of sufficient conditions for optimality in Chap. 6. For the linear systems considered in this section, the optimality of all the syntheses constructed here follows from Theorem 2.5.3.

2.6.2 A Hyperbolic Saddle

We now consider a system that has both a positive and negative eigenvalue:

$$\dot{x} = \begin{pmatrix} 0 & 1 \\ 1 & 0 \end{pmatrix} x + \begin{pmatrix} 0 \\ 1 \end{pmatrix} u, \qquad |u| \leq 1.$$

Again, the system is completely controllable,

$$K = (b, Ab) = \begin{pmatrix} 0 & 1 \\ 1 & 0 \end{pmatrix},$$

and the eigenvalues of A are $\mu_1 = -1$ and $\mu_2 = +1$. Hence optimal controls are bang-bang with at most one switching, and an extremal synthesis is sufficient for optimality.

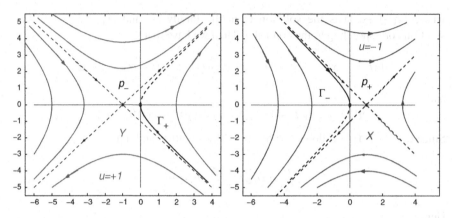

Fig. 2.2 Phase portrait for $u = +1$ (*left*) and for $u = -1$ (*right*)

As for the double integrator, geometric properties of the phase portrait of the uncontrolled system determine the structure of the overall synthesis of optimal controlled trajectories. The origin is a hyperbolic saddle for the system $\dot{z} = Az$, and the stable and unstable subspaces at the equilibria are spanned by the eigenvectors v_1 and v_2 of the eigenvalues μ_1 and μ_2, respectively,

$$v_1 = \begin{pmatrix} 1 \\ -1 \end{pmatrix}, \qquad v_2 = \begin{pmatrix} 1 \\ 1 \end{pmatrix}.$$

That is, if p is a multiple of v_1, then the solution $z(t)$ to the initial value problem $\dot{z} = Az$, $z(0) = p$, is given by $z(t) = e^{At}p = e^{-t}p$ and thus satisfies $\lim_{t \to \infty} z(t) = 0$, while for multiples of v_2 the solution is given by $z(t) = e^{At}p = e^{t}p$ and thus satisfies $\lim_{t \to -\infty} z(t) = 0$. The phase portraits for the controlled vector fields with $u = +1$ or $u = -1$ are simply shifted versions of the phase portrait of the homogeneous system $\dot{z} = Az$ along the x_1-axis and are shown in Fig. 2.2.

If, as above, we denote by X the vector field corresponding to the control $u \equiv -1$ and by Y the vector field corresponding to the control $u \equiv +1$, i.e.,

$$X(x) = Ax - b = \begin{pmatrix} x_2 \\ x_1 - 1 \end{pmatrix}, \qquad Y(x) = Ax + b = \begin{pmatrix} x_2 \\ x_1 + 1 \end{pmatrix},$$

then these vector fields now have a hyperbolic saddle at the points $p_+ = (+1, 0)$ and $p_- = (-1, 0)$ and the stable and unstable subspaces of the matrix A are translated to become lines through p_+ and p_-. Note that there again exist unique trajectories Γ_- of X and Γ_+ of Y that steer the system into the origin forward in time, shown as solid black curves in Fig. 2.2. Their continuations, which will not be part of the synthesis, are shown dashed. As with the double integrator, an extremal synthesis is then constructed by integrating X backward from points in Γ_+ and Y backward from points in Γ_-. However, it is now no longer possible to steer every point into the origin as it was the case with the double integrator, and the controllable set is bounded by the stable manifolds of the equilibria p_+ and p_-, that is, by the lines

$$E_+ = p_+ + \operatorname{lin span}\{v_1\} = \{x \in \mathbb{R}^2 : x_1 + x_2 = +1\}$$

and

$$E_- = p_- + \operatorname{lin span}\{v_1\} = \{x \in \mathbb{R}^2 : x_1 + x_2 = -1\}.$$

Clearly, for any admissible control u, we have that

$$\frac{d}{dt}(x_1 + x_2) = (x_1 + x_2) + u$$

and since $|u| \leq 1$, we always have $\frac{d}{dt}(x_1 + x_2) \leq 0$ at points (x_1, x_2) satisfying $x_1 + x_2 \leq -1$ and $\frac{d}{dt}(x_1 + x_2) \geq 0$ at points satisfying $x_1 + x_2 \geq 1$. Thus no point outside of

$$\mathscr{C} = \{(x_1, x_2) : -1 < x_1 + x_2 < 1\}$$

can be steered into the origin. On the other hand, if a point lies in $\mathscr{C}$, then it is clear from the phase portraits that there exists a unique bang-bang control with at most one switching that steers this initial condition into the origin. This family of controlled trajectories is illustrated in Fig. 2.3. By construction this family of controlled trajectories is an extremal synthesis, and hence it is optimal by Theorem 2.5.3.

This example illustrates the obvious fact that complete controllability does not allow one to freely steer the system into arbitrary points if constraints are imposed on the control. As seen in Example 2.4.1, if the eigenvalues are critical, i.e., lie on the imaginary axis, then the instability can be fully overcome by any kind of control action (of course, the control set needs to contain the origin in its interior). Generally, for unstable systems with eigenvalues with positive real parts a certain degree of instability can be overcome depending on the size of the control that is allowed. In Example 2.4.2, when there still existed a one-dimensional stable subspace for the system, it was this subspace (and the size on the control) that determined the controllable set. The next example shows what happens if the system is an unstable node without any stable trajectories at all. Even in this case, the control is still able to overcome some of the instabilities, and the controllable set still is open.

2.6.3   An Unstable Node

Consider the system

$$\dot{x} = \begin{pmatrix} 2 & 1 \\ 2 & 3 \end{pmatrix} x + \begin{pmatrix} 0 \\ 1 \end{pmatrix} u, \qquad u \in [-1, 1].$$

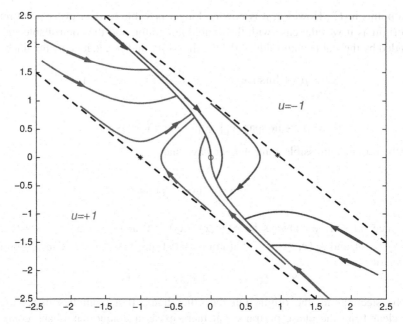

Fig. 2.3 Synthesis of optimal controlled trajectories for the time-optimal control problem to the origin for a hyperbolic saddle

As above, the system is completely controllable,

$$K = (b, Ab) = \begin{pmatrix} 0 & 1 \\ 1 & 3 \end{pmatrix},$$

with two real eigenvalues, $\mu_1 = 1$ and $\mu_2 = 4$, and as before, optimal controls are bang-bang with at most one switching and an extremal synthesis is optimal.

The uncontrolled system is an unstable node, and the phase portraits for the controlled vector fields X and Y corresponding to the constant controls $u = -1$ and $u = +1$, respectively, again are simply shifted versions along the x_1-axis of the phase portrait of $\dot{z} = Az$ shown in Fig. 2.4. In this case, the solutions along the eigenvectors do not play an important role, but instead the boundary of the controllable set is given by two specific trajectories Λ_+ and Λ_- of the vector fields X and Y: Λ_+ is the backward orbit of the trajectory of the vector field Y that passes through the equilibrium point $p_- = (-\frac{1}{4}, \frac{1}{2})$ of the vector field X at time 0 and converges to the equilibrium $p_+ = (\frac{1}{4}, -\frac{1}{2})$ of the vector field Y as $t \to -\infty$, and symmetrically, Λ_- is the backward orbit of the trajectory of the vector field X that passes through the equilibrium point p_+ of the vector field Y at time 0 and converges to the equilibrium p_- of the vector field X as $t \to -\infty$. The concatenation of these two curves with the equilibria p_+ and p_- forms a simple closed curve, and the controllable set $\mathscr{C}$ is the interior of this closed curve with $\mathscr{C}$ as its boundary. The optimal synthesis is constructed analogously as for the double integrator and the hyperbolic saddle by

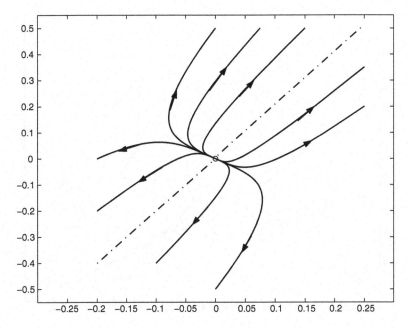

Fig. 2.4 Phase portrait of $\dot{z} = Az$ for an unstable node

integrating the vector fields X and Y backward from the unique trajectories Γ_- of X and Γ_+ of Y that steer the system into the origin forward in time, and it is illustrated in Fig. 2.5.

2.6.4   The Harmonic Oscillator

We close this section with an example of a matrix A that has complex eigenvalues. Because of the inherent oscillatory character of these systems, the number of switchings no longer can be bounded. We consider the harmonic oscillator. As before, $x(t)$ denotes the position of the object at time t, $\dot{x}(t)$ its velocity, and $u(t)$ the external force applied to the object, and we write the state as $x = (x_1, x_2)^T = (x, \dot{x})^T$. Now the dynamics takes the form

$$\dot{x} = \begin{pmatrix} 0 & 1 \\ -1 & 0 \end{pmatrix} x + \begin{pmatrix} 0 \\ 1 \end{pmatrix} u,$$

and the Hamiltonian H is given by

$$H = \lambda_0 + \lambda \left[\begin{pmatrix} 0 & 1 \\ -1 & 0 \end{pmatrix} x + \begin{pmatrix} 0 \\ 1 \end{pmatrix} u \right] = \lambda_0 + \lambda_1 x_2 + \lambda_2(-x_1 + u).$$

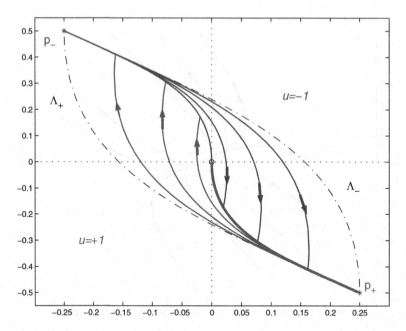

Fig. 2.5 Synthesis of optimal controlled trajectories for the time-optrimal control problem to the origin for an unstable node

Thus again, the minimum condition implies that

$$u(t) = \begin{cases} +1 & \text{if} \quad \lambda_2(t) < 0, \\ -1 & \text{if} \quad \lambda_2(t) > 0. \end{cases}$$

As for all linear systems, the adjoint equation is given by the system itself run backward, but written as a row vector

$$\dot{\lambda} = -\lambda \begin{pmatrix} 0 & 1 \\ -1 & 0 \end{pmatrix},$$

or

$$\dot{\lambda}_1 = \lambda_2, \qquad \dot{\lambda}_2 = -\lambda_1,$$

and thus

$$\ddot{\lambda}_2 \equiv \frac{d}{dt}(-\lambda_1) = -\lambda_2.$$

Hence, all solutions of the adjoint equation are integral curves of the harmonic oscillator. Thus again *optimal controls are bang-bang*, but now we cannot give an a priori bound on the number of switchings. In fact, depending on the initial condition, this number can be arbitrarily large. However, since switchings are the zeros of a solution to the harmonic oscillator, it follows that *all switchings* τ_n *are* *spaced exactly π units apart*, and the first switching τ_1 can take any value in the interval $(0, \pi]$. Analytically, any solution λ_2 of the adjoint equation is of the form $\lambda_2(t) = a\cos t + b\sin t$ for some constants a and b and therefore can be written in phase-angle form as

$$\lambda_2(t) = A\cos(t - \varphi)$$

with amplitude $A = \sqrt{a^2 + b^2}$ and phase $\varphi = \arctan\left(\frac{b}{a}\right)$.

With this information, as with the examples above, it is again straightforward to synthesize all possible extremals by analyzing the phase portraits of the systems corresponding to the constant controls $u \equiv +1$ and $u \equiv -1$ and then consider all possible concatenations that switch exactly π units of time apart. As before, let X and Y denote the vector fields corresponding to the controls $u \equiv -1$ and $u \equiv +1$, respectively. Integral curves of X are circles with center at the point $p_- = (-1, 0)$, and integral curves of Y are circles with center at $p_+ = (1, 0)$, both traversed clockwise. Exactly as in the case of the double integrator, among all these trajectories there are only two that steer the system into the origin directly, namely

$$\Gamma_+ : [-\pi, 0] \to \mathbb{R}^2, \quad t \mapsto (x_1(t), x_2(t)) = (1 - \cos(t), \sin(t)),$$

and

$$\Gamma_- : [-\pi, 0] \to \mathbb{R}^2, \quad t \mapsto (x_1(t), x_2(t)) = (-1 + \cos(t), -\sin(t)).$$

Note that Γ_- is the curve obtained by reflecting Γ_+ at the origin, and only these two semicircles are admissible extremal trajectories, since switchings must be spaced π units apart. Thus there cannot be any segment of an optimal X or Y trajectory longer than π. Any extremal control that steers the system into the origin needs to do so along either Γ_+ or Γ_- as final segment. The full family $\mathscr{F}$ is now constructed by picking a point $q_+(t) \in \Gamma_+$ (respectively $q_-(t) \in \Gamma_-$) for a time $t \in [-\pi, 0)$ and integrating the vector fields X and Y (respectively, Y and X) backward from $q_+(t)$ (respectively $q_-(t)$), switching between these vector fields at all times precisely π units apart. Thus, with the final time normalized to 0, the switchings occur at times t, $t - \pi$, $t - 2\pi$, ... and the curves where the switchings occur, the so-called *switching curves*, are obtained inductively by following the flow of X, respectively Y, for exactly π units of time starting with Γ_+ and Γ_-. Since integral curves are concentric circles centered at $p_\pm$, this generates a family of shifted semicircles of type Γ_+ below the positive x_1-axis and of type Γ_- above the negative x_1-axis as depicted in Fig. 2.6. In this figure, the curves Γ_+ and Γ_- are shown as solid curves since these are actually integral curves of X and Y, while all their translates are strictly switching curves that do not correspond to integral curves and are shown dashed. On all points on these translates, the controls switch between $+1$ and -1 and the corresponding trajectories cross the switching curves.

As with the double integrator, the family $\mathscr{F}$ covers the entire state space injectively, and for every initial condition (x_1^0, x_2^0) there exists a unique bang-bang extremal that steers the system into the origin. So again we have an extremal synthesis, and the control can be given as a feedback control. If we denote the switching locus by Υ and let G_+ denote the region below Υ in the (x_1, x_2)-plane

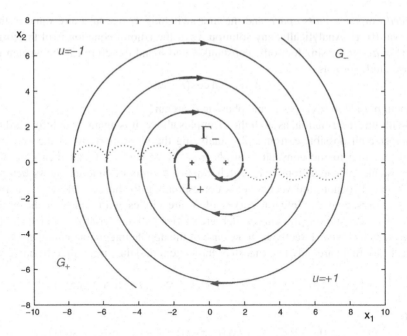

Fig. 2.6 Synthesis of optimal controlled trajectories for time-optimal control to the origin for the harmonic oscillator

and G_- the region above Υ, then the control is again a discontinuous feedback of the form

$$u_*(x) = \begin{cases} +1 & \text{for } x \in \Gamma_+ \cup G_+, \\ -1 & \text{for } x \in \Gamma_- \cup G_-. \end{cases}$$

As with the other examples considered in this section, by Theorem 2.5.3 the family $\mathscr{F}$ of controlled trajectories is an *optimal synthesis*.

We close this section with pointing out the special nature of the trajectories that end with the full semicircles Γ_+ and Γ_-. Let γ_+ and γ_- be the controlled trajectories in the field $\mathscr{F}$ that end at the origin at time 0 by following the full arcs Γ_+ and Γ_-, respectively, and have switchings at all negative integer multiples of π. These two extremal trajectories are *strictly abnormal*, i.e., the only way to satisfy the conditions of the maximum principle is with $\lambda_0 = 0$. Thus, the trajectories γ_+ and γ_- are examples of optimal trajectories whose extremals are abnormal.

Proposition 2.6.1. *The extremals corresponding to γ_+ and γ_- are unique (up to a positive multiple) and are strictly abnormal.*

Proof. Without loss of generality, we consider γ_-. Since the system is time-invariant, we can normalize the final time to be $T = 0$ and integrate backward. Then, as used already above, the parametrization of γ_- (respectively Γ_-) over the interval $[-\pi, 0]$ is given by

$$x_1(t) = -1 + \cos(t) \quad \text{and} \quad x_2(t) = -\sin(t).$$

The times $t = -\pi$ and $t = 0$ are switching times. Since λ itself is a solution of the harmonic oscillator, only multiples of $\sin(t)$ will satisfy this condition, and since the control is $u = -1$ on $[-\pi, 0]$, we must have $\lambda_2(t) = \alpha \sin(t)$ for some $\alpha < 0$. Hence $\lambda_1(t) = -\dot{\lambda}_2(t) = -\alpha \cos(t)$, and on $[-\pi, 0]$ the Hamiltonian H takes the form

$$
\begin{aligned}
H &= \lambda_0 + \lambda_1(t)x_2(t) + \lambda_2(t)(-x_1(t) + u(t)) \\
&= \lambda_0 + \alpha \cos(t)\sin(t) + \alpha \sin(t)(1 - \cos(t) - 1) \\
&= \lambda_0.
\end{aligned}
$$

But it follows from Theorem 2.5.1 that $H \equiv 0$, and thus we must have $\lambda_0 = 0$.   □

2.7   Extensions of the Model: Two Examples

The examples considered so far fall into well-established classes, linear-quadratic optimal control and time-optimal control for time-invariant linear systems. But the techniques that were used apply more generally, and as further illustration of how to use the conditions of the maximum principle, we shall solve a basic trading problem in economics and a classical example of a nonlinear system, the so-called moon landing problem, that will lead us to a discussion of general nonlinear control-affine systems in the next section.

2.7.1   An Economic Trading Model

We consider a simple model of a firm that buys and sells a product and has cash and the quantity of this product as its two assets; denote the values of these assets at time t by $x_1(t)$ and $x_2(t)$, respectively [139]. The initial values of the assets, $x_1(0)$ and $x_2(0)$, are given. If the company's reservation utility for the price of the product at the end of some planning period $[0, T]$ is denoted by π, then the firm's goal is to *maximize*

$$
C(u) = x_1(T) + \pi x_2(T).
$$

Ideally, the reservation utility would agree with the price at time T, but a priori this price is unknown. The control in the problem, represented by $u(t)$, is the rate of buying and selling the product at time t with $u(t) > 0$ considered a purchase and $u(t) < 0$ a sale. We assume that at any time, there are (self-imposed) limits on the amount of the product the company wants to buy or sell, say $m \leq u(t) \leq M$ with $m < 0$ and $M > 0$ given constants. If $p(t)$ denotes the price of the product at time t, then the effect of a trading operation on the company's assets is given by

$$
\dot{x}_1(t) = -\alpha x_2(t) - p(t)u(t), \qquad \dot{x}_2(t) = u(t),
$$

where $\alpha > 0$ is a constant associated with the cost of storing a unit of the product and the term $p(t)u(t)$ gives the cost of purchase or the revenue from sales at time t.

The dynamics now has the form $\dot{x} = Ax + B(t)u$, where A is time-invariant, but B is time-varying,

$$A = \begin{pmatrix} 0 & -\alpha \\ 0 & 0 \end{pmatrix} \quad \text{and} \quad B(t) = \begin{pmatrix} -p(t) \\ 1 \end{pmatrix}.$$

The Hamiltonian H for the problem is

$$H = H(t,\lambda,x,u) = \lambda_1(-\alpha x_2 - p(t)u) + \lambda_2 u = -\alpha \lambda_1 x_2 + (\lambda_2 - \lambda_1 p(t))u,$$

and the adjoint equations are given by

$$\dot{\lambda}_1 = 0, \qquad \dot{\lambda}_2 = \alpha \lambda_1,$$

with transversality conditions

$$\lambda_1(T) = -\lambda_0, \qquad \lambda_2(T) = -\lambda_0 \pi.$$

Notice the minus signs in the transversality conditions that arise, since in our formulation, we minimize the objective $J(u) = -C(u)$. If $\lambda_0 = 0$, then also $\lambda(t) \equiv 0$, contradicting the nontriviality of the multiplier, and thus we normalize $\lambda_0 = 1$. Hence

$$\lambda_1(t) \equiv -1 \quad \text{and} \quad \lambda_2(t) = \alpha(T-t) - \pi, \qquad 0 \le t \le T.$$

The minimization condition on the Hamiltonian therefore implies that the optimal control $u_*(t)$ satisfies

$$u_*(t) = \begin{cases} m & \text{if} \quad p(t) > \alpha(t-T) + \pi, \\ M & \text{if} \quad p(t) < \alpha(t-T) + \pi, \end{cases}$$

while it is not specified through the minimization condition if $p(t) = \alpha(t-T) + \pi$. In fact, if the price were to follow this linear relationship, then the minimization condition would be inconclusive, and this leads to the concept of singular controls that we shall describe in Sect. 2.8. Here we consider only the simpler case in which p is continuous and piecewise continuously differentiable with $\dot{p}(t) \ne \alpha$ everywhere. In this case, whenever $p(t) = \alpha(t-T) + \pi$, then p crosses the line $\ell(t) = \alpha(t-T) + \pi$, and at every such crossing a switch from m to M or vice versa occurs. Let us illustrate this solution with a particular case of price function as given below for the numerical values $T = 9$, $m = -1$, $M = 1$, and $\alpha = \frac{1}{3}$:

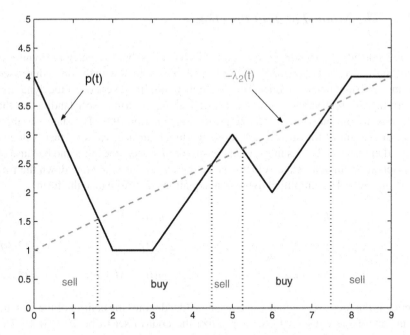

Fig. 2.7 An optimal trading strategy

$$p(t) = \begin{cases} -\frac{3}{2}t + 4 & \text{for} \quad 0 \le t \le 2, \\ 1 & \text{for} \quad 2 \le t \le 3, \\ t - 2 & \text{for} \quad 3 \le t \le 5, \\ -t + 8 & \text{for} \quad 5 \le t \le 6, \\ t - 4 & \text{for} \quad 6 \le t \le 8, \\ 4 & \text{for} \quad 8 \le t \le 9. \end{cases}$$

Choosing $\pi = 4$, we get that $\lambda_2(t) = -\frac{1}{3}t - 1$, and Fig. 2.7 illustrates the optimal buy–sell decisions for this price function. The optimal control for the problem thus is

$$u_*(t) = \begin{cases} m & \text{for} \quad 0 \le t \le 1.64, \\ M & \text{for} \quad 1.64 \le t \le 4.5, \\ m & \text{for} \quad 4.5 \le t \le 5.25, \\ M & \text{for} \quad 5.25 \le t \le 7.5, \\ m & \text{for} \quad 7.5 \le t \le 9. \end{cases}$$

2.7.2   The Moon-Landing Problem

We now consider a problem with a nonlinear dynamics, but for which the synthesis of optimal controlled trajectories can still easily be obtained with the procedure used for time-invariant linear systems. This is a highly simplified version of the dynamics underlying the real version of a spacecraft making a vertical soft landing on the surface of the moon while minimizing fuel consumption [95]. The state variables are h, the height of the space craft above the lunar surface; v, its vertical velocity oriented upward; and m, its mass. Fuel consumption lowers the mass and because of the orientation, increases the velocity, since the jets are used to slow down the free fall of the craft. The simplified dynamical equations therefore take the form

$$\dot{h} = v, \qquad\qquad\qquad h(0) = h_0, \qquad\qquad (2.29)$$

$$\dot{v} = -g + \frac{u}{m}, \qquad\qquad\qquad v(0) = v_0, \qquad\qquad (2.30)$$

$$\dot{m} = -ku, \qquad\qquad\qquad m(0) = M + F, \qquad\qquad (2.31)$$

where g is the moon's gravitational constant and u denotes the control of the system. By means of the constant k, we normalize the control set to be $U = [0,1]$. The coefficients M and F in the initial condition for m denote the mass of the spacecraft and the total mass of the fuel at the beginning of descent. The optimal control problem then becomes the following:

[ML]   For a free terminal time T, minimize the total amount of fuel used,

$$J(u) = \int_0^T u(t)dt,$$

over all piecewise continuous functions $u : [0,T] \to [0,1]$, subject to the dynamics (2.29)–(2.31) and terminal conditions

$$h(T) = 0 \qquad \text{and} \qquad v(T) = 0.$$

Clearly, an implicit assumption in the model is the state constraint $h \geq 0$, and obviously we also cannot allow that $h = 0$ at some intermediate time with negative velocity v. However, for the moment we ignore these constraints, and it will be seen that the optimal solution fulfills these obvious physical side conditions.

The Hamiltonian function for the moon-landing problem is given as

$$H = \lambda_0 u + \lambda_1 v + \lambda_2 \left(-g + \frac{u}{m}\right) - \lambda_3 ku = \lambda_1 v - \lambda_2 g + u \left(\lambda_0 - \frac{\lambda_2}{m} - \lambda_3 k\right).$$

If (x_*, u_*) is an optimal controlled trajectory defined over the interval $[0,T]$, then there exist a constant $\lambda_0 \geq 0$ and an adjoint variable $\lambda = (\lambda_1, \lambda_2, \lambda_3) : [0,T] \to$

$(\mathbb{R}^3)^*$ such that the following conditions are satisfied: (a) λ_0 and $\lambda(t)$ do not vanish simultaneously over $[0, T]$, (b) $\lambda(t)$ satisfies the adjoint equations

$$\dot{\lambda}_1 = 0, \qquad \dot{\lambda}_2 = -\lambda_1, \qquad \dot{\lambda}_3 = \lambda_2 \frac{u}{m^2},$$

with transversality condition $\lambda_3(T) = 0$, and (c) the control $u_*(t)$ minimizes the Hamiltonian H as a function of u over the control set $[0, 1]$ with minimum value 0.

Since the Hamiltonian H is linear in u, this minimum is determined by the sign of the function

$$\Phi(t) = \lambda_0 + \frac{\lambda_2(t)}{m(t)} - \lambda_3(t)k,$$

and we have that

$$u_*(t) = \begin{cases} 0 & \text{if} \quad \Phi(t) > 0, \\ \text{undefined} & \text{if} \quad \Phi(t) = 0, \\ 1 & \text{if} \quad \Phi(t) < 0. \end{cases}$$

Again Φ is the *switching function* of the problem.

For the time-optimal control problem, intuition would say that the optimal solution should be free fall ($u_* = 0$) followed, at the right moment, by a maximum thrust ($u_* = 1$) to slow down the craft to make a soft landing. This corresponds to a bang-bang control that has exactly one switching from $u = 0$ to $u = 1$. For this problem, this also is the minimum-fuel-consumption solution. To see this, we analyze the derivative of the switching function. It follows from the dynamics and adjoint equation that

$$\dot{\Phi}(t) = \frac{\dot{\lambda}_2(t)}{m(t)} - \lambda_2(t) \frac{\dot{m}(t)}{m(t)^2} - \dot{\lambda}_3(t)k = -\frac{\lambda_1}{m(t)}.$$

If $\lambda_1 = 0$, then the switching function Φ is constant. But Φ cannot vanish identically, since the condition that $H = \lambda_1 v - \lambda_2 g + u\Phi(t) \equiv 0$ then also gives that $\lambda_2 = 0$, which implies $\lambda_3 = 0$ as well and thus also $\lambda_0 = 0$ from $\Phi = 0$, contradicting the nontriviality of the multipliers. Clearly, Φ also cannot be positive, since v decreases along the control $u \equiv 0$, and thus we cannot meet the terminal condition $v = 0$. Hence, in this case, Φ must be negative, giving the constant control $u_*(t) \equiv 1$. This corresponds to braking with full thrust throughout, and clearly this is the optimal control for specific initial conditions. If $\lambda_1 \neq 0$, then the switching function is strictly monotone and thus has at most one zero. Again, the only choice that can satisfy the terminal condition is $\lambda_1 > 0$, and hence optimal controls must be bang-bang with at most one switching from $u = 0$ to $u = 1$.

Once this is known, a field of extremals can be constructed as before in the examples for linear systems. Suppose the control is given by $u_* \equiv 1$ on the interval $[\zeta, T]$. It then follows from the terminal conditions $h(T) = 0$ and $v(T) = 0$ that

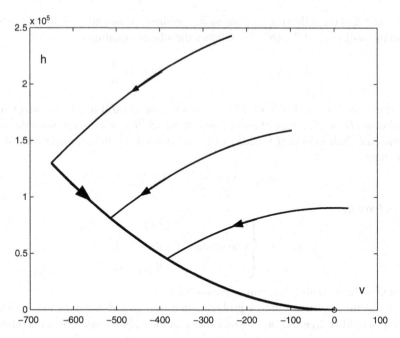

Fig. 2.8 Switching curve and optimal controlled trajectories near the final time T

$$h(\zeta) = -\frac{1}{2}g(T-\zeta)^2 - \frac{M+F}{k^2}\ln\left(1 - \frac{k(T-\zeta)}{M+F}\right) - \frac{T-\zeta}{k},$$

$$v(\zeta) = g(T-\zeta) + \frac{1}{k}\ln\left(1 - \frac{k(T-\zeta)}{M+F}\right).$$

Plotting $h(\zeta)$ against $v(\zeta)$, we get a curve $\mathscr{L}$, $\zeta \mapsto (h(\zeta), v(\zeta))$, that represents the set of all initial conditions (height and velocity pairs) that would result in a soft landing with full thrust $u_* \equiv 1$. Since there exists a restriction that the total amount of fuel will burn in time $\frac{F}{k}$ seconds, one further needs to restrict the curve to the ζ-values in the interval $[T - \frac{F}{k}, T]$. The first part of the trajectory simply is free fall ($u_* = 0$), and the initial portions of the equations are given by

$$h(t) = -\frac{1}{2}gt^2 + v_0t + h_0 \quad \text{and} \quad v(t) = -gt + v_0$$

so that

$$h(t) = h_0 - \frac{1}{2g}\left(v^2(t) - v_0^2\right), \qquad t \geq 0. \tag{2.32}$$

Once this parabola meets the curve $\mathscr{L}$, the thrusters need to be engaged at full force to make a soft landing. If the parabola does not meet the switching curve $\mathscr{L}$, a soft landing is impossible. Figure 2.8 illustrates the synthesis.

2.8 Singular Controls and Lie Derivatives

Both the linear time-optimal control problems in the plane and also the examples considered in the previous section lead to minimizing a Hamiltonian function that is linear in a scalar control u over a compact interval $[a,b]$. Clearly, this minimum is attained at $u = a$ if the function Φ multiplying u is positive and at $u = b$ if this function is negative. In the examples we have considered so far, it always turned out that optimal controls were *bang-bang*, i.e., consisted of a finite number of switchings between $u = a$ and $u = b$. We shall show in Sect. 3.6, that this is "always" the case for a time-invariant linear system whose control set is a compact polyhedron. More precisely, for these systems, it is always possible to find an optimal control that switches finitely many times between controls that take their values in one of the vertices of the control set; in addition, the number of switchings over a finite interval $[0,T]$ can be bounded. This no longer holds once the dynamics of the control system becomes nonlinear: optimal controls need not be bang-bang, and even when optimal controls switch only between $u = a$ and $u = b$, the number of switchings on a compact interval $[0,T]$ can be countably infinite. We shall see in the remaining sections of this chapter that these phenomena are linked with controls that arise when the function Φ multiplying u vanishes over some interval, so-called *singular controls*. We now develop geometric tools and techniques required for their analysis.

2.8.1 Time-Optimal Control for a Single-Input Control-Affine Nonlinear System

Again we use the time-optimal control problem as the vehicle to develop these tools, but now allow for nonlinearities in the state. We consider a time-invariant, single-input, control-affine system Σ of the form

$$\Sigma: \qquad \dot{x} = f(x) + g(x)u, \qquad f, g \in V^{\infty}(\Omega), \qquad x \in \Omega. \qquad (2.33)$$

Here, Ω is a domain (i.e., an open and connected subset) in $\mathbb{R}^n$, and $f : \Omega \to \mathbb{R}^n$ and $g : \Omega \to \mathbb{R}^n$ are two infinitely often continuously differentiable vector fields defined on Ω. We use $V^r(\Omega)$ to denote the set of all vector fields defined on Ω for which all components are $C^r(\Omega)$-functions, i.e., are defined and r times continuously differentiable on Ω. Clearly, the C^{∞} assumption is without loss of generality and can be replaced by requiring that the vector fields be sufficiently smooth, say $f, g \in V^r(\Omega)$, with r large enough for all the derivatives that arise to exist.

[NTOC] Given a time-invariant, single-input, control-affine control system Σ of the form (2.33), among all piecewise continuous (more generally, Lebesgue measurable) controls u that take values in the compact interval $[-1,1]$, $u : [0,T] \to [-1,1]$, find one that steers a given initial point $p \in \Omega$ into a target point $q \in \Omega$ in minimum time.

In the formulation of Sect. 2.2, we have that $M = \Omega$, $L(t,x,u) \equiv 1$, $f(t,x,u) = f(x) + g(x)u$, $\varphi \equiv 0$, and Ψ is given by $\Psi : [0,\infty) \times \Omega \to \Omega$, $(t,x) \mapsto \Psi(t,x) = x - q$, i.e., $N = \{q\}$. As with the linear system, since both initial and terminal points on the state are specified, the transversality conditions give no information about the multiplier λ, and the terminal value $\lambda(T)$ is free. But the transversality condition on the final time T implies that $H(T,\lambda_0,\lambda,x,u) = 0$, and since

$$H = \lambda_0 + \lambda\left(f(x) + g(x)u\right) = \lambda_0 + \langle \lambda, f(x) + g(x)u \rangle$$

is time-invariant, it follows that the Hamiltonian vanishes identically along any extremal. Equivalently, we have that

$$\langle \lambda(t), f(x_*(t)) + u_*(t)g(x_*(t)) \rangle \equiv \text{const} \leq 0.$$

We freely use the notation $\langle \cdot, \cdot \rangle$ for the inner product. In particular, note that this implies that $\lambda(t) \neq 0$, since otherwise also $\lambda_0 = 0$ contradicting the nontriviality condition of the maximum principle. The adjoint equation is given by

$$\dot{\lambda}(t) = -\lambda(t)\left(Df(x_*(t)) + u_*(t)Dg(x_*(t))\right),$$

where Df and Dg denote the matrices of the partial derivatives of the vector fields f and g, respectively, and the minimum condition implies that

$$u_*(t) = \begin{cases} +1 & \text{if} \quad \langle \lambda(t), g(x_*(t)) \rangle < 0, \\ -1 & \text{if} \quad \langle \lambda(t), g(x_*(t)) \rangle > 0. \end{cases}$$

Summarizing, we thus have the following result:

Theorem 2.8.1 (Maximum principle for problem (NTOC)). *Let (x_*, u_*) be a controlled trajectory defined over the interval $[0,T]$. If (x_*, u_*) minimizes the time of transfer from $p \in \Omega$ to $q \in \Omega$, then there exists a nontrivial solution $\lambda : [0,T] \to (\mathbb{R}^n)^*$ to the adjoint equation*

$$\dot{\lambda}(t) = -\lambda(t)\left(Df(x_*(t)) + u_*(t)Dg(x_*(t))\right) \tag{2.34}$$

such that

$$\langle \lambda(t), f(x_*(t)) + u_*(t)g(x_*(t)) \rangle \equiv \text{const} \leq 0$$

and the control u_* satisfies

$$u_*(t) = -\text{sgn}\,\langle \lambda(t), g(x_*(t)) \rangle.$$

2.8.2   The Switching Function and Singular Controls

Definition 2.8.1 (Switching function). Let Γ be an extremal lift for the problem [NTOC] consisting of a controlled trajectory (x_*, u_*) defined over the interval $[0,T]$ with corresponding adjoint vector $\lambda : [0,T] \to (\mathbb{R}^n)^*$. The function

$$\Phi_\Gamma(t) = \lambda(t)g(x_*(t)) = \langle \lambda(t), g(x_*(t)) \rangle \qquad (2.35)$$

is called the (corresponding) switching function.

We usually drop the subscript Γ in the notation if the extremal under consideration is understood. Clearly, properties of the switching function Φ determine the structure of the optimal controls. As long as Φ is not zero, the optimal control is simply given by

$$u_*(t) = -\mathrm{sgn}\ \Phi(t)$$

and thus takes its value in one of the vertices of the control set. A priori, the control is not determined by the minimum condition at times when $\Phi(t) = 0$. Naturally, if $\Phi(\tau) = 0$ and the derivative $\dot{\Phi}(\tau)$ exists and does not vanish, then the control switches between $u = -1$ and $u = +1$ with the order depending on the sign of $\dot{\Phi}(\tau)$. Such a time τ simply is a bang-bang junction, exactly as with linear systems. On the other hand, if $\Phi(t)$ were to vanish identically on an open interval I, then although the minimization property by itself gives no information about the control, in this case, also all the derivatives of $\Phi(t)$ must vanish, and this, except for some degenerate situations, generally does determine the control as well. Controls of this kind are the *singular* controls referred to above, while we refer to the constant controls $u = -1$ and $u = +1$ as *bang* controls. Strictly speaking, to be singular is not a property of the control, but of the extremal lift, since it clearly also depends on the multiplier λ defining the switching function.

Definition 2.8.2 (Singular controls and extremals). Let Γ be an extremal lift for the problem [NTOC] consisting of a controlled trajectory (x_*, u_*) defined over the interval $[0,T]$ with corresponding adjoint vector $\lambda : [0,T] \to (\mathbb{R}^n)^*$. We say that the control u is singular on an open interval $I \subset [0,T]$ if the switching function vanishes identically on I. The corresponding portion of the trajectory x defined over I is called a singular arc, and Γ a singular extremal (respectively, singular extremal lift).

Historically, this terminology has its origin in the following simple observation: in terms of the Hamiltonian H for problem [NTOC], the switching function can be expressed as

$$\Phi(t) = \frac{\partial H}{\partial u}(\lambda_0, \lambda(t), x_*(t), u_*(t)),$$

and thus the condition $\Phi(t) = 0$ formally is the first-order necessary condition for the Hamiltonian to have a minimum in the interior of the control set. For a general optimal control problem, extremal lifts are called singular, respectively nonsingular, over an open interval I if the first-order necessary condition

$$\frac{\partial H}{\partial u}(\lambda_0, \lambda(t), x_*(t), u_*(t)) = 0$$

is satisfied for $t \in I$ and if the matrix of the second-order partial derivatives,

$$\frac{\partial^2 H}{\partial u^2}(\lambda_0, \lambda(t), x_*(t), u_*(t)),$$

is singular, respectively nonsingular, on I. For problem [NTOC], this quantity is always zero, and thus any optimal control that takes values in the interior of the control set is necessarily singular. On the other hand, for example, for the linear-quadratic optimal control problem [LQ] considered earlier, this matrix is always positive definite and all extremals are nonsingular.

In order to solve the problem [NTOC], optimal controls need to be synthesized from bang and singular controls, the potential candidates for optimality, through an analysis of the zero set $\mathscr{Z}_\Phi$ of the switching function,

$$\mathscr{Z}_\Phi = \{t \in [0, T]: \ \Phi(t) = 0\}.$$

This, however, can become a very difficult problem, since a priori, all we know about $\mathscr{Z}_\Phi$ is that it is a closed set.

Proposition 2.8.1. *Given any closed subset $Z \subset \mathbb{R}^n$, there exists a nonnegative C^∞-function φ such that $Z = \{y \in \mathbb{R}^n : \varphi(y) = 0\}$.*

Proof. [108] Let $B = Z^c$, the complement of Z. Since $\mathbb{R}^n$ is second countable (see Appendix C), there exists a sequence of open balls B_i, $i \in \mathbb{N}$, such that $B = \cup_{i \in \mathbb{N}} B_i$. It is a standard calculus exercise to verify that the function Γ defined by

$$\Gamma(y) = \begin{cases} \exp\left(-\frac{1}{(y-1)^2}\right) & \text{for} \quad y < 1, \\ 0 & \text{for} \quad y \geq 1, \end{cases}$$

is C^∞: derivatives of arbitrary order exist and all derivatives at $y = 1$ from the left vanish. Let $D = B_r(p)$ be an open ball with radius r centered at p. Defining a radially symmetric function $\psi : D \to \mathbb{R}^n$ by

$$\psi(y) = \Gamma\left(\frac{\|y - p\|^2}{r^2}\right),$$

it then follows that ψ is nonnegative, $\psi \in C^\infty(D)$, and ψ vanishes identically outside of D. For each open ball B_i, $i \in \mathbb{N}$, let ψ_i be the correspondingly defined function. For a multi-index $\alpha = (\alpha_1, \ldots, \alpha_n)$, $\alpha_i \in \mathbb{N}$, let $|\alpha| = a_1 + \cdots + a_n$ and denote the corresponding partial derivatives of ψ_i by $D^\alpha \psi_i$,

$$D^\alpha \psi_i = \frac{\partial^{\alpha_1} \cdots \partial^{\alpha_n} \psi_i}{\partial y^{\alpha_1} \cdots \partial y^{\alpha_n}}.$$

Let

$$M_i = \sup_{|\alpha| \le i} \|D^\alpha \psi_i\|;$$

since all functions ψ_i have compact support and the summation is finite, all the numbers M_i are finite. Thus the series

$$\varphi(y) = \sum_{i=1}^{\infty} \frac{\psi_i}{2^i M_i}$$

converges uniformly (we have $\|\psi_i\| \le M_i$ for all $i \in \mathbb{N}$), and so do all its partial derivatives. For since also $\|D^\alpha \psi_i\| \le M_i$ for all $i \ge |\alpha|$, the termwise differentiated series

$$\sum_{i=1}^{\infty} \frac{D^\alpha \psi_i}{2^i M_i}$$

converges uniformly and its limit is the αth derivative of φ, $D^\alpha \varphi$. Thus φ is a C^∞-function that is positive on each ball B_i and vanishes identically outside $\cup_{i \in \mathbb{N}} B_i$, i.e., on Z. □

Thus, in principle, the zero set $\mathcal{Z}_\Phi$ of the switching function can be an arbitrary closed subset of the interval $[0, T]$, and a better understanding of this set is needed to solve the optimal control problem [NTOC]. In order to achieve this, we now analyze the derivatives of the switching function. Since both λ and the state x satisfy differential equations, the switching function Φ is differentiable, and we obtain

$$\begin{aligned}
\dot{\Phi}(t) &= \dot{\lambda}(t)g(x_*(t)) + \lambda(t)Dg(x_*(t))\dot{x}_*(t) \\
&= -\lambda(t)\left[Df(x_*(t)) + u_*(t)Dg(x_*(t))\right]g(x_*(t)) \\
&\quad + \lambda(t)Dg(x_*(t))\,f(x_*(t)) + u_*(t)g(x_*(t)) \\
&= \lambda(t)\,Dg(x_*(t))f(x_*(t)) - Df(x_*(t))g(x_*(t)). \quad (2.36)
\end{aligned}$$

The coefficients at $u_*(t)$ cancel, and thus the derivative of the switching function does not depend on the control u_*. Hence $\dot{\Phi}(t)$ is once more differentiable, and we can iterate this calculation to find higher-order derivatives. This very much is the approach pursued in older textbooks on the subject. However, brute force is not necessarily always a good strategy, and now it is of benefit to develop the proper formalism. The key is to observe that the tangent vector that multiplies λ in (2.36) is the coordinate expression of the *Lie bracket* of the vector fields f and g, and this quantity is of fundamental importance in the control of nonlinear systems. We therefore digress to give some of the background that not only is fundamental for nonlinear optimal control theory in general, but also provides us with an elegant and transparent scheme to carry out the required calculations.

2.8.3   Lie Derivatives and the Lie Bracket

As before, let Ω be a domain in $\mathbb{R}^n$ and denote the space of all infinitely often continuously differentiable functions on Ω by $C^\infty(\Omega)$. Also let $X : \Omega \to \mathbb{R}^n$ be a C^∞ vector field defined on Ω, $X \in V^\infty(\Omega)$. As before, the assumption $r = \infty$ is taken for simplicity of notation, and it suffices to have all functions and vector fields to be r-times continuously differentiable with the blanket assumption that r is large enough for all the required differentiations to be permissible. The vector field X can be viewed as defining a first-order differential operator from the space $C^\infty(\Omega)$ into $C^\infty(\Omega)$ by taking at every point $q \in \Omega$ the directional derivative of a function $\varphi \in C^\infty(\Omega)$ in the direction of the vector field $X(q)$, i.e.,

$$X : C^\infty(\Omega) \to C^\infty(\Omega), \quad \varphi \mapsto X\varphi,$$

defined by

$$(X\varphi)(q) = \nabla\varphi(q) \cdot X(q),$$

where $\nabla\varphi$ denotes the gradient of the function φ, as always written as a row vector. While this is a convenient notation, which we freely use, in order to distinguish the values of the vector field from its action when considered as an operator, it is more customary to denote this operator by L_X, i.e., $L_X(\varphi)(q) = (X\varphi)(q)$, and this function is called the *Lie derivative* of the function φ along the vector field X.

Definition 2.8.3 (Lie bracket). The Lie bracket of two vector fields X and Y defined on Ω is the operator defined by the commutator

$$[X,Y] = X \circ Y - Y \circ X = XY - YX.$$

Formally, this is a second-order differential operator. But in fact, all second-order terms cancel, and the Lie bracket defines another first-order differential operator. For if we denote the Hessian matrix of a function φ by $H(\varphi)$ and the action of this symmetric matrix on the vector fields X and Y by $H(\varphi)(X,Y)$, then we simply have that

$$
\begin{aligned}
[X,Y](\varphi) &= X(Y\varphi) - Y(X\varphi) = X(\nabla\varphi \cdot Y) - Y(\nabla\varphi \cdot X) \\
&= \nabla(\nabla\varphi \cdot Y) \cdot X - \nabla(\nabla\varphi \cdot X) \cdot Y \\
&= H(\varphi)(Y,X) + \nabla\varphi \cdot DY \cdot X - H(\varphi)(X,Y) - \nabla\varphi \cdot DX \cdot Y \\
&= \nabla\varphi \cdot (DY \cdot X - DX \cdot Y).
\end{aligned}
$$

This calculation verifies that if $X : \Omega \to \mathbb{R}^n$, $z \mapsto X(z)$, and $Y : \Omega \to \mathbb{R}^n$, $z \mapsto Y(z)$, are coordinates for these vector fields, then the coordinate expression for the Lie bracket is given by

$$[X,Y](z) = DY(z) \cdot X(z) - DX(z) \cdot Y(z). \tag{2.37}$$

This computation directly extends to calculating Lie brackets if we consider C^∞ vector fields as a module over $C^\infty(\Omega)$, i.e., multiply the vector fields by smooth functions.

Lemma 2.8.1. *Suppose α and β are smooth functions on Ω, α, $\beta \in C^\infty(\Omega)$, and X and Y are C^∞ vector fields on Ω. Then*

$$[\alpha X, \beta Y] = \alpha\beta[X,Y] + \alpha(L_X\beta)Y - \beta(L_Y\alpha)X.$$

Proof. This simply follows from the product rule:

$$\begin{aligned}
[aX, \beta Y] &= (\alpha X(\beta Y)) - (\beta Y(\alpha X)) \\
&= \alpha\{(X\beta)Y + \beta XY\} - \beta\{(Y\alpha)X + \alpha YX\} \\
&= \alpha\beta[X,Y] + \alpha(X\beta)Y - \beta(Y\alpha)X.
\end{aligned}$$

$\square$

A more important, and less obvious identity is the Jacobi identity.

Proposition 2.8.2. *For any C^∞ vector fields X, Y, and Z defined on Ω we have that*

$$[X,[Y,Z]] + [Y,[Z,X]] + [Z,[X,Y]] \equiv 0.$$

Proof. Again, computing as operators, we have that

$$\begin{aligned}
[X,[Y,Z]] &= X[Y,Z] - [Y,Z]X \\
&= X(YZ - ZY) - (YZ - ZY)X \\
&= XYZ - XZY - YZX + ZYX.
\end{aligned}$$

Adding the corresponding terms for the other brackets thus gives

$$\begin{aligned}
&[X,[Y,Z]] + [Y,[Z,X]] + [Z,[X,Y]] \\
&\equiv (XYZ - XZY - YZX + ZYX) + (YZX - YXZ - ZXY + XZY) \\
&\quad + (ZXY - ZYX - XYZ + YXZ),
\end{aligned}$$

and all terms cancel. $\square$

Note that the Jacobi identity can be written in the form

$$[X,[Y,Z]] = [[X,Y],Z] + [Y,[X,Z]],$$

and this simply states that taking the Lie bracket with X (i.e., the Lie derivative of a vector field along X) satisfies the product rule. These rules show that the vector fields, understood as differential operators, form a Lie algebra. A *Lie algebra* over

$\mathbb{R}$ is a real vector space $\mathfrak{G}$ together with a bilinear operator $[\cdot,\cdot] : \mathfrak{G} \times \mathfrak{G} \to \mathfrak{G}$ such that for all X, Y, and $Z \in \mathfrak{G}$ we have $[X,Y] = -[Y,X]$ and $[X,[Y,Z]] + [Y,[Z,X]] + [Z,[X,Y]] = 0$. Many of the essential concepts and computational tools that will be developed in Sect. 4.5 depend only on these general identities abstracted from the above properties of vector fields.

These notions allow us to restate the formula for the derivative of the switching function in a more general format, equally simple, but of great importance.

Theorem 2.8.2. *Let $Z : \Omega \to \mathbb{R}^n$ be a differentiable vector field defined on Ω and let (x,u) be a controlled trajectory defined over an interval I with trajectory in Ω. Let λ be a solution to the corresponding adjoint equation and define the function*

$$\Psi(t) = \langle \lambda(t),\, Z(x(t)) \rangle.$$

Then Ψ is differentiable with derivative given by

$$\dot{\Psi}(t) = \langle \lambda(t),\, [f + ug, Z](x(t)) \rangle.$$

Proof. This is the same calculation as above. Note that for any row vector $\lambda \in (\mathbb{R}^n)^*$, matrix $A \in \mathbb{R}^{n \times n}$, and column vector $x \in \mathbb{R}^n$ we have that $\langle \lambda, Ax \rangle = \lambda Ax = \langle \lambda A, x \rangle$. Thus, and dropping the argument t in the calculation, we have that

$$\dot{\Psi}(t) = \left\langle \dot{\lambda},\, Z(x) \right\rangle + \langle \lambda,\, DZ(x)\dot{x}_* \rangle$$

$$= -\langle \lambda\,(Df(x) + uDg(x)),\, Z(x) \rangle + \langle \lambda,\, DZ(x)\,(f(x) + ug(x)) \rangle$$

$$= \langle \lambda,\, DZ(x)f(x) - Df(x)Z(x) \rangle + u\langle \lambda,\, DZ(x)g(x) - Dg(x)Z(x) \rangle$$

$$= \langle \lambda,\, [f,Z](x) \rangle + u\langle \lambda(t),\, [g,Z](x) \rangle,$$

which, for simplicity of notation, we also write as $\dot{\Psi}(t) = \langle \lambda(t),\, [f + ug, Z](x(t)) \rangle$, noting that $u(t)$ simply is a real number under differentiation with respect to the state variables involved in the calculation of the Lie brackets. $\qquad\square$

2.8.4 The Order of a Singular Control and the Legendre–Clebsch Conditions

It follows from Theorem 2.8.2 that the first and second derivatives of the switching function $\Phi(t) = \langle \lambda(t), g(x(t)) \rangle$ are given by

$$\dot{\Phi}(t) = \langle \lambda(t),\, [f,g](x(t)) \rangle \qquad\qquad (2.38)$$

and

$$\ddot{\Phi}(t) = \langle \lambda(t),\, [f,[f,g]](x(t)) \rangle + u(t)\langle \lambda(t),\, [g,[f,g]](x(t)) \rangle. \qquad (2.39)$$

If now $\Gamma = ((x,u),\lambda)$ is an extremal lift for which the control is singular on an open interval I, then all derivatives of Φ vanish identically on I, so that we have

$$\langle \lambda(t), \ [f,g](x(t)) \rangle \equiv 0$$

and

$$\langle \lambda(t), \ [f,[f,g]](x(t)) \rangle + u(t) \langle \lambda(t), \ [g,[f,g]](x(t)) \rangle \equiv 0.$$

Clearly, at times t when $\langle \lambda(t), \ [g,[f,g]](x(t)) \rangle$ does not vanish, this equation determines the singular control, and this leads to the following definition:

Definition 2.8.4 (Order 1 singular control). Let $\Gamma = ((x,u),\lambda)$ be an extremal lift for the problem [NTOC] consisting of a controlled trajectory (x,u) defined over the interval $[0,T]$ with corresponding adjoint vector $\lambda : [0,T] \to (\mathbb{R}^n)^*$. If Γ is a singular extremal lift over an open interval I, then Γ, and also the control u, are said to be singular of order 1 over I if $\langle \lambda(t), \ [g,[f,g]](x(t)) \rangle$ does not vanish on the interval I.

We thus immediately have the following formula for the singular control in terms of the state and multiplier.

Proposition 2.8.3. *If* $\Gamma = ((x,u),\lambda)$ *is a singular extremal lift of order* 1 *over an open interval* I, *then the singular control is given by*

$$u_{\text{sing}}(t) = - \frac{\langle \lambda(t), \ [f,[f,g]](x(t)) \rangle}{\langle \lambda(t), \ [g,[f,g]](x(t)) \rangle}. \tag{2.40}$$

Note that this formula defines the singular control as a function of the state and the multiplier, and thus it depends on the extremal lift. In differential-geometric terms, it defines the singular control in the cotangent bundle (see Appendix C). Generally, it is not a feedback function in the state space. However, more can be said in low dimensions n of the state space and this will be pursued later on. Naturally, this formula in no way guarantees that the control bounds imposed in the problem are satisfied, and thus $u_{\text{sing}}(t)$ is admissible only if the values of this expression lie in the control set, the interval $[-1,1]$.

Similar to the Legendre condition in the calculus of variations, for singular controls a generalized version of the Legendre condition also is necessary for optimality. This result will be proven in Sect. 4.6.1.

Theorem 2.8.3 (Legendre–Clebsch condition). *Suppose the controlled trajectory* (x_*,u_*) *defined over the interval* $[0,T]$ *minimizes the time of transfer from* $p \in \Omega$ *to* $q \in \Omega$ *for problem [NTOC], and the control* u_* *is singular over an open interval* $I \subset [0,T]$. *Then there exists an extremal lift* $\Gamma = ((x_*,u_*),\lambda)$ *with the property that*

$$\langle \lambda(t), \ [g,[f,g]](x(t)) \rangle \leq 0 \quad \text{for all} \quad t \in I.$$

Definition 2.8.5 (Strengthened Legendre–Clebsch condition). Let $\Gamma = ((x,u),\lambda)$ be an extremal lift for the problem [NTOC] consisting of a controlled trajectory (x,u)

defined over the interval $[0,T]$ and corresponding adjoint vector $\lambda : [0,T] \to (\mathbb{R}^n)^*$ that is singular of order 1 over an open interval I. We say that the strengthened Legendre–Clebsch condition is satisfied along Γ over I if $\langle \lambda(t), [g, [f, g]](x(t)) \rangle$ is negative on I, and that it is violated if this expression is positive.

An important property of singular extremals that satisfy the strengthened Legendre–Clebsch condition is that if the singular control takes values in the interior of the control set, then at any time $t \in I$, it can be concatenated with either of the two bang controls $u = -1$ and $u = +1$ in the sense that this generates junctions that satisfy the conditions of the maximum principle. As before, let $X = f - g$ and $Y = f + g$ denote the corresponding vector fields. We write XS for a concatenation of a trajectory corresponding to the control $u = -1$ with a singular arc; i.e., for some $\varepsilon > 0$ the control is given by

$$ u(t) = \begin{cases} -1 & \text{for} \quad t \in (\tau - \varepsilon, \tau), \\ u_{\text{sing}}(t) & \text{for} \quad t \in [\tau, \tau + \varepsilon). \end{cases} $$

The time τ is called a junction time, and the corresponding point $x(\tau)$ a junction point. Similarly, concatenations of the type YS, SX, and SY are defined, and we use the symbol B to denote any one of X or Y.

Proposition 2.8.4. *Let $\Gamma = ((x, u), \lambda)$ be an extremal lift for the problem* [NTOC] *defined over the interval $[0,T]$ that is singular over an open interval I and suppose the strengthened Legendre–Clebsch condition is satisfied on I. If the singular control at the time $\tau \in I$ has a value in the open interval $(-1, 1)$, then there exists an $\varepsilon > 0$ such that any concatenation of the singular control with a bang control $u = -1$ or $u = +1$ at time τ satisfies the necessary conditions of the maximum principle; i.e., concatenations of the types BS and SB are allowed.*

Proof. It follows from Eq. (2.40) that the singular control is continuous if the strengthened Legendre–Clebsch condition is satisfied and so trivially are the constant controls $u = \pm 1$. For any control u that is continuous from the left $(-)$ or right $(+)$, the second derivative of the switching function at time τ is given by

$$ \ddot{\Phi}(\tau_{\pm}) = \langle \lambda(\tau), [f, [f, g]](x(\tau)) \rangle + u(\tau_{\pm}) \langle \lambda(\tau), [g, [f, g]](x(\tau)) \rangle, $$

and it vanishes identically along the singular control. If the strengthened Legendre–Clebsch condition is satisfied, then we have $\lambda(\tau)[g, [f, g]](x(\tau)) < 0$. By assumption, the singular control takes values in the interior of the control set, $u(\tau) \in (-1, 1)$, and thus we get for $u = -1$ that

$$ \ddot{\Phi}(\tau_{\pm}) = \langle \lambda(\tau), [X, [f, g]](x(\tau)) \rangle $$
$$ = \langle \lambda(\tau), [f, [f, g]](x(\tau)) \rangle - \langle \lambda(\tau), [g, [f, g]](x(\tau)) \rangle $$
$$ > \langle \lambda(\tau), [f, [f, g]](x(\tau)) \rangle + u(\tau_{\pm}) \langle \lambda(\tau), [g, [f, g]](x(\tau)) \rangle = 0, $$

and for $u = +1$ we have

$$\ddot{\Phi}(\tau_{\pm}) = \langle \lambda(\tau), \, [Y, [f, g]](x(\tau)) \rangle$$
$$= \langle \lambda(\tau), \, [f, [f, g]](x(\tau)) \rangle + \langle \lambda(\tau), \, [g, [f, g]](x(\tau)) \rangle$$
$$< \langle \lambda(\tau), \, [f, [f, g]](x(\tau)) \rangle + u(\tau_{\pm}) \langle \lambda(\tau), \, [g, [f, g]](x(\tau)) \rangle = 0.$$

For each control, these signs are consistent with both entry and exit from the singular arc. For example, if $u = -1$ on an interval $(\tau - \varepsilon, \tau)$, then the switching function has a local minimum at time $t = \tau$ with minimum value 0, and thus Φ is positive over this interval, consistent with the minimum condition of the maximum principle. □

The order of a singular control over an interval I need not be constant, since the function $\langle \lambda(t), \, [g, [f, g]](x(t)) \rangle$ may vanish on some portions of I. If these are isolated times, then typically at those times the local optimality status of the singular control changes from minimizing to maximizing, and the resulting subintervals simply need to be analyzed separately. A more degenerate situation would arise if $\langle \lambda(t), \, [g, [f, g]](x(t)) \rangle$ were to vanish identically on some subinterval $J \subset I$. In this case, many more relations need to be satisfied for the conditions to be consistent. Since we have both

$$\langle \lambda(t), \, [f, [f, g]](x(t)) \rangle \equiv 0 \qquad \text{and} \qquad \langle \lambda(t), \, [g, [f, g]](x(t)) \rangle \equiv 0 \qquad \text{for all } t \in J,$$

differentiating both identities, we get the following two equations on J:

$$0 \equiv \langle \lambda(t), \, [f, [f, [f, g]]](x(t)) \rangle + u(t) \langle \lambda(t), \, [g, [f, [f, g]]](x(t)) \rangle,$$
$$0 \equiv \langle \lambda(t), \, [f, [g, [f, g]]](x(t)) \rangle + u(t) \langle \lambda(t), \, [g, [g, [f, g]]](x(t)) \rangle.$$

Each condition by itself determines the control if the functions multiplying the control $u(t)$ are not zero. Since the pair $(1, u(t))$ is a nontrivial solution to this homogeneous system, however, we also need to have the compatibility condition

$$\langle \lambda(t), \, [f, [f, [f, g]]](x(t)) \rangle \langle \lambda(t), \, [g, [g, [f, g]]](x(t)) \rangle = \langle \lambda(t), \, [g, [f, [f, g]]](x(t)) \rangle^2,$$

where we use that by the Jacobi identity,

$$[g, [f, [f, g]]] = [f, [g, [f, g]]].$$

It is clear that these are increasingly more and more demanding requirements for the singular extremal to satisfy, and it seems plausible that "typically" these conditions should be difficult to satisfy, even in higher dimensions. This indeed is correct and can be made precise in the sense that "generically" singular extremals are of order 1, as shown by Bonnard and Chyba in [44, Sects. 8.3 and 8.5].

While this result, and also the results by Chitour, Jean and Trélat [71, 72] imply that we should not expect higher-order singular extremals for too many systems, this does not mean that these do not exist nor that these may not be of particular interest for some specific problem. One common way in which these higher-order singular extremals arise is that the control vector field g and the Lie bracket $[f,g]$ commute, i.e., that

$$[g,[f,g]] = 0.$$

In this case, the brackets $[f,[g,[f,g]]]$ and $[g,[g,[f,g]]]$ also are zero, and thus the calculation of the derivatives of the switching function simply continues as

$$\Phi^{(3)}(t) = \langle \lambda(t), \, [f,[f,[f,g]]](x(t)) \rangle = 0$$

and

$$\Phi^{(4)}(t) = \langle \lambda(t), \, [f,[f,[f,[f,g]]]](x(t)) \rangle + u(t) \langle \lambda(t), \, [g,[f,[f,[f,g]]]](x(t)) \rangle = 0.$$
$$(2.41)$$

This seems an adequate place to introduce a shorter notation for the iterated Lie brackets. It is common (for reasons that are connected with what is called the adjoint representation in Lie theory [256]) to think of taking the Lie bracket of a fixed vector field X with another vector field as a linear operator on the set of all smooth vector fields defined on Ω, $V^{\infty}(\Omega)$, and to denote it by $\mathrm{ad}X$,

$$\mathrm{ad}X \; : V^{\infty}(\Omega) \to V^{\infty}(\Omega), \quad Y \mapsto \mathrm{ad}_X(Y) = [X,Y].$$

The composition of these operators is then defined as

$$\mathrm{ad}_X^i = \mathrm{ad}_X^{i-1} \circ \mathrm{ad}_X,$$

so that, for example, we have

$$[f,[f,[f,[f,g]]]] = \mathrm{ad}_f^4(g).$$

In this notation, Eq. (2.41) can be written more compactly as

$$\Phi^{(4)}(t) = \left\langle \lambda(t), \, \mathrm{ad}_f^4(g)(x(t)) \right\rangle + u(t) \left\langle \lambda(t), \, [g,\mathrm{ad}_f^3(g)](x(t)) \right\rangle = 0.$$

Definition 2.8.6 (Higher-order singular control). Let Γ be an extremal lift for the problem [NTOC] consisting of a controlled trajectory (x,u) defined over the interval $[0,T]$ and corresponding adjoint vector $\lambda : [0,T] \to (\mathbb{R}^n)^*$ that is singular over an open interval I. The singular control is said to be of *intrinsic order k* over I if the following conditions are satisfied: *(1)* the first $2k-1$ derivatives of the switching function do not depend on the control u and vanish identically, i.e., for $i = 1,\ldots,2k-1$ we have that

$$\Phi^{(i)}(t) = \left\langle \lambda(t), \, \mathrm{ad}_f^i(g)(x(t)) \right\rangle \equiv 0,$$

and *(2)* $\left\langle \lambda(t), \mathrm{ad}_f^{2k}(g)(x(t)) \right\rangle$ does not vanish on I.

Theorem 2.8.4 (Generalized Legendre–Clebsch condition). *Suppose the controlled trajectory* (x_*, u_*) *defined over the interval* $[0,T]$ *minimizes the time of transfer from* $p \in \Omega$ *to* $q \in \Omega$ *for problem* [NTOC] *and the control* u_* *is singular of intrinsic order k over an open interval* $I \subset [0,T]$. *Then there exists an extremal lift* $\Gamma = ((x_*, u_*), \lambda)$ *with the property that*

$$(-1)^k \frac{\partial}{\partial u} \frac{d^{2k}}{dt^{2k}} \frac{\partial H}{\partial u} (\lambda_0, \lambda(t), x_*(t), u_*(t))$$

$$= (-1)^k \left\langle \lambda(t), [g, \mathrm{ad}_f^{2k-1}(g)](x(t)) \right\rangle \geq 0 \quad \text{for all} \quad t \in I.$$

This result is also known as the Kelley condition [131, 132, 262]. For a singular extremal of intrinsic order 2, it states that

$$\left\langle \lambda(t), [g, \mathrm{ad}_f^3(g)](x(t)) \right\rangle \geq 0 \quad \text{for all} \quad t \in I, \tag{2.42}$$

and a proof of this condition will be given in Sect. 4.6.2. The strengthened version of this condition has very interesting consequences.

Proposition 2.8.5. *Let* $\Gamma = ((x,u), \lambda)$ *be an extremal lift for the problem* [NTOC] *defined over the interval* $[0,T]$ *that is singular of intrinsic order 2 over an open interval I for which*

$$\left\langle \lambda(t), [g, \mathrm{ad}_f^3(g)](x(t)) \right\rangle > 0 \quad \text{for all} \quad t \in I.$$

Suppose that the singular control u takes values in the interior of the control set over the interval I. Then at no time $\tau \in I$ can the control u be concatenated with a bang control $u = -1$ or $u = +1$: concatenations of the types BS and SB violate the conditions of the maximum principle and are not optimal.

Proof. Without loss of generality, we consider a concatenation of the type SX at time $\tau \in I$. That is, we assume that for some $\varepsilon > 0$ the control is singular over the interval $(\tau - \varepsilon, \tau)$ and is given by $u = -1$ over the interval $(\tau, \tau + \varepsilon)$. Since the singular control is of order 2, the first three derivatives of the switching function do not depend on the control and thus are all continuous and given by $\Phi^{(i)}(t) = \left\langle \lambda(t), \mathrm{ad}_f^i(g)(x(t)) \right\rangle$, $i = 1,2,3$. The fourth derivative of Φ at τ from the right is thus given by

$$\Phi^{(4)}(\tau) = \left\langle \lambda(\tau), \mathrm{ad}_f^4(g)(x(\tau)) \right\rangle - \left\langle \lambda(\tau), [g, \mathrm{ad}_f^3(g)](x(\tau)) \right\rangle$$

$$< \left\langle \lambda(\tau), [f + u(\tau)g, \mathrm{ad}_f^3(g)](x(\tau)) \right\rangle = 0,$$

since the singular control $u(\tau)$ takes a value in $(-1,1)$. Thus the switching function has a local maximum for $t = \tau$ and is negative over the interval $(\tau, \tau + \varepsilon)$. But then the minimization property of the Hamiltonian implies that the control must be

$u = +1$. The analogous contradiction arises for concatenations of the type SY or for the order BS.   □

This result implies that an optimal singular arc of order 2 cannot be concatenated with a bang control. In fact, an optimal control needs to switch infinitely many times between the controls $u = -1$ and $u = +1$ on any interval $(\tau, \tau + \varepsilon)$ if a singular junction occurs at time τ. Corresponding trajectories are called *chattering arcs*. In Sect. 2.11 we shall give an example that shows that these can be optimal for the seemingly most innocent-looking system.

2.8.5   Multi-input Systems and the Goh Condition

We close this section with some comments about the multi-input case in which the dynamics takes the form

$$\dot{x} = f(x) + \sum_{i=1}^{m} g_i(x)u_i, \qquad x \in \Omega, \qquad u \in U. \tag{2.43}$$

As before, for simplicity, we assume that all vector fields are C^∞ on Ω. Clearly, now geometric properties of the control set $U \subset \mathbb{R}^m$ matter. If U is a compact polyhedron, then the Hamiltonian will be minimized at one of the vertices, and singular controls arise as the minimum is attained along one of the faces of the polyhedron. The situation that most closely resembles the structures for the single-input case above, and probably is the practically most important one, occurs when the control set is a multi-dimensional rectangle,

$$U = [\alpha_1, \beta_1] \times \cdots \times [\alpha_m, \beta_m].$$

In this case, the minimization of the Hamiltonian function

$$H = \lambda_0 + \left\langle \lambda, \ f(x) + \sum_{i=1}^{m} g_i(x)u_i \right\rangle = \lambda_0 + \langle \lambda, \ f(x)\rangle + \sum_{i=1}^{m} \Phi_i u_i$$

still splits into m scalar minimization problems as in the single-input case, and optimal controls satisfy

$$u_i(t) = \begin{cases} \alpha_i & \text{if} \quad \Phi_i(t) = \langle \lambda(t), \ g_i(x(t))\rangle < 0, \\ \beta_i & \text{if} \quad \Phi_i(t) = \langle \lambda(t), \ g(x(t))\rangle > 0. \end{cases}$$

As before, now the switching functions Φ_i, $i = 1, \ldots, m$, need to be analyzed to determine the optimal controls, and in principle, this follows the pattern discussed above. For example, Theorem 2.8.2 applies to give the derivatives of the switching functions as

$$\dot{\Phi}_i(t) = \left\langle \lambda(t), \left[f + \sum_{j=1}^{m} g_j u_j, g_i \right](x(t)) \right\rangle$$

$$= \langle \lambda(t), [f, g_i](x(t)) \rangle + \sum_{j \neq i} u_j(t) \langle \lambda(t), [g_j, g_i](x(t)) \rangle.$$

In contrast to the single-input case, now the derivative $\dot{\Phi}_i$ depends on the controls; on the controls other than u_i, that is. Hence, whether higher derivatives can be computed depends on the type of the controls, since these now need to be differentiated in time. Clearly, this is no issue for those components that are bang controls, but it needs to be checked if some of the controls are singular. All this leads to a much more elaborate analysis, which is best left for the particular problem under consideration. For example, if only one of the components is singular, with all other controls held constant, all the necessary conditions for optimality for the single-input control system are applicable. If more than one component is singular at the same time, the following result, the so-called *Goh condition*, [107] provides an extra necessary condition for optimality. This condition will be derived in Sect. 4.6.3.

Theorem 2.8.5 (Goh condition). *[107] Suppose the controlled trajectory (x_*, u_*) defined over the interval $[0, T]$ minimizes the time of transfer from $p \in \Omega$ to $q \in \Omega$ for the multi-input control system with dynamics given by Eq. (2.43) and a rectangular control set U. Suppose the ith and jth controls are simultaneously singular over an open interval $I \subset [0, T]$. Then there exists an extremal lift $\Gamma = ((x_*, u_*), \lambda)$ with the property that*

$$\langle \lambda(t), [g_i, g_j](x(t)) \rangle \equiv 0 \quad \text{for all} \quad t \in I.$$

2.9 Time-Optimal Control for Nonlinear Systems in the Plane

We use the time-optimal control problem [NTOC] in the plane as an instrument to provide a first illustration of the use of geometric methods and the Lie-derivative-based techniques introduced above in the analysis of optimal control problems. These results are due to H. Sussmann, who in a series of papers [230, 236–238], gave a complete solution for this optimal control problem in dimension 2. The two-dimensional problem allows for easy visualization of the results, yet the general problem quickly gets very difficult, both in dimension 2 and even more so in higher dimensions. While Sussmann's results and the monograph by Boscain and Piccoli [51] provide a comprehensive analysis of the time-optimal control problem for two-dimensional systems, only partial results about the structure of time-optimal controls in higher dimensions (mostly in $\mathbb{R}^3$ [54, 141, 210, 211] and some in $\mathbb{R}^4$ [221]) are currently known.

[TOC in $\mathbb{R}^2$]   Let Ω be an open and simply connected subset of $\mathbb{R}^2$ and let $f : \Omega \to \mathbb{R}^2$ and $g : \Omega \to \mathbb{R}^2$ be two C^∞ vector fields defined on Ω. For the control-affine system Σ with dynamics given by

$$\dot{x} = f(x) + g(x)u,$$

among all piecewise continuous (more generally, Lebesgue measurable) controls u, $u : [0, T] \to [-1, 1]$, find one that steers a given initial point $q_1 \in \Omega$ into a target point $q_2 \in \Omega$ (while remaining in Ω) in minimum time.

Our *aim* is to *determine the concatenation structure of optimal controls whose trajectories entirely lie in Ω*. More precisely, we are asking the question what can be said about time-optimal controlled trajectories that lie in Ω if certain assumptions are made on the vector fields f and g at some reference point $p \in \Omega$. We consider Ω to be a sufficiently small neighborhood of the point p, and thus by continuity, any inequality-type condition imposed on the values of f and g and/or their Lie brackets at p can also be assumed to hold in Ω. But we shall develop the arguments as much as possible semiglobally, i.e., state them in a way that they are valid for sets Ω that satisfy the required conditions throughout. It is natural to tackle this question by proceeding from the most general to increasingly more and more degenerate situations. That is, we first assume that the vectors $f(p)$ and $g(p)$ and other relevant Lie brackets are in general position, i.e., are linearly independent, and then proceed to consider more degenerate cases in which dependencies are allowed. In this spirit, throughout this section we make the following assumption:

(A0)   The vector fields f and g are linearly independent everywhere on $\Omega \subset \mathbb{R}^2$.

Under this assumption, in this and the next section, we fully determine the structure of time-optimal controlled trajectories that lie in Ω under generic conditions. These results are due to H. Sussmann, and in our presentation we follow his arguments that beautifully illustrate the use of geometric techniques in optimal control theory. In particular, as we proceed, it will become clear how these methods are needed to complement the first-order conditions of the Pontryagin maximum principle in order to arrive at deep and sharp results such as Proposition 2.9.5. Some of these results that we shall develop go well beyond the conditions of the maximum principle.

2.9.1 Optimal Bang-Bang Controls in the Simple Subcases

Lemma 2.9.1. *Any control corresponding to an abnormal extremal whose trajectory lies in Ω is constant equal to $u \equiv +1$ or $u \equiv -1$.*

Proof. Let $\Gamma = ((x, u), \lambda)$ be an extremal and suppose $\lambda_0 = 0$. If the switching function $\Phi(t) = \langle \lambda(t), g(x(t)) \rangle$ vanishes at some time τ, then it follows from

$$H(t) = \langle \lambda(t), \, f(x(t)) \rangle + u(t) \langle \lambda(t), \, g(x(t)) \rangle \equiv -\lambda_0$$

that we also must have $\langle \lambda(\tau), \, f(x(\tau)) \rangle = 0$. Hence $\lambda(\tau)$ vanishes against both $f(x(\tau))$ and $g(x(\tau))$. Since these two vectors are linearly independent, it follows that $\lambda(\tau) = 0$. But this contradicts the nontriviality of the multipliers. Hence there cannot be any zeros for the switching function, and thus the corresponding controls must be constant.   □

Having taken care of this special case, we henceforth assume that all extremals are normal and set $\lambda_0 = 1$. In particular, whenever τ is a switching time, it follows that

$$\langle \lambda(\tau), \, f(x(\tau)) \rangle = -1. \tag{2.44}$$

Using f and g as a basis, we can express any higher-order bracket of f and g as a linear combination of this basis. In particular, there exist smooth functions α and β, $\alpha, \beta \in C^\infty(\Omega)$, such that for all $x \in \Omega$ we have that

$$[f,g](x) = \alpha(x)f(x) + \beta(x)g(x). \tag{2.45}$$

We say an optimal controlled trajectory is of type XY if the corresponding control is bang-bang with at most one switching from $u = -1$ to $u = +1$ and use analogous labels for controlled trajectories that are concatenations of more segments. For example, a controlled trajectory of type $XYSY$ is a concatenation of an X-trajectory followed by a Y-trajectory, a singular arc, and one more Y-trajectory. However, we always allow for the possibility that some of the segments are absent and thus a specific trajectory of type $XYSY$ may simply be a concatenation of an X-trajectory with a single Y-trajectory.

Proposition 2.9.1. *If α does not vanish on Ω, then optimal controlled trajectories that lie in Ω are of type XY if α is positive and of type YX if α is negative. Corresponding optimal controls are bang-bang with at most one switching.*

Proof. Recall that as always, $X = f - g$ and $Y = f + g$. Let (x, u) be an optimal controlled trajectory that transfers a point $q_1 \in \Omega$ into the point $q_2 \in \Omega$ in minimum time with the trajectory x lying in Ω and let λ be an adjoint vector such that the conditions of the maximum principle are satisfied. If τ is a zero of the corresponding switching function, then we have that

$$\begin{aligned}
\Phi(\tau) &= \langle \lambda(\tau), \, [f,g](x(\tau)) \rangle \\
&= \alpha(x(\tau)) \cdot \langle \lambda(\tau), \, f(x(\tau)) \rangle + \beta(x(\tau)) \cdot \langle \lambda(\tau), \, g(x(\tau)) \rangle \\
&= -\alpha(x(\tau)).
\end{aligned}$$

Since α has constant sign on Ω, it follows that at every zero of Φ, the derivative of the switching function Φ is nonzero and has the same sign. But then Φ can have at most one zero, changing from positive to negative if $\alpha > 0$ and from negative to positive if $\alpha < 0$. Thus optimal controls must switch from $u = -1$ to $u = +1$ if $\alpha > 0$ and from $u = +1$ to $u = -1$ if $\alpha < 0$. This proves the result.   □

For **example**, for the harmonic oscillator of Sect. 2.6, we have that

$$[f,g](x) = \begin{pmatrix} -1 \\ 0 \end{pmatrix} = -\frac{1}{x_2}\begin{pmatrix} x_2 \\ -x_1 \end{pmatrix} - \frac{x_1}{x_2}\begin{pmatrix} 0 \\ 1 \end{pmatrix} = \alpha(x)Ax + \beta(x)g.$$

We can take as Ω either the upper or lower half-plane, $\Omega_+ = \{(x_1,x_2) : x_2 > 0\}$ or $\Omega_- = \{(x_1,x_2) : x_2 < 0\}$, and it follows that optimal trajectories that entirely lie in Ω_+ or Ω_- are bang-bang with at most one switching and that the switchings are from $u = +1$ to $u = -1$ in Ω_+ and from $u = -1$ to $u = +1$ in Ω_-. Also, note that the controls corresponding to the abnormal trajectories γ_+ and γ_- that lie in Ω_+ and Ω_- are constant. As this example shows, there clearly can be more switchings, but the trajectories need to leave and reenter the region Ω for this to be possible.

This proposition settles the local structure of time-optimal controlled trajectories near all points p where α does not vanish, i.e., where g and the Lie-bracket $[f,g]$ are in general position as well. Clearly, this does not suffice to settle the structure of optimal controls since there may and generally will exist some points where the vector fields g and $[f,g]$ are linearly dependent and the local structure near these points will need to be determined too. Proceeding from the general case to the more special ones, but still maintaining condition (A0), we now assume that $\alpha(p) = 0$. At the same time, however, we want for this to occur in as nondegenerate a scenario as possible. That is, no other equality relations that would matter should hold at p. In terms of *singularity theory*, after determining the structure of optimal controlled trajectories near points of *codimension* 0 (only two inequality relations are imposed, one in the form of assumption (A0), the other as $\alpha(p) \neq 0$, but no equality relations hold at the reference point), we now proceed to the analysis of the *codimension* 1 scenario when we allow for exactly one equality constraint, but otherwise again only impose inequality relations. More specifically we assume that

(A1)   The vector fields f and g are linearly independent everywhere on $\Omega \subset \mathbb{R}^2$ and there exists a point $p \in \Omega$ with $\alpha(p) = 0$, but the Lie derivatives of α along $X = f - g$ and $Y = f + g$ do not vanish on Ω,

$$(X\alpha)(x) = L_X(\alpha)(x) \neq 0, \quad (Y\alpha)(x) = L_Y(\alpha)(x) \neq 0 \qquad \text{for all } x \in \Omega.$$

Furthermore, we assume that the zero set $\mathscr{S} = \{x \in \Omega : \alpha(x) = 0\}$ is a curve (embedded one-dimensional submanifold) in Ω that divides Ω into two connected subregions $\Omega_+ = \{x \in \Omega : \alpha(x) > 0\}$ and $\Omega_- = \{x \in \Omega : \alpha(x) < 0\}$ so that $\Omega = \Omega_- \cup \mathscr{S} \cup \Omega_+$.

With the understanding that Ω is a sufficiently small neighborhood of p, this geometric assumption on the structure of the zero set of α simply follows from the implicit function theorem, since the assumption on the Lie derivatives implies that the gradient of α is non zero at p. On the other hand, several of the results below are valid as long as Ω has this geometric separation property, not just in small neighborhoods, and therefore we prefer to state the results as such.

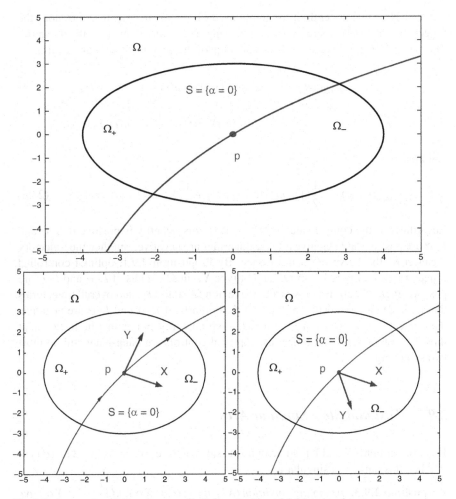

Fig. 2.9 Assumption (A1): subcases with (*left*) and without (*right*) a singular arc

Assumption (A1) by itself does not lead to a unique structure of time-optimal trajectories, but several subcases exist, since it matters to which side of $\mathscr{S}$ the vector fields X and Y point (see Fig. 2.9).

Proposition 2.9.2. *Assuming condition (A1), if $L_X(\alpha) = X\alpha$ and $L_Y(\alpha) = Y\alpha$ have the same sign on Ω, then optimal controlled trajectories that lie in Ω are of type YXY if the Lie derivatives are positive and of type XYX if they are negative. Optimal controls are bang-bang with at most two switchings.*

Proof. As above, let (x,u) be an optimal controlled trajectory that transfers a point $q_1 \in \Omega$ into the point $q_2 \in \Omega$ in minimum time with the trajectory x lying in Ω and let λ be an adjoint vector such that the conditions of the maximum principle are satisfied. Note that under these assumptions, the directional derivative of α along the trajectory x is strictly increasing or decreasing. For at any point $x(t)$, the dynamics $f(x(t)) + u(t)g(x(t))$ is a convex combination of the vectors $X(x(t))$ and $Y(x(t))$,

$$f(x(t)) + u(t)g(x(t)) = \frac{1}{2}(X + Y)(x(t)) + u(t)\frac{1}{2}(Y - X)(x(t))$$

$$= \frac{1}{2}(1 - u(t))X(x(t)) + \frac{1}{2}(1 + u(t))Y(x(t)),$$

and thus

$$\left(L_{f+ug}\alpha\right)(x(t)) = \frac{1}{2}(1 - u(t))L_X\alpha(x(t)) + \frac{1}{2}(1 + u(t))L_Y\alpha(x(t)). \qquad (2.46)$$

Regardless of the control value $u(t) \in [-1, 1]$, this quantity is positive if $L_X\alpha$ and $L_Y\alpha$ are positive and negative if these quantities are negative. But then the trajectory x can cross the curve $\mathscr{S}$ at most once. By Proposition 2.9.1, optimal controlled trajectories are of type YX in Ω_- and of type XY in Ω_+. Thus, if $L_X\alpha$ and $L_Y\alpha$ are positive, then trajectories move from the region Ω_- into Ω_+, and overall trajectories that lie in Ω can at most be of type YXY. Similarly, if $L_X\alpha$ and $L_Y\alpha$ are negative, then trajectories move from Ω_+ into Ω_-, and now trajectories that lie in Ω can at most be of type XYX. In either case, optimal controls are bang-bang with at most two switchings. $\qquad \square$

2.9.2 Fast and Slow Singular Arcs

If the vector fields X and Y point to opposite sides of the curve $\mathscr{S} = \{x \in \Omega : \alpha(x) = 0\}$, then this curve is a singular arc.

Proposition 2.9.3. *Assuming condition (A1), if $L_X(\alpha) = X\alpha$ and $L_Y(\alpha) = Y\alpha$ have opposite signs on Ω, then $\mathscr{S}$ is a singular arc. If $\Gamma = ((x, u), \lambda)$ is a corresponding singular extremal lift, then the strengthened Legendre–Clebsch condition is satisfied if $L_X\alpha$ is negative, and it is violated if $L_X\alpha$ is positive.*

Proof. In this case, X and Y always point to opposite sides of $\mathscr{S}$. Hence, at every point $x \in \mathscr{S}$ there exists a convex combination $u = u(x)$ such that the vector $f(x) + u(x)g(x)$ is tangent to $\mathscr{S}$ at x. This control is the unique solution to the equation $L_{f+ug}\alpha = 0$, i.e., solving from Eq. (2.46), we have that

$$u(x) = \frac{L_X\alpha(x) + L_Y\alpha(x)}{L_X\alpha(x) - L_Y\alpha(x)}.$$

Since $L_X\alpha$ and $L_Y\alpha$ have opposite signs, it follows that this value $u(x)$ lies strictly between -1 and $+1$, i.e., lies in the interior of the control set. In particular, it is admissible. Thus, if this control is optimal, then it must be singular.

We verify that the associated controlled trajectory through an initial condition $q \in \mathscr{S}$ is extremal by constructing a singular extremal lift $\Gamma = ((x, u), \lambda)$. Let $x = x(t)$ be the solution to the initial value problem

$$\dot{x} = f(x) + u(x)g(x), \qquad x(0) = q.$$

This solution exists over a maximal interval (t_-, t_+) with $t_- < 0 < t_+$. Let $\psi \in (\mathbb{R}^2)^*$ be a covector such that

$$\langle \psi, g(q) \rangle = 0 \qquad \text{and} \qquad \langle \psi, f(q) \rangle = -1$$

and let $\lambda = \lambda(t)$ be the solution of the corresponding adjoint equation

$$\dot{\lambda} = -\lambda \left(Df(x(t)) + u(x(t))Dg(x(t)) \right)$$

with initial condition $\lambda(0) = \psi$. This triple defines a singular extremal lift if the switching function $\Phi(t) = \langle \lambda(t), g(x(t)) \rangle$ vanishes identically on (t_-, t_+). But this is clear by construction: we have $\left(L_{f+ug}\alpha \right) x(t) \equiv 0$, and since $\alpha(q) = 0$, it follows that $\alpha(x(t)) \equiv 0$ on (t_-, t_+). Hence we get

$$\begin{aligned}
\dot{\Phi}(t) &= \langle \lambda(t), [f, g](x(t)) \rangle \\
&= \alpha(x(t)) \langle \lambda(t), f(x(t)) \rangle + \beta(x(t)) \cdot \langle \lambda(t), g(x(t)) \rangle \\
&= \beta(x(t)) \Phi(t).
\end{aligned}$$

But $\Phi(0) = \langle \psi, g(q) \rangle = 0$, and so the switching function vanishes identically. Hence Γ is a normal singular extremal lift.

It remains to check the Legendre–Clebsch condition. Using Lemma 2.8.1, it follows that

$$[g, [f, g]] = [g, \alpha f + \beta g] = (L_g \alpha) f - \alpha[f, g] + (L_g \beta)g.$$

Along the singular extremal, $\langle \lambda, g(x) \rangle \equiv 0$ and $\langle \lambda, [f, g](x) \rangle \equiv 0$, and thus, using Eq. (2.44), it also follows that $\langle \lambda(t), f(x(t)) \rangle \equiv -1$. Hence

$$\langle \lambda(t), [g, [f, g]](x(t)) \rangle = -L_g \alpha(x(t))$$

$$= -L_{\frac{1}{2}(Y-X)} \alpha(x(t)) = \frac{1}{2} (L_X \alpha - L_Y \alpha)(x(t)). \qquad (2.47)$$

This quantity has the same sign as $L_X \alpha$, and the strengthened Legendre–Clebsch condition is satisfied if $\langle \lambda(t), [g, [f, g]](x(t)) \rangle < 0$. Hence the result follows. $\qquad \square$

The Legendre–Clebsch condition distinguishes *fast* from *slow singular arcs*. On a set Ω in the plane where f and g are linearly independent, this can be seen with an instructive geometric argument by introducing a 1-form that measures the time along the trajectories. Differential forms provide a superior formalism for these computations, and for the sake of completeness, we provide the needed definitions and results. These are standard concepts from differential geometry and can be found in any text on the subject, such as, for example, [50, 256]. One-forms are simply linear functionals on the space of all tangent vectors; hence the space of 1-forms on Ω is a two-dimensional vector space as well. If we write $x = x_1 e_1 + x_2 e_2$, where $\{e_1, e_2\}$ is the canonical ordered basis for $\mathbb{R}^2$, we denote the corresponding dual basis by dx_1 and dx_2; that is, dx_i is the linear functional that satisfies

$$\langle dx_i, e_j \rangle = \begin{cases} 1 & \text{if } i = j, \\ 0 & \text{if } i \neq j. \end{cases}$$

Since f and g are linearly independent on Ω, there exists a unique 1-form ω on Ω that satisfies

$$\langle \omega(x), f(x) \rangle \equiv 1 \qquad \text{and} \qquad \langle \omega(x), g(x) \rangle \equiv 0 \qquad \text{for all } x \in \Omega. \qquad (2.48)$$

This form ω is easily computed: if f and g have the representations

$$f(x) = \begin{pmatrix} f_1(x_1, x_2) \\ f_2(x_1, x_2) \end{pmatrix} \qquad \text{and} \qquad g(x) = \begin{pmatrix} g_1(x_1, x_2) \\ g_2(x_1, x_2) \end{pmatrix},$$

then

$$\omega(x) = \frac{g_2(x)dx_1 - g_1(x)dx_2}{f_1(x)g_2(x) - f_2(x)g_1(x)} = \frac{g_2(x)dx_1 - g_1(x)dx_2}{\det(f(x), g(x))}, \qquad (2.49)$$

with $\det(f(x), g(x))$ denoting the determinant of the matrix

$$\begin{pmatrix} f_1 & g_1 \\ f_2 & g_2 \end{pmatrix}.$$

This determinant does not vanish on Ω, since the vector fields f and g are linearly independent. Depending on the sign of this determinant, the ordered basis $\mathscr{B} = \{f, g\}$ is said to be positively, respectively negatively, oriented.

Let (x, u) be a controlled trajectory defined over an interval $[t_0, t_1]$ with trajectory x lying in Ω. Then the line integral of ω along the curve $x(\cdot)$ is given by

$$\int_{x(\cdot)} \omega = \int_{t_0}^{t_1} \langle \omega(x(t)), \dot{x}(t) \rangle \, dt$$

$$= \int_{t_0}^{t_1} \langle \omega(x(t)), f(x(t)) \rangle \, dt + \int_{t_0}^{t_1} u(t) \langle \omega(x(t)), g(x(t)) \rangle \, dt$$

$$= \int_{t_0}^{t_1} dt = t_1 - t_0,$$

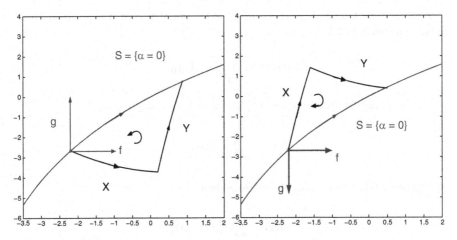

Fig. 2.10 Positively (*left*) and negatively (*right*) oriented vector fields f and g

so that ω measures the time along trajectories. For this reason, ω sometimes is called the clock form [44].

We now show how this differential form ω can be used to determine which type of trajectory is faster. Consider a point $q_1 \in \mathscr{S}$ and let (x, u) be the controlled trajectory that steers q_1 to another point $q_2 \in \mathscr{S}$ along the singular arc $\mathscr{S} \subset \Omega$ in time τ. If τ is small, then there exists a unique XY-trajectory that also steers q_1 into q_2 and lies in Ω. Simply consider the forward orbit of the X-trajectory that starts at q_1 and the backward orbit of the Y-trajectory that ends at q_2. Since X and Y point to opposite sides of $\mathscr{S}$, it follows that these two orbits intersect in some point $r \in \Omega$. Suppose it takes time s to go from q_1 to r along X and time t to go from r to q_2 along Y. If we denote the mapping that follows the flow of the vector field X for time s by Ψ_s^X, then we can write $r = \Psi_s^X(q_1)$, and analogously, for Y, we have that $q_2 = \Psi_t^Y(r)$. Overall, therefore,

$$q_2 = \Psi_t^Y(r) = \Psi_t^Y(\Psi_s^X(q_1)) = \left(\Psi_t^Y \circ \Psi_s^X\right)(q_1).$$

Stokes's theorem allows us to compare the time $s + t$ along the XY-trajectory with the time τ along the singular arc. Denote the closed curve consisting of the XY-trajectory concatenated with the singular arc run backward from q_2 to q_1 by Δ. The orientation of this closed curve matters in Stokes's theorem, and Δ has the same orientation as the ordered basis $\mathscr{B} = \{f, g\}$: since

$$\det(X(x), Y(x)) = \det(f(x) - g(x), \ f(x) + g(x)) = 2\det(f(x), g(x)),$$

the ordered basis $\{X, Y\}$ has the same orientation as $\{f, g\}$. But if $\{X, Y\}$ is positively oriented, the curve Δ is traversed counterclockwise, while it is traversed clockwise if $\{X, Y\}$ is oriented negatively (see Fig. 2.10).

Without loss of generality, we assume that the orientation of Δ is positive. Then Stokes's theorem [50, 256] gives that

$$s + t - \tau = \int_\Delta \omega = \int_R d\omega$$

where R denotes the region enclosed by Δ. For a 1-form ω given as

$$\omega(x) = \sum_{i=1}^n \xi_i(x) dx_i$$

with ξ_i smooth functions, the 2-form $d\omega$ is defined as

$$d\omega(x) = \sum_{i=1}^n d\xi_i(x) \wedge dx_i$$

with

$$d\xi_i(x) = \sum_{j=1}^n \frac{\partial \xi_i}{\partial x_j}(x) dx_j,$$

the differential of ξ_i. For the wedge product, $\wedge$, the rules of an alternating product apply, i.e., $dx_1 \wedge dx_2 = -dx_2 \wedge dx_1$ and $dx_i \wedge dx_i = 0$. In order to evaluate the area integral on the right, we need some facts about the actions of differential forms on vector fields. If ϕ is a smooth function defined on some domain $D \subset \mathbb{R}^n$, $\phi \in C^\infty(D)$, then the action of the 1-form $d\phi$ on a smooth vector field is simply taking the Lie derivative of ϕ along Z,

$$\langle d\phi(x), Z(x) \rangle = L_Z \phi(x).$$

For writing out the inner product in terms of the basis vectors, we have that

$$\langle d\phi(x), Z(x) \rangle = \left\langle \sum_{i=1}^n \frac{\partial \phi}{\partial x_i}(x) dx_i, \sum_{j=1}^n Z_j(x) e_j \right\rangle$$

$$= \sum_{i=1}^n \sum_{j=1}^n \frac{\partial \phi}{\partial x_i}(x) Z_j(x) \langle dx_i, e_j \rangle$$

$$= \sum_{i=1}^n \frac{\partial \phi}{\partial x_i}(x) Z_i(x) = L_Z \phi(x).$$

If ψ is another smooth function on D, $\psi \in C^\infty(D)$, then the action of the 2-form $d\phi \wedge d\psi$ on a pair of smooth vector fields f and g is defined as the alternating product

$$\langle d\phi(x) \wedge d\psi(x), (f(x), g(x)) \rangle$$

$$= \langle d\phi(x), f(x) \rangle \cdot \langle d\psi(x), g(x) \rangle - \langle d\phi(x), g(x) \rangle \cdot \langle d\psi(x), f(x) \rangle$$

$$= L_f \phi(x) \cdot L_g \psi(x) - L_g \phi(x) \cdot L_f \psi(x). \tag{2.50}$$

Note, in particular, that this gives 0 if $f = g$. These actions are then related to the Lie bracket through the following relation:

Lemma 2.9.2. [50] *Given any 1-form ω and smooth vector fields f and g defined on $D \subset \mathbb{R}^n$, it follows that*

$$\langle d\omega, (f,g) \rangle = L_f \langle \omega, g \rangle - L_g \langle \omega, f \rangle - \langle \omega, [f,g] \rangle.$$

Proof. It suffices to prove the Lemma if ω is of the form $\omega = \phi d\psi$, where ϕ and ψ are smooth functions on D. In this case, and dropping the argument x, we have that

$$L_f \langle \omega, g \rangle - L_g \langle \omega, f \rangle - \langle \omega, [f,g] \rangle$$
$$= L_f \langle \phi d\psi, g \rangle - L_g \langle \phi d\psi, f \rangle - \langle \phi d\psi, [f,g] \rangle$$
$$= L_f(\phi) \langle d\psi, g \rangle + \phi L_f \langle d\psi, g \rangle - L_g(\phi) \langle d\psi, f \rangle - \phi L_g \langle d\psi, f \rangle - \phi \langle d\psi, [f,g] \rangle$$
$$= L_f(\phi) L_g(\psi) + \phi L_f(L_g(\psi)) - L_g(\phi) L_f(\psi) - \phi L_g(L_f(\psi)) - \phi L_{[f,g]}\psi$$
$$= L_f(\phi) L_g(\psi) - L_g(\phi) L_f(\psi) + \phi \left\{ L_f(L_g(\psi)) - L_g(L_f(\psi)) - L_{[f,g]}\psi \right\}.$$

But

$$L_{[f,g]}\psi = L_f(L_g(\psi)) - L_g(L_f(\psi))$$

and thus since $d\omega = d\phi \wedge d\psi$, the result follows from Eq. (2.50). $\square$

For the 1-form ω defined by Eq. (2.48), we have $\langle \omega, f \rangle \equiv 1$ and $\langle \omega, g \rangle \equiv 0$, and thus the Lie derivatives of these functions vanish giving

$$\langle d\omega, (f,g) \rangle = -\langle \omega, [f,g] \rangle = -\langle \omega, \alpha f + \beta g \rangle$$
$$= -\alpha \langle \omega, f \rangle - \beta \langle \omega, g \rangle = -\alpha.$$

Furthermore,

$$\langle d\omega, (f,g) \rangle = \langle d\omega, (f_1 e_1 + f_2 e_2, \ g_1 e_1 + g_2 e_2) \rangle$$
$$= f_1 \langle d\omega, (e_1, \ g_1 e_1 + g_2 e_2) \rangle + f_2 \langle d\omega, (e_2, \ g_1 e_1 + g_2 e_2) \rangle$$
$$= f_1 g_1 \langle d\omega, (e_1, \ e_1) \rangle + f_1 g_2 \langle d\omega, (e_1, \ e_2) \rangle$$
$$\quad + f_2 g_1 \langle d\omega, (e_2, \ e_1) \rangle + f_2 g_2 \langle d\omega, (e_2, \ e_2) \rangle$$
$$= (f_1 g_2 - f_2 g_1) \langle d\omega, (e_1, \ e_2) \rangle$$
$$= \det(f(x), g(x)) \langle d\omega, (e_1, e_2) \rangle,$$

so that

$$\langle d\omega, (e_1, e_2) \rangle = -\frac{\alpha(x)}{\det(f(x), g(x))}.$$

Hence

$$\tau - (s+t) = -\int_R d\omega = \int_R \frac{\alpha(x)}{\det(f(x),g(x))}dx. \tag{2.51}$$

By construction, the region R lies entirely in Ω_+ or Ω_-, namely in Ω_+ if the Lie derivative $L_X\alpha$ is positive and in Ω_- if it is negative. In the first case, the integral is positive (recall that we assume that the basis $\mathcal{B} = \{f,g\}$ is positively oriented), and thus the singular arc takes longer than the XY-trajectory, while it does better in the second case when the region R lies in Ω_-. These conclusions are consistent with the strengthened Legendre–Clebsch condition. In carrying out this argument for YX-trajectories, the same consistency shows. This explicitly verifies that the Legendre–Clebsch condition distinguishes fast from slow singular arcs.

This calculation can also be used to show that *increasing the number of switchings along a bang-bang trajectory speeds up the time of transfer if the strengthened Legendre–Clebsch condition is satisfied.* Again, consider the XY-trajectory that steers q_1 into q_2 and is part of the curve Δ constructed above. Construct an $XYXY$-trajectory that connects q_1 with q_2 in Ω as follows: (i) Starting from q_1, follow the X-trajectory for time $s_1 < s$ and let r_1 denote the point reached, $r_1 = \Psi^X_{s_1}(q_1)$. (ii) At r_1, change to the Y-trajectory and follow it for time t_1 until the Y-trajectory again reaches the singular curve $\mathcal{S}$ in some point r_2, $r_2 = \Psi^Y_{t_1}(r_1) \in \mathcal{S}$. (iii) Here once more switch to the X-trajectory and follow it for time s_2 until it intersects the original Y-trajectory in the point r_3, $r_3 = \Psi^X_{s_2}(r_2)$. (iv) Then follow this Y-trajectory from r_3 into q_2, say $q_2 = \Psi^Y_{t_2}(r_3)$. Thus, overall,

$$q_2 = \left(\Psi^Y_{t_2} \circ \Psi^X_{s_2} \circ \Psi^Y_{t_1} \circ \Psi^X_{s_1}\right)(q_1).$$

Denote by $\Diamond$ the diamond-shaped curve that is obtained by concatenating the X-trajectory from r_1 to r first with the Y-trajectory from r to r_3, then with the X-trajectory run backward from r_3 to r_2, and finally the Y-trajectory run backward from r_2 to r_1 (see Fig. 2.11). Since we assume that the basis $\{f,g\}$ is positively oriented, the curve $\Diamond$ also is mathematically positively (counterclockwise) oriented. Let D denote the region enclosed by $\Diamond$. Using the 1-form ω, the difference in time between the original XY-trajectory and the newly constructed $XYXY$-trajectory can then be calculated as

$$(s+t) - (s_1 + t_1 + s_2 + t_2) = (s - s_1) + (t - t_2) - s_2 - t_1$$

$$= -\int_\Diamond \omega = -\int_D d\omega = -\int_D \frac{\alpha(x)}{\det(f(x),g(x))}dx.$$

By construction of $\Diamond$, the region D lies entirely in Ω_+ if $L_X\alpha > 0$ and in Ω_- if $L_X\alpha < 0$. Hence, the $XYXY$-trajectory steers q_1 into q_2 faster than the XY-trajectory does if $L_X\alpha < 0$, and it is slower if $L_X\alpha > 0$. Thus, *if the strengthened Legendre–Clebsch condition is satisfied,* i.e., for $L_X\alpha < 0$, *bang-bang trajectories with more switchings near the singular arc are faster, while they are slower if the strengthened*

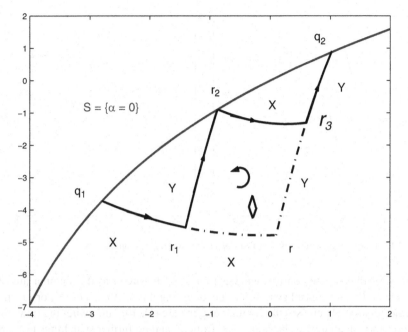

Fig. 2.11 Comparison of an $XYXY$-trajectory with an XY-trajectory

Legendre–Clebsch condition is violated. In either case, the singular arc can closely be approximated by bang-bang trajectories with an increasing number of switchings, and it is therefore to be expected that in the limit, optimal controls will follow the singular arc if the strengthened Legendre–Clebsch condition is satisfied, while they will avoid it, i.e., have as few switchings as possible, if it is violated. This indeed is the case.

Proposition 2.9.4. *Assuming condition (A1), if $L_X(\alpha) = X\alpha$ is negative and $L_Y(\alpha) = Y\alpha$ is positive on Ω, then optimal controlled trajectories that lie in Ω are of the type BSB, that is, are at most concatenations of a bang arc (X or Y) followed by a singular arc and possibly one more bang arc.*

Proof. Let (x,u) be an optimal controlled trajectory that transfers a point $q_1 \in \Omega$ into the point $q_2 \in \Omega$ in minimum time with the trajectory x lying in Ω and let λ be an adjoint vector such that the conditions of the maximum principle are satisfied. Once more, recall that by Proposition 2.9.1, optimal controlled trajectories are at most of type YX in Ω_- and of type XY in Ω_+.

Suppose $q_1 \in \Omega_-$. Initially, since $q_1 \notin \mathscr{S}$, the optimal control can be only $u = -1$ or $u = +1$. If the control starts with $u = -1$, then since $L_X(\alpha) < 0$, the trajectory moves away from $\mathscr{S} = \{x \in \Omega : \alpha(x) = 0\}$, and no junctions from X to Y are possible in Ω_-. Hence this trajectory simply is an X-arc, and the corresponding control is constant, given by $u \equiv -1$. On the other hand, if the control starts with $u = +1$, then the trajectory moves toward $\mathscr{S} = \{x \in \Omega : \alpha(x) = 0\}$. In this case, it

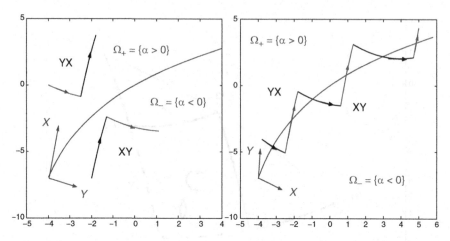

Fig. 2.12 Bang-bang switchings near fast (*left*) and slow (*right*) singular arcs

is possible that (a) the control switches to $u = -1$ before or as the singular arc $\mathscr{S}$ is reached, (b) the control switches to become singular as it reaches $\mathscr{S}$, or (c) this Y-trajectory simply crosses the singular arc. In case (a), after the switching time, the X-trajectory again moves the state away from $\mathscr{S}$ and no further switchings to Y are allowed in Ω_-. Hence in this case the trajectory is of type YX. In case (c), once the Y-trajectory enters the region Ω_+, switchings to X are no longer allowed and thus this trajectory simply is a Y-arc with constant control $u = +1$. The interesting case is (b). It follows from Proposition 2.8.4 that switchings onto and off the singular arc are extremal at any time. Thus, in this case, after following the singular arc for some time, the trajectory can leave $\mathscr{S}$ with the bang control $u = -1$ or $u = +1$. Using an X-trajectory, the system enters the region Ω_-, while it enters Ω_+ along a Y-trajectory. In any case, no more switchings are possible in Ω by Proposition 2.9.1. Thus overall, the structure is at most of type BSB. The analogous reasoning for an initial condition $q_1 \in \Omega_+$ shows the same concatenation structure to be valid.   □

2.9.3 Optimal Bang-Bang Trajectories near a Slow Singular Arc

What makes Proposition 2.9.4 work is that optimal bang-bang switchings in the regions Ω_- and Ω_+ move the system *away* from the singular arc $\mathscr{S}$ if $L_X(\alpha) = X\alpha$ is negative and $L_Y(\alpha) = Y\alpha$ is positive on Ω. The resulting synthesis of the type BSB is quite common around optimal singular arcs in small dimensions and will still be encountered several times throughout this text (e.g., Sects. 6.2 and 7.3). If, however, $L_X(\alpha) = X\alpha$ is positive and $L_Y(\alpha) = Y\alpha$ is negative on Ω, and in this case the singular arc is not optimal by the Legendre–Clebsch condition, or "slow," the opposite is true. Now optimal bang-bang junctions steer trajectories *toward* the singular arc $\mathscr{S}$ (see Fig. 2.12). In fact, in this case, there exist bang-bang extremals (i.e., bang-bang trajectories that satisfy the necessary conditions for optimality of the maximum principle) whose trajectories lie in Ω and have an arbitrarily large number of switchings. But as the geometric argument carried out above indicates, in this case, making more switchings slows down the trajectories, and thus none of these are optimal. This reasoning, however, is quite more intricate and goes well beyond a direct application of the conditions of the maximum principle, but involves the generalization of the concept of an *envelope* from the calculus of variations to the optimal control problem. We shall more generally develop this theory in Sect. 5.4, but here we include a self-contained proof of the result below due to Sussmann.

Proposition 2.9.5. [230, 236] *Let Ω be a domain on which condition (A1) is satisfied and where $L_X(\alpha) = X\alpha$ is positive and $L_Y(\alpha) = Y\alpha$ is negative. If Ω is taken sufficiently small, then optimal controls for trajectories that lie in Ω are bang-bang with at most one switching.*

Note that in contrast to the previous results, here we need to include the requirement that Ω be a small enough neighborhood of the reference point. This result does not hold in the more semiglobal setting without additional assumptions. The essential new concept involved in the proof of this result involves what are called conjugate points in [230]. However, for reasons that will be explained below, we prefer to use the terminology of *g-dependent points* instead.

Definition 2.9.1 (Variational vector field). Let $(x,u) : [0,T] \to \Omega \times U$ be an extremal controlled trajectory with multiplier λ. A variational vector field w along $\Gamma = ((x,u),\lambda)$ is a solution $w : [0,T] \to \mathbb{R}^2$ of the corresponding variational equation

$$\dot{w}(t) = \{Df(x(t)) + u(t)Dg(x(t))\} \cdot w(t). \tag{2.52}$$

The adjoint equation for the multiplier λ actually is the "adjoint" in the sense of linear differential equations to this variational equation (2.52). Thus, for any variational vector field w along Γ, the function $h : [0,T] \to \mathbb{R}, t \mapsto h(t) = \langle \lambda(t), w(t) \rangle$ is constant:

$$\dot{h}(t) = \left\langle \dot{\lambda}(t), w(t) \right\rangle + \langle \lambda(t), \dot{w}(t) \rangle = 0.$$

Suppose now that the switching function $\Phi(t) = \langle \lambda(t), g(x(t)) \rangle$ vanishes at times $t_1 < t_2$ and let w be the variational vector field that satisfies $w(t_1) = g(x(t_1))$. Since $\Phi(t_1) = 0$, it then follows that $\langle \lambda(t_2), w(t_2) \rangle = 0$. But $\Phi(t_2) = \langle \lambda(t_2), g(x(t_2)) \rangle = 0$ as well, and since $\lambda(t_2) \neq 0$, the vectors $g(x(t_2))$ and $w(t_2)$ must be linearly dependent. This leads to the following definition of g-dependent points in the plane.

Definition 2.9.2 (g-dependent). Let $(x,u) : [0,T] \to \Omega \times U$ be an extremal controlled trajectory with multiplier λ. Given times t_1 and t_2, $0 \leq t_1 < t_2 \leq T$, let $w(\cdot)$ be the variational vector field that satisfies $w(t_1) = g(x(t_1))$. We call the points $x(t_1)$ and $x(t_2)$ g-dependent (along $\Gamma = ((x,u),\lambda)$) if the vectors $g(x(t_2))$ and $w(t_2)$ are linearly dependent.

Thus, if $\Gamma = ((x,u),\lambda)$ is an extremal lift for which the control u switches at times $t_1 < t_2$, then the switching points $x(t_1)$ and $x(t_2)$ are g-dependent. As the example of time-optimal control for the harmonic oscillator shows, optimality of trajectories need not cease at g-dependent points. It does in the case that will be considered here, and thus the terminology of conjugate points is used in [230]. However, we generally prefer to restrict the terminology "conjugate point" to the case when optimality of trajectories ceases. We shall elaborate more on this in Sect. 6.1.

The key to the proof of Proposition 2.9.5 is to establish an inversion of g-dependent points around $\mathscr{S}$. For this calculation, a good choice of coordinates around $\mathscr{S} = \{x \in \Omega : \alpha(x) = 0\}$ is beneficial. The type of coordinates used here will also be needed in Sect. 2.10 and we therefore consider a slightly weaker version of assumption (A1). Let $p \in \Omega$ be a point at which (i) the vector fields f and g (and thus also X and Y) are linearly independent; (ii) $\alpha(p) = 0$, but the Lie derivative of α along X does not vanish, $L_X \alpha(p) \neq 0$; and (iii) the Lie derivative of α along g does not vanish, $L_g \alpha(p) \neq 0$. Conditions (i) and (ii) imply that the geometric properties of $\mathscr{S} = \{x \in \Omega : \alpha(x) = 0\}$ required in assumption (A1) are satisfied on a sufficiently small neighborhood of p. The third condition ensures that the vector field

$$S(x) = f(x) + \frac{L_f \alpha(x)}{L_g \alpha(x)} g(x) = f(x) + \frac{L_X \alpha(x) + L_Y \alpha(x)}{L_X \alpha(x) - L_Y \alpha(x)} g(x)$$

is well-defined near p. If the quotient $\frac{L_f \alpha(x)}{L_g \alpha(x)}$ lies between -1 and $+1$, then this is the singular vector field. But for the current reasoning it is not necessary that S correspond to a trajectory of the system, only that the integral curve of S through p be the curve $\mathscr{S}$. (This was shown in the proof of Proposition 2.9.3.) Let $[a,b]$ be an interval that contains 0 in its interior on which the solution to the initial value problem $\dot{y} = S(y)$, $y(0) = p$, exists. It then follows from a standard compactness argument that there exists an $\varepsilon > 0$ such that the solution $z = z(\cdot;s)$ to the initial value problem $\dot{z} = X(z)$, $z(0) = y(s)$, exists on the interval $[-\varepsilon, \varepsilon]$. Using the notation Ψ for the flow, we denote this solution by

$$\psi(s,t) = \Psi_t^X \circ \Psi_s^S(p).$$

If, in addition, the Lie derivative $L_X(\alpha)$ does not vanish at $y(t)$ for all $t \in [a,b]$, then the X-flow is everywhere transversal to the curve $\mathscr{S}$ and the map ψ is a diffeomorphism from some square $Q(\varepsilon) = (-\varepsilon, \varepsilon) \times (-\varepsilon, \varepsilon)$ onto some neighborhood $\psi(Q)$ of p. If we now choose this set $\psi(Q)$ as Ω,

$$\Omega = \{\psi(s,t) : -\varepsilon < s < \varepsilon, -\varepsilon < t < \varepsilon\},$$

then the times $(s,t) \in Q$ provide us with a good set of coordinates on Ω called *canonical coordinates of the second kind* in Lie theory (also, see Sects. 4.5 and 7.1). In these coordinates, the curve $\mathscr{S}$ corresponds to the s-axis, $\mathscr{S} \cong \{(s,t) \in Q : t = 0\}$, and integral curves of the vector field X are the vertical lines $s = \text{const}$. We call such a mapping $\psi : Q \to \Omega$ an X-*aligned chart of coordinates* centered at the point p (see Fig. 2.13).

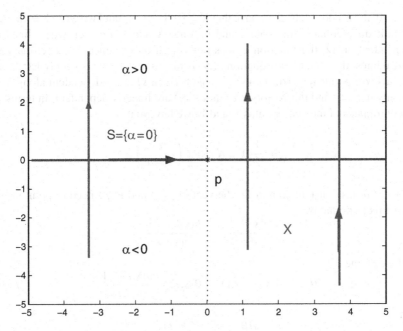

Fig. 2.13 An X-aligned coordinate chart

Definition 2.9.3 (X-aligned chart of coordinates). An X-aligned chart of coordinates (centered at p) is a diffeomorphism ψ,

$$\psi : Q(\varepsilon) \subset \mathbb{R}^2 \to \Omega, \quad (s,t) \mapsto \psi(s,t) = \Psi_t^X \circ \Psi_s^S(p),$$

such that X and Y are linearly independent everywhere on Ω, the set $\mathscr{S} = \{x \in \Omega : \alpha(x) = 0\}$ is the integral curve of the vector field S through p, and the Lie derivatives $L_X \alpha$ and $L_g \alpha$ are everywhere nonzero on Ω.

Lemma 2.9.3. [230] *Given an X-aligned chart of coordinates, $\Omega = \psi(Q(\varepsilon))$, for ε small enough, there exists a differentiable function*

$$\zeta : Q \to \mathbb{R}, \quad (s,t) \mapsto \zeta(s,t),$$

that satisfies

$$\zeta(s,0) = 0, \quad \frac{\partial \zeta}{\partial t}(s,0) = -1,$$

and is such that two points $q = \psi(s,t)$ and $q' = \psi(s',t')$ in Ω are g-dependent along an X-extremal if and only if $s' = s$ and $t' = \zeta(s,t)$. Thus ζ defines the mapping from q to its g-dependent point in this X-aligned chart of coordinates.

Proof. In these coordinates, we have that $X \cong (0,1)^T = \frac{\partial}{\partial t}$, and we write $Y \cong (a,b)^T$ for some differentiable functions a and b. Since X and Y are everywhere linearly independent on Ω, the function a does not vanish on Q. Since X-trajectories are vertical lines, the variational equation (2.52) along X-extremals is simply $\dot{w}(t) \equiv 0$, and thus two points $q = \psi(s,t)$ and $q' = \psi(s',t')$ in Ω are g-dependent along X if and only if $s = s'$ and the vectors $g(q)$ and $g(q')$ are linearly dependent. In terms of the coordinates of the vector fields X and Y, we have that

$$g = \frac{1}{2}(Y - X) \cong \frac{1}{2}\begin{pmatrix} a \\ b-1 \end{pmatrix},$$

and since a has constant sign in Ω, the vectors $g(q)$ and $g(q')$ need to point in the same direction; that is,

$$\frac{b(s,t) - 1}{a(s,t)} = \frac{b(s,t') - 1}{a(s,t')}.$$

If we define

$$\theta : Q \rightarrow \mathbb{R}, \quad (s,t) \mapsto \theta(s,t) = \frac{b(s,t) - 1}{a(s,t)},$$

then

$$\frac{\partial \theta}{\partial t}(s,t) = \frac{\xi(s,t)}{a^2(s,t)},$$

where

$$\xi(s,t) = \frac{\partial b}{\partial t}(s,t)a(s,t) - (b(s,t) - 1)\frac{\partial a}{\partial t}(s,t).$$

This expression relates to the determinant of $[f,g]$ and g: suppressing the arguments, we have that

$$[f,g] = [X,g] \cong Dg \cdot X = \frac{1}{2}\begin{pmatrix} \dfrac{\partial a}{\partial s} & \dfrac{\partial a}{\partial t} \\ \dfrac{\partial b}{\partial s} & \dfrac{\partial b}{\partial t} \end{pmatrix}\begin{pmatrix} 0 \\ 1 \end{pmatrix} = \frac{1}{2}\begin{pmatrix} \dfrac{\partial a}{\partial t} \\ \dfrac{\partial b}{\partial t} \end{pmatrix}$$

and thus

$$\det([f,g],g) \cong \frac{1}{4}\begin{vmatrix} \dfrac{\partial a}{\partial t} & a \\ \dfrac{\partial b}{\partial t} & b-1 \end{vmatrix} = -\frac{1}{4}\xi(s,t).$$

Expressing the Lie bracket $[f,g]$ in terms of f and g, we therefore get that

$$\xi(s,t) = -4\det([f,g],g) = -4\det(\alpha f + \beta g, g) = -4\alpha \det(f,g),$$

where α and the vector fields f and g are evaluated at the point $q = \psi(s,t) \in \Omega$. In particular, since α vanishes for $t = 0$ ($\psi(s,0) \in \mathscr{S}$), it follows that $\xi(s,0) \equiv 0$, and therefore t can be factored from $\xi(s,t)$, say

$$\xi(s,t) = t\tilde{\xi}(s,t).$$

Thus, with all functions and vector fields evaluated at $\psi(s,0) \in \mathscr{S}$, it follows that

$$\tilde{\xi}(s,0) = \frac{\partial \xi}{\partial t}(s,0) = -4 \frac{\partial}{\partial t}_{|t=0} \left(\alpha \det(f,g) \right) = -4 L_X \alpha \cdot \det(f,g) \neq 0.$$

Hence

$$\frac{\partial \theta}{\partial t}(s,0) = 0$$

and

$$\frac{\partial^2 \theta}{\partial t^2}(s,t) = \frac{\frac{\partial \xi}{\partial t}(s,t)a(s,t) - 2\frac{\partial a}{\partial t}(s,t)\xi(s,t)}{a^3(s,t)}$$

gives that

$$\frac{\partial^2 \theta}{\partial t^2}(s,0) = \frac{\tilde{\xi}(s,0)}{a^2(s,0)} \neq 0.$$

Overall, we therefore can write

$$\theta(s,t) = \theta(s,0) + t^2 \tilde{\theta}(s,t)$$

for some smooth function $\tilde{\theta} = \tilde{\theta}(s,t)$ that satisfies $\tilde{\theta}(s,0) \neq 0$ for all $s \in [-\varepsilon, \varepsilon]$. By shrinking ε further, if necessary, we may assume that $\tilde{\theta}(s,t)$ does not vanish on Q.

If one now expresses the difference

$$\delta(s,t,t') = \tilde{\theta}(s,t) - \tilde{\theta}(s,t')$$

as

$$\delta(s,t,t') = (t-t')\tilde{\delta}(s,t,t'),$$

then the equation $\theta(s,t) = \theta(s,t')$ is equivalent to

$$\begin{aligned}
0 &= t^2 \tilde{\theta}(s,t) - \left(t'\right)^2 \tilde{\theta}(s,t') \\
&= \left(t^2 - (t')^2\right) \tilde{\theta}(s,t') + t^2 \left(\tilde{\theta}(s,t) - \tilde{\theta}(s,t')\right) \\
&= (t-t') \left[(t+t') \tilde{\theta}(s,t') + t^2 \tilde{\delta}(s,t,t') \right],
\end{aligned}$$

and thus we need to solve the equation

$$\Delta(s,t,t') = (t+t') \tilde{\theta}(s,t') + t^2 \tilde{\delta}(s,t,t') = 0.$$

Clearly, $\Delta(0,0,0) = 0$ and

$$\frac{\partial \Delta}{\partial t'}(s,0,0) = \tilde{\theta}(s,0) \neq 0.$$

Hence, by the implicit function theorem, the equation $\Delta(s,t,t') = 0$ can be solved for t' near $(0,0,0)$ in terms of a differentiable function $t' = \zeta(s,t)$. Furthermore, for $t = 0$, we have that

$$0 = \Delta(s,0,\zeta(s,0)) = \zeta(s,0) \cdot \tilde{\theta}(s,\zeta(s,0)),$$

and since $\tilde{\theta}(s,t)$ does not vanish, it follows that $\zeta(s,0) \equiv 0$ for all $s \in [-\varepsilon,\varepsilon]$. Finally, differentiating $\Delta(s,t,\zeta(s,t))$ with respect to t and setting $t = 0$ gives

$$0 = \frac{\partial \Delta}{\partial t}(s,0,\zeta(s,0)) + \frac{\partial \Delta}{\partial t'}(s,0,\zeta(s,0))\frac{\partial \zeta}{\partial t}(s,0)$$

$$= \frac{\partial \Delta}{\partial t}(s,0,0) + \frac{\partial \Delta}{\partial t'}(s,0,0)\frac{\partial \zeta}{\partial t}(s,0).$$

But

$$\frac{\partial \Delta}{\partial t}(s,0,0) = \frac{\partial \Delta}{\partial t'}(s,0,0) = \tilde{\theta}(s,0) \neq 0,$$

and therefore

$$\frac{\partial \zeta}{\partial t}(s,0) = -1.$$

$\square$

We now prove Proposition 2.9.5: Let $\Omega = \psi(Q(\varepsilon))$ be an X-aligned chart of coordinates and suppose ε is small enough that there exists a differentiable function $\zeta : Q \to \mathbb{R}$, $(s,t) \mapsto \zeta(s,t)$, with the properties of Lemma 2.9.3. By making ε smaller if necessary, we also may assume that $\frac{\partial \zeta}{\partial t}$ is negative on Q. As before, $X \cong (0,1)^T = \frac{\partial}{\partial t}$ and we write $Y \cong (a,b)^T$ for some differentiable functions a and b. In these coordinates,

$$L_X \alpha \cong \frac{\partial \alpha}{\partial t}$$

and

$$L_Y \alpha \cong \frac{\partial \alpha}{\partial s} \cdot a + \frac{\partial \alpha}{\partial t} \cdot b.$$

Since the singular curve $\mathscr{S}$ is given by the s-axis, we have $\alpha(s,0) \equiv 0$ and therefore $\frac{\partial \alpha}{\partial s}(s,0) \equiv 0$ as well. Hence, at the reference point p, we get

$$\frac{L_Y \alpha(p)}{L_X \alpha(p)} \cong b(0,0), \tag{2.53}$$

and thus $b(0,0)$ is negative. By choosing ε small enough, we may assume that b is negative everywhere on Q. Similarly, without loss of generality we assume that $L_X\alpha > 0$ and $L_Y\alpha < 0$ on all of Ω.

Let $(\bar{x}, \bar{u})$ be a time-optimal YXY- trajectory that transfers a point $q_1 \in \Omega$ into $q_2 \in \Omega$ with the entire trajectory $\bar{x}$ lying in Ω. Denote the switching times by τ and τ', $\tau < \tau'$, and the corresponding junctions by r and r', respectively. The points r and r' are g-dependent along X, and thus if $r = \psi(s,t)$ and $r' = \psi(s',t')$, then $s' = s$ and $t' = \zeta(s,t)$. Note that $t < 0$ and $t' > 0$. (For by Proposition 2.9.1, YX-junctions need to lie in $\alpha \le 0$ and XY-junctions in $\alpha \ge 0$. Since $L_X\alpha > 0$ on Ω, we thus have $t \le 0$ and $t' \ge 0$. But $\zeta(s,0) = 0$, and thus neither can be zero, since otherwise $r = r'$.) The next lemma is one of the two key arguments in the construction, and it is only for this result that we need to make the neighborhood Ω small.

Lemma 2.9.4. *Let γ denote the restriction of the YXY-trajectory $\bar{x}$ to some small interval $[\tau - \varepsilon, \tau]$, where τ is the first switching time and let γ' be the image of this curve under the mapping $Z : (s,t) \mapsto (s, \zeta(s,t))$. For ε sufficiently small, the curve γ' is a trajectory of the system.*

Proof. It suffices to show that the tangent vector to the curve γ' at the point r' is a linear combination of $X(r')$ and $Y(r')$ with positive coefficients. For if this is the case, then by choosing the times sufficiently close to τ, at every point q' on the curve γ' there exists a continuous control $u(q') \in (-1,1)$ such that $f(q') + u(q')g(q')$ is tangent to γ'. After a suitable reparameterization, the curve thus becomes a trajectory of Σ.

This property, however, can be guaranteed only in a sufficiently small neighborhood of p. The tangent vector t' to the curve γ' at r' is the image of the vector $Y(r)$ under the differential of the mapping Z, i.e.,

$$
\begin{aligned}
t' &= \begin{pmatrix} 1 & 0 \\ \frac{\partial\zeta}{\partial s}(s,t) & \frac{\partial\zeta}{\partial t}(s,t) \end{pmatrix} \begin{pmatrix} a(s,t) \\ b(s,t) \end{pmatrix} = \begin{pmatrix} a(s,t) \\ \frac{\partial\zeta}{\partial s}(s,t)a(s,t) + \frac{\partial\zeta}{\partial t}(s,t)b(s,t) \end{pmatrix} \\
&= \frac{a(s,t)}{a(s,t')}\begin{pmatrix} a(s,t') \\ b(s,t') \end{pmatrix} + \left[\frac{\partial\zeta}{\partial s}(s,t)a(s,t) + \frac{\partial\zeta}{\partial t}(s,t)b(s,t) - \frac{a(s,t)}{a(s,t')}b(s,t')\right]\begin{pmatrix} 0 \\ 1 \end{pmatrix} \\
&= \frac{a(s,t)}{a(s,t')}Y(r') + b(s,t)\left[\frac{\partial\zeta}{\partial s}(s,t)\frac{a(s,t)}{b(s,t)} + \frac{\partial\zeta}{\partial t}(s,t) - \frac{a(s,t)}{a(s,t')}\frac{b(s,t')}{b(s,t)}\right]X(r').
\end{aligned}
$$

$$(2.54)$$

Since a has constant sign on Q, the quotient $\frac{a(s,t)}{a(s,t')}$ is positive. The function b is negative on Q, and by Lemma 2.9.3,

$$
\frac{\partial\zeta}{\partial s}(s,0)\frac{a(s,0)}{b(s,0)} + \frac{\partial\zeta}{\partial t}(s,0) \equiv -1.
$$

Hence, and once more by choosing the neighborhood Q small enough, we may assume that

$$\frac{\partial \zeta}{\partial s}(s,t)\frac{a(s,t)}{b(s,t)} + \frac{\partial \zeta}{\partial t}(s,t) < -\frac{1}{2} \qquad \text{for all} \qquad (s,t) \in Q.$$

Thus the coefficient at $X(r')$ is positive as well. $\square$

Remark 2.9.1. The construction of an X-aligned chart of coordinates $\Omega = \psi(Q(\varepsilon))$ does not require that $L_Y \alpha \neq 0$, and it is still applicable if $L_Y \alpha(p) = 0$, since then $L_g \alpha(p) = \frac{1}{2}L_X \alpha(p) > 0$. But in this case $b(0,0) = 0$, and thus the dominance argument above no longer can be made. For later reference, however, we already note here that such an argument is not needed at points where the Lie derivative of ζ along Y is positive,

$$L_Y \zeta(s,t) = \frac{\partial \zeta}{\partial s}(s,t)a(s,t) + \frac{\partial \zeta}{\partial t}(s,t)b(s,t) > 0,$$

and where $b(s,t')$ is negative. In this case, (2.54) directly gives that t' is a linear combination of $X(r')$ and $Y(r')$ with positive coefficients. This will allow us to deal with codimension-2 cases in the next section.

We now show that Lemma 2.54 precludes the optimality of the YXY-trajectory $\bar{x}$. In fact, the curve γ' is an envelope for the control system Σ, and the generalization of the theory of envelopes to optimal control shows that it cannot be optimal. We shall develop this theory for a general control problem in Sect. 5.4 but already here anticipate this argument with a direct calculation invoking the clock form ω introduced earlier.

Let Γ be the restriction of the YXY-trajectory to the interval $[\tau - \varepsilon, \tau']$ so that Γ is the concatenation of the curve γ with the X-trajectory that steers r into r'. Define another trajectory Γ' of Σ that steers the point $\bar{x}(\tau - \varepsilon)$ into r' by first following the X-trajectory from $\bar{x}(\tau - \varepsilon)$ to its g-dependent point on the curve γ' and then concatenating with the Σ-trajectory that corresponds to γ' (see Fig. 2.14).

Lemma 2.9.5. *The times along the trajectories Γ and Γ' are equal, $T(\Gamma) = T(\Gamma')$.*

Proof. The concatenation Υ of Γ with the curve Γ' run backward is a closed curve, and by Stokes's theorem, the difference in the times along these two trajectories is given by

$$T(\Gamma) - T(\Gamma') = \int_\Upsilon \omega = \int_R d\omega,$$

where R denotes the region enclosed by Υ. The coordinate expression for ω (see Eq. (2.49)) is given by

$$\omega = \frac{g_2 ds - g_1 dt}{\det(f,g)} = \frac{\frac{1}{2}(b-1)ds - \frac{1}{2}adt}{-\frac{1}{2}a} = dt + \frac{1-b(s,t)}{a(s,t)}ds$$

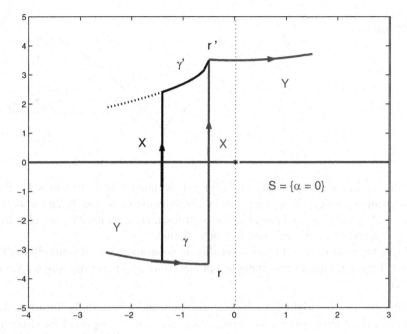

Fig. 2.14 Conjugate curve γ'

and thus, and using the notation θ from the proof of Lemma 2.9.3,

$$d\omega = \frac{\partial}{\partial t}\left(\frac{b(s,t)-1}{a(s,t)}\right)ds \wedge dt = \frac{\partial \theta}{\partial t}(s,t)ds \wedge dt.$$

Since Y is transversal to X, we can parameterize the curve γ as the graph of a function σ of s over some interval $[s_\varepsilon, s_\tau]$, say $\gamma: [s_\varepsilon, s_\tau] \to Q, s \mapsto \gamma(s) = (s, \sigma(s))$. Evaluating the double integral by integrating over the vertical segments therefore gives

$$\int_R d\omega = \int_{s_\varepsilon}^{s_\tau} \int_{\sigma(s)}^{\zeta(s,\sigma(s))} \frac{\partial \theta}{\partial t}(s,t)dtds$$

$$= \int_{s_\varepsilon}^{s_\tau} [\theta(s,\zeta(s,\sigma(s))) - \theta(s,\sigma(s))]\,ds.$$

But by construction, the points $(s,\sigma(s))$ on γ and $(s,\zeta(s,\sigma(s)))$ on γ' are g-dependent, and therefore for all $s \in [s_\varepsilon, s_\tau]$,

$$\theta(s,\zeta(s,\sigma(s))) = \theta(s,\sigma(s)).$$

Hence $\int_R d\omega = 0$ and thus $T(\Gamma) = T(\Gamma')$. □

Fig. 2.15 A variation along
a YXY-trajectory

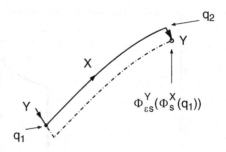

This precludes the optimality of Γ: for if Γ is time-optimal, then so is Γ'. But the dynamics along γ' is a strict convex combination of X and Y, and thus the control takes values in the interior of the control set. Hence it must be singular. But $\alpha(r') > 0$, and so this is not possible. Contradiction.

Since the roles of X and Y are reversible in our assumptions, it similarly can be shown that XYX-trajectories cannot be optimal either, and thus Proposition 2.9.5 is proven.   □

This proof is the original one by H. Sussmann, and it beautifully illustrates the underlying *geometric* aspects (i.e., conjugate points and envelopes) of the structure of optimal bang-bang trajectories near a slow singular arc. We shall return to this topic for a general n-dimensional system in Sect. 6.1.3 about transversal folds.

There exists an alternative, and in some sense more direct, *algebraic* approach that is based on a variation analogous to the one used in [209] for the three-dimensional case. Suppose again that Γ is a YXY-trajectory with the switching points given by q_1 and $q_2 = \Phi_s^X(q_1)$. It is geometrically clear (see Fig. 2.15), and not difficult to verify analytically, that there exist continuously differentiable positive functions $r = r(\varepsilon)$ and $t = t(\varepsilon)$ such that

$$\Phi_{\varepsilon s}^Y\left(\Phi_s^X(q_1)\right) = \Phi_{r(\varepsilon)}^X\left(\Phi_{t(\varepsilon)}^Y(q_1)\right).$$

The difference in time between these trajectories is then given by

$$\Delta(\varepsilon) = s(1 + \varepsilon) - t(\varepsilon) - s(\varepsilon),$$

and the YXY-trajectory is not time-optimal if $\Delta(\varepsilon) > 0$ for small $\varepsilon > 0$. Hence, the optimality of bang-bang trajectories with two switchings can be excluded by computing the Taylor expansion of Δ at $\varepsilon = 0$. It can be shown that the fact that q_1 and q_2 are g-dependent points is equivalent to $\Delta'(0) = 0$, and it thus becomes necessary to compute the second derivative $\Delta''(0)$. This, however, requires a good algebraic framework that is provided by a Lie-algebraic formalism that we shall establish only in Sect. 4.5. We shall return to this second approach in Sect. 7.3 when we analyze the corresponding situation—the structure of time-optimal bang-bang trajectories near a slow singular arc—in dimension three.

2.10   Input Symmetries and Codimension-2 Cases in the Plane

The results of the last section cover the local structure of time-optimal controlled trajectories near a point p in the plane under codimension-0 and some codimension-1 conditions. In order to classify the structure of time-optimal controlled trajectories for a generic time-invariant control-affine nonlinear system of the type [NTOC] in the plane, by Thom's transversality theorem [108] it is necessary to analyze all other possible codimension-1 and codimension-2 conditions. Other codimension-1 conditions arise if assumption (A0) is violated, i.e., if the vector fields f and g are linearly dependent at p; codimension-2 conditions arise if two independent equality relations are imposed. In this section, we still analyze those codimension-2 situations that arise if condition (A0) is met. These correspond to situations in which in addition to $\alpha(p)$, also one of the Lie derivatives of α along X or Y vanishes at p. The results of this and the previous section then collectively describe the structure of time-optimal controls near a reference point p where the vector fields $f(p)$ and $g(p)$ are linearly independent under otherwise generic conditions on the vector fields f and g.

2.10.1   Input Symmetries

In cases of higher codimensions, the number of possibilities increases significantly, and it now helps to use *input symmetries* and other invariances to reduce this number. Since the control set $U = [-1, 1]$ is invariant under a reflection at the origin, the problem [NTOC] remains unchanged if we use as control $v = -u$ instead. This transformation, however, changes the vector fields: g becomes $-g$ while f remains the same. Thus their Lie brackets and hence also the functions α and β and their Lie derivatives are affected. This allows us to normalize the signs of some of these functions.

Definition 2.10.1 (Input symmetry). An input symmetry is a linear transformation on the vector fields f and g that leaves the control system $\Sigma : \dot{x} = f(x) + ug(x)$, $u \in U$ (including the class of admissible controls), invariant.

Definition 2.10.2 (Reflection). For the system $\Sigma : \dot{x} = f(x) + ug(x)$, $|u| \leq 1$, define the reflection ρ by $\rho(f) = f$ and $\rho(g) = -g$, or equivalently, as the transformation that interchanges the vector fields X and Y,

$$\rho(X) = \rho(f - g) = \rho(f) - \rho(g) = f + g = Y$$

and

$$\rho(Y) = \rho(f + g) = \rho(f) + \rho(g) = f - g = X.$$

This definition naturally extends as a *homomorphism* to the Lie algebra generated by the vector fields f and g if we define

$$\rho\left([f,g]\right) = [\rho(f),\rho(g)]$$

and inductively extend this relation to higher-order Lie brackets. As before, we assume that Ω is a simply connected region of $\mathbb{R}^2$ and that f and g are linearly independent vector fields on Ω. Thus, all higher-order Lie brackets of f and g can be expressed as linear combinations of f and g with coefficients that are smooth functions of x. Suppose $[f,g](x) = \alpha(x)f(x) + \beta(x)g(x)$ and write

$$\rho\left([f,g]\right) = \rho(\alpha)\rho(f) + \rho(\beta)\rho(g).$$

The effects that an input symmetry has on the higher-order brackets and coordinate expressions can then easily be calculated through straightforward algebraic substitutions. We have that

$$[\rho(f),\rho(g)] = -[f,g] = -\alpha f - \beta g = -\alpha \rho(f) + \beta \rho(g)$$

and thus

$$\rho(\alpha) = -\alpha \qquad \text{and} \qquad \rho(\beta) = \beta. \tag{2.55}$$

Considering higher-order brackets, we arrive at analogous formulas for the Lie derivatives of α and β:

$$
\begin{aligned}
[X,[f,g]] = [X, \alpha f + \beta g] &= L_X(\alpha)f + \alpha[X,f] + L_X(\beta)g + \beta[X,g] \\
&= L_X(\alpha)f + L_X(\beta)g + (\alpha + \beta)[f,g] \\
&= (L_X(\alpha) + (\alpha + \beta)\alpha)f + (L_X(\beta) + (\alpha + \beta)\beta)g,
\end{aligned}
$$

and analogously

$$[Y,[f,g]] = (L_Y(\alpha) - (\alpha - \beta)\alpha)f + (L_Y(\beta) - (\alpha - \beta)\beta)g.$$

Applying the input symmetry ρ, we have that

$$
\begin{aligned}
\rho\left([X,[f,g]]\right) = [\rho(X),[\rho(f),\rho(g)]] &= -[Y,[f,g]] \\
&= -(L_Y(\alpha) - (\alpha - \beta)\alpha)f - (L_Y(\beta) - (\alpha - \beta)\beta)g \\
&= (-L_Y(\alpha) + (\rho(\alpha) + \rho(\beta))\rho(\alpha))\rho(f) \\
&\quad + (L_Y(\beta) + (\rho(\alpha) + \rho(\beta))\rho(\beta))\rho(g),
\end{aligned}
$$

and therefore

$$\rho\left(L_X(\alpha)\right) = -L_Y(\alpha) \qquad \text{and} \qquad \rho\left(L_X(\beta)\right) = L_Y(\beta).$$

Since $-L_Y(\alpha) = L_Y(-\alpha) = L_{\rho(X)}(\rho(\alpha))$, this relation can succinctly be expressed in the form

$$\rho\left(L_X(\alpha)\right) = L_{\rho(X)}(\rho(\alpha)). \tag{2.56}$$

Analogously, it follows that

$$\rho\left(L_Y(\alpha)\right) = L_{\rho(Y)}(\rho(\alpha)) = -L_X(\alpha)$$

and

$$\rho\left(L_Y(\beta)\right) = L_{\rho(Y)}(\rho(\beta)) = L_X(\beta).$$

Similarly, for higher-order derivatives we have that

$$\rho\left(L_X^2(\alpha)\right) = L_{\rho(X)}\left(L_{\rho(X)}(\rho(\alpha))\right) = L_Y\left(L_Y(-\alpha)\right) = -L_Y^2(\alpha)$$

and

$$\rho\left(L_Y^2(\alpha)\right) = L_{\rho(Y)}\left(L_{\rho(Y)}(\rho(\alpha))\right) = L_X\left(L_X(-\alpha)\right) = -L_X^2(\alpha),$$

and so on. Once more, *the effects that an input symmetry has on the vector fields f and g and their Lie brackets are easily obtained through straightforward algebraic substitutions.*

We briefly reconsider the results of the previous section with this point of view. If α is positive on some region Ω, we have shown that optimal controlled trajectories that lie in Ω have at most the structure XY. Applying the input symmetry ρ to the system changes the sign of α and interchanges X with Y. Thus, it directly follows that optimal controlled trajectories are at most of type YX if α is negative (see Proposition 2.9.1). On the other hand, in the codimension-1 situation (A1), the relevant conditions are all invariant under this input symmetry. For example, the singular arc is given by

$$S = f + \frac{L_X\alpha(x) + L_Y\alpha(x)}{L_X\alpha(x) - L_Y\alpha(x)}g = f + \frac{L_f\alpha}{L_g\alpha}g$$

and

$$\rho(S) = \rho(f) + \frac{\rho\left(L_f\alpha(x)\right)}{\rho\left(L_g\alpha(x)\right)}\rho(g) = f + \frac{-L_f\alpha(x)}{L_g\alpha(x)}(-g) = S.$$

Naturally, the strengthened Legendre–Clebsch condition (see Eq. (2.47),

$$\langle \lambda(t), \; [g, [f, g]](x(t)) \rangle = -L_g\alpha(x(t)),$$

is invariant under this input symmetry as well. In fact, the assumptions for each of the various codimension-1 cases considered in the last section are invariant under ρ. Still, this input symmetry is useful in the proof of Proposition 2.9.5, where we carried out the construction only for YXY-trajectories and merely claimed that the analogous construction excludes XYX-trajectories as well. Since ρ interchanges $L_X(\alpha)$ and $-L_Y(\alpha)$,

$$\rho\left(L_X(\alpha)\right) = -L_Y(\alpha) \qquad \text{and} \qquad \rho\left(L_Y(\alpha)\right) = -L_X(\alpha),$$

the assumptions of Proposition 2.9.5 are invariant under ρ, and thus, applying ρ, it immediately follows that XYX-trajectories cannot be optimal either. No further argument is necessary.

It is in the codimension-2 scenario, that input symmetries really become useful. We can limit our analysis to the case that one of the Lie derivatives of α with respect to X or Y vanishes, and without loss of generality, we shall consider the case when

$$L_X(\alpha)(p) \neq 0 \qquad \text{and} \qquad L_Y(\alpha)(p) = 0, \quad \text{while} \quad L_Y^2(\alpha)(p) \neq 0.$$

Using a second symmetry that optimal trajectories possess, we can in addition normalize the sign for the second Lie derivative $L_Y^2(\alpha)(p)$. Time-optimal trajectories are also invariant under *time reversal*. If (x_*, u_*) is a time-optimal trajectory for the system $\Sigma : \dot{x} = f(x) + ug(x)$, $|u| \leq 1$, defined over an interval $[0, T]$ that steers a point q_1 into q_2, then the pair (y_*, v_*) defined by $y_*(t) = x_*(T - t)$ and $v_*(t) = u_*(T - t)$ is a time-optimal trajectory that steers q_2 into q_1 for the system $\check{\Sigma} : \dot{y} = \check{f}(y) + v\check{g}(y)$, $|v| \leq 1$, where time has been reversed. Since

$$\dot{y}_*(t) = -\dot{x}_*(T - t) = -f(x_*(T - t)) - u_*(T - t)g(x_*(T - t))$$
$$= -f(y(t)) - v(t)g(y(t)),$$

this property can be expressed in terms of a second input symmetry that reverses the signs of the vector fields f and g.

Definition 2.10.3 (Time reversal). For the system $\Sigma : \dot{x} = f(x) + ug(x)$, $|u| \leq 1$, define time reversal τ by $\tau(f) = -f$ and $\tau(g) = -g$, or equivalently, by $\tau(X) = -X$ and $\tau(Y) = -Y$.

As above, we extend this definition to the Lie algebra generated by f and g and then calculate the relations it implies on the coordinates with respect to the basis in terms of f and g. Simple computations verify that

$$\tau(\alpha) = -\alpha, \qquad\qquad\qquad \tau(\beta) = -\beta,$$
$$\tau(L_X(\alpha)) = L_X(\alpha), \qquad\qquad \tau(L_Y(\alpha)) = L_Y(\alpha),$$
$$\tau(L_X^2(\alpha)) = -L_X^2(\alpha), \qquad\qquad \tau(L_Y^2(\alpha)) = -L_Y^2(\alpha),$$

and it is the last relation that, without loss of generality, allows us to assume that $L_Y^2(\alpha)$ is positive.

In a more abstract framework, the input symmetries generate a group $\mathscr{G} = \{\text{id}, \rho, \tau, \tau \circ \rho\}$ of idempotent elements (i.e., $\rho \circ \rho = \text{id}$, etc.) and using them, it is possible to reduce the number of codimension-2 scenarios by a factor of 4. It is to be expected that the mathematically more difficult scenarios arise when the Lie bracket configurations are invariant under this group of symmetries, and this will again happen for nongeneric codimension-3 situations. The codimension-2 cases, however, essentially can be fully analyzed based on the earlier codimension-1 results of Sect. 2.9 and some additional geometric considerations.

2.10.2 Saturating Singular Arcs

We now assume that

(A2)   the vector fields f and g are linearly independent everywhere on $\Omega \subset \mathbb{R}^2$ and there exists a point $p \in \Omega$ with $\alpha(p) = 0$, but the Lie derivative of α along X does not vanish on Ω,

$$\alpha(p) = 0, \qquad L_X(\alpha)(x) \neq 0 \quad \text{for all } x \in \Omega;$$

furthermore, the Lie derivative of α along Y vanishes at p, but the second Lie derivative of α along Y is positive on Ω,

$$L_Y(\alpha)(p) = 0, \qquad L_Y^2(\alpha)(x) > 0 \qquad \text{for all } x \in \Omega.$$

Note that $L_g \alpha(p) = \frac{1}{2} L_X(\alpha)(p) \neq 0$, and thus there exists an X-aligned chart of coordinates (centered at p), $\psi : Q(\varepsilon) \subset \mathbb{R}^2 \to \Omega = \psi(Q(\varepsilon))$, $(s,t) \mapsto \psi(s,t) = \Psi_t^X \circ \Psi_s^S(p)$. As above, in these coordinates $X \cong (0,1)^T = \frac{\partial}{\partial t}$, and we write $Y \cong (a,b)^T$ for some differentiable functions a and b. Since X and Y are everywhere linearly independent on Ω, the function a does not vanish on Q, and without loss of generality, we assume that a is positive on Ω. (If a is negative, then simply change s in the definition of the coordinates to $-s$.) Assumption (A2) also implies that the integral curve Υ of Y through the point p is tangent to the curve $S = \{x \in \Omega : \alpha(x) = 0\}$ at p and that the order of contact is 1, i.e., for r near 0,

$$\alpha(\Upsilon(r)) = \alpha(p) + L_Y \alpha(p) r + \frac{1}{2} L_Y^2 \alpha(p) r^2 + o(r^2)$$

$$= \frac{1}{2} L_Y^2 \alpha(p) r^2 + o(r^2).$$

Hence, except for the point p, the curve Υ lies in $\Omega_+ = \{x \in \Omega : \alpha(x) > 0\}$ and can be parameterized as the graph of a function of s. By choosing ε sufficiently small, we again can assume that this parameterization is defined on the full interval $[-\varepsilon, \varepsilon]$, say $\Upsilon : [-\varepsilon, \varepsilon] \to Q(\varepsilon)$, $s \mapsto (s, y(s))$, and $y'(0) = 0$. The geometry is illustrated in Fig. 2.16.

The point p is the beginning or end point of an admissible singular arc. The singular control at p is given by

$$u_{\text{sing}}(p) = \frac{L_X \alpha(p) + L_Y \alpha(p)}{L_X \alpha(p) - L_Y \alpha(p)} = +1,$$

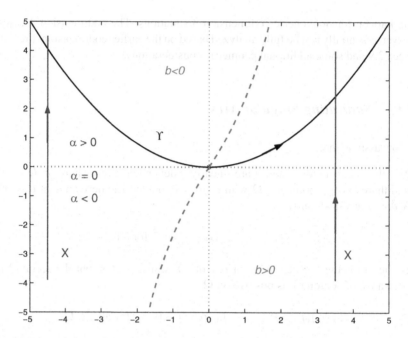

Fig. 2.16 Assumption (A2)

and thus the singular control saturates at p at its upper value. Since we normalized a to be positive, we have $L_Y \alpha(s,0) > 0$ for $s > 0$ and $L_Y \alpha(s,0) < 0$ for $s < 0$. Depending on the sign of $L_X \alpha$, the singular control is admissible for one side of the s-axis and inadmissible for the other. The singular arc itself is fast, (respectively, satisfies the strengthened Legendre–Clebsch condition), if $L_X(\alpha)$ is negative and it violates it if $L_X(\alpha)$ is positive. The structure of optimal controlled trajectories on $\Omega = \psi(Q(\varepsilon))$ thus depends on this sign, and we now analyze these two cases.

Proposition 2.10.1. *Let Ω be a domain on which condition (A2) is satisfied and suppose $L_X(\alpha) = X\alpha$ is positive everywhere on Ω. Then, for Ω sufficiently small, optimal controlled trajectories that entirely lie in Ω are of type XYXY.*

Proof. It follows from the results of Sect. 2.9 that except for p, every point in Ω has a neighborhood such that optimal controls for trajectories that lie entirely in this neighborhood are bang-bang. Thus optimal controlled trajectories are concatenations of X- and Y-arcs in Ω. It remains to establish the concatenation sequence. Recall that it follows from Proposition 2.9.1 that XY-junctions can lie only in $\{x \in \Omega : \alpha(x) \geq 0\}$ and YX-junctions must lie in $\{x \in \Omega : \alpha(x) \leq 0\}$.

As before, we consider an X-aligned chart of coordinates (centered at p), $\psi : Q(\varepsilon) \to \Omega = \psi(Q(\varepsilon))$. In this case, X and Y point to opposite sides of the s-axis for $s < 0$ and to the same side for $s > 0$ (see Fig. 2.16). Thus the set $\mathscr{S}_- = \{(s,0) : s < 0\}$ is a slow singular arc that saturates with $u_{\text{sing}}(p) = +1$ at p and $\mathscr{S}_+ = \{(s,0) : s > 0\}$ is inadmissible. Partition Q into three regions R_0, R_1, and R_2 according to the following specifications:

$$R_0 = \{(s,t) \in Q : t > y(s)\},$$
$$R_1 = \{(s,t) \in Q : s < 0, \, t < y(s)\},$$

and

$$R_2 = \{(s,t) \in Q : s > 0, \, t < y(s)\}.$$

Thus R_0 is the set above the integral curve Υ, and the region below this curve is divided further into its components in $\{s < 0\}$ and $\{s > 0\}$ with the boundaries given by the trajectory Υ and the negative t-axis, $\{(s,t) \in Q : s = 0, t < 0\}$. Since $X \cong (0,1)^T = \frac{\partial}{\partial t}$ is vertical, X-trajectories cross Υ into R_0. Once there, since $R_0 \subset \Omega_+$, at most one switching from X to Y can occur in R_0 and thus trajectories cannot leave R_0 forward in time as long as they are contained in Ω. If an optimal trajectory were to switch from X to Y on the curve Υ, then another junction with X is possible only at p followed possibly by one more switch to Y. It will follow from our argument below that no prior switchings can exist in this case, and overall, such a trajectory is at most of type $XYXY$.

The switchings in R_1 and R_2 can be analyzed with the tools developed in the proof of Proposition 2.9.5. By choosing ε small enough, we can assume that the function ζ constructed in Lemma 2.9.3 exists on $Q(\varepsilon)$ with the properties specified there. It was shown in the proof of Proposition 2.9.5 that YXY-trajectories are not optimal if the component b in the vector field Y is negative over the neighborhood $Q(\varepsilon)$, but under assumption (A2) the function b vanishes at p, and we first need to analyze its zero set in Q.

We first show that under assumption (A2), we have that

$$b(0,0) = 0 \quad \text{and} \quad \frac{\partial b}{\partial s}(s,0) > 0 \quad \text{for all } s \in [-\varepsilon, \varepsilon].$$

For recall from the proof of Proposition 2.9.5 that $L_X \alpha = \frac{\partial \alpha}{\partial t}$ and

$$L_Y \alpha(s,t) = \frac{\partial \alpha}{\partial s}(s,t) \cdot a(s,t) + \frac{\partial \alpha}{\partial t}(s,t) \cdot b(s,t).$$

The singular curve $\mathscr{S}$ is given by the s-axis, $\alpha(s,0) \equiv 0$, and therefore $\frac{\partial \alpha}{\partial s}(s,0) \equiv 0$. Hence we have along the s-axis that

$$L_Y \alpha(s,0) = L_X \alpha(s,0) \cdot b(s,0), \tag{2.57}$$

and thus, under assumption (A2), it follows that $b(0,0) = 0$. Differentiating Eq. (2.57) once more along the vector field Y, we get that

$$L_Y^2 \alpha(s,0) = L_Y L_X \alpha(s,0) \cdot b(s,0) + L_X \alpha(s,0) \cdot \left(\frac{\partial b}{\partial s}(s,0) \cdot a(s,0) + \frac{\partial b}{\partial t}(s,0) \cdot b(s,0) \right)$$

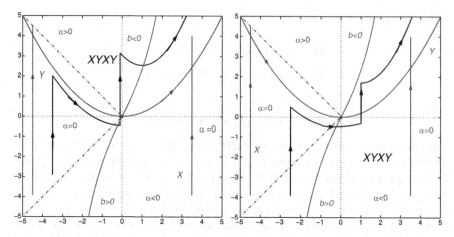

Fig. 2.17 Optimal $XYXY$-trajectories for $L_X\alpha > 0$

and therefore, upon evaluation at $p \cong (0,0)$,

$$L_Y^2\alpha(0,0) = L_X\alpha(0,0) \cdot \frac{\partial b}{\partial s}(0,0) \cdot a(0,0).$$

Hence, with our normalization of a to be positive, we get that $\frac{\partial b}{\partial s}(0,0) > 0$. In particular, $b(s,0)$ is negative for $s < 0$ and positive for $s > 0$. It follows from the implicit function theorem that the equation $b(s,t) = 0$ can be solved in terms of a differentiable function $s = \sigma(t)$, $\sigma(0) = 0$, in a neighborhood of the origin. By making ε smaller, if necessary, we may assume that the function σ is defined over the full interval $[-\varepsilon, \varepsilon]$. Furthermore, the graph of σ is transversal to the integral curve Υ of Y at p (see Figs. 2.16 and 2.17).

We also need to know the signs of the Lie derivative of the function ζ along the vector field Y. By definition,

$$L_Y\zeta(s,t) = \frac{\partial\zeta}{\partial s}(s,t)a(s,t) + \frac{\partial\zeta}{\partial t}(s,t)b(s,t),$$

and it follows from Lemma 2.9.3 that $\zeta(s,0) \equiv 0$ and $\frac{\partial\zeta}{\partial t}(s,0) \equiv -1$. In particular, all partial derivatives of ζ with respect to s vanish along $s = 0$. Hence we have that $L_Y\zeta(0,0) = 0$ and for $s < 0$,

$$L_Y\zeta(s,0) = \frac{\partial\zeta}{\partial s}(s,0)a(s,0) + \frac{\partial\zeta}{\partial t}(s,0)b(s,0) = -b(s,0) > 0.$$

Furthermore, with $b(0,0) = 0$, the second Lie derivative with respect to Y at the origin simplifies to

$$L_Y^2 \zeta(0,0) = \frac{\partial \zeta}{\partial t}(0,0)\frac{\partial b}{\partial s}(0,0)a(0,0) = -\frac{\partial b}{\partial s}(0,0) < 0.$$

Thus the zero set $Z = \{(s,t) : L_Y \zeta(s,t) = 0\}$ of the Lie derivative $L_Y \zeta$ near the origin is a one-dimensional embedded submanifold that is transversal to the singular curve $\mathscr{S} = \{(s,0) : |s| \le \varepsilon\}$.

Lemma 2.10.1. *YXY-trajectories that lie in the closure of R_1 are not optimal.*

Proof. Since the zero sets of b and $L_Y \zeta$ are transversal to Υ at p, it follows that there exists an open sector $V = \{(s,t) \in Q : s < 0,\ t < 2\omega|s|\}$ that lies entirely in the set $\{(s,t) \in Q : b(s,t) < 0,\ L_Y \zeta(s,t) > 0\}$. Since Y is tangent to the s-axis at p, by making ε smaller if necessary, we may assume that the curve Υ for $s < 0$ lies entirely in the smaller sector $W = \{(s,t) \in Q : s < 0,\ t < \omega|s|\}$ (see Fig. 2.17). Now consider a YXY-trajectory that lies in R_1 and suppose it has switchings at the points $(\tilde{s},\tilde{t})$ and $(\tilde{s},\tilde{t}')$, respectively. If this trajectory is optimal, then the two junctions are g-dependent along X and we have that $\tilde{t}' = \zeta(\tilde{s},\tilde{t})$. Since junctions of the type XY are optimal only in Ω_+, we have that $(\tilde{s},\tilde{t}') \in W$. It follows from Lemma 2.9.3 that

$$\tilde{t}' = \zeta(\tilde{s},\tilde{t}) = \frac{\partial \zeta}{\partial t}(\tilde{s},0)\tilde{t} + o(\tilde{t}) = -\tilde{t} + o(\tilde{t}).$$

(We have $\zeta(s,0) \equiv 0$, and thus all derivatives in s vanish identically.) But then, for ε small enough, the first junction point $(\tilde{s},\tilde{t})$ still must lie in the larger sector V where the Lie derivative $L_Y \zeta$ is positive, and by construction the second junction point lies in the region where b is negative. It follows from the remark following Lemma 2.9.4 that an envelope can be constructed, and thus this trajectory cannot be optimal. □

In particular, if there is an XY-junction on the curve Υ, then there could not have been a previous YX-junction. Hence, as claimed earlier, such a trajectory can be at most of type $XYXY$.

The remainder of the argument follows from a direct geometric analysis of X and Y trajectories. It is possible that optimal trajectories are of the type $XYXY$ in $\{s < 0\}$, but then the last switching must lie above Υ in R_0, and overall such a trajectory cannot switch any more. Trajectories that do not cross Υ in $\{s < 0\}$ are at most concatenations of type XY in $\{s < 0\}$. If they switch to X at $s = 0$, then once more, only one additional switch to Y in $\{t > 0\}$ is possible. If they cross $\{s = 0\}$ along Y, then it is possible to have a switch to X in the fourth quadrant $\{(s,t) \in Q : s > 0,\ t < 0\}$ and one more switch to Y in the first quadrant $\{(s,t) \in Q : s > 0,\ t > 0\}$ (cf., Fig. 2.17). In any case, an optimal controlled bang-bang trajectory that lies in Ω can have at most the concatenation sequence $XYXY$. □

Proposition 2.10.2. *Let Ω be a domain on which condition (A2) is satisfied and suppose $L_X(\alpha) = X\alpha$ is negative everywhere on Ω. Then, for Ω sufficiently small, optimal controlled trajectories that lie entirely in Ω are at most concatenations of type $XYSB$.*

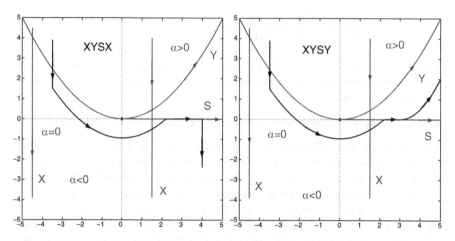

Fig. 2.18 Optimal $XYSB$-trajectories for $L_X \alpha < 0$

Proof. This is the easier case, and the result follows by a direct geometric reasoning from the codimension-1 scenarios. As above, consider an X-aligned chart of coordinates (centered at p), $\psi : Q(\varepsilon) \to \Omega = \psi(Q(\varepsilon))$. In this case X and Y point to the same sides of the s-axis for $s < 0$ and to opposite sides for $s > 0$. Furthermore, now $\mathscr{S}_+ = \{(s,0) : s > 0\}$ is a fast singular arc (that again saturates with $u_{\text{sing}}(p) = +1$ at p) and $\mathscr{S}_- = \{(s,0) : s < 0\}$ is inadmissible. By choosing Q small enough, it follows from Proposition 2.9.2 that optimal controlled trajectories that lie in $Q_- = \{(s,t) : s < 0\}$, the second and third quadrants, are at most of type XYX, and by Proposition 2.9.4, optimal controlled trajectories that lie in $Q_+ = \{(s,t) : s > 0\}$, the first and fourth quadrants, are at most of type BSB. However, overall, at most the concatenation sequence $XYSB$ is possible. For only Y-trajectories can cross from Q_- into Q_+, and XY-junctions are optimal only in Ω_+, the first and second quadrants, while YX-junctions are optimal only in Ω_-, the third and forth quadrants. Therefore, if a Y-trajectory crosses the t-axis at a positive value, then no further switching is possible, and such a trajectory can be at most of type XY. If it crosses for $t = 0$ and does not switch at p, the same is true. If it switches at p to a singular arc, then only SB is possible afterward, and this limits the concatenation sequence to $XYSB$. If a switch to X occurs at p, then again no further switches are possible, and such a trajectory is at most of type XYX. Finally, if the crossing happens for $t < 0$, then it is possible to switch to X in the forth quadrant (or also on the t-axis itself), and again in such a case we get at most XYX. If there is no switch to X, then the Y-trajectory may reach the singular arc and switch there, ending up with an SB concatenation. Overall, because of the directions of the vector fields X and Y near p, only concatenations of type $XYSB$ can be optimal (see Fig. 2.18).   □

Altogether, we have shown the following result:

Theorem 2.10.1. *Let p be a point where the vector fields $f(p)$ and $g(p)$ are linearly independent. Then, under generic conditions on the vector fields f and g, there exists* *a neighborhood Ω of p such that optimal controlled trajectories that lie entirely in Ω are concatenations of at most four pieces of either $X = f - g$, $Y = f + g$, or the singular arc S. At most one of these pieces can be a singular arc, and if there are four segments, then it must be the second or third leg in the concatenation sequence.* ∎

In all the examples considered here, there is a very simple relation between the number of X, Y, and singular segments in concatenation sequences that lie in a sufficiently small neighborhood Ω of some reference point p and what is called the codimension of the *Lie-bracket configuration* of the system $\Sigma = (f, g)$ at the point p that we still briefly want to point out. Loosely speaking, this Lie-bracket configuration consists of all the values of the vector fields f and g and their Lie brackets at p, and its *codimension* is given by the number of linearly independent "relevant" equality relations that hold between these vector fields at p. We are assuming that f and g are linearly independent on Ω and thus always can express the Lie bracket as $[f, g] = \alpha f + \beta g$ with some smooth functions α and β defined on Ω. In this case, the first "relevant" relation is that g and $[f, g]$ are linearly dependent at p, characterized by $\alpha(p) = 0$. If α does not vanish on Ω, the codimension-0 case, optimal controls are simply bang-bang with one switching, and the sign of α determines the order of the switchings. If α does vanish at p, higher-order terms in the Taylor expansion of α along the flows of X and Y at p matter, and depending on whether these Lie derivatives of α vanish at p, more degenerate scenarios arise. In the codimension-1 cases, characterized by the fact that both Lie derivatives of α along X and Y do not vanish at p, only three segments are possible. If we allow that one of the Lie derivatives vanishes, but again in a nondegenerate way, so that its second Lie derivative is nonzero, the codimension-2 case, this number increases to four. Overall, in each case we have the following simple relation:

Σ_p: The maximum number of concatenations of X, Y, and singular segments in time-optimal controlled trajectories that lie in a sufficiently small neighborhood Ω of some reference point p is given by

$$2 + \text{codim}(\Sigma_p) = \dim \Omega + \text{codim}(\Sigma_p).$$

This relation has also been verified for numerous cases of low codimension in dimensions 3 and 4 (e.g., see [210, 211, 221]). For example, the possible concatenation sequences BBB and BSB that arise in the codimension-1 cases in the plane are precisely the time-optimal concatenation sequences in the codimension-0 three-dimensional case (see Sect. 7.3), and the optimal sequences $BBBB$, $BBSB$, and $BSBB$ for the codimension-2 case in the plane are the optimal sequences for the codimension-1 cases in $\mathbb{R}^3$ (see Sect. 7.5) and the codimension-0 cases in $\mathbb{R}^4$. This is very much like the *unfolding of singularities* in the theory of differentiable mappings [108]. Thus, a *general classification of the concatenation sequences that optimal controlled trajectories* for planar systems *can have locally* in more degenerate cases *based on Lie-theoretic conditions* is not merely of intrinsic interest, but it also points to the structures of optimal solutions in higher dimensions. We shall return to this topic in Chap. 7. In the next section, we shall analyze another classical optimal control problem in which the codimension of the Lie-bracket configuration becomes infinite, and indeed, optimal trajectories require an infinite number of switchings on a finite interval and thus are no longer piecewise continuous.

Examples of these correspondences abound not only for the time-optimal control problem, but in general. For example, in Sect. 6.2, we shall consider a three-dimensional optimal control problem for a mathematical model for tumor anti-angiogenesis [160] in which, because of the presence of optimal saturating singular controls, the solution is fully characterized by the concatenation sequences determined here for the codimension-2 scenario. Indeed, the optimal concatenation structures encountered for the time-optimal control problem in the plane that were analyzed in the last two sections consistently reappear in optimal solutions for general optimal control problems in increasing dimensions.

2.11   Chattering Arcs: The Fuller Problem

The Fuller problem has its origin in electronics, arising in communication across a nonlinear channel [34, 35, 94]. In this section, we give a solution to this problem, an innocent-looking problem whose optimal controlled trajectories are chattering arcs for which the controls switch infinitely often on an arbitrarily small interval as the switchings accumulate at the final time. In particular, optimal controls are no longer piecewise continuous, but lie in the class of Lebesgue measurable functions. The reason for this behavior lies in the presence of an optimal singular arc of order 2.

[Fuller]   Given a point $p \in \mathbb{R}^2$, find a control (Lebesgue measurable function) with values in the interval $[-1, 1]$ that steers p into the origin under the dynamics

$$\dot{x}_1 = x_2, \quad \dot{x}_2 = u,$$

and minimizes the objective

$$J(u) = \frac{1}{2} \int_0^T x_1^2(t) dt.$$

The time T of transfer is finite, but otherwise free. Since the problem is time-invariant, we can arbitrarily shift the interval of definition for the control, and for this problem it is more convenient to normalize the terminal time to be 0. We thus consider the controls and trajectories to be defined over intervals $[-T, 0] \subset (-\infty, 0]$.

Theorem 2.11.1. Let $\zeta = \sqrt{\frac{\sqrt{33}-1}{24}} = 0.4446236\ldots$, the unique positive root of the equation $z^4 + \frac{1}{12}z^2 - \frac{1}{18} = 0$, and define

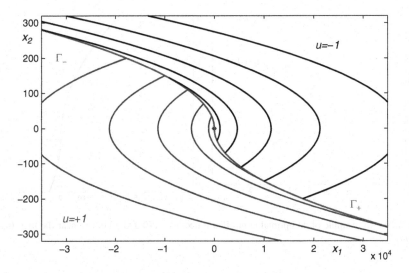

Fig. 2.19 Optimal synthesis for the Fuller problem

$$\Gamma_+ = \{(x_1, x_2) \in \mathbb{R}^2 : x_1 = \zeta x_2^2, \, x_2 < 0\},$$

$$\Gamma_- = \{(x_1, x_2) \in \mathbb{R}^2 : x_1 = -\zeta x_2^2, \, x_2 > 0\},$$

$$G_+ = \{(x_1, x_2) \in \mathbb{R}^2 : x_1 < -sgn(x_2)\zeta x_2^2\},$$

$$G_- = \{(x_1, x_2) \in \mathbb{R}^2 : x_1 > -sgn(x_2)\zeta x_2^2\}.$$

Then, the optimal control for the Fuller problem is given in feedback form as

$$u_*(x) = \begin{cases} +1 & \text{for} \quad x \in G_+ \cup \Gamma_+, \\ -1 & \text{for} \quad x \in G_- \cup \Gamma_-. \end{cases} \tag{2.58}$$

Corresponding trajectories cross the switching curves Γ_+ and Γ_- transversally, changing from $u = -1$ to $u = +1$ at points on Γ_+ and from $u = +1$ to $u = -1$ at points on Γ_-. These trajectories are chattering arcs with an infinite number of switchings that accumulate with a geometric progression at the final time $T = 0$.

Figures 2.18 and 2.19 depict the optimal synthesis for the Fuller problem. It looks very much like the synthesis for the double integrator, but with the significant difference that the switching curve $\Gamma = \Gamma_+ \cup \{(0,0)\} \cup \Gamma_-$ now is *not* a trajectory. Thus trajectories always cross Γ and cannot enter the origin along these curves.

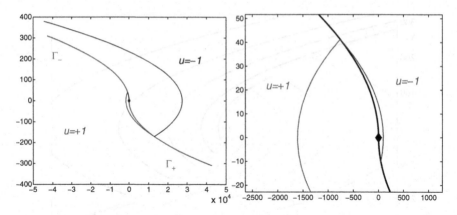

Fig. 2.20 An example of an optimal controlled trajectory (*left*) and a blowup near the final time (*right*)

2.11.1   The Fuller Problem as a Time-Optimal Control Problem in $\mathbb{R}^3$

The reason for the occurrence of the chattering controls is best understood if one embeds the Fuller problem into a time-optimal control problem of the form [NTOC] in $\mathbb{R}^3$ by adding the objective as a third variable, $\dot{x}_3 = \frac{1}{2}x_1^2$, i.e., the drift vector field f and control vector field g are given by

$$f(x) = \begin{pmatrix} x_2 \\ 0 \\ \frac{1}{2}x_1^2 \end{pmatrix} \quad \text{and} \quad g(x) = \begin{pmatrix} 0 \\ 1 \\ 0 \end{pmatrix}.$$

If one then considers the time-optimal control problem to the origin, the classical Fuller problem arises for initial conditions of the form $p = (x_1^0, x_2^0, -J(x_1^0, x_2^0))$, where $J(x_1^0, x_2^0)$ is the optimal value for the Fuller problem with initial condition (x_1^0, x_2^0). It will be seen that the solution to the Fuller problem is unique, and thus there exists exactly one control that steers p into the origin. Hence this control is the time-optimal one. In fact, the solutions to the Fuller problem are optimal abnormal extremals for this three-dimensional time-optimal control problem: the Hamiltonian for the Fuller problem is given by

$$H = \frac{1}{2}\lambda_0 x_1^2 + \lambda_1 x_2 + \lambda_2 u$$

with $\lambda_0 \geq 0$, while the Hamiltonian for the time-optimal control problem, where we change the notation for the multiplier to ψ in order to distinguish these two formulations, is given by

$$H = \psi_0 + \psi_1 x_2 + \psi_2 u + \frac{1}{2}\psi_3 x_1^2.$$

We shall see below that extremals for the Fuller problem cannot be abnormal ($\lambda_0 > 0$), and for the time-optimal control problem, ψ_3 is a constant that cannot vanish if $\psi_0 = 0$ with time-minimizing extremals corresponding to $\psi_3 > 0$ and maximizing ones to $\psi_3 < 0$. Normalizing $\psi_3 = 1$ and taking $\psi_0 = 0$, the conditions of the maximum principle for these two problems agree.

The Lie brackets of the vector fields f and g are easily computed as

$$[f,g](x) \equiv \begin{pmatrix} -1 \\ 0 \\ 0 \end{pmatrix}, \quad [f,[f,g]](x) = \begin{pmatrix} 0 \\ 0 \\ x_1 \end{pmatrix}, \quad \text{and} \quad [g,[f,g]](x) \equiv 0.$$

Since $[g,[f,g]]$ vanishes identically, so do the brackets $[f,[g,[f,g]]]$ and $[g,[g,[f,g]]]$, and singular controls are of higher order. The other relevant fourth- and fifth-order brackets are

$$\text{ad}_f^3\, g(x) = \begin{pmatrix} 0 \\ 0 \\ x_2 \end{pmatrix}, \quad \text{ad}_f^4\, g(x) \equiv 0, \quad \text{and} \quad [g,\text{ad}_f^3\, g](x) \equiv \begin{pmatrix} 0 \\ 0 \\ 1 \end{pmatrix}.$$

In particular,

$$\langle \psi, [g,\text{ad}_f^3\, g](x)\rangle = \psi_3 = 1 > 0,$$

and the Kelley condition for optimality of an order-2 singular arc is satisfied. The equation defining the singular control is

$$\Phi^{(4)}(t) = \psi \text{ad}_f^4\, g(x) + u\psi[g,\text{ad}_f^3\, g](x) = u$$

and thus the singular control is given by

$$u_{\text{sing}} \equiv 0.$$

The corresponding singular extremal Γ_F is therefore given by $u \equiv 0$, $x_1 \equiv x_2 \equiv 0$, with multipliers $\psi_0 = \psi_1 = \psi_2 = 0$ and $\psi_3 \equiv 1$. The classical Fuller problem can thus be interpreted as the problem of steering a point in $\mathbb{R}^3$ time-optimally into an order-2 singular arc that satisfies the Kelley condition. By Proposition 2.8.5, the singular control cannot be concatenated with a constant bang control without violating the necessary conditions of the maximum principle. This can be accomplished only by means of a chattering control.

2.11.2 Elementary Properties of Extremals

We now construct an extremal synthesis for the Fuller problem following an argument of Kupka [143]. (The optimality of this synthesis will be verified in Sects. 5.1 and 5.2.3 by means of two completely different arguments.) Let (x, u) be an optimal controlled trajectory that transfers p into the origin and minimizes the integral $\int_0^T x_1^2(t)dt$. By Theorem 2.2.1, there exist a constant $\lambda_0 \geq 0$ and an adjoint vector $\lambda = (\lambda_1, \lambda_2)$ such that (i) $(\lambda_0, \lambda_1, \lambda_2)$ do not vanish simultaneously, (ii) $\dot{\lambda}_1 = -\lambda_0 x_1$, $\dot{\lambda}_2 = -\lambda_1$, and (iii) the control minimizes the Hamiltonian $H = \frac{1}{2}\lambda_0 x_1^2 + \lambda_1 x_2 + \lambda_2 u$ over the interval $[-1, 1]$ with the minimum value identically zero.

Lemma 2.11.1. *Extremals of optimal controlled trajectories are normal.*

Proof. Suppose $\lambda_0 = 0$. The switching function Φ is given by the multiplier λ_2, and in this case $\dot{\lambda}_2 = 0$. Hence the corresponding control u is bang-bang with at most one switching ending with either $u = +1$ or $u = -1$. But this contradicts Proposition 2.8.5. For if we define a new control $\breve{u}$ by adding an interval $[T, T+\varepsilon]$ with control $\breve{u}(t) \equiv 0$ on this interval, then the value of the objective does not change under this extension, and thus $\breve{u}$ is optimal as well. But the final segment with $u = 0$ is a singular arc of order 2, and thus it cannot be concatenated optimally with a bang control. Contradiction. $\square$

We henceforth normalize $\lambda_0 = 1$. Then the derivatives of the switching function $\Phi = \lambda_2$ are given by

$$\dot{\Phi}(t) = -\lambda_1(t), \quad \ddot{\Phi}(t) = x_1(t), \quad \Phi^{(3)}(t) = x_2(t), \quad \Phi^{(4)}(t) = u(t),$$

and the minimum condition implies

$$u(t) = -\operatorname{sgn}\Phi(t).$$

In particular, the switching function is a solution to the nonsmooth differential equation $\Phi^{(4)}(t) = -\operatorname{sgn}\Phi(t)$. We start with some elementary properties of extremals.

Lemma 2.11.2. *Let $u = \pm 1$; then the functions*

$$I_{1,\pm} = x_1 - \frac{1}{2}ux_2^2 \quad \text{and} \quad I_{2,\pm} = -\lambda_1 - ux_1x_2 + \frac{1}{3}x_2^3$$

are first integrals for the extremals of the Fuller problem. That is, the functions $I_{1,\pm}$ and $I_{2,\pm}$ are constant along extremals for the controls $u = \pm 1$.

Proof. This follows by direct differentiation from the system and adjoint equations

$$\dot{l}_{1,\pm} = \dot{x}_1 - ux_2\dot{x}_2 = x_2 - u^2x_2 = 0,$$

$$\dot{l}_{2,\pm} = -\dot{\lambda}_1 - u\dot{x}_1x_2 - ux_1\dot{x}_2 + x_2^2\dot{x}_2 = x_1 - ux_2^2 - u^2x_1 + x_2^2u = (1 - u^2)x_1 = 0.$$

$\square$

Lemma 2.11.3. *Let* $\Gamma = ((x,u),\lambda)$ *be an extremal defined over the interval* $[-T,0]$. *If* $\tau < 0$ *is a switching time, then* τ *is an isolated zero of the switching function, and a bang-bang switch occurs at time* τ. *This switch is from* $u = +1$ *to* $u = -1$ *if* $x_2(\tau) > 0$ *and from* $u = -1$ *to* $u = +1$ *if* $x_2(\tau) < 0$.

Proof. Suppose $\Phi(\tau) = \lambda_2(\tau) = 0$. It is clear that a bang-bang switch occurs if $\dot{\lambda}_2(\tau) = -\lambda_1(\tau)$ does not vanish. If $\lambda_1(\tau) = 0$ as well, then the condition

$$0 = H(\tau) = \frac{1}{2}x_1^2(\tau) + \lambda_1(\tau)x_2(\tau) \tag{2.59}$$

implies that $x_1(\tau) = 0$, and thus, since the junction point is not the origin, we have $x_2(\tau) \neq 0$. But then $\Phi(\tau) = \dot{\Phi}(\tau) = \ddot{\Phi}(\tau) = 0$ and

$$\Phi^{(3)}(\tau) = x_2(\tau) \neq 0.$$

Thus the switching function changes from negative to positive if $x_2(\tau) > 0$ and from positive to negative if $x_2(\tau) > 0$ and the corresponding bang-bang switch occurs. For the case $\lambda_1(\tau) \neq 0$, the same structure follows, since $x_2(\tau)$ and $\lambda_1(\tau)$ have opposite signs by (2.59). $\square$

2.11.3 Symmetries of Extremals

The family of all extremals possesses two groups of symmetries, one continuous, the other discrete, which can be used very much to advantage in calculating the extremal synthesis. Without loss of generality, we define all extremals over the full interval $(-\infty, 0]$ with the terminal time T normalized to be $T = 0$. Let $\mathscr{G}_\alpha$ denote the multiplicative group of positive reals and define a 1-parameter group of scaling symmetries on $(-\infty, 0] \times [-1, 1] \times \mathbb{R}^2 \times (\mathbb{R}^2)^*$ by

$$\mathscr{G}_\alpha : (t, u, x_1, x_2, \lambda_1, \lambda_2) \mapsto \left(\frac{t}{\alpha}, \alpha^0 u, \alpha^2 x_1, \alpha x_2, \alpha^3 \lambda_1, \alpha^4 \lambda_2 \right).$$

Proposition 2.11.1. *Given an extremal lift* $\Gamma = ((x,u),\lambda)$ *for the Fuller problem and* $\alpha > 0$, *define* $\Gamma^\alpha = ((x^\alpha, u^\alpha), \lambda^\alpha)$ *as the controlled trajectory* (x^α, u^α) *and corresponding adjoint vector* λ^α *that are obtained under the action of the group* $\mathscr{G}_\alpha$ *on the variables; that is, by*

$$u^\alpha(t) = u\left(\frac{t}{\alpha}\right), \quad x_1^\alpha(t) = \alpha^2 x_1\left(\frac{t}{\alpha}\right), \quad x_2^\alpha(t) = \alpha x_2\left(\frac{t}{\alpha}\right),$$

and

$$\lambda_1^\alpha(t) = \alpha^3 \lambda_1\left(\frac{t}{\alpha}\right), \quad \lambda_2^\alpha(t) = \alpha^4 \lambda_2\left(\frac{t}{\alpha}\right).$$

Then Γ^α again is an extremal for the Fuller problem.

Proof. Consider the controlled trajectory (x, u) over the interval $[-\bar{t}, 0]$ with initial condition at time $-\bar{t}$ given by $(\bar{x}_1, \bar{x}_2)$. The rescaled control u^α, restricted to $[-\alpha\bar{t}, 0]$, then steers $(\bar{x}_1^\alpha, \bar{x}_2^\alpha) = (\alpha^2 \bar{x}_1, \alpha\bar{x}_2)$ into the origin with corresponding trajectory x^α. A direct calculation verifies that the adjoint equation is invariant under this transformation as well,

$$\dot{\lambda}_1^\alpha(t) = \alpha^3 \dot{\lambda}_1\left(\frac{t}{\alpha}\right)\frac{1}{\alpha} = -\alpha^2 x_1\left(\frac{t}{\alpha}\right) = -x_1^\alpha(t),$$

$$\dot{\lambda}_2^\alpha(t) = \alpha^4 \dot{\lambda}_2\left(\frac{t}{\alpha}\right)\frac{1}{\alpha} = -\alpha^3 \lambda_1\left(\frac{t}{\alpha}\right) = -\lambda_1^\alpha(t),$$

and also the Hamiltonian H remains unchanged:

$$H(\lambda^\alpha(t), x^\alpha(t), u^\alpha(t))$$

$$= \frac{1}{2}x_1^\alpha(t)^2 + \lambda_1^\alpha(t)x_2^\alpha(t) + \lambda_2^\alpha(t)u^\alpha(t)$$

$$= \frac{1}{2}\left[\alpha^2 x_1\left(\frac{t}{\alpha}\right)\right]^2 + \alpha^3 \lambda_1\left(\frac{t}{\alpha}\right)\alpha x_2\left(\frac{t}{\alpha}\right) + \alpha^4 \lambda_2\left(\frac{t}{\alpha}\right)u\left(\frac{t}{\alpha}\right)$$

$$= \alpha^4 H\left(\lambda\left(\frac{t}{\alpha}\right), x\left(\frac{t}{\alpha}\right), u\left(\frac{t}{\alpha}\right)\right) = 0.$$

Furthermore, by construction, the minimum condition on the control carries over from the extremal lift Γ. Hence Γ^α is an extremal as well.   □

A second symmetry is given by reflecting controlled trajectories and their multipliers at the origin, in mathematical terms, by the action of the discrete group S_2. Let $\mathscr{R}$ denote the reflection symmetry defined on $(-\infty, 0] \times [-1, 1] \times \mathbb{R}^2 \times (\mathbb{R}^2)^*$ by

$$\mathscr{R} : (t, u, x_1, x_2, \lambda_1, \lambda_2) \mapsto (t, -u, -x_1, -x_2, -\lambda_1, -\lambda_2).$$

Proposition 2.11.2. *Given an extremal lift $\Gamma = ((x, u), \lambda)$ for the Fuller problem, define $\check{\Gamma} = ((\check{x}, \check{u}), \check{\lambda})$ as the controlled trajectory $(\check{x}, \check{u})$ and corresponding adjoint vector $\check{\lambda}$ that are obtained under the action of $\mathscr{R}$, that is, by*

$$\check{u}(t) = -u(t), \quad \check{x}_1(t) = -x_1(t), \quad \check{x}_2(t) = -x_2(t),$$

and

$$\check{\lambda}_1(t) = -\lambda_1(t), \quad \check{\lambda}_2(t) = -\lambda_2(t).$$

Then $\check{\Gamma}$ again is an extremal for the Fuller problem.

Proof. It is clear that all the conditions of the maximum principle are invariant under this transformation.   □

2.11.4 A Synthesis of Invariant Extremals

Whenever a mathematical problem exhibits symmetries, it is a good strategy to seek solutions that obey these symmetries. In fact, there is one extremal that is invariant under the action of all symmetries $\mathscr{G}_\alpha$ for all $\alpha > 0$ and $\mathscr{R}$, namely the trivial solution for $u \equiv 0$ with $x \equiv 0$ and $\lambda \equiv 0$. (The nontriviality condition is satisfied by $\lambda_0 = 1$.) In some sense, this is responsible for the special properties of trajectories that need to steer the system into the origin. But there also exists a specific value α for which *all* extremals are invariant (as individual curves, not just as the whole family) under the actions of $\mathscr{R}$ and $\mathscr{G}_\alpha$. These are the optimal controlled trajectories for the Fuller problem, and we now calculate this value.

Let $\Gamma = ((x,u),\lambda)$ be an extremal for the Fuller problem and suppose $t_0 < 0$ is a switching time where the control switches from $u = +1$ to $u = -1$. Since the switchings are isolated, but must accumulate for $T = 0$, there exists a sequence $\{t_n\}_{n\in\mathbb{Z}}$ of switching times that converges to 0 as $n \to \infty$ and the control switches from $u = +1$ to $u = -1$ at even indices and from $u = -1$ to $u = +1$ at odd indices. Let $\check{\Gamma}_\alpha = ((\check{x}^\alpha,\check{u}^\alpha),\check{\lambda}^\alpha)$ denote the image of the extremal Γ under the combined action $\mathscr{A}_\alpha$ of the reflection $\mathscr{R}$ and the group $\mathscr{G}_\alpha$ for a fixed $\alpha > 0$, i.e., for all $t \leq 0$,

$$\check{u}^\alpha(t) = -u\left(\frac{t}{\alpha}\right), \quad \check{x}_1^\alpha(t) = -\alpha^2 x_1\left(\frac{t}{\alpha}\right), \quad \check{x}_2^\alpha(t) = -\alpha x_2\left(\frac{t}{\alpha}\right),$$

and

$$\check{\lambda}_1^\alpha(t) = -\alpha^3 \lambda_1\left(\frac{t}{\alpha}\right), \quad \check{\lambda}_2^\alpha(t) = -\alpha^4 \lambda_2\left(\frac{t}{\alpha}\right).$$

By Propositions 2.11.1 and 2.11.2, $\check{\Gamma}_\alpha = \mathscr{A}_\alpha(\Gamma)$ again is an extremal, but generally it will be different from Γ. If the extremals Γ and $\check{\Gamma}_\alpha$ are the same, i.e., if $\Gamma(t) = \check{\Gamma}_\alpha(t)$ for all $t \leq 0$, then the extremal is a *fixed point* under this transformation, and we say that it is invariant under this action. Note that if Γ is $\mathscr{A}_\alpha$-invariant, then it is also invariant under the action of any odd power α^{2k+1} for all $k \in \mathbb{Z}$. But there always exists a smallest $\alpha > 1$, and this number will be called the generator.

Proposition 2.11.3. *let $\Gamma = ((x,u),\lambda)$ be an extremal for the Fuller problem defined over the semi-infinite interval $(-\infty,0]$ with switching times $\{t_n\}_{n\in\mathbb{Z}}$ and suppose the control switches from $u = -1$ to $u = +1$ for even indices. If the extremal Γ is invariant under the combined action $\mathscr{A}_\alpha$ of the reflection $\mathscr{R}$ and the group $\mathscr{G}_\alpha$ with generator α, i.e., if $\Gamma(t) = \check{\Gamma}_\alpha(t)$ for all $t \leq 0$, then*

$$\alpha = \sqrt{\frac{1+2\zeta}{1-2\zeta}}, \qquad \text{where} \qquad \zeta = \sqrt{\frac{\sqrt{33}-1}{24}}.$$

The switching points lie on the curves

$$\Gamma_+ = \{(x_1,x_2) \in \mathbb{R}^2 : x_1 = \zeta x_2^2, \; x_2 < 0\}$$

and

$$\Gamma_- = \{(x_1,x_2) \in \mathbb{R}^2 : x_1 = -\zeta x_2^2, \; x_2 > 0\},$$

and switchings are from $X = f - g$ to $Y = f + g$ at points on Γ_+ and from Y to X at points on Γ_-.

Proof. The invariance condition and the choice of α as the generator imply that the switching times t_i follow a geometric progression, $t_{i-1} = \alpha t_i$, $i \in \mathbb{Z}$. Starting at the switching time t_0 and integrating the control $u = +1$ until the time $t_1 = \frac{t_0}{\alpha}$, using the first integral $I_{1,+}$, we obtain that

$$x_1(t_1) - \frac{1}{2}x_2^2(t_1) = x_1(t_0) - \frac{1}{2}x_2^2(t_0) \tag{2.60}$$

and thus

$$\frac{x_1(t_1)}{x_1(t_0)} - 1 = \frac{1}{2}\frac{x_2^2(t_1) - x_2^2(t_0)}{x_1(t_0)} = \frac{1}{2}\left(\frac{x_2^2(t_1)}{x_2^2(t_0)} - 1\right)\frac{x_2^2(t_0)}{x_1(t_0)}.$$

It follows from the invariance of the trajectory under the action of $\mathscr{G}_\alpha$ and $\mathscr{R}$ that

$$x_1(t_0) = \breve{x}_1^\alpha(t_0) = -\alpha^2 x_1\left(\frac{t_0}{\alpha}\right) = -\alpha^2 x_1(t_1)$$

and

$$x_2(t_0) = \breve{x}_2^\alpha(t_0) = -\alpha x_2\left(\frac{t_0}{\alpha}\right) = -\alpha x_2(t_1). \tag{2.61}$$

In particular, $x_i(t_1)$ and $x_i(t_0)$ have opposite signs at consecutive switchings for both $i = 1,2$. Hence

$$\frac{x_1(t_1)}{x_1(t_0)} = -\frac{1}{\alpha^2} \qquad \text{and} \qquad \frac{x_2(t_1)}{x_2(t_0)} = -\frac{1}{\alpha}. \tag{2.62}$$

But then we get for the XY-junction at time t_0 that

$$-\frac{1}{\alpha^2} - 1 = \frac{1}{2}\left(\frac{1}{\alpha^2} - 1\right)\frac{x_2^2(t_0)}{x_1(t_0)}$$

or equivalently,

$$x_1(t_0) = \frac{1}{2}\frac{\alpha^2 - 1}{\alpha^2 + 1}x_2^2(t_0).$$

Also, by Lemma 2.11.3, $x_2(t_0)$ is negative.

Similarly, if we integrate $u = -1$ between the switching times t_1 and t_2, then we get from $I_{1,-}$ that

$$x_1(t_2) + \frac{1}{2}x_2^2(t_2) = x_1(t_1) + \frac{1}{2}x_2^2(t_1),$$

which then leads to

$$\frac{x_1(t_2)}{x_1(t_1)} - 1 = \frac{1}{2}\frac{x_2^2(t_1) - x_2^2(t_2)}{x_1(t_1)} = \frac{1}{2}\left(1 - \frac{x_2^2(t_2)}{x_2^2(t_1)}\right)\frac{x_2^2(t_1)}{x_1(t_1)}.$$

Analogous to Eq. (2.62), we also have that

$$\frac{x_1(t_2)}{x_1(t_1)} = -\frac{1}{\alpha^2} \quad \text{and} \quad \frac{x_2(t_2)}{x_2(t_1)} = -\frac{1}{\alpha},$$

and so it follows that

$$-\frac{1}{\alpha^2} - 1 = \frac{1}{2}\left(1 - \frac{1}{\alpha^2}\right)\frac{x_2^2(t_1)}{x_1(t_1)}.$$

Hence

$$x_1(t_1) = -\frac{1}{2}\frac{\alpha^2 - 1}{\alpha^2 + 1}x_2^2(t_1),$$

and Lemma 2.11.3 now implies that $x_2(t_0)$ is positive. Setting

$$\zeta = \frac{1}{2}\frac{\alpha^2 - 1}{\alpha^2 + 1} \in \left(0, \frac{1}{2}\right), \tag{2.63}$$

the formulas for the switching curves follow.

It remains to calculate the value for ζ. For the switching times t_0 and t_1 we have that

$$x_1(t_0) = \zeta x_2(t_0)^2 \quad \text{and} \quad x_1(t_1) = -\zeta x_2(t_1)^2$$

and thus, once more using the first integral $I_{1,+}$, we get from Eq. (2.60) that

$$I_{1,+}(t_1) = \left(\zeta - \frac{1}{2}\right)x_2^2(t_1) = \left(-\zeta - \frac{1}{2}\right)x_2^2(t_0)$$

or equivalently, by Eq. (2.62),

$$\left(\zeta - \frac{1}{2}\right) = \left(-\zeta - \frac{1}{2}\right)\alpha^2. \tag{2.64}$$

Similarly, using the first integral $I_{2,+}$, we also have that

$$-\lambda_1(t_1) - x_1(t_1)x_2(t_1) + \frac{1}{3}x_2^3(t_1) = -\lambda_1(t_0) - x_1(t_0)x_2(t_0) + \frac{1}{3}x_2^3(t_0). \tag{2.65}$$

It follows from the condition $H(t) \equiv 0$ that at every switching time t_i we have that

$$0 = \frac{1}{2}x_1^2(t_i) + \lambda_1(t_i)x_2(t_i)$$

and thus

$$\lambda_1(t_i) = -\frac{1}{2}\frac{x_1^2(t_i)}{x_2(t_i)} = -\frac{1}{2}\zeta^2 x_2^3(t_i).$$

Hence Eq. (2.65) becomes

$$\left(\frac{1}{2}\zeta^2 - \zeta + \frac{1}{3}\right)x_2^3(t_1) = \left(\frac{1}{2}\zeta^2 + \zeta + \frac{1}{3}\right)x_2^3(t_0),$$

and thus, again by Eq. (2.62),

$$\left(\frac{1}{2}\zeta^2 - \zeta + \frac{1}{3}\right) = -\left(\frac{1}{2}\zeta^2 + \zeta + \frac{1}{3}\right)\alpha^3. \tag{2.66}$$

Solving Eqs. (2.64) and (2.66) for α and equating the resulting expressions gives the following relation on ζ:

$$\frac{\left(\frac{1}{2}\zeta^2 - \zeta + \frac{1}{3}\right)^2}{\left(\zeta - \frac{1}{2}\right)^3} = \frac{\left(\frac{1}{2}\zeta^2 + \zeta + \frac{1}{3}\right)^2}{\left(-\zeta - \frac{1}{2}\right)^3}. \tag{2.67}$$

This expressions simplifies to the equation

$$\zeta^4 + \frac{1}{12}\zeta^2 - \frac{1}{18} = 0,$$

which has a unique positive solution given by

$$\zeta = \sqrt{\frac{\sqrt{33} - 1}{24}}.$$

The formula for α follows from Eq. (2.63). □

These calculations prove that if there exist extremal controlled trajectories that are invariant under the combined action $\mathscr{A}$ defined by the composition of the group actions $\mathscr{R}$ and $\mathscr{G}_\alpha$ for some α, then the generator is given by

$$\alpha = \sqrt{\frac{1 + 2\zeta}{1 - 2\zeta}} = 4.1301599\ldots, \tag{2.68}$$

and the trajectories are those corresponding to the synthesis defined in Theorem 2.11.1. It is not difficult to reverse these computations and show that this construction indeed gives rise to a family of $\mathscr{A}_\alpha$-invariant extremals.

Proposition 2.11.4. *The synthesis $\mathscr{F}$ defined in Theorem 2.11.1 generates a family of $\mathscr{A}_\alpha$-invariant extremals.*

Proof. Let $p > 0$ and consider the point $\gamma(p) = (\zeta p^2, -p)$ on the switching curve Γ_+. We first calculate the total time T_p it takes for the controlled trajectory of the synthesis $\mathscr{F}$ that starts at the point $\gamma(p)$ to reach the origin. If we take $t_0 = -T_p$ as initial time t_0 for the trajectory, then the time of the next switching is $t_1 = \frac{t_0}{\alpha}$, and $\frac{x_2(t_1)}{x_2(t_0)} = -\frac{1}{\alpha}$. Since $\dot{x}_2 = 1$ over $[t_0, t_1]$, we have that

$$x_2(t_1) - x_2(t_0) = t_1 - t_0 = \left(\frac{1}{\alpha} - 1\right) t_0$$

and thus, dividing by $-x_2(t_0)$,

$$\left(1 - \frac{1}{\alpha}\right) \frac{t_0}{(-p)} = 1 - \frac{x_2(t_1)}{x_2(t_0)} = 1 + \frac{1}{\alpha},$$

which gives

$$T_p = -t_0 = \frac{1 + \frac{1}{\alpha}}{1 - \frac{1}{\alpha}} p = \frac{\alpha + 1}{\alpha - 1} p.$$

Given $\gamma(p)$, define a control u_p over the infinite interval $(-\infty, 0)$ to have the switching times $\{t_n\}_{n \in \mathbb{Z}}$ given by $t_0 = \frac{1+\alpha}{1-\alpha} p < 0$ and $t_i = \alpha^{-i} t_0$ with the controls alternating between $+1$ and -1 at the switching times and $u_p \equiv +1$ on the interval (t_0, t_1). Let $x_p = (x_1, x_2)^T$ be the corresponding trajectory. This is the controlled trajectory generated by the synthesis $\mathscr{F}$ through the point $\gamma(p)$. Define a solution $\lambda_p = (\lambda_1, \lambda_2)$ of the corresponding adjoint equation by taking as initial conditions at time t_0 the values

$$\lambda_1(t_0) = -\frac{1}{2} \frac{x_1^2(t_0)}{x_2(t_0)} = -\frac{1}{2} \zeta^2 p^3 \quad \text{and} \quad \lambda_2(t_0) = 0.$$

We claim that this defines an $\mathscr{A}$-invariant extremal $\Gamma_p = ((x_p, u_p), \lambda_p)$. This is fairly obvious by construction. Clearly, the control u_p is $\mathscr{A}$-invariant, and calculations invoking the first integral I_1 analogous to those carried out in the proof of Lemma 2.11.3 verify that the corresponding trajectory x_p is invariant as well. We have taken care to choose the correct initial condition for the multiplier, and the $\mathscr{A}$-invariance of the adjoint vector can be verified using the other first integral I_2. Finally, the fact that the Hamiltonian is identically zero simply follows from the fact that

$$H(t_0) = \frac{1}{2}x_1^2(t_0) + \lambda_1(t_0)x_2(t_0) + \lambda_2(t_0)u(t_0)$$

$$= -\frac{1}{2}(\zeta p^2)^2 - \frac{1}{2}\zeta^2 p^3(-p) + 0 = 0$$

and $\frac{d}{dt}H(t)$ vanishes, since λ is a solution to the corresponding adjoint equation. Since every controlled trajectory generated by the synthesis $\mathscr{F}$ is of this form, this proves the proposition.   □

It is easy to define a "patch" Ξ_0 of controlled trajectories that generates this synthesis. Simply take the value $p = 1$ and consider the point $(\zeta, -1) \in \Gamma_+$. The first return of this trajectory to the curve Γ_+ then is at

$$\bar{x}_2 = x_2(t_2) = \frac{1}{\alpha^2}x_2(t_0) = -\frac{1}{\alpha^2}.$$

Define the function $t_0 : [0, \infty) \to (-\infty, 0]$ by

$$t_0(p) = \frac{1+\alpha}{1-\alpha}p$$

and let

$$D_0 = \left\{ (t,p) : \frac{1}{\alpha^2} < p \le 1, \, t_0(p) \le t < t_1(p) = \frac{t_0(p)}{\alpha} \right\}$$

be the domain for a parametrization of the controlled trajectories of the Fuller synthesis $\mathscr{F}$,

$$\Xi_0 : D_0 \to \mathbb{R}^2 \backslash \{(0,0)\}, \quad (t,p) \mapsto (x_1(t,p), x_2(t,p)).$$

Then the iterates Ξ_n under the action defined by $\mathscr{A}$,

$$\Xi_n : D_0 \to \mathbb{R}^2 \backslash \{(0,0)\}, \quad (t,p) \mapsto (-1)^n \begin{pmatrix} x_1^{\alpha^n}(t,p) \\ x_2^{\alpha^n}(t,p) \end{pmatrix},$$

for all $n \in \mathbb{Z}$ cover the full state space, except for the origin.

We used invariance properties of the extremals to give a rather elegant and short construction of an extremal synthesis. By itself, however, this does not guarantee optimality. The optimality of this field will be verified in Sect. 5.2.3. In fact, there we shall give a rather elementary constructive argument that proves the optimality of this synthesis based on the parameterization of the patch Ξ_0.

It is also true that the extremals constructed here are the *only* extremals possible, but this argument is quite a bit more technical and involved (for example, see [34, 35]). Coupled with a standard result that guarantees the existence of optimal solutions for the Fuller problem, this indeed then proves the optimality of the synthesis constructed. But here we are interested rather in illustrating the use of invariance properties, a tremendously powerful tool in the analysis of nonlinear systems with symmetries. Our presentation here is based on ideas and arguments of Kupka. While this problem with its solution given by chattering arcs was considered an aberration for a long time, in his paper [143], Kupka has shown that this is far from the truth and that chattering extremals indeed are a generic phenomenon, i.e., are in some very precise mathematical sense "typical" in higher state-space dimensions.

Another important point that is made with this problem is that optimal controls in general need not be piecewise continuous for even the simplest-looking real analytic system. It is easy to see that an arbitrary measurable control u can be the solution of a time-optimal control problem for a system of the form $\dot{x} = f(x) + ug(x)$ with control set $U = [-1, 1]$ and some sufficiently "weird" smooth vector fields $f, g \in C^\infty$. But whether optimal controls can be that general if the vector fields are real analytic, or whether they then do have some regularity properties, as might be expected, still is an open problem for which only partial results exist. While the structure of these optimal chattering controls still is rather simple, nevertheless these are not piecewise continuous, but only Lebesgue measurable controls. And this is the correct class of controls to consider in any optimal control problem, since it allows for a reasonable theory of existence of optimal solutions (e.g., [33]). The main aim of this chapter was to illustrate how the conditions of the maximum principle can be used to solve problems, and for this the class of piecewise continuous controls is mostly adequate. But in order to proceed with the deeper theory, even if we shall not concern ourselves with existence theory, we shall need to allow for Lebesgue measurable controls. We shall see next that even for linear systems this is indispensable.

2.12   Notes

Linear-quadratic optimal control is a classical design principle in automatic control and is at the heart of many actual control schemes including autopilots on commercial aircraft, process control in chemical engineering, and many other regulation processes. There exist many excellent engineering textbooks that are fully devoted to this subject and its extensions, both as deterministic systems and in a stochastic (noisy) environment. For this reason, we included only the most fundamental results on this topic. We highly recommend the classical text by Kwakernaak and Sivan [144] to the interested reader. We used the textbook by Knowles [139] as a source for the introductory one-dimensional examples that allow for explicit integrations of the solutions.

Time-optimal control for linear systems also is a classical topic treated in depth in many of the textbooks from the 1960s and 1970s such as those by Lee and Marcus [147] and Athans and Falb [25]. We shall take up this topic in some more detail next.

The necessary conditions for optimality of singular controls presented in Sects. 2.8.4 and 2.8.5 represent only the culmination of the classical research on this topic that was carried out in the 1960s, e.g., [31, 104, 107, 121, 122, 131, 132, 169, 173, 178, 201]. We shall prove these results in Chap. 4, but using very different computational methods. Also, the lecture notes by H.W. Knobloch [137] provide an alternative approach to many of these conditions. A treatment of singular trajectories that proceeds beyond these classical developments and takes into account conjugate points is given by Bonnard and Kupka [48, 49], and for an in-depth analysis of singular trajectories, we highly recommend the monograph by Bonnard and Chyba [44]. Genericity properties of singular trajectories are developed in the work by Chitour, Jean, and Trélat [71–73].

There do not exist many textbooks that provide the differential-geometric framework that we employ in our treatment of optimal control. In fact, the early texts that give some of these foundations are in engineering, such as those by Isidori [120] and Nijmeijer and van der Schaft [176], but these texts focus on concepts from automatic control such as regulation and disturbance decoupling and do not address optimal control. The textbooks by Sontag [225] and Jurdjevic [126] address a more mathematical audience. While focused on the foundations of nonlinear systems theory (e.g., reachability and controllability, integral manifolds), these texts also include an introduction to optimal control problems, however largely motivated by linear-quadratic control problems.

The results that are included in the later sections of this chapter were for a long time only scattered in the research literature or some edited volumes such as [1, 6] and [4]. It is only more recently that some specialized monographs have been published that include these issues, such as those by Bonnard and Chyba [44], Boscain and Piccoli [51], and Bressan and Piccoli [56]. Among these, the book by Boscain and Piccoli is fully devoted to optimal control problems in the plane. We refer the reader to this text and Sussmann's original paper [236] for a complete analysis of generic systems. In his papers, Sussmann carries this analysis further, analyzing all cases of positive codimension for a nondegenerate dynamical system with smooth vector fields f and g in $C^\infty(\Omega)$ [236] and arbitrary real analytic vector fields f and g in $C^\omega(\Omega)$ [237]. In [238], it is then shown how these local results combine to provide a global solution to the problem in terms of a *regular synthesis*. These results are developed further by Boscain and Piccoli, who, more generally, analyze the time-optimal control problem and the structure of its optimal syntheses for systems on two-dimensional manifolds [51]. Much less is known in dimension three, and we shall pick up this topic in Chap. 7.

The Fuller problem is another classical optimal control problem. For quite some time, the structure of its solution was considered an aberration until I.A.K. Kupka showed that indeed this is a common phenomenon in higher dimensions [143]. As in the Fuller problem, it arises naturally if controlled trajectories need to follow or leave a locally optimal singular arc that is of order 2 and the singular controls take values in the interior of the control set. While this, in principle, is not a generic scenario, there are many interesting practical problems in which this happens. For example, in mathematical models for tumor anti-angiogenic treatments (see Sect. 6.2), there exists an optimal singular arc of order 1 that on addition of pharmacokinetic models for the drug action becomes of order 2, leading to optimal chattering connections [165]. Similarly, these phenomena arise in the control of autonomous underwater robots [74, 75]. The most comprehensive treatment of chattering arcs so far is given in the monograph by Zelikin and Borisov [262].

Chapter 3
Reachable Sets of Linear Time-Invariant Systems: From Convex Sets to the Bang-Bang Theorem

As a precursor to the proof of the maximum principle for a general nonlinear system, in this chapter we develop the classical results about the structure of the reachable set for linear time-invariant systems with bounded control sets and prove Theorem 2.5.3 of Chap. 2.

We always consider a system Σ of the form

$$\Sigma: \qquad \dot{x} = Ax + Bu, \qquad x(0) = p, \qquad u \in U, \tag{3.1}$$

where $x \in \mathbb{R}^n$, $u \in \mathbb{R}^m$, $A \in \mathbb{R}^{n \times n}$, $B \in \mathbb{R}^{n \times m}$, and the control set U is an arbitrary subset of $\mathbb{R}^m$. Since the system is time-invariant, without loss of generality we normalize the initial time to $t_0 = 0$. In order to obtain some of the deeper results of the theory, it now becomes necessary to consider Lebesgue measurable functions as controls, and we therefore take as the class $\mathscr{U}$ of admissible controls the class of all locally bounded Lebesgue measurable functions $u : [0, \infty) \to U$ that take values in the control set U almost everywhere (a.e.); that is, for any finite interval $[0, T] \subset [0, \infty)$ there exists a constant $M = M(T) < \infty$ such that $\|u(t)\| \leq M$ for all $t \in [0, T]$ with the possible exception of a Lebesgue null set.[1] With this choice of controls, the classical solution formula for a linear differential equation remains valid, and the solution $x(\cdot; p)$ to the initial value problem (3.1) is given as

$$x(t; p) = e^{At} p + \int_0^t e^{A(t-s)} Bu(s) ds. \tag{3.2}$$

Recall that the time-t-reachable set from p is the set of all points that can be reached from p by means of an admissible control defined on the interval $[0, t]$, i.e.,

---

[1] The technical aspects of measure theory and measurable functions are beyond the scope of this text. A brief summary of the basic theory is given in Appendix D, and more advanced results will be quoted from the literature. However, only few of these will be needed.

H. Schättler and U. Ledzewicz, *Geometric Optimal Control: Theory, Methods and Examples*, Interdisciplinary Applied Mathematics 38, DOI 10.1007/978-1-4614-3834-2_3, © Springer Science+Business Media, LLC 2012

$$\text{Reach}_{\Sigma,t}(p) = \left\{ q \in \mathbb{R}^n : \exists\, u \in \mathcal{U} \text{ so that } q = e^{At}p + \int_0^t e^{A(t-s)} Bu(s)ds \right\}.$$

In this chapter, we shall show that if the control set U is compact and convex, then so are the reachable sets $\text{Reach}_{\Sigma,t}(0)$ (Sect. 3.3), and we will use this fact to derive the celebrated *bang-bang theorem* for linear systems (Sect. 3.4). This theorem states that whenever a point q is reachable in time t by means of a Lebesgue measurable control that takes values in U, then it is also reachable in the same time t by means of a control that takes its values in the set of extreme points of U, a much smaller set. If U is a compact polyhedron, then in fact, q is reachable by means of a bang-bang control with a finite number of switchings (Sect. 3.6). But we start with two introductory sections in which we develop required background material. Properties of convex sets in $\mathbb{R}^n$ are essential in the analysis of reachable sets for linear systems and also in approximations to reachable sets of general nonlinear systems later on. We therefore fully develop these elementary but important facts in Sect. 3.1. Section 3.2 presents the useful notion of weak convergence in the Banach space $L^1(I)$ of integrable Lebesgue measurable functions on an interval I, which we shall also use on various occasions throughout the text. For this topic, some more advanced background from real analysis is required, and we need to refer to the literature for some of the results that will be used.

3.1   Elementary Theory of Convex Sets

We review some elementary but important facts about convex set in $\mathbb{R}^n$ that we will need in this chapter and also later on.

Definition 3.1.1 (Convex set). A set $E \subset \mathbb{R}^n$ is convex if whenever two points x and y lie in E, then the entire line segment connecting x with y lies in E:

$$x \in E,\ y \in E \implies \lambda x + (1-\lambda)y \in E \quad \text{for all } \lambda \in [0,1].$$

Recall that the closure of a set E, $\bar{E}$ or $\text{Clos}(E)$, consists of all accumulation points of E, and the interior of E, $\mathring{E}$, consists of all interior points (see Appendix A).

Proposition 3.1.1. *If $E \subset \mathbb{R}^n$ is convex, then so are its closure $\bar{E}$ and its interior $\mathring{E}$.*

Proof. These topological properties inherit from the convexity of E. Given two points x and y in the closure of E, pick sequences $\{x_n\}_{n \in \mathbb{N}} \subset E$ and $\{y_n\}_{n \in \mathbb{N}} \subset E$ such that $x_n \to x$ and $y_n \to y$. By the convexity of E, for all $\lambda \in [0,1]$ we have $\lambda x_n + (1-\lambda)y_n \in E$ and $\lambda x_n + (1-\lambda)y_n \to \lambda x + (1-\lambda)y$. Hence $\lambda x + (1-\lambda)y \in \bar{E}$ for all $\lambda \in [0,1]$. The fact that the interior of E is also convex immediately follows from the following lemma.

Lemma 3.1.1. *Let E be a convex set and let $x \in \mathring{E}$ and $y \in \bar{E}$. Then*

$$\lambda x + (1 - \lambda) y \in \mathring{E} \quad \text{for all} \quad \lambda \in (0, 1].$$

Proof. Choose $\varepsilon > 0$ such that $B_{2\varepsilon}(x) = \{z \in \mathbb{R}^n : \|z - x\| < 2\varepsilon\} \subset E$ and pick a sequence $\{y_n\}_{n \in \mathbb{N}} \subset E$ converging to y. Given $\lambda \in (0, 1]$, take n large enough that

$$\frac{1 - \lambda}{\lambda} \|y - y_n\| < \varepsilon.$$

Then for every point $z \in B_\varepsilon(x)$ we have that

$$\lambda z + (1 - \lambda) y = \lambda \left(z + \frac{1 - \lambda}{\lambda} (y - y_n) \right) + (1 - \lambda) y_n \in \lambda B_{2\varepsilon}(x) + (1 - \lambda) y_n \subset E,$$

and thus the neighborhood $\lambda B_\varepsilon(x) + (1 - \lambda) y$ of $\lambda x + (1 - \lambda) y$ lies in E. Hence $\lambda x + (1 - \lambda) y \in \mathring{E}$. $\square$

Definition 3.1.2 (Convex hull of a set). Given an arbitrary subset $S \subset \mathbb{R}^n$, the convex hull of S, $\mathrm{co}(S)$, is the smallest convex subset that contains S.

It is clear that an intersection of convex sets is again convex. Hence the convex hull of S is the intersection of all convex sets that contain S and as such is well-defined. The next result gives an analytic description of the convex hull in terms of convex combinations.

Proposition 3.1.2. *Given $S \subset \mathbb{R}^n$, the convex hull of S is given by*

$$\mathrm{co}(S) = \left\{ x \in \mathbb{R}^n : x = \sum_{i=0}^{k} \lambda_i x_i, \quad x_i \in S, \quad \lambda_i \geq 0, \quad \sum_{i=0}^{k} \lambda_i = 1, \quad k \in \mathbb{N}_0 \right\}.$$

Proof. Denote the set on the right by $\hat{S}$. This set is convex, for if x and y are points in $\hat{S}$, then without loss of generality, we may assume that x and y are written as convex combinations of the same set of vectors (by simply adding some terms with coefficients 0), say

$$x = \sum_{i=0}^{k} \lambda_i z_i \quad \text{and} \quad y = \sum_{i=0}^{k} \mu_i z_i$$

with $z_i \in S$, and all λ_i and μ_i nonnegative and summing to 1. But then, for any $\rho \in [0, 1]$,

$$\rho x + (1 - \rho) y = \sum_{i=0}^{k} (\rho \lambda_i + (1 - \rho) \mu_i) z_i$$

with $\rho \lambda_i + (1 - \rho)\mu_i \geq 0$ and

$$\sum_{i=0}^{k} \rho \lambda_i + (1 - \rho)\mu_i = \rho \left(\sum_{i=0}^{k} \lambda_i\right) + (1 - \rho) \left(\sum_{i=0}^{k} \mu_i\right) = 1.$$

Hence $\rho x + (1 - \rho)y \in \hat{S}$. On the other hand, any convex set that contains S also must contain any convex combinations of points from S, and thus it must contain $\hat{S}$. Hence $\hat{S}$ is the smallest convex set that contains S. $\qquad\square$

We next establish that a convex set has a well-defined dimension (which may be smaller than n). To this effect, we need the concept of an affine variety.

Definition 3.1.3 (Affine variety generated by a convex set). A set V of the form $V = p + W$, where p is a point in $\mathbb{R}^n$ and W is a subspace of $\mathbb{R}^n$, is called an affine variety, and the dimension of W is called the dimension of the affine variety. An $(n - 1)$-dimensional affine variety is called a hyperplane. Given an arbitrary subset E of $\mathbb{R}^n$, the affine variety generated by E, $\mathscr{A}(E)$, is the smallest affine variety that contains E.

Lemma 3.1.2. *Given any set $E \subset \mathbb{R}^n$, the affine variety generated by E is well-defined and is given by the set of all affine combinations of points from E, that is,*

$$\mathscr{A}(E) = \left\{x \in \mathbb{R}^n : x = \sum_{i=0}^{k} \lambda_i x_i, \quad x_i \in E, \quad \lambda_i \in \mathbb{R}, \quad \sum_{i=0}^{k} \lambda_i = 1, \quad k \in \mathbb{N}\right\}.$$

Proof. Points $p_0, \ldots, p_m$ are said to be affinely independent if (and only if) the vectors $p_1 - p_0, \ldots, p_m - p_0$ are linearly independent. (It is easy to see that this definition does not depend on the ordering of the points.) Thus, in $\mathbb{R}^n$ any set consisting of more than $n + 1$ points is affinely dependent. Choose m maximal so that there exist affinely independent points $p_0, \ldots, p_m$ from the set E and let W be the subspace spanned by the vectors $p_1 - p_0, \ldots, p_m - p_0$. Then $p_0 + W$ indeed is the smallest affine variety that contains E.

To see this, first note that any point $q \in E$ necessarily lies in $p_0 + W$. For if there existed a point $q \in E$ that lay outside $p_0 + W$, then the points $p_0, \ldots, p_m$ and q would be affinely independent, contradicting the maximality of m. Thus, $p_0 + W$ is an affine variety that contains E. On the other hand, since the vectors $p_1 - p_0, \ldots, p_m - p_0$ are linearly independent, any affine variety that contains E also must contain $p_0 + W$ and thus $p_0 + W \subset \mathscr{A}(E)$. Hence $\mathscr{A}(E) = p_0 + W$.

We now show that

$$p_0 + W = \left\{x \in \mathbb{R}^n : x = \sum_{i=0}^{k} \lambda_i x_i, \quad x_i \in E, \quad \lambda_i \in \mathbb{R}, \quad \sum_{i=0}^{k} \lambda_i = 1, \quad k \in \mathbb{N}\right\}.$$

Any point q in $p_0 + W$ can be written as

$$q = p_0 + \sum_{i=1}^{m} \mu_i (p_i - p_0)$$

for some $\mu_i \in \mathbb{R}$. Setting $\lambda_0 = 1 - \sum_{i=1}^{m} \mu_i$, and $\lambda_i = \mu_i$ for $i = 1, \ldots, m$, the point q can therefore be written as

$$q = \sum_{i=0}^{m} \lambda_i p_i \quad \text{with} \quad \sum_{i=0}^{m} \lambda_i = 1,$$

and thus

$$p_0 + W \subset \left\{ x \in \mathbb{R}^n : x = \sum_{i=0}^{k} \lambda_i x_i, \quad x_i \in E, \quad \lambda_i \in \mathbb{R}, \quad \sum_{i=0}^{k} \lambda_i = 1, \quad k \in \mathbb{N} \right\}.$$

Conversely, let $x_0, \ldots, x_k$ be arbitrary points in E, and λ_i real numbers that sum to 1. Since $x_i \in p_0 + W$, we can write

$$x_i = p_0 + \sum_{j=1}^{m} \mu_{ij} (p_j - p_0)$$

with some $\mu_{ij} \in \mathbb{R}$. Hence

$$\sum_{i=0}^{k} \lambda_i x_i = \sum_{i=0}^{k} \lambda_i \left(p_0 + \sum_{j=1}^{m} \mu_{ij} (p_j - p_0) \right)$$

$$= p_0 + \sum_{j=1}^{m} \left(\sum_{i=0}^{k} \lambda_i \mu_{ij} \right) (p_j - p_0) \in p_0 + W$$

and thus

$$\left\{ x \in \mathbb{R}^n : x = \sum_{i=0}^{k} \lambda_i x_i, \quad x_i \in E, \quad \lambda_i \in \mathbb{R}, \quad \sum_{i=0}^{k} \lambda_i = 1, \quad k \in \mathbb{N} \right\} \subset p_0 + W,$$

proving the reverse inclusion. □

Definition 3.1.4 (Dimension of a convex set). The dimension of a convex set E is defined as the dimension of the affine variety $\mathscr{A}(E)$ generated by E. A point $p \in E$ is called an internal point of E if p is an interior point of E relative to the affine variety $\mathscr{A}(E)$ generated by E.

The following result is the key to separation theorems about convex sets and is the finite-dimensional analogue to the Hahn–Banach theorem in infinite dimensions. In finite-dimensional spaces, its proof is elementary by induction on the dimension.

Theorem 3.1.1 (Hahn–Banach). *Let E be a nonempty open convex set in $\mathbb{R}^n$ and let V be an affine variety disjoint from E. Then there exists a hyperplane H that contains V and still is disjoint from E.*

Proof. The result is vacuous in $\mathbb{R}$: any affine variety disjoint from E is a point and thus is a hyperplane in $\mathbb{R}$. In order to emphasize the constructive nature of the argument, we also give the reasoning for the two-dimensional case. Suppose $E \subset \mathbb{R}^2$ and let V be a point not in E. (Nothing needs to be shown if V already is a line.) Without loss of generality we may assume that $V = \{0\}$. Denote by kE the set of all points that are of the form kx for some $x \in E$ and let

$$C = \bigcup_{k>0} kE.$$

Then C is an open convex cone: a set C is called a cone (with apex at 0) if whenever $x \in C$, then also $kx \in C$ for all $k > 0$. This property is obvious from the construction. To see that C is convex, consider arbitrary points y_1 and y_2 from C, say $y_1 = k_1 x_1$ and $y_2 = k_2 x_2$ with x_1 and x_2 from E, and let $\lambda \in [0,1]$. Then

$$\lambda y_1 + (1-\lambda) y_2 = (\lambda k_1 + (1-\lambda)k_2) \left(\frac{\lambda k_1 x_1}{\lambda k_1 + (1-\lambda)k_2} + \frac{(1-\lambda)k_2 x_2}{\lambda k_1 + (1-\lambda)k_2} \right) \in C.$$

Now let x be a boundary point from C, $x \in \partial C$, different from 0, and let H be the line through x. Then this hyperplane is disjoint from E: Since C is open, it is clear that x cannot lie in C. But $-x$ also cannot lie in C, since otherwise, by Lemma 3.1.1, $0 = \frac{1}{2}x + \frac{1}{2}(-x) \in C$. Hence the hyperplane H does not intersect C. This proves the result for $n = 2$.

Now suppose $n \geq 3$. Without loss of generality, again assume $0 \in V$ and let W be a subspace of maximal dimension that contains V and still is disjoint from E. Nothing needs to be shown in $\dim W = n - 1$. If the codimension of W is at least 2, then let U be a two-dimensional subspace of the orthogonal complement of W and consider the space $X = U \oplus W$. Denote the orthogonal projection in X along W by π, so that $\pi(W) = \{0\}$. The set $E' = \pi(E)$ is open in U, and since W and E are disjoint, does not contain 0. By the previous argument, there exists a line l through 0 that is disjoint from 0. But then l and W span a linear subspace disjoint from E of dimension strictly larger than $\dim W$. Contradiction. Thus $\dim W = n - 1$.   □

Definition 3.1.5 (Supporting hyperplane). let $E \subset \mathbb{R}^n$ be a convex set. A hyperplane $H = \{x \in \mathbb{R}^n : c^T x = a\}$ is called a supporting hyperplane to E if H intersects the boundary of E, $H \cap \partial E \neq \emptyset$, and E lies entirely to one side of H, i.e., $H \subset H^- = \{x \in \mathbb{R}^n : c^T x \leq a\}$ or $H \subset H^+ = \{x \in \mathbb{R}^n : c^T x \geq a\}$.

Corollary 3.1.1. *Let E be a nonempty convex set. Then, through every point p in the boundary of E, $p \in \partial E$, there exists a supporting hyperplane H.*

Proof. If the dimension of E is n, then the interior of E, $\mathring{E}$, is a nonempty open convex set, and by the Hahn–Banach theorem, through every boundary point p there exists a hyperplane H that is still disjoint from $\mathring{E}$, say $H = \{x \in \mathbb{R}^n : c^T x = a\}$. Then $\mathring{E}$ needs to lie to one side of the hyperplane H, say $c^T x > a$ for all $x \in \mathring{E}$. It follows from Lemma 3.1.1 that the closure of E is actually the same as the closure of the interior of E. Hence we have $c^T x \geq a$ for all $x \in \bar{E}$, and thus H is a supporting hyperplane to E at p. The result is trivially true if the dimension of E is smaller than n: in this case the affine variety generated by E is at most $(n-1)$-dimensional, and thus the entire set E is contained in a hyperplane H. This hyperplane is supporting. $\square$

Definition 3.1.6 (Separating hyperplane). Given two nonempty convex sets, E_1 and E_2, a hyperplane H is called separating if $E_1 \subset H^-$ and $E_2 \subset H^+$.

Proposition 3.1.3. *Let E be a nonempty closed and convex set and let $B = B_\rho(p) = \{y : \|y - p\| < \rho\}$ be an open ball disjoint from E. Then there exists a supporting hyperplane H to E that separates E from B.*

Proof. Let δ be the distance from p to E. Then $0 < \rho \leq \delta < \infty$ and there exists a point $x_* \in E$ such that $\delta = \|x_* - p\|$. This simply follows from the fact that a continuous function attains its minimum on a compact set (see Appendix A). Let $H = \{x \in \mathbb{R}^n : c^T x = a\}$ be the tangent plane to the sphere $\{x : \|x - p\| = \delta\}$ through x_*, and without loss of generality, assume that $B_\delta(p) \subset H^-$. Then E must lie in H^+: for if $c^T q < a$, then the segment connecting q with x_* lies in E and points on this segment close to x_* have a smaller distance to p than δ. Contradiction. Thus H is a supporting hyperplane to E that separates E from B. $\square$

Corollary 3.1.2. *Let E be a nonempty closed and convex set different from $\mathbb{R}^n$. Then there exist countably many supporting hyperplanes H_i, $i \in \mathbb{N}$, to E such that $E = \cap_{i=1}^{\infty} H_i^+$.*

Proof. The complement G of E, $G = E^c$, is a nonempty open set and can be written as a countable union of open balls B_i, $i \in \mathbb{N}$, $G = \cup_{i=1}^{\infty} B_i$. By Proposition 3.1.3, there exists a supporting hyperplane H_i to E such that $E \subset H_i^+$ and $B_i \subset \text{int}(H_i^-)$. Then $E \subset \cap_{i=1}^{\infty} H_i^+$ and $E^c = \cup_{i=1}^{\infty} B_i \subset \cup_{i=1}^{\infty} \text{int}(H_i^-) = \cup_{i=1}^{\infty}(H_i^+)^c = (\cap_{i=1}^{\infty} H_i^+)^c$. Hence $E = \cap_{i=1}^{\infty} H_i^+$. $\square$

Definition 3.1.7 (Extreme points). Let $E \subset \mathbb{R}^n$ be a convex set. A point $q \in E$ is said to be an extreme point of E if q cannot be written as a nontrivial convex combination of other points from E, i.e., if $q = \lambda x + (1 - \lambda)y$ with x and y from E and $\lambda \neq 0$ and $\lambda \neq 1$, then $x = y$. We denote the set of all extreme points of E by E^{ext}.

Clearly, extreme points are boundary points of E, but not every boundary point is an extreme point (see Fig. 3.1). For $E = \{x \in \mathbb{R}^n : \|x\|_2 = 1\}$, this is the case, but only the vertices are extreme points of $E = \{x \in \mathbb{R}^n : \|x\|_\infty = 1\}$. Also, as the

Fig. 3.1 Every point of a
circle is an extreme point, but
only the vertices are extreme
points of a rectangle

example of a closed hyperplane shows, closed convex sets need not have extreme
points at all. But if they are compact, this is the case, and we have the following
important result.

Theorem 3.1.2 (Krein–Milman). *Let E be a non-empty, compact and convex set
in $\mathbb{R}^n$. The set E^{ext} of extreme points of E is nonempty, and every point in E can be
written as a convex combination of at most $n + 1$ extreme points. In particular, E is
the convex hull of its extreme points, $E = \text{co}(E^{\text{ext}})$.*

For the proof we need the following auxiliary result, which is important enough
to be stated separately.

Proposition 3.1.4. *Let E be a nonempty compact and convex set. Then every
supporting hyperplane to E contains at least one extreme point.*

Proof. The proof is by induction on the dimension of E. If $\dim E = 1$, then the affine
variety generated by E is a line and E is a compact interval in it. In this case, each
of the two boundary points (relative to the affine variety) is an extreme point. So
now assume that the result is true for convex sets E of dimension at most $n - 1$ and
suppose E is a convex set of dimension n. Any supporting hyperplane $H = \{x \in \mathbb{R}^n :
c^T x = a\}$ for E contains at least one boundary point p. The intersection $E \cap H$ is a
nonempty compact and convex set of strictly smaller dimension. By the inductive
assumption there exists an extreme point q of $E \cap H$. But since H is a supporting
hyperplane, q actually is an extreme point of E. For suppose $q = \lambda x + (1 - \lambda)y$ with
x and y from E and $\lambda \neq 0$ and $\lambda \neq 1$. Since H is a supporting hyperplane, the set
E lies to one side of H, and we have, say, $c^T x \geq a$ and $c^T y \geq a$. But $c^T q = a$, since
$q \in H$, and so we must have $c^T x = a$ and $c^T y = a$, i.e., x and y lie in $E \cap H$. Since q is
an extreme point of $E \cap H$, it follows that $x = y$. Thus q is an extreme point of E. $\square$

Proof of the Krein–Milman Theorem: Once more, the proof is by induction on
the dimension of the convex set. The result is trivially true if $\dim E = 0$, in which
case E consists of a single point. It is also obvious if $\dim E = 1$, in which case E is
a compact interval in a line, and clearly every point in E is a convex combination of
the two boundary points that are the extreme points of E. So again assume that the
statement is correct for convex sets E of dimensions $\leq n - 1$ and let E be a convex
set of dimension n.

Let x be any point in E. By Corollary 3.1.1 and Proposition 3.1.4 there exists an
extreme point p, $p \in E^{\text{ext}}$. Consider the ray starting at p in the direction of x. Since
E is compact, there exists a last element $y \in E$ on this ray and $y \in \partial E$. Let H be a

supporting hyperplane to E through y. Then $E \cap H$ is a nonempty compact convex set of dimension strictly smaller than n. Hence, by the inductive assumption, y can be written as a convex combination of at most n extreme points q_i of $E \cap H$. But as shown above, extreme points of a supporting hyperplane are also extreme points of the set itself, and thus also $q_i \in E^{\text{ext}}$. Hence we have

$$y = \sum_{i=1}^{n} \lambda_i q_i, \qquad q_i \in E^{\text{ext}}, \qquad \lambda_i \geq 0, \qquad \sum_{i=1}^{n} \lambda_i = 1.$$

But x itself is a convex combination of p and y, and thus $x = \mu p + (1 - \mu) y$ for some $\mu \in [0, 1]$. Hence, altogether,

$$x = \mu p + (1 - \mu) \sum_{i=1}^{n} \lambda_i q_i,$$

and thus x is a convex combination of at most $n + 1$ extreme points from E. This proves the inductive step. $\square$

Corollary 3.1.3 (Carathéodory). *Given $S \subset \mathbb{R}^n$, the convex hull of S is given by*

$$\mathrm{co}(S) = \left\{ x \in \mathbb{R}^n : x = \sum_{i=0}^{n} \lambda_i x_i, \quad x_i \in S, \quad \lambda_i \geq 0, \quad \sum_{i=0}^{n} \lambda_i = 1 \right\}.$$

Proof. We already know that

$$\mathrm{co}(S) = \left\{ x \in \mathbb{R}^n : x = \sum_{i=0}^{k} \lambda_i x_i, \quad x_i \in S, \quad \lambda_i \geq 0, \quad \sum_{i=0}^{k} \lambda_i = 1, \quad k \in \mathbb{N}_0 \right\}$$

with the upper limit k in the summation free. The difference in the corollary is that the upper limit has been fixed to n, the dimension of the convex set. Every point $x \in \mathrm{co}(S)$ lies in the convex hull of a finite set of points $\{x_0, \ldots, x_k\}$ from S. But the convex hull of this finite set is compact and convex, and thus x can be written as a convex combination of at most $n + 1$ extreme points of $\mathrm{co}(\{x_0, \ldots, x_k\})$. But any extreme point of this set must be one of the x_i, $i = 0, \ldots, k$. Hence every point $x \in \mathrm{co}(S)$ can also be written as a convex combination of at most $n + 1$ points from S. $\square$

Corollary 3.1.4. *If the set S is compact, then so is its convex hull $\mathrm{co}(S)$.*

Proof. If S is bounded, say $\|x\| \leq C < \infty$ for all $x \in S$, and if y is some point in the convex hull of S,

$$y = \sum_{i=0}^{n} \lambda_i x_i, \quad x_i \in S, \quad \lambda_i \geq 0, \quad \sum_{i=0}^{n} \lambda_i = 1,$$

then

$$\|y\| \leq \sum_{i=0}^{n} \lambda_i \|x_i\| \leq \left(\sum_{i=0}^{n} \lambda_i \right) C = C,$$

and thus $co(S)$ is bounded. To show that $co(S)$ is closed, suppose $\{y_k\}_{k \in \mathbb{N}} \subset co(S)$ and $y_k \to y$. Write

$$y_k = \sum_{i=0}^{n} \lambda_i^k x_i^k, \quad x_i^k \in S, \quad \lambda_i^k \geq 0, \quad \sum_{i=0}^{n} \lambda_i^k = 1.$$

Since S is compact, there exist convergent subsequences, and without loss of generality, we may assume that $x_i^k \to x_i \in S$. But also the unit simplex

$$\Delta = \left\{ \lambda \in \mathbb{R}^{n+1} : \lambda_i \geq 0, \quad \sum_{i=0}^{n} \lambda_i = 1 \right\}$$

is compact, and thus, by taking convergent subsequences, we also may assume that $\lambda_i^k \to \lambda_i \geq 0$ and in the limit $\sum_{i=0}^{n} \lambda_i = 1$. Hence

$$y_k = \sum_{i=0}^{n} \lambda_i^k x_i^k \to \sum_{i=0}^{n} \lambda_i x_i \in co(S)$$

and thus $y \in co(S)$. □

We close this section with the following useful characterization of extreme points as lexicographically minimal elements of the set relative to some ordered basis $\mathscr{B}$ of $\mathbb{R}^n$ (see Fig. 3.2).

Definition 3.1.8 (Lexicographic ordering on $\mathbb{R}^n$). Given an ordered basis $\mathscr{B}$ consisting of n linearly independent vectors $(v_1, \ldots, v_n)$, every vector x has a unique coordinate representation $x = (x_1, \ldots, x_n)^T$, $x = \sum_{i=1}^{n} x_i v_i$. Given two vectors x and y of $\mathbb{R}^n$, $x \neq y$, let ι be the smallest index for which their coordinates differ. We say that x is lexicographically smaller than y, $x \prec y$, if $x_\iota < y_\iota$.

Proposition 3.1.5. *Let E be a nonempty compact and convex set in $\mathbb{R}^n$. Then a point $q \in E$ is an extreme point of E if and only if there exists a linear change of coordinates such that x is the lexicographically smallest element of E with respect to the new basis. Without loss of generality, the basis can always be chosen to be orthonormal.*

Proof. Clearly this condition is sufficient: suppose $x \prec y$ for all $y \in E$, $y \neq x$, and suppose x is a nontrivial convex combination of points from E, say

$$x = \lambda p + (1 - \mu)q, \qquad p, q \in E \quad \text{and} \quad \lambda \in (0, 1).$$

Fig. 3.2 p is the lexicographic minimum of the set E with respect to the basis whose coordinates are x_1 and x_2

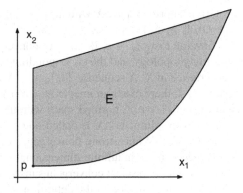

Since x is the lexicographic minimum over E, the first coordinates satisfy $x_1 \leq p_1$ and $x_1 \leq q_1$. But $x_1 = \lambda p_1 + (1-\mu)q_1$, and thus neither p_1 nor q_1 can be strictly larger. Hence $x_1 = p_1 = q_1$. But then, by the definition of the lexicographic order, the same argument applies to the second coordinate and inductively to each coordinate of x. Hence $x = p = q$, and x is an extreme point of E.

Conversely, suppose x is an extreme point of E. Then x is a boundary point of E and there exists a supporting hyperplane $H_1 = \{y \in \mathbb{R}^n : c_1^T y = a_1\}$ to E at x. Without loss of generality, suppose that $\|c_1\|_2 = 1$ and that E lies in the half-space $H_1^+ = \{y \in \mathbb{R}^n : c_1^T y \geq a_1\}$. Make a linear change of coordinates so that c_1 becomes the first basis vector in the new basis $\mathscr{B}_1$. Then, for any point $y \in E$, the corresponding coordinate is at least a_1, and it has this value for x. Thus, in the basis $\mathscr{B}_1$, x minimizes the first coordinate. If x is the unique minimizer, nothing more needs to be done and the basis $\mathscr{B}_1$ has the desired property. In general, the set of all minimizers of the first coordinate is given by $E \cap H_1$ and thus again is a nonempty compact and convex set. Since x is an extreme point of E, it trivially follows that x also is an extreme point of $E \cap H_1$. Let $H_2 = \{y \in \mathbb{R}^n : c_2^T y = a_2\}$ be a supporting hyperplane to $E \cap H_1$ at x. Without loss of generality, we may assume that c_2 is normalized, orthogonal to c_1, and that $E \cap H_1$ lies in the half-space $H_2^+ = \{y \in \mathbb{R}^n : c_2^T y \geq a_2\}$. Then, make another linear change of coordinates to construct a new basis $\mathscr{B}_2$ that retains the first basis vector c_1 and introduces c_2 as the second basis vector. As above, by construction the point x then minimizes the second coordinate over all points $y \in E$ that have the same first coordinate as x, i.e., $y \in E \cap H_1$. If necessary, iterate n times to arrive at a basis $\mathscr{B}_n$ that has the required property. □

3.2   Weak Convergence in $L^1(I)$

One of the main results about reachable sets of linear systems is that the time-t-reachable sets are compact and convex if the control set U is compact and convex. The proof of this result requires the important concept of weak convergence in the Banach space L^1 of Lebesgue measurable functions, and we briefly discuss this topic and some of the necessary background material.

Given a normed space X with norm $\|\cdot\|$, the norm induces a topology through the open balls $B_\varepsilon(x_0) = \{x \in X : \|x - x_0\| < \varepsilon\}$, and a sequence $\{x_n\}_{n \in \mathbb{N}} \subset X$ is said to converge to x, $x_n \to x$, if $\|x_n - x\|$ converges to 0 in $\mathbb{R}$. This topology is called the strong topology, and the corresponding concept of convergence is called *strong convergence* in X. A sequence $\{x_n\}_{n \in \mathbb{N}} \subset X$ is said to be a Cauchy sequence if for every $\varepsilon > 0$ there exists an integer $N = N(\varepsilon)$ such that for all $m, n \geq N(\varepsilon)$ we have that $\|x_n - x_m\| < \varepsilon$. A normed space with the property that every Cauchy sequence converges to a limit $x \in X$ is called complete, and a complete normed space is a *Banach space*. While strong convergence corresponds to the standard concept of convergence in $\mathbb{R}^n$, in infinite-dimensional spaces it often is much too stringent, i.e., it is too difficult to select convergent subsequences, a key argument in compactness considerations. This led to the definition of a weaker concept of convergence (one that makes it easier to find convergent subsequences) known as *weak convergence*.

Given a normed space X, a linear mapping $\ell : X \to \mathbb{R}$ is called a linear functional, and the vector space of all continuous linear functionals is called the dual space of X and is denoted by X'. It is not difficult to see that a linear functional ℓ is continuous if and only if it is bounded in the sense that there exists a constant $C < \infty$ such that

$$|\ell(x)| \leq C \|x\| \quad \text{for all} \quad x \in X. \tag{3.3}$$

[Clearly, this condition is sufficient: if $x_n \to x$, then $|\ell(x) - \ell(x_n)| = |\ell(x - x_n)| \leq C\|x - x_n\| \to 0$. On the other hand, if ℓ is not bounded, then there exists a sequence $\{x_n\}_{n \in \mathbb{N}} \subset S = \{x \in X : \|x\| = 1\}$ such that $|\ell(x_n)| \geq n$. Defining $y_n = \frac{x_n}{n}$, we have $y_n \to 0$, but $|\ell(y_n)| \geq 1$ for all $n \in \mathbb{N}$, and so $\ell(y_n) \not\to 0$.] In a finite-dimensional space, this condition is always satisfied: If $\mathscr{B} = \{v_1, \ldots, v_n\}$ is a basis and $x = \sum_{i=0}^n \alpha_i v_i$, then, with $C = \max_{i=1,\ldots,n} |\ell(v_i)|$,

$$|\ell(x)| = \left| \ell\left(\sum_{i=0}^n \alpha_i v_i \right) \right| \leq \sum_{i=0}^n |\alpha_i| \, |\ell(v_i)| \leq C \|x\|_1,$$

where $\|x\|_1$ denotes the ℓ_1-norm of the coordinate vector. Since all norms in a finite-dimensional space are equivalent (see Appendix A), this gives Eq. (3.3). Thus, the constant C is simply related to the maximum of a finite set of numbers arising for the basis vectors. Obviously, in infinite-dimensional spaces, this supremum no longer needs to be finite, and thus linear functionals are not automatically continuous. Often functionals involving derivatives will not be continuous. For a simple example, let X be the Banach space $C([0,1])$ of all continuous functions $f : [0,1] \to \mathbb{R}$ with the supremum norm and denote the subspace of polynomials by P. On P define a linear functional by the derivative evaluated at $t = 1$, i.e., $\ell : P \to \mathbb{R}$, $p \mapsto p'(1)$. Then for $p_n(t) = t^n$ we have $\|p_n\| = 1$ for all n, but $\ell(p_n) = n$, so this linear functional is not continuous.

Definition 3.2.1 (Weak convergence). Given a normed vector space X, a sequence $\{x_n\}_{n\in\mathbb{N}} \subset X$ is said to converge weakly to $x \in X$, $x_n \rightharpoonup x$, if for all continuous linear functionals ℓ, $\ell \in X'$, we have that $\lim_{n\to\infty} \ell(x_n) = \ell(x)$.

Clearly, if $\{x_n\}_{n\in\mathbb{N}} \subset X$ converges strongly to x, then

$$|\ell(x_n) - \ell(x)| \leq C\|x_n - x\| \to 0,$$

and so $\{x_n\}_{n\in\mathbb{N}} \subset X$ also converges weakly to x. In $\mathbb{R}^n$, these two concepts of convergence are equivalent, but no longer in infinite-dimensional spaces, and it then becomes easier to extract convergent subsequences under weak convergence.

Given a compact interval I, $L^1(I)$ denotes the normed space of all equivalence classes of Lebesgue measurable functions $f : I \to \mathbb{R}$ for which the norm

$$\|f\|_1 = \int_I |f|\, dt$$

is finite; here dt denotes integration against Lebesgue measure. Thus a sequence $\{f_n\}_{n\in\mathbb{N}} \subset L^1(I)$ converges strongly to $f \in L^1(I)$ if and only if $\lim_{n\to\infty} \|f_n - f\|_1 = 0$. It is well-known that the normed space $L^1(I)$ is complete, i.e., is a Banach space (see Appendix D). The space $L^\infty(I)$ consists of all equivalence classes of Lebesgue measurable functions $f : I \to \mathbb{R}$ that are (essentially) bounded, i.e., there exists a constant $C < \infty$ such that $|f| \leq C$ almost everywhere on I. When endowed with the norm

$$\|f\|_\infty = \inf\{C > 0 : |f| \leq C \text{ a.e. on } I\},$$

$L^\infty(I)$ also is a Banach space (see Appendix D). Note that if $h \in L^\infty(I)$, then the map $\ell : L^1(I) \to \mathbb{R}$ defined by

$$\ell(f) = \int_I hf\, dt \tag{3.4}$$

is a continuous linear functional on $L^1(I)$,

$$|\ell(f)| \leq \int_I |hf|\, dt \leq \|h\|_\infty \int_I |f|\, dt = \|h\|_\infty \cdot \|f\|_1.$$

Thus, every element $h \in L^\infty(I)$ gives rise to an element of the dual space of $L^1(I)$ by means of Eq. (3.4). In fact, these are *all* the continuous functionals on $L^1(I)$, i.e., every continuous linear functional ℓ on $L^1(I)$ is of the form given in Eq. (3.4) [59, Thm. 13.18]. Thus the dual space to $L^1(I)$ can be identified with $L^\infty(I)$,

$$\left(L^1(I)\right)' \simeq L^\infty(I).$$

Hence, for the space $L^1(I)$, the definition of weak convergence is equivalent to the following statement.

Definition 3.2.2 (Weak convergence in $L^1(I)$). A sequence $\{f_n\}_{n \in \mathbb{N}} \subset L^1(I)$ converges weakly to f in $L^1(I)$, $f_n \rightharpoonup f$, if and only if for all $h \in L^\infty(I)$,

$$\lim_{n \to \infty} \int_I h f_n \, dt = \int_I h f \, dt.$$

In our applications, the function f will be vector-valued, $f = (f_1, \ldots, f_n)^T$. The above definitions are then understood componentwise; for example,

$$\int_I f \, dt = \begin{pmatrix} \int_I f_1 \ dt \\ \vdots \\ \int_I f_n \ dt \end{pmatrix},$$

and we write $L^1(I; \mathbb{R}^n)$ for the Banach space of integrable vector-valued functions f. The dual space is then given by $L^\infty(I; \mathbb{R}^n)$, and the product becomes the inner product of vectors.

Definition 3.2.3 (Weakly sequentially compact). A subset $\mathscr{Q}$ of $L^1(I; \mathbb{R}^n)$ is said to be weakly sequentially compact if for any sequence $\{f_n\}_{n \in \mathbb{N}} \subset \mathscr{Q}$, there exists a subsequence $\{f_{n_k}\}_{k \in \mathbb{N}}$ that converges weakly to some limit $f \in \mathscr{Q}$.

Theorem 3.2.1. *Let $U \subset \mathbb{R}^m$ be compact and convex, I a compact interval, and let $\mathscr{U}$ denote the class of all Lebesgue measurable functions $u : I \to U$ defined on the interval I with values in the set U almost everywhere. Then $\mathscr{U}$ is weakly sequentially compact in $L^1(I; \mathbb{R}^m)$.*

This result is important in its own right and will also be used at various times later on in the context of nonlinear systems. For this reason, we include its proof. However, several advanced but nevertheless standard results from real analysis will need to be used that lie beyond the scope of our text, and for these we need to refer the reader to the abundant textbook literature on this topic (e.g., [59, 174, 257]).

Proof. Let $\{u_k\}_{k \in \mathbb{N}}$ be a sequence from $\mathscr{U}$. By changing the controls on a set of Lebesgue measure 0, if necessary, we may assume that the controls take values in U for all $t \in [0, T]$. The proof proceeds in two steps: First we show that there exists a Lebesgue measurable control $\bar{u} : I \to \mathbb{R}^m$ such that for all $h \in L^\infty(I; \mathbb{R}^m)$ we have that

$$\lim_{n \to \infty} \int_I \langle h, u_n \rangle \, dt = \int_I \langle h, \bar{u} \rangle \, dt.$$

Convexity of the control set is not required to prove this part, but is needed in the second step, in which it is shown that the values of $\bar{u}$ in fact lie in the control set U.

Step 1: It suffices to verify the limit property componentwise and thus for this step, we may assume that $m = 1$. Since the control set U is compact, all controls are uniformly bounded, say $|u_k(t)| \leq C < \infty$ for all $t \in [0,T]$. In particular, each control also is an element of the Hilbert space $L^2(I)$ of all square-integrable Lebesgue measurable functions f with the norm

$$\|f\|_2 = \sqrt{\int_I |f|^2 \, dt},$$

which is induced by the inner product

$$\langle f, g \rangle = \int_I fg \, dt.$$

There exists a countable complete orthonormal set $\{\Phi_i\}$ for $L^2(I)$ [257, Thm. 8.24], and every control u_k can be represented in $L^2(I)$ as a convergent Fourier series [257, Thm. 8.27],

$$u_k = \sum_{i=1}^{\infty} c_{k,i} \Phi_i \quad \text{with} \quad \sum_{i=1}^{\infty} |c_{k,i}|^2 = \int_I |u_k|^2 \, dt \leq C^2 \mu(I)$$

and $\mu(I)$ the length of the interval I. In particular, $|c_{k,i}| \leq C\sqrt{\mu(I)}$ for all i and k, and thus all the sequences $\{c_{k,i}\}_{k \in \mathbb{N}}$ are bounded. Hence there exists a sequence $\{k_r\}_{r \in \mathbb{N}}$ such that for each i the limits

$$\gamma_i = \lim_{r \to \infty} c_{k_r,i} \quad \text{exist and satisfy} \quad \sum_{i=1}^{\infty} |\gamma_i|^2 \leq C^2 \mu(I).$$

[Since $\{c_{k,1}\}_{k \in \mathbb{N}}$ is bounded, there exists a subsequence $\{k_r^{(1)}\}_{r \in \mathbb{N}}$ such that $\gamma_1 = \lim_{r \to \infty} c_{k_r^{(1)},1}$. Then pick another convergent subsequence $\{k_r^{(2)}\}_{r \in \mathbb{N}}$ of $\{k_r^{(1)}\}_{r \in \mathbb{N}}$ such that $\gamma_2 = \lim_{r \to \infty} c_{k_r^{(2)},2}$. It follows inductively that there exist subsequences $\{k_r^{(i)}\}_{r \in \mathbb{N}}$ of $\{k_r^{(i-1)}\}_{r \in \mathbb{N}}$ such that $\gamma_i = \lim_{r \to \infty} c_{k_r^{(i)},i}$ exists for all $i \in \mathbb{N}$. But then the diagonal sequence $\{k_r^{(r)}\}_{r \in \mathbb{N}}$ has the desired property: given any index i, the sequence $\{k_r^{(r)}\}_{r \geq i}$ is a subsequence of $\{k_r^{(i)}\}_{r \in \mathbb{N}}$ and thus also $\gamma_i = \lim_{r \to \infty} c_{k_r^{(r)},i}$. Furthermore, for any positive integer N, we have that

$$\sum_{i=1}^{N} |\gamma_i|^2 = \lim_{r \to \infty} \sum_{i=1}^{N} \left|c_{k_r^{(r)},i}\right|^2 \leq \lim_{r \to \infty} \sum_{i=1}^{\infty} \left|c_{k_r^{(r)},i}\right|^2 \leq C^2 \mu(I),$$

implying the second condition.] Hence, by the Riesz–Fischer theorem [257, Thm. 8.30], there exists a measurable function $\bar{u} : I \to \mathbb{R}^m$ such that

$$\bar{u} = \sum_{i=1}^{\infty} \gamma_i \Phi_i.$$

This function $\bar{u}$ is the desired limit. Let $h \in L^\infty(I)$ and suppose $|h(t)| \le M$ on I. The function h also lies in $L^2(I)$ and thus can be represented by its Fourier series as $h = \sum_{i=1}^\infty \eta_i \Phi_i$. For any finite sum, $p = \sum_{i=1}^N \eta_i \Phi_i$, we have that

$$
\lim_{r\to\infty} \int_I p u_{k_r} dt = \lim_{r\to\infty} \int_I \left(\sum_{i=1}^N \eta_i \Phi_i \right) u_{k_r} dt = \sum_{i=1}^N \eta_i \left(\lim_{r\to\infty} \int_I \Phi_i u_{k_r} dt \right)
$$

$$
= \sum_{i=1}^N \eta_i \left(\lim_{r\to\infty} c_{k_r,i} \right) = \sum_{i=1}^N \eta_i \gamma_i = \sum_{i=1}^N \eta_i \int_I \Phi_i \bar{u}\, dt = \int_I p \bar{u}\, dt.
$$

Hence

$$
\left| \int_I h u_{k_r} dt - \int_I h \bar{u}\, dt \right| \le \left| \int_I p\,(u_{k_r} - \bar{u})\, dt \right| + \left| \int_I (h - p)\,(u_{k_r} - \bar{u})\, dt \right|,
$$

and the first term converges to 0 as $r \to \infty$. Given $\varepsilon > 0$, choose a positive integer N such that $\int_I |h - p|^2\, dt < \varepsilon^2$. Using Hölder's inequality (see Proposition D.3.1 in Appendix D), the second term can then be bounded in the form

$$
\left| \int_I (h - p)\,(u_{k_r} - \bar{u})\, dt \right|^2 \le \left| \int_I (h - p)^2\, dt \right| \cdot \left| \int_I (u_{k_r} - \bar{u})^2\, dt \right|
$$

$$
\le \varepsilon^2 \| u_{k_r} - \bar{u} \|^2 \le 2C^2 \mu(I) \varepsilon^2
$$

and thus also converges to 0. Overall, this proves the first step:

$$
\lim_{r\to\infty} \int_I h u_{k_r} dt = \int_I h \bar{u}\, dt.
$$

Step 2: Since U is closed and convex, by Corollary 3.1.2 there exist (at most) countably many hyperplanes $H_i = \{x \in \mathbb{R}^n : c_i^T x = a_i\}$, $i \in \mathbb{N}$, such that $U = \cap_{i=1}^\infty H_i^+$, $H_i^+ = \{x \in \mathbb{R}^n : c_i^T x \ge a_i\}$. Let $E_i = \{t \in I : c_i^T \bar{u}(t) < a_i\}$ and denote the characteristic function of the set E_i by χ_i. By step 1, we therefore have that

$$
0 \ge \int_{E_i} \left(c_i^T \bar{u}(t) - a_i \right) dt = \int_I \chi_i \left(c_i^T \bar{u}(t) - a_i \right) dt = \lim_{r\to\infty} \int_I \chi_i \left(c_i^T u_{k_r}(t) - a_i \right) dt \ge 0.
$$

Hence E_i must be a null set, and thus $c_i^T \bar{u}(t) \ge a_i$ a.e. on I. Since this holds for all i, the values of $\bar{u}$ lie in U a.e., and thus $\bar{u} \in \mathscr{U}$. $\qquad\qquad\square$

3.3 Topological Properties of Reachable Sets

We use Theorem 3.2.1 to establish basic topological properties of reachable sets for linear systems.

Theorem 3.3.1. *If the control set U is compact and convex, then the time-t-reachable set $\mathrm{Reach}_{\Sigma,t}(0)$ is closed.*

Proof. Let $\{p_k\}_{k\in\mathbb{N}} \subset \mathrm{Reach}_{\Sigma,t}(0)$ be a sequence of points in the time-t-reachable set, say

$$p_k = \int_0^t e^{A(t-s)} B u_k(s)\,ds$$

for some admissible control $u_k(\cdot)$, and suppose $p_k \to p$. Since $\mathcal{U}$ is weakly sequentially compact in $L^1(I)$, there exists a subsequence $\{u_{k_r}\}_{r\in\mathbb{N}}$ such that u_{k_r} converges weakly to some admissible control $u \in \mathcal{U}$. The vector function $s \mapsto e^{A(t-s)}B$ is continuous, hence bounded, and thus lies in $L^\infty(I;\mathbb{R}^n)$. Therefore

$$p = \lim_{r\to\infty} p_{k_r} = \lim_{r\to\infty} \int_0^t e^{A(t-s)} B u_{k_r}(s)\,ds = \int_0^t e^{A(t-s)} B u(s)\,ds \in \mathrm{Reach}_{\Sigma,t}(0)$$

and thus p also is reachable from the origin in time t.   □

Corollary 3.3.1. *If the control set U is compact and convex, then the time-t-reachable set $\mathrm{Reach}_{\Sigma,t}(0)$ is compact and convex.*

Proof. It is clear from Eq. (3.2) that the reachable set is bounded. Hence, by Theorem 3.3.1, it is compact. Furthermore, if $u_1 : [0,t] \to U$ and $u_2 : [0,t] \to U$ are admissible controls that steer 0 into p_1 and 0 into p_2, respectively, then the convex combination $u_\lambda = \lambda u_1 + (1-\lambda)u_2$ also is admissible and steers 0 into $\lambda p_1 + (1-\lambda)p_2$,

$$\lambda p_1 + (1-\lambda)p_2 = \lambda \int_0^t e^{A(t-s)} B u_1(s)\,ds + (1-\lambda)\int_0^t e^{A(t-s)} B u_2(s)\,ds$$

$$= \int_0^t e^{A(t-s)} B u_\lambda(s)\,ds.$$

This verifies the corollary.   □

It is a remarkable result that convexity of the control set is not needed for the time-t-reachable sets to be convex. By Lyapunov's theorem on the range of a vector-valued measure, time-t-reachable sets are *always* convex. The reason essentially is that the class of Lebesgue measurable sets is so large that if u_1 steers 0 into p_1 and u_2 steers 0 into p_2 in time t, then these convex combinations can be realized by "switching" between the admissible controls u_1 and u_2 in time over Lebesgue measurable sets.

Theorem 3.3.2. *For any control set $U \subset \mathbb{R}^m$, the time-t-reachable set* $\text{Reach}_{\Sigma,t}(0)$ *is convex.*

The proof of this theorem is a direct consequence of the following interesting result, which gives an idea about the richness of the class of Lebesgue measurable sets. It holds under more general assumptions on the integrand f, but this version suffices for our purpose [70, 117].

Proposition 3.3.1. *Let $f \in L^\infty(I, \mathbb{R}^n)$ be a bounded vector-valued Lebesgue measurable function and let w be a scalar Lebesgue measurable function defined on I with values in $[0,1]$, $w : I \to [0,1]$. Then there exists a Lebesgue measurable subset E of I such that*

$$\int_I fw\, ds = \int_E f\, ds.$$

Proof. [70] Let $\mathcal{D}$ denote the weakly sequentially compact set of all Lebesgue measurable functions $v : I \to [0,1]$. Define a linear operator $T : \mathcal{D} \to \mathbb{R}^n$ by

$$T(v) = \int_I fv\, ds$$

and let $q = \int_I fw\, ds$. Denote the inverse image of the point q under the operator T by $\mathcal{S}$, $\mathcal{S} = T^{-1}(\{q\})$. This set $\mathcal{S}$ is convex: if $T(v_1) = q$ and $T(v_2) = q$, then for any $\lambda \in [0,1]$, the function $\lambda v_1 + (1-\lambda)v_2$ also lies in $\mathcal{D}$ and

$$T(\lambda v_1 + (1-\lambda)v_2) = \lambda T(v_1) + (1-\lambda)T(v_2) = q.$$

Furthermore, the operator T is continuous in the weak topology: if $\{v_n\}_{n \in \mathbb{N}} \subset \mathcal{D}$ converges weakly to some $v \in \mathcal{D}$, $v_n \rightharpoonup v$, then

$$\lim_{n \to \infty} T(v_n) = \lim_{n \to \infty} \int_I fv_n\, ds = \lim_{n \to \infty} \int_I fv\, ds = T(v).$$

Hence $\mathcal{S}$ is a weakly sequentially closed subset of $\mathcal{D}$. Altogether, $\mathcal{S}$ thus is a nonempty weakly compact and convex subset of $L^1(I)$.

It follows from an infinite-dimensional version of the Krein–Milman theorem [117] that $\mathcal{S}$ has at least one extreme point, say $\varpi \in \mathcal{D}^{\text{ext}}$. We *claim* that such an extreme point ϖ must have values in the set of extreme points of $[0,1]$, i.e., in $\{0,1\}$, almost everywhere. If we then take E as the set where ϖ takes the value 1, we get that

$$\int_I f w \, ds = q = \int_I f \varpi \, ds = \int_E f \, ds,$$

which concludes the proof.

While this claim may seem intuitively clear, this is the difficult part of the argument. We shall conclude the proof with an inductive argument on the dimension n. Suppose $P = \{s \in I : 0 < \varpi(s) < 1\}$ has positive measure. Then there exists an $\varepsilon > 0$ such that the set $A_\varepsilon = \{s \in I : \varepsilon < \varpi(s) < 1 - \varepsilon\}$ has positive measure (otherwise $P = \bigcup_{n \in \mathbb{N}} A_{\frac{1}{n}}$ is a null set). By choosing a Lebesgue measurable subset $A \subset A_\varepsilon$, if necessary, in addition we may assume that

$$\int_A \|f\|_\infty \, ds < \varepsilon.$$

(Suppose $I = [a,b]$ and let $m = \|f\|_\infty = \max_{i=1,\ldots,n} \|f_i\|_\infty$. If we partition $[a,b]$ into a finite number of intervals I_j of length less than $\frac{\varepsilon}{m}$, then at least one of the sets $A_\varepsilon \cap I_j$ has positive measure, and we may take any of these for A.) Write A as the disjoint union of two Lebesgue measurable sets F_1 and F_2 of positive measure, $A = F_1 \cup F_2$, $F_1 \cap F_2 = \varnothing$.

We first consider the scalar case, $n = 1$. Let

$$\alpha = \int_{F_1} f \, ds \quad \text{and} \quad \beta = \int_{F_2} f \, ds,$$

so that $|\alpha| < \varepsilon$, $|\beta| < \varepsilon$ and define a Lebesgue measurable function h as

$$h(s) = \begin{cases} \alpha & \text{if } s \in F_2, \\ -\beta & \text{if } s \in F_1, \\ 0 & \text{if } s \notin A. \end{cases}$$

In the special case in which both α and β are zero, set the values of h to $+1$ and -1, respectively. By construction,

$$\int_I f(\varpi \pm h) \, ds = \int_I f \varpi ds \pm \left(\alpha \int_{F_2} f \, ds - \beta \int_{F_1} f \, ds \right) = q \pm 0 = q,$$

and so both $\varpi + h$ and $\varpi - h$ lie in $\mathscr{S}$. But they differ from ϖ on a set of positive measure, and thus

$$\varpi = \frac{1}{2}(\varpi + h) + \frac{1}{2}(\varpi - h)$$

is a nontrivial convex combination of elements in $\mathscr{S}$. Thus ϖ is not an extreme point. Contradiction. This proves the proposition in the scalar case.

Now inductively assume that the proposition is correct for dimensions less than n and suppose $f \in L^\infty(I;\mathbb{R}^n)$. By the inductive assumption, there exist Lebesgue measurable subsets G_1 of F_1 and G_2 of F_2 such that for $i = 1,\ldots,n-1$, we have that

$$\frac{1}{2}\int_{F_1} f_i \, ds = \int_{G_1} f_i \, ds \quad \text{and} \quad \frac{1}{2}\int_{F_2} f_i \, ds = \int_{G_2} f_i \, ds.$$

For $j = 1,2$ define Lebesgue measurable functions h_j by

$$h_j(s) = \begin{cases} 1 & \text{if} \quad s \in G_j, \\ -1 & \text{if} \quad s \in F_j \backslash G_j, \\ 0 & \text{if} \quad s \notin F_j. \end{cases}$$

Note that the functions h_j vanish outside of F_j. Then, for each $i = 1,\ldots,n-1$, we have that

$$\int_{F_1} f_i h_1 \, ds = \int_{G_1} f_i \, ds - \int_{F_1 \backslash G_1} f_i \, ds = 2\int_{G_1} f_i \, ds - \int_{F_1} f_i \, ds = 0,$$

and analogously,

$$\int_{F_2} f_i h_2 \, ds = 0.$$

Since h_2 vanishes on F_1 and h_1 vanishes on F_2, we have that

$$\int_{F_1} f_i h_2 \, ds = 0 = \int_{F_1} f_i h_2 \, ds$$

for $i = 1,\ldots,n-1$. Thus it follows for $j = 1,2$, and $i = 1,\ldots,n-1$, that

$$\int_A f_i h_j \, ds = 0.$$

Since h_j vanishes outside of A, also

$$\int_I f_i h_j \, ds = 0.$$

As in the scalar case, for the last component f_n now define

$$\alpha = \int_{F_1} f_n h_1 \, ds \quad \text{and} \quad \beta = \int_{F_2} f_n h_2 \, ds,$$

so that $|\alpha| < \varepsilon$, $|\beta| < \varepsilon$, and define a Lebesgue measurable function h by

$$h = \alpha h_2 - \beta h_1.$$

If both α and β are zero, take $h = h_1 - h_2$. Then, as above,

$$\int_I f_n h \, ds = \alpha \int_{F_2} f_n h_2 \, ds - \beta \int_{F_1} f_n h_1 \, ds = 0,$$

and thus

$$\int_I f h \, ds = 0.$$

This provides the same contradiction as in the scalar case: the functions $\varpi + h$ and $\varpi - h$ still lie in $\mathscr{Q}$ and

$$\int_I f(\varpi \pm h) \, ds = \int_I f \varpi \, ds \pm \int_I f h \, ds = q,$$

i.e., both $\varpi + h$ and $\varpi - h$ still lie in $\mathscr{S}$. These functions differ from ϖ on a set of positive measure and $\varpi = \frac{1}{2}(\varpi + h) + \frac{1}{2}(\varpi - h)$ is a nontrivial convex combination of elements in $\mathscr{S}$. Contradiction. This concludes the inductive argument.   □

Proof of Theorem 3.3.2: Let $u_1 : [0,t] \to U$ and $u_2 : [0,t] \to U$ be two admissible controls that steer 0 into p_1 and 0 into p_2, respectively. Given $\lambda \in (0,1)$, take the function w in Proposition 3.3.1 constant equal to λ, $w(s) \equiv \lambda$. It follows that there exists a Lebesgue measurable subset E_λ of the interval $[0,t]$ such that

$$\lambda (p_1 - p_2) = \int_{[0,t]} e^{A(t-s)} B(u_1(s) - u_2(s)) w(s) \, ds = \int_{E_\lambda} e^{A(t-s)} B(u_1(s) - u_2(s)) \, ds.$$

The control u_λ defined as

$$u_\lambda(t) = \begin{cases} u_1(t) & \text{if } t \in E_\lambda, \\ u_2(t) & \text{if } t \notin E_\lambda, \end{cases}$$

is admissible, i.e., is Lebesgue measurable, bounded, takes values in the control set U almost everywhere, and we have that

$$\lambda p_1 + (1-\lambda) p_2 = \lambda (p_1 - p_2) + p_2$$

$$= \int_{E_\lambda} e^{A(t-s)} B(u_1(s) - u_2(s)) \, ds + \int_0^T e^{A(t-s)} B u_2(s) \, ds$$

$$= \int_0^T e^{A(t-s)} B u_\lambda(s) \, ds \in \text{Reach}_{\Sigma,t}(0).$$

Thus any convex combination lies in $\text{Reach}_{\Sigma,t}(0)$.   □

3.4   The General Bang-Bang Theorem

As above, let Σ be a linear time-invariant system with a control set $U \subset \mathbb{R}^m$. We assume that U is nonempty convex and compact and denote by $\mathscr{U}$ the class of all Lebesgue measurable functions that take values in U. Thus $\mathscr{U}$ is a nonempty, weakly sequentially compact, and convex subset of $L^1(I; \mathbb{R}^m)$. It follows from the infinite-dimensional version of the Krein–Milman theorem that $\mathscr{U}$ has extreme points, $\mathscr{U}^{\text{ext}} \neq \emptyset$, and that $\mathscr{U} = co\,(\mathscr{U}^{\text{ext}})$. Similarly, by Theorem 3.1.2, the set U^{ext} of extreme points of $U \subset \mathbb{R}^m$ is nonempty and U is the convex hull of the set U^{ext}. Denote by $\mathscr{U}_{\text{ext}}$ the class of all Lebesgue measurable functions $u : [0, \infty) \to U^{\text{ext}}$ that take values in U^{ext} almost everywhere and let Σ^{ext} be the same linear time-invariant system, but with control set U^{ext} and admissible class of controls given by $\mathscr{U}_{\text{ext}}$,

$$\Sigma^{\text{ext}}: \qquad \dot{x} = Ax + Bu, \qquad x(0) = p, \qquad u \in U^{\text{ext}}. \tag{3.5}$$

It is easy to see that $\mathscr{U}_{\text{ext}} \subset \mathscr{U}^{\text{ext}}$: if $v \in \mathscr{U}_{\text{ext}}$ and $v = \lambda u_1 + (1 - \lambda)u_2$ with $u_1, u_2 \in \mathscr{U}$ and $\lambda \in (0, 1)$, then a.e. in $[0, t]$, the point $v(t)$ is an extreme point of U, and thus $u_1(t) = u_2(t) = v(t)$. Hence $v = u_1 = u_2$, and thus v is an extreme point of $\mathscr{U}$, i.e., $v \in \mathscr{U}^{\text{ext}}$. The converse, however, is not so clear, and thus we consider the system Σ^{ext} with control set $\mathscr{U}_{\text{ext}}$. The following theorem states that any point q that is reachable in time T at all can also be reached in the same time T by means of a control that takes values in the set U^{ext} of extreme points of the control set, i.e., with a control in $\mathscr{U}_{\text{ext}}$. This is the celebrated bang-bang theorem for linear systems.

Theorem 3.4.1 (General bang-bang theorem). *For every $T > 0$, the time-T-reachable sets for the systems Σ and Σ^{ext} are equal,*

$$\text{Reach}_{\Sigma, T}(0) = \text{Reach}_{\Sigma^{\text{ext}}, T}(0). \tag{3.6}$$

Since $\mathscr{U}_{\text{ext}} \subset \mathscr{U}$, it is clear that

$$\text{Reach}_{\Sigma^{\text{ext}}, T}(0) \subset \text{Reach}_{\Sigma, T}(0).$$

By Corollary 3.3.1, the reachable set $\text{Reach}_{\Sigma, T}(0)$ is compact and convex, and thus by the Krein–Milman theorem, it is the convex hull of its extreme points,

$$\text{Reach}_{\Sigma, T}(0) = co\left\{ (\text{Reach}_{\Sigma, T}(0))^{\text{ext}} \right\}.$$

We shall show in Proposition 3.4.1 below that

$$(\text{Reach}_{\Sigma, T}(0))^{\text{ext}} \subset \text{Reach}_{\Sigma^{\text{ext}}, T}(0).$$

By Theorem 3.3.2, $\text{Reach}_{\Sigma^{ext}, T}(0)$ is convex, and thus it follows that

$$co\left[(\text{Reach}_{\Sigma, T}(0))^{\text{ext}} \right] \subset \text{Reach}_{\Sigma^{\text{ext}}, T}(0).$$

Hence Eq. (3.6) holds. The bang-bang theorem therefore follows from the following result:

Proposition 3.4.1. *Let p be an extreme point of the time-T-reachable set. Then there exists a control $v \in \mathcal{U}_{\text{ext}}$ such that*

$$p = \int_{[0,T]} e^{A(T-s)} Bv(s)ds, \qquad \text{i.e., } p \in \text{Reach}_{\Sigma \text{ext},T}(0).$$

If the columns of the matrix B are linearly independent, then this control v is unique and is the only control that steers 0 into p.

Proof. Make a linear change of coordinates so that p becomes the first element with respect to the induced lexicographic ordering $\prec$ in a new orthonormal basis $\mathcal{B} = (h_1, \ldots, h_n)$. Let $u \in \mathcal{U}$ be an admissible control that steers 0 into p,

$$p = \int_{[0,T]} e^{A(T-s)} Bu(s)ds.$$

If necessary, change u on a set of measure zero so that $u(t) \in U$ for all $t \in [0,T]$. Denote by E_t the compact and convex set $E_t = \{e^{A(T-t)} Bv : v \in U\}$ and let w_t be the lexicographically first element of E_t with respect to the basis $\mathcal{B}$. By Proposition 3.1.5, w_t is an extreme point of E_t. It is a highly nontrivial technical issue to establish that the function w is actually Lebesgue measurable, and we only indicate the reasoning. The following Lemma is still rather elementary:

Lemma 3.4.1. *The graph of the function $w : [0,T] \to \mathbb{R}^n$ is Borel measurable.*

Proof. The Borel σ-algebra is the smallest σ-algebra that contains all open (and closed) sets (see Appendix D). Let $F = \{(t,x) : t \in [0,T], x \in E_t\}$; F is the continuous image of the compact set $[0,T] \times U$ under the mapping

$$(t,u) \mapsto \left(t, e^{A(T-t)} Bu\right)$$

and thus is compact, hence a Borel set. Let F_1 be the set of all points (t,x) such that x minimizes the first coordinate over the set E_t, i.e.,

$$F_1 = \{(t,x) \in F : \langle h_1, x \rangle \leq \langle h_1, y \rangle \text{ for all } y \in E_t\}.$$

In general, the set F_1 need no longer be compact, since the point where the minimum is attained can jump. But F_1 is Borel measurable: the set

$$G = \{(t,x,y) \in [0,T] \times \mathbb{R}^n \times \mathbb{R}^n : x \in E_t, \, y \in E_t\}$$

is the continuous image of $[0,1] \times U \times U$ under the mapping

$$(t,u,v) \longmapsto \left(t, e^{A(T-t)} Bu, e^{A(T-t)} Bv\right)$$

and thus is compact. Then

$$G_{1,n} = \left\{ (t,x,y) \in G : \langle h_1, y \rangle \leq \langle h_1, x \rangle - \frac{1}{n} \right\}$$

is a closed subset of a compact set and hence is compact as well. Let π denote the projection $\pi : (t,x,y) \mapsto (t,y)$. Since projections of compact sets are compact, the set $\pi(G_{1,n})$ is also compact and thus is a Borel set. But a point $(t,x) \in F$ does not lie in F_1 if and only if there exists a point $(t,y) \in F$ such that $\langle h_1, y \rangle < \langle h_1, x \rangle$ and thus

$$F_1 = F \cap \left(\bigcup_{n \in \mathbb{N}} \pi(G_{1,n}) \right)^c$$

is a Borel set. It is not difficult to iterate this construction. For example,

$$F_2 = \{ (t,x) \in F_1 : \langle h_2, x \rangle \leq \langle h_2, y \rangle \text{ for all } y \text{ such that } (t,y) \in F_1 \},$$

and again the set

$$G_{2,n} = \left\{ (t,x,y) \in G : \langle h_1, y \rangle = \langle h_1, x \rangle, \ \langle h_2, y \rangle \leq \langle h_2, x \rangle - \frac{1}{n} \right\}$$

is compact, and we also have that

$$F_2 = F_1 \cap \left(\bigcup_{n \in \mathbb{N}} \pi(G_{2,n}) \right)^c.$$

Proceeding with a finite induction, the set F_n is Borel measurable. But F_n is the graph of the function w, and thus this proves the lemma.   □

This reasoning is necessary, since in general, it is not true that the projection of a Borel measurable set is Borel measurable. But the following rather deep result holds.

Proposition 3.4.2. *Let B be a Borel measurable subset of $\mathbb{R}^n$ and let $f : \mathbb{R}^n \to \mathbb{R}^k$ be a continuous mapping. Then the image $f(B)$ is Lebesgue measurable in $\mathbb{R}^k$.* ∎

Corollary 3.4.1. *If the graph of a function $w : [0,T] \to \mathbb{R}^n$ is Borel measurable in $\mathbb{R} \times \mathbb{R}^n$, then the function w itself is Lebesgue measurable in $\mathbb{R}^n$.* □

This result establishes the measurability of the function w constructed as the lexicographic minimum over the sets E_t, a highly nontrivial technical issue. We refer the reader to the textbook [59] for this proof (see Exercise 11:6.2 on p. 496). The rest of the argument is rather straightforward.

Fig. 3.3 Lift of an extreme
point $\bar{v}(t) \in \bar{U}$ to an extreme
point $v(t) \in U$

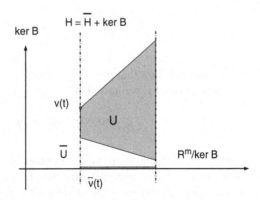

Lemma 3.4.2. *There exists a Lebesgue measurable control $v(t)$ with values in U^{ext} such that $w_t = e^{A(T-t)}Bv(t)$.*

Proof. If the columns of B are linearly independent, we can simply solve the equation $w_t = e^{A(T-t)}Bv(t)$ for $v(t)$ in terms of the pseudo-inverse as

$$v(t) = (B^T B)^{-1} B^T e^{A(t-T)} w_t,$$

and this solution is unique. By construction, the values of v are extreme points: if $v(t) = \lambda v_1(t) + (1-\lambda)v_2(t)$ for some $\lambda \in (0,1)$, then

$$\begin{aligned}
w_t &= e^{A(T-t)}Bv(t) \\
&= \lambda e^{A(T-t)}Bv_1(t) + (1-\lambda)e^{A(T-t)}Bv_2(t) \\
&= \lambda w_1(t) + (1-\lambda)w_2(t),
\end{aligned}$$

and $w_1(t)$ and $w_2(t)$ lie in the set E_t. Since w_t is an extreme point of E_t, we have $w_1(t) = w_2(t)$, and since the columns of B are linearly independent, this is possible only if $v_1(t) = v_2(t)$. Thus v takes values in the set U^{ext}.

If the columns of B are not linearly independent, we need to factor out the kernel of B, $\ker B$. In the factor space, there exists a unique control $\bar{v}(t)$, and this control takes values in the extreme points of the set $\bar{U} = U/\ker B$. But extreme points of $\bar{U}$ can be lifted to extreme points of U: if $\bar{u}$ is an extreme point of $\bar{U}$, pick a supporting hyperplane $\bar{H}$ to $\bar{U}$ at $\bar{u}$. Then $\bar{H}$ lifts to the supporting hyperplane $H = \bar{H} + \ker B$ to U (see Fig. 3.3), and thus H contains a least one extreme point $v(t)$. This lift can be done so that measurability is preserved. $\square$

It remains to show that the new control $v \in \mathscr{U}_{\text{ext}}$ steers the system into the same point p. For the moment, set

$$q = \int_{[0,T]} e^{A(T-t)} Bv(t)dt \in \text{Reach}_{\Sigma^{\text{ext}},T}(0).$$

We then need to show that $p = q$. By the choice of w as lexicographic minimum and the construction of v, we have that

$$\left\langle h_1, e^{A(T-t)} Bv(t) \right\rangle \leq \left\langle h_1, e^{A(T-t)} Bu(t) \right\rangle \qquad (3.7)$$

for all $t \in [0,T]$. Integrating over $[0,T]$ gives $\langle h_1, q \rangle \leq \langle h_1, p \rangle$. But by assumption, p was the lexicographic minimum, and therefore we must have $\langle h_1, q \rangle = \langle h_1, p \rangle$. Thus equality holds almost everywhere in Eq. (3.7), i.e.,

$$\left\langle h_1, e^{A(T-t)} Bv(t) \right\rangle = \left\langle h_1, e^{A(T-t)} Bu(t) \right\rangle \qquad \text{a.e.}$$

Now simply iterate this argument. We then also have that

$$\left\langle h_2, e^{A(T-t)} Bv(t) \right\rangle \leq \left\langle h_2, e^{A(T-t)} Bu(t) \right\rangle \qquad (3.8)$$

for all $t \in [0,T]$ and thus $\langle h_2, q \rangle \leq \langle h_2, p \rangle$. As above, this implies $\langle h_2, q \rangle = \langle h_2, p \rangle$, and again equality holds almost everywhere in Eq. (3.8), i.e.,

$$\left\langle h_2, e^{A(T-t)} Bv(t) \right\rangle = \left\langle h_2, e^{A(T-t)} Bu(t) \right\rangle \qquad \text{a.e.}$$

Inductively, it follows that

$$\left\langle h_i, e^{A(T-t)} Bv(t) \right\rangle = \left\langle h_i, e^{A(T-t)} Bu(t) \right\rangle \qquad \text{a.e.}$$

for $i = 1,\ldots,n$. The set $\mathscr{B} = (h_1,\ldots,h_n)$ is an ordered basis, and thus $p = q \in \text{Reach}_{\Sigma^{\text{ext}},T}(0)$. $\qquad \square$

Note that the general bang-bang theorem merely states that if a point q is reachable in time T at all, then it can also be reached in the same time T by means of a Lebesgue measurable control $v \in \mathscr{U}_{\text{ext}}$ that takes values in the set U^{ext} of extreme points of the control set U. This does not guarantee that the control v has any additional regularity properties such as being piecewise continuous, a practically desirable feature. For the case that U is a compact polyhedron, i.e., when U is the convex hull of a finite set of points, then indeed piecewise constant controls suffice. But this argument uses entirely different methods based on the characterization of boundary trajectories given in Theorem 2.5.3. We establish this result next.

3.5 Boundary Trajectories and Small-Time Local Controllability

The time-t-reachable sets for linear systems with compact control set are compact and convex. Hence boundary points can be characterized in terms of supporting hyperplanes. This directly leads to the following analytical characterization of boundary points.

Theorem 3.5.1. *Let Σ be a linear time-invariant system with a nonempty compact and convex control set U and let u be an admissible control that steers the origin into a point p in time T. Then p is a boundary point of the time-T-reachable set, $p \in \partial \operatorname{Reach}_{\Sigma,T}(0)$, if and only if there exists a nontrivial solution $\lambda : [0,T] \to (\mathbb{R}^n)^*$ to the adjoint equation $\dot\lambda = -\lambda A$ such that*

$$\lambda(t)Bu(t) = \min_{v \in U} \lambda(t)Bv \quad \text{a.e. on } [0,T]. \tag{3.9}$$

Proof. Since $\operatorname{Reach}_{\Sigma,T}(0)$ is compact and convex, a point $p \in \operatorname{Reach}_{\Sigma,T}(0)$ is a boundary point if and only if there exists a supporting hyperplane to $\operatorname{Reach}_{\Sigma,T}(0)$ at p. If u is an admissible control that steers 0 into p, then this is equivalent to the existence of a nonzero row vector $\eta \in (\mathbb{R}^n)^*$ such that

$$\left\langle \eta, \int_0^T e^{A(T-t)}Bu(t)dt \right\rangle \le \left\langle \eta, \int_0^T e^{A(T-t)}Bv(t)dt \right\rangle$$

for all admissible controls $v \in \mathscr{U}$. Equivalently,

$$\int_0^T \eta e^{A(T-t)}Bu(t)dt \le \int_0^T \eta e^{A(T-t)}Bv(t)dt, \tag{3.10}$$

and this holds if and only if

$$\eta e^{A(T-t)}Bu(t)dt \le \eta e^{A(T-t)}Bv \quad \text{for all } v \in U. \tag{3.11}$$

Clearly, Eq. (3.11) is sufficient for Eq. (3.10) to hold (simply integrate). Conversely, if Eq. (3.10) holds, consider a time $t \in [0,T)$ and choose $h > 0$ small enough that $t + h < T$. By taking as control

$$v(s) = \begin{cases} u(s) & \text{if } s \notin [t, t+h], \\ v & \text{if } s \in [t, t+h], \end{cases}$$

for some fixed value $v \in U$, it follows that

$$\frac{1}{h} \int\limits_t^{t+h} \eta e^{A(T-s)} Bu(s) \, ds \le \frac{1}{h} \int\limits_t^{t+h} \eta e^{A(T-s)} Bv \, ds.$$

Taking the limit as $h \to 0$, we obtain that

$$\eta e^{A(T-t)} Bu(t) dt \le \eta e^{A(T-t)} Bv \quad \text{a.e.}$$

For the right-hand side, this is simply the fundamental theorem of calculus; for the left-hand side, however, this is not that immediate, since the control u is only Lebesgue measurable. But it is valid as well by Lebesgue's differentiation theorem (see Appendix D, [257, Thm. 7.2]). Hence the theorem follows with $\lambda(t) = \eta e^{A(T-t)}$. $\square$

Definition 3.5.1 (Small-time locally controllable). A system Σ is said to be small-time locally controllable from a point p if there exists an $\varepsilon > 0$ such that p is an interior point of the time-t-reachable set from p for all times $t \in (0, \varepsilon)$,

$$p \in \text{int} \left(\text{Reach}_{\Sigma, t}(p) \right), \quad t \in (0, \varepsilon).$$

A necessary condition for $\Sigma : \dot{x} = Ax + Bu$, $u \in U$, to be small-time locally controllable is that it must be completely controllable. Otherwise the reachable set lies in a lower-dimensional affine subspace and does not have interior points. Clearly, in such a case, one can redefine the state space as the controllable subspace, and thus, without loss of generality, we can always assume that the underlying system is completely controllable.

The obstruction to small-time local controllability is that the dynamics forces the system to move away from p, as illustrated in the simple one-dimensional example $\dot{x} = u$ with control set U given by $U = [1, 2]$ (see Fig. 3.4). However, if 0 lies in the interior of the control set, this is not possible, and the proposition below gives a simple sufficient condition for small-time local controllability. It is not too difficult to give a complete characterization of small-time locally controllable time-invariant linear systems, but for this we refer the reader to the literature [225].

Proposition 3.5.1. *A completely controllable linear time-invariant system $\Sigma : \dot{x} = Ax + Bu$, $u \in U$, for which 0 is an interior point of the control set U, $0 \in \text{int}(U)$, is small-time locally controllable from the origin.*

Proof. By assumption, U contains a ball $B_\delta(0) = \{u \in \mathbb{R}^m : \|u\|_2 \le \delta\}$. Pick $\varepsilon > 0$ such that for all $t \in [0, T]$ we have that $\left\| e^{A(T-t)} Bv \right\|_2 < \delta$ whenever $\|v\|_2 < \varepsilon$. Then the controls $u(t) = e^{A(T-t)} Bv$ are admissible, and thus

$$W(T)v = \left(\int\limits_0^T e^{A(T-t)} BB^T e^{A^T(T-t)} dt \right) v = \int\limits_0^T e^{A(T-t)} Bu(t) dt \in \text{Reach}_{\Sigma, T}(0),$$

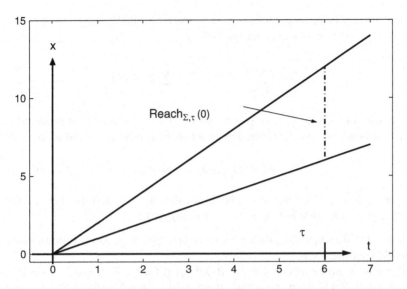

Fig. 3.4 For $\Sigma : \dot{x} = u \in [1,2]$ we have that $Reach_{\Sigma,t}(0) = [t, 2t]$

i.e., $W(T)B_\varepsilon(0) \subset Reach_{\Sigma,T}(0)$. Since (A,B) is completely controllable, the matrix $W(T)$ is positive definite (see Sect. 2.5), and thus $W(T)B_\varepsilon(0)$ contains a neighborhood of 0.   □

Theorem 3.5.2. *Let $\Sigma : \dot{x} = Ax + Bu$, $u \in U$, be small-time locally controllable from the origin. Then, a point p in the time-T-reachable set from the origin can be reached in time T time-optimally if and only if $p \in \partial\, Reach_{\Sigma,T}(0)$.*

Proof. We first show that if p lies in the interior of the time-T-reachable set from the origin, then it can in fact also be reached in shorter time. This is a direct consequence of the lemma below, which, for reasons of further applications, we formulate a bit more generally. Note that its proof does not use linearity of the system, but only the fact that the reachable sets are convex. It thus holds equally for nonlinear systems that have this property.

Lemma 3.5.1. *Let $\gamma : (a,b) \to \mathbb{R}^n$ be a curve such that for some $t \in (a,b)$ the point $\gamma(t)$ lies in the interior of the time-t-reachable set, $\gamma(t) \in \text{int}\,(Reach_{\Sigma,t}(0))$. Then for some $\varepsilon > 0$ we have $\gamma(s) \in Reach_{\Sigma,s}(0)$ for all $s \in [t-\varepsilon, t]$.*

Proof of the Lemma. Since $\gamma(t)$ is an interior point of the time-t-reachable set, there exist $n+1$ affinely independent points $q_0, \ldots, q_n$ of the time-t-reachable set such that $\gamma(t)$ becomes the barycenter of the simplex generated by the points q_i, i.e.,

$$\gamma(t) = \frac{1}{n+1}(q_0 + \cdots + q_n).$$

Let $u_i : [0,t] \to U$ be admissible controls that steer 0 into the points q_i and denote the corresponding trajectories by γ_i, $i = 0, 1, \ldots, n$. Then, the points $\gamma_i(s)$, $i = 0, 1, \ldots, n$, are still affinely independent for $s \in [t - \varepsilon, t]$ and $\varepsilon > 0$ small enough. By the implicit function theorem, there exist continuous functions ξ_i, $i = 0, 1, \ldots, n$, such that

$$\gamma(s) = \sum_{i=0}^{n} \xi_i(s)\gamma_i(s), \qquad \sum_{i=0}^{n} \xi_i(s) \equiv 1.$$

These functions satisfy $\xi_i(t) = \frac{1}{n+1}$ for all $i = 0, 1, \ldots, n$, and thus remain positive for $\varepsilon > 0$ small enough. Thus this defines a convex combination, and we have that

$$\gamma(s) \in \mathrm{co}\{\gamma_0(s), \ldots, \gamma_n(s)\}.$$

But the curves $\gamma_i(s)$ lie in the reachable set $\mathrm{Reach}_{\Sigma,s}(0)$, and this set is convex. Hence $\gamma(s)$ is reachable in time s, $\gamma(s) \in \mathrm{Reach}_{\Sigma,s}(0)$. $\square$

If $p \in int\,(\mathrm{Reach}_{\Sigma,T}(0))$, then we simply take $\gamma(t) \equiv p$, and it follows that p is reachable in time $T - \varepsilon$.

Conversely, now suppose that $p \in \mathrm{Reach}_{\Sigma,t}(0)$ for $t < T$. Since Σ is small-time locally controllable from the origin, there exists a neighborhood W of 0 that is contained in the reachable set for time $T - t$. (If Σ is small-time locally controllable from the origin, then it follows that 0 is an interior point of the time-t-reachable set from the origin for all times $t > 0$.) Suppose $u : [0, t] \to U$ is an admissible control that steers 0 to p in time t. The same control u steers any other point x to $e^{At}x + p$. If we now apply this control over the interval $[T - t, T]$ to the points in $W \subset \mathrm{Reach}_{\Sigma,T-t}(0)$, then it follows that the neighborhood $e^{At}W + p$ of p lies in $\mathrm{Reach}_{\Sigma,T}(0)$. Hence p is an interior point. $\square$

Corollary 3.5.1. *Let $\Sigma : \dot{x} = Ax + Bu$, $u \in U$, be small-time locally controllable from the origin. Then an admissible control $u \in \mathcal{U}$ is time-optimal if and only if there exists a nontrivial solution $\lambda : [0, T] \to (\mathbb{R}^n)^*$ to the adjoint equation $\dot{\lambda} = -\lambda A$ such that*

$$\lambda(t)Bu(t) = \min_{v \in U} \lambda(t)Bv \quad \text{a.e. on } [0, T].$$

Thus, for small-time locally controllable linear time-invariant systems, the conditions of the maximum principle are both necessary and sufficient for time optimality. This applies to the examples considered in Sect. 2.6 and proves the optimality of the syntheses constructed there. Compactness and convexity of the reachable sets is the key to this result.

3.6   The Bang-Bang Theorem for Compact Polyhedra

We now use the maximum principle, respectively, Theorem 3.5.1, to improve on the conditions of the general bang-bang theorem for compact polyhedra U, i.e., control sets that are the convex hull of a finite set, $U = \mathrm{co}\{u_0, u_1, \ldots, u_r\}$.

Theorem 3.6.1 (Bang-bang theorem for compact polyhedra). *Let $U = \mathrm{co}(F)$ be a compact polyhedron and consider the linear time-invariant system $\Sigma : \dot{x} = Ax + Bu$, $u \in U$. Then, for every $T > 0$, there exists an integer $N = N(T)$ such that whenever a point q is reachable from p in time $\tau \leq T$, $q \in \mathrm{Reach}_{\Sigma,\tau}(p)$, then q can also be reached from p in time τ by means of a piecewise constant control u that has values in F and has at most $N(T)$ switchings.*

Proof. We first carry out some normalizations that will simplify the notation.

(i) Note that the set $BU = \{Bu : u \in U\}$ also is a compact polyhedron in $\mathbb{R}^n$, and in fact, it is the convex hull of the set BF, $BU = B\mathrm{co}(F) = \mathrm{co}(BF)$. Thus, we may as well assume that $B = \mathrm{Id}$ and $F \subset \mathbb{R}^n$.

(ii) By shifting the initial condition if necessary, we may also assume that $u_0 = 0$. For if u steers p into q in time τ, then

$$q = e^{A\tau}p + \int_0^\tau e^{A(\tau-t)}u(t)dt$$

$$= e^{A\tau}\left(p + \int_0^\tau e^{-At}u_0 dt\right) + \int_0^\tau e^{A(\tau-t)}\left(u(t) - u_0\right)dt$$

$$\in \mathrm{Reach}_{\Sigma^*,\tau}\left(p + \int_0^\tau e^{-At}u_0 dt\right),$$

and hence q is reachable in time τ with the control $u(\cdot) - u_0$ from the point $p^* = p + \int_0^\tau e^{-At}u_0 dt$. Thus, with a new control set $U^* = U - u_0$, we can consider the system

$$\Sigma^* : \quad \dot{x} = Ax + u, \quad u \in U^* = \mathrm{co}\{0, u_1, \ldots, u_r\}, \quad x(0) = p^*.$$

(iii) Since admissible controls are convex combinations of the points u_i, the controllable subspace $\mathscr{C}^*(A, I)$ for the system Σ^* is given by

$$\mathscr{C}^*(A, I) = \mathrm{lin\ span}\left\{A^i u_j : i = 0, 1, \ldots, n - 1, j = 1, \ldots, r\right\}.$$

If the system is not completely controllable, then replace $\mathbb{R}^n$ by the affine variety $e^{A\tau}p^* + \mathscr{C}^*(A, I)$, and relative to this space, the reachable set will have a nonempty interior. Therefore, without loss of generality, we may assume that the system is completely controllable in the sense that $\mathscr{C}^*(A, I) = \mathbb{R}^n$.

(iv) Finally, simplifying the notation, we drop the $*$'s and we also take the origin as initial point. This only causes a shift of the reachable set.

We now proceed with the actual proof of the result. Suppose $q \in \mathrm{Reach}_{\Sigma,\tau}(0)$. The strategy is to reduce the proof to the consideration of boundary points and then use their characterization in Theorem 3.5.1 to prove the bang-bang property with a bound on the number of switchings.

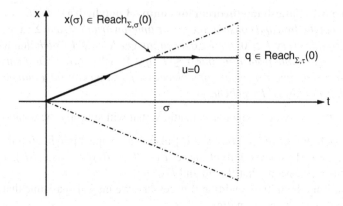

Fig. 3.5 $x(\sigma) \in Reach_{\Sigma,\sigma}(0)$

Step 1: Reduction to boundary points. Let $\bar{x}$ denote the solution to the linear equation $\dot{x} = Ax$ (that is, for the control $u = 0$) with terminal condition $\bar{x}(\tau) = q$ and let

$$\sigma = \inf\{t \le \tau : \bar{x}(s) \in Reach_{\Sigma,s}(0) \text{ for all } s \in [t, \tau]\}.$$

Since $\bar{x}(\tau) \in Reach_{\Sigma,\tau}(0)$, this set of times is nonempty, and thus σ is well-defined (see Fig. 3.5). We first want to show that σ actually is a minimum, i.e., $\bar{x}(\sigma) \in Reach_{\Sigma,\sigma}(0)$. Pick a sequence of times $\{t_j\}_{j \in \mathbb{N}}$ that monotonically decreases to σ. Since $\bar{x}(t_j) \in Reach_{\Sigma,t_j}(0)$, there exist controls u_j such that

$$\bar{x}(t_j) = \int_0^{t_j} e^{A(t_j - s)} u_j(s) ds.$$

Extend all the controls u_j to the full interval $[0, \tau]$ by choosing the control value to be 0 on the interval $(t_j, \tau]$. Since U is compact and convex, the class of admissible controls is weakly sequentially compact in $L^1([0, t]; \mathbb{R}^m)$, and we can select a weakly convergent subsequence. Without loss of generality, assume that $u_j \rightharpoonup u$. Denote the characteristic function of the interval $[0, t]$ by $\chi_{[0,t]}(s)$. Then we have for all $t \in [0, \tau]$ that

$$\int_0^t e^{A(t-s)} u_j(s) ds = \int_0^\tau \chi_{[0,t]}(s) e^{A(t-s)} u_j(s) ds$$

$$\rightarrow \int_0^\tau \chi_{[0,t]}(s) e^{A(t-s)} u(s) ds = \int_0^t e^{A(t-s)} u(s) ds.$$

Writing

$$\int_0^\sigma e^{A(\sigma-s)}u(s)ds - \bar{x}(t_j) = \int_0^\sigma e^{A(\sigma-s)}u(s)ds - \int_0^{t_j} e^{A(t_j-s)}u_j(s)ds$$

$$= \int_0^\sigma e^{A(\sigma-s)}\left(u(s)-u_j(s)\right)ds$$

$$+ \left(\mathrm{Id} - e^{A(t_j-\sigma)}\right)\int_0^\sigma e^{A(\sigma-s)}u_j(s)ds$$

$$- e^{A(t_j-\sigma)}\int_\sigma^{t_j} e^{A(\sigma-s)}u_j(s)ds,$$

each of these terms converges to 0: the first one by the weak convergence of u_j to u, the second one since $e^{A(t_j-\sigma)} \to \mathrm{Id}$ as $t_j \to \sigma$, and the last one since the integrand remains bounded. Thus we have that

$$\bar{x}(\sigma) = \lim_{j\to\infty} \bar{x}(t_j) = \int_0^\sigma e^{A(\sigma-s)}u(s)ds \in \mathrm{Reach}_{\Sigma,\sigma}(0)$$

and $\bar{x}(\sigma)$ lies in the time-σ-reachable set. It then follows from Lemma 3.5.1 that this point actually lies in the boundary of the time-σ-reachable set, $\bar{x}(\sigma) \in \partial \mathrm{Reach}_{\Sigma,\sigma}(0)$.

Step 2:   Using the conditions of the maximum principle, we now prove that 0 can be steered into the boundary point $\bar{x}(\sigma)$ in time σ by a control with a finite number of switchings and that a bound on the number of switchings can be given. Using Step 1, it then follows that any point in the time-τ-reachable set can be reached with one more switching, and this proves the theorem.

Let u be an admissible control that steers the origin into $\bar{x}(\sigma)$ in time σ. By Theorem 3.5.1 there exists a nonzero row vector η such that

$$\eta e^{A(\sigma-t)}u(t) = \min_{v\in U} \eta e^{A(\sigma-t)}v \quad \text{a.e. on } [0,\sigma].$$

For the controls $u_i \in F$, $i = 1,\ldots,r$, define functions ψ_i as

$$\psi_i(t) = \eta e^{A(\sigma-t)}u_i$$

and set $\psi_0(t) \equiv 0$ (corresponding to $u_0 = 0$). It follows from the normalization undertaken in (iii) that not all of the functions ψ_i vanish identically. (Otherwise, η is a nonzero vector orthogonal to the controllable subspace $\mathscr{C}^*(A,I)$, contradicting complete controllability.) The function $\lambda(t) = \eta e^{A(\sigma - t)}$ is a solution to the adjoint equation $\dot{\lambda}(t) = -\lambda(t)A$, and thus all the scalar functions ψ_i are solutions to one and the same homogeneous nth-order linear differential equation with constant coefficients of the form

$$\varphi^{(n)} = b_{n-1}\varphi^{(n-1)} + b_{n-2}\varphi^{(n-2)} + \cdots + b_0\varphi, \tag{3.12}$$

with the coefficients depending only on the characteristic polynomial of A (cf., Proposition 2.5.1).

Lemma 3.6.1. *There exists a positive constant $C < \infty$ (depending only on the length T of the interval and the system Σ) such that any nontrivial solution φ of Eq. (3.12) has at most $n-1$ zeros in an interval of length C.*

Proof of the Lemma. Let $z = (\varphi, \dot{\varphi}, \ldots, \varphi^{(n-1)})^T$ and write the nth-order equation (3.12) as a first-order system of the form $\dot{z} = Bz$, where B is the companion matrix,

$$\dot{z} = \begin{pmatrix} 0 & 1 & 0 & 0 & \cdots & 0 & 0 \\ 0 & 0 & 1 & 0 & \cdots & 0 & 0 \\ 0 & 0 & 0 & 1 & \cdots & 0 & 0 \\ \vdots & \vdots & \vdots & \vdots & \ddots & \vdots & \vdots \\ 0 & 0 & 0 & 0 & \cdots & 1 & 0 \\ 0 & 0 & 0 & 0 & \cdots & 0 & 1 \\ b_0 & b_1 & b_2 & b_3 & \cdots & b_{n-2} & b_{n-1} \end{pmatrix} z.$$

Since φ is a nontrivial solution, the vector $z(t)$ is nonzero for all t. Multiplying z by a positive constant does not effect the zeros, and thus we may normalize z such that $\|z(0)\|_2 = 1$. The solution is given by $z(t) = e^{Bt}z_0$, and if M denotes the least upper bound matrix norm of B,

$$M = \text{lub}_2(B) = \max_{\|z\|_2=1} \|Bz\|_2,$$

then we have that

$$\|z(t)\|_2 \le \text{lub}_2\left(e^{Bt}\right) \|z_0\|_2 \le e^{Mt}.$$

Hence the derivative is bounded over the interval $[0, T]$,

$$\|\dot{z}\|_2 \le Me^{MT} \quad \text{for all } t \in [0, T].$$

Since $\|z(0)\|_2 = 1$, at least one component, say the ith, satisfies $\left|\varphi^{(i)}\right| \ge \frac{1}{\sqrt{n}}$. Since all derivatives are bounded by Me^{MT}, it follows that the ith component cannot have a zero in the interval $[0, \frac{1}{2\sqrt{n}Me^{MT}}]$. But then φ itself can have at most $n-1$ zeros in the interval $[0, \frac{1}{2\sqrt{n}Me^{MT}}]$. (Otherwise, no matter what i is, the ith derivative would need to have a zero as well.) This proves the Lemma. $\qquad \square$

We now prove the result by induction on the cardinality of the set F. If $r = 0$, i.e., if there is only one point in the control set, the result is trivial, and we therefore may assume it to be true for sets F consisting of r points and now consider the case $F = \{0, u_1, \ldots, u_r\}$. Recall that u is an admissible control that steers 0 into $\bar{x}(\sigma) \in \partial \text{Reach}_{\Sigma, \sigma}(0)$ in time σ. By Theorem 3.5.1,

$$\lambda e^{A(\sigma-t)} u(t) = \min\{0, \psi_1(t), \ldots, \psi_r(t)\}.$$

Let ψ_k be a function that does not vanish identically. It follows from Lemma 3.6.1 that there exists a time C such that ψ_k has at most $n-1$ zeroes in any interval of length C. Hence there exists a constant $v = v(T)$ such that ψ_k has at most $v(T)$ zeros in the interval $[0, T]$. Enumerate these zeros as $0 = t_0 < t_1 < \cdots < t_v < t_{v+1} = T$. Then the function ψ_k has constant sign over the open intervals (t_i, t_{i+1}), $i = 0, 1, \ldots, r$. If ψ_k is negative on (t_i, t_{i+1}), then on this interval, we have

$$\min\{0, \psi_1(t), \ldots, \psi_r(t)\} = \min\{\psi_1(t), \ldots, \psi_r(t)\},$$

and if ψ_k is positive, then we have

$$\min\{0, \psi_1(t), \ldots, \psi_r(t)\} = \min\{0, \psi_1(t), \ldots, \psi_{k-1}(t), \psi_{k+1}(t), \ldots, \psi_r(t)\}.$$

In either case, on the subinterval (t_i, t_{i+1}) we can drop one of the controls and thus have reduced the problem to one with a subset of F as controls that has cardinality $r - 1$. By the inductive assumption, the control u is bang-bang on the interval (t_i, t_{i+1}) with a bound on the number of switchings. Suppose there are at most μ switchings in each of the intervals. Then overall there are at most $(v + 1)\mu$ switchings. This proves the theorem. $\qquad\square$

It follows from the proof that the number of switchings in the interval $[0, T]$ can be bounded by a constant C times the length of the interval, CT. The example of the harmonic oscillator (Sect. 2.6.4) shows that this bound CT cannot be improved upon. This result, which is due to Brunovsky [60], gives a significant improvement over the conclusions of the general bang-bang theorem and is the basis for the construction of piecewise defined optimal feedback controls [60] as illustrated in Sect. 2.6.

3.7 Notes

The results presented in this chapter form the core of historical developments in control theory in the 1960s and can be found in many textbooks from that area, such as, for example, [25, 117, 147]. We included them in this text because of their historical significance and the importance of convexity properties in the general theory to be developed next. We made the exposition as elementary as possible to be accessible to a reader (with possibly an engineering or economics background) who is familiar with basic real analysis, but not the more advanced functional-analytic results (such as the Banach–Alaoglu theorem) that often are used to give quick and elegant, but abstract and not very intuitive proofs of these results. While there is a need for technical constructions involving measurability, the arguments presented here hold these to a minimum. The proofs of both the general bang-bang theorem and the bang-bang theorem with a bound on the number of switchings presented here are based on lecture notes from a course on the topic given by H. Sussmann at Rutgers University in 1983.

Chapter 4
The High-Order Maximum Principle: From Approximations of Reachable Sets to High-Order Necessary Conditions for Optimality

In this chapter, we prove the Pontryagin maximum principle. The proof we present follows arguments by Hector Sussmann [244, 247, 248], but in a smooth setting. It is somewhat technical, but provides a uniform treatment of *first- and high-order variations*. As a result, we not only prove Theorem 2.2.1, but obtain a general high-order version of the maximum principle (e.g., see [140]) from which we then derive the high-order necessary conditions for optimality that were introduced in Sect. 2.8.

We consider a general nonlinear time-invariant control system Σ. Recall that a *control system* consists of a differential equation that describes its *dynamics* and a class of *admissible controls* that specifies the input functions that can be used to influence this dynamics. We write the dynamics in the form $\dot{x} = f(x, u)$, where the state variable x takes values in the *state-space* M and the control variable u takes values in a *control set* U. As before, we take U as a subset of $\mathbb{R}^m$, but otherwise arbitrary. In this chapter, we consider both the cases in which the state space M is an open subset of $\mathbb{R}^n$ and those in which it is a finite-dimensional manifold. While much of our interest and treatment of applications is for problems on $\mathbb{R}^n$, it is the language of differential geometry that clearly brings out the role of the objects (multipliers as cotangent vectors) and the constructions (differentials and pullbacks of maps), and for this reason we include this more general—and also more natural—formulation. Because of the importance of the underlying subject, we include a somewhat more comprehensive introduction to manifolds in Appendix C. Admissible controls are now always locally bounded Lebesgue measurable functions $u : I \mapsto U$ defined on some interval $I \subset \mathbb{R}$ with values in U. Overall, the setup remains unchanged under this modification: Given an admissible control u, a time-varying ordinary differential equation $\dot{x} = f(x, u(t))$ is obtained whose solution curve x is called the corresponding trajectory. (As before, we shall impose conditions that guarantee the existence and uniqueness of solutions.) The pair (x, u) is a *controlled trajectory* of the system. We say that a point $q \in M$ is *reachable* from an initial point $p \in M$ in time t if there exists a controlled trajectory (x, u) defined over the interval $[0, t]$ that starts at p, $x(0) = p$, and passes through q at time t, $x(t) = q$. The set of all points that are reachable from p in time t is the

H. Schättler and U. Ledzewicz, *Geometric Optimal Control: Theory, Methods and Examples*, Interdisciplinary Applied Mathematics 38,
DOI 10.1007/978-1-4614-3834-2_4, © Springer Science+Business Media, LLC 2012

time-t-reachable set, Reach$_{\Sigma,t}(p)$, and the union of all these sets for $t > 0$ is the *reachable set* from p, Reach$_\Sigma(p)$. In the literature, the term attainable set is also used for these fundamental objects, whose study goes back to the early paper [203] by E. Roxin (cf. also [204, 205]).

Our aim is to study the collection of all controlled trajectories as a whole and to characterize special trajectories that minimize some objective over a subset of trajectories. The latter is the typical situation in *optimal control problems*: Given an initial point $p \in M$ and a subset $N \subset M$ of terminal conditions, among all controlled trajectories that start at p and end in N, find those that minimize some cost functional J. We shall always assume that the objective is of the form

$$J(u) = \int_0^T L(x,u)dt + \varphi(x(T))$$

with the Lagrangian L describing the running cost of the controlled trajectory (x,u) and φ defining a penalty term at the endpoint. The necessary conditions for optimality of a controlled reference trajectory $\Gamma = (\bar{x}, \bar{u})$ given in the Pontryagin maximum principle (Theorem 2.2.1) are a direct corollary of a natural separation property. For example, suppose J is given in Mayer form as a pure penalty term $J(u) = \varphi(x(T))$ with φ a differentiable function defined on N. If Γ is optimal, then no point in the set $A = \{q \in N : \varphi(q) < J(\bar{u})\}$ can be reachable from p. That is, the reachable set Reach$_\Sigma(p)$ and A must be disjoint, Reach$_\Sigma(p) \cap A = \emptyset$. At the same time, the optimal controlled trajectory steers p into a point q in the closure of both of these sets. The empty intersection property therefore implies that the *cone of decrease* for the objective at q, which consists of all tangent vectors v to N at q for which $\nabla\varphi(q)v < 0$, can be separated from a suitably constructed approximating cone to the set of all points that are reachable in N from p at q. We shall see that the maximum principle is a direct consequence of the study of (suitably defined) *approximating cones for the reachable sets*. Necessary conditions for optimality will be recast as geometric conditions on reachable sets, and the geometric content of the maximum principle is that of a separation theorem.

Our main aim therefore is to develop criteria that allow us to determine whether a reachable point q is an interior point or a boundary point of the reachable set. These techniques also are of interest for the question about small-time local controllability, i.e., whether $p \in \text{int}(\text{Reach}_{\Sigma,\leq T}(p))$ for small time T. For linear time-invariant systems, given a compact and convex control set, the time-t-reachable set is compact and convex, and we thus have a characterization of boundary points in terms of supporting hyperplanes (Theorem 3.5.1). For a general nonlinear system, however, reachable sets no longer need to be convex or closed, and thus characterizations of boundary points are more difficult to give. By approximating the reachable set with a suitable convex cone at q, it is still possible to give necessary conditions for q to be a boundary point. These approximations will be developed as part of a general framework to construct and compute approximating directions to the reachable set at a point q. The essential ingredient of this framework is the definition of so-called *point variations*. This class of variations is modeled after the classical *needle variations* made by Pontryagin, Boltyansky, Gamkrelidze, and Mishchenko [193], but it also allows for the inclusion of so-called high-order variations. The needle variations themselves are a generalization of the variations Weierstrass made in the proof of his necessary condition for optimality in the calculus of variations. Given a reference trajectory $\Gamma = (\bar{x}, \bar{u})$ that steers p into q in time T, at every point $\bar{x}(\bar{t})$ on the reference trajectory, these variations generate a cone $C_{\bar{t}}$ of directions that then is moved to the terminal point q to generate a convex approximating cone $\mathcal{K}$ to the reachable set at q. The aim of this construction is to be able to distinguish between interior and boundary points of the reachable set. For this reason, we want that if every vector is an approximating direction to the reachable set at q, i.e., if $\mathcal{K} = \mathbb{R}^n$, then q should be an interior point of the reachable set. By negation, if q lies in the boundary of the reachable set, $q \in \partial \operatorname{Reach}_{\Sigma}(p)$, then $\mathcal{K}$ cannot be the full space $\mathbb{R}^n$ and thus there exist nontrivial supporting hyperplanes to $\mathcal{K}$. However, care needs to be exercised in the definition of an approximating cone for this to be true. We present in Sect. 4.1 an adequate definition that goes back to Boltyansky's original proof of the maximum principle. The necessary conditions for optimality of the maximum principle then directly follow from classical separation theorems about convex sets. In its purest form, *the maximum principle* indeed *gives necessary conditions for trajectories to lie in the boundary of the reachable set from a point.*

After establishing the concept of *approximating cone* to be used and the necessary separation results in Sect. 4.1, in Sect. 4.2 we first give the classical construction by Pontryagin et al. [193] to prove the maximum principle for systems on open sets in $\mathbb{R}^n$ with bounded Lebesgue measurable functions as the class of admissible controls. We use this section to introduce the main ideas of the construction with fewer technical details. These ideas, however, are more general, and by placing the construction on a *manifold*, the geometric content of the construction becomes clear. We also give some typical examples of control systems on manifolds in Sect. 4.3. A brief introduction to differentiable manifolds and the main concepts from differential geometry that will be used (tangent and cotangent vectors, integral curves, differentials, and pullbacks of mappings) is included in Appendix C, but we need to refer the reader to the literature for a deeper treatment of this subject. Besides the setting on manifolds, in Sect. 4.4 also a more general class of *point variations* will be considered that allows us to include high-order variations into the construction and thus this generates a larger approximating cone to the reachable set. This leads to a geometric form of the high-order maximum principle, Theorem 4.4.1 [140]. In Sect. 4.6 we present some results from Lie algebra on exponential representations of flows of vector fields that provide an efficient computational technique to compute variational vectors generated by high-order variations. We then use these constructions to derive the Legendre–Clebsch condition as well as the Kelley and Goh conditions formulated in Chap. 2.

The constructions and computations presented in this chapter admittedly are technical. But they form the foundation for the techniques of optimal control theory, which, as the argument is made by Sussmann and Willems in their review article [245], *"although deeply rooted in the classical calculus of variations in its problems and results, provides superior techniques which contrary to a still widespread misconception go well beyond just incorporating inequality constraints into the classical framework."*

4.1 Boltyansky Approximating Cones

In this section we give the results on convex cones that will be needed to translate geometric approximations of the reachable set (to be carried out in Sects. 4.2 and 4.4) into the analytical conditions of the maximum principle. Since these approximations will be local, it suffices to consider the case where the underlying state space is $\mathbb{R}^n$ and all our cones have apex at the origin.

Definition 4.1.1 (Cone). A cone C in $\mathbb{R}^n$ is a set with the property that $rC \subset C$ for all $r > 0$.

Lemma 4.1.1. *A cone C is convex if and only if $rC + sC \subset C$ for all $r, s > 0$.*

Proof. If C is convex and x and y are arbitrary points from C, then for any positive numbers r and s,

$$rx + sy = (r+s) \left(\frac{r}{r+s} x + \frac{s}{r+s} x \right) \in (r+s)C \subset C;$$

the converse is obvious.   □

There exists a large number of definitions that in one way or another express the notion of approximating a set with "tangent" vectors. They can differ widely in their technical assumptions, and we refer the interested reader to the book by Aubin and Frankowska [26] for a survey and comparison of some of these concepts. Our goal is to approximate the reachable set at a point q with a convex cone C in such a way that if the full space $\mathbb{R}^n$ is an approximating cone at q, then q is an interior point to the reachable set. The most typical and straightforward way of defining an approximating cone to a set S at q, the so-called *Bouligand tangent cone*, is to simply take the cone of tangent vectors at q: a vector v is tangent to a set S at a point q if there exist an $\varepsilon > 0$ and a differentiable curve $\gamma : [0, \varepsilon] \to S$, $\gamma(0) = q$, with tangent vector $\dot{\gamma}(0) = v$. However, this concept is far too general, and it does not serve our purpose.

Example. Let $A = \{(x, y) \in \mathbb{R}^2 : x \leq 0 \text{ or } y \leq 0 \text{ or } y \geq x^2\}$. Any line ℓ passing through the origin has a segment that has $(0, 0)$ in its relative interior, which entirely lies in A. Thus the set of all tangent vectors to A at $(0, 0)$ is $\mathbb{R}^2$, but $(0, 0)$ is not an interior point of A (see Fig. 4.1).

What fails in the example is that the approximation of the set by the tangent line $t v_\varepsilon$ becomes increasingly worse in t for vectors $v_\varepsilon = (1, \varepsilon)$ as $\varepsilon \to 0$, and it is not possible to approximate the set *uniformly* over all tangent lines. This is what makes the difference, and the following somewhat technical condition that goes back to Boltyansky makes such a notion of uniform approximation precise. We denote by $D_r^n(x_0)$ the closed ball (disk) with radius r around x_0 in $\mathbb{R}^n$, $D_r^n(x_0) = \{x \in \mathbb{R}^n : \|x - x_0\|_2 \leq r\}$, while we write $B_r^n(x_0)$ for the open ball. We also write $\mathbb{R}_+^n = \{x \in \mathbb{R}^n : x_i \geq 0 \text{ for all } i = 1, \ldots, n\}$ for the positive octant and $Q_\delta^n = \{x \in \mathbb{R}^n : 0 \leq x_i \leq \delta \text{ for } i = 1, \ldots, n\}$ for the cube of side length δ in $\mathbb{R}_+^n$.

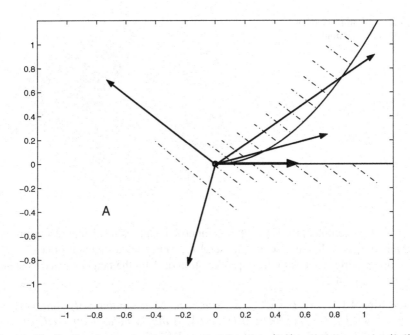

Fig. 4.1 The Bouligand approximating cone to A at the origin $(0,0)$ is the full space, but $(0,0)$ is not an interior point of A

Definition 4.1.2 (Boltyansky approximating cone). Let A be a subset of $\mathbb{R}^n$ and let q be a point in the closure of A. A cone $C \subset \mathbb{R}^n$ is a Boltyansky approximating cone for A at q if for any finite set of vectors $v_1, \ldots, v_k$ from C, $k \in \mathbb{N}$, and every $\varepsilon > 0$ there exist vectors $v'_1, \ldots, v'_k$ in $\mathbb{R}^n$ satisfying $\|v_i - v'_i\| < \varepsilon$ for $i = 1, \ldots, k$, and a continuous map Ξ defined on Q^k_δ for some $\delta > 0$ with values in $A \cup \{q\}$, $\Xi : Q^k_\delta \to A \cup \{q\}$, such that

$$\Xi(z_1, \ldots, z_k) = q + z_1 v'_1 + \cdots + z_n v'_k + o(\|z\|) \quad \text{as } z \to 0. \tag{4.1}$$

Here, and in the following, we use the standard Landau notation $o(\|z\|)$ to denote functions ρ that have the property that $\lim_{z \to 0} \frac{\rho(z)}{\|z\|} = 0$; we will also use $O(\|z\|)$ to denote functions ρ which remain bounded after division by $\|z\|$, i.e., $\limsup_{z \to 0} \frac{\rho(z)}{\|z\|}$ is finite. Generally, these functions correspond to higher-order terms whose specific form is irrelevant for the argument. From now on, approximating cones will always be understood in the sense of this definition.

Postulating that the map Ξ in Definition 4.1.2 approximates only the nearby vectors v'_i, although clearly more general, is not intuitive. This is a purely technical aspect that will be convenient later on. It allows us to combine approximating directions obtained from point variations made at the same time into an approximating map. This simple trick avoids serious problems one encounters when trying to do

Fig. 4.2 Tangent plane to an
embedded submanifold

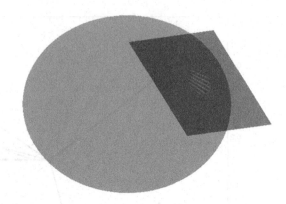

this (see, for example, [36, 37]). Little geometric insight is lost if the reader prefers
to think of v_i' as v_i. In the case that the set A is an embedded submanifold of $\mathbb{R}^n$, the
definition readily reduces to the approximation of A by the tangent space as stated
below (see Fig. 4.2).

Proposition 4.1.1. *Let $F : \mathbb{R}^n \to \mathbb{R}^m$, $m < n$, be a continuously differentiable
mapping and let $A = \{x \in \mathbb{R}^n : F(x) = 0\}$. If the Jacobian matrix $DF(q)$ is of
full rank m at a point $q \in A$, then the kernel of $DF(q)$,*

$$C = \ker DF(q) = \{v \in \mathbb{R}^n : DF(q)v = 0\},$$

*is a Boltyansky approximating cone to A at q. This $(n-m)$-dimensional subspace is
also called the* tangent space *to A at q and is denoted by $T_q A$.*

This is a classical result, sometimes also referred to as the Lusternik theorem.
Here we only indicate the main idea of the proof; the full proof is included in
Appendix C, Proposition C.1.2. Given a fixed finite collection of vectors $v_1, \ldots, v_k$
from C, define a linear functional ℓ as $\ell : \mathbb{R}^k \to \mathbb{R}$, $z \mapsto \ell(z) = \sum_{i=1}^{k} z_i v_i$. There is
no need to introduce the approximating vectors v_i', and we simply take $v_i' = v_i$.
Furthermore, since C is a subspace, if $v \in C$, then also $-v \in C$. Hence, the
approximating map naturally needs to be defined not only on a cube Q_δ^k, but on a
full neighborhood $B_\delta^k(0)$ of zero, $\Xi : B_\delta^k(0) \to A$, $z \mapsto \Xi(z)$. Thus one needs to show
that there exist a neighborhood $B_\delta^k(0)$ and a continuous function $r = r(z)$ defined on
$B_\delta^k(0)$ that is of order $o(\|z\|)$ as $z \to 0$ such that $\Xi(z) = q + \ell(z) + r(z) \in A$. This
function r can be computed by solving the equation $F(z) = 0$ near q using a quasi-
Newton algorithm [228], and the superlinear convergence of the procedure implies
that r is of the desired order. If F is twice continuously differentiable, the Newton
algorithm can be used, and then its quadratic convergence implies that the remainder
$r = r(z)$ actually is of order $O(\|z\|^2)$ as $z \to 0$.

Equation (4.1) requires a uniform approximation over the positive octant $\mathbb{R}_+^n$. As
a result, it is possible to take linear combinations with positive coefficients, and thus
the definition automatically extends to the convex hull as well.

Lemma 4.1.2. *If C is a Boltyansky approximating cone for a set A at q, then so is its convex hull* $\mathrm{co}(C)$. ∎

The proof of this lemma simply consists in combining the definitions of the convex hull with the notion of a Boltyansky approximating cone and is left as an exercise for the reader. From now on, without loss of generality, we always assume that Boltyansky approximating cones are convex. The essential result is the following, which states that a Boltyansky approximating cone provides the desired topological properties.

Theorem 4.1.1. *If all of $\mathbb{R}^n$ is a Boltyansky approximating cone for a set A at q, then q is an interior point of $A \cup \{q\}$.*

Proof. Without loss of generality, we assume that $q = 0$. Let $e_1,\ldots,e_n$ be the canonical basis for $\mathbb{R}^n$ and set $e_0 = -(e_1 + \cdots + e_n)$. The points $e_0,e_1,\ldots,e_n$, are affinely independent, and we denote by S the convex hull of these points. The set S contains the origin in its interior, and every point $s \in S$ can then be written in a unique way as a convex combination of the points $e_0,\ldots,e_n$, $s = z_0 e_0 + \cdots + z_n e_n$, with $z_i \geq 0$ for all $i = 0,1,\ldots,n$, and $z_0 + \cdots + z_n = 1$. Let $\mathbf{z} = (z_0,\ldots,z_n)$ denote this vector of affine coordinates.

By assumption, the full space $\mathbb{R}^n$ is an approximating cone for A at 0 and thus, given any $\varepsilon > 0$, there exist vectors $e_0',\ldots,e_n'$ in $\mathbb{R}^n$ such that $||e_i - e_i'|| < \varepsilon$ for all $i = 0,\ldots,n$, a positive number $\delta > 0$, and a continuous map $\Xi : Q_\delta^{n+1} \to A \cup \{0\}$ such that

$$\Xi(\mathbf{z}) = z_0 e_0' + \cdots + z_n e_n' + o(||z||) \quad \text{as } z \to 0.$$

Choose an $\alpha > 0$ such that the closed ball $D_\alpha^n(0)$ lies in S (see Fig. 4.3), and for r sufficiently small, define a map $\Phi_r : S \to \mathbb{R}^n$ by

$$\Phi_r(s) = \frac{1}{r}\Xi(r\mathbf{z}).$$

For ε and r small enough, we then have for all $s \in S$ that

$$||\Phi_r(s) - s|| = \left\| \sum_{i=0}^{n} z_i(e_i' - e_i) + \frac{o(r)}{r} \right\|$$

$$\leq \sum_{i=0}^{n} z_i \left\| e_i' - e_i \right\| + \frac{o(r)}{r} \leq \varepsilon + \frac{o(r)}{r} \leq \frac{\alpha}{2}.$$

Since we can choose α arbitrarily small, this inequality states that the map $\Phi_r : S \to \mathbb{R}^n$ can be made to be arbitrarily close to the identity map $\mathrm{id} : S \to S$. But the identity map trivially covers a neighborhood of the origin, and it thus is to be expected that a continuous map close enough to the identity will have the same property. This is indeed correct and essentially goes back to a result in algebraic topology that the identity map on a sphere is not homotopic to a constant. An

Fig. 4.3 Neighborhood
$D_\alpha^n(0)$

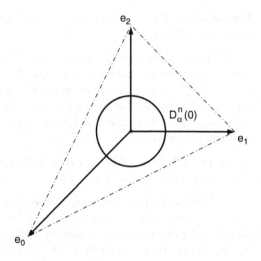

equivalent, and quicker way to see this is to invoke the Brouwer fixed-point theorem, which can be derived from this result. (For an elegant and simple exposition of these results, we refer the reader to Milnor's notes [184].)

Theorem 4.1.2 (Brouwer fixed point theorem). *Let* $D \subset \mathbb{R}^n$ *be a nonempty compact and convex set and let* $f : D \to D$ *be a continuous map that maps* D *into itself. Then* f *has a fixed point in* D. ■

If f is a scalar function, $f : [a,b] \to [a,b]$, then this is nothing but the mean value theorem. For in this case, nothing needs to be shown if either $f(a) = a$ or $f(b) = b$. Otherwise, the function $g : [a,b] \to [a,b]$ defined by $g(x) = f(x) - x$ is positive at $x = a$ and negative at $x = b$ and hence needs to have a zero in (a,b). But this simple geometric reasoning does not generalize to higher dimensions.

For $q \in D_{\frac{\alpha}{2}}^n(0)$ fixed, define the continuous map

$$F : S \times D_{\frac{\alpha}{2}}^n(0) \to D_\alpha^n(0),$$

$$(s,q) \mapsto F(s,q) = q - \Phi_r(s) + s.$$

It follows from the above estimate that F has values in $D_\alpha^n(0)$, and thus for q fixed, the mapping $F(\cdot,q)$ maps $D_\alpha^n(0) \subset S$ into itself. By the Brouwer fixed-point theorem, F has a fixed point μ, $\mu = \mu(q)$, in $D_\alpha^n(0) \subset S$. But $F(\mu,q) = \mu$ is equivalent to $\Phi_r(\mu) = q$, and thus every point $q \in D_{\frac{\alpha}{2}}^n(0)$ is the image of some point $\mu \in S$ under Φ_r. Hence the closed ball $D_{\frac{\alpha}{2}}^n(0)$ lies in the image of $\Phi_r(S)$, and $A \cup \{0\}$ contains the neighborhood $rB_{\frac{\alpha}{2}}^n(0)$ of zero. □

The next theorem will allow us to translate the geometric conditions about reachable sets into necessary conditions for optimality in the optimal control problem.

Fig. 4.4 Strongly transversal
cones C_1 and C_2

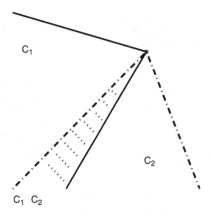

Definition 4.1.3 (Transversal cones). [247] Two convex cones C_1 and C_2 in $\mathbb{R}^n$ are said to be transversal if $C_1 - C_2 = \{x - y : x \in C_1, \; y \in C_2\} = \mathbb{R}^n$. They are strongly transversal if C_1 and C_2 are transversal and if in addition, $C_1 \cap C_2 \neq \{0\}$, i.e., C_1 and C_2 intersect at least along a half-line.

We give a simple sufficient condition for convex cones to be strongly transversal (see Fig. 4.4) that will suffice for the purpose of this text.

Definition 4.1.4 (Separation of convex cones). Let C_1 and C_2 be convex cones in $\mathbb{R}^n$. The cones C_1 and C_2 are separated if there exists a nontrivial linear functional λ such that

$$\langle \lambda, x \rangle \geq 0 \quad \text{for all } x \in C_1 \quad \text{and} \quad \langle \lambda, y \rangle \leq 0 \quad \text{for all } y \in C_2.$$

Proposition 4.1.2. *Let C_1 and C_2 be two nonempty convex cones that are not separated and suppose C_2 is not a linear subspace. Then C_1 and C_2 are strongly transversal.*

Proof. We first show that C_1 and C_2 are transversal. It is easy to verify that $C_1 - C_2$ is a convex cone. Hence, if $C_1 - C_2 \neq \mathbb{R}^n$, then $C_1 - C_2$ necessarily needs to lie in a half-space, and thus there exists a nontrivial linear functional λ such that $\langle \lambda, z \rangle \geq 0$ for all $z \in C_1 - C_2$. Since cones are invariant under positive scalings, this implies that

$$\langle \lambda, x \rangle \geq r \langle \lambda, y \rangle \quad \text{for all } x \in C_1, \; y \in C_2 \text{ and } r > 0.$$

Taking the limit $r \to 0$, we obtain $\langle \lambda, x \rangle \geq 0$ for all $x \in C_1$. Similarly, we also have that

$$r \langle \lambda, x \rangle \geq \langle \lambda, y \rangle \quad \text{for all } x \in C_1, y \in C_2 \text{ and } r > 0,$$

and thus it follows that $\langle \lambda, y \rangle \leq 0$ for all $y \in C_2$. Hence C_1 and C_2 are separated. Contradiction.

Fig. 4.5 Geometric
interpretation of
Theorem 4.1.3

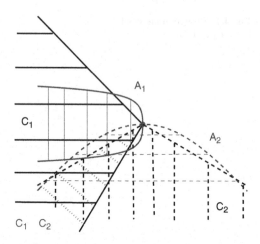

Fig. 4.5 Geometric interpretation of Theorem 4.1.3

But then C_1 and C_2 intersect at least along a nontrivial half-line. For suppose $C_1 \cap C_2 \subset \{0\}$ and pick any nonzero vector $z \in C_2$. Since C_1 and C_2 are transversal, there exist vectors $x \in C_1$ and $y \in C_2$ such that $z = x - y$. Thus $z + y = x \in C_1 \cap C_2 = \{0\}$, and so $z = -y$. Hence $-z$ also lies in C_2. But this holds for any $z \in C_2$, and so C_2 is a linear subspace. Contradiction. Thus C_1 and C_2 are strongly transversal.   □

Theorem 4.1.3. *Let A_1 and A_2 be subsets of $\mathbb{R}^n$ and suppose q is a point in the intersection of the closures of these sets. Let C_1 and C_2 be convex approximating cones for A_1 and A_2 at q, respectively, that are strongly transversal. Then q is a limit point of $A_1 \cap A_2$, i.e., every neighborhood V of q contains a point $\bar{q} \in A_1 \cap A_2$, $\bar{q} \neq q$.*

The geometric contents of this theorem is visualized in Fig. 4.5.

Proof. Without loss of generality, we again assume that $q = 0$ and let $e_1, \ldots, e_n$ be the canonical basis for $\mathbb{R}^n$. As in the proof of Theorem 4.1.1, set $e_0 = -(e_1 + \cdots + e_n)$. Since C_1 and C_2 are transversal, we can write every vector as a difference of vectors from C_1 and C_2, say

$$e_i = f_i - g_i, \quad f_i \in C_1, \quad g_i \in C_2.$$

Since there exists a nonzero vector $z \in C_1 \cap C_2$, we also have for any $m > 0$ that

$$e_i = (f_i + mz) - (g_i + mz)$$

with $f_i + mz \in C_1$ and $g_i + mz \in C_2$. The convex hull of the vectors $\{f_i + mz : i = 0, \ldots, n\}$ simply translates the convex hull of the vectors $\{f_i : 0 = 1, \ldots, n\}$ in $\mathbb{R}^n$ by mz. But this convex hull is a compact set, and by choosing m large enough, we can guarantee that it does not contain $q = 0$. Equivalently, without loss of generality, we may assume that the original vectors f_i are such that 0 does not lie in the convex hull of the vectors $\{f_0, \ldots, f_n\}$.

Once more, let S be the convex hull of $e_0, e_1, \ldots, e_n$, i.e., the corresponding n-dimensional unit simplex in $\mathbb{R}^n$. Recall that elements $s \in S$ can be written in a unique way as a convex combination in the form $s = z_0 e_0 + \cdots + z_n e_n$ with $z_i \geq 0$ for all $i = 0, 1, \ldots, n$, and $z_0 + \cdots + z_n = 1$, and again let $\mathbf{z} = (z_0, \ldots, z_n)$ denote this vector of affine coordinates. Since C_1 and C_2 are approximating cones for A_1 and A_2 at 0, given $\varepsilon > 0$, there exist vectors $f_0', \ldots, f_n'$ and $g_0', \ldots, g_n'$ such that $\|f_i - f_i'\| < \frac{\varepsilon}{2}$ and $\|g_i - g_i'\| < \frac{\varepsilon}{2}$ for all $i = 0, \ldots, n$, and continuous maps $\Xi_j : Q_\delta^{n+1} \to A_j \cup \{0\}$, $j = 0, 1$, such that

$$\Xi_1(\mathbf{z}) = z_0 f_0' + \cdots + z_n f_n' + o(\|z\|)$$

and

$$\Xi_2(\mathbf{z}) = z_0 g_0' + \cdots + z_n g_n' + o(\|z\|)$$

as $z \to 0$. Let $D_\alpha^n(0)$ be a ball contained in the interior of the unit simplex S, and for small $r > 0$ now define a map $\Phi_r : S \to \mathbb{R}^n$ by

$$\Phi_r(s) = \frac{1}{r} \Xi_1(r\mathbf{z}) - \frac{1}{r} \Xi_2(r\mathbf{z}).$$

Then, again for sufficiently small ε and r, we have that

$$\|\Phi_r(s) - s\| = \left\| \sum_{i=0}^{n} z_i (f_i' - g_i') + \frac{o(r)}{r} - \sum_{i=0}^{n} z_i e_i \right\|$$

$$\leq \left\| \sum_{i=0}^{n} z_i (f_i' - f_i) - \sum_{i=0}^{n} z_i (g_i' - g_i) + \frac{o(r)}{r} \right\|$$

$$\leq \sum_{i=0}^{n} z_i \|f_i' - f_i\| + \sum_{i=0}^{n} z_i \|g_i' - g_i\| + \frac{o(r)}{r}$$

$$\leq \varepsilon + \frac{o(r)}{r} \leq \frac{\alpha}{2}.$$

Using the Brouwer fixed-point theorem, it now follows, exactly as in the proof of Theorem 4.1.1, that the image $\Phi_r(S)$ contains the closed ball $D_{\frac{\alpha}{2}}^n(0)$. Especially, $0 \in \Phi_r(S)$, and so there exists a point $\mathbf{z}^r = (z_0^r, \ldots, z_n^r) \in S$ such that $\Xi_1(r\mathbf{z}^r) = \Xi_2(r\mathbf{z}^r)$. But for r small enough, $\Xi_1(r\mathbf{z}^r) \neq 0$. [For we have that

$$\frac{\Xi_1(r\mathbf{z})}{r} = z_0 f_0' + \cdots + z_n f_n' + \frac{o(r)}{r}$$

and the point $z_0 f_0' + \cdots + z_n f_n'$ lies in the convex hull of the vectors $\{f_0', \ldots, f_n'\}$. By choosing ε small enough, we can guarantee that this compact set still has positive distance to the origin and thus by also choosing r small enough, $\frac{\Xi_1(r\mathbf{z})}{r}$ has positive distance to the origin.] Hence $\bar{q} = \Xi_1(r\mathbf{z}^r) = \Xi_2(r\mathbf{z}^r) \in A_1 \cap A_2$, $\bar{q} \neq 0$, and thus for every sufficiently small r there exists a nonzero point $\bar{q} \in A_1 \cap A_2$ that lies in the neighborhood $rB_{\frac{\alpha}{2}}^n(0)$ of $q = 0$. $\square$

4.2   Proof of the Pontryagin Maximum Principle

We now prove the Pontryagin maximum principle (Theorem 2.2.1). Recall that we view a control systems as a collection of time-dependent vector fields defined on some state space M indexed by a class $\mathscr{U}$ of admissible control functions. In this section, we still assume that the state space M is an open and connected subset of $\mathbb{R}^n$, but controls are Lebesgue measurable functions. More precisely, we consider a control system $\Sigma = (M, U, f, \mathscr{U})$ consisting of the following objects:

1. The *state space* M is an open and connected subset of $\mathbb{R}^n$.
2. The *control set* U is an arbitrary subset of $\mathbb{R}^m$.
3. The *dynamics* or dynamical law is given by a function $f : M \times U \to \mathbb{R}^n$, $(x, u) \mapsto f(x, u)$, that is continuous and continuously differentiable in x for every $u \in U$ fixed with continuous partial derivatives $\frac{\partial f}{\partial x}(x, u)$.
4. The class $\mathscr{U}$ of *admissible controls* is given by all locally bounded Lebesgue measurable functions with values in the control set U. Thus, if η is an admissible control defined over a compact interval I, then η takes values in a compact subset of U almost everywhere on I.

Given an admissible control $u : J \to U$ defined on some open interval J, we obtain a time-dependent differential equation $\dot{x} = f(x, u(t))$ whose right-hand side now is only Lebesgue measurable in t. As in the case when the right-hand side is continuous in t, it still follows under our assumptions that solutions to initial value problems exist, are unique, and are continuously differentiable functions of the initial conditions (t_0, x_0). But these solutions $x : J \to \mathbb{R}^n$, $t \mapsto x(t)$, now are no longer continuously differentiable, but only absolutely continuous curves that pass through x_0 at time t_0 and satisfy the differential equation almost everywhere.

Definition 4.2.1 (Absolutely continuous). A continuous curve $\xi : J \to \mathbb{R}^n$, $t \mapsto \xi(t)$, is said to be absolutely continuous, $\xi \in AC(J; \mathbb{R}^n)$, if there exists a Lebesgue measurable integrable function $v : J \to \mathbb{R}^n$, $t \mapsto v(t)$, $v \in L^1(J; \mathbb{R}^n)$, such that for some $t_0 \in J$,

$$\xi(t) = \int_{[t_0, t]} v(s) ds,$$

with ds denoting integration against Lebesgue measure.

In fact, for any function $v \in L^1(J; \mathbb{R}^n)$, the components of such a curve ξ define σ-additive set functions that are absolutely continuous against Lebesgue measure,[1] whence this terminology. It can be shown (the Radon–Nikodym theorem [174]) that an absolutely continuous curve ξ is differentiable almost everywhere in I with derivative given by v. The function v is called the Radon–Nikodym derivative and is also denoted by $\dot{\xi}$.

---

[1] A σ-additive set function v is said to be absolutely continuous with respect to some measure μ if whenever E is a measurable set for which $\mu(E) = 0$, then also $v(E) = 0$.

Under our assumptions on the control system, for any admissible control u defined on an open interval J, the function $(t,x) \mapsto F(t,x) = f(x,u(t))$ is Lebesgue measurable in t and continuously differentiable in x. Furthermore, restricted to a compact subinterval $I \subset J$, the control u takes values in a compact subset of U (without loss of generality, we may assume that this holds everywhere on I), and restricted to a compact subset K of M, the continuous functions $f(x,u)$ and $\frac{\partial f}{\partial x}(x,u)$ are bounded. Hence it follows that both F and $\frac{\partial F}{\partial x}$ are bounded for $t \in I$ and $x \in K$. Thus the so-called C^1-Carathéodory conditions are satisfied (see Appendix D), and as in the case that the right-hand side of the differential equation is continuous in time, it follows that for any initial condition $(t_0,x_0) \in J \times M$, there exists a unique solution $x(\cdot;t_0,x_0)$ of the initial value problem $\dot{x} = f(x,u(t))$, $x(t_0) = x_0$, defined on a maximal interval of definition, $(\tau_-(t_0,x_0), \tau_+(t_0,x_0))$. This solution is the *corresponding trajectory*, and as before, we refer to the pair (x,u) as a *controlled trajectory*. Furthermore, $x(\cdot;t_0,x_0)$ is a continuously differentiable function of t_0 and x_0, and the partial derivatives $\frac{\partial x}{\partial t_0}$ and $\frac{\partial x}{\partial x_0}$ are solutions to the *variational equation* along $x(\cdot;t_0,x_0)$,

$$\dot{z} = \frac{\partial f}{\partial x}(x(t;t_0,x_0),u(t))z, \tag{4.2}$$

with the appropriate initial condition. Recall that the variational equation can formally be obtained by simply differentiating the differential equation $\dot{x}(t;t_0,x_0) = f(x(t;t_0,x_0),u(t))$ with respect to t_0 and x_0. Similarly, the initial conditions for the variational equation follow by differentiating the identity $x(t_0;t_0,x_0) \equiv x_0$. Since (4.2) is a time-varying linear differential equation whose matrix is bounded, these solutions exist on the full interval $(\tau_-(t_0,x_0), \tau_+(t_0,x_0))$ on which $x(\cdot;t_0,x_0)$ is defined (see Appendices B and D).

We close these introductory comments with a brief outline of the actual proof. We assume as given a controlled reference trajectory $(\bar{x},\bar{u})$ defined over an interval $[0,T]$ that steers $\bar{p} = \bar{x}(0)$ into $\bar{q} = \bar{x}(T)$. The proof is divided into four main steps:

Step 1: Generation of approximating directions by means of variations (Sect. 4.2.1). At every time $\bar{t} \in [0,T]$, variations $\mathcal{V}$ are made that generate curves $\zeta^{\mathcal{V}} : [0,\varepsilon] \to \mathbb{R}^n$, $\alpha \mapsto \zeta^{\mathcal{V}}(\alpha)$, anchored at $\zeta^{\mathcal{V}}(0) = \bar{x}(\bar{t})$. Integrating the variational equation along the controlled reference trajectory $(\bar{x},\bar{u})$ maps these curves $\zeta^{\mathcal{V}}$ into curves $\kappa^{\mathcal{V}} : [0,\varepsilon] \to \mathbb{R}^n$, $\alpha \mapsto \kappa^{\mathcal{V}}(\alpha)$, that lie in the reachable set from $\bar{p}$, $\text{Reach}_\Sigma(\bar{p})$, and are anchored at $\bar{q}$. The derivatives of these curves at $\alpha = 0$, $\dot{\kappa}^{\mathcal{V}}(0)$, define tangent vectors to the reachable set $\text{Reach}_\Sigma(\bar{p})$ at $\bar{q}$.

Step 2: Construction of an approximating cone (Sect. 4.2.2). The tangent vectors computed in step 1 are combined into an approximating map Ξ to create a Boltyansky approximating cone $\mathcal{K}$ to $\text{Reach}_\Sigma(\bar{p})$ at $\bar{q}$.

Step 3: Necessary conditions for boundary points (Sect. 4.2.3). If $\bar{q}$ is a boundary point of the reachable set $\text{Reach}_\Sigma(\bar{p})$, then by Theorem 4.1.1, this approximating cone cannot be the full space $\mathbb{R}^n$, and thus there exists a supporting hyperplane to $\mathscr{K}$. This translates into analytical necessary conditions for $\bar{q} \in \partial \operatorname{Reach}_\Sigma(\bar{p})$.

Step 4: Necessary conditions for optimal control (Sect. 4.2.4). By augmenting the dynamics with an extra variable that keeps track of the running cost in the objective function, optimality of the controlled trajectory $(\bar{x}, \bar{u})$ is described in terms of an empty intersection property between the reachable set $\operatorname{Reach}_\Sigma(\bar{p})$ and the set A of all terminal conditions that if reachable, would give a better value for the objective. The approximating cone to the set A at $\bar{q}$ is not a half-space, and thus by Theorem 4.1.3, these two sets can be separated at $\bar{q}$ by a hyperplane. The analytical conditions of the Pontryagin maximum principle follow from this separation property.

4.2.1   Tangent Vectors to the Reachable Set

Let $\bar{x}(\cdot)$ be a controlled trajectory defined on $[0, T]$ corresponding to an admissible control $\bar{u}(\cdot)$ with initial point $\bar{x}(0) = \bar{p}$ and terminal point $\bar{x}(T) = \bar{q}$. Henceforth we call $\bar{x}(\cdot)$ the *reference trajectory* and $\bar{u}(\cdot)$ the *reference control*. Our goal is to construct a (reasonably large) approximating cone for the reachable set $\operatorname{Reach}_\Sigma(\bar{p})$ from $\bar{p}$ at $\bar{q}$. In a first step we simply construct curves $\kappa^{\mathscr{V}}$,

$$\kappa^{\mathscr{V}} : [0, \varepsilon] \to \operatorname{Reach}_\Sigma(\bar{p}), \quad \alpha \mapsto \kappa^{\mathscr{V}}(\alpha),$$

that lie in the reachable set. The mechanism for doing so is to make variations in the reference control that change the reference trajectory infinitesimally as a function of a one-dimensional parameter α.

Definition 4.2.2 (Tangent vector to the reachable set). We say that a nonzero vector $v \in \mathbb{R}^n$ is a tangent vector to the reachable set from $\bar{p}$ at $\bar{q}$ if there exists a curve $\kappa^{\mathscr{V}} : [0, \varepsilon] \to \operatorname{Reach}_\Sigma(\bar{p})$, $\alpha \mapsto \kappa^{\mathscr{V}}(\alpha)$, that is differentiable from the right at $\alpha = 0$ and satisfies

$$v = \left(\frac{d}{d\alpha}_{|\alpha=0} \right) \kappa^{\mathscr{V}}(\alpha) = \dot{\kappa}^{\mathscr{V}}(0).$$

The following two variations generate the only curves needed for the proof of the Pontryagin maximum principle:

(1) Variation $\mathscr{V}_1$ (CUT): For any $\bar{t} \in (0, T]$, we modify the reference control $\bar{u}(\cdot)$ by deleting the portion that lies over the interval $(\bar{t} - \alpha, \bar{t}]$. This defines a 1-parameter family of admissible controls as

$$u_\alpha^{\mathscr{V}_1}(t) = \begin{cases} \bar{u}(t) & \text{if } 0 \leq t < \bar{t} - \alpha, \\ \bar{u}(t + \alpha) & \text{if } \bar{t} - \alpha \leq t \leq T - \alpha, \end{cases}$$

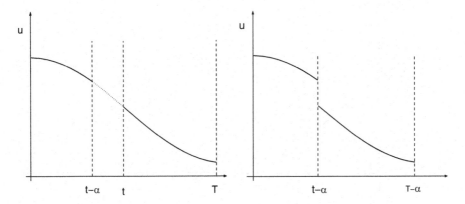

Fig. 4.6 Variation CUT

(see Fig. 4.6), and we denote the corresponding trajectories by $x_\alpha^{\mathcal{V}_1}(\cdot)$. For α small enough, the trajectory will exist on the full interval $[0, T - \alpha]$, and we denote the endpoint by $\kappa^{\mathcal{V}_1}(\alpha)$,

$$\kappa^{\mathcal{V}_1}(\alpha) = x_\alpha^{\mathcal{V}_1}(T - \alpha). \tag{4.3}$$

Since $\bar{x}$ is a solution to the differential equation $\dot{x} = f(x, \bar{u}(t))$, the curve

$$\zeta^{\mathcal{V}_1} : [0, \varepsilon] \to \mathbb{R}^n, \qquad \alpha \mapsto \zeta^{\mathcal{V}_1}(\alpha) = \bar{x}(\bar{t} - \alpha) \tag{4.4}$$

is differentiable at zero for almost every $\bar{t} \in (0, T]$ with derivative

$$\dot{\zeta}^{\mathcal{V}_1}(0) = -f(\bar{x}(\bar{t}), \bar{u}(\bar{t})).$$

This curve $\zeta^{\mathcal{V}_1}$ has the advantage that its tangent vector can be easily calculated, but it is not a curve in the reachable set. Rather, this curve is anchored at the point $\bar{x}(\bar{t})$ on the reference trajectory. The curve $\kappa^{\mathcal{V}_1}(\alpha)$ in the reachable set will then be the image of $\zeta^{\mathcal{V}_1}(\alpha)$ under the flow of the solutions to the differential equation $\dot{x} = f(x, \bar{u}(t))$. We shall show how this allows us to compute $\dot{\kappa}_1^{\mathcal{V}}(0)$ after the second variation has been defined.

(2) Variation $\mathcal{V}_2$ (PASTE): For any $\bar{t} \in [0, T]$ and an arbitrary admissible control value $v \in U$, insert a short constant segment given by $u(t) \equiv v$ for $\bar{t} \leq t < \bar{t} + \alpha$. The control corresponding to this variation is given by

$$u_\alpha^{\mathcal{V}_2}(t) = \begin{cases} \bar{u}(t) & \text{if } 0 \leq t < \bar{t}, \\ v & \text{if } \bar{t} \leq t < \bar{t} + \alpha, \\ \bar{u}(t - \alpha) & \text{if } \bar{t} + \alpha \leq t \leq T + \alpha, \end{cases}$$

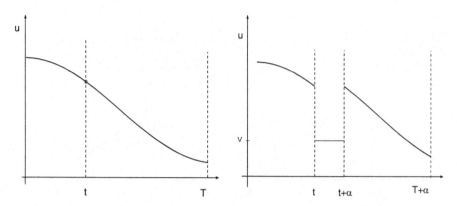

Fig. 4.7 Variation PASTE

(see Fig. 4.7), and as above, we denote the corresponding trajectory by $x_{\alpha}^{\gamma_2}$.

Given $\bar{t}$, it follows from local existence of solutions to the differential equation and continuous dependence on initial conditions that there exists an $\varepsilon > 0$ such that the trajectory $x_{\alpha}^{\gamma_2}$ will exist on the full interval $[\bar{t}, T + \alpha]$ for all $\alpha \in [0, \varepsilon]$. If we define

$$\zeta^{\gamma_2}(\alpha) = x_{\alpha}^{\gamma_2}(\bar{t} + \alpha), \tag{4.5}$$

then this curve is always differentiable from the right at 0 and

$$\dot{\zeta}^{\gamma_2}(0) = f(\bar{x}(\bar{t}), v).$$

Once again, the curve $\kappa^{\gamma_2}(\alpha)$ in the reachable set is defined as the endpoint of the trajectories,

$$\kappa^{\gamma_2}(\alpha) = x_{\alpha}^{\gamma_2}(T + \alpha), \tag{4.6}$$

and it is the image of $\zeta^{\gamma_2}(\alpha)$ under the flow of the solutions to the differential equation $\dot{x} = f(x, \bar{u}(t))$.

We now show how to calculate the tangent vectors $\dot{\kappa}_1^{\gamma_i}(0)$ for $i = 1, 2$. For this we need to embed the controlled reference trajectory into a parameterized field of controlled trajectories.

Proposition 4.2.1. *Given a controlled trajectory $(\bar{x}(\cdot), \bar{u}(\cdot))$ defined on the interval $[0, T]$, there exists a neighborhood P of $\bar{p}$ such that for every $p \in P$ the solution $x(\cdot; p)$ to the differential equation*

$$\dot{x} = f(x, \bar{u}(t)), \quad x(0) = p,$$

exists on the full interval $[0, T]$ and is continuously differentiable with respect to the initial condition p. The partial derivative of the general solution with respect to the parameter p, $\Theta(t) = \frac{\partial x}{\partial p}(t; p)$, is the fundamental solution of the variational

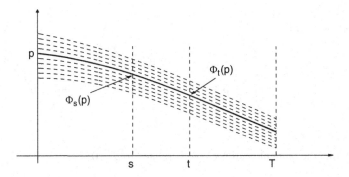

Fig. 4.8 Canonical embedding of the controlled reference trajectory $(\bar{x}, \bar{u})$

equation, that is,

$$\dot{\Theta}(t) = \frac{\partial f}{\partial x}(x(t;p), \bar{u}(t))\Theta(t), \qquad \Theta(0) = \mathrm{Id}. \tag{4.7}$$

Proof. This simply is a consequence of the general results about solutions to an ordinary differential equation alluded to above: By extending $\bar{u}(\cdot)$ to an admissible control on some open interval that contains $[0, T]$, we may assume that the reference trajectory $\bar{x}(\cdot)$ is defined on some interval $[-\varepsilon, T + \varepsilon]$, $\varepsilon > 0$. It follows from our general setup that the time-dependent vector field $(t, x) \mapsto f(x, \bar{u}(t))$ on M satisfies the C^1-Carathéodory conditions (see Appendix D). Hence there exists a neighborhood P of $\bar{p}$ such that for every $p \in P$ the differential equation $\dot{x} = f(x, \bar{u}(t))$ has a unique solution $x(\cdot; p)$ that satisfies $x(0) = p$, and by choosing P small enough, this solution exists on the interval $[0, T]$. Furthermore, the general solution $x(t; p)$ is continuous in (t, p), differentiable in p for fixed t with jointly continuous derivative $\frac{\partial x}{\partial p}(t; p)$ that can be calculated as the solution at time t to the variational equation (4.7). As solution to a linear time-varying ordinary differential equation whose matrix is given by a bounded Lebesgue measurable function, this solution exists on the full interval $[0, T]$. $\qquad \square$

Let $P_t = \{x(t; p) : p \in P\}$, $0 \leq t \leq T$, be the image of the neighborhood P under the flow of the solutions of this differential equation. In particular, $P_0 = P$. We also denote this flow by

$$\Phi_t : P_0 \to P_t, \qquad p \mapsto \Phi_t(p) = x(t; p),$$

and more generally, let $\Phi_{t,s}$ be the mapping that moves points from time s to time t along this flow,

$$\Phi_{t,s} : P_s \to P_t, \qquad \Phi_{t,s} = \Phi_t \circ (\Phi_s)^{-1}. \tag{4.8}$$

We call this flow $\Phi_{t,s}$ the *canonical embedding* of the controlled reference trajectory $(\bar{x}, \bar{u})$ into a local field of controlled trajectories (see Fig. 4.8). For both variations $\mathscr{V}_1$ and $\mathscr{V}_2$, the curve $\kappa^{\mathscr{V}}(\alpha)$ is given by $\kappa^{\mathscr{V}}(\alpha) = \Phi_{T,\bar{t}}(\zeta^{\mathscr{V}}(\alpha))$, and by

Fig. 4.9 The curve $\kappa^{\mathscr{V}}(\cdot)$ is the image of the curve $\zeta^{\mathscr{V}}(\cdot)$ under the flow $\Phi_{\bar{t},T}$

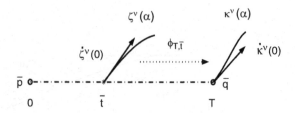

the chain rule, the tangent vector $\dot{\kappa}^{\mathscr{V}}(0)$ is related to $\dot{\zeta}^{\mathscr{V}}(0)$ as

$$\dot{\kappa}^{\mathscr{V}}(0) = \left(\Phi_{T,\bar{t}}\right)_* \dot{\zeta}^{\mathscr{V}}(0),$$

where $\left(\Phi_{T,\bar{t}}\right)_*$ denotes the derivative, also called the *differential* of the flow map (see Appendix C). In our case here, this differential is simply computed as the solution to the variational equation (4.7).

Proposition 4.2.2. *For the variations CUT and PASTE, the tangent vector $\dot{\kappa}^{\mathscr{V}}(0)$ is given by the value $\theta(T)$ of the solution θ to the variational equation with initial condition $\dot{\zeta}^{\mathscr{V}}(0)$ at time $\bar{t}$,*

$$\dot{\theta}(t) = \frac{\partial f}{\partial x}(\bar{x}(t), \bar{u}(t))\theta(t), \qquad \theta(\bar{t}) = \dot{\zeta}^{\mathscr{V}}(0). \tag{4.9}$$

We say the vector $\dot{\zeta}^{\mathscr{V}}(0)$ is moved from time $\bar{t}$ into the vector $\dot{\kappa}^{\mathscr{V}}(0)$ at time T along the flow of the canonical embedding (see Fig. 4.9).

Proof. The flow $\Phi_{\bar{t}}$ is a diffeomorphism, and thus there exists a differentiable curve $\pi^{\mathscr{V}} : [0, \varepsilon] \to P_0$, $\alpha \mapsto \pi^{\mathscr{V}}(\alpha)$, that parameterizes the initial conditions corresponding to the curve $\zeta^{\mathscr{V}}(\cdot)$, i.e.,

$$\zeta^{\mathscr{V}}(\alpha) = \Phi_{\bar{t}}\left(\pi^{\mathscr{V}}(\alpha)\right) = x(\bar{t}; \pi^{\mathscr{V}}(\alpha)).$$

Since $\pi^{\mathscr{V}}(0) = \bar{p}$, we therefore have that

$$\dot{\zeta}^{\mathscr{V}}(0) = \frac{\partial x}{\partial p}(\bar{t}; \bar{p})\dot{\pi}^{\mathscr{V}}(0).$$

By Proposition 4.2.1, the derivative of $x(\bar{t}; \bar{p})$ in the direction of the vector $\dot{\pi}^{\mathscr{V}}(0)$ at $\bar{p}$ can be calculated as the solution at time $\bar{t}$ of the variational equation (4.9) with initial condition $\theta(0) = \dot{\pi}^{\mathscr{V}}(0)$. If we denote the solutions of the matrix-valued initial value problem

$$\dot{\Psi}(t) = \frac{\partial f}{\partial x}(\bar{x}(t), \bar{u}(t))\Psi(t), \qquad \Psi(r) = \mathrm{Id},$$

by $\Psi(t;r)$, then we simply have that $\dot{\zeta}^{\mathscr{V}}(0) = \Psi(\bar{t};0)\dot{\pi}^{\mathscr{V}}(0)$. Analogously, the curve $\kappa^{\mathscr{V}}(\cdot)$ is the image of the curve $\pi^{\mathscr{V}}(\cdot)$ under the flow Φ_T,

$$\kappa^{\mathscr{V}}(\alpha) = \Phi_{T,\bar{t}}\left(\zeta^{\mathscr{V}}(\alpha)\right) = \Phi_T\left((\Phi_{\bar{t}})^{-1}\left(\zeta^{\mathscr{V}}(\alpha)\right)\right) = \Phi_T\left(\pi^{\mathscr{V}}(\alpha)\right) = x(T;\pi^{\mathscr{V}}(\alpha)),$$

and thus also

$$\dot{\kappa}^{\mathscr{V}}(0) = \frac{\partial x}{\partial p}(T;\bar{p})\dot{\pi}^{\mathscr{V}}(0) = \Psi(T,0)\dot{\pi}^{\mathscr{V}}(0).$$

Hence

$$\dot{\kappa}^{\mathscr{V}}(0) = \Psi(T,0)\dot{\pi}^{\mathscr{V}}(0) = \Psi(T,0)\Psi(\bar{t},0)^{-1}\dot{\zeta}^{\mathscr{V}}(0)$$
$$= \Psi(T,0)\Psi(0,\bar{t})\dot{\zeta}^{\mathscr{V}}(0) = \Psi(T,\bar{t})\dot{\zeta}^{\mathscr{V}}(0),$$

and thus $\dot{\kappa}^{\mathscr{V}}(0)$ is the value of the solution θ of the variational equation (4.9) at time T that has the initial condition $\dot{\zeta}^{\mathscr{V}}(0)$ at time $\bar{t}$. $\square$

4.2.2 Construction of an Approximating Cone

The variations CUT and PASTE, carried out for all $\bar{t} \in [0,T]$ and all possible values v from the control set U, thus generate a collection of tangent vectors $\dot{\kappa}^{\mathscr{V}}(0)$ to the reachable set $Reach_\Sigma(\bar{p})$ from $\bar{p}$ at $\bar{q}$. As we have seen in Sect. 4.1, this in itself is not sufficient to obtain a Boltyansky approximating cone, but we need to be able to combine different variations into an approximating map. We now show that this can be done for all these vectors.

Theorem 4.2.1. Let $\mathscr{K}$ be the convex hull of all the vectors $\dot{\kappa}^{\mathscr{V}}(0)$ that are generated by the variations CUT and PASTE along the reference trajectory, i.e., all vectors of the form $\left(\Phi_{T,\bar{t}}\right)_*\dot{\zeta}^{\mathscr{V}}(0)$, where $\dot{\zeta}^{\mathscr{V}}(0)$ is given by $f(\bar{x}(\bar{t}),v)$ for all possible values $v \in U$ and all times $\bar{t} \in [0,T]$ and by $-f(\bar{x}(\bar{t}),\bar{u}(\bar{t}))$ for almost every $\bar{t} \in [0,T]$. Then $\mathscr{K}$ is a Boltyansky approximating cone to $Reach_\Sigma(\bar{p})$ at $\bar{q}$.

Proof. By Lemma 4.1.2, it suffices to show that the prescribed collection of vectors $\left(\Phi_{T,\bar{t}}\right)_*\dot{\zeta}^{\mathscr{V}}(0)$ is a Boltyansky approximating cone.

In a first step, suppose that $v_1,\ldots,v_r$, is a finite collection of vectors from $\mathscr{K}$ that are of the form

$$v_i = \left(\Phi_{T,t_i}\right)_* w_i \tag{4.10}$$

with $w_i = \dot{\zeta}^{\mathscr{V}_i}(0)$ for variations $\mathscr{V}_1,\ldots,\mathscr{V}_r$, all of the type CUT or PASTE, but for *distinct* times t_i, say $t_1 < \cdots < t_r$. For sufficiently small $\delta > 0$, define the so-called endpoint mapping $\Xi : Q_\delta^r \to Reach_\Sigma(\bar{p})$ by

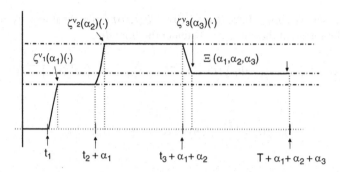

Fig. 4.10 An illustration of the definition of the endpoint mapping $\Xi(\alpha_1,\ldots,\alpha_r)$. (The flow of the canonical embedding is indicated by the *horizontal dash-dotted lines*)

$$\Xi(\alpha_1,\ldots,\alpha_r) = \Phi_{T,t_r} \circ \zeta^{\mathcal{V}_r}(\alpha_r) \circ \Phi_{t_r,t_{r-1}} \circ \cdots \circ \Phi_{t_2,t_1} \circ \zeta^{\mathcal{V}_1}(\alpha_1) \circ \Phi_{t_1,0}(\bar{p}), \quad (4.11)$$

where analogously to Eqs. (4.4) and (4.5), the mappings $\zeta^{\mathcal{V}_i}(\alpha_i)$ describe the effects of the variations $\mathcal{V}_i$, $i = 1,\ldots,r$, made at time t_i acting on the point at which the variation is anchored.

As the name indicates, this mapping simply describes the endpoints of the combined successive trajectories x_α corresponding to the admissible controls u_α defined by the variations (see Fig. 4.10). The first variation $\mathcal{V}_1$ is made at time t_1 and generates the curve $\zeta^{\mathcal{V}_1}(\alpha_1)$ anchored at $\bar{x}(t_1)$. After the first variation has been made, the trajectory follows the flow Φ_{t_2,t_1} of the canonical embedding of the reference controlled trajectory until time t_2, when the second variation is made. This second variation, however, is no longer made from the reference trajectory, but from a point on a nearby trajectory of the canonical embedding that is determined by the effects of the first variation. This step is then repeated finitely many times. In this construction, we need to carefully track the errors that are made as we move from one variation to the next in order to show that the higher-order term indeed is of order $o(|\alpha|)$ as $\alpha \to 0$, $\alpha = (\alpha_1,\ldots,\alpha_r)$, $|\alpha| = \alpha_1 + \cdots + \alpha_r$.

We show that Ξ is an approximating map for the vectors v_i, i.e., that we can take $v_i' = v_i$ in Definition 4.1.2. It follows from our general regularity assumptions that Ξ is continuous. The proof that Ξ is an approximating map in the sense of the property (4.1) is by induction on r, the number of variations made. Let $r = 1$. Since the first variation is always made from the reference trajectory, setting $\bar{t} = t_1$, the map Ξ simply reduces to the curve $\kappa^{\mathcal{V}}(\alpha)$ constructed above: with the curve $\zeta^{\mathcal{V}}(\alpha)$ anchored at $\Phi_{\bar{t},0}(\bar{p}) = \bar{x}(\bar{t})$, it follows that

$$\Xi(\alpha) = \Phi_{T,\bar{t}} \circ \zeta^{\mathcal{V}}(\alpha) \circ \Phi_{\bar{t},0}(\bar{p}) = \Phi_{T,\bar{t}}\left(\zeta^{\mathcal{V}}(\alpha)\right) = \kappa^{\mathcal{V}}(\alpha).$$

In particular, the values of Ξ thus lie in $\text{Reach}_\Sigma(\bar{p})$. By Taylor's theorem, we have that

$$\Xi(\alpha) = \Xi(0) + \Xi'(0)\alpha + o(\alpha) = \bar{x}(T) + \dot{\kappa}^{\mathscr{V}}(0)\alpha + o(\alpha)$$

$$= \bar{q} + \alpha\left(\Phi_{T,\bar{t}}\right)_* \dot{\zeta}^{\mathscr{V}}(0) + o(\alpha) = \bar{q} + \alpha\left(\Phi_{T,\bar{t}}\right)_* w + o(\alpha)$$

$$= \bar{q} + \alpha v + o(\alpha)$$

and thus Ξ is an approximating map.

We now proceed with the inductive step. In order to simplify the notation, we formulate it only for $r = 2$. The general argument is exactly the same, only with more cumbersome indices. Hence, we assume that two vectors v_1 and v_2 are generated by two variations $\mathscr{V}_1$ and $\mathscr{V}_2$ of the type CUT or PASTE at times $t_1 < t_2$, i.e., for $i = 1, 2$,

$$v_i = \dot{\kappa}^{\mathscr{V}_i}(0) = \left(\Phi_{T,t_i}\right)_* \dot{\zeta}^{\mathscr{V}_i}(0) = \left(\Phi_{T,t_i}\right)_* w_i.$$

We need to show that for $|\alpha| = \alpha_1 + \alpha_2$ small enough, the mapping

$$\Xi(\alpha_1, \alpha_2) = \Phi_{T,t_2} \circ \zeta^{\mathscr{V}_2}(\alpha_2) \circ \Phi_{t_2,t_1} \circ \zeta^{\mathscr{V}_1}(\alpha_1) \circ \Phi_{t_1,0}(\bar{p})$$

takes values in the reachable set and that it has a linear approximation of the form

$$\Xi(\alpha_1, \alpha_2) = \bar{q} + \alpha_1 v_1 + \alpha_2 v_2 + o(\alpha_1 + \alpha_2).$$

We first verify that the image lies in the reachable set. This actually requires that we choose $|\alpha|$ small. For only then is the control determined by the variation admissible. This control is easily written down. For example, say variation $\mathscr{V}_1$ is of the type CUT for time α_1 and variation $\mathscr{V}_2$ inserts the constant control $v \in U$ for time α_2. Then the corresponding control u_α is given by

$$u_\alpha(t) = \begin{cases} \bar{u}(t) & \text{if } 0 \le t < t_1 - \alpha_1, \\ \bar{u}(t + \alpha_1) & \text{if } t_1 - \alpha_1 \le t < t_2 - \alpha_1, \\ v & \text{if } t_2 - \alpha_1 \le t < t_2 - \alpha_1 + \alpha_2, \\ \bar{u}(t + \alpha_1 - \alpha_2) & \text{if } t_2 - \alpha_1 + \alpha_2 \le t \le T - \alpha_1 + \alpha_2. \end{cases}$$

In general, however, restrictions on the parameters α_i need to be imposed to the effect that *the two variations do not interfere with each other*. For example, if variation $\mathscr{V}_1$ is of the type PASTE and variation $\mathscr{V}_2$ is of the type CUT, we need to have that $t_1 + \alpha_1 < t_2 - \alpha_2$. Since $t_1 < t_2$, this condition can always be satisfied by taking $|\alpha| = \alpha_1 + \alpha_2$ small enough. Also, for $|\alpha|$ small enough, the corresponding solution x_α to the dynamics will exist on the full interval, and thus $\Xi(\alpha_1, \alpha_2)$ defines a point in the reachable set. Hence Ξ takes values in $\text{Reach}_\Sigma(\bar{p})$.

We now calculate the derivative of Ξ. The first variation $\mathscr{V}_1$ produces the curve

$$\zeta^{\mathscr{V}_1}(\alpha_1) \circ \Phi_{t_1,0}(\bar{p}) = \bar{x}(t_1) + \alpha_1 \dot{\zeta}^{\mathscr{V}_1}(0) + o(\alpha_1),$$

which then is moved along the flow of the canonical embedding from time t_1 to time t_2. The flow Φ_{t_2,t_1} is a diffeomorphism, and using Taylor's theorem, it follows that

$$\Phi_{t_2,t_1} \circ \zeta^{\mathscr{V}_1}(\alpha_1) \circ \Phi_{t_1,0}(\bar{p}) = \Phi_{t_2,t_1} \left[\bar{x}(t_1) + \alpha_1 \dot{\zeta}^{\mathscr{V}_1}(0) + o(\alpha_1)\right]$$

$$= \Phi_{t_2,t_1}(\bar{x}(t_1)) + (\Phi_{t_2,t_1})_* \left[\alpha_1 \dot{\zeta}^{\mathscr{V}_1}(0) + o(\alpha_1)\right] + o(\alpha_1)$$

$$= \bar{x}(t_2) + \alpha_1 (\Phi_{t_2,t_1})_* \dot{\zeta}^{\mathscr{V}_1}(0) + o(\alpha_1). \tag{4.12}$$

Here we use that $\Phi_{t_2,t_1}(\bar{x}(t_1)) = \bar{x}(t_2)$, since the canonical embedding reduces to the reference controlled trajectory for $p = \bar{p}$. Now the second variation $\mathscr{V}_2$ is made, but starting from the point $x = \bar{x}(t_2) + \alpha_1 (\Phi_{t_2,t_1})_* \dot{\zeta}^{\mathscr{V}_1}(0) + o(\alpha_1)$, *not* from the point $y = \bar{x}(\bar{t}_2)$ on the reference trajectory as in our earlier calculations. We thus need to compare the effects of a variation $\mathscr{V}$ made at the two different starting points x and y. We therefore now include the anchor point in our notation and let $\zeta^{\mathscr{V}}(\alpha;x)$ denote the curve in α generated by a variation (CUT or PASTE) starting at x, $\zeta^{\mathscr{V}}(0;x) = x$. By continuous dependence of the solutions of a differential equation on initial conditions, $\zeta^{\mathscr{V}}(\alpha;\cdot)$ is differentiable in x and

$$\zeta^{\mathscr{V}}(\alpha;x) - \zeta^{\mathscr{V}}(\alpha;y) = \int_0^1 \frac{d}{ds}\left(\zeta^{\mathscr{V}}(\alpha;y+s(x-y))\right) ds$$

$$= \left(\int_0^1 \frac{\partial \zeta^{\mathscr{V}}}{\partial x}(\alpha;y+s(x-y))ds\right)(x-y)$$

$$= x-y + \left(\int_0^1 \left[\frac{\partial \zeta^{\mathscr{V}}}{\partial x}(\alpha;y+s(x-y))-1\right]ds\right)(x-y)$$

$$= x-y+r(x,y;\alpha)(x-y), \tag{4.13}$$

where the last line defines r. By definition, r is continuous. Since $\zeta^{\mathscr{V}}(0;x) = x$, we have $r(x,y;0) \equiv 0$, and thus uniformly in x and y over a compact neighborhood of the point $\bar{x}(\bar{t})$ on the reference trajectory, we have that $\lim_{\alpha\to0+} \tilde{r}(x,y;\alpha) = 0$. Thus the function r is of order $o(1)$ as $\alpha \to 0$. Applying Eq. (4.13) to Eq. (4.12), we thus get that

$$\zeta^{\mathscr{V}_2}(\alpha_2) \circ \Phi_{t_2,t_1} \circ \zeta^{\mathscr{V}_1}(\alpha_1) \circ \Phi_{t_1,0}(\bar{p})$$

$$= \zeta^{\mathscr{V}_2}\left(\alpha_2;\bar{x}(t_2) + \alpha_1 (\Phi_{t_2,t_1})_* \dot{\zeta}^{\mathscr{V}_1}(0) + o(\alpha_1)\right) = \zeta^{\mathscr{V}_2}(\alpha_2;x)$$

$$= \zeta^{\mathscr{V}_2}(\alpha_2;y) + x-y+r(x,y,\alpha)(x-y)$$

$$= \zeta^{\mathscr{V}_2}(\alpha_2;\bar{x}(t_2)) + \alpha_1 (\Phi_{t_2,t_1})_* \dot{\zeta}^{\mathscr{V}_1}(0) + o(\alpha_1) + o(1)\cdot O(\alpha_1).$$

The error made in this calculation by switching back from the point x (on the trajectory reached after the first variation) to the point y (on the reference trajectory) is thus of order $O(\alpha_1)$, and it is propagated with order $o(1)$ as $\alpha_2 \to 0$. Since we still have that

$$\frac{o(1) \cdot O(\alpha_1)}{\alpha_1 + \alpha_2} \to 0 \qquad \text{as } \alpha_1 + \alpha_2 \to 0,$$

it follows that these remainders together are of order $o(\alpha_1 + \alpha_2)$. Furthermore, for the second variation $\mathscr{V}_2$ acting on the point $\bar{x}(t_2)$ of the reference trajectory, we have that

$$\zeta^{\mathscr{V}_2}(\alpha_2; \bar{x}(t_2)) = \bar{x}(t_2) + \alpha_2 \dot{\zeta}^{\mathscr{V}_2}(0) + o(\alpha_2).$$

Combining these equations yields

$$\zeta^{\mathscr{V}_2}(\alpha_2) \circ \Phi_{t_2,t_1} \circ \zeta^{\mathscr{V}_1}(\alpha_1) \circ \Phi_{t_1,0}(\bar{p})$$
$$= \bar{x}(\bar{t}_2) + \alpha_2 \dot{\zeta}^{\mathscr{V}_2}(0) + \alpha_1 (\Phi_{t_2,t_1})_* \dot{\zeta}^{\mathscr{V}_1}(0) + o(\alpha_1 + \alpha_2).$$

Finally, moving all the vectors with the flow Φ_{T,t_2} to the terminal point, we obtain

$$\Xi(\alpha_1, \alpha_2) = \Phi_{T,t_2} \left[\bar{x}(\bar{t}_2) + \alpha_2 \dot{\zeta}^{\mathscr{V}_2}(0) + \alpha_1 (\Phi_{t_2,t_1})_* \dot{\zeta}^{\mathscr{V}_1}(0) + o(\alpha_1 + \alpha_2) \right]$$
$$= \Phi_{T,t_2}(\bar{x}(\bar{t}_2)) + \alpha_2 (\Phi_{T,t_2})_* \left[\dot{\zeta}^{\mathscr{V}_2}(0) \right] + \alpha_1 (\Phi_{T,t_2})_* \left[(\Phi_{t_2,t_1})_* \dot{\zeta}^{\mathscr{V}_1}(0) \right]$$
$$+ o(\alpha_1 + \alpha_2)$$
$$= \bar{q} + \alpha_1 \dot{\kappa}^{\mathscr{V}_1}(0) + \alpha_2 \dot{\kappa}^{\mathscr{V}_2}(0) + o(\alpha_1 + \alpha_2)$$
$$= \bar{q} + \alpha_1 v_1 + \alpha_2 v_2 + o(\alpha_1 + \alpha_2).$$

Thus Ξ is an approximating map for $v_1 = \dot{\kappa}^{\mathscr{V}_1}(0)$ and $v_2 = \dot{\kappa}^{\mathscr{V}_2}(0)$. This concludes the main step of the proof. Summarizing, *we have shown that the endpoint mapping* $\Xi = \Xi(\alpha_1, \ldots, \alpha_r)$ *of trajectories defined through a finite number of variations of types CUT and PASTE that are made at different times is an approximating map for the corresponding tangent vectors to the reachable set.*

We still need to take care of the situation in which variations are made at the same time. In this case, it is possible that these variations interfere with each other, and then the construction just given no longer works. But both variations CUT and PASTE are stable in the sense that given $\bar{t}$, they can also be made at sufficiently many times τ near $\bar{t}$. More generally, if $\mathscr{V}_1$ and $\mathscr{V}_2$ are two variations made at the same time $\bar{t}$ that generate the tangent vectors $v_i = \dot{\kappa}^{\mathscr{V}_i}(0) = (\Phi_{T,\bar{t}})_* \dot{\zeta}_0^{\mathscr{V}_i}(\bar{x}(\bar{t}))$, it is possible to separate the times if the *variations can be approximated by other variations made at nearby points.* Specifically, suppose *Knobloch's condition* (K) [137] is satisfied:

[K] there exist times $\{t_{i,n}\}_{n \in \mathbb{N}} \subset [0, T]$, $t_{i,n} \neq \bar{t}$, that converge to $\bar{t}$, $\lim_{n \to \infty} t_{i,n} = \bar{t}$, and variations $\mathscr{V}_{i,n}$ made at $t_{i,n}$ such that

$$w_{i,n} = \dot{\zeta}^{\mathscr{V}_{i,n}}(0; \bar{x}(t_{i,n})) \to w_i = \dot{\zeta}^{\mathscr{V}_i}(0; \bar{x}(\bar{t})).$$

If this condition is satisfied, then we have that

$$v_{i,n} = \dot{\kappa}^{\mathscr{V}_{i,n}}(0) = \left(\Phi_{T,t_{i,n}}\right)_* w_{i,n} \to v_i = \dot{\kappa}^{\mathscr{V}_i}(0) = \left(\Phi_{T,\bar{t}}\right)_* w_i,$$

and given $\varepsilon > 0$, by choosing n and m large enough, we can make the norms $\|v_1 - v_{1,n}\|$ and $\|v_2 - v_{2,m}\|$ less than ε while at the same time making sure that $t_{1,n} \neq t_{2,m}$. By the previous step of the proof, the endpoint map Ξ corresponding to the variations $\mathscr{V}_{1,n}$ and $\mathscr{V}_{2,m}$ made at the times $t_{1,n}$ and $t_{2,m}$ then still is an approximating map for the original tangent vectors v_1 and v_2. It is exactly for this reason that the map Ψ in the definition of an approximating cone (Definition 4.1.2) is only required to approximate vectors v_i' that are close to the tangent vectors v_i. Thus, we also have an approximating map for variations made at the same time if these variations are stable in the sense that property (K) is valid. Clearly, this argument extends to any finite number of variations.

It thus remains only to verify that the variations CUT and PASTE have the stability property (K). For the variations PASTE this is immediate: if $\mathscr{V}$ denotes the variation PASTE made at $\bar{t} \in [0,T]$ by inserting a constant control with value $v \in U$, then this variation can be made at any time $t_n \in [0,T]$ where it generates the vector $w_n = f(\bar{x}(t_n),v)$, and by continuity $w_n = f(\bar{x}(t_n),v) \to f(\bar{x}(\bar{t}),v) = w$ whenever $t_n \to \bar{t}$. For the variation CUT, using some more specific results about Lebesgue measurable functions, it can be shown that property (K) still is valid almost everywhere in $[0,T]$.

Proposition 4.2.3. *For almost every $\bar{t} \in [0,T]$, there exists a sequence of times $\{t_n\}_{n \in \mathbb{N}} \subset [0,T]$, $t_n \neq \bar{t}$, that converge to $\bar{t}$, $\bar{t} = \lim_{n \to \infty} t_n$, with the property that the variation CUT made at time t_n generates the tangent vector $w_n = -f(\bar{x}(t_n)),\bar{u}(t_n))$ and that $f(\bar{x}(\bar{t})),\bar{u}(\bar{t})) = \lim_{n \to \infty} f(\bar{x}(t_n)),\bar{u}(t_n))$.*

Proof. The variation CUT can be defined for any time $\bar{t} \in (0,T]$. But the reference control $\bar{u}$ is only measurable, and thus the function $\alpha \mapsto \zeta^{\mathscr{V}_1}(\alpha) = \bar{x}(\bar{t} - \alpha)$ need not be differentiable in α at $\alpha = 0$ for all $\bar{t} \in (0,T]$. However, it follows from Lebesgue's differentiation theorem [174] that there exists a measurable set $F \subset [0,T]$ of full measure such that $\dot{\zeta}^{\mathscr{V}_1}(0)$ exists for all $\bar{t} \in F$ and is given by $\dot{\zeta}^{\mathscr{V}_1}(0) = -f(\bar{x}(\bar{t})),\bar{u}(\bar{t}))$. Let $\bar{\phi} : [0,T] \to \mathbb{R}^n$, $t \mapsto f(\bar{x}(t),\bar{u}(t))$. Given $n \in \mathbb{N}$, by Lusin's theorem [174] there exists a closed subset $F_n \subset F$ of measure greater than $T - \frac{1}{n}$ such that the restriction of $\bar{\phi}$ to F_n is continuous. In particular, whenever $\{t_n\} \subset F_n$ and $t_n \to \bar{t} \in F_n$, then $f(\bar{x}(t_n)),\bar{u}(t_n)) \to f(\bar{x}(\bar{t})),\bar{u}(\bar{t}))$. Property (K) is therefore valid for any time $\bar{t} \in F_n$ that is a limit point of the set F_n (i.e., there exists a sequence of times $\{t_n\} \subset F_n$, $t_n \neq \bar{t}$, and $t_n \to \bar{t}$). But almost every time in F_n is a limit point of F_n. For if $\tau \in F_n$ is not a limit point of F_n, then there exists a neighborhood of τ that does not contain another point from F_n, and thus the set of times $\tau \in F_n$ that are not limit points of F_n is discrete and hence at most countable. Thus, if $\tilde{F}_n$ denotes the set of times $\tau \in F_n$ that are limit points of F_n, then its measure is still greater than $T - \frac{1}{n}$. Consequently, $\tilde{F} = \cup_{n=1}^{\infty} \tilde{F}_n \subset F$ is a set of full measure, and for every $\bar{t} \in \tilde{F}$ property (K) holds. $\square$

This completes the proof of Theorem 4.2.1. $\square$

The last step in the proof is the reason for the somewhat odd Definition 4.1.2 of an approximating cone being able to approximate nearby vectors. This definition has the advantage that we do not need to address the highly nontrivial issue of whether and when variations made at the same time $\bar{t}$ can be combined. For more general variations than CUT and PASTE, as we shall consider them later on, this is not necessarily true in general (see, for instance, [36, 137]).

4.2.3   Boundary Trajectories

Theorem 4.2.1 immediately gives us necessary conditions for the point $\bar{q}$ to lie in the boundary of the reachable set. For if $\bar{q} \in \partial \operatorname{Reach}_{\Sigma}(\bar{p})$, then by Theorem 4.1.1 the approximating cone $\mathscr{K}$ constructed above cannot be the full space $\mathbb{R}^n$. Hence there exists a nonzero covector $\bar{\lambda} \in (\mathbb{R}^n)^*$ such that

$$\langle \bar{\lambda}, v \rangle \geq 0 \quad \text{for all } v \in \mathscr{K}. \tag{4.14}$$

In the construction of $\mathscr{K}$, the tangent vectors $w = \zeta^{\gamma}(0)$ generated at the point $\bar{x}(\bar{t})$ are moved forward to the endpoint $\bar{q} = \bar{x}(T)$ along the reference trajectory. Rather than moving all these tangent vectors forward to the terminal point, instead we can move the covector $\bar{\lambda}$ backward along the reference trajectory to $\bar{x}(\bar{t})$. This has the obvious advantage that overall, only one object needs to be moved.

Proposition 4.2.4. *Let $\theta : [0,T] \to \mathbb{R}^n$, $t \mapsto \theta(t)$, be a solution to the variational equation*

$$\dot{\theta}(t) = \frac{\partial f}{\partial x}(\bar{x}(t), \bar{u}(t))\theta(t),$$

and let $\lambda : [0,T] \to (\mathbb{R}^n)^*$, $t \mapsto \lambda(t)$, be a solution to its adjoint equation,

$$\dot{\lambda}(t) = -\lambda(t)\frac{\partial f}{\partial x}(\bar{x}(t), \bar{u}(t)). \tag{4.15}$$

Then the function $h : [0,T] \to \mathbb{R}$, $t \mapsto \langle \lambda(t), \theta(t) \rangle = \lambda(t)\theta(t)$, is constant.

Proof. This is an immediate consequence of the fact that the second equation is the adjoint of the first:

$$\frac{d}{dt}\langle \lambda(t), \theta(t) \rangle = \dot{\lambda}(t)\theta(t) + \lambda(t)\dot{\theta}(t)$$

$$= -\lambda(t)\frac{\partial f}{\partial x}(\bar{x}(t), \bar{u}(t))\theta(t) + \lambda(t)\frac{\partial f}{\partial x}(\bar{x}(t), \bar{u}(t))\theta(t) = 0. \qquad \square$$

Recall that a tangent vector $v = \kappa^{\gamma}(0)$ to the reachable set at $\bar{q}$ is obtained from the vector $w = \zeta^{\gamma}(0)$ as the solution $\theta(T)$ of the variational equation at time T with initial condition w at time t (Proposition 4.2.2). If we therefore choose $\lambda(\cdot)$ as the solution to the adjoint equation (4.15) with terminal condition $\lambda(T) = \bar{\lambda}$, then we have for all $t \in [0,T]$ that

$$\langle \bar{\lambda}, v \rangle = \langle \lambda(T), \theta(T) \rangle = \langle \lambda(t), \theta(t) \rangle = \langle \lambda(t), w \rangle.$$

Substituting for w the results of the variations CUT and PASTE, we therefore obtain that

$$\langle \lambda(t), f(\bar{x}(t), v) \rangle \geq 0 \qquad \text{for all } t \in [0, T] \text{ and all } v \in U$$

and

$$-\langle \lambda(t), f(\bar{x}(t), \bar{u}(t)) \rangle \geq 0 \qquad \text{for almost every } t \in [0, T].$$

Defining the Hamiltonian H as

$$H : (\mathbb{R}^n)^* \times \mathbb{R}^n \times U \to \mathbb{R},$$

$$(\lambda, x, u) \mapsto H(\lambda, x, u) = \langle \lambda, f(x, u) \rangle,$$

we obtain the following result:

Theorem 4.2.2 (Maximum principle for boundary trajectories). *Let $\Gamma = (\bar{x}(\cdot), \bar{u}(\cdot))$ be an admissible controlled trajectory defined on $[0, T]$ with initial point $\bar{x}(0) = \bar{p}$ and terminal point $\bar{x}(T) = \bar{q}$. If $\bar{q}$ lies in the boundary of the reachable set from $\bar{p}$, $\bar{q} \in \partial \operatorname{Reach}_\Sigma(\bar{p})$, then there exists a nontrivial solution $\lambda : [0, T] \to (\mathbb{R}^n)^*$, $t \mapsto \lambda(t)$, to the adjoint equation along Γ,*

$$\dot{\lambda}(t) = -\lambda(t)\frac{\partial f}{\partial x}(\bar{x}(t), \bar{u}(t)) = -\frac{\partial H}{\partial x}(\lambda(t), \bar{x}(t), \bar{u}(t)), \qquad \lambda(T) \neq 0, \quad (4.16)$$

such that almost everywhere on $[0, T]$

$$0 = H(\lambda(t), \bar{x}(t), \bar{u}(t)) = \min_{u \in U} H(\lambda(t), \bar{x}(t), u). \qquad (4.17)$$

4.2.4 Necessary Conditions for Optimality

The necessary conditions for optimality of the Pontryagin maximum principle, Theorem 2.2.1, also follow from Theorem 4.2.1. The additional ingredient now is an *objective* that will be minimized over possibly a subset of all controlled trajectories. Let $L : M \times U \to \mathbb{R}$, $(x, u) \mapsto L(x, u)$, be a function that has the same regularity properties as postulated for the dynamics, i.e., is continuous on $M \times U$ and, for each $u \in U$ fixed, is continuously differentiable in x with partial derivative $\frac{\partial L}{\partial x}(x, u)$ continuous on $M \times U$. Furthermore, let $\Psi : \mathbb{R}^n \to \mathbb{R}^{n-k}$, $x \mapsto \Psi(x)$, be a continuously differentiable mapping with the property that the gradients of the components are linearly independent on the set $N = \{x \in M : \Psi(x) = 0\}$. Thus the set N is a k-dimensional embedded C^1-submanifold of the state space M, the terminal manifold. Let $\varphi : N \to \mathbb{R}$ be a continuously differentiable function defined on N. For a controlled trajectory (x, u) defined on the interval $[0, T]$ with terminal value $x(T)$ in N, define its cost as

$$\mathscr{J}(u) = \int_0^T L(x(t), u(t))dt + \varphi(x(T)).$$

We then consider the following optimal control problem:

[OC] Given a control system Σ, minimize $\mathscr{J}(u)$ over all controlled trajectories (x, u) defined over some compact interval $[0, T]$ with fixed initial value $x(0) = \bar{p}$ and terminal value $x(T) \in N$. The final time T is free.

It is possible to express optimality of a controlled trajectory in terms of the reachable set for a system in which the state space has been augmented with the cost as extra variable. Let $\widetilde{\Sigma}$ be the control system with state space $\widetilde{M} = M \times \mathbb{R}$, the same control set U, and same class $\mathscr{U}$ of admissible controls, but dynamical law given by

$$\widetilde{f} : \widetilde{M} \times U \to \mathbb{R}^{n+1}, \qquad (x, y, u) \mapsto \widetilde{f}(x, y, u) = \begin{pmatrix} f(x, u) \\ L(x, u) \end{pmatrix}$$

with initial condition $\tilde{p} = (\bar{p}, 0)^T$. Thus the differential equations become

$$\dot{x} = f(x, u(t)), \qquad x(0) = \bar{p},$$
$$\dot{y} = L(x, u(t)), \qquad y(0) = 0.$$

The terminal manifold remains the same, but is now considered an embedded submanifold $\widetilde{N} = N \times \mathbb{R}$ of $\widetilde{M}$. Hence, to each admissible controlled trajectory (x, u) of Σ defined over the interval $[0, T]$, there corresponds a unique controlled trajectory (x, y, u) of $\widetilde{\Sigma}$ with y given as

$$y(t) = \int_0^t L(x(s), u(s))ds,$$

and the value of the corresponding objective is $\mathscr{J}(u) = y(T) + \varphi(x(T))$. Define a function $\omega : \widetilde{N} = N \times \mathbb{R} \to \mathbb{R}$, $(x, y) \mapsto \omega(x, y) = \varphi(x) + y$, such that

$$\mathscr{J}(u) = \omega(x(T), y(T)).$$

Given a controlled reference trajectory $\Gamma = (\bar{x}, \bar{u})$ defined over the interval $[0, T]$ that steers $\bar{p}$ into $\bar{q}$, let $\bar{\omega} = \omega(\bar{q}, \bar{y}(T))$ be the corresponding value of the objective and define

$$\widetilde{A} = \{(x, y) \in N \times \mathbb{R} : \omega(x, y) = \varphi(x) + y < \bar{\omega}\}.$$

Thus, this is the set of all possible terminal points that *in principle* (i.e., if they were reachable), would give a better value for the objective. An approximating cone to this set is easily constructed:

Proposition 4.2.5. *The set*

$$\widetilde{\mathscr{J}} = \{(v,\alpha) \in \mathbb{R}^n \times \mathbb{R}: \; D\Psi(\bar{q})v = 0, \; \langle\nabla\varphi(\bar{q}),v\rangle + \alpha < 0\}$$

is a Boltyansky approximating cone for $\widetilde{A}$ at $\widetilde{q} = (\bar{q},\bar{\omega})$.

It is clear that $\widetilde{\mathscr{J}}$ is a nonempty convex cone that is not a subspace. For $(0,-1) \in \widetilde{\mathscr{J}}$ and $(0,1) \notin \widetilde{\mathscr{J}}$.

Proof. In order to show that $\widetilde{\mathscr{J}}$ is approximating, suppose $(v_1,\alpha_1),\dots,(v_r,\alpha_r) \in \widetilde{\mathscr{J}}$. It follows from Proposition 4.1.1 that the null space of $D\Psi(\bar{q})$, the tangent space to N at $\bar{q}$, is an approximating cone for N. Thus there exists a continuous map Ξ defined on a full neighborhood $B_\delta^r(0)$ of 0, $\Xi: B_\delta^r(0) \to N$, such that

$$\Xi(z_1,\dots,z_r) = \bar{q} + \sum_{i=1}^{r} z_i v_i + o(\|z\|).$$

Define the augmented map $\widetilde{\Xi}: B_\delta^r(0) \to N \times \mathbb{R}$ by

$$\widetilde{\Xi}(z_1,\dots,z_r) = \left(\Xi(z_1,\dots,z_r), \; \bar{\omega} + \sum_{i=1}^{r} \alpha_i z_i\right).$$

Clearly, $\widetilde{\Xi}$ is continuous, and it satisfies

$$\widetilde{\Xi}(z_1,\dots,z_r) = (\bar{q},\bar{\omega}) + \sum_{i=1}^{r} z_i (v_i,\alpha_i) + o(\|z\|),$$

i.e., Eq. (4.1) is satisfied with $(v_i',\alpha_i') = (v_i,\alpha_i)$. Therefore, all that remains to be shown is that for sufficiently small positive coordinates z_i, $z \in Q_\delta^r$, the mapping $\widetilde{\Xi}$ takes values in the set $\widetilde{A}$. We have that

$$\omega\left(\widetilde{\Xi}(z_1,\dots,z_r)\right) = \varphi(\Xi(z_1,\dots,z_r)) + \bar{\omega} + \sum_{i=1}^{r} \alpha_i z_i$$

$$= \varphi\left(\bar{q} + \sum_{i=1}^{r} z_i v_i + o(\|z\|)\right) + \bar{\omega} + \sum_{i=1}^{r} \alpha_i z_i$$

$$= \varphi(\bar{q}) + \left\langle d\varphi(\bar{q}), \sum_{i=1}^{r} v_i z_i \right\rangle + o(\|z\|) + \bar{\omega} + \sum_{i=1}^{r} \alpha_i z_i$$

$$= \varphi(\bar{q}) + \bar{\omega} + \sum_{i=1}^{r} [\langle d\varphi(\bar{q}),v_i\rangle + \alpha_i] z_i + o(\|z\|)$$

$$< \varphi(\bar{q}) + \bar{\omega},$$

where the last inequality follows from the facts that $z_i > 0$ and that by definition of $\widetilde{\mathscr{J}}$, we have $\langle d\varphi(\bar{q}),v_i\rangle + \alpha_i < 0$. Thus the linear term is negative, and for $\|z\|$

sufficiently small it dominates the remainder $o(\|z\|)$. Hence $\widetilde{\Xi}$ takes values in $\widetilde{A}$, and for $\delta > 0$ small enough the map $\widetilde{\Xi} : Q_\delta^r \to \widetilde{A}$ is approximating. $\qquad\square$

If $\Gamma = (\bar{x}, \bar{u})$ is optimal, then no point in $\widetilde{A}$ can be reachable for the augmented system $\widetilde{\Sigma}$, and thus the reachable set $\mathit{Reach}_{\widetilde{\Sigma}}(\tilde{p})$ must be disjoint from $\widetilde{A}$:

$$\mathit{Reach}_{\widetilde{\Sigma}}(\tilde{p}) \cap \widetilde{A} = \emptyset.$$

In particular, $\tilde{q} = (\bar{q}, \bar{y}(T))$ is an isolated point of $\mathit{Reach}_{\widetilde{\Sigma}}(\tilde{p}) \cap (\widetilde{A} \cup \{\tilde{q}\})$. Thus, if $\mathscr{K}$ denotes the approximating cone to $\mathit{Reach}_{\widetilde{\Sigma}}(\tilde{p})$ at $\tilde{q}$ constructed in Theorem 4.2.1, then it follows from Theorem 4.1.3 that $\mathscr{K}$ and $\mathscr{J}$ cannot be strongly transversal. Since $\mathscr{J}$ is not a linear subspace, it follows from Proposition 4.1.2 that $\mathscr{K}$ can be separated from $\mathscr{J}$. Hence there exists a nonzero covector $\tilde{\lambda} \in (\mathbb{R}^{n+1})^*$ such that

$$\langle \tilde{\lambda}, \tilde{v} \rangle \geq 0 \quad \text{for all} \quad \tilde{v} \in \mathscr{K} \tag{4.18}$$

and

$$\langle \tilde{\lambda}, \tilde{j} \rangle \leq 0 \quad \text{for all} \quad \tilde{j} \in \mathscr{J}. \tag{4.19}$$

We again use the adjoint equation to move the covector $\tilde{\lambda}$ back along the controlled reference trajectory Γ. As in Sect. 4.2.3, condition (4.18) implies that there exists a nontrivial solution to the adjoint equation, which we also call $\tilde{\lambda}$, such that we have almost everywhere on $[0, T]$,

$$0 \equiv \langle \tilde{\lambda}(t), \tilde{f}(\bar{x}(t), \bar{y}(t), \bar{u}(t)) \rangle = \min_{u \in U} \langle \tilde{\lambda}(t), \tilde{f}(\bar{x}(t), \bar{y}(t), u) \rangle.$$

If we write $\tilde{\lambda} = (\lambda, \lambda_0)$, then the adjoint equation for the augmented system reads

$$\left(\dot{\lambda}(t), \dot{\lambda}_0(t) \right) = -(\lambda(t), \lambda_0(t)) \begin{pmatrix} \dfrac{\partial f}{\partial x}(\bar{x}(t), \bar{u}(t)) & 0 \\[2mm] \dfrac{\partial L}{\partial x}(\bar{x}(t), \bar{u}(t)) & 0 \end{pmatrix},$$

and thus

$$\dot{\lambda}(t) = -\lambda(t)\frac{\partial f}{\partial x}(\bar{x}(t), \bar{u}(t)) - \lambda_0(t)\frac{\partial L}{\partial x}(\bar{x}(t), \bar{u}(t))$$

and $\dot{\lambda}_0(t) \equiv 0$, such that λ_0 is constant. If we now define the Hamiltonian H as

$$H : \mathbb{R} \times (\mathbb{R}^n)^* \times M \times U \to \mathbb{R},$$

$$(\lambda_0, \lambda, x, u) \mapsto H(\lambda_0, \lambda, x, u) = \lambda_0 L(x, u) + \langle \lambda, f(x, u) \rangle,$$

then, exactly as in Theorem 4.2.2, the adjoint equation for λ can be rewritten as

$$\dot{\lambda}(t) = -\frac{\partial H}{\partial x}(\lambda_0, \lambda(t), \bar{x}(t), \bar{u}(t)),$$

and the minimum condition becomes

$$0 = H(\lambda_0, \lambda(t), \bar{x}(t), \bar{u}(t)) = \min_{u \in U} H(\lambda_0, \lambda(t), \bar{x}(t), u).$$

Furthermore, since $\tilde{\lambda}$ is nonzero, we also have the nontriviality condition $(\lambda_0, \lambda(t)) \neq 0$ for all $t \in [0, T]$.

For the optimal control problem, condition (4.19) in addition gives that

$$\langle \lambda(T), v \rangle + \lambda_0 \alpha \leq 0 \quad \text{for all} \quad (v, \alpha) \in \widetilde{\mathscr{J}}.$$

Taking $(0, -1) \in \widetilde{\mathscr{J}}$, it follows that $\lambda_0 \geq 0$. Furthermore, in $\widetilde{\mathscr{J}}$ we can take the limit as $\alpha \nearrow -\langle \nabla \varphi(\bar{q}), v \rangle$ from below, and this implies

$$\langle \lambda(T), v \rangle \leq \lambda_0 \langle \nabla \varphi(\bar{q}), v \rangle \quad \text{for all} \quad v \in \ker D\Psi(\bar{q}).$$

But $\ker D\Psi(\bar{q})$ is a linear subspace, and thus with v also $-v$ lies in $\ker D\Psi(\bar{q})$. Hence this inequality must be an equality. Thus, the restriction of $\lambda(T)$ to $\ker D\Psi(\bar{q})$, i.e., to the tangent space of N at $\bar{q}$, is given by $\lambda_0 \nabla \varphi(\bar{q})$. This is the *transversality condition* of the maximum principle. Summarizing, we have proven the following theorem:

Theorem 4.2.3 (Maximum principle for optimal control problems). *[193] Suppose the controlled trajectory $\Gamma = (\bar{x}(\cdot), \bar{u}(\cdot))$ defined on $[0, T]$ with initial point $\bar{x}(0) = \bar{p}$ and terminal point $\bar{x}(T) = \bar{q}$ is optimal for the optimal control problem* [OC]. *Then there exist a constant $\lambda_0 \geq 0$ and a solution $\lambda : [0, T] \rightarrow (\mathbb{R}^n)^*$ of the adjoint equation along Γ,*

$$\dot{\lambda}(t) = -\lambda_0 \frac{\partial L}{\partial x}(\bar{x}(t), \bar{u}(t)) - \lambda(t)\frac{\partial f}{\partial x}(\bar{x}(t), \bar{u}(t)) = -\frac{\partial H}{\partial x}(\lambda_0, \lambda(t), \bar{x}(t), \bar{u}(t)),$$

that satisfies the terminal condition

$$\langle \lambda(T), v \rangle = \lambda_0 \langle \nabla \varphi(\bar{q}), v \rangle \quad \text{for all} \quad v \in \ker D\Psi(\bar{q})$$

such that

$$(\lambda_0, \lambda(t)) \neq 0 \quad \text{for all} \quad t \in [0, T],$$

and almost everywhere on $[0, T]$ we have that

$$0 = H(\lambda_0, \lambda(t), \bar{x}(t), \bar{u}(t)) = \min_{u \in U} H(\lambda_0, \lambda(t), \bar{x}(t), u).$$

4.3   Control Systems on Manifolds: Definition and Examples

The results developed so far are adequate for problems that are naturally formulated on all of $\mathbb{R}^n$ or some open subset of it. For example, many models in biomedical applications fall into this category (see, for example, [155, 156, 160]). Mechanical systems, on the other hand, are restricted in their movements. A robotic arm is of a fixed length and may only be able to rotate. For such a system, $\mathbb{R}^n$ provides an inadequate model for the state space. At a minimum, additional constraints need to be imposed that reflect the physical limitations of the underlying system. Often these problems then are more naturally (and without constraints) formulated on a manifold. Even for problems on $\mathbb{R}^n$, not only do differential-geometric concepts provide an excellent framework that, for example, illuminates the geometric meaning of the multipliers in the Pontryagin maximum principle, but they also enrich the analysis of these problems with tools and techniques that we shall use later in the derivation of high-order necessary conditions for optimality. We therefore extend our previous definition of a control system to allow the state space to be a differentiable manifold. A rather self-contained but brief introduction to differentiable manifolds and its main definitions and concepts that will be used (tangent and cotangent vectors, integral curves, differentials, and pullbacks of mappings) is given in Appendix C. But we need to refer the reader to the rich literature for a deeper treatment of this subject (e.g., [50, 102, 256]).

In our presentation, we do not strive for utmost generality, but instead our aim is to generalize the key ideas of the classical construction to a reasonably broad framework. Therefore we still employ a set of simplifying assumptions that lead to what sometimes is referred to as the *smooth maximum principle* for time-invariant systems, but now on a differentiable manifold M.

Definition 4.3.1. A control system is a 4-tuple $\Sigma = (M, U, f, \mathscr{U})$ consisting of the following objects:

1. The *state space* M is a connected n-dimensional differentiable manifold[2] of class C^ℓ with $\ell \geq 1$.
2. The *control set* U is an arbitrary subset of $\mathbb{R}^m$.
3. The *dynamics* or dynamical law assigns to every point $(q, u) \in M \times U$ a tangent vector $f(q, u)$ to the manifold M at the point q, $f(q, u) \in T_q M$. We assume that this map has the following regularity property: for every coordinate chart $\phi : V \to \mathbb{R}^n$, $q \mapsto x = \phi(q)$, on M with domain V, the coordinate representation f_ϕ of f on $\phi(V)$ is continuous on $\phi(V) \times U$, continuously differentiable in x for each fixed $u \in U$, and the partial derivatives $\frac{\partial f_\phi}{\partial x}(x, u)$ are continuous on $\phi(V) \times U$.
4. The class $\mathscr{U}$ of *admissible controls* is given by all locally bounded Lebesgue measurable functions.

---

[2]It will always be assumed that M is second countable and Hausdorff. A manifold M is second countable if its topology has a countable basis, and it is Hausdorff if for any two points p and q in M there exist open neighborhoods U of p and V of q that are disjoint. The interested reader is referred to [81] or any other textbook on topology for this background material.

Given an admissible control u defined on an interval I, for every coordinate chart $\phi : V \to \mathbb{R}^n$ on M, the map $F_\phi : I \times V \to \mathbb{R}^n$, $(t,x) \mapsto F_\phi(t,x) = f_\phi(x,u(t))$, satisfies the C^1-Carathéodory conditions on $I \times V$, and thus the results about existence of solutions and smooth dependence on initial conditions are valid for each coordinate chart. It is a standard (but somewhat tedious) procedure to verify that it is possible to patch together different coordinate neighborhoods. It follows that the time-dependent vector field $(t,q) \mapsto f(q,u(t))$ defined on $I \times M$ satisfies the C^1-Caratheodory conditions, and thus for every initial condition $(t_0,p_0) \in I \times M$ the initial value problem $\dot{q} = f(q,u(t))$, $q(t_0) = p_0$, has a unique absolutely continuous solution $x = x(t;t_0,p_0)$ that is defined on a maximal open subinterval that contains t_0. As before, this solution is a continuously differentiable function of the initial time t_0 and initial point p_0. We shall not go into the details regarding coordinate charts, and for ease of notation we use the same label x for points on the manifold and their coordinate representations, $x = x(q)$. We thus also write $f(x,u)$ for both the map on the manifold and its coordinate representation f_ϕ. As before, we call x the trajectory corresponding to the control u, and the pair (x,u) a controlled trajectory. The various definitions of reachability all carry over verbatim.

We give three examples of control systems on manifolds that should convince the reader that this is a natural and important generalization. The first one is the classical problem of geodesics on a sphere, the second one a standard model for the control of a rigid body such as a satellite in space, and the third one is about the control of a kinematically redundant robotic manipulator.

4.3.1 Shortest Paths on a Sphere

Given two points p and q on the unit sphere $S^2 = \{z \in \mathbb{R}^3 : \|z\|_2 = 1\}$ in $\mathbb{R}^3$, what is the shortest "path" that connects p to q and lies on S^2? Without going into details, let us simply informally assume that paths are (piecewise) continuously differentiable curves that lie on S^2. More generally, solutions to this problem on an arbitrary manifold are called *geodesics*. On the sphere, the solution is geometrically obvious: the points p, q and the origin in $\mathbb{R}^3$ define a plane P whose intersection with S^2 defines a so-called great circle $C = P \cap \mathbb{R}^3$, i.e., a circle that lies on S^2 and has radius $r = 1$. The points p and q divide this great circle into two disjoint segments that connect p with q, and the shorter of the two is the solution to the problem (see Fig. 4.11).

Mathematically, this is the optimal control problem to minimize the arc length over all piecewise continuously differentiable curves in $\mathbb{R}^3$ that connect p with q and lie on S^2. Clearly, one possibility would be to choose $\mathbb{R}^3$ as state space and to enforce the constraint that all curves must lie on S^2. But this is not necessary: in this new setup, we simply take the manifold $M = S^2$ as the state space, the control set U is given by the unit circle in $\mathbb{R}^2$, $U = \{(u_1,u_2)^T : u_1^2 + u_2^2 = 1\}$, and the dynamics simply assigns to every point $q \in S^2$ and control vector $u \in U$ the tangent vector $f(q,u) \in T_q S^2$ that is defined by u in the tangent plane to S^2 at q.

Fig. 4.11 Geodesic curve on S^2

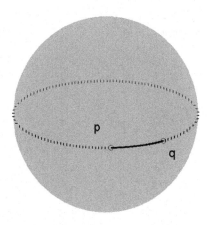

While the manifold S^2 is embedded into $\mathbb{R}^3$, it is a two-dimensional object. The coordinates $\phi : V \subset S^2 \to \mathbb{R}^2$, $q \mapsto x = \phi(q)$, relate the optimal control problem defined on S^2 to an optimal control problem defined on $\mathbb{R}^2$. But these correspondences are never valid globally, and the problem is not equivalent to an optimal control problem on $\mathbb{R}^2$. For example, there clearly exists a distinguished set of coordinates for S^2 defined by spherical coordinates,

$$z_1 = \sin\theta\cos\phi, \qquad z_2 = \sin\theta\sin\phi, \qquad z_3 = \cos\theta,$$

where $\theta \in [0,\pi]$ denotes the angle between the vector $z \in S^2$ and the positive z_3-axis and $\phi \in [0,2\pi]$ is the angle between the projection of z into the (z_1,z_2)-plane and the positive z_1-axis (see Fig. 4.12). Thus $\theta =$ const corresponds to a circle of constant latitude ("parallel"), while $\phi =$ const defines a circle of constant longitude ("meridian") that connects the north pole $N = (0,0,1)^T$ with the south pole $S = (0,0,-1)^T$. This coordinate transform, which actually defines the inverse mapping $x = (\theta,\phi) \mapsto z \in S^2$, is an analytic diffeomorphism if the angle θ is restricted to the open interval $(0,\pi)$, but it has singularities at the north and south poles; the angle φ can be varied over $[0,2\pi]$. Thus this provides a coordinate neighborhood for $S^2/\{N,S\}$.

These coordinates can be used to give a traditional, but somewhat incomplete, description of the optimal control problem [64]. Using the obvious symmetries of the problem, without loss of generality we may assume that the great circle $C = P \cap \mathbb{R}^3$ is the equator $E = \{z \in S^2 : z_3 = 0\}$ and that $p = (1,0,0)$. The second point q also lies on E and has coordinates $\theta = \frac{\pi}{2}$ and $\phi = \phi_f \leq \pi$, where we pick the shorter segment to define the coordinate. The line element ds for a curve defined in spherical coordinates on S^2 is given by

$$(ds)^2 = (d\theta)^2 + \cos^2\theta\,(d\phi)^2.$$

Fig. 4.12 Spherical
coordinates

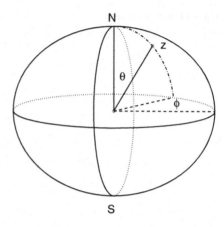

If one anticipates (using heuristic arguments) that the optimal curve can be described as a function of ϕ in this coordinate patch, then the problem can be formulated mathematically as to minimize the integral

$$J(u) = \int_0^{\phi_f} \sqrt{w^2 + \cos^2 \theta} \, d\phi$$

over all differentiable curves $\theta : [0, \phi_f] \to (0, \pi)$, $\varphi \mapsto \theta(\phi)$, subject to the dynamics

$$\frac{d\theta}{d\phi} = w, \qquad \theta(0) = \frac{\pi}{2},$$

and terminal condition $\theta(\phi_f) = \frac{\pi}{2}$. In this formulation, the derivative becomes a one-dimensional unrestricted control. While this engineering-type formulation turns out to be good enough for this particular case, clearly, the formulation on manifolds is more accurate, more general, and more elegant.

4.3.2   Control of a Rigid Body

A *rigid body* is a mechanical object for which the distance between any two of its points remains constant during motion and the orientation is preserved; that is, a right-handed coordinate system won't be changed into a left-handed one. For example, satellites are rigid bodies that need to be controlled in space to maintain an appropriate antenna orientation to enable efficient communication with specific locations on Earth. In addition, nowadays often a primary source of energy for satellites comes from solar cells, and thus the orientation of the satellite also should try to maximize the exposure of its solar cells to the sun.

Fig. 4.13 A circular satellite path

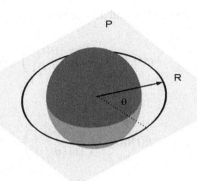

Motions of a rigid body are described with respect to two coordinate frames, one a *body frame* that is fixed on the rigid body, the other an *inertial frame* that is fixed in space. For the example of a satellite, we can think of the latter as the standard orthonormal coordinate system $\{e_1, e_2, e_3\}$ in $\mathbb{R}^3$ with origin at the center of the Earth and mutually orthogonal directions e_i with say two endpoints located somewhere on the equator and the third the North Pole. Similarly, a body frame is given by another orthornormal coordinate system $\{f_1, f_2, f_3\}$ that has its origin in the center of the satellite and one of its coordinate directions aligned with the antenna to be directed. Thus these coordinates describe the points on the satellite and do not change with time, while its coordinates relative to the inertial frame describe the motion of the object. The latter consist of the position of the center of the satellite and its orientation. This naturally leads to a manifold M as the state space for this system. We briefly discuss simplified versions of these dynamics.

Denote the position of the center of the satellite by p and, for simplicity, assume that the satellite nominally moves on a circular orbit in a fixed plane P at a distance R from the center of the Earth (see Fig. 4.13). Generally, there exist small deviations of the actual distance from the desired value R, and a stabilizing control law needs to be implemented. (We shall discuss this topic in the context of perturbation feedback control in Sect. 5.3). The underlying dynamics is simply Newton's second law, but it needs to be expressed in polar coordinates. If we write the position p of the satellite's center in the plane P as

$$p(t) = R(t) \begin{pmatrix} \cos(\theta(t)) \\ \sin(\theta(t)) \end{pmatrix},$$

with θ denoting the angle relative to some reference value $\theta = 0$, then the motion of the satellite in a gravitational field of strength $\left(\frac{\omega}{R}\right)^2$ obeys the differential equations

$$\ddot{R} = R\dot{\theta}^2 - \left(\frac{\omega}{R}\right)^2, \qquad R\ddot{\theta} = -2\dot{\theta}\dot{R}.$$

For the first and second derivatives of $p(t)$ are given by

$$\dot{p}(t) = \dot{R}(t) \begin{pmatrix} \cos(\theta(t)) \\ \sin(\theta(t)) \end{pmatrix} + R(t)\dot{\theta}(t) \begin{pmatrix} -\sin(\theta(t)) \\ \cos(\theta(t)) \end{pmatrix}$$

and

$$\ddot{p}(t) = \left(\ddot{R}(t) - R(t)\dot{\theta}(t)^2\right) \begin{pmatrix} \cos(\theta(t)) \\ \sin(\theta(t)) \end{pmatrix} + \left(R(t)\ddot{\theta}(t) + 2\dot{R}(t)\dot{\theta}(t)\right) \begin{pmatrix} -\sin(\theta(t)) \\ \cos(\theta(t)) \end{pmatrix},$$

and thus the normal centrifugal force on the satellite, which is opposed by the gravitational force $-\left(\frac{\omega}{R}\right)^2$, is given by $\ddot{R}(t) - R(t)\dot{\theta}(t)^2$, and the tangential force becomes $R(t)\ddot{\theta}(t) + 2\dot{R}(t)\dot{\theta}(t)$. Hence, a simple model for the orbit of the satellite becomes

$$\ddot{R} = \dot{R}\dot{\theta}^2 - \left(\frac{\omega}{R}\right)^2 + u, \qquad\qquad \ddot{\theta} = -2\dot{\theta}\frac{\dot{R}}{R} + \frac{1}{R}v$$

where u and v denote the effects of controls that influence the normal and tangential motions of the satellite, respectively.

The orientation of the satellite at the point $p(t)$ is determined by the coordinate frame $\{f_1, f_2, f_3\} = \{f_1(t), f_2(t), f_3(t)\}$. If we write these three vectors as the columns of a matrix Q,

$$Q = Q(t) = (f_1(t), \ f_2(t), \ f_3(t)),$$

this defines an orthogonal matrix Q, $Q^T Q = \mathrm{Id}$, and without loss of generality, we are assuming that the orientation is positive, i.e., $\det(Q) = +1$. Then Q is an element of the special orthogonal group $SO(3)$, a Lie group (see Appendix C). The orientation of the satellite over time therefore is described by a curve $t \mapsto Q(t) \in SO(3)$ in this manifold. Its tangent vector can be identified with a skew-symmetric matrix, i.e., an element in the Lie algebra $\mathfrak{so}(3)$, which represents the tangent space to this manifold. In order to see this, consider a fixed, but arbitrary, point on the satellite. Denote its coordinates in the body frame by $\mathfrak{x}$ (the German letter x) and let $\mathfrak{y}$ (the German letter y) denote the coordinates of the same point, but relative to the inertial frame $\{e_1, e_2, e_3\}$ that has been translated to the center of the satellite. Then the coordinate transformation simply becomes $\mathfrak{y}(t) = Q(t)\mathfrak{x}(t)$, and the corresponding velocities relate as $\dot{\mathfrak{y}}(t) = \dot{Q}(t)\mathfrak{x}(t) + Q(t)\dot{\mathfrak{x}}(t)$. However, the coordinates of a given point on the satellite are always the same in the body frame, and thus $\dot{\mathfrak{x}}(t) \equiv 0$. Hence the velocity is given by

$$\dot{\mathfrak{y}}(t) = \dot{Q}(t)\mathfrak{x}(t) = \dot{Q}(t)Q^{-1}(t)\mathfrak{y}(t),$$

and the matrix $S(t) = \dot{Q}(t)Q^{-1}(t)$ is skew-symmetric. This is easily seen by differentiating the identity $Q(t)^T Q(t) = \mathrm{Id}$ which gives $\dot{Q}(t)^T Q(t) + Q(t)^T \dot{Q}(t) = 0$ and thus

$$\dot{Q}(t)Q^{-1}(t) = -Q^{-T}(t)\dot{Q}(t)^T = -\left(\dot{Q}(t)Q^{-1}(t)\right)^T.$$

Any matrix $S \in \mathfrak{so}(3)$ is of the form

$$S(\omega) = \begin{pmatrix} 0 & \omega_3 & -\omega_2 \\ -\omega_3 & 0 & \omega_1 \\ \omega_2 & -\omega_1 & 0 \end{pmatrix}$$

for some vector $\omega = (\omega_1, \omega_2, \omega_3)$. Physically, ω gives the angular velocities relative to the body frame, and the matrix $S(\omega)$ gives the instantaneous angular velocities as seen from the inertial frame. The angular accelerations are then of the form

$$J\dot{\omega} = S(\omega)J\omega$$

with J a positive definite symmetric matrix also called the moment of inertia tensor related to these moments. The matrix J will be diagonal if the axes of the body frame coincide with the rigid body's so-called principal axes. If one now considers the problem to control the satellite's orientation through changes in the angular accelerations around two of the satellite's axes, say $\dot{\omega}_1$ and $\dot{\omega}_2$, then the corresponding dynamics becomes

$$\dot{Q} = S(\omega)Q, \quad J\dot{\omega} = S(\omega)J\omega + \begin{pmatrix} 1 \\ 0 \\ 0 \end{pmatrix} u + \begin{pmatrix} 0 \\ 1 \\ 0 \end{pmatrix} v,$$

with u and v denoting the corresponding angular accelerations.

For example, if the satellite is in a geostationary orbit over a base point, it is of interest to have the antenna always directed toward this point. If this direction is taken as the third axis, then one would want to control the orientation of the satellite through the angular accelerations around the other two axes in such a way that disturbances do not affect the desired orientation. It clearly is always possible to counter steer for deviations in the first two angular accelerations, but the question becomes whether it is possible to decouple disturbances in the angular velocities $\dot{\omega}_3$ from the third row of the orientation R. This indeed can be done, and we refer the interested reader to the papers [188, 200].

The manifold point of view offers distinct advantages in describing the dynamics. The orientation of the satellite relative to an inertial reference is described by an orthonormal frame, and mathematically this becomes an element of the Lie group $SO(3)$. This provides a far superior model for the dynamics and clearly brings out the geometric content of the model.

Fig. 4.14 Schematic diagram
of a planar robot

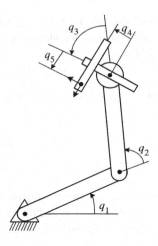

4.3.3   Trajectory Planning for Redundant Robotic Manipulators

Redundant manipulators are robotic configurations that possess more degrees of freedom than are strictly necessary to perform a required task. Redundancy offers several practical advantages. For example, it gives increased dexterity for obstacle avoidance and allows for the generation of trajectories that optimize dynamic performance. One type of redundant manipulators are so-called macro–micro configurations. In many industrial applications, tasks often require high dynamic motion over a large work space. Examples include routing and trimming of large panels and ultrasonic scanning for nondestructive defect inspection. In order to cover this large work envelope, typically large manipulators are used. However, because of the massive castings used as the links, these robots lack the performance required for dynamic trajectories, such as tight radius turns at high speed and high-G accelerations at the start and stop of motions. To overcome these problems, one proposed solution is to use a macro–micro manipulator configuration in which a short-stroke micromanipulator with high dynamic performance characteristics is coupled to a large-stroke macromanipulator. Overall, this creates a redundant robot with improved dynamic performance while retaining the work envelope of the large manipulator.

Figure 4.14 gives an example of a simple planar robot that combines three rotary links with two translational links. The macro configuration consists of the two rotational links of lengths L_1 and L_2, respectively, whose positions are characterized by the two angles q_1 and q_2. The micro configuration consists of two rigid shafts that are mounted to the end of the second link. The first of these shafts can still be rotated, and q_3 denotes the angle relative to the orientation of the second link. The second shaft is mounted rigidly such that it is perpendicular to the first shaft, and it cannot be rotated. But both shafts have mechanisms (small motors) that allow for forward or backward movements. The variable q_4 denotes the displacement of the second shaft relative to the center position of the first one, and $q_5 \geq 0$ denotes the displacement of the tool tip relative to the first shaft. The position of the endpoint of the distal fifth link of the manipulator contains the tool tip, and it is this point that generally needs to be controlled. Forward kinematic equations describe its position in $\mathbb{R}^2$ and can easily be written down for this simple model using elementary geometry:

$$x = L_1 \cos(q_1) + L_2 \cos(q_1 + q_2) + q_4 \cos(q_1 + q_2 + q_3) - q_5 \sin(q_1 + q_2 + q_3),$$

$$y = L_1 \sin(q_1) + L_2 \sin(q_1 + q_2) + q_4 \sin(q_1 + q_2 + q_3) + q_5 \cos(q_1 + q_2 + q_3).$$

Thus, the point (x,y) describes the position of the tool tip for this planar 5-axis mechanism. In many practical robotic applications, it is not just the position of the end-effector that is of interest, but also the orientation of the tool. For example, imagine that the tool corresponds to some cutting or welding process. Therefore, generally not only the position of the end-effector, but also its orientation is of interest. For the example here, the orientation can simply be described by the total angle

$$\theta = q_1 + q_2 + q_3.$$

The combined variables $(x,y) \in \mathbb{R}^2$ and the orientation $\theta \in S^1$ are called the *end-effector coordinates*. Note that the circle S^1 is isomorphic to $SO(2)$, the group of all orthogonal 2×2 matrices Q that have determinant $+1$, the special orthogonal group in $\mathbb{R}^2$, by means of the mapping

$$\theta \mapsto \begin{pmatrix} \cos\theta & -\sin\theta \\ \sin\theta & \cos\theta \end{pmatrix}$$

(see Appendix C).

The *task-space path plan* then consists of a C^2 curve in $\mathbb{R}^2 \times SO(2)$ describing the position and orientation of the end-effector and a time parametrization of the path. More generally, for a general-purpose robot in $\mathbb{R}^3$, the task-space path plan typically is given by a C^2 curve in $\mathbb{R}^3 \times SO(3)$ and a time parametrization of the path. If we denote the path plan by $x : [a,b] \to \mathbb{R}^2 \times SO(2), t \mapsto x(t)$, then the role of the trajectory planner is to develop a continuous function $q : [a,b] \to \mathbb{R}^5, t \mapsto q(t)$, that describes the joint positions as a function of time so that the tool tip follows the defined path plan, i.e.,

$$x(t) = f(q(t)),$$

where f is the forward kinematics of the robot. For a redundant manipulator, we have that $\dim(q) > \dim(x)$, and thus the robot has extra degrees of freedom, and typically there is an infinite set of solutions to this equation. This leads to the notion of a *self-motion manifold*. The self-motion manifold is defined for a fixed point x in the task space as the disjoint union of all feasible joint configurations q that result in the same end-effector position and orientation [63]. If the Jacobian matrix of the forward kinematics f at the point x, $Df(x)$, has linearly independent rows, then this is a possibly disjoint union of embedded submanifolds. Thus any point on the self-motion manifold of a particular task point x generates the same end-effector position and orientation, and movement on the self-motion manifold keeps the end-effector coordinates stationary in the task space.

For our case, this self-motion manifold can easily be calculated, and it is a two-dimensional torus $T^2 = S^1 \times S^1$. Clearly, we have $\dim(x) = \dim(\mathbb{R}^2 \times SO(2)) = 3$ and $\dim(q) = 5$, and thus there exist two degrees of freedom. Intuitively, it is clear that the variables (q_3, q_4, q_5) provide a nonredundant mechanism of joints in the task space, and thus the free variables are the two angles $(z_1, z_2) = (q_1, q_2)$. For the planar model considered here, it is not difficult to give a parameterization of the joints in terms of these self-motion variables: clearly,

$$q_1 = z_1, \qquad q_2 = z_2, \qquad q_3 = \theta - z_1 - z_2,$$

and

$$\begin{pmatrix} q_4 \\ q_5 \end{pmatrix} = \begin{pmatrix} \cos\theta & \sin\theta \\ -\sin\theta & \cos\theta \end{pmatrix} \begin{pmatrix} x - L_1\cos(z_1) - L_2\cos(z_1 + z_2) \\ y - L_1\sin(z_1) - L_2\sin(z_1 + z_2) \end{pmatrix}$$

$$= \begin{pmatrix} x\cos\theta + y\sin\theta - L_1\cos(z_1 - \theta) - L_2\cos(z_1 + z_2 - \theta) \\ -x\sin\theta + y\cos\theta - L_1\sin(z_1 - \theta) - L_2\sin(z_1 + z_2 - \theta) \end{pmatrix}.$$

Coupling the self-motion manifolds with the movement along a prescribed path $t \mapsto x(t)$ for the end-effector coordinates (possibly for small time intervals and assuming that no singular configurations arise along the path) generates a foliation structure of the interval with a constant manifold, the 2-dimensional torus T^2 in this particular example. An optimal control problem then consists in determining a path $t \mapsto z(t)$ for which $z(t)$ lies in the self-motion manifold for the point $x(t)$ that minimizes some objective function associated with the corresponding curve $t \mapsto q(t)$ in the joint space. Constraints on the control are then given by the mechanical restrictions of the robot. For example, in [100] the problem of minimizing the jerk, i.e., the derivative of the acceleration, along a prescribed trajectory plan is considered and analyzed as an optimal control problem for a redundant manipulator. This is the problem to minimize the integral

$$J(u) = \int_0^T \left\| q^{(3)}(t) \right\|^2 dt$$

over all paths $t \mapsto q(t)$ that lie in the foliation of self-motion manifolds defined by an a priori prescribed task-space path plan $t \mapsto x(t)$ subject to existing control constraints.

4.4 The High-Order Maximum Principle

We generalize the construction given in Sect. 4.2 to a class of variations that includes as special cases the needle variations CUT and PASTE. The procedure thus generates larger approximating cones $\mathscr{K}$ and hence leads to improved necessary conditions for optimality. In particular, the framework presented here allows for a uniform treatment of both first- and high-order variations, and we shall use it in Sect. 4.6 to derive well-known high-order necessary conditions for optimality, including the Legendre–Clebsch and Goh conditions. Concepts from differential geometry and Lie algebra provide the framework to formulate and evaluate these more complicated (high-order) variations, and it is for this reason that we formalize the construction for systems on manifolds. The structure of the proof, which is based on lecture notes by H. Sussmann, is identical to the procedure followed in Sect. 4.2.

4.4.1 Embeddings and Point Variations

As before, let $\bar{x}(\cdot)$ be a trajectory defined on $[0,T]$ corresponding to the admissible control $\bar{u}(\cdot)$ with initial point $\bar{x}(0) = \bar{p}$ and terminal point $\bar{x}(T) = \bar{q}$, the *controlled reference trajectory*. An important aspect of the classical construction is that variational vectors $\zeta^{\gamma}(0)$ that are generated at a point $\bar{x}(\bar{t})$ need to be transported to the endpoint $\bar{q}$. This was achieved by means of the solutions of the dynamics for the reference control $\bar{u}(\cdot)$. It is not necessary to use this specific flow, but any embedding of the reference trajectory into a smooth family of trajectories can be used.

Definition 4.4.1 (Embedding of a controlled trajectory). An embedding E of the reference controlled trajectory $\Gamma = (\bar{x}(\cdot), \bar{u}(\cdot))$ defined on $[0,T]$ is a family of controlled trajectories

$$\left\{ \Gamma^{E}(\cdot) = (x^{E}(\cdot;p), u^{E}(\cdot;p)) : u^{E}(\cdot;p) \in \mathscr{U}, \ p \in N_0^{E} \right\}$$

with the following properties:

1. N_0^{E} is a neighborhood of $\bar{p}$, and for every point $p \in N_0^{E}$ the pair $(x^{E}(\cdot;p), u^{E}(\cdot;p))$ is a controlled trajectory defined over the full interval $[0,T]$;
2. for $p = \bar{p}$, the controlled trajectory $(x^{E}(\cdot;\bar{p}), u^{E}(\cdot;\bar{p}))$ is given by the controlled reference trajectory,

$$(x^{E}(t;\bar{p}), u^{E}(t;\bar{p})) = (\bar{x}(t), \bar{u}(t)) \quad \text{for all } t \in [0,T];$$

3. for each $t \in [0,T]$ the flow defined by the embedding

$$\Phi_t^{E}(\cdot) : N_0^{E} \to M, \ p \mapsto \Phi_t^{E}(p) = x^{E}(t;p),$$

is a C^1-diffeomorphism from N_0^E onto some neighborhood N_t^E of $\bar{x}(t)$, and the partial derivatives of $\Phi_t^E(\cdot)$ with respect to the initial condition p are continuous jointly in (t, p).

Under the technical assumptions made in Sect. 4.3, as was the case on an open subset M of $\mathbb{R}^n$, also on manifolds it is always possible to construct an embedding of the controlled reference trajectory $(\bar{x}(\cdot), \bar{u}(\cdot))$ using the flow induced by the reference control $\bar{u}$ (analogous to Proposition 4.2.1). The same control is used for all controlled trajectories, $u^E(t; p) = \bar{u}(t)$ for all $t \in [0, T]$, and $x^E(t; p)$ is the corresponding trajectory with initial condition $x^E(0; p) = p$. We again call this the *canonical embedding* of the controlled reference trajectory. While this is a natural way to construct an embedding, it is not required that the embedding be given in this form. In fact, only the properties postulated in Definition 4.4.1 are needed in the construction, not actual properties of the reference vector field $(t, x) \mapsto f(x, \bar{u}(t))$. It is only the flow that determines how to move vectors and covectors along the reference trajectory. Examples can be given of vector fields that do not satisfy the C^1-Carathéodory conditions, but for which an embedding of the desired smoothness properties still exists. The reason for this lies in the fact that flows of vector fields can be "nicer" than the actual vector fields that generate them. Simply put, *flows have smoothing effects*. Generalizations along these lines are given in the research of H. Sussmann (e.g., see [244, 247, 248]).

We now proceed to define *point variations*. Point variations include the needle variations CUT and PASTE defined earlier and, more generally, are a generalization of the variations Weierstrass made in the proof of his side condition in the calculus of variations. They are the fundamental tool for calculating approximating vectors to the reachable set. The definition below goes back to H. Sussmann, and it formalizes a general mechanism for the construction of approximating curves $\kappa^{\mathcal{V}}$ to the reachable set at $\bar{q}$. As before, the procedure is to change the reference trajectory infinitesimally as a function of a one-dimensional parameter α, i.e., by making variations in the control. The guiding principle behind the construction is to generate as many tangent vectors as possible that still can be combined into an approximating cone $\mathcal{K}$. The framework presented here applies directly to so-called high-order variations and thus allows for the inclusion of the corresponding tangent vectors (see also [137, 140]). Recall that the variations CUT and PASTE can *always* be made, regardless of the structure of the reference control or trajectory. If the reference control has special properties, then this structure may be used to make additional variations. For example, if the reference control takes values in the interior of the control set, point variations can be defined that both increase and decrease the values of the reference control, while this is not possible if the values are at their upper or lower limits. In this way, the Legendre–Clebsch condition will be derived in Sect. 4.6.

Definition 4.4.2 (Point variation). Given an embedding E of a controlled reference trajectory $\Gamma = (\bar{x}(\cdot), \bar{u}(\cdot))$ defined on $[0, T]$, a point variation $\mathcal{V}$ along E is a 1-parameter family of variations characterized by a 7-tuple,

$$\mathscr{V} = \{\bar{t}, \varepsilon, t_1(\cdot), t_2(\cdot), \Delta_+(\cdot), \xi(\cdot, \cdot), \eta(\cdot, \cdot)\},$$

consisting of

1. a time $\bar{t} \in [0, T]$ that determines the anchor point for the variation on the controlled reference trajectory, $\bar{x}(\bar{t})$, and a positive number $\varepsilon > 0$ that defines the domain $[0, \varepsilon]$ for the 1-parameter family of variations,
2. continuous functions $t_1 : [0, \varepsilon] \to [0, T]$, $t_2 : [0, \varepsilon] \to [0, T]$, and $\Delta_+ : [0, \varepsilon] \to [0, \infty)$ that satisfy

$$t_1(0) = t_2(0) = \bar{t} \quad \text{and} \quad \Delta_+(0) = 0,$$

3. and a family of admissible controlled trajectories $(\xi(\alpha, \cdot), \eta(\alpha, \cdot))$, parameterized by $\alpha \in [0, \varepsilon]$ and defined on the interval $[0, \Delta_+(\alpha)]$,

$$\mathscr{D} = \{(\alpha, s) : 0 \le \alpha \le \varepsilon, 0 \le s \le \Delta_+(\alpha)\},$$

such that $\xi : \mathscr{D} \to M$, $(\alpha, s) \mapsto \xi(\alpha, s)$, is continuous in both variables and satisfies the initial condition

$$\xi(\alpha, 0) = \bar{x}(t_1(\alpha)) \quad \text{for all} \quad \alpha \in [0, \varepsilon].$$

For ε sufficiently small, for each $\alpha \in [0, \varepsilon]$, this one-parameter family of variations gives rise to a new controlled trajectory $(x_\alpha^{\mathscr{V}}(\cdot), u_\alpha^{\mathscr{V}}(\cdot))$, which we now define. As with the variations CUT and PASTE, it is advantageous to parameterize curves as images of a curve $\pi^{\mathscr{V}}(\cdot)$ in the neighborhood N_0^E of $\bar{p}$ introduced in the embedding.

Lemma 4.4.1. *For ε sufficiently small, there exists a continuously differentiable curve $\pi^{\mathscr{V}} : [0, \varepsilon] \to N_0^E$, $\alpha \mapsto \pi^{\mathscr{V}}(\alpha)$, such that $z = \pi^{\mathscr{V}}(\alpha)$ is the unique solution to the equation*

$$x^E(t_2(\alpha); z) = \xi(\alpha, \Delta_+(\alpha)) \tag{4.20}$$

in N_0^E.

Proof. The set $V = \{(t, q) : 0 \le t \le T, \ q \in N_t^E\}$ is the image of $[0, T] \times N_0^E$ under the flow of the embedding $(t, p) \mapsto (t, \Phi_t^E(p)) = (t, x^E(t; p))$. This flow is a C^1-diffeomorphism, and thus V is a neighborhood of $\{(t, \bar{x}(t)) : 0 \le t \le T\}$. Since $\lim_{\alpha \to 0} t_2(\alpha) = \bar{t}$ and

$$\lim_{\alpha \to 0} \xi(\alpha, \Delta_+(\alpha)) = \xi(0, 0) = \bar{x}(t_1(0)) = \bar{x}(\bar{t}),$$

for ε sufficiently small, the points $(t_2(\alpha), \xi(\alpha, \Delta_+(\alpha))$ lie in V and thus $\xi(\alpha, \Delta_+(\alpha)) \in N_{t_2(\alpha)}^E$. But then there exists a unique point $z = \pi^{\mathscr{V}}(\alpha) \in N_0^E$ such that

$$\xi(\alpha, \Delta_+(\alpha)) = \Phi_{t_2(\alpha)}^E(z) = x^E(t_2(\alpha); z).$$

Since the flow $\Phi^E_{t_2(\alpha)} : N^E_0 \to N^E_{t_2(\alpha)}$ is a diffeomorphism, by the implicit function theorem this solution $\pi^{\mathcal{V}}$ is continuously differentiable as a function of α.   □

The point variation $\mathcal{V}$ gives rise to the following 1-parameter family $u^{\mathcal{V}}_\alpha$, $0 \leq \alpha \leq \varepsilon$, of admissible controls:

$$u^{\mathcal{V}}_\alpha(t) = \begin{cases} \bar{u}(t) & \text{if } 0 \leq t \leq t_1(\alpha), \\ \eta(\alpha, t - t_1(\alpha)) & \text{if } t_1(\alpha) \leq t \leq t_1(\alpha) + \Delta_+(\alpha) = t_2(\alpha) + \Delta(\alpha), \\ u^E(t - \Delta(\alpha); \pi^{\mathcal{V}}(\alpha)) & \text{if } t_2(\alpha) + \Delta(\alpha) \leq t \leq T + \Delta(\alpha), \end{cases}$$

with corresponding trajectory given by

$$x^{\mathcal{V}}_\alpha(t) = \begin{cases} \bar{x}(t) & \text{if } 0 \leq t \leq t_1(\alpha), \\ \xi(\alpha, t - t_1(\alpha)) & \text{if } t_1(\alpha) \leq t \leq t_1(\alpha) + \Delta_+(\alpha) = t_2(\alpha) + \Delta(\alpha), \\ x^E(t - \Delta(\alpha); \pi^{\mathcal{V}}(\alpha)) & \text{if } t_2(\alpha) + \Delta(\alpha) \leq t \leq T + \Delta(\alpha). \end{cases}$$

Definition 4.4.3 (Controlled trajectories $(x^{\mathcal{V}}_\alpha(\cdot), u^{\mathcal{V}}_\alpha(\cdot))$ generated by a point variation). We call the 1-parameter family $(x^{\mathcal{V}}_\alpha(\cdot), u^{\mathcal{V}}_\alpha(\cdot))$, $0 \leq \alpha \leq \varepsilon$, the controlled trajectories generated by the variation $\mathcal{V}$.

Thus, using the variation $\mathcal{V}$, a modification of the controlled reference trajectory is defined as follows: $\bar{t}$ denotes the time at which the variation is made and $\bar{x}(\bar{t})$ is the point on the controlled reference trajectory where the variation is anchored. All the actual computations will be made at $\bar{t}$. The functions $t_1(\alpha)$ and $t_2(\alpha)$, respectively, denote the times at which a change from the reference trajectory is initiated and when the return to the reference trajectory is made in the form of following the trajectory of the embedding. Thus the control is initially given by $\bar{u}(t)$ on the interval $[0, t_1(\alpha)]$ and then is again given by the restriction of the control $u^E(\cdot; \pi^{\mathcal{V}}(\alpha))$ of the embedding to the interval $[t_2(\alpha), T]$, but shifted by the time $\Delta(\alpha)$. Overall, the time $\Delta_-(\alpha) = t_2(\alpha) - t_1(\alpha)$ has been "taken away" from the reference control. Note that while typically $t_2(\alpha) > t_1(\alpha)$, this is not required. The piece of the reference trajectory over the interval $(t_1(\alpha), t_2(\alpha))$ is replaced by the trajectory $\xi(\alpha, \cdot)$ corresponding to the admissible control $\eta(\alpha, \cdot)$, which is used for a time interval of length $\Delta_+(\alpha)$. Hence $\xi(\alpha, \Delta_+(\alpha))$ is the point to which the variation corresponding to a fixed value α steers the system in time $t_1(\alpha) + \Delta_+(\alpha) = t_2(\alpha) + \Delta(\alpha)$. At this point, the trajectory switches back to the "reference" trajectory in the sense that from now on, it follows the trajectory of the embedding of Γ that passes through this point $\xi(\alpha, \Delta_+(\alpha))$ at time $t_2(\alpha)$. But note that we do *not* just pick up the trajectory of the embedding that passes through $\xi(\alpha, \Delta_+(\alpha))$ at time $t_1(\alpha) + \Delta_+(\alpha)$, the total time that has elapsed since the beginning of the variation. Rather, the point $\xi(\alpha, \Delta_+(\alpha))$ is taken as initial condition for the trajectory of the embedding at time $t_2(\alpha)$. This corresponds to still using the full control $u^E(\cdot; \pi^{\mathcal{V}}(\alpha)) \restriction [t_2(\alpha), T]$ over the interval $[t_2(\alpha) + \Delta(\alpha), T + \Delta(\alpha)]$

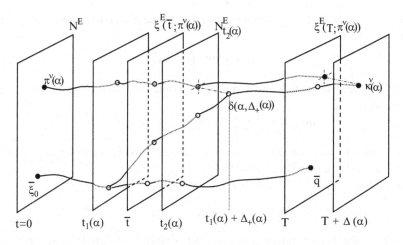

Fig. 4.15 Point variations

and allows us to make variations in the terminal time (see Fig. 4.15). The overall net change of time in the variation is therefore given by $\Delta(\alpha) = \Delta_+(\alpha) - \Delta_-(\alpha)$. In particular, variations for which $\Delta(\alpha) \equiv 0$ are "equal-time" variations and generate curves that lie in $\mathrm{Reach}_{\Sigma, T}(\bar{p})$.

Summarizing, the family of controlled trajectories obtained from the point variation $\mathcal{V}$ is given by

$$\Gamma^{\mathcal{V}} = \left\{ \Gamma_{\alpha}^{\mathcal{V}} = \left(x_{\alpha}^{\mathcal{V}}(\cdot), u_{\alpha}^{\mathcal{V}}(\cdot) \right), \, 0 \le \alpha \le \varepsilon \right\},$$

and for each $\alpha \in [0, \varepsilon]$, the terminal point $x_{\alpha}^{\mathcal{V}}(T + \Delta(\alpha))$ lies in the reachable set $\mathrm{Reach}_{\Sigma}(\bar{p})$. However, our assumptions do not guarantee that the curve

$$\kappa^{\mathcal{V}} : [0, \varepsilon] \to \mathrm{Reach}_{\Sigma}(\bar{p}),$$

$$\alpha \mapsto \kappa^{\mathcal{V}}(\alpha) = x_{\alpha}^{\mathcal{V}}(T + \Delta(\alpha)) = x^E(T; \pi^{\mathcal{V}}(\alpha)), \qquad (4.21)$$

is differentiable, and even if the derivative of this curve $\kappa^{\mathcal{V}}$ at $\alpha = 0$ exists, it may be zero. As before, a nonzero vector

$$\dot{\kappa}^{\mathcal{V}}(0) = \left(\frac{d}{d\alpha}_{|\alpha=0} \right) \kappa^{\mathcal{V}}(\alpha) \neq 0,$$

will be called *a tangent vector to the reachable set* $\mathrm{Reach}_{\Sigma}(\bar{p})$ at $\bar{q}$. As with the variations CUT and PASTE introduced earlier, $\dot{\kappa}^{\mathcal{V}}(0)$ will be computed as the image of the tangent vector $\zeta^{\mathcal{V}}(0)$, where $\zeta^{\mathcal{V}}(\cdot)$ is the curve defined as

$$\zeta^{\mathcal{V}} : [0, \varepsilon] \to M, \qquad \alpha \mapsto \zeta^{\mathcal{V}}(\alpha) = x^E(\bar{t}; \pi^{\mathcal{V}}(\alpha)), \qquad (4.22)$$

with $\kappa^{\mathcal{V}}(\alpha)$ the image of $\zeta^{\mathcal{V}}(\alpha)$ under the flow of the embedding from time $\bar{t}$ to the terminal time T,

$$\kappa^{\mathcal{V}}(\alpha) = \Phi_{T,\bar{t}}^{E}\left(\zeta^{\mathcal{V}}(\alpha)\right),\qquad(4.23)$$

and $\Phi_{t,s}^{E}$ is the mapping that moves points from time s to time t along the flow of the embedding,

$$\Phi_{t,s}^{E}: N_s^E \to N_t^E,\qquad \Phi_{t,s}^{E} = \Phi_t^E \circ (\Phi_s^E)^{-1}.$$

It is easy to see that the variations CUT and PASTE are examples of point variations in the sense of this definition:

Variation $\mathcal{V}_1$ (CUT): For $\bar{t} \in (0,T]$ define

$$t_1(\alpha) = \bar{t} - \alpha,\qquad t_2(\alpha) = \bar{t},\qquad \text{and}\qquad \Delta_+(\alpha) = 0.$$

Since no controlled trajectory is inserted, we simply have $\xi(\alpha,0) = \bar{x}(t_1(\alpha))$ and do not need to specify η. Undoing all the definitions, we see that the curve $\zeta^{\mathcal{V}_1}(\alpha)$ is given by

$$\zeta^{\mathcal{V}_1}(\alpha) = x^E(\bar{t}; \pi^{\mathcal{V}_1}(\alpha)) = x^E(t_2(\alpha); \pi^{\mathcal{V}_1}(\alpha))$$
$$= \xi(\alpha,\Delta_+(\alpha)) = \xi(\alpha,0) = \bar{x}(t_1(\alpha)) = \bar{x}(\bar{t}-\alpha).$$

Variation $\mathcal{V}_2$ (PASTE): Pick a value $v \in U$, and for $\bar{t} \in [0,T]$ let

$$t_1(\alpha) = t_2(\alpha) = \bar{t},\qquad \Delta_+(\alpha) = \alpha.$$

The control η is constant, given by $\eta(\alpha,s) \equiv v$ for all $s \in [0,\alpha]$, and $\xi(\alpha,\cdot)$ is the corresponding trajectory. By choosing ε small enough, this solution exists on $[0,\varepsilon]$. Note that for $\alpha_1 \leq \alpha_2$, the trajectory $\xi(\alpha_1,\cdot)$ is just the restriction of the trajectory $\xi(\alpha_2,\cdot)$ to the interval $[0,\alpha_1]$. Hence the curve $\zeta^{\mathcal{V}_2}(\alpha)$ is given by the value at time α of the inserted trajectory,

$$\zeta^{\mathcal{V}_2}(\alpha) = x^E(\bar{t}; \pi^{\mathcal{V}_2}(\alpha)) = x^E(t_2(\alpha); \pi^{\mathcal{V}_2}(\alpha)) = \xi(\alpha,\alpha) = x_{\alpha}^{\mathcal{V}_2}(\bar{t}+\alpha).$$

4.4.2   Variational Vector and Covector Fields

We have seen for the variations CUT and PASTE that it was easy to compute the tangent vectors $\dot{\zeta}^{\mathcal{V}}(0)$ at the anchor point $\bar{x}(\bar{t})$, and then the tangent vectors $\dot{\kappa}^{\mathcal{V}}(0)$ to the reachable set at $\bar{q}$ were computed as images under the differential of the flow $\Phi_{\bar{t},T}^{E}$. This leads to the concept of a variational vector field along a controlled reference trajectory as a curve in the *tangent bundle* of M (see Appendix C). For our purpose, we do not really need a formal definition of the tangent bundle TM as a manifold, and hence merely state that it simply consists of all the pairs (q,θ), where q is a point on M and θ is a tangent vector to M at q, $\theta \in T_qM$.

Definition 4.4.4 (Variational vector field). Given a controlled trajectory $\Gamma = (\bar{x}(\cdot), \bar{u}(\cdot))$ defined on $[0,T]$ and an embedding E of Γ, a variational vector field along (Γ, E) is a map $\Theta : [0,T] \to TM, t \mapsto (\bar{x}(t), \theta(t))$, i.e., $\theta(t) \in T_{\bar{x}(t)}M$ for all t, such that with $\left(\Phi_{t,s}^E\right)_*$ denoting the differential of the flow $\Phi_{t,s}^E = \Phi_t^E \circ \left(\Phi_s^E\right)^{-1}$, we have that

$$\theta(t) = \left(\Phi_{t,s}^E\right)_* \theta(s) \quad \text{for all} \quad s,t \in [0,T].$$

Thus, a variational vector field along (Γ, E) is obtained by moving the vector $\theta(0)$ along the reference trajectory using the differential of the flow. Given any $\bar{t} \in [0,T]$ and any $\bar{\theta} \in T_{\bar{x}(t)}M$, there exists a unique variational vector field that satisfies $\theta(\bar{t}) = \bar{\theta}$, namely $\theta(t) = \left(\Phi_{t,\bar{t}}^E\right)_* \bar{\theta}$. As before, for the canonical embedding, a variational vector field simply is a solution to the variational equation (4.9) over the interval $[0,T]$, and this solution is uniquely determined by its value at one specific time.

As we have seen in the formulation of necessary conditions for optimality in Sect. 4.2.3, it is equally important to move the multiplier λ back along the flow of the controlled reference trajectory. Geometrically, multipliers are covectors and correspond to linear functionals on the tangent spaces. Hence this naturally leads to the dual notion of a variational covector field as a curve in the *cotangent bundle* T^*M (see Appendix C). Again, for our purpose it suffices to think of the cotangent bundle as the collection of all pairs (q, λ), where q is a point on M and λ is a cotangent vector to M at q, $\lambda \in T_q^*M$. Variational covector fields are defined in terms of the adjoint of the differential of the flow, $\left(\Phi_{t,s}^E\right)_*$, also called the pullback of the flow $\Phi_{t,s}^E$. Recall that if $A \in \mathbb{R}^{n \times m}$ defines the linear mapping $L : \mathbb{R}^m \to \mathbb{R}^n, x \mapsto L(x) = Ax$, then the adjoint of A, A^*, is the unique linear mapping $L^* : \mathbb{R}^n \to \mathbb{R}^m, y \mapsto L^*(y) = A^*y$, such that $\langle y, Ax \rangle = \langle A^*y, x \rangle$ for all $x \in \mathbb{R}^m$ and all $y \in \mathbb{R}^n$. For the inner product, we simply get $\langle y, Ax \rangle = y^T Ax = (A^T y)^T x = \langle A^T y, x \rangle$, and thus $A^* = A^T$. If we write y as a row vector, $y \in (\mathbb{R}^n)^*$, then this simplifies to $\langle y, Ax \rangle = yAx = \langle yA, x \rangle$, and it is not necessary to take the transpose. Consistent with our notation of writing multipliers as row vectors, we therefore let the pullback act on the right. That is, $\lambda(s) \left(\Phi_{s,t}^E\right)^*$ denotes the covector obtained from $\lambda(s)$ by moving $\lambda(s)$ along the flow of the embedding from time s to time t. For a variational covector field, this covector must be the same as $\lambda(t)$.

Definition 4.4.5 (Variational covector field). Given a controlled trajectory $\Gamma = (\bar{x}(\cdot), \bar{u}(\cdot))$ defined on $[0,T]$ and an embedding E of Γ, a variational covector field along (Γ, E) is a map $\Lambda(\cdot) : [0,T] \to T^*M, t \mapsto (\bar{x}(t), \lambda(t))$, i.e., $\lambda(t) \in T_{\bar{x}(t)}^*(M)$ for all t, such that

$$\lambda(s) \left(\Phi_{s,t}^E\right)^* = \lambda(t) \quad \text{for all} \quad s,t \in [0,T].$$

The following simple but very important fact immediately follows from these definitions.

Proposition 4.4.1. *If θ is a variational vector field along (Γ, E) and λ is a variational covector field along (Γ, E), then the function $h : [0,T] \to \mathbb{R}$, $t \mapsto \langle \lambda(t), \theta(t) \rangle$, is constant.*

Proof. For all $s, t \in [0,T]$ we have that

$$\langle \lambda(t), \theta(t) \rangle = \langle \lambda(s) \left(\Phi_{s,t}^E \right)^{\star}, \theta(t) \rangle = \langle \lambda(s), \left(\Phi_{s,t}^E \right)_{\star} \theta(t) \rangle = \langle \lambda(s), \theta(s) \rangle. \qquad \square$$

Proposition 4.4.2. *For the canonical embedding E, variational vector fields are solutions to the variational equation*

$$\dot{\theta}(t) = \frac{\partial f}{\partial x}(\bar{x}(t), \bar{u}(t)) \theta(t),$$

and variational covector fields are solutions to the corresponding adjoint equation

$$\dot{\lambda}(t) = -\lambda(t) \frac{\partial f}{\partial x}(\bar{x}(t), \bar{u}(t)).$$

Proof. The result for variational vector fields has already been shown in Proposition 4.2.2. To compute the pullback of the flow, observe that for all variational vector fields θ we have that

$$0 \equiv \frac{d}{dt} \langle \lambda(t), \theta(t) \rangle = \langle \dot{\lambda}(t), \theta(t) \rangle + \langle \lambda(t), \dot{\theta}(t) \rangle$$

$$= \langle \dot{\lambda}(t), \theta(t) \rangle + \left\langle \lambda(t), \frac{\partial f}{\partial x}(\bar{x}(t), \bar{u}(t)) \theta(t) \right\rangle$$

$$= \langle \dot{\lambda}(t), \theta(t) \rangle + \left\langle \lambda(t) \frac{\partial f}{\partial x}(\bar{x}(t), \bar{u}(t)), \theta(t) \right\rangle$$

$$= \left\langle \dot{\lambda}(t) + \lambda(t) \frac{\partial f}{\partial x}(\bar{x}(t), \bar{u}(t)), \theta(t) \right\rangle.$$

Note once more that there is no need to take transposes, since we write covectors as row vectors. Hence the covector $\dot{\lambda}(t) + \lambda(t) \frac{\partial f}{\partial x}(\bar{x}(t), \bar{u}(t)) \in T^*_{\bar{x}(t)}M$ is orthogonal to every tangent vector $\theta \in T_{\bar{x}(t)}M$. But this is possible only for the zero vector, and thus we have that $\dot{\lambda}(t) = -\lambda(t) \frac{\partial f}{\partial x}(\bar{x}(t), \bar{u}(t))$. $\qquad \square$

4.4.3 C^1-Extendable Variations

So far, we have only generated curves $\kappa^{\mathscr{V}}(\cdot)$ in the reachable set $Reach_{\Sigma}(\bar{p})$ and calculated their tangent vectors $\dot{\kappa}^{\mathscr{V}}(0)$. We need to be able to combine finitely many tangent vectors generated by different point variations into an approximating map. In order to do so, Definition 4.4.2 given above is not strong enough. As for the variations CUT and PASTE, we also need these variations to be *stable* in the sense that they *can be made at nearby points as well*. We therefore require that these variations, or rather the corresponding flows of trajectories, extend into a neighborhood of $\bar{x}(\bar{t})$ such that it is possible to apply these maps not only at the points on the reference trajectory, but also at points in a full neighborhood of it. As in the classical construction given earlier, the reason for this requirement simply is that subsequent variations will not be made from the reference trajectory, but only from nearby trajectories defined by the embedding.

Definition 4.4.6 (C^1-Extendable point variation). A point variation $\mathscr{V}$ is called C^1-extendable if there exist a neighborhood W of $\bar{x}(\bar{t})$ and a map $\tilde{\xi}$ with values in M defined on the set

$$\tilde{D} = \{(z;\alpha,s) : z \in W,\ 0 \le \alpha \le \varepsilon,\ 0 \le s \le \Delta_+(\alpha)\}$$

such that the following conditions are satisfied:

1. For (z,α) fixed, $\tilde{\xi}(z;\alpha,\cdot)$ is an admissible controlled trajectory of the system Σ that starts at the point z, $\tilde{\xi}(z;\alpha,0) = z$.
2. If the initial point z corresponds to the initial point of the variation $\mathscr{V}$ for α, i.e., if $z = \bar{x}(t_1(\alpha))$, then $\tilde{\xi}(z;\alpha,\cdot)$ is given by the trajectory $\xi(\alpha,\cdot)$ of the variation, i.e.,

$$\tilde{\xi}(\bar{x}(t_1(\alpha));\alpha,s) = \xi(\alpha,s).$$

3. For (α,s) fixed, $\tilde{\xi}(\cdot;\alpha,s)$ is continuously differentiable on W and both $\tilde{\xi}$ and $\frac{\partial\tilde{\xi}}{\partial z}$ are continuous as functions of all variables on $\tilde{D}$.

Proposition 4.4.3. *The variations CUT and PASTE are C^1-extendable.*

Proof. This is trivial for the variation CUT: in this case $\Delta_+(\alpha) \equiv 0$, and we can simply define $\tilde{\xi}(z;\alpha,0) \equiv z$. All requirements are met, since from the definition of a point variation we have that $\bar{x}(t_1(\alpha)) = \xi(\alpha,0)$ and thus

$$\tilde{\xi}(\bar{x}(t_1(\alpha));\alpha,0) = \bar{x}(t_1(\alpha)) = \xi(\alpha,0).$$

For the variation PASTE, we choose as $\tilde{\xi}$ the obvious map that extends the definition of ξ as a solution to the dynamics with the constant control $v \in U$. Thus we define $\tilde{\xi}(z;\alpha,s)$ as the value at time s of the trajectory corresponding to the constant control v that starts at z at time 0. If W is chosen as a sufficiently small neighborhood of $\bar{x}(\bar{t})$, then it follows from the continuous dependence of the solution on initial conditions that this solution exists on the interval $[0,\Delta_+(\alpha)]$ for all $\alpha \in [0,\varepsilon)$, and by construction, the map $\tilde{\xi}$ agrees with ξ for initial points $z = \bar{x}(t_1(\alpha))$. The required smoothness properties of $\tilde{\xi}$ are again immediate consequences of the C^1-dependence of the solution of an ordinary differential equation on initial conditions. $\square$

We then use the function $\tilde{\xi}$ to extend the mapping that assigns to the anchor point $\bar{x}(\bar{t})$ the point $\zeta^{\mathscr{V}}(\alpha) = x^E(\bar{t};\pi^{\mathscr{V}}(\alpha))$ generated by the variation

Fig. 4.16 Illustration of the definition of the mapping $\Psi^{\mathcal{Y}}(\alpha;z)$; the trajectories of the embedding are represented by *horizontal* lines

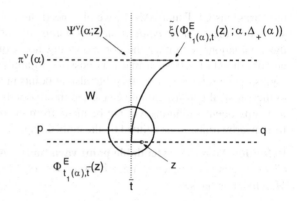

$$\bar{x}(\bar{t}) \mapsto \zeta^{\mathcal{Y}}(\alpha) = x^E(\bar{t};\pi^{\mathcal{Y}}(\alpha))$$

into the neighborhood W of $\bar{x}(\bar{t})$: define

$$\Psi^{\mathcal{Y}} : [0,\varepsilon] \times W \to M, \qquad (\alpha,z) \mapsto \Psi^{\mathcal{Y}}(\alpha;z),$$

by

$$\Psi^{\mathcal{Y}}(\alpha;z) = \Phi^E_{\bar{t},t_2(\alpha)} \circ \tilde{\xi}\left(\Phi^E_{t_1(\alpha),\bar{t}}(z);\alpha,\Delta_+(\alpha)\right). \tag{4.24}$$

Geometrically, starting at the point $z \in W$, the flow of the embedding is used to move z from time $\bar{t}$ to time $t_1(\alpha)$. Then the extension of the variation corresponding to α is run for the full time $\Delta_+(\alpha)$. The resulting point $\tilde{\xi}(\Phi^E_{t_1(\alpha),\bar{t}}(z),\alpha,\Delta_+(\alpha))$ is then moved back to time $\bar{t}$ by picking up the trajectory of the embedding that passes through $\tilde{\xi}(\Phi^E_{t_1(\alpha),\bar{t}}(z),\alpha,\Delta_+(\alpha))$ at time $t_2(\alpha)$ (see Fig. 4.16).

From these definitions, it follows that

$$\Psi^{\mathcal{Y}}(\alpha;\bar{x}(\bar{t})) = \Phi^E_{\bar{t},t_2(\alpha)} \circ \tilde{\xi}\left(\Phi^E_{t_1(\alpha),\bar{t}}(\bar{x}(\bar{t}));\alpha,\Delta_+(\alpha)\right)$$
$$= \Phi^E_{\bar{t},t_2(\alpha)} \circ \tilde{\xi}\left(\bar{x}(t_1(\alpha)));\alpha,\Delta_+(\alpha)\right)$$
$$= \Phi^E_{\bar{t},t_2(\alpha)} \circ \xi(\alpha,\Delta_+(\alpha))$$
$$= \Phi^E_{\bar{t},t_2(\alpha)} \circ x^E(t_2(\alpha);\pi^{\mathcal{Y}}(\alpha))$$
$$= x^E(\bar{t};\pi^{\mathcal{Y}}(\alpha)) = \zeta^{\mathcal{Y}}(\alpha),$$

and for $\alpha = 0$,

$$\Psi^{\mathcal{Y}}(0;z) \equiv \Phi^E_{\bar{t},t_2(0)} \circ \tilde{\xi}\left(\Phi^E_{t_1(0),\bar{t}}(z);0,\Delta_+(0)\right)$$
$$= \Phi^E_{\bar{t},\bar{t}} \circ \tilde{\xi}\left(\Phi^E_{\bar{t},\bar{t}}(z);0,0\right) = \tilde{\xi}(z;0,0) = z,$$

i.e., $\Psi^{\mathcal{Y}}(0;\cdot) = \mathrm{id}$, the identity map.

Proposition 4.4.4. *The function* $\Psi^{\gamma} = \Psi^{\gamma}(\alpha;z)$ *is continuous on* $[0,\varepsilon] \times W$, *continuously differentiable in* z *for fixed* α *and the partial derivative* $\frac{\partial \Psi^{\gamma}}{\partial z}$ *is continuous on* $[0,\varepsilon] \times W$. *For* $x,y \in W$ *and* $\alpha \in [0,\varepsilon]$, *define a function* $r = r(x,y,\alpha)$ *by*

$$\Psi^{\gamma}(\alpha;x) - \Psi^{\gamma}(\alpha;y) = x - y + r(x,y,\alpha)(x-y). \tag{4.25}$$

Then, for any convex neighborhood $W' \subset W$ *of* $\bar{x}(\bar{t})$ *with compact closure in* W, *there exists a function* $C : [0,\varepsilon] \to \mathbb{R}_+$ *such that* $\|r(x,y,\alpha)\| \leq C(\alpha)$ *for all* x *and* y *in* W *and* $\lim_{\alpha \to 0+} C(\alpha) = 0$.

Proof. Since the flows $\Phi^E_{s,t}$ are diffeomorphisms, and since $\tilde{\xi}$ is continuously differentiable on W, the smoothness properties of Ψ^{γ} are a direct consequence of our assumptions and the corresponding properties of the flows. Let W' be a convex neighborhood $W' \subset W$ of $\bar{x}(\bar{t})$ that has compact closure in W. We need to show that the remainder r defined by Eq. (4.25) is of order $o(1)$ uniformly for $x,y \in W'$. On a convex neighborhood, this remainder can be expressed explicitly with the following calculation already made earlier:

$$\begin{aligned}
\Psi^{\gamma}(\alpha;x) - \Psi^{\gamma}(\alpha;y) &= \int_0^1 \frac{d}{ds}\left(\Psi^{\gamma}(\alpha;y+s(x-y))\right) ds \\
&= \left(\int_0^1 \frac{\partial \Psi^{\gamma}}{\partial x}(\alpha;y+s(x-y))ds\right)(x-y) \\
&= x - y + \left(\int_0^1 \left[\frac{\partial \Psi^{\gamma}}{\partial x}(\alpha;y+s(x-y)) - 1\right] ds\right)(x-y) \\
&= x - y + r(x,y,\alpha)(x-y),
\end{aligned}$$

with the last line defining r. It therefore suffices to show that $\|r(x,y,\alpha)\| \leq C(\alpha)$ with $C(\alpha) \to 0$ as $\alpha \searrow 0$. It follows from the explicit form we just computed that r is continuous for x,y in the compact closure of W'. Let $C(\alpha) = \sup_{x,y \in W'} r(x,y,\alpha)$. Since $\Psi^{\gamma}(0;z) = z$, we have that $r(x,y,0) \equiv 0$. A continuous function on a compact set is uniformly continuous, and this therefore implies that $\lim_{\alpha \to 0+} C(\alpha) = 0$. $\quad\square$

4.4.4 Construction of an Approximating Cone

We are now ready to define the collection of tangent vectors that can be combined into an approximating map. As in the classical construction, the ability to make variations at nearby points allows us to combine variations made at the same time. We therefore consider only variational vectors that satisfy Knobloch's condition [K] and lie in the following sets: let

$Q(\bar{t})$ be the set of all tangent vectors $\dot{\zeta}^{\gamma}(0)$ generated by C^1-extendable point variations at time $\bar{t}$ for which this derivative exists and is nonzero, and let

P($\bar{t}$) be the set of all vectors $w \in T_{\bar{x}(\bar{t})}M$ for which there exist a sequence $\{t_n\}_{n \in \mathbb{N}}$, $t_n \neq \bar{t}$, t_n converging to $\bar{t}$, and a sequence of vectors $\{w_n\}_{n \in \mathbb{N}}$, $w_n \in Q(t_n)$, such that $w_n \to w$ as $n \to \infty$.

As in Sect. 4.2, the vectors in P($\bar{t}$) correspond to approximating directions that can be generated not only at some specific time $\bar{t}$, but also at nearby points on the reference trajectory. We have already seen in the proof of Theorem 4.2.1 that this is true for all $\bar{t} \in [0, T]$ for the tangent vectors generated by the variations PASTE, and for the tangent vectors generated by the variations CUT this still holds a.e. on $[0, T]$.

Moving the vectors in P($\bar{t}$) forward to the terminal point $\bar{q}$ by means of the embedding, we generate tangent directions to the reachable set at $\bar{q}$. Let

$$P^*(\bar{t}) = \left\{ \left(\Phi_{T,\bar{t}}^E \right)_* w : w \in P(\bar{t}) \right\}$$

and let $\mathscr{K}$ be the convex hull generated by all possible vectors in $P^*(\bar{t})$,

$$\mathscr{K} = \mathrm{co} \left(\bigcup_{0 \leq \bar{t} \leq T} P^*(\bar{t}) \right).$$

Then we have the main result of this construction:

Theorem 4.4.1. $\mathscr{K}$ is a Boltyansky approximating cone to $\mathrm{Reach}_\Sigma(\bar{p})$ at $\bar{q}$.

Proof. This proof is the same as for Theorem 4.2.1, only using the more general concepts formulated above. We therefore only indicate the main steps.

As before, it suffices to show that $\bigcup_{0 \leq \bar{t} \leq T} P^*(\bar{t})$ is a Boltyansky approximating cone and for this, given any vectors $v_1, \ldots, v_r$, $v_i \in P^*(t_i)$ for some t_i, not necessarily distinct, we need to construct an approximating map.

The key step consists in showing that the endpoint mapping for vectors v_i of the form $v_i = (\Phi_{T,t_i}^E)_* w_i$ with $w_i \in Q(t_i)$ for distinct times t_i, say $t_1 < \cdots < t_r$, is an approximating map for the vectors v_i. Let $\mathscr{V}_i$ be C^1-extendable point variations that generate the vectors w_i. For sufficiently small $\delta > 0$ define $\Xi : Q_\delta^r(0) \to \mathrm{Reach}_\Sigma(\bar{p})$ by

$$\Xi(\alpha_1, \ldots, \alpha_r) = \Phi_{T,t_r}^E \circ \Psi^{\mathscr{V}_r}(\alpha_r; \cdot) \circ \Phi_{t_r,t_{r-1}}^E \circ \cdots \circ \Phi_{t_2,t_1}^E \circ \Psi^{\mathscr{V}_1}(\alpha_1; \cdot) \circ \Phi_{t_1,0}^E(\bar{p}). \tag{4.26}$$

As before, it follows from our general regularity assumptions that Ξ is continuous, and the proof that Ξ is an approximating map is by induction on r, the number of point variations made. Let $r = 1$. Since the first variation is made from the reference trajectory, we have $\Psi^{\mathscr{V}}(\alpha; \bar{x}(\bar{t})) = \zeta^{\mathscr{V}}(\alpha)$, and thus, if we set $\bar{t} = t_1$, the map Ξ reduces to the curve $\kappa^{\mathscr{V}}(\alpha)$,

$$\Xi(\alpha) = \Phi_{T,\tilde{t}}^{E} \circ \Psi^{\mathcal{V}}(\alpha; \cdot) \circ \Phi_{\tilde{t},0}^{E}(\bar{p}) = \Phi_{T,\tilde{t}}^{E} \circ \Psi^{\mathcal{V}}(\alpha; \bar{x}(\tilde{t}))$$

$$= \Phi_{T,\tilde{t}}^{E}\left(\zeta^{\mathcal{V}}(\alpha)\right) = \kappa^{\mathcal{V}}(\alpha) \in \text{Reach}_{\Sigma}(\bar{p}).$$

Furthermore, by Taylor's theorem we have that

$$\Xi(\alpha) = \Xi(0) + \Xi'(0)\alpha + o(\alpha) = \bar{x}(T) + \dot{\kappa}^{\mathcal{V}}(0)\alpha + o(\alpha)$$

$$= \bar{q} + \alpha\left(\Phi_{T,\tilde{t}}^{E}\right)_{*}\dot{\zeta}^{\mathcal{V}}(0) + o(\alpha) = \bar{q} + \alpha\left(\Phi_{T,\tilde{t}}^{E}\right)_{*}w + o(\alpha)$$

$$= \bar{q} + \alpha v + o(\alpha),$$

and thus Ξ is an approximating map.

Again, for reasons of notational simplicity, we formulate the inductive step only for $r = 2$. It is assumed that the vectors w_1 and w_2 are generated by C^1-extendable variations $\mathcal{V}_1$ and $\mathcal{V}_2$ at times $t_1 < t_2$,

$$v_i = \dot{\kappa}^{\mathcal{V}_i}(0) = \left(\Phi_{T,t_i}^{E}\right)_{*}\dot{\zeta}^{\mathcal{V}_i}(0) = \left(\Phi_{T,t_i}^{E}\right)_{*}w_i \qquad \text{for } i = 1,2.$$

As before, we need to show that the map

$$\Xi = \Phi_{T,t_2}^{E} \circ \Psi^{\mathcal{V}_2}(\alpha_2; \cdot) \circ \Phi_{t_2,t_1}^{E} \circ \Psi^{\mathcal{V}_1}(\alpha_1; \cdot) \circ \Phi_{t_1,0}^{E}(\bar{p})$$

is approximating, i.e., takes values in $\text{Reach}_{\Sigma}(\bar{p})$ and that its linear approximation is of the form

$$\Xi(\alpha_1, \alpha_2) = \bar{q} + \alpha_1 v_1 + \alpha_2 v_2 + o(\alpha_1 + \alpha_2).$$

We first show that the image lies in the reachable set. Let $t_i^{(1)}$, $t_i^{(2)}$, Δ_{+}^{i}, and ξ_i, $i = 1,2$, be the functions t_1, t_2, Δ_{+}, and ξ in the definition of a point variation for the variations $\mathcal{V}_1$ and $\mathcal{V}_2$, respectively, and also let $\tilde{\xi}_2$ be the extension of ξ_2. Disentangling the definitions of the construction (cf. Eq. (4.22) and (4.20)), we have that

$$\Phi_{t_2,t_1}^{E} \circ \Psi^{\mathcal{V}_1}(\alpha_1; \cdot) \circ \Phi_{t_1,0}^{E}(\bar{p}) = \Phi_{t_2,t_1}^{E} \circ \Psi^{\mathcal{V}_1}(\alpha_1; \bar{x}(t_1)) = \Phi_{t_2,t_1}^{E} \circ x^E(t_1; \pi^{\mathcal{V}}(\alpha_1))$$

$$= \Phi_{t_2,t_1}^{E} \circ \Phi_{t_1,t_1^{(2)}(\alpha_1)}^{E} \circ x^E(t_1^{(2)}(\alpha_1); \pi^{\mathcal{V}}(\alpha_1))$$

$$= \Phi_{t_2,t_1^{(2)}(\alpha_1)}^{E} \circ \xi_1(\alpha_1, \Delta_{+}^{1}(\alpha_1)).$$

This is the actual state into which the corresponding control steers the initial point $\bar{p}$ in time t_2, and let us simply denote it by $y = y(\alpha_1)$. Since y no longer lies on the reference trajectory, now the extension $\tilde{\xi}_2$ needs to be used to carry out the second variation. We have

$$\Xi(\alpha_1, \alpha_2) = \Phi^E_{T,t_2} \circ \Psi^{\mathscr{V}_2}(\alpha_2; \cdot) \circ \Phi^E_{t_2,t_1} \circ \Psi^{\mathscr{V}_1}(\alpha_1; \cdot) \circ \Phi^E_{t_1,0}(\bar{p})$$

$$= \Phi^E_{T,t_2} \circ \Psi^{\mathscr{V}_2}(\alpha_2; y(\alpha_1))$$

$$= \Phi^E_{T,t_2} \circ \Phi^E_{t_2,t_2^{(2)}(\alpha_2)} \circ \tilde{\xi}\left(\Phi^E_{t_2^{(1)}(\alpha_2),t_2}(y(\alpha_1)); \alpha_2, \Delta_+(\alpha_2)\right)$$

$$= \Phi^E_{T,t_2^{(2)}(\alpha_2)} \circ \tilde{\xi}\left(\Phi^E_{t_2^{(1)}(\alpha_2),t_2}(y(\alpha_1)); \alpha_2, \Delta_+(\alpha_2)\right).$$

It is part of the definition of a C^1-extendable variation that the curves $\tilde{\xi}(z; \alpha_2, \cdot)$ are admissible controlled trajectories. Since the times t_i are all distinct, for sufficiently small $\|\alpha\|$ we can guarantee that $t_1^{(2)}(\alpha_1) < t_2^{(1)}(\alpha_2)$, and thus this defines a controlled trajectory of the system. Hence $\Xi(\alpha_1, \alpha_2)$ is the endpoint of an admissible controlled trajectory and thus lies in $\text{Reach}_\Sigma(\bar{p})$.

It remains to show that the map is approximating. This is the same argument as for the variations CUT and PASTE. From the inductive beginning we have that

$$\Psi^{\mathscr{V}_1}(\alpha_1; \cdot) \circ \Phi^E_{t_1,0}(\bar{p}) = \Psi^{\mathscr{V}_1}(\alpha_1; \bar{x}(t_1)) = \zeta^{\mathscr{V}_1}(\alpha_1) = \bar{x}(t_1) + \alpha_1 \dot{\zeta}^{\mathscr{V}_1}(0) + o(\alpha_1).$$

$$(4.27)$$

Applying the diffeomorphism $\Phi^E_{t_2,t_1}$ to Eq. (4.27), we get, again using Taylor's theorem, that

$$\Phi^E_{t_2,t_1} \circ \Psi^{\mathscr{V}_1}(\alpha_1; \cdot) \circ \Phi^E_{t_1,0}(\bar{p}) = \Phi^E_{t_2,t_1}\left(\Psi^{\mathscr{V}_1}(\alpha_1; \bar{x}(t_1))\right)$$

$$= \Phi^E_{t_2,t_1}\left[\bar{x}(t_1) + \alpha_1 \dot{\zeta}^{\mathscr{V}_1}(0) + o(\alpha_1)\right]$$

$$= \Phi^E_{t_2,t_1}(\bar{x}(t_1)) + \left(\Phi^E_{t_2,t_1}\right)_* \left[\alpha_1 \dot{\zeta}^{\mathscr{V}_1}(0) + o(\alpha_1)\right] + o(\alpha_1)$$

$$= \bar{x}(t_2) + \alpha_1 \left(\Phi^E_{t_2,t_1}\right)_* \dot{\zeta}^{\mathscr{V}_1}(0) + o(\alpha_1). \qquad (4.28)$$

Now the second variation $\mathscr{V}_2$ is made, but starting from this point, *not* from the point $\bar{x}(\bar{t}_2)$ on the reference trajectory. Proposition 4.4.4 provides the required control over the error made in this process. For α_1 sufficiently small, the point $\bar{x}(t_2) + \alpha_1 \left(\Phi^E_{t_2,t_1}\right)_* \dot{\zeta}^{\mathscr{V}_1}(0) + o(\alpha_1)$ lies in a prescribed compact neighborhood of $\bar{x}(t_2)$ where the proposition will apply. Taking $x = \bar{x}(t_2) + \alpha_1 \left(\Phi^E_{t_2,t_1}\right)_* \dot{\zeta}^{\mathscr{V}_1}(0) + o(\alpha_1)$ and $y = \bar{x}(t_2)$ in Proposition 4.4.4, it follows that

$$\Psi^{\mathscr{V}_2}(\alpha_2; \cdot) \circ \Phi^E_{t_2,t_1} \circ \Psi^{\mathscr{V}_1}(\alpha_1; \cdot) \circ \Phi^E_{t_1,0}(\bar{p})$$

$$= \Psi^{\mathscr{V}_2}\left(\alpha_2; \bar{x}(t_2) + \alpha_1 \left(\Phi^E_{t_2,t_1}\right)_* \dot{\zeta}^{\mathscr{V}_1}(0) + o(\alpha_1)\right)$$

$$= \Psi^{\mathscr{V}_2}(\alpha_2; \bar{x}(t_2)) + \alpha_1 \left(\Phi^E_{t_2,t_1}\right)_* \dot{\zeta}^{\mathscr{V}_1}(0) + o(\alpha_1) + o(1) \cdot O(\alpha_1).$$

The additional remainder in this equation is generated by the term

$$C(\alpha_2) \left\| \alpha_1 \left(\Phi^E_{t_2,t_1} \right)_* \dot\zeta^{\gamma_1}(0) + o(\alpha_1) \right\|$$

arising in the bound in Proposition 4.4.4. The error made by switching back to the reference trajectory from the new trajectory reached after the first variation is thus of order $O(\alpha_1)$, and it is propagated only with order $o(1)$ as $\alpha_2 \to 0$. Since we still have that

$$\frac{o(1) \cdot O(\alpha_1)}{\alpha_1 + \alpha_2} \to 0 \qquad \text{as } \alpha_1 + \alpha_2 \to 0,$$

it follows that these remainders together are of order $o(\alpha_1 + \alpha_2)$. Furthermore, for the second variation $\mathscr{V}_2$ we also have that

$$\Psi^{\gamma_2}(\alpha_2; \bar{x}(t_2)) = \zeta^{\gamma_2}(\alpha_2) = \bar{x}(t_2) + \alpha_2 \dot\zeta^{\gamma_2}(0) + o(\alpha_2).$$

Combining these equations yields

$$\Psi^{\gamma_2}(\alpha_2; \cdot) \circ \Phi^E_{t_2,t_1} \circ \Psi^{\gamma_1}(\alpha_1; \cdot) \circ \Phi^E_{t_1,0}(\bar{p})$$
$$= \bar{x}(\bar{t}_2) + \alpha_2 \dot\zeta^{\gamma_2}(0) + \alpha_1 \left(\Phi^E_{t_2,t_1} \right)_* \dot\zeta^{\gamma_1}(0) + o(\alpha_1 + \alpha_2).$$

Finally, moving all the vectors with the flow Φ^E_{T,t_2} to the terminal point, we obtain

$$\Xi(\alpha_1, \alpha_2) = \Phi^E_{T,t_2} \left[\bar{x}(\bar{t}_2) + \alpha_2 \dot\zeta^{\gamma_2}(0) + \alpha_1 \left(\Phi^E_{t_2,t_1} \right)_* \dot\zeta^{\gamma_1}(0) + o(\alpha_1 + \alpha_2) \right]$$
$$= \Phi^E_{T,t_2} (\bar{x}(\bar{t}_2)) + \alpha_2 \left(\Phi^E_{T,t_2} \right)_* \left[\dot\zeta^{\gamma_2}(0) \right]$$
$$\quad + \alpha_1 \left(\Phi^E_{T,t_2} \right)_* \left[\left(\Phi^E_{t_2,t_1} \right)_* \dot\zeta^{\gamma_1}(0) \right] + o(\alpha_1 + \alpha_2)$$
$$= \bar{q} + \alpha_1 \dot\kappa^{\gamma_1}(0) + \alpha_2 \dot\kappa^{\gamma_2}(0) + o(\alpha_1 + \alpha_2)$$
$$= \bar{q} + \alpha_1 v_1 + \alpha_2 v_2 + o(\alpha_1 + \alpha_2).$$

Thus Ξ is an approximating map for $v_1 = \dot\kappa^{\gamma_1}(0)$ and $v_2 = \dot\kappa^{\gamma_2}(0)$. This concludes the main step of the proof.

Now consider the general case and suppose $v_1, \ldots, v_r$ are arbitrary vectors from $\bigcup_{0 \le \bar{t} \le T} P^*(\bar{t})$. There exist times $t_i \in [0, T]$, not necessarily distinct, and vectors $w_i \in P(t_i)$ such that $v_i = (\Phi^E_{T,t_i})_* w_i$. By definition of $P(t_i)$, there exist sequences $\{t_{i,n}\}_{n \in \mathbb{N}}$, $t_{i,n} \ne t_i$, $t_{i,n} \to t_i$, and vectors $w_{i,n} \in Q(t_{i,n})$ converging to w_i. Given $\varepsilon > 0$, choose integers $n_1, \ldots, n_r$ such that the times t_{i,n_i} are all distinct and such that $\|w_i - w_{i,n_i}\| < \varepsilon$. It follows from the previous step of the proof that there exists an approximating map Ξ for the vectors $v'_i = (\Phi^E_{T,t_i})_* w_{i,n_i}$. But this, by Definition 4.1.2, defines an approximating map for the vectors $v_1, \ldots, v_r$. Hence $\bigcup_{0 \le \bar{t} \le T} P^*(\bar{t})$, and thus also $\mathscr{K}$, is an approximating cone. $\qquad \square$

Theorem 4.4.1 once more immediately gives necessary conditions for the point $\bar{q}$ to lie in the boundary of the reachable set. In this case $\mathcal{K}$ cannot be the full tangent space $T_{\bar{q}}M$, and thus, as in the classical case, there exists a nonzero covector $\bar{\lambda} \in T_{\bar{q}}^*M$ such that

$$\langle \bar{\lambda}, v \rangle \geq 0 \quad \text{for all } v \in \mathcal{K}.$$

Moving the covector $\bar{\lambda}$ backward along the controlled reference trajectory, we get the following result:

Corollary 4.4.1. *If* $\bar{q} \in \partial \operatorname{Reach}_{\Sigma}(\bar{p})$, *then there exists a nontrivial variational covector field* $\lambda : [0, T] \to T^*M$ *such that*

$$\langle \lambda(t), w \rangle \geq 0 \quad \text{for all } w \in P(t). \tag{4.29}$$

Abstract necessary conditions for optimality follow analogously to the derivation given in Sect. 4.2.4. However, in order to get concrete conditions that go beyond the conditions of the Pontryagin maximum principle (Theorem 4.2.3), it will be necessary to include additional and substantially different variations into this framework. This will be the topic of the remaining parts of this chapter. However, we first need to establish the algebraic framework that allows us to compute these high-order variations efficiently.

4.5   Exponential Representations of Flows

Exponential representations of flows of vector fields, coupled with important Lie-algebraic formulas such as the Baker–Campbell–Hausdorff formula [256], provide a powerful set of tools for computations involving nonlinear control systems. We develop them here in a generality that is sufficient for our purpose. In Chap. 7, this framework will also be used to make explicit computations for nonlinear systems based on properties of the vector fields in the dynamics and their Lie brackets in so-called *canonical coordinates*, a "good" set of coordinates that allows us to make explicit computations in the construction of small-time reachable sets.

Let $\Omega \subset \mathbb{R}^n$ be an open set, and without loss of generality, assume that $X : \Omega \to \mathbb{R}^n$ is a C^{∞} vector field[3] defined on Ω. For every point $p \in \Omega$, there exist a neighborhood V of p and an $\varepsilon > 0$ such that the initial value problem $\dot{x} = X(x)$, $x(0) = q_0 \in V$, has a unique solution defined for $|t| < \varepsilon$. So far, we have denoted this solution by $x(t; q_0)$ or, if we wanted to emphasize the flow aspects, by $\Phi_t^X(q_0)$. The most important property of solutions derived from their uniqueness is the *semigroup property*: if s and t are times such that the corresponding solutions are defined, then

$$\Phi_{s+t}^X(q_0) = \Phi_s^X(\Phi_t^X(q_0)).$$

---

[3]It obviously suffices for the vector fields to be of class C^r with r large enough that all derivatives exist that need to be taken.

Fig. 4.17 $p = q_0 \exp(t_1 X)$
$\exp(t_2 Y) \exp(-t_3 X) \exp(t_4 Y)$
$\exp(-t_5 X)$

$q_5 = q_4 \exp(-t_5 X)$ $\qquad$ $q_4 = q_3 \exp(t_4 Y)$

$q_3 = q_2 \exp(-t_3 X)$

q_0 $\qquad\qquad\qquad$ $q_2 = q_1 \exp(t_2 Y)$

$q_1 = q_0 \exp(t_1 X)$

Many of the formal properties of solutions to differential equations follow from this relation, the functional equation for the exponential. For this reason, even for nonlinear vector fields, exponentials give rise to a powerful calculus to work with solutions of differential equations.

Definition 4.5.1 (Exponential representation of solutions to differential equations). Given a C^1-vector field X defined on some open set $\Omega \subset \mathbb{R}^n$, $X : \Omega \to \mathbb{R}^n$, we also denote the solution to the initial value problem $\dot{x} = X(x)$, $x(0) = q_0$, at time t by

$$q_t = q_0 e^{tX} = q_0 \exp(tX) \sim \Phi_t^X(q_0).$$

Thus $q_t = q_0 e^{tX}$ simply is another notation for the point at time t on the integral curve of the vector field X that starts at q_0 at time 0. For the beginning, we also include the more standard classical notation $\Phi_t^X(q_0)$ relating the two with the symbol $\sim$. However, there is one significant difference: in the exponential notation we let *the diffeomorphisms act on the right*. Thus, the standard formulation $f(x)$ for the value of the function f at the point x would take the uncommon form xf. While the chain rule is the reason to write compositions of functions in the standard form $f(x)$, for many other computations (as we shall see below), this uncommon formulation becomes more useful. It is quite regularly used in Lie algebra (for example, see [123]), and it significantly simplifies the formal calculations we shall be using. For example, if X and Y are two vector fields, then the order of the flows in the exponential notation

$$q_0 e^{sX} e^{tY} = q_0 \exp(sX) \exp(tY) \sim \Phi_t^Y(\Phi_s^X(q_0))$$

is much more natural for the integral curve of Y that starts at the point $q_0 e^{sX}$ than the common notation, which will become highly cumbersome as additional concatenations are added (see Fig. 4.17).

The main advantage of the exponential notation for the flow is that it equally applies to the differential and pullbacks of the flow. In fact, the same notation will be used for the flow of the original vector field and for any of its extensions that are generated on arbitrary tensor fields. For example, a vector Y transported from q_0 to $q_t = q_0 e^{tX}$ along the flow of the vector field X is denoted simply by $q_0 Y e^{tX}$. As was shown in Proposition 4.4.2, this vector is the solution at time t of the variational equation along the integral curve $s \mapsto q_0 e^{sX}$ with initial condition $Y(q_0)$ at the point q_0; the traditional notation for this vector used so far is

$$q_0 Y e^{tX} = q_0 Y \exp(tX) \sim \left(\Phi_t^X\right)_* Y(q_0).$$

The true advantage of the exponential notation becomes clear in the following important computation:

Proposition 4.5.1. *Given two C^1-vector fields X and Y defined on an open subset $\Omega \subset \mathbb{R}^n$, for $q \in \Omega$ and sufficiently small t let*

$$V(t) = q_0 e^{tX} Y e^{-tX} = q_0 \exp(tX) Y \exp(-tX) \sim \left(\Phi_{-t}^X\right)_* Y(\Phi_t^X(q_0)),$$

i.e., $V(t)$ is the vector obtained by moving the vector field Y evaluated at the point $q_0 e^{tX}$ back to the initial point q_0 along the flow of the vector field X. The vector-valued curve V is continuously differentiable and

$$\dot{V}(t) = \frac{dV}{dt}(t) = q_0 e^{tX} [X, Y] e^{-tX} \sim \left(\Phi_{-t}^X\right)_* [X, Y](\Phi_t^X(q_0)) \qquad (4.30)$$

with $[X, Y]$ the Lie bracket of the vector fields X and Y. In particular,

$$\dot{V}(0) = q_0 [X, Y] \sim [X, Y](q_0).$$

Proof. We first carry out the proof in the standard notation and then show how these calculations simplify using the formal calculus of exponential representations. Note, however, that it is the proof given here that justifies these formal exponential computations.

Write $Z(t) = Y(\Phi_t^X(q_0))$ for the vector field Y evaluated along the solution $\Phi_t^X(q_0)$ to the initial value problem $\dot{x} = X(x)$, $x(0) = q_0$ (see Fig. 4.18). By definition, the vector-valued function $V(t)$ is the solution to the variational equation along X at time $s = 0$ with initial condition given by $Z(t)$ at time $s = t$. This is a time-varying linear ODE, and its solution can be expressed in terms of the so-called fundamental matrix (see Sect. 2.1). If we denote the fundamental matrix of the variational equation along X by $\Psi(s, t)$, then $V(t) = \Psi(0, t) Z(t)$. (Recall that $\Psi(s, t)$ is the inverse to $\Psi(t, s)$.) Differentiating, we get

$$\dot{V}(t) = \left(\frac{d\Psi}{dt}(0, t)\right) \cdot Z(t) + \Psi(0, t) \cdot \frac{dZ}{dt}(t).$$

Fig. 4.18 $Z(t) = q_0 \exp(tX)Y$
$\sim Y(\Phi_t^X(q_0))$

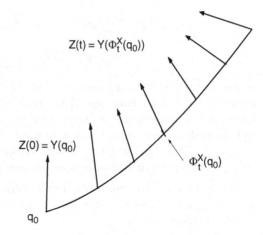

By the chain rule,

$$\frac{dZ}{dt}(t) = DY(\Phi_t^X(q_0)) \cdot X(\Phi_t^X(q_0)),$$

and by definition of the fundamental matrix,

$$\frac{\partial \Psi}{\partial s}(s,t) = DX(\Phi_t^X(q_0)) \cdot \Psi(s,t).$$

Differentiating the identity $\Psi(t,0)\Psi(0,t) \equiv \mathrm{Id}$ yields

$$DX(\Phi_t^X(q_0))\Psi(t,0) \cdot \Psi(0,t) + \Psi(t,0) \cdot \frac{d\Psi}{dt}(0,t) \equiv 0,$$

which gives

$$\frac{d\Psi}{dt}(0,t) = -\Psi(t,0)^{-1}DX(\Phi_t^X(q_0)) \cdot \mathrm{Id} = -\Psi(0,t)DX(\Phi_t^X(q_0)).$$

Substituting this relation into the equation for $\dot{V}(t)$, we therefore get that

$$\dot{V}(t) = -\Psi(0,t)DX(\Phi_t^X(q_0)) \cdot Z(t) + \Psi(0,t)DY(\Phi_t^X(q_0)) \cdot X(\Phi_t^X(q_0))$$

$$= \Psi(0,t)(DY \cdot X - DX \cdot Y)(\Phi_t^X(q_0))$$

$$= \Psi(0,t)[X,Y](\Phi_t^X(q_0)) = (\Phi_{-t}^X)_* [X,Y](\Phi_t^X(q_0)),$$

verifying Eq. (4.30).

Using exponential notation, this derivation simply follows by differentiating $V(t) = q_0 e^{tX} Y e^{-tX}$ using the product rule and interpreting the vector fields as differential operators, $[X,Y] = XY - YX$,

$$\frac{d}{dt}\left(q_0 e^{tX} Y e^{-tX}\right) = q_0\left(e^{tX}X\right) Y e^{-tX} - q_0 e^{tX}Y\left(e^{-tX}X\right)$$

$$= q_0 e^{tX}\left(XY - YX\right)e^{-tX} = q_0 e^{tZ}[Y,X]e^{-tZ}.$$

In this computation, we are also using that the vector field X commutes with its own flow, i.e., that we have $\left(q_0 e^{tX}\right)X = (q_0 X)e^{tX}$. Here the expression on the left represents the vector field X evaluated at the point $q_0 e^{tX}$, while the expression on the right formally is the vector $q_0 X$ moved from time $s=0$ to time $s=t$ along the flow of the vector field X. But this gives the same vector. Comparing the two computations above should convince the reader of the efficiency of the exponential notation. $\square$

If X and Y are C^∞-vector fields, then this procedure can be iterated. Recall that $\operatorname{ad}X(Y) = [X,Y]$ and high-order brackets are defined by $\operatorname{ad}^n X(Y) = [X,\operatorname{ad}^{n-1} X(Y)]$. It follows inductively that

$$\frac{d^n}{dt^n}_{|t=0}\left(q_0 e^{tX} Y e^{-tX}\right) = q_0 \operatorname{ad}^n X(Y) \sim \operatorname{ad}^n X(Y)(q_0),$$

and we therefore obtain the following Taylor series expansion:

Corollary 4.5.1 [256]. *For two C^∞-vector fields X and Y, we have the Taylor series expansion*

$$V(t) = q_0 e^{tX} Y e^{-tX} = q_0 \exp(tX)Y\exp(-tX) = \sum_{n=0}^{\infty}\left(\frac{t^n}{n!}\right) q_0 \operatorname{ad}^n X(Y).$$

Note that this series simply is an exponential of the operator ad, and in standard notation it is written in the form

$$e^{t\operatorname{ad}X}Y = \sum_{n=0}^{\infty}\left(\frac{t^n}{n!}\right)\operatorname{ad}^n X(Y) = Y + t[X,Y] + \frac{t^2}{2!}[X,[X,Y]] + \cdots . \qquad (4.31)$$

We henceforth use both $X(q_0)$ and $q_0 X$, and we freely switch between $V(t) = q_0 e^{tX} Y e^{-tX}$ and the more common notation $V(t) = (e^{t\operatorname{ad}X}Y)(q_0)$. By Corollary 4.5.1, for small $|t|$, the vector $(e^{t\operatorname{ad}X}Y)(q_0)$ can be approximated by the terms in this series when all the vector fields are evaluated at q_0.

The representation (4.31) is fundamental for many explicit computations in the context of nonlinear systems. As one example that we shall use extensively, we give an asymptotic infinite product expansion for an exponential Lie series[4] due to Sussmann [234]. However, we limit the derivation to the formal aspects and refer the reader to Sussmann's paper and textbooks on Lie algebra [52, 123] for the algebraic definitions and constructions. For later use, we need the expansion including all Lie

---

[4]This expansion is in terms of exponentials of elements in a Philip Hall basis [52] of the Lie subalgebra $L(\mathscr{X})$ of the free associative algebra generated by a family of noncommutative indeterminates $\mathscr{X} = \{X_1,\ldots,X_r\}$. We refer the interested reader to [234] for a proof of these algebraic constructions and to [52] for a definition of a Hall basis. We do not need these in our text.

brackets of order ≤ 5. This expansion will be used in Sects. 4.6.1 and 4.6.2 to derive the Legendre–Clebsch and Kelley conditions for optimality of singular controls.

Proposition 4.5.2 [209]. *Let X and Y be C^∞-vector fields and let v and w be Lebesgue integrable functions of t on $[0,T]$; set*

$$V(t) = \int_0^t v(s)ds, \qquad W(t) = \int_0^t w(s)ds, \qquad \text{and} \qquad U(t) = \int_0^t w(s)V(s)ds,$$

and let $S(t)$ be the solution of the initial value problem $\dot{S} = S(vX + wY)$, $S(0) = \mathrm{Id}$. Then S has an asymptotic product expansion of the form

$$S(t) = \cdots \exp\left(\frac{1}{2}\int_0^t wV^2 U ds \cdot [[X,Y],[X,[X,Y]]]\right)$$

$$\exp\left(\int_0^t wVWU ds \cdot [[X,Y],[Y,[X,Y]]]\right)$$

$$\exp\left(\frac{1}{24}\int_0^t wV^4 ds \cdot \mathrm{ad}^4 X(Y)\right) \exp\left(\frac{1}{6}\int_0^t wV^3 W ds \cdot [Y,\mathrm{ad}^3 X(Y)]\right)$$

$$\exp\left(\frac{1}{4}\int_0^t wV^2 W^2 ds \cdot [Y,[Y,[X,[X,Y]]]]\right)$$

$$\exp\left(\frac{1}{6}\int_0^t wVW^3 ds \cdot [Y,[Y,[Y,[X,Y]]]]\right)$$

$$\exp\left(\frac{1}{2}\int_0^t wVW^2 ds \cdot [Y,[Y,[X,Y]]]\right)$$

$$\exp\left(\frac{1}{2}\int_0^t wV^2 W ds \cdot [Y,[X,[X,Y]]]\right)$$

$$\exp\left(\frac{1}{6}\int_0^t wV^3 ds \cdot [X,[X,[X,Y]]]\right)$$

$$\exp\left(\int_0^t wVW ds \cdot [Y,[X,Y]]\right) \exp\left(\frac{1}{2}\int_0^t wV^2 ds \cdot [X,[X,Y]]\right)$$

$$\exp\left(\int_0^t wV ds \cdot [X,Y]\right) \exp(W \cdot Y) \exp(V \cdot X), \tag{4.32}$$

with the remainder indicated by the dots an infinite product of exponentials of iterated Lie brackets of X and Y of orders 6 and higher.

For small $|t|$, this representation gives rise to an asymptotic product expansion for the solution $z(\cdot)$ of the time-varying differential equation

$$\dot{z} = v(t)X(z) + w(t)Y(z), \qquad z(0) = q_0,$$

which will be used frequently in the subsequent sections. For the moment, we simply establish the algebraic formula.

Proof. Make an ansatz for the solution S as a product of the form $S = S_1 e^{VX}$. Differentiating in t, we obtain that

$$\dot{S} = \dot{S}_1 e^{VX} + S_1 e^{VX}(vX) = \dot{S}_1 e^{VX} + S(vX).$$

Solving this equation for $\dot{S}_1$ and using the asymptotic series (4.31) gives

$$\dot{S}_1 = (\dot{S} - S(vX))e^{-VX} = S(wY)e^{-VX} = S_1 e^{VX}(wY)e^{-VX} = S_1 e^{V \operatorname{ad} X}(wY)$$

$$= S_1 \left(wY + wV[X,Y] + \frac{1}{2}wV^2[X,[X,Y]] \right.$$

$$\left. + \frac{1}{6}wV^3 \operatorname{ad}^3 X(Y) + \frac{1}{24}wV^4 \operatorname{ad}^4 X(Y) \cdots \right).$$

In our formal computation, we keep track of all iterated brackets of X and Y up to order 5 and indicate higher-order brackets with dots. Now write $S_1 = S_2 e^{WY}$ and repeat this step to get

$$\dot{S}_2 = (\dot{S}_1 - S_1(wY))e^{-WY}$$

$$= S_2 e^{WY} \left(wV[X,Y] + \frac{1}{2}wV^2[X,[X,Y]] + \frac{1}{6}wV^3 \operatorname{ad}^3 X(Y) \right.$$

$$\left. + \frac{1}{24}wV^4 \operatorname{ad}^4 X(Y) \cdots \right) e^{-WY}$$

$$= S_2 e^{W \operatorname{ad} Y}$$

$$\left(wV[X,Y] + \frac{1}{2}wV^2[X,[X,Y]] + \frac{1}{6}wV^3 \operatorname{ad}^3 X(Y) + \frac{1}{24}wV^4 \operatorname{ad}^4 X(Y) \cdots \right)$$

$$= S_2 \left(wV[X,Y] + \frac{1}{2}wV^2[X,[X,Y]] + wVW[Y,[X,Y]] + \frac{1}{6}wV^3 \operatorname{ad}^3 X(Y) \right.$$

$$+ \frac{1}{2}wV^2 W[Y,[X,[X,Y]]] + \frac{1}{2}wVW^2[Y,[Y,[X,Y]]] + \frac{1}{24}wV^4 \operatorname{ad}^4 X(Y)$$

$$+ \frac{1}{6}wV^3 W[Y, \operatorname{ad}^3 X(Y)] + \frac{1}{4}wV^2 W^2[Y,[Y,[X,[X,Y]]]]$$

$$\left. + \frac{1}{6}wVW^3[Y,[Y,[Y,[X,Y]]]] + \cdots \right).$$

One more time, let $S_2 = S_3 e^{U[X,Y]}$ and solve for $\dot{S}_3$ as

$$\dot{S}_3 = (\dot{S}_2 - S_2(wV[X,Y]))e^{-U[X,Y]}$$

$$= S_3 e^{U[X,Y]} \left(\frac{1}{2}wV^2[X,[X,Y]] + wVW[Y,[X,Y]] + \text{terms from above} \right) e^{-U[X,Y]}$$

$$= S_3 e^{U \operatorname{ad}[X,Y]} \left(\frac{1}{2} wV^2 [X,[X,Y]] + wVW[Y,[X,Y]] + \text{terms from above} \right)$$

$$= S_3 \left(\frac{1}{2} wV^2 [X,[X,Y]] + wVW[Y,[X,Y]] \right.$$

$$+ \frac{1}{6} wV^3 \operatorname{ad}^3 X(Y) + \frac{1}{2} wV^2 W[Y,[X,[X,Y]]] + \frac{1}{2} wVW^2 [Y,[Y,[X,Y]]]$$

$$+ \frac{1}{24} wV^4 \operatorname{ad}^4 X(Y) + \frac{1}{6} wV^3 W[Y, \operatorname{ad}^3 X(Y)]$$

$$+ \frac{1}{4} wV^2 W^2 [Y,[Y,[X,[X,Y]]]] + \frac{1}{6} wVW^3 [Y,[Y,[Y,[X,Y]]]]$$

$$\left. + \frac{1}{2} wV^2 U[[X,Y],[X,[X,Y]]] + wVWU[[X,Y],[Y,[X,Y]]] \cdots \right).$$

This equation can be solved formally by integrating to obtain

$$S_3(t) = \exp \left(\left(\frac{1}{2} \int_0^t wV^2 ds \right) [X,[X,Y]] + \left(\int_0^t wVW ds \right) [Y,[X,Y]] \right.$$

$$\left. + \text{integrals of all other terms above} \right).$$

The derivation will be completed by writing the exponential of the sum as a product of exponentials. By definition, $S(t) = S_3(t) e^{U[X,Y]} e^{WY} e^{VX}$, and the calculation done so far implies that for any vector fields X and Y we have that

$$e^{X+Y} = e^{r(X,Y)} e^Y e^X,$$

where the remainder $r(X,Y)$ consists of commutator brackets of order at least 2. Thus, when we write $S_3(t)$ as a product of exponentials, then the commutator terms here are all brackets of order at least 6 and do not contribute to the terms we want to calculate. Hence, modulo terms of higher order, we can simply write $S_3(t)$ as the *product* of all the exponentials of the terms in the sum. This gives the expansion (4.32). □

We give some immediate applications of this product expansion.

Corollary 4.5.2 (Commutator formula).

$$q_0 e^{rX} e^{sY} = q_0 \cdots \exp \left(\frac{1}{2} rs^2 [Y,[X,Y]] \right) \exp \left(\frac{1}{2} r^2 s [X,[X,Y]] \right)$$

$$\exp (rs[X,Y]) \exp (sY) \exp (rX). \tag{4.33}$$

Proof. For an interval $I \subset \mathbb{R}$, denote the characteristic function of I by $\chi_I(t)$, i.e., $\chi_I(t) = 1$ if $t \in I$ and $\chi_I(t) = 0$ if $t \notin I$. Let $v = \chi_{[0,r]}$, $w = \chi_{(r,r+s]}$ and set $T = s + r$. Then elementary integrations show that $V(T) = r$, $W(T) = s$, $U(T) = rs$, $\frac{1}{2}\int_0^T w(t)V(t)^2 dt = \frac{1}{2}r^2 s$ and $\int_0^T w(t)V(t)W(t)dt = \frac{1}{2}rs^2$. $\square$

Corollary 4.5.3 (Expansion of the sum).

$$q_0 \exp(tX + tY) = q_0 \cdots \exp\left(\frac{1}{8}t^4[Y,[Y,[X,Y]]]\right) \exp\left(\frac{1}{8}t^4[Y,[X,[X,Y]]]\right)$$

$$\exp\left(\frac{1}{24}t^4[X,[X,[X,Y]]]\right)$$

$$\exp\left(\frac{1}{3}t^3[Y,[X,Y]]\right) \exp\left(\frac{1}{6}t^3[X,[X,Y]]\right)$$

$$\exp\left(\frac{1}{2}t^2[X,Y]\right) \exp(tY) \exp(sX).$$

Proof. Here we take $v = w = \chi_{[0,t]}$. Hence $V(s) = W(s) = s$, and thus for any $i, j \in \mathbb{N}$, we get $\int_0^t w(s)V(s)^i W(s)^j ds = \int_0^t s^{i+j}ds = \frac{1}{i+j+1}t^{i+j+1}$. $\square$

Corollary 4.5.4.

$$q_0 \exp(sX + tY) = q_0 \cdots \exp\left(\frac{1}{8}st^3[Y,[Y,[X,Y]]]\right) \exp\left(\frac{1}{8}s^2t^2[Y,[X,[X,Y]]]\right)$$

$$\exp\left(\frac{1}{24}s^3t[X,[X,[X,Y]]]\right)$$

$$\exp\left(\frac{1}{3}st^2[Y,[X,Y]]\right) \exp\left(\frac{1}{6}s^2t[X,[X,Y]]\right)$$

$$\exp\left(\frac{1}{2}st[X,Y]\right) \exp(tY) \exp(sX).$$

Proof. For the vector field $\tilde{X} = \frac{s}{t}X$ we have $q_0 \exp(sX + tY) = q_0 \exp(t\tilde{X} + tY)$, and the formula from Corollary 4.5.3 applies to give this result. $\square$

Corollary 4.5.5 [52] (Baker–Campbell–Hausdorff formula).

$$q_0 \exp(sX) \exp(tY) = q_0 \exp\left(sX + tY + \frac{1}{2}st[X,Y]\right.$$

$$\left. + \frac{1}{12}s^2t[X,[X,Y]] - \frac{1}{12}st^2[Y,[X,Y]] \cdots\right). \quad (4.34)$$

Proof. Using Corollaries 4.5.2 and 4.5.4, modulo commutator brackets of order ≥ 4 we obtain the following product expansion:

$$q_0 \exp\left(sX + tY + \frac{1}{2}st[X,Y] + \frac{1}{12}s^2t[X,[X,Y]] - \frac{1}{12}st^2[Y,[X,Y]]\right)$$

$$= q_0 \cdots \exp\left(\frac{1}{4}s^2t[X,[X,Y]] + \frac{1}{4}st^2[Y,[X,Y]]\right)$$

$$\exp\left(\frac{1}{2}st[X,Y] + \frac{1}{12}s^2t[X,[X,Y]] - \frac{1}{12}st^2[Y,[X,Y]]\right)\exp(sX+tY)$$

$$= q_0 \cdots \exp\left(\frac{1}{3}s^2t[X,[X,Y]]\right)\exp\left(\frac{1}{6}st^2[Y,[X,Y]]\right)\exp\left(\frac{1}{2}st[X,Y]\right)\exp(sX+tY)$$

$$= q_0 \cdots \exp\left(\frac{1}{3}s^2t[X,[X,Y]]\right)\exp\left(\frac{1}{6}st^2[Y,[X,Y]]\right)\exp\left(\frac{1}{2}st[X,Y]\right)\cdots$$

$$\exp\left(-\frac{1}{3}s^2t[X,[X,Y]]\right)\exp\left(-\frac{1}{6}st^2[Y,[X,Y]]\right)$$

$$\exp\left(-\frac{1}{2}st[X,Y]\right)\exp(sX)\exp(tY)$$

$$= q_0 \exp(R(s,t))\exp(sX)\exp(tY),$$

where $R(s,t)$ denotes the remainder terms in this product expansion consisting of exponentials of iterated Lie brackets of order ≥ 4. By Corollary 4.5.2, commuting these exponentials only adds higher-order terms, and thus

$$q_0 \exp(R(s,t))\exp(sX)\exp(tY) = q_0 \exp(sX)\exp(tY)\exp\left(\tilde{R}(s,t)\right),$$

where $\tilde{R}(s,t)$ also is a product of exponentials of iterated Lie brackets of order ≥ 4. Hence

$$q_0 \exp(sX)\exp(tY) = q_0 \exp\left(sX + tY + \frac{1}{2}st[X,Y] + \frac{1}{12}s^2t[X,[X,Y]]\right.$$

$$\left. - \frac{1}{12}st^2[Y,[X,Y]]\right)\exp\left(-\tilde{R}(,st)\right)$$

$$= q_0 \exp\left(sX + tY + \frac{1}{2}st[X,Y] + \frac{1}{12}s^2t[X,[X,Y]] - \frac{1}{12}st^2[Y,[X,Y]] + \cdots\right),$$

with the dots once more representing iterated Lie brackets of orders ≥ 4. $\square$

4.6 High-Order Necessary Conditions for Optimality

The necessary conditions of the maximum principle were derived solely using the variations CUT and PASTE. These variations can always be made, and hence they apply to every situation. If the controlled reference trajectory has special properties, it is possible to make additional variations that take advantage of these properties. For example, if the control takes values in the interior of the control set, we can vary the control in a full neighborhood of its values, which cannot be done if the control value lies on the boundary of the control set. The Legendre–Clebsch condition is an example of a second-order necessary condition for optimality that arises from a specific such variation. In this section, we prove the most important high-order necessary conditions for optimality of singular controls: the Legendre–Clebsch and Kelley conditions for single-input control systems and the Goh–condition for multi-input control systems. These results are more generally valid for nonlinear control systems, but we restrict our presentation to the class of control-affine systems,

$$\Sigma : \qquad \dot{x} = f(x) + \sum_{i=1}^{m} g_i(x) u_i.$$

While this indeed is the most important class of control systems in applications, our motivation here also is that it allows us to use the algebraic formalism introduced above to compute the tangent vectors $\zeta^{\mathcal{V}}(0)$ that arise as more complicated C^1-extendable variations $\mathcal{V}$ are made.

4.6.1 The Legendre–Clebsch Condition

The Legendre–Clebsch condition is a second-order necessary condition of optimality for controls that take values in the interior of the control set. It is a rather classical result in the optimal control literature (for example, see [64, 137, 140] and the references therein), but in its proof often the questionable normalization $\bar{u} \equiv 0$ is made, which a priori assumes that the corresponding control is smooth. Here we set up a general high-order point variation $\mathcal{V}$ that does not make such an assumption and use Proposition 4.5.2 to compute the leading term in $\zeta^{\mathcal{V}}(\alpha)$. The argument is a simplified version of the reasoning by M. Kawski and H. Sussmann from [130], where the leading term of a variation is computed in great generality.

For simplicity of notation, we consider a single-input control-linear system of the form

$$\Sigma : \qquad \dot{x} = f(x) + u g(x), \quad |u| \leq 1,$$

and we henceforth assume that $\Gamma = (\bar{x}(\cdot), \bar{u}(\cdot))$ is a controlled reference trajectory defined on $[0, T]$ with initial point $\bar{x}(0) = \bar{p}$ and terminal point $\bar{x}(T) = \bar{q}$. We recall from real analysis [257] that a time $\bar{t}$ is a *Lebesgue point* of an integrable function $\bar{u}$ if

$$\lim_{\alpha \to 0} \frac{1}{2\alpha} \int_{\bar{t}-\alpha}^{\bar{t}+\alpha} |\bar{u}(t) - \bar{u}(\bar{t})| \, dt = 0.$$

It is clear that all points $\bar{t}$ where $\bar{u}$ is continuous are Lebesgue points, and more generally, it follows from Lebesgue's differentiation theorem (see Appendix D, [257,

Thm. 7.2]) that almost every time $\bar{t} \in I$ of a locally bounded Lebesgue measurable function $\bar{u}$ is a Lebesgue point.

Theorem 4.6.1 [130]. *Let $\bar{t} \in I$ be a Lebesgue point of the reference control $\bar{u}$ and suppose that $|u(\bar{t})| < 1$. Then the vector $-[g,[f,g]](\bar{x}(\bar{t}))$ lies in the sets $P(\bar{t})$ and $Q(\bar{t})$.*

Theorem 4.6.1 immediately applies to boundary trajectories and the time-optimal control problem, and we get the following corollary:

Corollary 4.6.1 (Legendre–Clebsch condition). *Suppose $\bar{q}$ lies on the boundary of the reachable set, $\bar{q} \in \partial \operatorname{Reach}_\Sigma(\bar{p})$, and let $I \subset [0,T]$ be an open interval on which the reference control $\bar{u}$ takes values in the interior of the control set a.e. Then there exists a nontrivial solution $\lambda : [0,T] \to (\mathbb{R}^n)^*, t \mapsto \lambda(t)$, to the adjoint equation along Γ such that the conditions of Theorem 4.2.2 are satisfied and in addition, for all $t \in I$ we have that*

$$\langle \lambda(t), [f,g](\bar{x}(t)) \rangle \equiv 0$$

and

$$\langle \lambda(t), [g,[f,g]](\bar{x}(t)) \rangle \leq 0. \tag{4.35}$$

Proof of the Corollary. The first condition is an immediate consequence of Theorem 4.2.2. For if the values of $\bar{u}$ lie in the interior of the control set a.e. on I, then the minimization condition of Theorem 4.2.2 implies that the switching function $\Phi(t) = \langle \lambda(t), g(\bar{x}(t)) \rangle$ vanishes identically on I and so then does its derivative $\dot{\Phi}(t) = \langle \lambda(t), [f,g](\bar{x}(t)) \rangle$. It is not difficult to set up point variations that show that the vectors $\pm[f,g](\bar{x}(\bar{t}))$ lie in the sets $P(\bar{t})$ and $Q(\bar{t})$ for a.e. $\bar{t} \in I$, but this is not necessary. The inequality (4.35) on the second-order bracket $[g,[f,g]]$ is new, and it follows from Theorem 4.2.2 that there exists a multiplier λ such that this inequality holds a.e. on I. Since the function $t \mapsto \langle \lambda(t), [g,[f,g]](\bar{x}(t)) \rangle$ is continuous, it is valid on all of I. $\square$

We prove the theorem in several steps and begin with defining the structure of a general point variation $\mathcal{V}$ that also will be used in the proof of the other high-order necessary conditions. Again, we consider the canonical embedding of the controlled reference trajectory Γ. Given $\bar{t} \in I$, choose $\varepsilon > 0$ (sufficiently small) and for $0 \leq \alpha \leq \varepsilon$ replace the reference control $\bar{u}$ over the interval $[\bar{t}, \bar{t}+\alpha]$ with another, suitably chosen, admissible control $\eta(\alpha, \cdot)$. Since the control set is bounded, for sufficiently small $\varepsilon > 0$ these controlled trajectories will all exist over the full interval $[\bar{t}, \bar{t}+\alpha]$. In the notation of Definition 4.4.2, we have that $t_1(\alpha) = \bar{t}, t_2(\alpha) = \bar{t}+\alpha, \Delta_+(\alpha) = \alpha$, and the control of the variation is given by

$$u_\alpha^{\mathcal{V}}(t) = \begin{cases} \bar{u}(t) & \text{if } 0 \leq t \leq \bar{t}, \\ \eta(\alpha, t-\bar{t}) & \text{if } \bar{t} < t \leq \bar{t}+\alpha, \\ \bar{u}(t) & \text{if } \bar{t}+\alpha \leq t \leq T. \end{cases}$$

Thus, the controlled trajectory $(x_\alpha^{\mathscr{V}}, u_\alpha^{\mathscr{V}})$ follows the controlled reference trajectory up to time $t_1(\alpha) = \bar{t}$ and then switches to another admissible controlled trajectory $(\xi(\alpha, \cdot), \eta(\alpha, \cdot))$ (whose specific form is still to be determined) for an interval of length α until at time $\bar{t} + \alpha$ the system reaches the point $\xi(\alpha, \alpha)$. Recall that the curve $\pi^{\mathscr{V}}(\alpha)$ is defined as the unique solution z to the equation $x^E(\bar{t} + \alpha; z) = \xi(\alpha, \alpha)$, and the curve $\zeta^{\mathscr{V}}(\alpha)$, whose tangent vector we need to compute, is the image of this curve $\pi^{\mathscr{V}}(\alpha)$ under the flow of the canonical embedding from time $t = 0$ until time $t = \bar{t}$, i.e.,

$$\zeta^{\mathscr{V}}(\alpha) = x^E(\bar{t}; \pi^{\mathscr{V}}(\alpha)) = \Phi_{\bar{t}}^E(\pi^{\mathscr{V}}(\alpha)) = \left(\Phi_{\bar{t}, \bar{t}+\alpha}^E\right)(\xi(\alpha, \alpha)).$$

The curve $\zeta^{\mathscr{V}}(\alpha)$ is obtained by integrating the dynamics corresponding to the reference control $\bar{u}$ backward from $\xi(\alpha, \alpha)$ over the interval $[\bar{t}, \bar{t} + \alpha]$.

This construction can be expressed more succinctly using exponential notations. Let $S = f(x) + \bar{u}(t)g(x)$ denote the time-varying vector field defined by the reference control $\bar{u}$ and let $\rho(\alpha, \cdot)$ denote the difference between the control $\eta(\alpha, \cdot)$ of the variation and the reference control over $[\bar{t}, \bar{t} + \alpha]$, i.e.,

$$\eta(\alpha, t) = \bar{u}(\bar{t} + t) + \rho(\alpha, t) \qquad \text{for} \quad 0 \le t \le \alpha.$$

Then the curve $\zeta^{\mathscr{V}}(\alpha)$ can be expressed in the form

$$\zeta^{\mathscr{V}}(\alpha) = \bar{x}(\bar{t}) \exp(\alpha(S + \rho g)) \exp(-\alpha S). \tag{4.36}$$

(Starting at $\bar{x}(\bar{t})$, first the vector field $S + \rho g$ defined through the variation is integrated forward for time α and then the reference vector field S is integrated backward for time α.) Proposition 4.5.2 allows us to compute the leading term in an asymptotic expansion for $\zeta^{\mathscr{V}}(\alpha)$ near $\alpha = 0$. We write the dynamics in the form $v(t)f(x) + w(t)g(x)$ with v defined over $[0, 2\alpha]$ by

$$v(t) = \begin{cases} 1 & \text{if } 0 \le t \le \alpha, \\ -1 & \text{if } \alpha < t \le 2\alpha, \end{cases}$$

and we express $w(t)$ in the form $w(t) = \overline{w}(t) + h(t)$ with $\overline{w}(t)$ the contribution of the reference control and $h(t)$ the contribution from the variation, i.e.,

$$\overline{w}(t) = \begin{cases} \bar{u}(\bar{t} + t) & \text{if } 0 \le t \le \alpha, \\ -\bar{u}(\bar{t} + 2\alpha - t) & \text{if } \alpha < t \le 2\alpha, \end{cases} \quad \text{and} \quad h(t) = \begin{cases} \rho(\alpha, t) & \text{if } 0 \le t \le \alpha, \\ 0 & \text{if } \alpha < t \le 2\alpha. \end{cases}$$

The minus signs in the definitions of v and $\overline{w}$ over the interval $[\alpha, 2\alpha]$ express the fact that the reference control is integrated *backward*. Then, by Proposition 4.5.2, the curve $\zeta^{\mathscr{V}}(\alpha)$ has an asymptotic expansion of the form

$$\zeta^{\mathcal{V}}(\alpha) = \bar{x}(\bar{t}) \cdots \exp\left(\left(\int_0^{2\alpha} wVW ds\right)[g,[f,g]]\right)$$

$$\exp\left(\left(\frac{1}{2}\int_0^{2\alpha} wV^2 ds\right)[f,[f,g]]\right) \exp\left(\left(\int_0^{2\alpha} wV ds\right)[f,g]\right)$$

$$\exp\left(\left(\int_0^{2\alpha} w ds\right)g\right) \exp\left(\left(\int_0^{2\alpha} v ds\right)f\right), \tag{4.37}$$

where the dots represent a remainder term that is of order $o(\alpha^3)$ as $\alpha \to 0$, and for $0 \le t \le 2\alpha$, we have that

$$V(t) = \int_0^t v(s)ds = \begin{cases} t & \text{if } 0 \le t \le \alpha, \\ -t+2\alpha & \text{if } \alpha \le t \le 2\alpha, \end{cases} \quad \text{and} \quad W(t) = \int_0^t w(s)ds.$$

We also write

$$\overline{W}(t) = \int_0^t \overline{w}(s)ds \quad \text{and} \quad H(t) = \int_0^t h(s)ds = \begin{cases} \int_0^t \rho(\alpha,s)ds & \text{if } 0 \le t \le \alpha, \\ \int_0^\alpha \rho(\alpha,s)ds & \text{if } \alpha \le t \le 2\alpha, \end{cases}$$

such that $W(t) = \overline{W}(t) + H(t)$.

   Our *aim* is to determine the controls $\eta(\alpha,\cdot)$, respectively the increments $\rho(\alpha,\cdot)$, such that the coefficients at the vector fields f, g, $[f,g]$, and $[f,]f,g]]$ in the expansion (4.37) vanish, while the coefficient at $[g,[f,g]]$ is nonzero. Obviously, $V(2\alpha) = 0$, and thus the coefficient at the vector field f vanishes.

Lemma 4.6.1. *For $k = 0,1,\dots$ we have that*

$$\int_0^{2\alpha} w(s)V(s)^k ds = \int_0^\alpha s^k \rho(\alpha,s)ds.$$

Proof. We have

$$\int_0^{2\alpha} w(s)V(s)^k ds = \int_0^{2\alpha} \overline{w}(s)V(s)^k ds + \int_0^{2\alpha} h(s)V(s)^k ds.$$

Because of the telescoping nature of the variation, the integral that includes the reference $\overline{w}$ vanishes:

$$\int_0^{2\alpha} \overline{w}(s)V(s)^k ds = \int_0^\alpha \bar{u}(\bar{t}+s)s^k ds + \int_\alpha^{2\alpha} -\bar{u}(\bar{t}+2\alpha-s)(-s+2\alpha)^k ds$$

$$= \int_0^\alpha \bar{u}(\bar{t}+s)s^k ds + \int_\alpha^0 \bar{u}(\bar{t}+r)r^k dr = 0.$$

Hence,

$$\int_0^{2\alpha} w(s)V(s)^k ds = \int_0^{2\alpha} h(s)V(s)^k ds = \int_0^{\alpha} s^k \rho(\alpha,s)ds.$$ □

In particular, $W(2\alpha) = \int_0^\alpha \rho(\alpha,s)ds$, and we will choose the controls of the variations such that

$$\int_0^\alpha \rho(\alpha,s)ds = 0 \qquad \text{for all } \alpha \in [0,\varepsilon].$$

Thus the coefficient at the vector field g vanishes and also $H(t) \equiv 0$ on $[\alpha, 2\alpha]$. The coefficients at the Lie brackets $[f,g]$ and $[f,[f,g]]$ in Eq. (4.37) are given by

$$\int_0^{2\alpha} w(s)V(s)ds = \int_0^{\alpha} s\rho(\alpha,s)ds$$

and

$$\frac{1}{2}\int_0^{2\alpha} w(s)V(s)^2 ds = \frac{1}{2}\int_0^{\alpha} s^2 \rho(\alpha,s)ds,$$

and it is easy to choose $\rho(\alpha,\cdot)$ such that these terms vanish as well.

In order to bring out the structure of the variation more clearly, and also to minimize the technical aspects of the construction, let us first assume that the reference control $\bar{u}$ is continuous from the right at $\bar{t}$. Since $|\bar{u}(\bar{t})| < 1$, in this case there exist positive constants ε and c such that for all $t \in [\bar{t}, \bar{t}+\varepsilon]$ we have that

$$-1 + 2c \le \bar{u}(t) \le 1 - 2c,$$

and thus any continuous variation that satisfies $|\rho(\alpha,t)| \le c$ defines an admissible control. Specifically, we can choose the following function:

Proposition 4.6.1. *Let*

$$Q(t) = 20t^3 - 30t^2 + 12t - 1,$$

and for $0 \le s \le \alpha$, define

$$\rho(\alpha,s) = cQ\left(\frac{s}{\alpha}\right).$$

Then $\eta(\alpha,\cdot) = \bar{u}(\bar{t}+\cdot) + \rho(\alpha,\cdot)$ is an admissible control,

$$\int_0^\alpha s^i \rho(\alpha,s)ds = 0 \qquad \text{for } i = 0,1,2, \tag{4.38}$$

and

$$\int_0^{2\alpha} w(s)V(s)W(s)ds = -\omega\alpha^3 + o(\alpha^3) \tag{4.39}$$

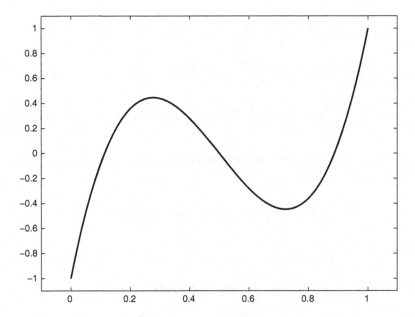

Fig. 4.19 Polynomial Q of the variation $\mathscr{V}$ for the Legendre–Clebsch condition

with

$$\omega = \frac{1}{2}c^2 \int_0^1 \left(\int_0^t Q(s)ds \right)^2 dt > 0.$$

This polynomial Q defines the variation that generates the Legendre–Clebsch condition. Its graph is shown in Fig. 4.19. The maximum of $|Q(t)|$ on the interval $[0,1]$ is taken on at its endpoints and is given by 1. Also note the symmetry of the graph with respect to the point $(\frac{1}{2},0)$, i.e., $Q(1-t) = -Q(t)$. This symmetry is preserved in similar higher-order variations.

Proof. The polynomial Q is the unique cubic polynomial that satisfies

$$\int_0^1 t^i Q(t)dt = 0 \qquad \text{for } i = 0,1,2 \qquad \text{and} \qquad \int_0^1 t^3 Q(t)dt = \frac{1}{140}$$

and is easily computed as the solution of a 4×4 system of linear equations. The normalization has been made such that $\max\{|Q(t)| : 0 \le t \le 1\} = 1$. Hence it is clear that $\eta(\alpha,\cdot)$ is an admissible control and Eq. (4.38) holds. We need to compute the integral $\int_0^{2\alpha} w(s)V(s)W(s)ds$.

Recall that $W = \bar{W} + H$ and H vanishes identically on $[\alpha,2\alpha]$. Thus,

$$\int_0^{2\alpha} w(s)V(s)W(s)ds = \frac{1}{2}V(s)W(s)^2\Big|_0^{2\alpha} - \frac{1}{2}\int_0^{2\alpha} v(s)W(s)^2 ds$$

$$= -\frac{1}{2}\left(\int_0^\alpha W(s)^2 ds + \int_\alpha^{2\alpha} -W(s)^2 ds\right)$$

$$= -\frac{1}{2}\left(\int_0^\alpha \overline{W}(s)^2 ds + 2\int_0^\alpha \overline{W}(s)H(s)ds + \int_0^\alpha H(s)^2 ds - \int_\alpha^{2\alpha} \overline{W}(s)^2 ds\right).$$

Analogously to the earlier computation above in the proof of Lemma 4.6.1, the $\overline{W}^2$ integrals cancel: we have

$$\overline{W}(t) = \int_0^t \bar{u}(\bar{t}+s)ds \quad \text{for} \quad 0 \le t \le \alpha,$$

and for $\alpha \le t \le 2\alpha$,

$$\overline{W}(t) = \int_0^\alpha \bar{u}(\bar{t}+s)ds + \int_\alpha^t -\bar{u}(\bar{t}+2\alpha-s)ds$$

$$= \int_0^\alpha \bar{u}(\bar{t}+s)ds + \int_\alpha^{2\alpha-t} \bar{u}(\bar{t}+r)dr = \int_0^{2\alpha-t} \bar{u}(\bar{t}+s)ds. \qquad (4.40)$$

Hence

$$\int_\alpha^{2\alpha} \overline{W}(t)^2 dt = \int_\alpha^{2\alpha}\left(\int_0^{2\alpha-t} \bar{u}(\bar{t}+s)ds\right)^2 dt = -\int_\alpha^0\left(\int_0^r \bar{u}(\bar{t}+s)ds\right)^2 dr$$

$$= \int_0^\alpha\left(\int_0^r \bar{u}(\bar{t}+s)ds\right)^2 dr = \int_0^\alpha \overline{W}(r)^2 dr,$$

and overall, we have that

$$\int_0^{2\alpha} w(s)V(s)W(s)ds = -\frac{1}{2}\int_0^\alpha H(s)^2 ds - \int_0^\alpha \overline{W}(s)H(s)ds. \qquad (4.41)$$

Furthermore,

$$\int_0^\alpha H(s)^2 ds = \alpha\int_0^1 H(\alpha t)^2 dt = \alpha\int_0^1\left(\int_0^{\alpha t} cQ\left(\frac{r}{\alpha}\right)dr\right)^2 dt$$

$$= \alpha^3 c^2\int_0^1\left(\int_0^t Q(s)ds\right)^2 dt = 2\omega\alpha^3$$

with

$$\omega = \frac{1}{2}c^2\int_0^1\left(\int_0^t Q(s)ds\right)^2 dt > 0.$$

The second term in Eq. (4.41) is of higher order, for since $\bar{u}$ is continuous from the right at $\bar{t}$, we have that

$$\overline{W}(t) = \int_0^t \bar{u}(\bar{t}+s)ds = \bar{u}(\bar{t})t + o(t)$$

and thus also

$$\int_0^t \overline{W}(s)ds = \frac{1}{2}\bar{u}(\bar{t})t^2 + o(t^2).$$

Integrating by parts, we get that

$$\int_0^\alpha \overline{W}(s)H(s)ds = \left(\int_0^t \overline{W}(s)ds\right)H(t)\Big|_0^\alpha - \int_0^\alpha \left(\int_0^s \overline{W}(t)dt\right)h(s)ds$$

$$= -\int_0^\alpha \left(\frac{1}{2}\bar{u}(\bar{t})s^2 + o(s^2)\right)cQ\left(\frac{s}{\alpha}\right)ds$$

$$= -\frac{1}{2}\bar{u}(\bar{t})\alpha^3 c\int_0^1 t^2 Q(t)dt + o\left(\alpha^3\right) = o\left(\alpha^3\right),$$

since $\int_0^1 t^2 Q(t)dt = 0$. Overall, we thus have that

$$\int_0^{2\alpha} w(s)V(s)W(s)ds = -\omega\alpha^3 + o\left(\alpha^3\right).$$

This concludes the proof of the proposition.   □

Essentially, with some technical adjustments that account for measurability, this argument remains valid if $\bar{t}$ is a Lebesgue point for the reference control $\bar{u}$. Since $\bar{u}(\bar{t})$ lies in the open interval $(-1,1)$, it is still possible to pick small positive constants c such that $-1 + 2c \leq \bar{u}(\bar{t}) \leq 1 - 2c$, but now these inequalities need not hold everywhere on the interval $[\bar{t}, \bar{t} + \alpha]$. We therefore define the "good" set G_α and the "bad" set B_α as

$$G_\alpha = \{t \in [0,\alpha] : |\bar{u}(\bar{t}+t) - \bar{u}(\bar{t})| \leq c\} \quad \text{and} \quad B_\alpha = G_\alpha^c = [0,\alpha]\backslash G_\alpha.$$

We will now choose $\rho(\alpha,\cdot)$ as a measurable function that vanishes on B_α with $|\rho(\alpha,t)| \leq c$ for $t \in G_\alpha$. Hence, as above, the control $\eta(\alpha,t) = \bar{u}(\bar{t}+t) + \rho(\alpha,t)$ is admissible, and we need to show that Eqs. (4.38) and (4.39) remain valid.

On the set B_α, we have that $|\bar{u}(\bar{t}+t) - \bar{u}(\bar{t})| > c$, and since $\bar{t}$ is a Lebesgue point, it follows that

$$c\mu(B_\alpha) \leq \int_{[\bar{t},\bar{t}+\alpha]} |\bar{u}(t) - \bar{u}(\bar{t})|\,d\mu = o(\alpha).$$

Hence the Lebesgue measure $\mu(B_\alpha)$ of the bad set is of order $o(\alpha)$; equivalently, $\mu(G_\alpha) = \alpha - o(\alpha)$. If we define $\widehat{G}_\alpha = \{s \in [0,1] : \alpha s \in G_\alpha\}$, the set G_α rescaled as a subset of $[0,1]$, then it follows that

$$\lim_{\alpha \to 0} \mu\left(\widehat{G}_\alpha\right) = \lim_{\alpha \to 0}\left(1 - \frac{o(\alpha)}{\alpha}\right) = 1,$$

and thus, in the limit $\alpha \to 0$, this set has full measure; without loss of generality, we set $\widehat{G}_0 = [0,1]$. For $\alpha \geq 0$, define cubic polynomials $Q_\alpha(s) = a_0^\alpha + a_1^\alpha s + a_2^\alpha s^2 + a_3^\alpha s^3$ on $[0,1]$ such that

$$\int_{\widehat{G}_\alpha} s^i Q_\alpha(s)ds = 0 \qquad \text{for } i = 0,1,2 \qquad \text{and} \qquad \int_{\widehat{G}_\alpha} s^3 Q_\alpha(s)ds = \frac{1}{140}.$$

As before, the coefficients of the polynomials Q_α are the unique solutions to a system of linear equations of the form $M(\alpha)a^\alpha = b$, where $a^\alpha = \left(a_0^\alpha, a_1^\alpha, a_2^\alpha, a_3^\alpha\right)^T$, $b = (0,0,0,\frac{1}{140})^T$, and $M(\alpha)$ is the 4×4 matrix with entries

$$m_{ij}(\alpha) = \int_{\widehat{G}_\alpha} s^{i+j-2}d\mu.$$

In the limit $\alpha \to 0$, these entries converge to the matrix $M(0)$ with entries

$$m_{ij}(0) = \int_{[0,1]} s^{i+j-2}d\mu = \frac{1}{i+j-1}.$$

This matrix is nonsingular, and thus for small enough α and $i = 0,1,2$, and 3, the coefficients a_i^α are uniquely determined and as $\alpha \to 0$ converge to the coefficients a_i of the polynomial $Q(t) = 20t^3 - 30t^2 + 12t - 1$ considered above. In particular, we may assume that $\max_{[0,1]}|Q_\alpha(t)| \leq 2$ for all $\alpha \in [0,\varepsilon]$. Hence, if we now define

$$\rho(\alpha,t) = \begin{cases} \frac{c}{2}Q_\alpha\left(\frac{t}{\alpha}\right) & \text{if } t \in G_\alpha, \\ 0 & \text{if } t \in B_\alpha, \end{cases} \tag{4.42}$$

then it follows for $i = 0,1$, and 2 that

$$\int_0^\alpha t^i \rho(\alpha,t)dt = \frac{c}{2}\int_{G_\alpha} t^i Q_\alpha\left(\frac{t}{\alpha}\right)dt = \frac{c}{2}\alpha^{i+1}\int_{\widehat{G}_\alpha} s^i Q_\alpha(s)\,ds = 0.$$

Furthermore,

$$\int_0^\alpha H(t)^2 dt = \alpha \int_0^1 H(\alpha s)^2 ds = \alpha \int_0^1 \left(\int_{[0,\alpha s]\cap G_\alpha} \frac{c}{2}Q_\alpha\left(\frac{r}{\alpha}\right)dr\right)^2 ds$$

$$= \alpha^3 \frac{c^2}{4}\int_0^1 \left(\int_{[0,s]\cap G_\alpha} Q_\alpha(t)dt\right)^2 ds$$

$$= \alpha^3 \frac{c^2}{4}\left[\int_0^1 \left(\int_0^s Q(t)dt\right)^2 ds + o(1)\right],$$

where we once more use that $\mu(B_\alpha)$ is of order $o(\alpha)$ as $\alpha \to 0$. Thus, taking

$$\omega = \frac{c^2}{8} \int_0^1 \left(\int_0^s Q(t)dt \right)^2 ds > 0,$$

we again have that

$$\int_0^\alpha H(t)^2 dt = 2\omega\alpha^3 + o(\alpha^3).$$

The second term in Eq. (4.41) remains of higher order $o(\alpha^3)$: the condition that $\bar{t}$ is a Lebesgue point still implies that

$$\overline{W}(t) = \int_0^t \bar{\eta}(\bar{t}+s)ds = \bar{\eta}(\bar{t})t + o(t),$$

and the rest of the calculation is as above. Thus we have the same conditions as in Proposition 4.6.1.

These properties imply that the asymptotic expansion for the curve

$$\zeta^{\mathcal{V}}(\alpha) = \bar{x}(\bar{t}) \exp(\alpha(S + \rho g)) \exp(-\alpha S)$$

has the form

$$\zeta^{\mathcal{V}}(\alpha) = \bar{x}(\bar{t}) \exp(-\omega\alpha^3 [g, [f, g]] + o(\alpha^3)).$$

Replacing α with $\beta = \omega\alpha^3$, the resulting curve is differentiable at $\beta = 0$ and we have

$$\dot{\zeta}^{\mathcal{V}}(0) = -[g, [f, g]] (\bar{x}(\bar{t})),$$

verifying that this vector lies in $P(\bar{t})$.

Furthermore, since $\bar{t}$ is a Lebesgue point, it also follows that for every small enough α, the interval $[\bar{t}, \bar{t} + \alpha]$ must contain a set of positive measure where $|\bar{u}(t)| < 1$. (Otherwise, for some positive constant k we have that $\int_{[\bar{t},\bar{t}+\alpha]} |\bar{u}(t) - \bar{u}(\bar{t})| d\mu \geq k\alpha$, contradicting the definition of a Lebesgue point.) Since almost every time t is a Lebesgue point, it follows that there exists a sequence of times $\{\bar{t}_n\}_{n\in\mathbb{N}}$ that converges to $\bar{t}$ from the right such that all times $\bar{t}_n$ are Lebesgue points of $\bar{u}$ with $|\bar{u}(\bar{t}_n)| < 1$. Hence Knobloch's condition (K) is satisfied, and we have that $-[g, [f, g]] (\bar{x}(\bar{t})) \in Q(\bar{t})$. This concludes the proof of Theorem 4.6.1.

4.6.2   The Kelley Condition

The Legendre–Clebsch condition is nontrivial if the singular control is of order 1, that is, if

$$\langle \lambda(t), [g, [f, g]] (\bar{x}(t)) \rangle \neq 0.$$

If this term vanishes on an interval, then the singular control is of higher order, and these computations can be extended, provided this order is *intrinsic*, i.e., the vector fields that arise in the derivatives of the switching function at the control vanish identically (see Sect. 2.8). We already have seen in Chap. 2 that singular controls that are of intrinsic order 2 lead to chattering concatenations (cf., the Fuller problem). This phenomenon also arises naturally in mathematical models for cancer treatments involving anti-angiogenic therapy. For a wide class of these models, optimal controls are characterized by an optimal singular arc of order 1 if dosage and concentration of the drugs are identified; e.g., see [160]. If a standard linear pharmacokinetic model is added that describes the concentration in the plasma, this singular arc remains optimal, but its order increases to 2 [165]. These problems will be analyzed in a companion volume to this text [166]. We therefore include a proof of the Kelley condition for optimality of singular controls of intrinsic order 2. We again consider a system of the form

$$\Sigma : \qquad \dot{x} = f(x) + u g(x), \quad |u| \leq 1.$$

As above, let $\Gamma = (\bar{x}(\cdot), \bar{u}(\cdot))$ be a controlled reference trajectory defined on $[0, T]$ with initial point $\bar{x}(0) = \bar{p}$ and terminal point $\bar{x}(T) = \bar{q}$.

Theorem 4.6.2. *Suppose $[g, [f, g]]$ vanishes identically in a neighborhood of the reference trajectory $\bar{x}$. Let $\bar{t} \in I$ be a Lebesgue point of the reference control $\bar{u}$ and suppose that $|u(\bar{t})| < 1$. Then the vector $[g, \mathrm{ad}^3 f(g)](\bar{x}(\bar{t}))$ lies in the sets $P(\bar{t})$ and $Q(\bar{t})$.*

Corollary 4.6.2 (Kelley condition). *Suppose $\bar{q}$ lies in the boundary of the reachable set, $\bar{q} \in \partial \operatorname{Reach}_{\Sigma}(\bar{p})$, and let $I \subset [0, T]$ be an open interval on which the reference control $\bar{u}$ takes values in the interior of the control set. If the vector field $[g, [f, g]]$ vanishes identically (in a neighborhood of the controlled reference trajectory), then there exists a nontrivial solution $\lambda : [0, T] \to (\mathbb{R}^n)^*$, $t \mapsto \lambda(t)$, to the adjoint equation along Γ such that the conditions of Theorem 4.2.2 are satisfied and in addition, for all $t \in I$ we have that*

$$\langle \lambda(t), \mathrm{ad}^i f(g)(\bar{x}(t)) \rangle \equiv 0 \qquad \text{for } i = 1, 2, 3,$$

and

$$\langle \lambda(t), [g, \mathrm{ad}^3 f(g)](\bar{x}(t)) \rangle \geq 0.$$

The proof of the corollary is as above. As in the case of the Legendre–Clebsch condition, the first relation is an immediate consequence of Theorem 4.2.2 and the fact that $[g, [f, g]] \equiv 0$. In this case, the first three derivatives of the switching function $\Phi(t) = \langle \lambda(t), g(\bar{x}(t)) \rangle$ are given by

$$\Phi^{(i)}(t) = \langle \lambda(t), \mathrm{ad}^i f(g)(\bar{x}(t)) \rangle$$

and vanish identically on I. The inequality on the Lie bracket $[g, \mathrm{ad}^3 f(g)]$ is new, and by Theorem 4.2.2 there exists a multiplier λ such that this inequality holds a.e. on I. Since the function $t \mapsto \langle \lambda(t), [g, \mathrm{ad}^3 f(g)](\bar{x}(t)) \rangle$ is continuous, it thus is valid on all of I.

We now prove the theorem using the same structure and notation for the variation $\mathcal{V}$ as in the proof of Theorem 4.6.1. Here we need the full asymptotic product expansion computed in Proposition 4.5.2 with all Lie brackets of order ≤ 5. However, since $[g, [f, g]] \equiv 0$, several fourth- and fifth-order brackets vanish, leading to a significant simplification for this expression. Clearly, any brackets with $[g, [f, g]]$ such as $[f, [g, [f, g]]]$ vanish identically. But also other identities are induced by the Jacobi condition. For example, $[g, [f, [f, g]]] = [f, [g, [f, g]]] \equiv 0$. Furthermore,

$$[g, [f, \mathrm{ad}^2 f(g)]] + [f, [\mathrm{ad}^2 f(g), g]] + [\mathrm{ad}^2 f(g), [g, f]] \equiv 0.$$

Since $[\mathrm{ad}^2 f(g), g] = -[g, [f, [f, g]]] \equiv 0$, we get that

$$[g, \mathrm{ad}^3 f(g)] = -[[f, g], [f, [f, g]]],$$

and the corresponding exponentials in Eq. (4.32) can be combined. We therefore obtain the following reduced formula for $\zeta^{\mathcal{V}}(\alpha)$:

$$\zeta^{\mathcal{V}}(\alpha) = \bar{x}(\bar{t}) \exp\left(\alpha(S + \rho g)\right) \exp\left(-\alpha S\right)$$

$$= \bar{x}(\bar{t}) \cdots \exp\left(\left(\frac{1}{6}\int_0^{2\alpha} wV^3 W ds - \frac{1}{2}\int_0^{2\alpha} wV^2 U ds\right) \cdot [g, \mathrm{ad}^3 f(g)]\right)$$

$$\exp\left(\frac{1}{4!}\int_0^{2\alpha} wV^4 ds \cdot \mathrm{ad}^4 f(g)\right) \exp\left(\frac{1}{3!}\int_0^{2\alpha} wV^3 ds \cdot \mathrm{ad}^3 f(g)\right)$$

$$\exp\left(\frac{1}{2}\int_0^{2\alpha} wV^2 ds \cdot \mathrm{ad}^2 f(g)\right) \exp\left(\int_0^{2\alpha} wV ds \cdot \mathrm{ad} f(g)\right)$$

$$\exp\left(\int_0^{2\alpha} w\, ds \cdot g\right), \tag{4.43}$$

where the dots now represent a remainder term of order $o(\alpha^5)$ and we already use that $V(2\alpha) = 0$, so that the term at f vanishes.

The technical aspects of carrying out the construction for a time $\bar{t}$ that is a Lebesgue point are the same as in the proof above, and thus without loss of generality, we simply assume that the control $\bar{u}$ is continuous at $\bar{t}$ from the right. As above, let ε and c be positive constants such that for all $t \in [\bar{t}, \bar{t} + \varepsilon]$ we have that $-1 + 2c \leq \bar{u}(t) \leq 1 - 2c$.

Proposition 4.6.2. *Let*

$$Q(t) = 252t^5 - 630t^4 + 560t^3 - 210t^2 + 30t - 1,$$

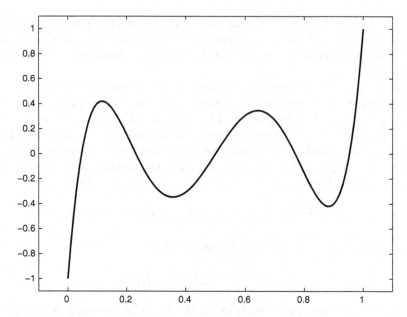

Fig. 4.20 Polynomial Q of the variation $\mathscr{V}$ for the Kelley condition

and for $0 \leq s \leq \alpha$, define

$$\rho(\alpha,s) = cQ\left(\frac{s}{\alpha}\right).$$

Then $\eta(\alpha,\cdot) = \bar{u}(\bar{t}+\cdot)+\rho(\alpha,\cdot)$ is an admissible control,

$$\int_0^\alpha s^i \rho(\alpha,s)ds = 0 \qquad\qquad \text{for } i = 0,\ldots,4,$$

and

$$\frac{1}{6}\int_0^{2\alpha} w(s)V(s)^3 W(s)ds = -\frac{1}{4}\psi_1 c^2 \alpha^5 + o\left(\alpha^5\right), \qquad (4.44)$$

$$\frac{1}{2}\int_0^{2\alpha} w(s)V(s)^2 U(s)ds = -\frac{1}{4}\psi_2 c^2 \alpha^5 + o\left(\alpha^5\right), \qquad (4.45)$$

with

$$\psi_1 = \int_0^1 t^2\left(\int_0^t Q(s)ds\right)^2 dt < \psi_2 = \int_0^1 \left(\int_0^t sQ(s)ds\right)^2 dt.$$

The graph of this polynomial Q is shown in Fig. 4.20. As in the case of the Legendre–Clebsch condition, the normalization has been chosen so that $\max\{|Q(t)| : 0 \leq t \leq 1\} = 1$, and this maximum of $|Q(t)|$ is taken on at both endpoints. The symmetry with respect to the point $(\frac{1}{2},0)$, i.e., $Q(1-t) = -Q(t)$, is preserved.

Proof. Here Q is the unique fifth-order polynomial defined on the interval $[0,1]$ that satisfies

$$\int_0^1 t^i Q(t)dt = 0 \qquad \text{for } i = 0,\ldots,4 \qquad \text{and} \qquad \int_0^1 t^5 Q(t)dt = \frac{1}{2772};$$

$\eta(\alpha,\cdot)$ is an admissible control, and by Lemma 4.6.1 the coefficients in the expansion at the Lie brackets $\text{ad}^i f(g)$, $i = 1,\ldots 4$, vanish. Hence we are left with only

$$\zeta^{\gamma}(\alpha) = \bar{x}(\bar{t})\cdots\exp\left(\left(\frac{1}{6}\int_0^{2\alpha} wV^3W ds - \frac{1}{2}\int_0^{2\alpha} wV^2U ds\right)\cdot[g, \text{ad}^3 f(g)]\right).$$

$$(4.46)$$

We compute these two integrals separately. We again write $W(s) = \overline{W}(s) + H(s)$ and recall that $H(s) \equiv 0$ on $[\alpha, 2\alpha]$. Thus, analogously to the earlier computations, we get that

$$\int_0^{2\alpha} w(s)V(s)^3 W(s)ds = \frac{1}{2}V(s)^3 W(s)^2\Big|_0^{2\alpha} - \frac{3}{2}\int_0^{2\alpha} v(s)V(s)^2 W(s)^2 ds$$

$$= -\frac{3}{2}\left(\int_0^{\alpha} s^2 W(s)^2 ds + \int_{\alpha}^{2\alpha} -(2\alpha - s)^2 W(s)^2 ds\right)$$

$$= -\frac{3}{2}\left(\int_0^{\alpha} s^2\overline{W}(s)^2 ds + 2\int_0^{\alpha} s^2\overline{W}(s)H(s)ds + \int_0^{\alpha} s^2 H(s)^2 ds\right.$$

$$\left. - \int_{\alpha}^{2\alpha} (2\alpha - s)^2\overline{W}(s)^2 ds\right).$$

Once more, the $\overline{W}^2$ integrals cancel: using Eq. (4.40), we get that

$$\int_{\alpha}^{2\alpha} (2\alpha - s)^2\overline{W}(s)^2 ds = \int_{\alpha}^{2\alpha} (2\alpha - s)^2\left(\int_0^{2\alpha - s} \bar{u}(\bar{t}+t)dt\right)^2 ds$$

$$= \int_0^{\alpha} r^2\left(\int_0^r \bar{u}(\bar{t}+t)dt\right)^2 dr = \int_0^{\alpha} r^2\overline{W}(r)^2 dr.$$

Hence

$$\frac{1}{6}\int_0^{2\alpha} w(s)V(s)^3 W(s)ds = -\frac{1}{4}\int_0^{\alpha} s^2 H(s)^2 ds - \frac{1}{2}\int_0^{\alpha} s^2\overline{W}(s)H(s)ds.$$

The first integral equals

$$\int_0^\alpha s^2 H(s)^2 ds = \alpha \int_0^1 (\alpha t)^2 H(\alpha t)^2 dt = \alpha^3 \int_0^1 t^2 \left(\int_0^{\alpha t} cQ\left(\frac{r}{\alpha}\right) dr \right)^2 dt$$

$$= \alpha^5 c^2 \int_0^1 t^2 \left(\int_0^t Q(s) ds \right)^2 dt,$$

and we set

$$\psi_1 = \int_0^1 t^2 \left(\int_0^t Q(s) ds \right)^2 dt > 0.$$

The second integral is of order $o\left(\alpha^5\right)$: as before,

$$\overline{W}(t) = \bar{u}(\bar{t})t + o(t)$$

and

$$\int_0^t \overline{W}(s) ds = \frac{1}{2}\bar{u}(\bar{t})t^2 + o(t^2).$$

Integrating by parts, we get that

$$\int_0^\alpha s^2 \overline{W}(s) H(s) ds = \left(\int_0^t s^2 \overline{W}(s) ds \right) H(t) \Big|_0^\alpha - \int_0^\alpha \left(\int_0^s t^2 \overline{W}(t) dt \right) h(s) ds$$

$$= - \int_0^\alpha \left(\frac{1}{2}\bar{u}(\bar{t})s^4 + o(s^4) \right) cQ\left(\frac{s}{\alpha}\right) ds$$

$$= -\frac{1}{2}\bar{u}(\bar{t})\alpha^5 c \int_0^1 t^4 Q(t) dt + o\left(\alpha^5\right) = o\left(\alpha^5\right),$$

where we use that $\int_0^1 t^4 Q(t) dt = 0$. Overall, we therefore have that

$$\frac{1}{6} \int_0^{2\alpha} w(s) V(s)^3 W(s) ds = -\frac{1}{4}\psi_1 c^2 \alpha^5 + o\left(\alpha^5\right),$$

verifying Eq. (4.44).

We similarly evaluate the second integral in Eq. (4.46): using that

$$\frac{d}{ds}\left(\frac{1}{2}U(s)^2 \right) = w(s) V(s) U(s),$$

we obtain

$$\int_0^{2\alpha} w(s)V(s)^2 U(s)ds = \frac{1}{2}V(s)U(s)^2\Big|_0^{2\alpha} - \frac{1}{2}\int_0^{2\alpha} v(s)U(s)^2 ds$$

$$= -\frac{1}{2}\int_0^{\alpha} U(s)^2 ds + \frac{1}{2}\int_\alpha^{2\alpha} U(s)^2 ds.$$

Once again, writing $w(s) = \overline{w}(s) + h(s)$, define

$$\overline{U}(s) = \int_0^s \overline{w}(t)V(t)dt \quad \text{and} \quad K(s) = \int_0^s h(t)V(t)dt.$$

For $s \in [\alpha, 2\alpha]$ we have that

$$\overline{U}(s) = \int_0^{\alpha} \bar{u}(\bar{t}+t)t\,dt + \int_\alpha^s -\bar{u}(\bar{t}+2\alpha-t)(2\alpha-t)\,dt$$

$$= \int_0^{\alpha} \bar{u}(\bar{t}+t)t\,dt + \int_\alpha^{2\alpha-s} \bar{u}(\bar{t}+r)r\,dr = \int_0^{2\alpha-s} t\bar{u}(\bar{t}+t)\,dt,$$

and thus

$$\overline{U}(s) = \int_0^s \overline{w}(t)V(t)dt = \begin{cases} \int_0^s t\bar{u}(\bar{t}+t)dt & \text{if } 0 \le s \le \alpha, \\ \int_0^{2\alpha-s} t\bar{u}(\bar{t}+t)dt & \text{if } \alpha \le s \le 2\alpha. \end{cases}$$

In particular, $\overline{U}(2\alpha) = 0$, and as with $\overline{W}$, we have the symmetry

$$\overline{U}(\alpha - s) = \overline{U}(\alpha + s) \quad \text{for} \quad 0 \le s \le \alpha.$$

Furthermore, since $\int_0^s tp(\alpha,t)dt = 0$, we also obtain that

$$K(s) = \int_0^s h(t)V(t)dt = \begin{cases} \int_0^s tp(\alpha,t)dt & \text{if } 0 \le s \le \alpha, \\ 0 & \text{if } \alpha \le s \le 2\alpha. \end{cases}$$

Hence, overall,

$$\int_0^{2\alpha} w(s)V(s)^2 U(s)ds = -\frac{1}{2}\int_0^{\alpha} \left(\overline{U}(s) + K(s)\right)^2 ds + \frac{1}{2}\int_\alpha^{2\alpha} \overline{U}(s)^2 ds$$

$$= -\frac{1}{2}\int_0^{\alpha} \overline{U}(s)^2 ds + \frac{1}{2}\int_\alpha^{2\alpha} \overline{U}(s)^2 ds - \int_0^{\alpha} \overline{U}(s)K(s)ds$$

$$- \frac{1}{2}\int_0^{\alpha} K(s)^2 ds.$$

As with the first integral, the $\overline{U}$-integrals cancel, the mixed term is of higher order, and the K^2-integral gives the dominant term. Specifically,

$$\int_\alpha^{2\alpha} \overline{U}(s)^2 ds = \int_\alpha^{2\alpha} \left(\int_0^{2\alpha-s} t\bar{u}(\bar{t}+t)dt \right)^2 ds$$

$$= \int_0^\alpha \left(\int_0^r t\bar{u}(\bar{t}+t)dt \right)^2 dr = \int_0^\alpha \overline{U}(r)^2 dr,$$

$$\int_0^\alpha \overline{U}(s)K(s)ds = \left(\int_0^t \overline{U}(s)ds \right) K(t) \Big|_0^\alpha - \int_0^\alpha \left(\int_0^s \overline{U}(t)dt \right) h(s)V(s)ds$$

$$= -\int_0^\alpha \left(\int_0^s \frac{1}{2}\bar{u}(\bar{t})t^2 + o(t^2)dt \right) cQ\left(\frac{s}{\alpha}\right) s\, ds$$

$$= -\frac{1}{2}\bar{u}(\bar{t}) \int_0^\alpha \left(\frac{1}{3}s^4 cQ\left(\frac{s}{\alpha}\right) + o(s^4) \right) ds$$

$$= -\frac{1}{6}\bar{u}(\bar{t})\alpha^5 c \int_0^1 t^4 Q(t)dt + o\left(\alpha^5\right) = o\left(\alpha^5\right),$$

and

$$\int_0^\alpha K(s)^2 ds = \alpha \int_0^1 K(\alpha t)^2 dt = \alpha^3 \int_0^1 \left(\int_0^{\alpha t} rcQ\left(\frac{r}{\alpha}\right) dr \right)^2 dt$$

$$= \alpha^5 c^2 \int_0^1 \left(\int_0^t sQ(s)ds \right)^2 dt = \psi_2 c^2 \alpha^5$$

with

$$\psi_2 = \int_0^1 \left(\int_0^t sQ(s)ds \right)^2 dt.$$

Hence

$$\frac{1}{2}\int_0^{2\alpha} w(s)V(s)^2 U(s)ds = -\frac{1}{4}\psi_2 c^2 \alpha^5 + o\left(\alpha^5\right),$$

verifying Eq. (4.45).

An explicit computation shows that

$$\psi_2 - \psi_1 = \int_0^1 \left(\int_0^t sQ(s)ds \right)^2 dt - \int_0^1 t^2 \left(\int_0^t Q(s)ds \right)^2 dt > 0,$$

and the result follows.   □

Thus, the expansion of $\zeta^\gamma(\alpha)$ takes the form

$$\zeta^{\mathscr{V}}(\alpha) = \bar{x}(\bar{t}) \cdots \exp\left(\left(\frac{1}{4}(\psi_2 - \psi_1)c^2\alpha^5 + o\left(\alpha^5\right)\right) \cdot [g, \mathrm{ad}^3 f(g)]\right),$$

and this proves that $[g, \mathrm{ad}^3 f(g)](\bar{x}(\bar{t})) \in P(\bar{t})$. The rest of the argument is as for the Legendre–Clebsch condition. This concludes the proof of Theorem 4.6.2. □

These computations generalize to controls that are singular of intrinsic order k, and we briefly recall these statements from Sect. 2.8. If Γ is an extremal lift for a controlled trajectory $(\bar{x}, \bar{u})$ defined over the interval $[0, T]$ with corresponding adjoint vector $\lambda : [0, T] \to (\mathbb{R}^n)^*$, then the corresponding switching function is $\Phi(t) = \langle \lambda(t), g(x(t)) \rangle$, and if the control is singular on an interval I, then $\Phi(t)$ and all its derivatives vanish identically on I. In the case analyzed in this section, $[g, f, g]] \equiv 0$ and thus

$$\Phi^{(i)}(t) = \langle \lambda(t), \mathrm{ad}^i f(g)(x(t)) \rangle \equiv 0$$

for $i = 1, 2$, and 3, and

$$\Phi^{(4)}(t) = \langle \lambda(t), [f + ug, \mathrm{ad}^3 f(g)](x(t)) \rangle$$

gives rise to the Kelley condition. This procedure can be continued as long as all the relevant Lie brackets vanish identically. In such a case, because of Lie-algebraic identities, the control can appear for the first time only in an even derivative. (We warn the reader that contrary to what is being claimed in several textbooks and research papers, this need not be the case for singular controls in general.) A singular control u is said to be of *intrinsic order* k on I if u takes on values in the interior of the control set and if all the Lie brackets that would arise at the control variable u when the switching function is differentiated at most $2k - 1$ times vanish identically, while $\langle \lambda(t), \mathrm{ad}_f^{2k}(g)(x(t)) \rangle$ does not vanish on I. In this case, the first $2k - 1$ derivatives of the switching function are given by

$$\Phi^{(i)}(t) = \langle \lambda(t), \mathrm{ad}^i f(g)(x(t)) \rangle, \quad \text{for} \quad i = 1, \ldots, 2k - 1,$$

and vanish identically, and

$$\Phi^{(2k)}(t) = \left\langle \lambda(t), [f + ug, \mathrm{ad}^{2k-1} f(g)](x(t)) \right\rangle$$

can be solved for u. It then is a necessary condition for the trajectory to lie on the boundary of the reachable set or be time-optimal that

$$(-1)^k \left\langle \left\langle \lambda(t), [g, \mathrm{ad}^{2k-1} f(g)](x(t)) \right\rangle \right\rangle \geq 0. \tag{4.47}$$

Since the switching function Φ can also be expressed in the form

$$\Phi(t) = \langle \lambda(t), g(x(t)) \rangle = \frac{\partial H}{\partial u}(\lambda_0, \lambda(t), x_*(t), u_*(t)),$$

in its classical formulation [64] the *generalized Legendre–Clebsch* or *Kelley condition* can be stated as follows:

$$(-1)^k \frac{\partial}{\partial u} \frac{d^{2k}}{dt^{2k}} \frac{\partial H}{\partial u}(\lambda_0, \lambda(t), x_*(t), u_*(t)) \geq 0 \quad \text{for all} \quad t \in I.$$

A proof of this condition can be given by constructing variations analogous to those made above, but it requires a deeper look into the Lie-algebraic relations that enter the asymptotic expansion (4.32) and will not be pursued in this text.

4.6.3 The Goh Condition for Multi-input Systems

The results derived above also apply to multi-input control-affine systems. For simplicity, suppose the controls take values in a compact interval, $u_i \in [\alpha_i, \beta_i]$. If only one of the controls is singular and the other controls are locally constant except for a discrete set of switching times, then the results above directly apply if we simply incorporate all these controls into the reference drift vector field f. Even if more than one control is singular, we can just limit the variations to one control at a time, and the same necessary conditions for optimality are valid. This could be seen by doing the analogous computations within the framework presented or by simply noting that the exponential formalism extends to the case of time-varying vector fields. This gets more technical and is known as the *chronological calculus* of A. Agrachev and R. Gamkrelidze [11, 12], but the computations made above extend to this setting and the results remain true (e.g., see [130]). However, if more than one control is singular at the same time, there is an important additional necessary condition for optimality that arises, the so-called Goh condition. We close this section with a derivation of this condition.

For simplicity of notation, we consider a system with two inputs of the form

$$\Sigma : \dot{x} = f(x) + u_1 g_1(x) + u_2 g_1(x), \qquad |u_1| \leq 1, |u_2| \leq 1,$$

and, as always, assume that $\Gamma = (\bar{x}(\cdot), \bar{u}(\cdot))$ is a controlled reference trajectory defined on $[0, T]$ with initial point $\bar{x}(0) = \bar{p}$ and terminal point $\bar{x}(T) = \bar{q}$.

Theorem 4.6.3. *Let $\bar{t} \in I$ be a Lebesgue point for both reference controls $\bar{u}_1$ and $\bar{u}_2$ and suppose that $|u_i(\bar{t})| < 1$ for $i = 1, 2$. Then the vectors $\pm[g_1, g_2]](\bar{x}(\bar{t}))$ lie in the sets $P(\bar{t})$ and $Q(\bar{t})$.*

Corollary 4.6.3 (Goh condition). *Suppose $\bar{q}$ lies on the boundary of the reachable set, $\bar{q} \in \partial \mathrm{Reach}_\Sigma(\bar{p})$, and let $I \subset [0,T]$ be an open interval on which the reference controls $\bar{u}_1$ and $\bar{u}_2$ take values in the interior of the control set. Then there exists a nontrivial solution $\lambda : [0,T] \to (\mathbb{R}^n)^*$, $t \mapsto \lambda(t)$, to the adjoint equation along Γ such that the conditions of Theorem 4.2.2 are satisfied and in addition, for all $t \in I$ we have that*

$$\langle \lambda(t), [g_1, g_2](\bar{x}(t)) \rangle \equiv 0.$$

Proof of Theorem. We need a version of the asymptotic product expansion in Proposition 4.5.2 with three terms, but only up to second-order brackets. Let $S(t)$ be the solution of the initial value problem

$$\dot{S} = S(uX + vY + wZ), \quad S(0) = \mathrm{Id}.$$

and denote the integrals over the integrable functions u, v, and w by the corresponding capital letters U, V, and W. Making the ansatz $S = S_1 e^{UX}$ and differentiating in t leads to

$$\dot{S}_1 = (\dot{S} - S(uX))e^{-UX} = S_1 e^{UX}(vY + wZ)e^{-UX} = S_1 e^{U \, \mathrm{ad} X}(vY + wZ)$$

$$= S_1(vY + wZ + vU[X,Y] + wU[X,Z] + \cdots).$$

Writing $S_1 = S_2 e^{VY}$ and repeating this step one more time gives

$$\dot{S}_2 = (\dot{S}_1 - S_1(vY))e^{-VY} = S_2 e^{V \, \mathrm{ad} Y}(wZ + vU[X,Y] + wU[X,Z] + \cdots)$$

$$= S_2(wZ + vU[X,Y] + wU[X,Z] + wV[Y,Z] + \cdots).$$

As in the proof of Proposition 4.5.2, this equation can be solved formally by integrating and then, modulo higher-order terms, expanding the sum into a product to obtain

$$S_2(t) = \cdots \exp\left(\int_0^t wV ds \cdot [Y,Z]\right) \exp\left(\int_0^t wU ds \cdot [X,Z]\right) \exp\left(\int_0^t vU ds \cdot [X,Y]\right)$$

$$\exp(W \cdot Z) \exp(V \cdot Y) \exp(U \cdot X).$$

We now use this formula to compute variational vectors

$$\zeta^\gamma(\alpha) = \bar{x}(\bar{t}) \exp(\alpha(S + \rho_1 g_1 + \rho_2 g_2)) \exp(-\alpha S),$$

where we use the same setup as before: $S = f + \bar{u}_1 g_1 + \bar{u}_2 g_2$ denotes the controlled reference vector field and the terms $\rho_i g_i$ describe the variations to be made. Given $\bar{t} \in I$ and $\varepsilon > 0$ small enough, for $0 \le \alpha \le \varepsilon$ replace the reference controls $\bar{u}_1$ and $\bar{u}_2$ over the interval $[\bar{t}, \bar{t} + \alpha]$ with other, suitably chosen admissible controls $\eta_1(\alpha, \cdot)$ and $\eta_2(\alpha, \cdot)$. We again take $t_1(\alpha) = \bar{t}$, $t_2(\alpha) = \bar{t} + \alpha$ and $\Delta_+(\alpha) = \alpha$, and for $i = 1, 2$, the controls of the variation are given by

$$u_{i,\alpha}^{\gamma}(t) = \begin{cases} \bar{u}_i(t) & \text{if } 0 \leq t \leq \bar{t}, \\ \eta_i(\alpha, t - \bar{t}) & \text{if } \bar{t} < t \leq \bar{t} + \alpha, \\ \bar{u}_i(t) & \text{if } \bar{t} + \alpha \leq t \leq T. \end{cases}$$

The curve $\zeta^{\gamma}(\alpha)$ once more is obtained by integrating the dynamics corresponding to the reference controls $\bar{u}_1$ and $\bar{u}_2$ backward from the endpoint $\xi(\alpha, \alpha)$ of the variation over the interval $[\bar{t}, \bar{t} + \alpha]$. As before, let $\rho_i(\alpha, \cdot)$ denote the difference between the control $\eta_i(\alpha, \cdot)$ of the variation and the reference control $\bar{u}_i$ over $[\bar{t}, \bar{t} + \alpha]$, i.e., $\eta_i(\alpha, t) = \bar{u}_i(\bar{t} + t) + \rho_i(\alpha, t)$ for $0 \leq t \leq \alpha$. Writing the dynamics in the form $\dot{S} = S(uf + vg_1 + wg_2)$, the functions u, v, and w defined on $[0, 2\alpha]$ are given by

$$u(t) = \begin{cases} 1 & \text{if } 0 \leq t \leq \alpha, \\ -1 & \text{if } \alpha \leq t \leq 2\alpha, \end{cases}$$

and $v(t) = \bar{v}(t) + h_1(t)$ and $w(t) = \bar{w}(t) + h_2(t)$, where $\bar{v}(t)$ and $\bar{w}(t)$ denote the contributions of the reference controls and $h_i(t)$ denotes the contributions from the variation, i.e.,

$$\bar{v}(t) = \begin{cases} \bar{u}_1(\bar{t} + t) & \text{if } 0 \leq t \leq \alpha, \\ -\bar{u}_1(\bar{t} + 2\alpha - t) & \text{if } \alpha < t \leq 2\alpha, \end{cases} \qquad h_1(t) = \begin{cases} \rho_1(\alpha, t) & \text{if } 0 \leq t \leq \alpha, \\ 0 & \text{if } \alpha < t \leq 2\alpha, \end{cases}$$

and

$$\bar{w}(t) = \begin{cases} \bar{u}_2(\bar{t} + t) & \text{if } 0 \leq t \leq \alpha, \\ -\bar{u}_2(\bar{t} + 2\alpha - t) & \text{if } \alpha < t \leq 2\alpha, \end{cases} \qquad h_2(t) = \begin{cases} \rho_2(\alpha, t) & \text{if } 0 \leq t \leq \alpha, \\ 0 & \text{if } \alpha < t \leq 2\alpha. \end{cases}$$

Modulo the same technical argument about measurability that was used in the proof of the Legendre–Clebsch condition, without loss of generality we may assume that the reference controls $\bar{u}_1$ and $\bar{u}_2$ are continuous from the right at $\bar{t}$. Then there exist positive constants ε, c_1, and c_2 such that for $i = 1$ and 2 we have for all $t \in [\bar{t}, \bar{t} + \varepsilon]$ that

$$-1 + 2c_i \leq \bar{u}_i(t) \leq 1 - 2c_i,$$

and thus measurable variations ρ_i with values in the intervals $[-c_i, c_i]$ will give rise to an admissible control. As before, we define the functions ρ_i in terms of polynomials Q_i defined on the interval $[0, 1]$ as

$$\rho_i(\alpha, s) = c_i Q_i\left(\frac{s}{\alpha}\right) \qquad \text{for } 0 \leq s \leq \alpha,$$

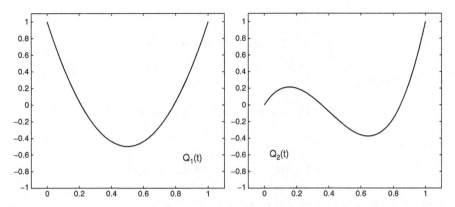

Fig. 4.21 Polynomials Q_1 (*left*) and Q_2 (*right*) for the variation $\mathcal{V}$ to derive the Goh condition

where $\max\{|Q_i(t)| : 0 \le t \le 1\} = 1$. The first polynomial Q_1 is chosen as the unique quadratic polynomial that satisfies

$$\int_0^1 t^i Q_1(t)dt = 0 \qquad \text{for } i = 0, 1, \qquad \text{and} \qquad \int_0^1 t^2 Q_1(t)dt = \frac{1}{30},$$

i.e.,

$$Q_1(t) = 6t^2 - 6t + 1.$$

However, totally symmetric variations in the two controls do not work, and we break the symmetry by choosing Q_2 as the cubic polynomial

$$Q_2(t) = 10t^3 - 12t^2 + 3t.$$

This polynomial satisfies for $i = 0, 1$ that

$$\int_0^1 t^i Q_2(t)dt = 0 \quad \text{and} \quad \int_0^1 Q_2(t) \cdot \left(\int_0^t Q_1(s)ds \right) dt = \frac{1}{140} \ne 0.$$

The graphs of these polynomials are shown in Fig. 4.21.

The telescoping structure of the variations is as before, and the calculations made in the proof of Lemma 4.6.1 immediately imply that we have

$$U(2\alpha) = 0, \quad V(2\alpha) = 0, \quad W(2\alpha) = 0,$$

as well as

$$\int_0^{2\alpha} v(s)U(s)ds = \int_0^{\alpha} s\rho_1(\alpha, s)ds = 0$$

and

$$\int_0^{2\alpha} w(s)U(s)ds = \int_0^\alpha sp_2(\alpha,s)ds = 0.$$

Therefore, the expansion for

$$\zeta^{\mathcal{V}}(\alpha) = \bar{x}(\bar{t})\exp(\alpha(S + p_1 g_1 + p_2 g_2))\exp(-\alpha S)$$

reduces to

$$\zeta^{\mathcal{V}}(\alpha) = \bar{x}(\bar{t})\cdots\exp\left(\int_0^t wVds\cdot[g_1,g_2]\right),$$

which gives the leading term. We have

$$\int_0^{2\alpha} wVds = \int_0^{2\alpha}(\overline{w}+h_2)(\overline{V}+H_1)ds$$

$$= \int_0^{2\alpha}\overline{w}\overline{V}ds + \int_0^\alpha h_2\overline{V}ds + \int_0^\alpha \overline{w}H_1 ds + \int_0^\alpha h_2 H_1 ds.$$

As before, the first integral, which involves only the reference trajectory, will be zero, the second and third integrals will be of higher order, and the last term will give the dominant term: using Eq. (4.40) to represent $\overline{V}$ on the interval $[\alpha, 2\alpha]$, we have that

$$\int_0^{2\alpha}\overline{w}(t)\overline{V}(t)dt = \int_0^\alpha \bar{u}_2(\bar{t}+t)\left(\int_0^t \bar{u}_1(\bar{t}+s)ds\right)dt$$

$$- \int_\alpha^{2\alpha}\bar{u}_2(\bar{t}+2\alpha-t)\left(\int_0^{2\alpha-t}\bar{u}_1(\bar{t}+s)ds\right)dt$$

$$= \int_0^\alpha \bar{u}_2(\bar{t}+t)\left(\int_0^t \bar{u}_1(\bar{t}+s)ds\right)dt$$

$$+ \int_\alpha^0 \bar{u}_2(\bar{t}+r)\left(\int_0^r \bar{u}_1(\bar{t}+s)ds\right)dr = 0.$$

Furthermore,

$$\int_0^\alpha h_2(s)\overline{V}(s)ds = \int_0^\alpha p_2(\alpha,s)(\bar{u}_1(\bar{t})s + o(s))ds$$

$$= \bar{u}_1(\bar{t})\int_0^\alpha sp_2(\alpha,s)ds + o(\alpha^2) = o(\alpha^2),$$

$$\int_0^\alpha \overline{w}(s)H_1(s)ds = \overline{W}(s)H_1(s)\big|_0^\alpha - \int_0^\alpha \overline{W}(s)h_1(s)ds$$

$$= -\int_0^\alpha(\bar{u}_2(\bar{t})s + o(s))p_1(\alpha,s)ds = o(\alpha^2),$$

and

$$\int_0^\alpha h_2(s)H_1(s)ds = \int_0^\alpha c_1 Q_2\left(\frac{s}{\alpha}\right)\left(\int_0^s c_2 Q_1\left(\frac{r}{\alpha}\right)dr\right)ds$$

$$= \alpha^2 c_1 c_2 \int_0^1 Q_2(t)\left(\int_0^t Q_1(\ell)d\ell\right)dt = \frac{\alpha^2 c_1 c_2}{140} > 0.$$

If we were to use $Q_1 = Q_2 = Q$, it would actually hold that

$$\int_0^1 Q(t)\left(\int_0^t Q(\ell)d\ell\right)dt = 0,$$

and thus this term would drop out. For this reason, we needed to pick two polynomials such that one is not orthogonal to the integral of the other. After normalizing the parameter α, we thus get the Lie bracket $[g_1,g_2](\bar{x}(\bar{t}))$ as the variational vector,

$$\zeta^\nu(0) = [g_1,g_2](\bar{x}(\bar{t})).$$

Replacing Q_2 with its negative, the negative of this value can be realized as well, and thus both $[g_1,g_2](\bar{x}(\bar{t}))$ and $-[g_1,g_2](\bar{x}(\bar{t}))$ lie in the sets $P(\bar{t})$. It then follows as for the Legendre–Clebsch condition that these vectors also lie in $Q(\bar{t})$.   □

4.7 Notes

In an article commemorating the 300th anniversary of the brachistochrone problem [245], Hector Sussmann and Jan Willems view this problem as the birth of optimal control. Indeed, the brachistochrone problem is the first problem to deal with dynamic behavior, asking for an optimal selection of a curve. H. Sussmann and J. Willems make a compelling argument for "the superiority of the optimal control method for the brachistochrone problem" [245, p. 44], which, in modern optimal control theory, would be considered a minimum-time problem. Viewing the history of the calculus of variations as a "search for the simplest and most general statement of the necessary conditions for optimality, ... this statement is provided by the maximum principle of optimal control" [245, p. 35].

The proof of the maximum principle presented in this chapter is based on lecture notes of one of the authors for a course on optimal control that Hector Sussmann taught at Rutgers University in 1983. The underlying approach is the original one of the group of mathematicians around Pontryagin [193], especially Boltyansky's [42], and the proof given in Sect. 4.2 reproduces their arguments for the case of the needle variations CUT and PASTE (also, see [44, 56]). These needle variations have their historic origin in the variations made by Weierstrass in the proof of his side condition in the calculus of variations. The differential-geometric language and concepts that are used here clearly bring out the role of the various components in the proof: variations as a means to generate tangent vectors to the reachable set, the embedding of the controlled reference trajectory as a means to move or transport vectors and covectors along the controlled reference trajectory, and the multipliers as separating hyperplanes between suitably constructed approximating cones to the reachable set and lower level sets of the objective. It especially becomes clear in this construction that there is no need to distinguish between first- and higher-order maximum principles [140]. Clearly, any point variation can be reparameterized to give a first-order tangent vector.

Naturally, there also exist other approaches to prove the maximum principle. We only briefly mention those that are based on ideas from optimization theory and variational analysis. In the first approach, which goes back to Dubovitskii and Milyutin [106, 185], constraints are treated separately and simple approximating cones give rise to an Euler–Lagrange-type equation. From it one can easily obtain a weak version of the maximum principle in which the minimization condition on the Hamiltonian is replaced by the corresponding necessary conditions for optimality. However, it is not all evident in these constructions, and indeed quite cumbersome, to obtain the full version of the maximum principle. This approach, however, is convenient to give extensions of the maximum principle for situations in which the equality constraints (including the dynamics) are no longer regular. In this way, one can give additional and stronger necessary conditions for abnormal extremals. This research goes back to the work of Avakov [27–29] and has found numerous extensions, for example in the work of Arutyunov [23,24], Tretyakov and Brezhneva [251], and ourselves [149, 150, 153, 154]. In particular, the paper [152] gives a high-order version of the concept of an approximating cone for these problems. Proofs of the maximum principle that follow variational arguments, but in a nonsmooth setting, are given by Ioffe and Tikhomirov [119], Clarke [76–78], and Vinter [254]. For an in-depth account of variational analysis and generalized differentiation we refer the reader to the monographs by Aubin and Frankowska [26] and B. Mordukhovich [186, 187]. The proof given in this chapter has been extended to a nonsmooth setting in the work of H. Sussmann, e.g., [244, 247, 248], and provides the most general version of the conditions of the maximum principle proven so far.

Exponential representations of solutions to differential equations are classical in differential geometry in the context of canonical coordinates, geodesics, and the exponential mapping [50, 256], but their use in optimal control is not widespread. They have been consistently employed by one of the authors in his work on time-optimal control for nonlinear systems in small dimensions in the late 1980s, and some of this work will be presented in Chap. 7. These techniques also were used in several engineering problems in the early nineties, especially for path planning in robotics. But overall, these techniques have not achieved the prominence they deserve. More general versions of exponential representations for time-varying vector fields are given in the research of A.A. Agrachev and R.V. Gamkrelidze [11, 12] and are known as chronological calculus. We refer the reader to the advanced research monograph by Agrachev and Sachkov [15] for these generalizations and applications to high-order necessary and sufficient conditions for optimality that are based on the formalism of symplectic geometry [13, 14].

Our treatment of high-order necessary conditions for optimality uses the product expansion in Proposition 4.5.2 as it was derived in the thesis of one of the authors in 1986 [209]. The construction in the proof of the Legendre–Clebsch condition is a detailed version of the argument by Kawski and Sussmann in their paper [130]. In this paper, in great generality the leading term of a variation is computed using Lie-algebraic computations based on a shuffle-product. These become of interest with the use of highly oscillatory variations as they are considered in the research on local controllability of M. Kawski [128, 129]. The variations that we give to prove the Kelley and Goh conditions are our own way to prove these necessary conditions for optimality. They present a much more streamlined proof of these classical results than can be found in the literature (e.g., [64]). Especially for the Goh-condition [107], we are not aware of a simpler treatment.

For some historical comments on the development of the conditions of the maximum principle, especially its connections with Hestenes's approach [118], we refer the reader to the review article by Pesch and Plail [194].

Chapter 5
The Method of Characteristics: A Geometric Approach to Sufficient Conditions for a Local Minimum

So far, our focus has been on necessary conditions for optimality. The conditions of the Pontryagin maximum principle, Theorem 2.2.1, collectively form the first-order necessary conditions for optimality of a controlled trajectory (aside from the much stronger minimum condition on the Hamiltonian that generalizes the Weierstrass condition of the calculus of variations). Clearly, as in ordinary calculus, first-order conditions by themselves are no guarantee that even a local extremum is attained. High-order tests, based on second- and increasingly higher-order derivatives, like the Legendre–Clebsch conditions for singular controls, can be used to restrict the class of candidates for optimality further, but in the end, *sufficient conditions* need to be provided that at least guarantee some kind of local optimality. These will be the topic of the next two chapters of our text.

Once more, we consider the optimal control problem [OC] formulated in Sect. 2.2. The topic now is to study the behavior of the optimal value of the solutions as a function of the initial data, the initial time t_0, and the initial value x_0. As before, we denote the trajectory corresponding to an admissible control that satisfies this initial condition by x, and the optimal control problem under consideration therefore is to

[OC] minimize the functional

$$\mathscr{J}(u;t_0,x_0) = \int_{t_0}^{T} L(s,x(s),u(s))ds + \varphi(T,x(T))$$

over all admissible controls $u \in \mathscr{U}$ for which the corresponding trajectory x,

$$\dot{x}(t) = f(t,x(t),u(t)), \quad x(t_0) = x_0,$$

satisfies the terminal constraint

$$(T,x(T)) \in N = \{(t,x) \in \mathbb{R} \times \mathbb{R}^n : \Psi(t,x) = 0\}.$$

H. Schättler and U. Ledzewicz, *Geometric Optimal Control: Theory, Methods and Examples*, Interdisciplinary Applied Mathematics 38, DOI 10.1007/978-1-4614-3834-2_5, © Springer Science+Business Media, LLC 2012

We have explicitly included the dependence of the value of the objective, $\mathscr{J}$, on the initial condition (t_0, x_0) in this notation. Throughout this chapter, we make the following regularity assumptions on the data: the dynamics f and the Lagrangian L are continuous in all variables, and for some positive integer ℓ or $\ell = \infty$, these functions are ℓ-times continuously differentiable in x and u. If these functions are real-analytic, we say they are of class C^ω. We use $C^{0,\ell,\ell}$ and analogous notations to denote the smoothness properties of various functions and mappings. The penalty term φ and the functions Ψ defining the constraint are assumed to be ℓ-times continuously differentiable in both variables, t and x, and we simply write $\varphi \in C^\ell$ if the smoothness assumptions are the same for all variables. Furthermore, we always assume that the Jacobian matrix of the equations Ψ defining the terminal constraint, $D\Psi$, is of full rank on $N = \{(t,x) \in \mathbb{R} \times \mathbb{R}^n : \Psi(t,x) = 0\}$. Thus N is an embedded C^ℓ-submanifold of $\mathbb{R} \times \mathbb{R}^n$.

Time-dependent and time-independent models. In this formulation, we again have included an explicit dependence of the data on the time t. In most practical problems, these functions actually are time-invariant or autonomous. However, if the problem is one of regulating a system over a fixed finite time interval $[0, T]$ (e.g., the linear-quadratic regulator or models for cancer treatments over a prescribed therapy horizon), even if all other data are time-invariant, the optimal control clearly will depend not only on the state x, but also on the time t, or, more intuitively, the time that is left until the terminal time T. We therefore call the optimal control problem [OC] *time-dependent* if any one of the functions f, L, φ, Ψ depends on t. In particular, in any optimal control problem over a fixed finite time-interval $[0, T]$, one of the functions defining N is given by $\psi(t,x) = t - T$, and thus the problem automatically becomes time-dependent regardless of whether the dynamics and objective depend on t. We call the problem [OC] *time-independent* if none of the data depend on t. For example, the time-optimal control problem to a point as considered in Chap. 2 is time-independent. It is sometimes convenient to normalize the final times in these problems, but this does not constitute a fixed terminal time. We use the terminology time-varying and time-invariant, respectively, to indicate whether the dynamics f and the Lagrangian L depend on t or not. The theory presented in this chapter will be developed mostly for the time-dependent formulation, but we also include time-independent versions. At times, this is as simple as setting $x' = (t,x)$. In other situations, this formal substitution to make the problem time-independent is somewhat inadequate. The reason is that the trivial dynamics $\dot{t} = 1$ introduces a distinguished monotone variable that a general time-invariant system does not have. Once flows of solutions to differential equations are considered, however, such a variable will be needed, but there is no need to duplicate it if it already exists in the formulation. And if one formally does so, this leads to rather awkward formulations. For this reason, and at the expense of some minor duplications, we consider both formulations in this chapter, but with emphasis on the time-dependent problem.

The main idea in studying sufficient conditions for optimality is to consider the value V of the optimal control problem as a function of the initial conditions. In this context, it is customary and more convenient, although somewhat ambiguous, to denote the initial time by t and the initial value by x so that the value function reads

$$V(t,x) = \inf_{u \in \mathscr{U}} \mathscr{J}(u;t,x),$$

where the infimum is taken over all admissible controls $u \in \mathscr{U}$ whose corresponding trajectories start at the point x at time t and satisfy all other requirements of the optimal control problem. It is not difficult to see—and this is known as Bellman's *dynamic programming principle*—that if the function V is differentiable at (t,x) with gradient $(\frac{\partial V}{\partial t}(t,x), \frac{\partial V}{\partial x}(t,x))$, then for all $u \in U$ the following inequality is satisfied:

$$\frac{\partial V}{\partial t}(t,x) + \frac{\partial V}{\partial x}(t,x)f(t,x,u) + L(t,x,u) \geq 0.$$

Furthermore, if u_* is an optimal control for the initial condition (t,x) that is continuous at the initial time t, then equality holds for $u = u_*(t)$. Thus, in this case the value V satisfies the following first-order linear partial differential equation that is coupled with the optimal control u_* through a minimum condition,

$$\frac{\partial V}{\partial t}(t,x) + \min_{u \in U}\left\{ \frac{\partial V}{\partial x}(t,x)f(t,x,u) + L(t,x,u) \right\} \equiv 0,$$

the so-called **Hamilton–Jacobi–Bellman** equation [32]. The importance of this equation, however, is not as a necessary condition for optimality, but it lies in its significance as a sufficient condition for optimality. Indeed, if the pair (V,u_*) is a solution to this equation (in the sense that V is a continuously differentiable function and $u_* = u_*(t,x)$ is an admissible control for which the minimum is realized), then u_* is an optimal control. This also, as will be shown below, is easily seen. However, for various reasons, this result does not provide a satisfactory sufficient condition for optimality. While V turns out to be C^1 for some problems, the common scenario is that the value function V has singularities and that it will not be differentiable on lower-dimensional submanifolds of the state space. This fact indeed severely impedes a straightforward application of the dynamic programming principle, and essentially, the study of sufficient conditions for optimality becomes equivalent to analyzing the singularities of solutions to its associated Hamilton–Jacobi–Bellman equation, a rather difficult problem.

In this chapter, we adapt the **method of characteristics**, the classical solution procedure for first-order partial differential equations [124], to construct the value function associated with a parameterized family of extremals. The so-called *characteristic equations* for the Hamilton–Jacobi–Bellman equation are given by the dynamics of the system and the adjoint equations of the maximum principle [33]; the minimum condition in the maximum principle becomes the minimum condition in the Hamilton–Jacobi–Bellman equation. This construction clearly brings out the relationships between the necessary conditions of the maximum principle and the sufficiency of the dynamic programming principle, and it provides the generalization of the concept of a field of extremals from the calculus of variations to optimal control theory.

After outlining the main ideas of dynamic programming in Sect. 5.1, Sect. 5.2 presents the construction of the value function associated with a parameterized family of extremals. While our constructions may be global in some situations—and we shall use them to give a straightforward and short proof of the optimality of the synthesis for the Fuller problem constructed in Sect. 2.11—with the conditions that need to be imposed, these constructions are inherently local in nature and will be used to compute "patches" of the value function that then need to be "glued together" to obtain a global solution. However, the global aspects of the construction, known as a *regular synthesis* of optimal controlled trajectories [41, 196], will be postponed until Chap. 6. Here, we develop the local aspects and use the procedure to give sufficient conditions for a local minimum of a controlled trajectory (Sect. 5.3). These results form the core of what is known as *perturbation feedback control* in the engineering literature [64], and here a geometric approach will be taken to rigorously derive the underlying formulas. Geometric arguments are also crucial in Sects. 5.4 and 5.5, where we investigate the behavior of a parameterized flow of extremals near singularities, that is, when the corresponding trajectories overlap. This brings us back to *conjugate points*, and one main tool in their investigation will be the generalization of the theory of *envelopes* from the calculus of variations to the optimal control problem. It will be seen that very different local syntheses arise depending on the type of the conjugate point: while the local geometry of extremal trajectories near a fold singularity is exactly the same as for the family of catenaries in the problem of minimum surfaces of revolution in the calculus of variations (Sect. 1.3), a cusp singularity generates a cut-locus of extremals akin to the intersection of the catenaries with the Goldschmidt extremals (Sect. 1.7). Throughout this chapter, we assume that the control u is continuous, a natural condition if only one "patch" of the field of extremals is considered. In Chap. 6, the results will then be extended to situations in which various patches will be glued together.

5.1   The Value Function and the Hamilton–Jacobi–Bellman Equation

We present the main ideas of Bellman's principle of optimality and illustrate it with some classical examples. In this section, we denote the initial conditions by (t,x), and in order to keep the notation unambiguous, we use the Greek letters η and ξ for admissible controls and their corresponding trajectories. Thus, if $\eta : [t,T] \to U$ is an admissible control, then ξ is the solution to the dynamics $\dot{\xi}(s) = f(s, \xi(s), \eta(s))$ with initial condition $\xi(t) = x$. In subsequent sections, we shall return to the more intuitive notation for the controls and trajectories as u and x. We first consider the time-dependent formulation.

Definition 5.1.1 (Value function).   Given an initial condition (t,x), denote by $\mathscr{U}_{(t,x)}$ the set of all admissible controls η defined over some interval $[t,T]$, $\eta : [t,T] \to U$, for which the corresponding trajectory ξ exists on the full interval $[t,T]$ and steers the system into the terminal manifold N. The value $V = V(t,x)$ of the optimal control problem [OC] is defined as

$$V(t,x) = \inf_{u \in \mathscr{U}_{(t,x)}} \mathscr{J}(u;t,x), \tag{5.1}$$

i.e., as the infimum of the values of the cost functional taken over all controlled trajectories with initial condition (t,x). If the set $\mathscr{U}_{(t,x)}$ is empty, we define $V(t,x) = +\infty$. The function $V : G \to \mathbb{R}$, $(t,x) \mapsto V(t,x)$, defined on the set G of all possible initial conditions, is called the *value function* of the optimal control problem [OC].

Proposition 5.1.1 (Dynamic programming principle). *If the value V is differentiable at (t,x), then we have for every control value u in the control set, $u \in U$, that*

$$\frac{\partial V}{\partial t}(t,x) + \frac{\partial V}{\partial x}(t,x)f(t,x,u) + L(t,x,u) \geq 0.$$

If $\eta_* : [t,T] \to U$ is an optimal control that is continuous at the initial time t with value $u_* = \lim_{s \searrow t} \eta(s)$, then

$$\frac{\partial V}{\partial t}(t,x) + \frac{\partial V}{\partial x}(t,x)f(t,x,u_*) + L(t,x,u_*) = 0.$$

Overall, we therefore have that

$$\frac{\partial V}{\partial t}(t,x) + \min_{u \in U}\left\{\frac{\partial V}{\partial x}(t,x)f(t,x,u) + L(t,x,u)\right\} \equiv 0. \tag{5.2}$$

Proof. Let $u \in U$ be an admissible control value and for $h > 0$, let $\eta_u : [t,t+h] \to U$ denote the admissible control that is constant on the interval $[t,t+h]$ given by $\eta_u(s) \equiv u$. For h sufficiently small, the corresponding trajectory ξ_u exists on the interval $[t,t+h]$. If $\eta : [t+h,T] \to U$ is any control that is admissible for the initial condition $(t+h, \xi_u(t+h))$, then the concatenation $\eta_u * \eta : [t,T] \to U$, defined by

$$(\eta_u * \eta)(s) = \begin{cases} u & \text{for } t \leq s \leq t+h, \\ \eta(s) & \text{for } t+h < s \leq T, \end{cases}$$

is admissible for the initial condition (t,x), $\eta_u * \eta \in \mathscr{U}_{(t,x)}$, and therefore, by definition of the value function, we have that

$$V(t,x) \leq \int_t^{t+h} L(s,\xi_u(s),u)ds + V(t+h,\xi_u(t+h)). \tag{5.3}$$

Hence

$$\frac{V(t+h,\xi_u(t+h)) - V(t,x)}{h} + \frac{1}{h}\int_t^{t+h} L(s,\xi_u(s),u)ds \geq 0.$$

Since V is differentiable at (t,x), taking the limit $h \to 0$, $h > 0$, it follows that

$$\frac{\partial V}{\partial t}(t,x) + \frac{\partial V}{\partial x}(t,x)f(t,x,u) + L(t,x,u) \geq 0.$$

Furthermore, if η_* is an optimal control, then equality holds in Eq. (5.3), and thus in the limit $h \to 0+$, equality holds with $u = u_*$.   □

Equation (5.2) is called the *Hamilton–Jacobi–Bellman (HJB) equation*, and its solutions are essential for any kind of theory of sufficient conditions for optimality. Note that the function to be minimized in the HJB equation is the same expression that is obtained if in the definition of the Hamiltonian H for the control system, $H = \lambda_0 L(t,x,u) + \lambda f(t,x,u)$, we normalize $\lambda_0 = 1$ and replace the multiplier λ by $\frac{\partial V}{\partial x}(t,x)$. In fact, as we shall see in this chapter, *the identification of the adjoint variable λ with this gradient is the key relation that connects necessary to sufficient conditions for optimality.*

Definition 5.1.2 (Admissible feedback controls). Let G be a region in (t,x)-space. We call a feedback control $u : G \to U$, $(t,x) \mapsto u(t,x)$, admissible (on G) for the control problem [OC] if for every initial condition $(t,x) \in G$, the initial value problem

$$\dot{\xi} = f(s,\xi,u(s,\xi)), \quad \xi(t) = x, \tag{5.4}$$

has a unique solution $\xi : [t,T] \to \mathbb{R}^n$ (forward in time) for which the corresponding open-loop control $\eta : [t,T] \to U$, $\eta(s) = u(s,\xi(s))$, satisfies the regularity properties postulated in the definition of the class $\mathcal{U}$ of admissible controls.

If the feedback control u is continuous, standard results on ODEs guarantee the existence and uniqueness of solutions to the initial value problem (5.4); if, however, this feedback control is discontinuous—and as we already have seen in Chap. 2, this is the more typical scenario in optimal control problems—then these standard results are not enough to clarify the existence of solutions. Rather than going into the intricacies as to when solutions to ordinary differential equations with discontinuous right-hand sides exist, in the formulation adopted here we simply require the existence and uniqueness of solutions to Eq. (5.4), while at the same time, demanding that the open-loop control that gives rise to this controlled trajectory be admissible. This indeed typically holds for *optimal* feedback controls and will be satisfied for all the examples considered in this text.

Definition 5.1.3 (Classical solution to the Hamilton–Jacobi–Bellman equation). Let G be a region in (t,x)-space that contains the terminal manifold N in its boundary. We call the pair (V,u_*) a classical solution to the Hamilton–Jacobi–Bellman equation on G if (i) $V : G \to \mathbb{R}$ is continuously differentiable on G and extends continuously onto N, (ii) u_* is an admissible feedback control, (iii) we have

$$\frac{\partial V}{\partial t}(t,x) + \min_{u \in U}\left\{\frac{\partial V}{\partial x}(t,x)f(t,x,u) + L(t,x,u)\right\} \equiv 0,$$

with equality holding for the feedback control $u_* = u_*(t,x)$, and (iv) the boundary condition $V(t,x) = \varphi(t,x)$ holds for all $(t,x) \in N$.

Theorem 5.1.1. *If (V,u_*) is a classical solution to the Hamilton–Jacobi–Bellman equation on G, then the control u_* is optimal with respect to any other admissible control η for which the graph of the corresponding controlled trajectory ξ lies in G and V is the corresponding minimal value when taken over this class of controls. In particular, if a classical solution (V,u_*) exists on the full space, then u_* is an optimal control and V is the value function for the problem.*

Proof. Let $\eta : [t,T] \to U$ be any admissible control for initial condition $(t,x) \in G$, $\eta \in \mathcal{U}_{(t,x)}$, with corresponding trajectory ξ. By assumption, the graph of ξ lies in G for $s \in [t,T)$ and the function V is differentiable along the graph of ξ. Since ξ is an absolutely continuous curve, we have a.e. on $[t,T)$ that

$$\frac{d}{ds}V(s,\xi(s)) = \frac{\partial V}{\partial t}(s,\xi(s)) + \frac{\partial V}{\partial x}(s,\xi(s))f(s,\xi(s),\eta(s)).$$

It thus follows from the Hamilton–Jacobi–Bellman equation that

$$\frac{d}{ds}V(s,\xi(s)) \geq -L(s,\xi(s),\eta(s)).$$

Integrating this inequality from t to some time $T - \varepsilon$ and then taking the limit as $\varepsilon \to 0$ therefore yields

$$V(T,\xi(T)) - V(t,x) \geq -\int_t^T L(s,\xi(s),\eta(s))ds.$$

The boundary condition for V states that $V(T,\xi(T)) = \varphi(T,\xi(T))$, and thus it follows that

$$V(t,x) \leq \int_t^T L(s,\xi(s),\eta(s))ds + \varphi(T,\xi(T)) = \mathscr{J}(\eta;t,x).$$

Furthermore, for the control η_*, $\eta_*(s) = u_*(s,\xi_*(s))$, we have equality and thus $V(t,x) = \mathscr{J}(\eta_*;t,x)$. Hence V is the value function, $V(t,x) = \min_{\eta \in \mathcal{U}_{(t,x)}} \mathscr{J}(\eta;t,x)$. This proves the theorem.   □

More generally, this argument shows that any differentiable function W that satisfies

$$\frac{\partial W}{\partial t}(t,x) + \frac{\partial W}{\partial x}(t,x)f(t,x,u) + L(t,x,u) \geq 0 \qquad \text{for all} \quad u \in U$$

with boundary condition

$$W(T,x) \leq \varphi(T,x) \qquad \text{for} \quad (T,x) \in N$$

provides a lower bound for the value function. Such a function is called a lower value.

The solution of optimal control problems is thus closely related to finding solutions to the Hamilton–Jacobi–Bellman equation (5.2). It is the coupling of two aspects, first-order PDE and optimization problem, that makes the solution of HJB equations a challenge. One possible approach is to first solve the minimization problem for u and "define" the control as a "function" of the state and the gradient $\frac{\partial V}{\partial x}$, $u = u(t, x, \frac{\partial V}{\partial x})$, and then substitute the resulting relation into the partial differential equation. Clearly, there exist various obstacles to this approach: the point where the minimum is attained need not be unique, and even if it is, (e.g., if the Hamiltonian of the associated control problem is strictly convex in u), then the resulting PDE typically becomes highly nonlinear and difficult to solve. In some special cases, often related to the existence of symmetries of the underlying optimal control problem, an explicit solution to the HJB equation becomes possible, and below we shall give two examples in which this procedure does work, one straightforward (the linear-quadratic regulator), the other somewhat involved (the Fuller problem). For a general problem, however, it is often preferable to retain the structure of the HJB equation as a first-order linear PDE and reduce its solution to a system of ordinary differential equations in the so-called *method of characteristics*. For the optimal control problem, these characteristic equations are the dynamics in the state space and the adjoint equations on the multipliers. In the next section, we shall show how the method of characteristics can be used in a methodical way in connection with an analysis of extremals to construct the value function for an optimal control problem and thus to solve HJB equations. But we start with two classical examples in which underlying symmetries of the problem allow an explicit solution.

Example 5.1.1 (Linear-quadratic regulator).   Recall that the linear-quadratic regulator problem [LQ] (see Sect. 2.1) is the optimal control problem to minimize a quadratic objective of the form

$$J(\eta; 0, x) = \frac{1}{2} \int_0^T \left[\xi^T(t) Q(t) \xi(t) + \eta^T(t) R(t) \eta(t) \right] dt + \frac{1}{2} \xi^T(T) S_T \xi(T)$$

over all locally bounded Lebesgue measurable functions $\eta : [0, T] \to \mathbb{R}^m$ subject to the linear dynamics

$$\dot{\xi}(t) = A(t)\xi(t) + B(t)\eta(t), \qquad \xi(0) = x.$$

The entries of the matrices $A(\cdot)$, $B(\cdot)$, $R(\cdot)$, and $Q(\cdot)$ all are continuous functions on the interval $[0, T]$, and the matrices $R(\cdot)$ and $Q(\cdot)$ are symmetric; $R(\cdot)$ is positive definite and $Q(\cdot)$ positive semidefinite; S_T is a constant positive definite matrix. Since the time horizon is specified, even if dynamics and Lagrangian are time-invariant, this is a time-dependent problem, and the Hamilton–Jacobi–Bellman equation needs to be considered in the form Eq. (5.2),

$$\frac{\partial V}{\partial t}(t,x) + \min_{u \in \mathbb{R}^m} \left\{ \frac{\partial V}{\partial x}(t,x)(A(t)x + B(t)u) + \frac{1}{2} \left(x^T Q(t)x + u^T R(t)u \right) \right\} \equiv 0,$$

with boundary condition

$$V(T,x) = \frac{1}{2} x^T S_T x.$$

Since the matrix $R(t)$ is positive definite, the function to be minimized is strictly convex in u. Hence the minimization problem has a unique solution given by the stationary point, i.e.,

$$\frac{\partial V}{\partial x}(t,x)B(t) + u^T R(t) = 0,$$

or equivalently,

$$u = -R^{-1}(t)B^T(t) \left(\frac{\partial V}{\partial x}(t,x) \right)^T.$$

Substituting this expression back into the HJB equation, we obtain a now nonlinear first-order partial differential equation,

$$\frac{\partial V}{\partial t}(t,x) + \frac{\partial V}{\partial x}(t,x)A(t)x + \frac{1}{2}x^T Q(t)x$$
$$- \frac{1}{2}\frac{\partial V}{\partial x}(t,x)B(t)R^{-1}(t)B^T(t) \left(\frac{\partial V}{\partial x}(t,x) \right)^T \equiv 0. \qquad (5.5)$$

While there is little hope of explicitly solving nonlinear partial differential equations in general, this equation is separable. Since $\frac{\partial V}{\partial x}(t,x)A(t)x \in \mathbb{R}$, we can rewrite the linear term as

$$\frac{\partial V}{\partial x}(t,x)A(t) = \frac{1}{2} \left(\frac{\partial V}{\partial x}(t,x)A(t)x + x^T A^T(t) \left(\frac{\partial V}{\partial x}(t,x) \right)^T \right),$$

and thus, because of the quadratic nature of the underlying optimal control problem, Eq. (5.5) possesses a symmetry that can be exploited by considering a quadratic function V of the form

$$V(t,x) = \frac{1}{2}x^T S(t)x$$

with a symmetric matrix S. This ansatz separates the variables t and x, and Eq. (5.5) reduces to

$$\frac{1}{2}x^T \left[\dot{S}(t) + S(t)A(t) + A^T(t)S(t) - S(t)B(t)R^{-1}(t)B^T(t)S(t) + Q(t) \right]x \equiv 0,$$

and the boundary condition can be satisfied by choosing $S(T) = S_T$. Thus $V(t,x) = \frac{1}{2}x^T S(t)x$ is a classical solution to the Hamilton–Jacobi–Bellman equation if and only if the matrix $S(t)$ is a solution to the Riccati terminal value problem

$$\dot{S} + SA(t) + A^T(t)S - SB(t)R^{-1}(t)B^T(t)S + Q(t) = 0, \quad S(T) = S_T,$$

and the minimizing control is given by the linear feedback law

$$u(t,x) = -R^{-1}(t)B^T(t)S(t)x.$$

We already know that the solution to this Riccati equation exists over the full interval $[0,T]$ (Sect. 2.1), and thus this defines an admissible feedback control, and the pair (V,u_*) is a classical solution of the HJB equation for the linear-quadratic regulator. Note that even if the dynamics and Lagrangian are time-invariant, the optimal control depends on the time t via the solution S of the Riccati equation and is a time-varying linear feedback. This clearly attests to the time-dependent nature of the problem formulation.

Other problems, such as, for example, the time-optimal control problems to a point considered in Sects. 2.6 and 2.9, are time-independent. Then the Hamilton–Jacobi–Bellman equation is naturally defined in the state space, i.e., as a function of x alone. The connections between the time-dependent and time-independent formulations are easily made by means of the identification $x' = (t,x)$ when the trivial equation $\dot{t} = 1$ is added to the dynamics. Writing

$$f'(x',u) = \begin{pmatrix} 1 \\ f(x',u) \end{pmatrix}, \tag{5.6}$$

Eq. (5.2) reads

$$\min_{u\in U}\left\{\frac{\partial V}{\partial x'}(x')f(x',u) + L(x',u)\right\} \equiv 0,$$

and deleting the primes, this defines the time-independent or *autonomous version of the Hamilton–Jacobi–Bellman equation*,

$$\min_{u\in U}\left\{\frac{\partial V}{\partial x}(x)f(x,u) + L(x,u)\right\} \equiv 0. \tag{5.7}$$

Example 5.1.2 (Fuller problem). We once more consider the problem (see Sect. 2.11) of finding a Lebesgue measurable function with values in the interval $[-1,1]$ that steers a point $x = (x_1,x_2) \in \mathbb{R}^2$ into the origin under the dynamics $\dot{x}_1 = x_2$, $\dot{x}_2 = u$, and minimizes the objective $J(u) = \frac{1}{2}\int_0^T x_1^2(t)dt$. This problem is time-invariant, and the corresponding HJB equation for $V = V(x)$ reads

$$\min_{|u|\leq 1}\left\{x_2\frac{\partial V}{\partial x_1} + u\frac{\partial V}{\partial x_2} + \frac{1}{2}x_1^2\right\} \equiv 0$$

with boundary condition $V(0,0) = 0$. The minimization condition gives

$$u(x) = -\text{sgn}\left(\frac{\partial V}{\partial x_2}(x)\right),$$

and thus the HJB equation reduces to the following first-order nonlinear PDE:

$$x_2 \frac{\partial V}{\partial x_1}(x_1, x_2) - \left|\frac{\partial V}{\partial x_2}(x_1, x_2)\right| + \frac{1}{2}x_1^2 \equiv 0, \qquad V(0,0) = 0. \qquad (5.8)$$

We shall show here (and once more by different and more direct means in the next section) that the cost function associated with the extremal synthesis constructed in Sect. 2.11 indeed is a classical solution $V \in C(\mathbb{R}^2)$ that is continuously differentiable away from the origin. By Theorem 5.1.1, this proves the global optimality of this synthesis.

Recall (Theorem 2.11.1) that the feedback control u_* is given by

$$u_*(x) = \begin{cases} +1 & \text{for } x \in G_+ \cup \Gamma_+, \\ -1 & \text{for } x \in G_- \cup \Gamma_-, \end{cases}$$

where with $\zeta = \sqrt{\frac{\sqrt{33}-1}{24}} = 0.4446236\ldots$ the unique positive root of the equation $z^4 + \frac{1}{12}z^2 - \frac{1}{18} = 0$,

$$\Gamma_+ = \{(x_1, x_2) \in \mathbb{R}^2 : x_1 = \zeta x_2^2, \ x_2 < 0\},$$
$$\Gamma_- = \{(x_1, x_2) \in \mathbb{R}^2 : x_1 = -\zeta x_2^2, \ x_2 > 0\},$$

and

$$G_+ = \{(x_1, x_2) \in \mathbb{R}^2 : x_1 < -\text{sgn}(x_2)\zeta x_2^2\},$$
$$G_- = \{(x_1, x_2) \in \mathbb{R}^2 : x_1 > -\text{sgn}(x_2)\zeta x_2^2\}.$$

Proposition 5.1.2 [259]. *The Hamilton–Jacobi–Bellman equation for the Fuller problem has a classical solution (V, u_*) with $V \in C(\mathbb{R}^2) \cap C^1(\mathbb{R}^2 \setminus \{0,0\})$ given by*

$$V(x) = \begin{cases} V_+(x) & \text{for } x \in G_+ \cup \Gamma_+, \\ V_-(x) & \text{for } x \in G_- \cup \Gamma_-, \end{cases}$$

where

$$V_+(x) = -\frac{1}{15}x_2^5 + \frac{1}{3}x_1 x_2^3 - \frac{1}{2}x_1^2 x_2 + A\left(\frac{1}{2}x_2^2 - x_1\right)^{5/2}$$

and

$$V_-(x) = V_+(-x) = \frac{1}{15}x_2^5 + \frac{1}{3}x_1 x_2^3 + \frac{1}{2}x_1^2 x_2 + A\left(\frac{1}{2}x_2^2 + x_1\right)^{5/2}$$

with

$$A = \frac{2}{5} \frac{\frac{1}{3} + \zeta + \frac{1}{2}\zeta^2}{\left(\frac{1}{2} + \zeta\right)^{3/2}} = 0.382\ldots .$$

Proof. Define the function $V = V(x)$ as the value associated with the extremal synthesis $\mathfrak{S}$ for the Fuller problem constructed in Theorem 2.11.1, i.e., $V(x)$ is the value of the objective for the extremal controlled trajectory that starts at the point $x = (x_1, x_2)$ at time 0. We shall show that this function is given by the formulas above and that it is a continuously differentiable solution to the Hamilton–Jacobi–Bellman equation for the Fuller problem away from the origin.

We follow the argument of Wonham [259] that utilizes the symmetries of the extremal synthesis $\mathfrak{S}$. We have seen in Proposition 2.11.1 that for any $\alpha > 0$, extremal controlled trajectories of the Fuller problem are invariant under the scaling symmetry $\mathscr{G}_\alpha$ defined by

$$t^\alpha = \frac{t}{\alpha}, \qquad \eta^\alpha(t) = \eta(t^\alpha), \qquad \xi_1^\alpha(t) = \alpha^2 \xi_1(t^\alpha), \quad \text{and} \quad \xi_2^\alpha(t) = \alpha \xi_2(t^\alpha).$$

This scaling symmetry extends to the value V: if η is an extremal control for the initial condition (x_1, x_2) defined over the interval $[0, T]$, then η^α is an extremal control for initial condition $(\alpha^2 x_1, \alpha x_2)$ defined over the interval $[0, \alpha T]$ and

$$V\left(\alpha^2 x_1, \alpha x_2\right) = \frac{1}{2} \int_0^{\alpha T} \xi_1^\alpha(t)^2 \, dt = \frac{1}{2} \int_0^{\alpha T} \left[\alpha^2 \xi_1\left(\frac{t}{\alpha}\right)\right]^2 dt$$

$$= \alpha^5 \frac{1}{2} \int_0^T \xi_1(s)^2 \, ds = \alpha^5 V(x_1, x_2).$$

Hence, the value V of the extremal synthesis satisfies

$$V\left(\alpha^2 x_1, \alpha x_2\right) = \alpha^5 V(x_1, x_2), \qquad \alpha > 0.$$

Functions that have this property are easily computed [191]: differentiating with respect to α and evaluating at $\alpha = 1$ yields the first-order linear PDE

$$2x_1 \frac{\partial V}{\partial x_1}(x_1, x_2) + x_2 \frac{\partial V}{\partial x_2}(x_1, x_2) = 5V(x_1, x_2). \tag{5.9}$$

Clearly, the function $V(x_1, x_2) = x_2^5$ is a particular solution, and the expression $Y = x_1/x_2^2$ is a functional invariant to the homogeneous equation

$$2x_1 \frac{\partial Y}{\partial x_1}(x_1, x_2) + x_2 \frac{\partial Y}{\partial x_2}(x_1, x_2) = 0.$$

The general solution to Eq. (5.9) is therefore of the form $V(x_1, x_2) = x_2^5 \Psi\left(\frac{x_1}{x_2^2}\right)$, where Ψ is an arbitrary continuously differentiable function $\Psi : \mathbb{R} \to \mathbb{R}$ [124].

Substituting this general form for the function V into Eq. (5.8) gives

$$x_2^4 \Psi'' \left(\frac{x_1}{x_2^2}\right) - \left|5x_2^4 \Psi \left(\frac{x_1}{x_2^2}\right) + x_2^5 \Psi' \left(\frac{x_1}{x_2^2}\right)\left(-2\frac{x_1}{x_2^3}\right)\right| + \frac{1}{2}x_1^2 \equiv 0,$$

or equivalently, after dividing by x_2^4, in the variable $y = x_1/x_2^2$ we obtain

$$\Psi'(y) - \left|5\Psi(y) - 2y\Psi'(y)\right| + \frac{1}{2}y^2 \equiv 0.$$

Taking out the absolute values with positive and negative signs gives two first-order linear differential equations,

$$(1+2y)\Psi'_+(y) - 5\Psi(y) + \frac{1}{2}y^2 \equiv 0$$

and

$$(1-2y)\Psi'_-(y) + 5\Psi(y) + \frac{1}{2}y^2 \equiv 0,$$

which can be solved explicitly in the form

$$\Psi_+(y) = -\frac{1}{15} + \frac{1}{3}y - \frac{1}{2}y^2 + A_+ \left(\frac{1}{2}-y\right)^{5/2}$$

and

$$\Psi_-(y) = \frac{1}{15} + \frac{1}{3}y + \frac{1}{2}y^2 + A_- \left(\frac{1}{2}+y\right)^{5/2},$$

where A_+ and A_- are constants of integration. In the original variables, this gives the functions

$$V_+(x) = -\frac{1}{15}x_2^5 + \frac{1}{3}x_1 x_2^3 - \frac{1}{2}x_1^2 x_2 + A_+ \left(\frac{1}{2}x_2^2 - x_1\right)^{5/2}$$

and

$$V_-(x) = \frac{1}{15}x_2^5 + \frac{1}{3}x_1 x_2^3 + \frac{1}{2}x_1^2 x_2 + A_- \left(\frac{1}{2}x_2^2 + x_1\right)^{5/2}.$$

The function V_+ is well-defined and continuously differentiable for $x_1 < \frac{1}{2}x_2^2$, while V_- is well-defined and continuously differentiable for $x_1 > -\frac{1}{2}x_2^2$. Thus V_+ and V_- are C^1 in open neighborhoods of the domains $G_+ \cup \Gamma_+$ and $G_- \cup \Gamma_-$ on which the feedback control u_* is defined by $u_* = +1$ and $u_* = -1$, respectively. Setting $A_+ = A_- = A$, the combined function V possesses the reflection symmetry

$$V_-(x) = V_+(-x) \qquad \text{and} \qquad V_+(x) = V_-(-x),$$

which is also present in the extremal synthesis. Making this choice, it therefore suffices to show that V is continuously differentiable on one of the switching curves $\Gamma_\pm$. Without loss of generality, we consider $\Gamma_+ = \{(x_1, x_2) \in \mathbb{R}^2 : x_1 = \zeta x_2^2, \ x_2 < 0\}$.

The optimal control law satisfies the relation

$$u_*(x) = -\operatorname{sgn}\left(\frac{\partial V}{\partial x_2}(x)\right),$$

and therefore this partial derivative needs to vanish on the switching curve Γ_+. The choice of A ensures this condition. For we have that

$$\left(\frac{\partial V_+}{\partial x_2}(x)\right)_{|\Gamma_+} = \left(-\frac{1}{3} + \zeta - \frac{1}{2}\zeta^2 - \frac{5}{2}A\left(\frac{1}{2} - \zeta\right)^{\frac{3}{2}}\right)x_2^4$$

and

$$\left(\frac{\partial V_-}{\partial x_2}(x)\right)_{|\Gamma_+} = \left(\frac{1}{3} + \zeta + \frac{1}{2}\zeta^2 - \frac{5}{2}A\left(\frac{1}{2} + \zeta\right)^{\frac{3}{2}}\right)x_2^4.$$

(Note that $\sqrt{x_2^2} = -x_2$, since x_2 is negative on Γ_+.) Thus, by definition of A,

$$\left(\frac{\partial V_-}{\partial x_2}(x)\right)_{|\Gamma_+} = 0.$$

Furthermore, recall from Sect. 2.11, Eq. (2.67), that ζ is defined as the unique positive solution of the following relation:

$$\frac{\left(\frac{1}{3} + \zeta + \frac{1}{2}\zeta^2\right)^2}{\left(\frac{1}{2} + \zeta\right)^3} = \frac{\left(\frac{1}{3} - \zeta + \frac{1}{2}\zeta^2\right)^2}{\left(\frac{1}{2} - \zeta\right)^3}.$$

Taking the square root, and noting that $\frac{1}{3} - \zeta + \frac{1}{2}\zeta^2 < 0$, it follows that

$$\frac{5}{2}A = \frac{\frac{1}{3} + \zeta + \frac{1}{2}\zeta^2}{\left(\frac{1}{2} + \zeta\right)^{3/2}} = \frac{-\frac{1}{3} + \zeta - \frac{1}{2}\zeta^2}{\left(\frac{1}{2} - \zeta\right)^{3/2}}$$

and thus also

$$\left(\frac{\partial V_+}{\partial x_2}(x)\right)_{|\Gamma_+} = 0$$

as well. The partial derivatives with respect to x_1 along Γ_+ are then given by

$$\left(\frac{\partial V_+}{\partial x_1}(x)\right)_{|\Gamma_+} = \left(\frac{1}{3} - \zeta + \frac{5}{2}A\left(\frac{1}{2} - \zeta\right)^{\frac{3}{2}}\right)x_2^3 = -\frac{1}{2}\zeta^2 x_2^3$$

and

$$\left(\frac{\partial V_-}{\partial x_1}(x)\right)_{\mid \Gamma_+} = \left(\frac{1}{3}+\zeta-\frac{5}{2}A\left(\frac{1}{2}+\zeta\right)^{\frac{3}{2}}\right)x_2^3 = -\frac{1}{2}\zeta^2 x_2^3.$$

Hence the gradient of V is continuous on Γ_+ and given by

$$\left(\frac{\partial V}{\partial x_1}(x), \frac{\partial V}{\partial x_2}(x)\right)_{\mid \Gamma_+} = \left(-\frac{1}{2}\zeta^2 x_2^3, 0\right). \qquad (5.10)$$

It remains to verify that the two branches V_+ and V_- combine to a continuous function on Γ_+. We have that

$$
\begin{aligned}
V_+(x)_{\mid \Gamma_+} &= \left(-\frac{1}{15}+\frac{1}{3}\zeta-\frac{1}{2}\zeta^2 - A\left(\frac{1}{2}-\zeta\right)^{5/2}\right)x_2^5 \\
&= \left(-\frac{1}{15}+\frac{1}{3}\zeta-\frac{1}{2}\zeta^2 - \frac{2}{5}\left(-\frac{1}{3}+\zeta-\frac{1}{2}\zeta^2\right)\left(\frac{1}{2}-\zeta\right)\right)x_2^5 \\
&= -\frac{1}{5}\zeta^3 x_2^5
\end{aligned}
$$

and

$$
\begin{aligned}
V_-(x)_{\mid \Gamma_+} &= \left(\frac{1}{15}+\frac{1}{3}\zeta+\frac{1}{2}\zeta^2 - A\left(\frac{1}{2}+\zeta\right)^{5/2}\right)x_2^5 \\
&= \left(\frac{1}{15}+\frac{1}{3}\zeta+\frac{1}{2}\zeta^2 - \frac{2}{5}\left(\frac{1}{3}+\zeta+\frac{1}{2}\zeta^2\right)\left(\frac{1}{2}+\zeta\right)\right)x_2^5 \\
&= -\frac{1}{5}\zeta^3 x_2^5.
\end{aligned}
$$

Hence V is continuous on Γ_+. This verifies that V is continuous and continuously differentiable away from the origin. $\qquad \square$

This is one of few examples in which a nonlinear HJB equation can be solved explicitly, albeit with considerable effort. For a general problem, however, the obstacle to applying Theorem 5.1.1 lies more in the fact that solutions V to the Hamilton–Jacobi–Bellman equation rarely are differentiable everywhere, but typically have singularities along lower-dimensional sets. Rather than being the exception, this is the norm, and can be seen in even the simplest problems.

Example 5.1.3 (Time-optimal control for the double integrator). Consider the time-optimal control problem to the origin for the linear system $\dot{x}_1 = x_2$, $\dot{x}_2 = u$, with control set $|u| \leq 1$ (see Sect. 2.6.1). Similar to the Fuller problem, the optimal control $u_*(x)$ is a feedback control given in the form

$$u_*(x) = \begin{cases} +1 & \text{for } x \in G_+ \cup \Gamma_+, \\ -1 & \text{for } x \in G_- \cup \Gamma_-, \end{cases}$$

where now

$$\Gamma_+ = \left\{ (x_1, x_2) \in \mathbb{R}^2 : x_1 = \frac{1}{2}x_2^2,\ x_2 < 0 \right\},$$

$$\Gamma_- = \left\{ (x_1, x_2) \in \mathbb{R}^2 : x_1 = -\frac{1}{2}x_2^2,\ x_2 > 0 \right\},$$

and

$$G_+ = \left\{ (x_1, x_2) \in \mathbb{R}^2 : x_1 < -\operatorname{sgn}(x_2)\frac{1}{2}x_2^2 \right\},$$

$$G_- = \left\{ (x_1, x_2) \in \mathbb{R}^2 : x_1 > -\operatorname{sgn}(x_2)\frac{1}{2}x_2^2 \right\}.$$

While trajectories cross the switching curves Γ_+ and Γ_- transversally for the Fuller problem (and we shall see in Sect. 6.1 in general that this easily verifiable geometric property in fact ensures that the value function V remains continuously differentiable at the switching curves), here the switching curves are trajectories of the system and the optimal trajectories follow these (lower-dimensional) curves. This causes a loss of differentiability of the value function. For this example, it is elementary to compute the value by integrating the time along the trajectories, and the result is

$$V(x) = \begin{cases} V_+(x) & \text{for } x \in G_+ \cup \Gamma_+, \\[2ex] V_-(x) & \text{for } x \in G_- \cup \Gamma_-, \end{cases}$$

with

$$V_+(x) = 2\sqrt{\frac{1}{2}x_2^2 - x_1} - x_2$$

and

$$V_-(x) = 2\sqrt{\frac{1}{2}x_2^2 + x_1} + x_2.$$

As with the Fuller problem, V has the reflection symmetry

$$V_-(x) = V_+(-x) \qquad \text{and} \qquad V_+(x) = V_-(-x)$$

and we need to consider only one switching curve, say Γ_+. It is clear that V is continuous on Γ_+,

$$V_+(x)_{|\Gamma_+} = -x_2 = V_-(x)_{|\Gamma_+}.$$

Furthermore, the partial derivatives of V_- extend as continuously differentiable functions onto the switching surface Γ_+, which is transversal to the flow,

$$\left(\frac{\partial V_-}{\partial x_1}(x), \frac{\partial V_-}{\partial x_2}(x) \right)_{|\Gamma_+} = \left(-\frac{1}{x_2}, 0 \right).$$

But the partial derivatives of the tangential part V_+ have a singularity on Γ_+ with both derivatives diverging to $-\infty$ as $x \to \Gamma_+$ from within G_+. So the value function is no longer differentiable on the switching curves Γ_+ and Γ_-.

As already indicated, the singularity in the derivatives of the value function is caused by the drop in dimension of the surface that the optimal controlled trajectories follow, a rather commonplace phenomenon in optimal control problems. The nondifferentiability of solutions V to the Hamilton–Jacobi–Bellman equation along lower-dimensional sets B invalidates the simple argument used in the proof of Theorem 5.1.1: if η is an arbitrary control, then in principle, it is possible that the set of times t when the corresponding trajectory ξ lies in this set B can be arbitrarily complicated (recall that the zero set of a C^1 function defined on an interval $[0,T]$ can be any closed subset of $[0,T]$, Proposition 2.8.1), and one can no longer carry out the differentiation along the trajectory ξ. Clearly, there is no problem if there are only finitely many times when this trajectory crosses B, but for an arbitrary control η this cannot be asserted in general. Thus the concept of a classical solution generally is not adequate for the theory, and it needed to be revised.

This led to the definition of a *viscosity solution* to the Hamilton–Jacobi–Bellman equation [80, 96] where differentiability of the solution V is replaced by two inequalities that do not require the existence of derivatives of the solution: Consider a general first-order partial differential equation of the form

$$H(z, V, DV) = 0,$$

where z lies in some region Ω in $\mathbb{R}^k$, $V : \Omega \to \mathbb{R}$ is the desired solution, and DV denotes the gradient of V. Define the *subdifferential* $D^-V(z_0)$ of V at a point $z_0 \in \Omega$ as

$$D^-V(z_0) = \left\{ \lambda \in (\mathbb{R}^k)^* : \lim_{z \to z_0} \inf \frac{V(z) - V(z_0) - \lambda(z - z_0)}{\|z - z_0\|} \geq 0 \right\}$$

and the *superdifferential* $D^+V(z_0)$ of V at a point $z_0 \in \Omega$ as

$$D^+V(z_0) = \left\{ \lambda \in (\mathbb{R}^k)^* : \lim_{z \to z_0} \sup \frac{V(z) - V(z_0) - \lambda(z - z_0)}{\|z - z_0\|} \leq 0 \right\}.$$

Definition 5.1.4 (Viscosity solution). Let $V : \Omega \to \mathbb{R}$ be a continuous function. Then V is called a viscosity subsolution of $H(z, V, DV) = 0$ if

$$H(z, V(z), \lambda) \geq 0$$

for all $z \in \Omega$ and all $\lambda \in D^- V(z)$ and a viscosity supersolution of $H(z,V,DV) = 0$ if

$$H(z,V(z),\lambda) \leq 0$$

for all $z \in \Omega$ and all $\lambda \in D^+ V(z)$. If V is both a viscosity subsolution and a viscosity supersolution of $H(z,V,DV) = 0$, then V is called a *viscosity solution* of $H(z,V,DV) = 0$.

It can be shown that the value function to problem [OC] indeed is a viscosity solution to the HJB equation [30]. Results that establish existence and uniqueness of viscosity solutions to the Hamilton–Jacobi–Bellman equation thus characterize the value function as this unique viscosity solution. By now there exist numerous texts that develop the theory of viscosity solutions in the optimal control context, for example [30,96], and in our presentation here we do not pursue these theoretical developments. Rather, we are interested in a *constructive method* to define these solutions, not necessarily by means of explicit formulas, but in a way that provides adequate geometric insights that make it possible to prove the optimality of the underlying controls. Examples 5.1.1 and 5.1.2 above made use of existing symmetries of the problem to give an analytic solution to the HJB equation. While often elegant and convenient, a more direct approach is needed for the general case. The seemingly pedestrian approach of example 5.1.3, where we simply evaluated the cost function along a family of extremal controlled trajectories, provides such an approach. Of course, in a general method the point is to use such a construction to actually *prove* the optimality of an extremal synthesis that has been constructed. The *method of characteristics*, the classical method to solve first-order partial differential equations [124], reduces this computation of a solution to the integration of a pair of ordinary differential equations, the so-called characteristic equations. It has a natural generalization to the optimal control setting and provides such an approach. We develop this method and show that it indeed is an efficient technique to derive sufficient conditions for strong local optimality by constructing a local field of extremals around a reference trajectory. Here we focus on the local aspects of these constructions and defer a development of the global aspects, which are based on Boltyansky's concept of a regular synthesis [41], to the next chapter.

5.2   Parameterized Families of Extremals and the Shadow-Price Lemma

We now formalize the method of characteristics for an optimal control problem. The simple idea is to parameterize extremals by integrating the system and adjoint equation backward from the terminal manifold while maintaining the minimum condition of the maximum principle and then to investigate the mapping properties of the corresponding family of controlled trajectories. If this flow covers the state space (or if time-dependent, the product of time and state space) injectively, then the objective evaluated along this family of trajectories, also sometimes called the *cost-to-go function*, will be the desired solution to the Hamilton–Jacobi–Bellman equation. This procedure not only provides a general method to construct solutions to the HJB equation, but also allows us to investigate singularities in the solutions by relating them to singularities in the parametrization. The construction itself clearly brings out the relationships between the necessary conditions of the maximum principle, Theorem 2.2.1, and the dynamic programming principle, Theorem 5.1.1.

5.2.1 Parameterized Families of Extremals

As we have seen, for the time-optimal control problems in the plane and for the Fuller problem, solutions to optimal control problems typically consist of various "patches" that need to be glued together to form the full solution. In this section, we develop the theory that applies to *one* such patch, but analyze how these patches fit together only in Chap. 6. For example, within our construction, bang-bang controls give rise to several patches and the control is constant on each of them.

Throughout this section, we assume that the *controls depend in a sufficiently smooth way on nearby points*. More specifically, with p a parameter, we assume that the control $u = u(t,p)$ lies in $C^{0,r}$ with r a positive integer, $r = \infty$ or $r = \omega$. Thus, the controls are continuous and, for fixed t, depend r-times continuously differentiably on the parameter p with the partial derivatives $\frac{\partial u}{\partial p}(t,p)$ continuous in (t,p).

Definition 5.2.1 (C^r-Parameterized family of controlled trajectories). Given an open subset P of $\mathbb{R}^d$ with $0 \leq d \leq n$, let

$$t_- : P \to \mathbb{R}, \ p \mapsto t_-(p), \qquad \text{and} \qquad t_+ : P \to \mathbb{R}, \ p \mapsto t_+(p),$$

be two r-times continuously differentiable functions, $t_\pm \in C^r(P)$, that satisfy $t_-(p) < t_+(p)$ for all $p \in P$. We call t_- and t_+ the initial and terminal times of the parametrization and define its domain as

$$D = \{(t,p) : p \in P, t_-(p) \leq t \leq t_+(p)\}.$$

Let $\xi_- : P \to \mathbb{R}^n, \ p \mapsto \xi_-(p)$, and $\xi_+ : P \to \mathbb{R}^n, \ p \mapsto \xi_+(p)$, be r-times continuously differentiable functions, $\xi_\pm \in C^r(P)$. A d-dimensional, C^r-parameterized family $\mathscr{T}$ of controlled trajectories with domain D, initial conditions ξ_-, and terminal conditions ξ_+ consists of:

1. admissible controls, $u : D \to U, \ (t,p) \mapsto u(t,p)$, that are continuous and r-times continuously differentiable in p on D, ($u \in C^{0,r}(D)$), and
2. corresponding trajectories $x : D \to \mathbb{R}^n, \ (t,p) \mapsto x(t,p)$, i.e., solutions of the dynamics

$$\dot{x}(t,p) = f(t,x(t,p),u(t,p)), \tag{5.11}$$

Fig. 5.1 The flow F of a parameterized family of controlled trajectories

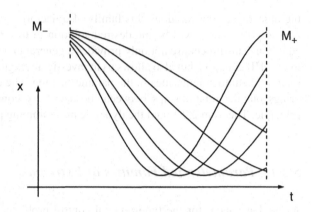

that exist over the full interval $[t_-(p), t_+(p)]$ and satisfy the initial condition $x(t_-(p), p) = \xi_-(p)$ and terminal condition $x(t_+(p), p) = \xi_+(p)$.

We shall be considering both time-dependent and time-independent formulations and always wish to separate the time t and the state x in our notation. It is convenient, however, to have a common notation for the associated flows (see Fig. 5.1).

Definition 5.2.2 (Flow of controlled trajectories). Let $\mathscr{T}$ be a C^r-parameterized family of controlled trajectories. For a time-dependent optimal control problem, we define the associated flow as the map

$$F : D \to \mathbb{R} \times \mathbb{R}^n, \quad (t, p) \longmapsto F(t, p) = \begin{pmatrix} t \\ x(t, p) \end{pmatrix},$$

i.e., in terms of the graphs of the corresponding trajectories. For a time-independent optimal control problem, we define the associated flow as the flow of the trajectories,

$$F : D \to \mathbb{R}^n, \quad (t, p) \longmapsto F(t, p) = x(t, p).$$

We say the flow F is a $C^{1,r}$-mapping on some open set $Q \subset D$ if the restriction of F to Q is continuously differentiable in (t, p) and r times differentiable in p with derivatives that are jointly continuous in (t, p). If $F \in C^{1,r}(Q)$ is injective and the Jacobian matrix $DF(t, p)$ is nonsingular everywhere on Q, then, by the inverse function theorem (see Appendix A, Theorem A.3.1), the mapping $F(t, p)$ has an inverse of class $C^{1,r}$, and we say that F is a $C^{1,r}$-diffeomorphism onto its image $F(Q)$.

The boundary sections

$$M_- = \{(t, p) : p \in P, t = t_-(p)\} \quad \text{and} \quad M_+ = \{(t, p) : p \in P, t = t_+(p)\}$$

of a C^r-parameterized family of controlled trajectories are the graphs of the functions t_- and t_+, $M_- = \mathfrak{gr}(t_-)$ and $M_+ = \mathfrak{gr}(t_+)$. We call the images of these sections under the flow F, $N_{\pm} = F(M_{\pm})$, the *source*, respectively the *target*, of the parametrization. Thus

$$N_{\pm} = \{(t,x) : t = t_{\pm}(p), \ x = \xi_{\pm}(p), \quad p \in P\}$$

in the time-dependent case and

$$N_{\pm} = \{x : x = \xi_{\pm}(p), \ p \in P\}$$

for a time-independent problem. In the construction, generally one of these is specified and the trajectories are defined as the solutions of the associated initial (or terminal) value problem. The other then simply is defined by the flow of these solutions. If $\mathscr{T}$ is a C^r-parameterized family of controlled trajectories with source N_-, then (with the obvious modifications to the definition) we also allow that $t_+(p) \equiv +\infty$, and for a family with target N_+ we may have that $t_-(p) \equiv -\infty$. We want to consider both the cases in which trajectories are integrated forward in time (families of controlled trajectories with source N_-) and backward in time (families of controlled trajectories with target N_+).

The controls $u(t,p)$ in the parameterized family are assumed to be continuous and r-times continuously differentiable in p, $u \in C^{0,r}(D)$. The parameter set P is open, but the domain D is closed at the endpoints of the intervals. Hence this assumption is understood in the sense that $u(t,p)$ extends as a $C^{0,r}$-function onto an open neighborhood of D. For instance, for constant (bang) controls or singular controls defined in feedback form, this is automatic and these are the only cases considered in this text.

It thus follows from the classical results about solutions to ODEs (see Appendix B) that the trajectories $x(t,p)$ and their time derivatives $\dot{x}(t,p)$ are r-times continuously differentiable in p and that these derivatives are continuous jointly in (t,p) in an open neighborhood of D, i.e., $x \in C^{1,r}(D)$. These partial derivatives can be calculated as solutions to the corresponding variational equations:

$$\frac{d}{dt}\left(\frac{\partial x}{\partial p}(t,p)\right) = \frac{\partial^2 x}{\partial t \partial p}(t,p) = \frac{\partial \dot{x}}{\partial p}(t,p) = \frac{\partial}{\partial p}\left(f(t,x(t,p),u(t,p))\right)$$

$$= f_x(t,x(t,p),u(t,p))\frac{\partial x}{\partial p}(t,p) + f_u(t,x(t,p),u(t,p))\frac{\partial u}{\partial p}(t,p).$$

$$(5.12)$$

In particular, the flow F is a $C^{1,r}$-mapping. Section 5.3 deals with the question when it is injective, i.e., a $C^{1,r}$-diffeomorphism.

In order to calculate the value of the objective along a parameterized family of controlled trajectories, we need to add an extra function γ that describes this cost at the source N_- or at the target N_+.

Definition 5.2.3 (C^r-Parameterized family of controlled trajectories with cost γ). Suppose $\mathscr{T}$ is a d-dimensional, C^r-parameterized family $\mathscr{T}$ of controlled trajectories with domain D and initial and terminal values ξ_- and ξ_+. Given an r-times continuously differentiable function $\gamma_- : P \to \mathbb{R}$, $p \mapsto \gamma_-(p)$ (respectively, $\gamma_+ : P \to \mathbb{R}$, $p \mapsto \gamma_+(p)$), we define the cost or cost-to-go function associated with $\mathscr{T}$ as

$$C(t,p) = \gamma_-(p) - \int_{t_-(p)}^{t} L(s,x(s,p),u(s,p))\,ds$$

(respectively as

$$C(t,p) = \int_{t}^{t_+(p)} L(s,x(s,p),u(s,p))\,ds + \gamma_+(p)$$

when the terminal value is specified), and call $\mathscr{T}$ a C^r-parameterized family of controlled trajectories with cost γ.

The functions γ_+ and γ_- propagate the cost along trajectories from patch to patch, and $C(t,p)$ represents the value of the objective for the control $u = u(\cdot,p)$ if the initial condition at time t is given by $x(t,p)$. This specification is equally valid for a time-dependent or time-independent problem. Since the value of the optimal cost on the terminal manifold N is specified by the penalty term φ in the objective, integrating trajectories backward in time is the typical procedure. For syntheses where trajectories can be successively integrated backward from the terminal manifold, these functions are easily computed.

Example 5.2.1 (Time-optimal control to the origin for the double integrator). Allowing for parameterized families of various dimensions d enables us to build a synthesis in an inductive way. For this optimal control problem the switching curves Γ_+ and Γ_- can be considered to be zero-dimensional "parameterized families," which, integrating backward from the target $N_+ = \{0\}$ and dropping the zero-dimensional parameter $p = 0$, can be described as

$$\Gamma_+ : P = \{0\} \quad t_- = -\infty,\, t_+ = 0, \quad D = (-\infty,0] \times \{0\},$$

$$\xi_+ = 0, \quad u(t) \equiv +1, \quad x(t) = \begin{pmatrix} \frac{1}{2}t^2 \\ t \end{pmatrix},$$

and

$$\Gamma_- : P = \{0\} \quad t_- = -\infty,\, t_+ = 0, \quad D = (-\infty,0] \times \{0\},$$

$$\xi_+ = 0, \quad u(t) \equiv -1, \quad x(t) = \begin{pmatrix} -\frac{1}{2}t^2 \\ -t \end{pmatrix}.$$

The full families of trajectories that cover the regions

$$G_+ = \left\{ (x_1, x_2) \in \mathbb{R}^2 : x_1 < -\operatorname{sgn}(x_2)\frac{1}{2}x_2^2 \right\}$$

and

$$G_- = \left\{ (x_1, x_2) \in \mathbb{R}^2 : x_1 > -\operatorname{sgn}(x_2)\frac{1}{2}x_2^2 \right\}$$

then form one-dimensional parameterized families $\mathscr{T}_+$ and $\mathscr{T}_-$ given by

$$\mathscr{T}_+ : P = (0, \infty), \quad t_-(p) = -\infty, \ t_+(p) = -p, \quad D = \{(t, p) : t \le -p < 0\},$$

$$\xi_+(p) = \begin{pmatrix} -\frac{1}{2}p^2 \\ p \end{pmatrix}, \quad u(t,p) \equiv +1, \quad x(t,p) = \begin{pmatrix} \frac{1}{2}t^2 + 2pt + p^2 \\ t + 2p \end{pmatrix},$$

and

$$\mathscr{T}_- : P = (0, \infty), \quad t_-(p) = -\infty, \ t_+(p) = -p, \quad D = \{(t, p) : t \le -p < 0\},$$

$$\xi_+(p) = \begin{pmatrix} \frac{1}{2}p^2 \\ -p \end{pmatrix}, \quad u(t,p) \equiv -1, \quad x(t,p) = \begin{pmatrix} -\frac{1}{2}t^2 - 2pt - p^2 \\ -t - 2p \end{pmatrix}.$$

This problem is time-independent, and in this case there is always freedom to choose either the initial or terminal time arbitrarily. In the parametrization above, we have chosen $t_+(p)$ so that the variable t in $x(t,p)$ is equal to the negative of the total time it takes to steer the initial condition $x(t,p)$ into the origin along the concatenated optimal trajectory, i.e., $C(t,p) = -t$. The cost functions on the targets of the parameterizations therefore are given by $\gamma_+ = 0$ for the zero-dimensional strata $N = \{0\}$, and $\gamma_+(p) = p$ for the two one-dimensional families $\mathscr{T}_\pm$. All these parameterizations are real analytic, C^ω (see Fig. 5.2).

In a general problem, the trajectories $x(\cdot, p)$ are defined as the solutions to the initial or terminal value problem for the dynamics (5.11), and an explicit integration is not required. For parameterized families as in Example 5.1.1, the cost functions $\gamma_\pm$ are easily obtained through backward integration from the targets. Naturally, such a direct integration approach becomes much less convenient for problems like the Fuller problem when the switching curves Γ_+ and Γ_- are no longer trajectories of the system. However, in order to find the correct functions γ that propagate the cost, it is not always necessary to integrate the cost along the trajectories in the family. These functions can sometimes be found in a much easier and more elegant way from necessary conditions for optimality (see Example 5.1.2 below). Clearly, we are interested in *optimal* controlled trajectories, and thus we now consider families of controlled trajectories that satisfy the conditions of the maximum principle, i.e., are extremals. Once more, we assume that u is of class $C^{0,r}$, which naturally extends to the state as $x \in C^{1,r}$. However, since the adjoint is defined in terms of data that are differentiated in the state x, we generally obtain only one degree less of

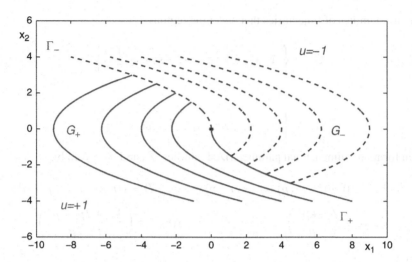

Fig. 5.2 Synthesis of optimal controlled trajectories for the time-optimal control problem to the origin for the double integrator

smoothness for the multipliers, $\lambda \in C^{1,r-1}$. The following definition is essential for the construction of solutions to the Hamilton–Jacobi–Bellman equation by means of the method of characteristics. As always,

$$H = H(t, \lambda_0, \lambda, x, u) = \lambda_0 L(t, x, u) + \lambda f(t, x, u).$$

Definition 5.2.4 (C^r-Parameterized family of extremals). As before, let P be an open subset of $\mathbb{R}^d$ with $0 \le d \le n$, let t_- and t_+, $t_\pm \in C^r(P)$, be the initial and terminal times for the parametrization, and let $D = \{(t, p) : p \in P, t_-(p) \le t \le t_+(p)\}$. A d-dimensional, C^r-parameterized family $\mathscr{E}$ of extremals (or extremal lifts) with domain D consists of

1. a C^r-parameterized family $\mathscr{T}$ of controlled trajectories (x, u) with domain D, initial and terminal conditions ξ_- and ξ_+, and cost γ_- (respectively, γ_+):

$$\dot{x}(t, p) = f(t, x(t, p), u(t, p)), \qquad x(t_\pm(p), p) = \xi_\pm(p);$$

2. a nonnegative multiplier $\lambda_0 \in C^{r-1}(P)$ and costate $\lambda : D \to (\mathbb{R}^n)^*$, $\lambda = \lambda(t, p)$, such that $(\lambda_0(p), \lambda(t, p)) \ne (0, 0)$ for all $(t, p) \in D$ and the adjoint equation

$$\dot{\lambda}(t, p) = -\lambda_0(p) L_x(t, x(t, p), u(t, p)) - \lambda(t, p) f_x(t, x(t, p), u(t, p)),$$

is satisfied on the interval $[\tau_-(p), \tau_+(p)]$ with boundary condition $\lambda_-(p) = \lambda(\tau_-(p), p)$ (respectively, $\lambda_+(p) = \lambda(\tau_+(p), p)$) given by an $(r-1)$-times continuously differentiable function of p, such that the following conditions are satisfied:

3. defining $h(t,p) = H(t, \lambda_0(p), \lambda(t,p), x(t,p), u(t,p))$, the controls $u = u(t,p)$ solve the minimization problem

$$h(t,p) = \min_{v \in U} H(t, \lambda_0(p), \lambda(t,p), x(t,p), v);$$

4. with $h_\pm(p) = h(t_\pm(p), p))$, the following transversality condition holds at the source (respectively, target):

$$\lambda_\pm(p) \frac{\partial \xi_\pm}{\partial p}(p) = \lambda_0(p) \frac{\partial \gamma_\pm}{\partial p}(p) + h_\pm(p) \frac{\partial t_\pm}{\partial p}(p); \qquad (5.13)$$

5. if the target N_+ is a part of the terminal manifold N, $N_+ \subset N$, then, setting $T(p) = t_+(p)$, with $\xi_+(p) = x(T(p), p)$ we have that $\gamma_+(p) = \varphi(T(p), \xi_+(p))$; furthermore, there exists an $(r-1)$-times continuously differentiable multiplier $v : P \to (\mathbb{R}^{n+1-k})^*$ such that the following transversality conditions are satisfied:

$$\lambda(T(p), p) = \lambda_0(p) \frac{\partial \varphi}{\partial x}(T(p), \xi_+(p)) + v(p) \frac{\partial \Psi}{\partial x}(T(p), \xi_+(p)), \qquad (5.14)$$

$$-h(T(p), p) = \lambda_0(p) \frac{\partial \varphi}{\partial t}(T(p), \xi_+(p)) + v(p) \frac{\partial \Psi}{\partial t}(T(p), \xi_+(p)). \qquad (5.15)$$

This definition merely formalizes that all controlled trajectories in the family $\mathscr{E}$ satisfy the conditions of the maximum principle while some smoothness properties are satisfied by the parametrization and natural geometric regularity assumptions are met at the terminal manifold N. Note that it has not been assumed that the parametrization $\mathscr{E}$ of extremals covers the state space injectively, and one of our objectives is to use this framework to analyze the geometry of the flow of the associated controlled trajectories as injectivity becomes lost (e.g., conjugate points). The degree r in the definition denotes the smoothness of the parametrization of the controls in the parameter p, $u \in C^{0,r}$, and as already explained, this implies that $x \in C^{1,r}$. The condition $\lambda \in C^{1,r-1}$ is ensured by requiring that the multipliers λ_0 and the boundary values $\lambda_\pm(p)$, respectively v, be $(r-1)$-times continuously differentiable. In particular, for a C^1-parameterized family of extremals, only continuity in p is required. If the data defining the problem [OC] possess an additional degree of differentiability in x and if the multiplier λ_0 and the function $\lambda_\pm(p)$ are r-times continuously differentiable with respect to p, then it follows that $\lambda(t,p) \in C^{1,r}$ as well. In particular, this is true if $r = \infty$ or $r = \omega$, as will be the case in almost all examples considered in this chapter. In such a case, we call $\mathscr{E}$ a *nicely C^r-parameterized family of extremals*. Also, if $\lambda_0(p) > 0$ for all $p \in P$, then all extremals are normal, and by dividing by $\lambda_0(p)$, we may assume that $\lambda_0(p) \equiv 1$, and we call such a family *normal*.

Example 5.2.1 (continued). By construction, the families $\mathscr{T}_\pm$ of controlled trajectories defined earlier are real analytic families of extremals: All extremals are normal, $\lambda_0(p) \equiv 1$, and the adjoint variables $\lambda(t,p)$ are the solutions to the equations $\dot{\lambda}_1 = 0$ and $\dot{\lambda}_2 = -\lambda_1$ with terminal conditions

$$\lambda_\pm(p) = \lambda(t_\pm(p),p) = \left(-\frac{\varepsilon}{p},0\right),$$

where $\varepsilon = +1$ for $\mathscr{T}_+$ and $\varepsilon = -1$ for $\mathscr{T}_-$. Recall that the multiplier λ_2 is the switching function for this problem (see Sect. 2.6.1), and thus it must vanish on the switching curves $\Gamma_\pm$. The value of $\lambda_1(t_\pm(p),p)$ is then determined by the fact that the Hamiltonian H vanishes identically, i.e.,

$$0 = H = 1 + \lambda_\pm(p)\begin{pmatrix} 0 & 1 \\ 0 & 0 \end{pmatrix}\xi_\pm(p)$$

$$= 1 + \lambda_1(t_\pm(p),p)\xi_2(t_\pm(p),p)$$

$$= 1 + \lambda_1(t_\pm(p),p)p\varepsilon,$$

so that $\lambda_1(t_\pm(p),p) = -\frac{\varepsilon}{p}$. With this specification, the controls satisfy the minimum condition by construction. Furthermore, $t_+(p) = -p$, $\gamma_+(p) = p$, and $\xi_+(p) = \varepsilon\left(-\frac{1}{2}p^2\right)$, which gives

$$\lambda_\pm(p)\frac{\partial \xi_\pm}{\partial p}(p) = \left(-\frac{\varepsilon}{p},0\right)\begin{pmatrix} -p \\ 1 \end{pmatrix}\varepsilon = \varepsilon^2 = 1 = \frac{\partial \gamma_\pm}{\partial p}(p),$$

i.e., the transversality condition (5.13) is satisfied. (Note that $h \equiv 0$ and $\lambda_0(p) \equiv 1$.) Since $p \neq 0$, these parameterizations are real-analytic.

We shall see below in Lemma 5.2.2 that it is this transversality condition that ensures the proper relationship between the multiplier λ and the cost γ. Essentially, this condition is the propagation of the transversality condition on the multiplier λ from the terminal constraint along the parameterized family of extremals.

Lemma 5.2.1. *For the optimal control problem [OC], if the target N_+ is a part of the terminal manifold N, $N_+ \subset N$, then condition (5.13) follows from the transversality conditions of the maximum principle.*

Proof. Given a C^r-parameterized family $\mathscr{E}$ of extremals with target $N_+ \subset N$, we have that $\xi_+(p) = x(T(p),p)$, $\gamma_+(p) = \varphi(T(p),\xi_+(p))$, and

$$\lambda(T(p),p) = \lambda_0(p)\varphi_x(T(p),\xi_+(p)) + v(p)\Psi_x(T(p),\xi_+(p)).$$

In the following calculation, all functions are evaluated at their proper argument, i.e.,

$$T = T(p), \qquad x = x(T(p), p), \qquad u = u(T(p), p),$$
$$\xi_+ = \xi_+(p), \qquad \gamma_+ = \gamma_+(p),$$
$$\varphi = \varphi(T(p), \xi_+(p)), \qquad f - f(T(p), x(T(p), p), u(T(p), p))$$

But it is more convenient—and the reasoning becomes much more transparent—if these arguments are omitted. Differentiating with respect to p (and denoting the partial derivatives of the data ϕ and Ψ by subscripts) gives

$$\frac{\partial \xi_+}{\partial p} = \frac{\partial x}{\partial t} \frac{\partial T}{\partial p} + \frac{\partial x}{\partial p} = f \frac{\partial T}{\partial p} + \frac{\partial x}{\partial p}$$

and

$$\frac{\partial \gamma_+}{\partial p} = \varphi_t \frac{\partial T}{\partial p} + \varphi_x \frac{\partial \xi_+}{\partial p} = \varphi_t \frac{\partial T}{\partial p} + \varphi_x \left(f \frac{\partial T}{\partial p} + \frac{\partial x}{\partial p} \right).$$

At the terminal point we have that $\Psi(T(p), x(T(p), p)) \equiv 0$ and therefore

$$\Psi_t \frac{\partial T}{\partial p} + \Psi_x \left(f \frac{\partial T}{\partial p} + \frac{\partial x}{\partial p} \right) \equiv 0$$

as well. Hence

$$\lambda \frac{\partial \xi_\pm}{\partial p} = \lambda \left(f \frac{\partial T}{\partial p} + \frac{\partial x}{\partial p} \right) = (\lambda_0 \varphi_x + v\Psi_x) \left(f \frac{\partial T}{\partial p} + \frac{\partial x}{\partial p} \right)$$
$$= \lambda_0 \left(\frac{\partial \gamma_+}{\partial p} - \varphi_t \frac{\partial T}{\partial p} \right) - v\Psi_t \frac{\partial T}{\partial p}$$
$$= \lambda_0 \frac{\partial \gamma_+}{\partial p} - (\lambda_0 \varphi_t + v\Psi_t) \frac{\partial T}{\partial p}$$
$$= \lambda_0 \frac{\partial \gamma_+}{\partial p} + H \frac{\partial T}{\partial p},$$

and condition (5.13) is satisfied. □

5.2.2 The Shadow-Price Lemma and Solutions to the Hamilton–Jacobi–Bellman Equation

The following result contains the key relation in making the step from necessary to sufficient conditions for optimality. In economic models, the dynamics of the optimal control problem often describes an actual production process, and the objective represents its cost. For these applications, the lemma below relates changes in the production, $\frac{\partial x}{\partial p}$, to the changes in the actual cost, $\frac{\partial C}{\partial p}$, and it is the multiplier λ

that assigns a price to these changes. For this reason, the multipliers are also known as shadow prices in economics. We retain this terminology.

Lemma 5.2.2 (Shadow-price Lemma). *Let $\mathscr{E}$ be a C^1-parameterized family of extremal lifts with domain D. Then for all $(t, p) \in D$,*

$$\lambda_0(p)\frac{\partial C}{\partial p}(t, p) = \lambda(t, p)\frac{\partial x}{\partial p}(t, p). \tag{5.16}$$

Proof. The parameterized cost $C = C(t, p)$ and trajectories $x = x(t, p)$ are continuously differentiable in p on D, and for fixed p, both sides of Eq. (5.16) are continuously differentiable functions of t. It therefore suffices to show that for p fixed, both sides have identical t-derivatives and the same value for $t = t_\pm(p)$.

(i) Both sides agree on the target (respectively, source) for $t = t_\pm(p)$: it follows from the definition of the parameterized cost (Definition 5.2.3) that

$$\frac{\partial C}{\partial p}(t_\pm(p), p) = L(t_\pm(p), \xi_\pm(p), u(t_\pm(p), p))\frac{\partial t_\pm}{\partial p}(p) + \frac{\partial \gamma_\pm}{\partial p}(p).$$

Furthermore, $\xi_\pm(p) \equiv x(t_\pm(p), p)$ implies that

$$\frac{\partial \xi_\pm}{\partial p}(p) = \frac{\partial x}{\partial t}(t_\pm(p), p)\frac{\partial t_\pm}{\partial p}(p) + \frac{\partial x}{\partial p}(t_\pm(p), p)$$

$$= f(t_\pm(p), \xi_\pm(p), u(\tau_\pm(p), p))\frac{\partial t_\pm}{\partial p}(p) + \frac{\partial x}{\partial p}(t_\pm(p), p).$$

Hence, and once more dropping the arguments in the calculation,

$$\lambda_0(p)\frac{\partial C}{\partial p}(t_\pm(p), p) = \lambda_0 L\frac{\partial t_\pm}{\partial p} + \lambda_0\frac{\partial \gamma_+}{\partial p}$$

$$= (H - \lambda f)\frac{\partial t_\pm}{\partial p}(p) + \lambda_0\frac{\partial \gamma_+}{\partial p}$$

$$= H\frac{\partial t_\pm}{\partial p} + \lambda_0\frac{\partial \gamma_+}{\partial p} - \lambda\frac{\partial \xi_\pm}{\partial p} + \lambda\frac{\partial x}{\partial p}$$

$$= \lambda(t_\pm(p), p)\frac{\partial x}{\partial p}(\tau_\pm(p), p),$$

where we use the transversality condition (5.13) in the last relation.

(ii) Both sides have the same time derivatives over the interval $[t_-(p), t_+(p)]$: using the adjoint equation and the variational equation (5.12), we get that

$$\frac{d}{dt}\left\{\lambda(t,p)\frac{\partial x}{\partial p}(t,p)\right\}$$

$$=\dot\lambda(t,p)\frac{\partial x}{\partial p}(t,p)+\lambda(t,p)\frac{\partial^2 x}{\partial t\partial p}(t,p)$$

$$=(-\lambda_0 L_x-\lambda f_x)\frac{\partial x}{\partial p}+\lambda\left(f_x\frac{\partial x}{\partial p}+f_u\frac{\partial u}{\partial p}\right)$$

$$=-\lambda_0\left(L_x\frac{\partial x}{\partial p}+L_u\frac{\partial u}{\partial p}\right)+H_u\frac{\partial u}{\partial p}$$

$$=\lambda_0(p)\frac{\partial^2 C}{\partial t\partial p}(t,p)+H_u(t,\lambda_0(p),\lambda(t,p),x(t,p),u(t,p))\frac{\partial u}{\partial p}(t,p).$$

Hence the proof will be completed by verifying that

$$H_u(t,\lambda_0(p),\lambda(t,p),x(t,p),u(t,p))\frac{\partial u}{\partial p}(t,p)\equiv 0 \qquad \text{on } D. \qquad (5.17)$$

To see this, fix a point (t,p) in the interior of D. Then, for q in some neighborhood of p, also the points (t,q) lie in the interior of D, and the control values $v=u(t,q)$ are admissible. Hence it follows from the minimization property of the extremal control $u(t,p)$ that the function

$$h(q)=\lambda_0(p)L(t,x(t,p),u(t,q))+\lambda(t,p)f(t,x(t,p),u(t,q))$$

has a local minimum at $q=p$. Since this function is differentiable in q, we have as necessary condition that

$$0=\frac{\partial h}{\partial p}(p)=H_u(t,\lambda_0(p),\lambda(t,p),x(t,p),u(t,p))\frac{\partial u}{\partial p}(t,p).$$

This verifies that

$$\frac{d}{dt}\left\{\lambda(t,p)\frac{\partial x}{\partial p}(t,p)\right\}=\lambda_0(p)\frac{\partial^2 C}{\partial t\partial p}(t,p).$$

Together (i) and (ii) prove Eq. (5.16). □

Note that this computation does not require smoothness properties on the multipliers $\lambda_0(p)$ and $v(p)$ and that the shadow-price lemma is also valid for abnormal extremals, albeit in the degenerate form $\lambda(t,p)\frac{\partial x}{\partial p}(t,p)=0$.

The following observation shows that the transversality condition (5.13) needed for the shadow-price lemma to be valid propagates between source and target of the parametrization. This will later allow us to concatenate different parameterized families and will be taken up again in Sect. 6.1.

Corollary 5.2.1. *Let ξ be a C^r-parameterized family of extremals with domain $D =$ $\{(t,p) : p \in P, t_-(p) \leq t \leq t_+(p)\}$. Given any continuously differentiable function $\tau : P \to \mathbb{R}$, $p \mapsto \tau(p)$, that satisfies $t_-(p) \leq \tau(p) \leq t_+(p)$, let $\xi(p) = x(\tau(p),p)$, $\gamma(p) = C(\tau(p),p)$, $\lambda(p) = \lambda(\tau(p),p)$, and $h(p) = h(\tau(p),p))$. Then we have that*

$$\lambda(p)\frac{\partial \xi}{\partial p}(p) = \lambda_0(p)\frac{\partial \gamma}{\partial p}(p) + h(p)\frac{\partial \tau}{\partial p}(p).$$

In particular, the transversality condition (5.13) propagates between source and target.

Proof. Essentially, this is the same calculation as in step (i) of the proof of the shadow-price lemma. For notational simplicity, we once more drop the arguments. We have that

$$\frac{\partial \xi}{\partial p} = \dot{x}\frac{\partial \tau}{\partial p} + \frac{\partial x}{\partial p} = f\frac{\partial \tau}{\partial p} + \frac{\partial x}{\partial p} \quad \text{and} \quad \frac{\partial \gamma}{\partial p} = -L\frac{\partial \tau}{\partial p} + \frac{\partial C}{\partial p}.$$

Hence the shadow-price lemma gives

$$\lambda\frac{\partial \xi}{\partial p} - \lambda_0\frac{\partial \gamma}{\partial p} - h\frac{\partial \tau}{\partial p} = \lambda\frac{\partial x}{\partial p} - (h - \lambda f)\frac{\partial \tau}{\partial p} - \lambda_0\frac{\partial \gamma}{\partial p}$$

$$= \lambda_0\left(\frac{\partial C}{\partial p} - L\frac{\partial \tau}{\partial p}\right) - \lambda_0\frac{\partial \gamma}{\partial p} = 0.$$

$\square$

For normal extremals, we now show that the shadow-price lemma implies that the associated cost is a classical solution to the Hamilton–Jacobi–Bellman equation on a region G of the state space if the corresponding family of trajectories covers G injectively. We consider the time-dependent formulation, i.e., the flow F is given by $F(t,p) = (t,x(t,p))$. If F is an injective $C^{1,r}$-map from some open set $Q \subset D$ onto a region $G \subset \mathbb{R} \times \mathbb{R}^n$, then F is a $C^{1,r}$-diffeomorphism if and only if the Jacobian matrix $\frac{\partial x}{\partial p}$ is nonsingular on D. In this case, for t fixed, the map $F(t,\cdot) : p \mapsto x(t,p)$ is a C^r-diffeomorphism and its inverse $F^{-1}(t,\cdot) : x \mapsto F^{-1}(t,x)$ also is r-times continuously differentiable in x.

Theorem 5.2.1. *Let $\mathcal{E}$ be a C^r-parameterized family of normal extremals for a time-dependent optimal control problem and suppose the restriction of its flow F to some open set $Q \subset D$ is a $C^{1,r}$-diffeomorphism onto an open subset $G \subset \mathbb{R} \times \mathbb{R}^n$ of the (t,x)-space. Then the value $V^{\mathcal{E}}$ of the parameterized family $\mathcal{E}$,*

$$V^{\mathcal{E}} : G \to \mathbb{R}, \quad V^{\mathcal{E}} = C \circ F^{-1},$$

is continuously differentiable in (t,x) and r-times continuously differentiable in x for fixed t. The function

$$u_* : G \to \mathbb{R}, \quad u_* = u \circ F^{-1},$$

is an admissible feedback control that is continuous and r-times continuously differentiable in x for fixed t. Together, the pair $(V^{\mathscr{E}}, u_*)$ is a classical solution of the Hamilton–Jacobi–Bellman equation

$$\frac{\partial V}{\partial t}(t,x) + \min_{u \in U} \left\{ \frac{\partial V}{\partial x}(t,x) f(t,x,u) + L(t,x,u) \right\} \equiv 0$$

on G. Furthermore, the following identities hold in the parameter space on Q:

$$\frac{\partial V^{\mathscr{E}}}{\partial t}(t,x(t,p)) = -H(t,\lambda(t,p),x(t,p),u(t,p)), \tag{5.18}$$

$$\frac{\partial V^{\mathscr{E}}}{\partial x}(t,x(t,p)) = \lambda(t,p). \tag{5.19}$$

If $\mathscr{E}$ is nicely C^r-parameterized, then $V^{\mathscr{E}}$ is $(r+1)$-times continuously differentiable in x on G, and we also have that

$$\frac{\partial^2 V^{\mathscr{E}}}{\partial x^2}(t,x(t,p)) = \frac{\partial \lambda^T}{\partial p}(t,p) \left(\frac{\partial x}{\partial p}(t,p) \right)^{-1}. \tag{5.20}$$

Proof. Since $F \upharpoonright Q : Q \to G$ is a $C^{1,r}$-diffeomorphism, the function $V^{\mathscr{E}}$ and the control u_* are well-defined, and the stated smoothness properties carry over from the parametrization. Furthermore, the differential equation

$$\dot{x} = f(t,x,u_*(t,x)), \quad x(t_0) = x(t_0,p_0) = x_0,$$

has the unique solution $\tilde{x}(t) = x(t,p_0)$ with corresponding open-loop control

$$u_{(t_0,x_0)}(t) = u_*(t,x(t,p_0)) = u(t,p_0).$$

Hence the feedback control u_* is admissible.

Since $C = V^{\mathscr{E}} \circ F$, we have that

$$\frac{\partial C}{\partial p}(t,p) = \frac{\partial V^{\mathscr{E}}}{\partial x}(t,x(t,p)) \frac{\partial x}{\partial p}(t,p).$$

In view of Lemma 5.2.2 and the fact that $\frac{\partial x}{\partial p}$ is nonsingular, Eq. (5.19) follows; furthermore,

$$-L(t,x(t,p),u(t,p)) = \frac{\partial C}{\partial t}(t,p)$$

$$= \frac{\partial V^{\mathscr{E}}}{\partial t}(t,x(t,p)) + \frac{\partial V^{\mathscr{E}}}{\partial x}(t,x(t,p))\dot{x}(t,p)$$

$$= \frac{\partial V^{\mathscr{E}}}{\partial t}(t,x(t,p)) + \lambda(t,p)f(t,x(t,p),u(t,p))$$

gives Eq. (5.18). But then the minimum condition in the definition of extremals implies that the pair $(V^{\mathscr{E}},u_*)$ solves the Hamilton–Jacobi–Bellman equation: for $(t,x) = (t,x(t,p)) \in G$ and an arbitrary control value $v \in U$ we have that

$$\frac{\partial V^{\mathscr{E}}}{\partial t}(t,x) + \frac{\partial V^{\mathscr{E}}}{\partial x}(t,x)f(t,x,v) + L(t,x,v)$$

$$= \frac{\partial V^{\mathscr{E}}}{\partial t}(t,x(t,p)) + \frac{\partial V^{\mathscr{E}}}{\partial x}(t,x(t,p))f(t,x(t,p),v) + L(t,x(t,p),v)$$

$$= \frac{\partial V^{\mathscr{E}}}{\partial t}(t,x(t,p)) + \lambda(t,p)f(t,x(t,p),v) + L(t,x(t,p),v)$$

$$= \frac{\partial V^{\mathscr{E}}}{\partial t}(t,x(t,p)) + H(t,\lambda(t,p),x(t,p),v)$$

$$\geq \frac{\partial V^{\mathscr{E}}}{\partial t}(t,x(t,p)) + H(t,\lambda(t,p),x(t,p),u(t,p)) = 0$$

with equality for $v = u(t,p)$.

If $\mathscr{E}$ is nicely C^r-parameterized, then in addition, λ also is C^r in p, and thus, since on G we have that $\frac{\partial V^{\mathscr{E}}}{\partial x} = \lambda \circ F^{-1}$, it follows that $\frac{\partial V^{\mathscr{E}}}{\partial x}$ is still r-times continuously differentiable in x. Differentiating the column vector $\lambda^T(t,p) = \left(\frac{\partial V^{\mathscr{E}}}{\partial x}\right)^T(t,x(t,p))$, we get that

$$\frac{\partial \lambda^T}{\partial p}(t,p) = \frac{\partial^2 V^{\mathscr{E}}}{\partial x^2}(t,x(t,p))\frac{\partial x}{\partial p}(t,p),$$

where consistent with our notation, $\frac{\partial \lambda^T}{\partial p}$ is the matrix of partial derivatives of the column vector λ^T. $\square$

If the problem is time-independent, then we simply use the relation (5.6) to relate the dynamics,

$$f'(x',u) = \begin{pmatrix} 1 \\ f(x',u) \end{pmatrix}.$$

In this case, the flow map is given by $F(t,p) = x(t,p)$, and the relation (5.19) reads

$$\frac{\partial V^{\mathscr{E}}}{\partial x}(x(t,p)) = \lambda(t,p). \tag{5.21}$$

Note that here the value function is independent of t and this is consistent with the fact that the Hamiltonian H vanishes identically since the dynamics is time-invariant and the terminal time T is free.

Definition 5.2.5 (Local field of extremals). A C^r-parameterized local field of extremals, $\mathscr{F}$, is a C^r-parameterized family of normal extremals for which the associated flow $F : D \to \mathbb{R} \times \mathbb{R}^n$, $(t,p) \mapsto F(t,p)$, is a $C^{1,r}$-diffeomorphism from the interior of the set D, $\mathring{D} = \{(t,p) : p \in P, t_-(p) < t < t_+(p)\}$, onto a region $G = F(\mathring{D})$.

We do not require that the flow F be a diffeomorphism on the source or target of the parametrization. If these are codimension-1 embedded submanifolds, and if the flow F is transversal to them, then the flow extends as a $C^{1,r}$-diffeomorphism onto a neighborhood of the full closed domain D, and this is a common scenario along switching surfaces (see Sect. 6.1). However, if the target parameterizes a section N_T of the terminal manifold N, $N_T \subset N$, and N is of codimension greater than 1, then the flow F is not a diffeomorphism for the terminal time $t = t_+(p) = T(p)$. In this case, regardless of the dimension of N, the value function $V^{\mathscr{E}}$ constructed in Theorem 5.2.1 has a well-defined continuous extension to N_T, since the terminal value of the cost,

$$C(T(p),p) = \varphi(T(p),\xi_+(p)) = \varphi(F(T(p),p)),$$

depends only on the terminal point $F(T(p),p)$, but not on the parameter p itself. Thus, even if parameterizations for $p_1 \neq p_2$ have the same terminal point, $(T,x) = F(T(p_1),p_1) = F(T(p_2),p_2)$, this nevertheless gives rise to a unique specification of the value as

$$V^{\mathscr{E}}(T,x) = \varphi(F(T(p),p)).$$

Hence, we can extend the definition of $V^{\mathscr{E}} = C \circ F^{-1}$ to the target $N_T \subset N$ of the parameterized family by taking any of the preimages of F in the parametrization of the target through $D_T = \{(T,p) : T = T(p), p \in P\}$.

Combining Theorems 5.1.1 and 5.2.1 therefore gives the following result about optimality (see Fig. 5.3):

Corollary 5.2.2. *Let $\mathscr{F}$ be a C^r-parameterized local field of extremals with target N_T in the terminal manifold N, $N_T \subset N$, and assume that its associated flow covers a domain G. Then, given any initial condition $(t_0,x_0) \in G$, $x_0 = x(t_0,p_0)$, the open-loop control $\bar{u}(t) = u(t,p_0)$, $t_0 \leq t \leq T(p_0)$, is optimal when compared with any other admissible control u for which the corresponding trajectory x (respectively, its graph) lies in G, i.e., $\mathscr{J}(\bar{u}) \leq \mathscr{J}(u)$.* ∎

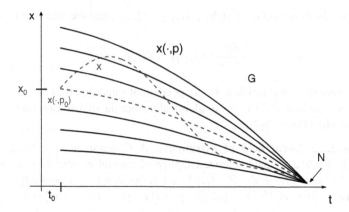

Fig. 5.3 A relative minimum: $\mathscr{J}\left(u_{(t_0,x_0)}\right) \leq \mathscr{J}(v)$

Corollary 5.2.2 is tailored to the formulation of sufficient conditions for local minima, which we shall pursue in the next section. For questions about global optimality, however, it is generally necessary to consider various local fields of extremals and then glue them together, a topic we shall develop in general only in Chap. 6. For a specific problem, however, this often is easily accomplished once the extremal trajectories have been constructed. Here we shall still show how these general constructions can rather easily (with considerably less technical effort than the calculation given in Sect. 5.1) be used to prove the optimality of the extremal synthesis for the Fuller problem constructed in Sect. 2.11.

5.2.3 The Fuller Problem Revisited

In contrast to the time-optimal control problem to the origin for the double integrator, in this problem the switching curves Γ_+ and Γ_- are not trajectories. However, as before, they form the source and target for two real analytic families $\mathscr{E}_\pm$ of normal extremals given by the trajectories for the constant controls $u = \pm 1$ (see Fig. 5.4).

It follows from the analysis of extremals in Sect. 2.11 that the switching curves Γ_+ and Γ_- can be parameterized as

$$\Gamma_+ : P = (-\infty, 0) \to \mathbb{R}^2, \qquad p \mapsto \begin{pmatrix} \zeta p^2 \\ p \end{pmatrix},$$

and

$$\Gamma_- : P = (0, \infty) \to \mathbb{R}^2, \qquad p \mapsto \begin{pmatrix} -\zeta p^2 \\ p \end{pmatrix},$$

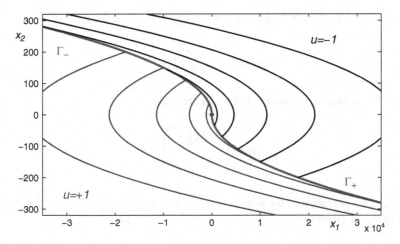

Fig. 5.4 Synthesis of optimal controlled trajectories for the Fuller problem

with $\zeta = \sqrt{\frac{\sqrt{33}-1}{24}}$. We again normalize the parameterizations so that all trajectories reach the origin at the final time $T = 0$. If a trajectory starts at the point $(\zeta p^2, p)^T \in \Gamma_+$ at time t_0, then it follows from Eq. (2.61) in the proof of Proposition 2.11.3 that the next switching is at time $t_1 = \frac{t_0}{\alpha}$ and that $x_2(t_0) + \alpha x_2(t_1) = 0$, where

$$\alpha = \sqrt{\frac{1+2\zeta}{1-2\zeta}} = 4.1302\ldots .$$

Starting at Γ_+, the control is given by $u \equiv +1$ on $[t_0, t_1] \subset (-\infty, 0)$, and we therefore also get that

$$x_2(t_1) - x_2(t_0) = t_1 - t_0 = \left(\frac{1}{\alpha} - 1\right) t_0.$$

Hence

$$t_0 = \frac{x_2(t_1) - x_2(t_0)}{\frac{1}{\alpha} - 1} = \frac{-\frac{1}{\alpha} - 1}{\frac{1}{\alpha} - 1} x_2(t_0) = \frac{\alpha + 1}{\alpha - 1} p.$$

We therefore define a family $\mathscr{E}_+$ of normal extremals that correspond to the control $u(t, p) \equiv +1$ over the domain

$$D_+ = \{(t, p):\ p < 0,\ t_-(p) \le t \le t_+(p)\}$$

with the functions t_+ and t_- given by

$$t_-(p) = \frac{\alpha + 1}{\alpha - 1} p \qquad \text{and} \qquad t_+(p) = \frac{1}{\alpha} \cdot \frac{\alpha + 1}{\alpha - 1} p.$$

The corresponding controlled trajectories $x(t,p)$ start at the point

$$\xi_-(p) = \begin{pmatrix} \zeta p^2 \\ p \end{pmatrix} \in \Gamma_+$$

at time $t_-(p)$ and then at time $t_+(p)$ reach the point

$$\xi_+(p) = \begin{pmatrix} -\zeta\left(\frac{p}{\alpha}\right)^2 \\ -\frac{p}{\alpha} \end{pmatrix} \in \Gamma_-.$$

An explicit integration of the trajectories is not required, but of course it can easily be done here, yielding

$$x_2(t,p) = t - \frac{2}{\alpha-1}p \quad \text{and} \quad x_1(t,p) = \frac{1}{2}x_2(t,p)^2 + \left(\zeta - \frac{1}{2}\right)p^2.$$

The multipliers $\lambda(t,p)$ are the solutions to the initial value problems

$$\dot{\lambda}_1(t,p) = -x_1(t,p), \qquad \lambda_1(t_-(p),p) = -\frac{1}{2}\zeta^2 p^3,$$

and

$$\dot{\lambda}_2(t,p) = -\lambda_1(t,p), \qquad \lambda_2(t_-(p),p) = 0.$$

As for the double integrator, $\lambda_2(t_-(p),p)$ vanishes, since the initial condition lies on the switching curve Γ_+, and λ_2 once more is the switching function of the problem. The value for $\lambda_1(t_-(p),p)$ then again is simply computed from the condition that

$$0 = H = \frac{1}{2}x_1^2 + \lambda_1 x_2 + \lambda_2 = \frac{1}{2}\zeta^2 p^4 + \lambda_1 p.$$

The multipliers λ could easily be computed explicitly, but there is no need to do so. It follows from the symmetries of the synthesis (see the proof of Proposition 2.11.3) that λ at the target is given by

$$\lambda_1(t_+(p),p) = -\frac{1}{2}\frac{x_1(t_+(p),p)^2}{x_2(t_+(p),p)} = -\frac{1}{2}\zeta^2 x_2(t_+(p),p)^3 \quad \text{and} \quad \lambda_2(t_+(p),p) = 0.$$

It remains to define the cost γ_- at the source Γ_+. This, in fact, does *not* require that one evaluate the objective along the trajectories, but it can be done much more elegantly by means of the shadow-price lemma. We shall prove in Sect. 6.1 (Theorem 6.1.1) that the value of a parameterized family of extremals remains continuously differentiable at a switching curve Γ if the two respective flows of trajectories cross Γ transversally. This is the case along Γ_+ and Γ_-. Already anticipating this result, the value of the cost $V_+ = V^{\mathcal{E}_+}$ of the parameterized family $\mathcal{E}_+$ for the source γ_- on the switching curve Γ_+, $\gamma_-(p) = V_+(\zeta p^2, p)$, can therefore be obtained by integrating the differential dV_+ of V_+ along the curve

$$Z : [p,0] \to \Gamma_+, \qquad s \mapsto (\zeta s^2, s).$$

By the shadow-price lemma, this gradient is given by the multiplier λ. Therefore, and simply postulating that

$$\nabla V_+(\zeta s^2, s) = \left(-\frac{1}{2}\zeta^2 s^3, 0\right), \tag{5.22}$$

we calculate the cost at the source as

$$
\begin{aligned}
\gamma_-(p) &= V_+(\zeta p^2, p) - V_+(0,0) = -\int_Z dV_+ \\
&= \int_0^p \frac{\partial V_+}{\partial x_1}(\zeta s^2, s)dx_1(s) + \frac{\partial V_+}{\partial x_2}(\zeta s^2, s)dx_2(s) \\
&= \int_0^p \left(-\frac{1}{2}\zeta^2 s^3 \cdot 2\zeta s + 0\right) ds = -\frac{1}{5}\zeta^3 p^5. \tag{5.23}
\end{aligned}
$$

With this specification, the transversality condition (5.13) is satisfied:

$$\lambda(t_-(p), p)\frac{\partial \xi_-}{\partial p}(p) = \left(-\frac{1}{2}\zeta^2 p^3, 0\right) \cdot \binom{2\zeta p}{1} = -\zeta^3 p^4 = \frac{\partial \gamma_-}{\partial p}(p).$$

Hence, $\mathcal{E}_+$ is a real-analytic parameterized family of normal extremals for the Fuller problem. Furthermore, since the control is constant, it simply follows from the uniqueness of solutions to ordinary differential equations that the corresponding trajectories do not intersect, and thus the corresponding flow map, here given by $F : D_+ \to G_+ \cup \Gamma_+$, $(t, p) \mapsto x(t, p)$, is a diffeomorphism (that extends into a neighborhood of the two switching curves Γ_+ and Γ_-). Thus $\mathcal{E}_+$ is a C^ω-field of normal extremals. We want to emphasize that *all these parameterizations are by means of simple calculations that directly follow from the analysis of extremals* carried out in Sect. 2.11.

Analogously, a real-analytic field $\mathcal{E}_-$ of extremal trajectories is constructed that corresponds to the control $u(t, p) \equiv -1$ and starts on the switching curve Γ_-. In this case the domain is described as

$$D_- = \{(t, p) : p > 0, \, t_-(p) \le t \le t_+(p)\}$$

with the functions t_+ and t_- now given by

$$t_-(p) = \frac{1+\alpha}{1-\alpha}p \quad \text{and} \quad t_+(p) = \frac{1}{\alpha} \cdot \frac{1+\alpha}{1-\alpha}p.$$

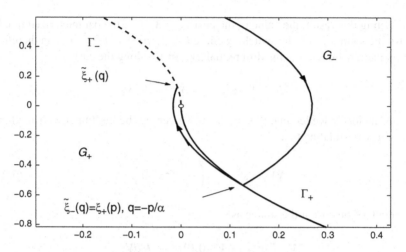

Fig. 5.5   The two parameterizations for a junction point on Γ_+

The initial conditions for the trajectories $x(t,p)$ on Γ_- are given by $\xi_-(p) = (-\zeta p^2, p)^T$, for the multipliers λ by $\lambda(t_-(p),p) = (-\frac{1}{2}\zeta^2 p^3, 0)$, and the cost γ_- at the source is $\gamma_-(p) = \frac{1}{5}\zeta^3 p^5$.

If we denote the flows induced by these two parameterized families of extremal trajectories by F_+ and F_-, respectively, then it follows from Theorem 5.2.1 that the values $V_+ = V^{\mathcal{E}_+} = C \circ F_+^{-1}$ and $V_- = V^{\mathcal{E}_-} = C \circ F_-^{-1}$ are continuously differentiable solutions to the Hamilton–Jacobi–Bellman equation for the Fuller problem on the images $G_+ = F_+(D_+)$ and $G_- = F_-(D_-)$, respectively. We now show that these two functions V_+ and V_- combine to a continuous function on the switching curves Γ_+ and Γ_-.

Proposition 5.2.1. *The values V_+ and V_- are continuous on the switching curves Γ_+ and Γ_-.*

Proof. Without loss of generality, we consider the switching curve Γ_+. It needs to be shown that the source cost of the parameterized field $\mathcal{E}_+$ is the same as the computed target cost for the parameterized field $\mathcal{E}_-$ at the junction on Γ_+. This indeed is a direct consequence of the construction, but we need to account for the fact that the same trajectory is described by different parameters in these two parameterizations (see Fig. 5.5).

We fix a parameter $p > 0$ in the field $\mathcal{E}_-$. The source cost γ_- for $\mathcal{E}_-$ on the switching curve Γ_- is defined as $\gamma_-(p) = \frac{1}{5}\zeta^3 p^5$, and it represents the value of the objective for the controlled trajectory in the field with initial point $\xi_-(p) = (-\zeta p^2, p)^T \in \Gamma_-$. The target cost for $\mathcal{E}_-$ on Γ_+, denoted by $\gamma_+(p)$, then is computed along the parameterized family $\mathcal{E}_-$ as

$$\gamma_+(p) = \gamma_-(p) - \int_{t_-(p)}^{t_+(p)} \frac{1}{2}x_1^2(t,p)dt.$$

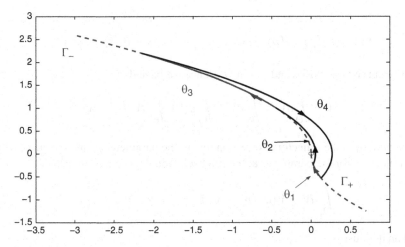

Fig. 5.6 The closed curve Θ

This is the cost of the objective when the initial point is given by the corresponding target $\xi_+(p)$, hence the minus sign at the integral. Note that it follows from the properties of the parameterized family that the target point is given by

$$\xi_+(p) = \begin{pmatrix} \varsigma \left(\frac{p}{\alpha}\right)^2 \\ -\frac{p}{\alpha} \end{pmatrix} \in \Gamma_+.$$

This point now becomes the source of a trajectory in the field $\mathscr{E}_+$. However, in $\mathscr{E}_+$, this point is parameterized in the form $\begin{pmatrix} \varsigma q^2 \\ q \end{pmatrix}$ for some $q < 0$. Hence the parameters relate as $q = -\frac{p}{\alpha}$. Using a tilde to distinguish the source cost for $\mathscr{E}_+$ from the analogous term for $\mathscr{E}_-$, this source cost $\tilde{\gamma}_-$ was therefore defined as $\tilde{\gamma}_-(q) = -\frac{1}{5}\varsigma^3 q^5$. Hence, continuity at the junction is equivalent to

$$\gamma_+(p) = \tilde{\gamma}_-\left(-\frac{p}{\alpha}\right). \tag{5.24}$$

In principle, this relation can be verified by an explicit computation. But a more elegant, and also more general, argument can be made using the theory developed so far. We work with the field $\mathscr{E}_-$, but in order not to clutter the notation, we drop the subscript for the associated objects. Thus V denotes the value associated with the field $\mathscr{E}_-$, $V = C \circ F_-^{-1}$, and so on. Since the corresponding parameterized trajectories $x(t,p)$ and parameterized cost $C(t,p)$ extend into an open neighborhood of the domain D, the associated value V extends as a continuously differentiable function into a simply connected open neighborhood $\tilde{G}$ of $G = F(D)$. For small $\varepsilon > 0$, consider the closed curve $\Theta = \Theta(\varepsilon)$ that lies in $\tilde{G}$ and is obtained by concatenating the following four smooth curves, $\Theta = \theta_1 * \theta_2 * \theta_3 * \theta_4$ (see Fig. 5.6):

$$\theta_1 : \quad [-p, -\varepsilon] \to \Gamma_+, \qquad\qquad s \mapsto \theta_1(s) = \xi_+(-s),$$

$$\theta_2 : \quad [-t_+(\varepsilon), -t_-(\varepsilon)] \to G_-, \qquad s \mapsto \theta_2(s) = x(-s, \varepsilon),$$

$$\theta_3 : \quad [\varepsilon, p] \to \Gamma_-, \qquad\qquad s \mapsto \theta_3(s) = \xi_-(s),$$
$$\theta_4 : \quad [t_-(p), t_+(p)] \to G_-, \qquad s \mapsto \theta_4(s) = x(s, p).$$

The curve Θ is closed and lies in $\tilde{G}$. Hence we have that

$$0 = \int_\Theta dV = \int_{\theta_1} dV + \int_{\theta_2} dV + \int_{\theta_3} dV + \int_{\theta_4} dV.$$

The curve θ_4 is the trajectory corresponding to the parameter p, and the curve θ_2 is the trajectory for the parameter ε, but run backward. Hence we have that

$$\int_{\theta_4} dV = V(\xi_+(p)) - V(\xi_-(p)) = \gamma_+(p) - \gamma_-(p)$$

and analogously

$$\int_{\theta_2} dV = V(\xi_-(\varepsilon)) - V(\xi_+(\varepsilon)) = \gamma_-(\varepsilon) - \gamma_+(\varepsilon) = \int_{t_-(\varepsilon)}^{t_+(\varepsilon)} \frac{1}{2} x_1(s, \varepsilon)^2 ds.$$

As $\varepsilon \to 0$, the length of the interval $[t_-(\varepsilon), t_+(\varepsilon)]$ converges to 0,

$$t_+(\varepsilon) - t_-(\varepsilon) = \left(\frac{1}{\alpha} - 1\right) \cdot \frac{1+\alpha}{1-\alpha}\varepsilon = \frac{1+\alpha}{\alpha}\varepsilon \to 0,$$

and $x_1(s, \varepsilon)$ remains bounded on this interval as $\varepsilon \to 0$. Hence $\int_{\theta_2} dV \to 0$ as $\varepsilon \to 0$.

The limits of θ_1 and θ_3 as $\varepsilon \to 0$ are well-defined and parameterize the sections on the switching curves Γ_+ and Γ_- that connect the origin with the points $\xi_+(p)$ and $\xi_-(p)$, respectively. It follows from Theorem 5.2.1 that the gradient of the value V associated with the family $\mathcal{E}_-$ is given by the multiplier λ of the parameterized family $\mathcal{E}_-$ (see Eq. (5.21)). Hence, and consistent with our original definition, it follows for the portion along Γ_- that

$$\lim_{\varepsilon \to 0} \int_{\theta_3} dV = \int_0^P \nabla V(\zeta s^2, s) d\xi_-(s)$$

$$= \int_0^P \lambda_1(t_-(s), s) d\xi_{1,-}(s) + \lambda_2(t_-(s), s) d\xi_{2,-}(s) ds$$

$$= \int_0^P \left(-\frac{1}{2}\zeta^2 s^3 \cdot (-2\zeta s) + 0\right) ds = \frac{1}{5}\zeta^3 p^5 = \gamma_-(p).$$

Analogously, the gradient of the value V along the switching curve Γ_+ is given by the multiplier λ at the terminal time $t_+(p)$. For $s \in [-p, -\varepsilon]$ the terminal points in the target are given by $\xi_+(-s) = \left(\zeta\left(\frac{s}{\alpha}\right)^2, \frac{s}{\alpha}\right)^T \in \Gamma_+$, and from the properties of the synthesis we have that

$$\lambda_1(t_+(-s),-s) = -\frac{1}{2}\zeta^2\xi_{2,+}(t_+(-s),-s)^3 = -\frac{1}{2}\zeta^2\left(\frac{s}{\alpha}\right)^3$$

and $\lambda_2(t_+(-s),-s) = 0$. Hence

$$\lim_{\varepsilon \to 0}\int_{\theta_1} dV = \int_{-p}^0 \nabla V\left(\zeta\left(-\frac{s}{\alpha}\right)^2, \frac{s}{\alpha}\right) d\xi_+(-s)$$

$$= \int_{-p}^0 \lambda_1(t_+(-s),-s)d\xi_{1,+}(-s) + \lambda_2(t_+(-s),-s)d\xi_{2,+}(-s)$$

$$= \int_{-p}^0 \left(-\frac{1}{2}\zeta^2\left(\frac{s}{\alpha}\right)^3 \cdot 2\zeta\left(\frac{s}{\alpha}\right)\frac{1}{\alpha} + 0\right) ds$$

$$= -\frac{1}{5}\zeta^3\left(\frac{s}{\alpha}\right)^5\bigg|_{-p}^0 = -\frac{1}{5}\zeta^3\left(\frac{p}{\alpha}\right)^5 = -\tilde{\gamma}_-\left(-\frac{p}{\alpha}\right).$$

Thus, overall we get that

$$\gamma_+(p) = \gamma_-(p) + \int_{\theta_4} dV = \left(\lim_{\varepsilon \to 0}\int_{\theta_3} dV\right) + \int_{\theta_4} dV$$

$$= -\lim_{\varepsilon \to 0}\left(\int_{\theta_1} dV + \int_{\theta_2} dV\right) = \tilde{\gamma}_-\left(-\frac{p}{\alpha}\right).$$

Hence the values join to a continuous function along Γ_+.   □

Thus the combined value function V,

$$V(x) = \begin{cases} V_+(x) & \text{if } x \in G_+ \cup \Gamma_+, \\ V_-(x) & \text{if } x \in G_- \cup \Gamma_-, \end{cases}$$

is well-defined and is continuous on $\mathbb{R}^2$. In fact, this function is continuously differentiable away from the origin, and this is an immediate consequence of the fact that the two flows corresponding to $\mathcal{E}_+$ and $\mathcal{E}_-$ cross the switching surfaces transversally. This result will be proven in general in Sect. 6.1 (Theorem 6.1.1). Hence the value V constructed above is a continuous function that is continuously differentiable away from the origin and solves the HJB equation. By Theorem 5.1.1, this proves the global optimality of the synthesis. Obviously, there is no need, nor reason, to actually calculate the value function V explicitly. *Having the synthesis of extremals* that was constructed in Sect. 2.11, *the method of characteristics immediately, and with little further ado, implies the optimality of the corresponding controlled trajectories.* This is the idea behind the concept of a regular synthesis to be developed further in Chap. 6.

5.3   Neighboring Extremals and Sufficient Conditions for a Local Minimum

We now use the method of characteristics to formulate sufficient conditions for a local minimum. We are interested in conditions that establish the optimality of a controlled reference trajectory $\bar{\Gamma} = (\bar{x}, \bar{u})$ over other controlled trajectories $\Gamma = (x, u)$ that have the property that the trajectories $\bar{x}$ and x are "close" to each other in the state space. When the terminal manifold in problem [OC] is of lower dimension, then the classical notion of a strong local minimum is somewhat too restrictive and inconvenient to handle. Therefore, here we adopt the following definition of a relative minimum.

Definition 5.3.1 (Relative minimum). Let $\bar{\Gamma} = (\bar{x}, \bar{u})$ be a controlled trajectory for the optimal control problem [OC] and let $\mathscr{T}$ be a family of controlled trajectories that contains $\bar{\Gamma}$. We say that $\bar{\Gamma}$ is a relative minimum over $\mathscr{T}$ if $\mathscr{J}(\bar{u}) \leq \mathscr{J}(u)$ for all controls u such that the corresponding controlled trajectory (x, u) lies in $\mathscr{T}$.

Clearly, this becomes a meaningful statement only if the class $\mathscr{T}$ is reasonably large. If $\mathscr{T}$ consists of all trajectories that steer the initial condition of $\bar{x}$ into N, then $\bar{\Gamma}$ is globally optimal. Typically, and this will be the case throughout this chapter, if $\bar{\Gamma}$ is defined over an interval $[\tau, T]$, then $\mathscr{T}$ will be the class of all controlled trajectories (x, u) for which the trajectories x (respectively, their graphs, in the time-dependent case) lie in some open neighborhood G of the restriction of the reference trajectory $\bar{x}$ (respectively, its graph) to the semiopen interval $[\tau, T)$. Such families arise naturally if the controlled reference trajectory $\bar{\Gamma}$ is embedded into a local field of extremals. Since the corresponding flow F of trajectories cannot be injective at the terminal manifold when N is of lower dimension, the final point is excluded (see also Fig. 5.3).

Definition 5.3.2 (Relative minimum over a domain G). We say that a controlled trajectory $\bar{\Gamma} = (\bar{x}, \bar{u})$ defined over a compact interval $[\tau, T]$ provides a relative minimum over a domain G if the restriction of the trajectory $\bar{x}$ (respectively, its graph) to the open interval (τ, T) is contained in the interior of G and if for any other controlled trajectory (x, u) that lies in G (respectively, has its graph in G) and steers $\bar{x}(\tau)$ into the terminal manifold N, the value for the cost is not better than for Γ, $\mathscr{J}(\bar{u}) \leq \mathscr{J}(u)$.

It is not required that the corresponding controls remain close as well. As in the calculus of variations, being able to embed the controlled reference trajectory into a local field gives rise to sufficient conditions for a relative minimum. We shall construct these embeddings in two steps: (i) set up a parameterized family of extremals that includes the reference extremal and (ii) develop conditions that guarantee that the associated flow F of controlled trajectories is injective around the reference trajectory. It then follows from Corollary 5.2.2 that the reference controlled trajectory is a relative minimum with respect to all controlled trajectories (x, u) for which x lies in the region G covered injectively by the flow associated with the trajectories in the parameterized family.

In this section, we consider the case that the reference control lies in the interior of the control set; we shall analyze bang-bang controls later in Sect. 6.1. We develop sufficient conditions for local optimality that involve second-order terms of the data, and we henceforth assume that the general regularity assumptions made in the beginning of the chapter are in effect with $\ell \geq 2$. Furthermore, since the questions addressed in this section typically arise in the context of regulating a system over a prescribed interval, *we consider only the time-dependent formulation*. As a side comment, the brachistochrone problem and the problem of minimum surfaces of revolution in the classical calculus of variations are initially formulated for problems for general curves in the plane. But once the mathematical model is written in terms of graphs of functions, this then matches the framework used here. The results that we present now are equally classical in the optimal control literature and date back to the early control literature from the sixties. They can be found in a more engineering orientated, but less rigorous formulation, for example, in [64, Chap. 6]. Our presentation here is mathematically precise and emphasizes the geometric properties of the underlying flow of extremals. For simplicity of exposition, we consider only the case that the terminal time T is fixed. In this case, the underlying geometric properties come out quite cleanly, whereas the formulas become a bit tedious if the final time T is defined implicitly through one of the constraints that define the terminal manifold N. We refer the reader to [64, Chap. 6] for a formal discussion of this situation, but restrict our presentation to the following optimal control problem:

[OC_T] For a *fixed terminal time* T, minimize the functional

$$\mathscr{J}(u;t_0,x_0) = \int_{t_0}^{T} L(s,x(s),u(s))ds + \varphi(x(T))$$

over all admissible controls $u \in \mathscr{U}$ for which the corresponding trajectory x,

$$\dot{x}(t) = f(t,x(t),u(t)), \qquad x(t_0) = x_0,$$

satisfies the terminal constraint $\Psi(x(T)) = 0$.

5.3.1 A Canonical Parameterized Family of Extremals

In this section we construct a canonical embedding for an extremal whose control takes values in the interior of the control set provided the matrix of the second derivatives $\frac{\partial^2 H}{\partial u^2}$ is nonsingular along the reference extremal (also, see Sect. 2.8).

Definition 5.3.3 (Nonsingular extremal). A normal extremal $\Lambda = ((x, u), \lambda)$ consisting of a controlled trajectory $\Gamma = (x, u)$ defined over an interval $[\tau, T]$ with corresponding multiplier λ is said to be nonsingular if for all $t \in [\tau, T]$ we have that

$$\frac{\partial H}{\partial u}(t, \lambda(t), x(t), u(t)) \equiv 0$$

and the matrix

$$\frac{\partial^2 H}{\partial u^2}(t, \lambda(t), x(t), u(t))$$

is positive definite. As in the calculus of variations, in this case we say that the *strengthened Legendre condition* is satisfied along the extremal Λ.

If the control $u(t)$ of a normal extremal Λ lies in the interior of the control set, then it is an immediate consequence of the actual minimization condition of the maximum principle that $\frac{\partial H}{\partial u}(t, \lambda(t), x(t), u(t)) \equiv 0$ and that the matrix $\frac{\partial^2 H}{\partial u^2}(t, \lambda(t), x(t), u(t))$ is positive semidefinite. We refer to this property also as the *Legendre condition*.

Theorem 5.3.1. *Let $\Lambda = ((\bar{x}, \bar{u}), \bar{\lambda})$ be a nonsingular extremal for problem [OC_T] defined over an interval $[\tau, T]$ and suppose*

1. *for every $t \in [\tau, T]$ the control $\bar{u}(t)$ lies in the interior of the control set, $\bar{u}(t) \in \text{int}(U)$, and*
2. *$\bar{u}(t)$ is the unique minimizer of the function $v \mapsto H(t, \bar{\lambda}(t), \bar{x}(t), v)$ over the control set U.*

Then there exist a canonical, nicely $C^{\ell-1}$-parameterized family of nonsingular extremals $\mathcal{E}$ with domain $D = \{(t, p) : p \in P, t_-(p) \le t \le T\}$ and a parameter value $p_0 \in P$ such that for all $t \in [\tau, T]$ we have that

$$x(t, p_0) \equiv \bar{x}(t), \quad u(t, p_0) \equiv \bar{u}(t), \quad \text{and} \quad \lambda(t, p_0) \equiv \bar{\lambda}(t).$$

We say that the reference extremal Λ is embedded into the family $\mathcal{E}$.

Condition (2) that $\bar{u}(t)$ is the unique minimizer is important in this construction. It holds, for instance, if the Hamiltonian H is strictly convex in u. If it is not satisfied, the local construction below can still be carried out, but it is not guaranteed that the resulting controlled trajectories are extremal, since the minimizing control might switch in any neighborhood of p_0. In such a case, the smoothness properties that we require will not be satisfied, and different parameterized families of extremals will need to be patched together.

Proof. Let ρ denote the lift of the graph of the controlled trajectory $\bar{x}$ into the cotangent bundle defined by the extremal Λ, i.e., $\rho : [\tau, T] \to [\tau, T] \times \mathbb{R}^n \times (\mathbb{R}^n)^*, t \mapsto \rho(t) = (t, \bar{x}(t), \bar{\lambda}(t))$. Since the strengthened Legendre condition is satisfied along ρ, it follows from the implicit function theorem that for every time $t \in [\tau, T]$ there

Fig. 5.7 A tubular neighborhood

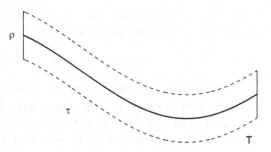

exist neighborhoods V_t of $\rho(t)$ and W_t of $\bar{u}(t)$, which, without loss of generality, we take in the forms

$$V_t = (t - \varepsilon, t + \varepsilon) \times B_\varepsilon(\bar{x}(t)) \times B_\varepsilon(\bar{\lambda}(t)) \subset \mathbb{R} \times \mathbb{R}^n \times (\mathbb{R}^n)^*$$

and $W_t = B_\delta(\bar{u}(t)) \subset \text{int}(U)$, such that for all $(s, x, \lambda) \in V_t$, the equation $H_u(s, \lambda, x, u) = 0$ has a unique solution $u = u(s, x, \lambda) \in W_t$. This solution is continuous, $(\ell - 1)$-times continuously differentiable in x and λ, and satisfies

$$u(s, \bar{x}(s), \bar{\lambda}(s)) = \bar{u}(s) \quad \text{for all} \quad s \in (t - \varepsilon, t + \varepsilon).$$

The parameters ε and δ generally depend on the time t, $\varepsilon = \varepsilon(t)$, and $\delta = \delta(t)$, and the stated properties are valid by choosing ε small enough. Since the curve ρ is compact, there exists a finite subcover of this curve with neighborhoods V_{t_i}, $i = 1, \ldots, k$. If two of these neighborhoods overlap, $V_{t_1} \cap V_{t_2} \neq \emptyset$, then there exists a section over the interval

$$(\sigma_1, \sigma_2) = (t_1 - \varepsilon(t_1), t_1 + \varepsilon(t_1)) \cap (t_2 - \varepsilon(t_2), t_2 + \varepsilon(t_2))$$

that is common to both neighborhoods, and over this section both functions $u_1 = u_1(t, x, \lambda)$ and $u_2 = u_2(t, x, \lambda)$ are solutions of the equation $H_u(t, \lambda, x, u) = 0$ that satisfy $u(s, \bar{x}(s), \bar{\lambda}(s)) = \bar{u}(s)$ for $t \in (\sigma_1, \sigma_2)$. Since the solutions in the neighborhoods V_{t_1} and V_{t_2} are unique, it follows that $u_1(t, x, \lambda) = u_2(t, x, \lambda)$ for all $(t, x, \lambda) \in V_{t_1} \cap V_{t_2}$. This allows us to define a function $u = u(s, x, \lambda)$ in a neighborhood of the curve ρ by setting $u(s, x, \lambda) = u_i(s, x, \lambda)$ whenever $(s, x, \lambda) \in V_{t_i}$. Since there are only finitely many neighborhoods, there exists a positive ε (now independent of t) such that this function exists, is continuous, and is $(\ell - 1)$-times continuously differentiable in x and λ on a tubular neighborhood

$$V = \left\{ (t, x, \lambda) : \ \tau \leq t \leq T, \ \|x - \bar{x}(t)\| < \varepsilon, \ \|\lambda - \bar{\lambda}(t)\| < \varepsilon \right\}$$

of the curve ρ (see Fig. 5.7).

By choosing ε sufficiently small, we can also guarantee that for all $(t,x,\lambda) \in V$, the control $u(t,x,\lambda)$ is the unique minimizer of the function

$$\eta : U \to \mathbb{R}, \quad v \mapsto \eta(v) = H(t,\lambda,x,v).$$

For by assumption, this holds along the curve ρ, and by construction, $u(t,x,\lambda)$ is a stationary point of η. Since the matrix $\frac{\partial^2 H}{\partial u^2}(t,\lambda,x,u(t,x,\lambda))$ is positive definite along the compact curve ρ, by continuity this condition is preserved for small enough ε, and thus $u(t,x,\lambda)$ is a strict local minimizer for the function η. By assumption, the control $\bar{u}(t)$ is the unique minimizer of the function $v \mapsto H(t,\bar{\lambda}(t),\bar{x}(t),v)$ over the control set U everywhere on the compact curve ρ. Since the minimum value is a continuous function, for ε small enough the function η still has a unique minimum for all points $(t,x,\lambda) \in V$, and this minimum lies in the interior of the control set, $\mathrm{int}(U)$. Once more, this is a consequence of the compactness of ρ. Hence $u(t,x,\lambda)$ is this minimum of η over the control set U. Note that this need not hold any longer if there exist a time $t \in [\tau,T]$ and a control value $v \in U$ such that $H(t,\bar{\lambda}(t),\bar{x}(t),\bar{u}(t)) = H(t,\bar{\lambda}(t),\bar{x}(t),v)$. In this case, it is possible that control values near v are better than the local solution $u(t,x,\lambda)$ near the point $(t,\bar{\lambda}(t),\bar{x}(t))$. For this reason assumption (2) is needed.

We define the flow of extremals by parameterizing the terminal points for the states and multipliers. Since N is an embedded submanifold, there exists a canonical parametrization of extremals in terms of the final points for the state x in the manifold N and the normal vectors to the manifold N at x, the so-called *normal bundle*: Let $\xi_0 = \bar{x}(T) \in N$ and let $v_0 \in (\mathbb{R}^{n-k})^*$ be the covector in the terminal condition for the multiplier $\bar{\lambda}(T)$, $\bar{\lambda}(T) = \varphi_x(\xi_0) + v_0 \Psi_x(\xi_0)$. Let P_1 be an open neighborhood of the origin in $\mathbb{R}^k$ and let $\xi : P_1 \to N$, $y \mapsto \xi(y)$, be a C^ℓ-coordinate chart for N centered at ξ_0, i.e., ξ is a C^ℓ-diffeomorphism such that $\xi(0) = \xi_0$. By our standing assumption, the columns of the matrix $\Psi_x^T(\xi(y))$ are linearly independent and therefore form a basis for the normal space to N at $\xi(y)$. Hence the vectors v^T provide coordinates for this normal space. Thus, if P_2 is an open neighborhood of v_0^T in $\mathbb{R}^{n-k}$ and $P = P_1 \times P_2$, then the mapping

$$\omega : P \to N \times \left(\mathbb{R}^{n-k}\right), \qquad p = (y,v^T) \mapsto \omega(p) = (\xi(y), v^T), \qquad (5.25)$$

defines a C^ℓ-coordinate chart for the normal bundle to N at the point (ξ_0, v_0^T). Define the *endpoint mapping for the states and costates*, $\Omega : P \to N \times (\mathbb{R}^n)^*$, as

$$p = (y,v^T) \mapsto \Omega(p) = (\xi(y), \varphi_x(\xi(y)) + v\Psi_x(\xi(y))) = (x(T,p), \lambda(T,p)). \quad (5.26)$$

Note that $\Omega(p_1) = \Omega(p_2)$ if and only if $\omega(p_1) = \omega(p_2)$, and thus Ω is a $C^{\ell-1}$-diffeomorphism onto its n-dimensional image. We then define the trajectories $x = x(t,p)$ and costates $\lambda = \lambda(t,p)$ as the solutions to the combined dynamics

$$\dot{x} = f(t,x,u(t,x,\lambda)), \qquad (5.27)$$

$$\dot{\lambda} = -L_x(t,x,u(t,x,\lambda)) - \lambda f_x(t,x,u(t,x,\lambda)), \qquad (5.28)$$

with terminal values $(x(T,p),\lambda(T,p)) = \Omega(p)$. The control $u(t,p)$ is given by

$$u(t,p) = u(t,x(t,p),\lambda(t,p)). \tag{5.29}$$

By taking ε in the definition of the neighborhood V small enough, it follows from the continuous dependence of solutions to ordinary differential equations on initial conditions and parameters that $x(\cdot,p)$ and $\lambda(\cdot,p)$ exist and are $(\ell-1)$-times continuously differentiable in the parameter p over some interval $[t_-(p),T]$ with $t_- : P \to (-\infty,T]$, $p \mapsto t_-(p)$, a smooth function that satisfies $t_-(p_0) = \tau$. This therefore defines a nicely $C^{\ell-1}$-parameterized family of nonsingular extremals $\mathscr{E}$ that reduces to the reference extremal Λ for $p_0 = (0,v_0^T)$. $\qquad\square$

We henceforth take this canonical, nicely $C^{\ell-1}$-parameterized family of nonsingular extremals $\mathscr{E}$ as given and denote its domain by $D = \{(t,p) : p \in P, t_-(p) \leq t \leq T\}$ with $t_-(p_0) = \tau$. For the parameterization of the terminal states and multipliers, we also write $\xi(p)$ and $v(p)$, but always assume that the mapping $\omega : P \to N \times \mathbb{R}^{n-k}$, $p \mapsto (\xi(p),v^T(p))$, is a diffeomorphism.

If the matrix $DF(t,p_0)$ is nonsingular on the interval $[\tau,T]$, then for every time t in this interval there exists a neighborhood O_t of (t,p_0), $O_t \subset D$, such that the restriction of the flow map F to O_t is a $C^{\ell-1}$-diffeomorphism. If we thus define

$$G = \bigcup_{\tau \leq t < T} F(O_t), \tag{5.30}$$

then G is an open set that except for the terminal point $\bar{x}(T)$, contains the controlled reference trajectory $\bar{\Gamma} = (\bar{x},\bar{u})$. By Corollary 5.2.2, $\bar{\Gamma}$ is a relative minimum over the domain G.

Corollary 5.3.1. *The canonical, nicely $C^{\ell-1}$-parameterized family of nonsingular extremals $\mathscr{E}$ constructed in Theorem 5.3.1 defines a* local field *around the reference trajectory if and only if the matrix $\frac{\partial x}{\partial p}(t,p_0)$ is nonsingular on the interval $[\tau,T]$.* ■

5.3.2 Perturbation Feedback Control and Regularity of the Flow F

Perturbation feedback control is probably the most important practical application of optimal control theory. It is widely used, for example, in autopilots on commercial aircraft, chemical process control, and many other control schemes that regulate a system around some predetermined set point. Mathematically, it is based on formulas that establish the local injectivity of the flow F. In our slightly different, but mathematically more classical viewpoint, the *neighboring extremals* considered in the engineering literature are the members of the canonical parameterized family of extremals constructed above, and the conditions that are imposed guarantee that this construction actually provides a local embedding of the reference extremal into a *field of extremals*. Chapter 6 of the textbook by Bryson and Ho [64] provides an engineering-type exposition of calculations that indeed provide sufficient conditions for the matrix $\frac{\partial x}{\partial p}(t, p_0)$ to be nonsingular along a reference extremal. As we have seen, the boundary conditions at time T for $x(T, p)$ and $\lambda(T, p)$ define an n-dimensional manifold. The combined flow in the cotangent bundle, the combined state-multiplier space, is a $C^{\ell-1}$-diffeomorphism in (x, λ)-space, since together (x, λ) are solutions to the differential equations (5.27) and (5.28), and thus the resulting flow map Φ_t, which maps the terminal conditions to the solutions at time t, is one-to-one. Hence the image of the flow for a given time t always is an n-dimensional manifold. *Local optimality* of the reference trajectory follows if one can ensure that *the projection* $\pi : (x, \lambda) \mapsto x$ *is one-to-one away from the terminal time* T. For if this is the case, then the flow locally covers an n-dimensional set in the state space, hence a neighborhood of the reference trajectory. For problem [OC$_T$] (and even in the case that the terminal time becomes part of the constraints and is defined implicitly) there exist classical results that characterize the local regularity of the flow along a reference extremal in terms of solutions to Riccati equations [64, Chap. 6]. In the engineering literature, these calculations, which are collectively known as perturbation feedback control, are usually treated formally without any relations to the underlying geometric properties they represent. It are these geometric facts that we want to elucidate.

In the subsequent calculations, x, u, and λ and their partial derivatives are evaluated at (t, p), partial derivatives of f are evaluated along the controlled trajectories of the family, $(t, x(t, p), u(t, p))$, and all partials of H are evaluated along the full extremals, $(t, \lambda(t, p), x(t, p), u(t, p))$. For notational clarity, however, we drop these arguments. Recall that the matrix $\frac{\partial x}{\partial p}(t, p)$ of the partial derivatives with respect to the parameter p is the solution of the variational equation of the dynamics, i.e.,

$$\frac{d}{dt}\left(\frac{\partial x}{\partial p}\right) = f_x \frac{\partial x}{\partial p} + f_u \frac{\partial u}{\partial p}$$

(see Eq. (5.12)). In order to eliminate $\frac{\partial u}{\partial p}$, we differentiate the identity

$$H_u^T(t, \lambda(t, p), x(t, p), u(t, p)) \equiv 0$$

with respect to p. The notation has been set up for column vectors, and differentiating the m-dimensional column vector H_u^T with respect to x, we get an $m \times n$ matrix whose row vectors are the x-gradients of the components of H_u^T. We denote this matrix by H_{ux}. In particular, under our general differentiability assumptions, the mixed partials are equal and we have that $H_{xu} = H_{ux}^T$. However, the multiplier λ is written as a row vector, and in order to be consistent in our notation, we need to differentiate H_u^T with respect to the column vector λ^T. For example, $H_{u\lambda^T}$ is the $m \times n$ matrix whose row vectors are the partial derivatives of the components of H_u^T with respect to λ^T. Since $H_u^T = f_u^T \lambda^T + L_u^T$, we simply have that $H_{u\lambda^T} = f_u^T$, etc.

Differentiating $H_u^T = 0$ thus gives

$$0 = H_{u\lambda^T}\frac{\partial\lambda^T}{\partial p} + H_{ux}\frac{\partial x}{\partial p} + H_{uu}\frac{\partial u}{\partial p} = f_u^T\frac{\partial\lambda^T}{\partial p} + H_{ux}\frac{\partial x}{\partial p} + H_{uu}\frac{\partial u}{\partial p}.$$

Along a nonsingular extremal, the matrix H_{uu} is nonsingular, and it therefore follows that

$$\frac{\partial u}{\partial p} = -H_{uu}^{-1}\left(H_{ux}\frac{\partial x}{\partial p} + f_u^T\frac{\partial\lambda^T}{\partial p}\right). \tag{5.31}$$

Substituting this relation into the variational equation for x, we obtain

$$\frac{d}{dt}\left(\frac{\partial x}{\partial p}\right) = \left(f_x - f_u H_{uu}^{-1}H_{ux}\right)\frac{\partial x}{\partial p} - f_u H_{uu}^{-1}f_u^T\frac{\partial\lambda^T}{\partial p}.$$

The equation for the partial derivative $\frac{\partial\lambda^T}{\partial p}$ follows by differentiating the adjoint equation:

$$\frac{d}{dt}\left(\frac{\partial\lambda^T}{\partial p}\right) = \frac{\partial^2\lambda^T}{\partial t\partial p} = \frac{\partial}{\partial p}\left(\dot{\lambda}^T\right) = \frac{\partial}{\partial p}\left(-\frac{\partial H}{\partial x}\right)^T$$

$$= -H_{xx}\frac{\partial x}{\partial p} - H_{x\lambda^T}\frac{\partial\lambda^T}{\partial p} - H_{xu}\frac{\partial u}{\partial p} \tag{5.32}$$

$$= -\left(H_{xx} - H_{xu}H_{uu}^{-1}H_{ux}\right)\frac{\partial x}{\partial p} - \left(f_x - f_u H_{uu}^{-1}H_{ux}\right)^T\frac{\partial\lambda^T}{\partial p}.$$

Hence, we have the following fact:

Proposition 5.3.1. *Given a C^r-parameterized family of nonsingular extremals, the matrices $\frac{\partial x}{\partial p}(t,p)$ and $\frac{\partial\lambda^T}{\partial p}(t,p)$ are solutions to the homogeneous linear matrix differential equation*

$$\begin{pmatrix} \frac{d}{dt}\left(\frac{\partial x}{\partial p}\right) \\ \frac{d}{dt}\left(\frac{\partial\lambda^T}{\partial p}\right) \end{pmatrix} = \begin{pmatrix} f_x - f_u H_{uu}^{-1}H_{ux} & -f_u H_{uu}^{-1}f_u^T \\ -\left(H_{xx} - H_{xu}H_{uu}^{-1}H_{ux}\right) & -\left(f_x - f_u H_{uu}^{-1}H_{ux}\right)^T \end{pmatrix}\begin{pmatrix} \frac{\partial x}{\partial p} \\ \frac{\partial\lambda^T}{\partial p} \end{pmatrix}. \tag{5.33}$$

We have already encountered these equations in Sect. 2.4. However, extensions of those results are needed to deal with terminal manifolds N. We briefly dispose of the simpler problem without terminal constraints.

(a) Neighboring extremals without terminal constraints.

In this case, the parameter set P is simply a neighborhood of the terminal value $\xi_0 = \bar{x}(T)$ of the controlled reference trajectory. Using $\xi(p) = p$, the terminal conditions for the variational equations (5.33) are given by

$$\frac{\partial x}{\partial p}(T, p) = \text{Id} \quad \text{and} \quad \frac{\partial \lambda^T}{\partial p}(T, p) = \varphi_{xx}(p).$$

Since $\frac{\partial x}{\partial p}(T, p)$ is nonsingular, Proposition 2.4.1 directly applies to give the following result:

Theorem 5.3.2. *Let $\mathscr{E}$ be a nicely C^1-parameterized family of nonsingular extremals with domain $D = \{(t, p) : p \in P, t_-(p) \le t \le T\}$ for the optimal control problem [OC_T] without terminal constraints. Then the matrix $\frac{\partial x}{\partial p}(t, p)$ is nonsingular over the interval $[\tau, T]$ if and only if the solution $S = S(t, p)$ to the Riccati differential equation*

$$\dot{S} + Sf_x + f_x^T S + H_{xx} - (Sf_u + H_{xu})H_{uu}^{-1}(H_{ux} + f_u^T S) \equiv 0 \qquad (5.34)$$

with terminal condition

$$S(T, p) = \varphi_{xx}(p) \qquad (5.35)$$

exists over the full interval $[\tau, T]$. ■

The Riccati equation

$$\dot{S} + S\left(f_x - f_u H_{uu}^{-1} H_{ux}\right) + \left(f_x - f_u H_{uu}^{-1} H_{ux}\right)^T S - Sf_u H_{uu}^{-1} f_u^T S + H_{xx} - H_{xu}H_{uu}^{-1}H_{ux} \equiv 0$$

of Proposition 2.4.1 can be written more concisely in the form (5.34) given here.

Proposition 5.3.2. *If $\frac{\partial x}{\partial p}(t, p_0)$ is nonsingular for all times $t \in [t_-(p_0), T]$, then there exists a neighborhood P of p_0 such that the restriction of the flow F to $[t_-(p_0), T] \times P$ is a $C^{1,\ell-1}$-diffeomorphism. In this case, the reference controlled trajectory is a strong local minimum for the optimal control problem [OC_T].*

Proof. Since the flow F along the trajectory $x(\cdot, p_0)$ is regular, it follows from the inverse function theorem that for every time $s \in [t_-(p_0), T]$ there exists a neighborhood D_s of (s, p_0) such that the restriction of F to D_s is a $C^{1,\ell-1}$-diffeomorphism. Without loss of generality, we may take D_s in the form $D_s = I_s \times P_s$, where I_s is an open interval and P_s an open neighborhood of p_0. The sets $\{D_s : s \in [t_-(p_0), T]\}$ form an open cover of the compact set $[t_-(p_0), T] \times \{p_0\}$, and thus there exists a finite subcover $\{D_{s_i} : s_i \in [t_-(p_0), T], \ i = 1, \ldots, r\}$. Let $P = \bigcap_{i=1}^{r} P_{s_i}$. Then the map F is a $C^{1,\ell-1}$-diffeomorphism on $D = \{(t, p) : p \in P, t_-(p_0) \le t \le T\}$. For suppose $F(s_1, p_1) = F(s_2, p_2)$. Since the flow map F is defined in terms of the graphs of the trajectories, we have $s_1 = s_2$, and this time lies in one of the intervals I_{s_i}. But $F \upharpoonright I_{s_i} \times P$ is a $C^{1,\ell-1}$-diffeomorphism, and since both p_1 and p_2 lie in P, we have $p_1 = p_2$ as well. Thus F is injective on D. Furthermore, F has a differentiable inverse by the inverse function theorem.   □

The controlled reference trajectory $(\bar{x}, \bar{u}) = (x(\cdot, p_0), u(\cdot, p_0))$ thus gives a strong local minimum for the problem $[OC_T]$ in the following classical form: there exists an $\varepsilon > 0$ such that for any other admissible controlled trajectory (x, u) with the same initial condition, $x(\tau) = \bar{x}(\tau)$, that satisfies $\|x(t) - \bar{x}(t)\| < \varepsilon$ for all $t \in [\bar{t}_0, T]$, we have that $\mathscr{J}(\bar{u}) \le \mathscr{J}(u)$. Summarizing, we thus have proven the following result:

Corollary 5.3.2. *Consider the optimal control problem $[OC_T]$ without terminal constraints. Let $\Lambda = ((\bar{x}, \bar{u}), \lambda)$ be a nonsingular extremal defined over $[\tau, T]$ and suppose that (i) for every $t \in [\tau, T]$ the control $\bar{u}(t)$ lies in the interior of the control set, $\bar{u}(t) \in \text{int}(U)$, (ii) $\bar{u}(t)$ is the unique minimizer of the function $v \mapsto H(t, \bar{\lambda}(t), \bar{x}(t), v)$ over the control set U, and, (iii) along Λ, the Riccati equation (5.34) with terminal condition (5.35) has a solution over the full interval $[\tau, T]$. Then $(\bar{x}, \bar{u})$ is a strong local minimum.*   ∎

Thus, for this problem, the existence of a solution to Eqs. (5.34) and (5.35) on the compact interval $[\tau, T]$ is the generalization of the *strengthened Jacobi condition* from the calculus of variations. As in the calculus of variations, we shall see in the next section that the existence of a solution on the half-open interval $(\tau, T]$ is a necessary condition for a local minimum. While the solution $S = S(t, p)$ exists, it follows from Proposition 2.4.1 that $\frac{\partial x}{\partial p}(t, p)$ and $\frac{\partial \lambda^T}{\partial p}(t, p)$ are related as

$$\frac{\partial \lambda^T}{\partial p}(t, p) = S(t, p) \frac{\partial x}{\partial p}(t, p),$$

and thus the Hessian matrix of the corresponding value $V^{\mathscr{E}}$ is given by (cf. Eq. (5.20))

$$\frac{\partial V^{\mathscr{E}}}{\partial x^2}(t, x(t, p)) = \frac{\partial \lambda^T}{\partial p}(t, p) \left(\frac{\partial x}{\partial p}(t, p) \right)^{-1} = S(t, p).$$

However, these relations need to be modified when terminal constraints are imposed.

(b) Neighboring extremals with terminal constraints $N = \{x \in \mathbb{R}^n : \Psi(x) = 0\}$.

In this case, the flow map F is singular at the terminal time $t = T$, and since $\frac{\partial x}{\partial p}(T, p) = \frac{\partial \xi}{\partial p}(p)$, the image of $\frac{\partial x}{\partial p}(T, p)$ is given by the tangent space of N at $\xi(p)$. It is the second part of the parameter, the multiplier v, that desingularizes the flow F around the terminal manifold. Differentiating the transversality condition (5.14) for $\lambda(T, p)$ with respect to p, and taking the transpose, we have that

$$\frac{\partial \lambda^T}{\partial p}(T, p) = [\varphi_{xx}(\xi(p)) + v(p)\Psi_{xx}(\xi(p))] \frac{\partial \xi}{\partial p}(p) + \Psi_x^T(\xi(p)) \frac{\partial v^T}{\partial p}(p). \quad (5.36)$$

In this equation, the notation $v\Psi_{xx}$ is a convenient shortcut for $v\Psi_{xx} = \sum_{i=1}^{n-k} v_i \frac{\partial^2 \psi_i}{\partial x^2}$ with $\frac{\partial^2 \psi_i}{\partial x^2}$ denoting the Hessian matrices of the functions ψ_i that define the terminal constraint, $\Psi = (\psi_1, \dots, \psi_{n-k})^T$. Thus, the first term in Eq. (5.36) is the sum of $1 + n - k$ vectors, each of which is obtained by an $n \times n$ matrix acting on the tangent vector $\frac{\partial \xi}{\partial p}(p)$ to N at $\xi(p)$. Recall that the multiplier $v = v(p)$ depends only on the extremal, but not on time. Equation (5.36) therefore points to the following ansatz for the solutions to the variational equations:

$$\frac{\partial \lambda^T}{\partial p}(t,p) = S(t,p)\frac{\partial x}{\partial p}(t,p) + R^T(t,p)\frac{\partial v^T}{\partial p}(p).$$

Substituting this relation into the differential equations (5.33) for $\frac{\partial x}{\partial p}$ and $\frac{\partial \lambda^T}{\partial p}$, differential equations for S and R can be derived that indeed generate this relation. In fact, for S this gives the same Riccati equation (5.34) as in the case without terminal constraints, but with modified terminal condition

$$S(T,p) = \varphi_{xx}(\xi(p)) + v(p)\Psi_{xx}(\xi(p)). \tag{5.37}$$

Assuming that this solution exists over the full interval $[\tau, T]$, the matrix $R = R(t,p) \in \mathbb{R}^{(n-k)\times n}$ can then simply be computed as the solution to the linear matrix differential equation

$$\dot{R} = R\left(-f_x + f_u H_{uu}^{-1} H_{ux} + f_u H_{uu}^{-1} f_u^T S\right) \tag{5.38}$$

with terminal condition

$$R(T,p) = \Psi_x(\xi(p)). \tag{5.39}$$

Lemma 5.3.1. *If the solution $S = S(t,p)$ to the Riccati equation (5.34) with terminal condition (5.37) exists over the full interval $[\tau, T]$ and R denotes the solution to the linear ODE (5.38) with boundary condition (5.39), then for all $t \in [\tau, T]$, we have that*

$$\frac{\partial \lambda^T}{\partial p}(t,p) = S(t,p)\frac{\partial x}{\partial p}(t,p) + R^T(t,p)\frac{\partial v^T}{\partial p}(p). \tag{5.40}$$

Proof. In order to simplify the notation, we use the abbreviations

$$\tilde{A} = f_x - f_u H_{uu}^{-1} H_{ux}, \quad \tilde{B} = f_u H_{uu}^{-1} f_u^T, \quad \text{and} \quad \tilde{C} = H_{xx} - H_{xu} H_{uu}^{-1} H_{ux}, \tag{5.41}$$

where, as always, all the partial derivatives are evaluated along the extremals

$$t \mapsto (t, \lambda(t,p), x(t,p), u(t,p)).$$

Note that S, $\tilde{B}$, and $\tilde{C}$ are symmetric matrices. Define

$$\Delta(t,p) = \frac{\partial \lambda^T}{\partial p}(t,p) - S(t,p)\frac{\partial x}{\partial p}(t,p) - R^T(t,p)\frac{\partial v^T}{\partial p}(p).$$

It follows from the specifications of the terminal values that $\Delta(T,p) = 0$. Using Eq. (5.33), we furthermore get that

$$
\begin{aligned}
\dot{\Delta}(t,p) &= \frac{d}{dt}\left(\frac{\partial \lambda^T}{\partial p}\right) - \dot{S}\frac{\partial x}{\partial p} - S\frac{d}{dt}\left(\frac{\partial x}{\partial p}\right) - \dot{R}^T\frac{\partial v^T}{\partial p} \\
&= -\tilde{A}^T\frac{\partial \lambda^T}{\partial p} - \tilde{C}\frac{\partial x}{\partial p} - \left(-S\tilde{A} - \tilde{A}^T S + S\tilde{B}S - \tilde{C}\right)\frac{\partial x}{\partial p} \\
&\quad - S\left(f_x\frac{\partial x}{\partial p} + f_u\frac{\partial u}{\partial p}\right) - \left(-\tilde{A} + \tilde{B}S\right)^T R^T\frac{\partial v^T}{\partial p} \\
&= -\tilde{A}^T\frac{\partial \lambda^T}{\partial p} + S\tilde{A}\frac{\partial x}{\partial p} - \left(-\tilde{A} + \tilde{B}S\right)^T\left(S\frac{\partial x}{\partial p} + R^T\frac{\partial v^T}{\partial p}\right) \\
&\quad - S\left(f_x\frac{\partial x}{\partial p} + f_u\frac{\partial u}{\partial p}\right).
\end{aligned}
$$

Since all extremals are nonsingular, Eq. (5.31) implies that

$$
\begin{aligned}
\dot{\Delta}&(t,p) \\
&= -\tilde{A}^T\frac{\partial \lambda^T}{\partial p} + S\tilde{A}\frac{\partial x}{\partial p} - \left(-\tilde{A} + \tilde{B}S\right)^T\left(S\frac{\partial x}{\partial p} + R^T\frac{\partial v^T}{\partial p}\right) - S\left(\tilde{A}\frac{\partial x}{\partial p} - \tilde{B}\frac{\partial \lambda^T}{\partial p}\right) \\
&= \left(-\tilde{A} + \tilde{B}S\right)^T\left(\frac{\partial \lambda^T}{\partial p} - S\frac{\partial x}{\partial p} - R^T\frac{\partial v^T}{\partial p}\right) \\
&= \left(-\tilde{A} + \tilde{B}S\right)^T\Delta
\end{aligned}
$$

and thus $\Delta(t,p) \equiv 0$. $\qquad\qquad\qquad\qquad\qquad\qquad\qquad\qquad\qquad\qquad\qquad\qquad$ $\square$

Lemma 5.3.2. *Let* $Q = Q(t,p) \in \mathbb{R}^{(n-k)\times(n-k)}$ *be the integral of*

$$\dot{Q} = Rf_u H_{uu}^{-1} f_u^T R^T, \qquad Q(T,p) = 0. \tag{5.42}$$

Then

$$R(t,p)\frac{\partial x}{\partial p}(t,p) + Q(t,p)\frac{\partial v^T}{\partial p}(p) \equiv 0. \tag{5.43}$$

Proof. Define

$$\Theta(t,p) = R(t,p)\frac{\partial x}{\partial p}(t,p) + Q(t,p)\frac{\partial v^T}{\partial p}(p).$$

It follows from $Q(T,p) = 0$ that

$$\Theta(T,p) = \Psi_x(\xi(p))\frac{\partial \xi}{\partial p}(p) + 0 = 0,$$

where in the last step, we use the fact that the columns of $\frac{\partial \xi}{\partial p}(p)$ are tangent to N at $\xi(p)$, while the rows of $\Psi_x(\xi(p))$ are a basis for the normal space to N at $\xi(p)$. Furthermore,

$$\dot{\Theta}(t,p) = \dot{R}\frac{\partial x}{\partial p} + R\frac{d}{dt}\left(\frac{\partial x}{\partial p}\right) + \dot{Q}\frac{\partial v^T}{\partial p}$$

$$= R\left(-\tilde{A} + \tilde{B}S\right)\frac{\partial x}{\partial p} + R\left(\tilde{A}\frac{\partial x}{\partial p} - \tilde{B}\frac{\partial \lambda^T}{\partial p}\right) + R\tilde{B}R^T\frac{\partial v^T}{\partial p}$$

$$= R\tilde{B}\left(S\frac{\partial x}{\partial p} + R^T\frac{\partial v^T}{\partial p} - \frac{\partial \lambda^T}{\partial p}\right) \equiv 0.$$

Thus $\Theta(t,p) \equiv 0$.   □

Corollary 5.3.3. *The matrix $\frac{\partial x}{\partial p}(t,p)$ is nonsingular if and only if $Q(t,p)$ is nonsingular.*

Proof. The matrix $R(t,p) \in \mathbb{R}^{(n-k)\times n}$ is the solution of the homogeneous linear matrix differential equation (5.38). Since the rows of the matrix $R(T,p) = \Psi_x(\xi(p))$ are linearly independent, it follows that the rows of $R(t,p)$ are linearly independent for all times t and thus $R(t,p)$ is of full rank $n-k$ everywhere. Thus, if $\frac{\partial x}{\partial p}(t,p)$ is nonsingular, then the rank of the product $R(t,p)\frac{\partial x}{\partial p}(t,p)$ is equal to $n-k$. By Eq. (5.43), this then also is the rank of the product $Q(t,p)\frac{\partial v^T}{\partial p}(p)$. Hence both $Q(t,p)$ and $\frac{\partial v^T}{\partial p}(p)$ must be of full rank $n-k$.

Conversely, suppose $Q(t,p)$ is nonsingular. If $z \in \ker(\frac{\partial x}{\partial p}(t,p))$, then by Eq. (5.43), we have that $\frac{\partial v^T}{\partial p}(p)z = 0$ and thus by Eq. (5.40) also $\frac{\partial \lambda^T}{\partial p}(t,p)z = 0$. Hence the vector functions $s \to \frac{\partial x}{\partial p}(s,p)z$ and $s \to \frac{\partial \lambda^T}{\partial p}(s,p)z$ are solutions to the homogeneous linear differential equation (5.33) that vanish for $s = t$, and thus these functions vanish identically. In particular, at the terminal time T, we have that $\frac{\partial \xi}{\partial p}(p)z = \frac{\partial x}{\partial p}(T,p)z = 0$. But the mapping $\omega : p \mapsto \omega(p) = (\xi(p), v(p)^T)$ is a diffeomorphism, and therefore

$$\left(\frac{\partial \xi}{\partial p}(p), \frac{\partial v^T}{\partial p}(p)\right)z = 0$$

implies that $z = 0$. Hence $\frac{\partial x}{\partial p}(t,p)$ is nonsingular.   □

Summarizing the construction, we have the following result:

Theorem 5.3.3. *Let $\mathscr{E}$ be the canonical, nicely $C^{\ell-1}$-parameterized family of nonsingular extremals constructed in Theorem 5.3.1 with domain $D = \{(t,p) : p \in P, t_-(p) \leq t \leq T\}$ for the optimal control problem [OC$_T$] with terminal constraints given by $N = \{x \in \mathbb{R}^n : \Psi(x) = 0\}$. Suppose the solution $S = S(t,p)$ to the Riccati differential equation*

$$\dot{S} + Sf_x + f_x^T S + H_{xx} - (Sf_u + H_{xu})H_{uu}^{-1}(H_{ux} + f_u^T S) \equiv 0,$$

with terminal condition

$$S(T,p) = \varphi_{xx}(\xi(p)) + \nu(p)\Psi_{xx}(\xi(p)),$$

exists over the full interval $[\tau,T]$ and let $R = R(t,p)$ and $Q = Q(t,p)$ be the solutions to the terminal value problems

$$\dot{R} = R\left(-f_x + f_u H_{uu}^{-1}H_{ux} + f_u H_{uu}^{-1}f_u^T S\right), \qquad R(T,p) = \Psi_x(\xi(p)),$$

and

$$\dot{Q} = Rf_u H_{uu}^{-1}f_u^T R^T, \qquad Q(T,p) = 0,$$

over the interval $[\tau,T]$. If $Q(t,p)$ is negative definite over the interval $[\tau,T)$, then the matrix $\frac{\partial x}{\partial p}(t,p)$ is nonsingular over the interval $[\tau,T)$ as well. In this case, there exists a domain G in (t,x)-space that, with the exception of the terminal point $\bar{x}(T)$, contains the graph of the controlled reference trajectory $\bar{\Gamma} = (\bar{x},\bar{u})$ and $\bar{\Gamma}$ is a relative minimum over the set G. ∎

The last statement of the theorem follows from Corollary 5.2.2. As before, the Hessian matrix of the corresponding value $V^{\mathscr{E}}$ is given by

$$\frac{\partial V^{\mathscr{E}}}{\partial x^2}(t,x(t,p)) = \frac{\partial \lambda^T}{\partial p}(t,p)\left(\frac{\partial x}{\partial p}(t,p)\right)^{-1}.$$

However, for $t < T$, in this case it follows from Eqs. (5.40) and (5.43) that

$$\frac{\partial V^{\mathscr{E}}}{\partial x^2}(t,x(t,p)) = S(t,p) + R^T(t,p)\frac{\partial \nu^T}{\partial p}(p)\left(\frac{\partial x}{\partial p}(t,p)\right)^{-1}$$

$$= S(t,p) - R^T(t,p)Q(t,p)^{-1}R(t,p). \tag{5.44}$$

This matrix is the so-called Schur complement of Q for the matrix $\begin{pmatrix} S & R^T \\ R & Q \end{pmatrix}$. The function

$$Z(t,p) = \frac{\partial \lambda^T}{\partial p}(t,p)\left(\frac{\partial x}{\partial p}(t,p)\right)^{-1}$$

satisfies the same Riccati differential equation as S. For if $\frac{\partial x}{\partial p}(t,p)$ is nonsingular, then it always holds that (see Eq. (5.32))

$$
\begin{aligned}
\dot{Z} &= \left[\frac{d}{dt}\left(\frac{\partial \lambda^T}{\partial p}\right)\right]\left(\frac{\partial x}{\partial p}\right)^{-1} + \left(\frac{\partial \lambda^T}{\partial p}\right)\left[\frac{d}{dt}\left(\frac{\partial x}{\partial p}\right)^{-1}\right] \\
&= \left[\frac{d}{dt}\left(\frac{\partial \lambda^T}{\partial p}\right)\right]\left(\frac{\partial x}{\partial p}\right)^{-1} - \left(\frac{\partial \lambda^T}{\partial p}\right)\left(\frac{\partial x}{\partial p}\right)^{-1}\left[\frac{d}{dt}\left(\frac{\partial x}{\partial p}\right)\right]\left(\frac{\partial x}{\partial p}\right)^{-1} \\
&= \left(-H_{xx}\frac{\partial x}{\partial p} - f_x^T\frac{\partial \lambda^T}{\partial p} - H_{xu}\frac{\partial u}{\partial p}\right)\left(\frac{\partial x}{\partial p}\right)^{-1} - Z\left(f_x\frac{\partial x}{\partial p} + f_u\frac{\partial u}{\partial p}\right)\left(\frac{\partial x}{\partial p}\right)^{-1} \\
&= -Z f_x - f_x^T Z - H_{xx} - (Z f_u + H_{xu})\left(\frac{\partial u}{\partial p}\right)\left(\frac{\partial x}{\partial p}\right)^{-1}
\end{aligned}
\tag{5.45}
$$

and along a nonsingular extremal we have that

$$
\left(\frac{\partial u}{\partial p}\right)\left(\frac{\partial x}{\partial p}\right)^{-1} = -H_{uu}^{-1}\left(H_{ux} + f_u^T Z\right).
$$

Naturally, in this case $Z(t,p)$ has a singularity as $t \to T$ and Eq. (5.44) resolves the behavior near the terminal time. The general formulation (5.45) will also come in useful for the case of bang-bang controls considered in Sect. 6.1 when $\frac{\partial u}{\partial p} \equiv 0$.

The condition that $Q(t,p)$ be nonsingular for all $t < T$, also called a normality condition in [64], has an intuitive and natural control engineering interpretation. This is a controllability assumption on the linearized time-varying system

$$
\dot{y} = \frac{\partial f}{\partial x}(t, x(t,p), u(t,p))y + \frac{\partial f}{\partial u}(t, x(t,p), u(t,p))v.
$$

This is best seen through the connection with solutions to the adjoint equation for this system.

Lemma 5.3.3. *The matrix $Q(\tau, p)$ is singular if and only if there exists a nontrivial solution $\mu = \mu(t,p)$ of the linear adjoint equation*

$$
\dot{\mu} = -\mu\frac{\partial f}{\partial x}(t, x(t,p), u(t,p))
$$

with terminal condition $\mu(T,p)$ that is perpendicular to N at $\xi(p) = x(T,p)$ such that

$$
\mu(t,p)\frac{\partial f}{\partial u}(t, x(t,p), u(t,p)) \equiv 0
$$

on the interval $[\tau, T]$.

Proof. The matrix $Q = Q(\tau, p)$ is negative semi definite: for any vector $z \in \mathbb{R}^{n-k}$ we have that

$$z^T Q(t,p)z = -\int_\tau^T z^T \dot{Q} z\, ds = -\int_\tau^T z^T R f_u H_{uu}^{-1} f_u^T R^T z\, ds \le 0.$$

Since the matrix H_{uu} is positive definite, $Q(\tau, p)$ is singular if and only if there exists a non-zero vector $z \in \mathbb{R}^{n-k}$ such that

$$z^T R(t,p) f_u(t, x(t,p), u(t,p)) \equiv 0 \qquad \text{on} \quad [\tau, T].$$

In this case, $\mu(t, p) = z^T R(t, p)$ is a nontrivial solution of the linear adjoint equation

$$\dot{\mu} = z^T \dot{R} = z^T R \left(-f_x + f_u H_{uu}^{-1} H_{ux} + f_u H_{uu}^{-1} f_u^T S \right) = -z^T R f_x = -\mu f_x$$

that satisfies $\mu(T, p) \perp N$ and $\mu f_u = z^T R f_u \equiv 0$. Conversely, if such a solution exists, then μ is also a solution to the ODE

$$\dot{\mu} = \mu \left(-f_x + f_u H_{uu}^{-1} H_{ux} + f_u H_{uu}^{-1} f_u^T S \right).$$

Since $\mu(T, p)$ is nontrivial and orthogonal to N, there exists a nonzero vector $z \in \mathbb{R}^{n-k}$ such that $\mu(T) = z^T \Psi_x(\xi(p))$, and thus μ is given by $\mu(t) = z^T R(t, p)$. Thus $z^T Q(\tau, p)z = 0$ and $Q(\tau, p)$ is singular. □

This relates the regularity of Q to a classical characterization for the controllability of a time-varying linear system (e.g., see [127, 138]).

Proposition 5.3.3. *A time-varying linear system $\dot{y} = A(t)y + B(t)v$ is controllable over an interval $[\tau, T]$ if and only if for every nontrivial solution μ of the adjoint equation $\dot{\mu} = -\mu A(t)$, the function $\mu(t)B(t)$ does not vanish identically on the interval $[\tau, T]$. It is completely controllable if this holds for any subinterval $[\tau, T]$.*

Proof. If there exists a nontrivial solution μ of the adjoint equation $\dot{\mu} = -\mu A(t)$ for which the function $\mu(t)B(t)$ vanishes identically on the interval $[\tau, T]$, then for any solution y of the dynamics we have that

$$\frac{d}{dt}(\mu y) = \dot{\mu} y + \mu \dot{y} = -\mu A(t)y + \mu (A(t)y + B(t)v) \equiv 0.$$

It thus follows, for arbitrary initial conditions y_τ and terminal conditions y_T, that $\mu(\tau)y_\tau = \mu(T)y_T$. In particular, initial points y_τ for which $\mu(\tau)y_\tau \ne 0$ cannot be steered into $y_T = 0$ at time T, and the system is not controllable over $[\tau, T]$.

Conversely, suppose that $\mu(t)B(t)$ does not vanish identically on the interval $[\tau, T]$ for any nontrivial solution μ of the adjoint equation $\dot{\mu} = -\mu A(t)$. We explicitly construct a control that steers a given initial point y_τ into a specified terminal point y_T over the interval $[\tau, T]$. If we denote the fundamental matrix of the homogeneous time-varying linear system $\dot{y} = A(t)y$ by $\Phi(t,s)$, then we need to find a control v such that

$$y_T = \Phi(T,\tau)y_\tau + \int_\tau^T \Phi(T,s)B(s)v(s)ds.$$

In terms of the fundamental matrix Φ, solutions to the adjoint equation are simply given in the form $\mu(s) = \bar{\mu}\Phi(T,s)$ with $\bar{\mu}$ some n-dimensional row vector. This is a consequence of the fact that $\Phi(T,s)$ is the inverse matrix of $\Phi(s,T)$. For using the formula for the differentiation of the inverse of a matrix,

$$\frac{d}{dt}X^{-1}(t) = -X^{-1}(t)\dot{X}(t)X^{-1}(t),$$

it follows that

$$\frac{\partial}{\partial s}\Phi(T,s) = -\Phi(T,s)\frac{\partial\Phi}{\partial t}(s,T)\Phi(T,s)$$

$$= -\Phi(T,s)A(s)\Phi(s,T)\Phi(T,s) = -\Phi(T,s)A(s).$$

By assumption, for any $\bar{\mu} \neq 0$, the function $\mu(s)B(s) = \bar{\mu}\Phi(T,s)B(s)$ does not vanish identically on $[\tau,T]$, and therefore

$$\bar{\mu}\left(\int_\tau^T \Phi(T,s)B(s)B^T(s)\Phi(T,s)^T ds\right)\bar{\mu}^T > 0.$$

Hence, the matrix

$$W(\tau,T) = \int_\tau^T \Phi(T,s)B(s)B^T(s)\Phi(T,s)^T ds$$

is positive definite. Thus, we can simply take

$$v(s) = B^T(s)\Phi(T,s)^T \bar{v},$$

where $\bar{v}$ is the solution to the linear equation

$$y_T - \Phi(T,\tau)y_\tau = W(\tau,T)\bar{v}.$$

This control steers y_τ into y_T over the interval $[\tau,T]$. Hence the linear system is completely controllable.   □

Thus, the fact that $Q(t,p)$ needs to be nonsingular for all $t \in [\tau,T]$ is related to a complete controllability condition on the linearized system. However, the situation here is different in the sense that the problem requires only that the system be steered into the terminal manifold N, and for this reason it is not required

Fig. 5.8 Foliation of the
flow F

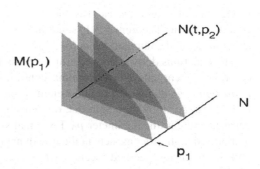

that the linearization be completely controllable; we need only that the system is
fully controllable with respect to the normal directions at the terminal manifold.
This is the significance of the terminal constraint $\mu(T) \perp N$; controllability of the
linearization along tangent directions to N is not required. In fact, depending on the
values of the parameter, two different solutions $x(\cdot, p_1)$ and $x(\cdot, p_2)$ may or may not
end at the same terminal point.

The assumption that $Q(\cdot, p)$ is nonsingular enforces a regular geometric structure
in the form of a foliation that clarifies these relations. Going back to the canonical, nicely C^1-parameterized family of nonsingular extremals $\mathcal{E}$ constructed in
Theorem 5.3.1, the parameter set P, $P = P_1 \times P_2$, is the direct product of an open
neighborhood P_1 of the origin in $\mathbb{R}^k$ that is the domain of a coordinate chart for
the terminal manifold N and an open neighborhood P_2 of v_0^T in $\mathbb{R}^{n-k}$ that defines
coordinates for the normal space and is used to parameterize the terminal conditions
on the multiplier λ. For $p_1 \in P_1$ fixed, the controlled trajectories

$$(t, p_2) \mapsto (t, x(t, p_1, p_2)), \qquad p_2 \in P_2, \ t < T,$$

form $(n - k + 1)$-dimensional integral submanifolds $M(p_1)$ of graphs of controlled
trajectories all of which steer the system into the point $(T, \xi(p_1))$ at the terminal
time T. On the other hand, if we freeze a parameter $p_2 \in P_2$, then the image of the
terminal manifold N under the flow F for a fixed time t,

$$(t, p_1) \mapsto (t, x(t, p_1, p_2)), \qquad p_1 \in P_1, \ t < T,$$

defines a k-dimensional submanifold $N(t, p_2)$ that is transversal to all the manifolds
$M(p_1)$ at the point $(t, x(t; p_1, p_2))$. Mathematically, such a decomposition is called
a *foliation*, with the surfaces $M(p_1)$ the *leaves of the foliation* and the manifolds
$N(t, p_2)$ the *transversal sections* (see Fig. 5.8).

Thus there exist both an intuitive control engineering interpretation and an
elegant geometric picture underlying the formal computations of neighboring
extremals done in [64, Chap. 6]. These interpretations can be extended to the case
that the terminal time is defined implicitly through one of the constraints in Ψ,
but a third equation that models the time evolution is required, and the notation
becomes somewhat cumbersome. Also, if one wants to make the constructions

mathematically rigorous, as was the case for the model considered here with condition (ii) in Theorem 5.3.1, some minor extra assumptions need to be made about the formal computations in [64].

The equations derived above form the basis for the linearization of a nonlinear control system around a locally optimal controlled reference trajectory, the so-called *perturbation feedback control* in the engineering literature [64, Sect. 6.4]. The nominal path is given by the controlled reference trajectory $(\bar{x}, \bar{u}) = (x(\cdot, p_0), u(\cdot, p_0))$ corresponding to the parameter p_0. For a real system, because of disturbances and unmodeled high-order aspects in the true dynamics, generally, at time t, the system will not be in its specified location $\bar{x}(t) = x(t, p_0)$, but is expected to have some small deviation from the actual position at time t given by $x \neq \bar{x}(t)$. If x lies in the region covered by the flow F of the parameterized family of extremals, then there exists a parameter $p \in P$ such that $x = x(t, p)$. The corresponding optimal solution therefore is given by $(x(\cdot, p), u(\cdot, p))$. Since the family is C^1-parameterized, a first-order Taylor expansion for this control around the reference value is given by

$$u(t, p) = u(t, p_0) + \frac{\partial u}{\partial p}(t, p_0)(p - p_0) + o(\|p - p_0\|),$$

and by Eq. (5.31) we have that

$$\frac{\partial u}{\partial p}(t, p_0) = -H_{uu}^{-1}\left(H_{ux}\frac{\partial x}{\partial p}(t, p_0) + f_u^T \frac{\partial \lambda^T}{\partial p}(t, p_0)\right),$$

with the partial derivatives of H all evaluated along the reference extremal for parameter p_0. It follows from Eqs. (5.40) and (5.43) that

$$\frac{\partial \lambda^T}{\partial p}(t, p_0) = \left(S(t, p_0) - R^T(t, p_0)Q(t, p_0)^{-1}R(t, p_0)\right)\frac{\partial x}{\partial p}(t, p_0),$$

and thus we have that

$$\frac{\partial u}{\partial p}(t, p_0) = -H_{uu}^{-1}\left(H_{ux} + f_u^T\left(-R^T Q^{-1}R\right)\right)\frac{\partial x}{\partial p}(t, p_0).$$

If we denote the deviation of the state from the reference trajectory by

$$\Delta x(t) = x(t, p) - x(t, p_0)$$

and the deviation of the reference control by

$$\Delta u(t) = u(t, p) - u(t, p_0),$$

then overall, this gives

$$\Delta u(t) = -H_{uu}^{-1}\left(H_{ux} + f_u^T\left(S - R^T Q^{-1}R\right)\right)\Delta x(t) + o(\|\Delta x(t)\|). \qquad (5.46)$$

This equation provides the linearization of the optimal control $u(t,p)$ around the nominal control $u(t,p_0)$ as a time-varying linear feedback control of the deviation from the nominal trajectory, information that typically is readily available with today's sensor technology. For a nonlinear control system with reasonable local stability properties around the reference, such a control scheme generally is highly effective, and it has the advantage that it requires computations only along one trajectory, the reference controlled trajectory, since all the matrices S, R, Q and all the partial derivatives of H are evaluated along the controlled reference trajectory for parameter p_0. Naturally, if the deviation becomes too large, this control law no longer is applicable. Even with the best technology, if turbulence occurs, the pilot needs to turn off the autopilot and return to work, since a linearization-based control scheme no longer is able to handle deviations of such a magnitude.

We note only that is not difficult to show that this perturbation feedback control scheme can also be derived as the solution to a linear-quadratic optimal control problem (but with terminal constraints), as would seem obvious from the structure involving the solution of a Riccati differential equation. This is the natural extension of the accessory problem of the calculus of variations to the control theory setting. However, as in that case, and in contrast to the classical formulation considered in Sect. 2.1, it is no longer guaranteed that a solution to the Riccati differential equation exists for all times. This will be our next topic. We still reiterate that some adjustments to the construction become necessary if the final time T is included in the terminal constraint. In this case, additional quantities that account for the time evolution near the terminal manifold become necessary, but otherwise our arguments carry over almost verbatim. In particular, the matrices S, R, and Q satisfy the same differential equations. We refer the interested reader to [64, Sects. 6.5 and 6.6].

SUMMARY: In this section, for problem [OC_T] we showed how to embed a nonsingular extremal $\Lambda = ((\bar{x}, \bar{u}), \lambda)$ in a canonical way into a nicely $C^{\ell-1}$-parameterized family of nonsingular extremals $\mathscr{E}$ with domain $D = \{(t,p) : p \in P, t_-(p) \leq t \leq T\}$ with Λ the extremal for parameter $p_0 \in P$ (Theorem 5.3.1). Then, for various formulations of the general optimal control problem [OC_T], we established conditions under which the members of $\mathscr{E}$ are locally optimal (e.g., Theorems 5.3.2 and 5.3.3). All these criteria are based on sufficient conditions for the corresponding flow F,

$$F : D \to \mathbb{R} \times \mathbb{R}^n, \qquad (t,p) \longmapsto F(t,p) = \begin{pmatrix} t \\ x(t,p) \end{pmatrix},$$

to be a diffeomorphism away from the terminal manifold N, and as such, these are all related to the fact that the Jacobian matrix $\frac{\partial x}{\partial p}(t,p)$ is nonsingular.

5.4 Fold Singularities and Conjugate Points

In the last two sections of this chapter, we analyze the behavior of the flow of trajectories F when this flow becomes singular. Throughout, we assume as given a C^2-parameterized family of normal extremals $\mathscr{E}$ with domain

$$D = \{(t,p) : p \in P, t_-(p) \le t \le t_+(p)\}.$$

We call the member of this family that corresponds to the parameter value $p_0 \in P$ the reference controlled trajectory and denote it by $\Lambda = ((\bar{x}, \bar{u}), \bar{\lambda})$. All our considerations here will be local, and we always assume that P is a sufficiently small neighborhood of p_0 that will be shrunk whenever necessary so that local properties of the construction are valid on all of P.

We study the behavior of the flow F of the family $\mathscr{E}$ and its associated value function near a singular point, but consider only the two least-degenerate scenarios, so-called *fold* points in this section, and the case of a *simple cusp* point in the next section. The analysis of more degenerate singularities very quickly becomes exceedingly difficult, and the geometric properties of the corresponding value functions in these cases are still largely unknown. The two cases analyzed here are commonly encountered in many low-dimensional optimal control problems (related to the fact that these are the only generic singularities for two-dimensional maps [258]) and often form determining structures in the optimal solutions. The canonical example for fold singularities is given by the envelope of the extremals for the minimum surfaces of revolution in the classical problem of the calculus of variations (see Sects. 1.3 and 1.7), and we shall show that the local geometry of the flow F near a fold point is exactly the same. In particular, local optimality of the extremals in the family ceases at fold points. In fact, we here carry out an argument using *envelopes* that generalizes this classical concept from the calculus of variations to the optimal control problem and mimics this theory.

Our aim is to study singularities in the flow $F = F(t,p)$ of a parameterized family of extremals. In the time-dependent case, the differential of the flow is given by

$$DF(t,p) = \begin{pmatrix} 1 & 0 \\ \frac{\partial x}{\partial t}(t,p) & \frac{\partial x}{\partial p}(t,p) \end{pmatrix},$$

and therefore singularities of the flow correspond to rank deficiencies of the matrix $\frac{\partial x}{\partial p}(t,p)$. In the time-independent case, since the dimension of the parameter space P is reduced by one, the full Jacobian also contains the time derivative $\frac{\partial x}{\partial t}(t,p)$. Still, in the flow F the variable t has a distinguished role, and we are interested in the singularities that arise in time, i.e., conjugate points. Therefore, here we disregard singularities that are caused when $\frac{\partial x}{\partial t} \in \text{lin span} \left(\frac{\partial x}{\partial p} \right)$ and $\frac{\partial x}{\partial p}$ is of full rank.

Fig. 5.9 Assumption (A)

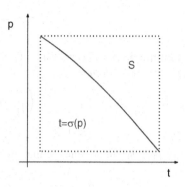

Definition 5.4.1 (Singular points). Let $\mathcal{E}$ a be k-dimensional, C^2-parameterized family of normal extremals, $k \le n$. We call a point (t_0, p_0) in the interior of D, $(t_0, p_0) \in \text{int}(D)$, a singular point if the matrix $\frac{\partial x}{\partial p}(t_0, p_0) \in \mathbb{R}^{n \times k}$ does not have full rank k. A singular point is called a corank-$\ell > 0$ singular point if the matrix $\frac{\partial x}{\partial p}(t_0, p_0)$ has rank $k - \ell$.

The $k \times k$ matrix $\frac{\partial x}{\partial p}(t, p)^T \frac{\partial x}{\partial p}(t, p)$ is positive semidefinite, and the corank of a singular point is the dimension of the kernel or null space of this matrix. In particular, at a corank-1 singular point (t_0, p_0), this kernel is spanned by any eigenvector for the algebraically simple eigenvalue 0. Let

$$\Delta(t, p) = \det \left[\left(\frac{\partial x}{\partial p}(t, p) \right)^T \frac{\partial x}{\partial p}(t, p) \right].$$

Definition 5.4.2 (t-Regular singular points). We call a singular point $(t_0, p_0) \in \text{int}(D)$ t-regular if

$$\frac{\partial \Delta}{\partial t}(t_0, p_0) \neq 0.$$

If (t_0, p_0) is a t-regular singular point, then by the implicit function theorem, there exists an open neighborhood $W = (t_0 - \varepsilon, t_0 + \varepsilon) \times B_\delta(p_0)$ of (t_0, p_0) with the property that the equation $\Delta(t, p) = 0$ has a unique solution on W given in the form $t = \sigma(p)$ with a continuously differentiable function $\sigma : B_\delta(p_0) \to (t_0 - \varepsilon, t_0 + \varepsilon)$, $p \mapsto \sigma(p)$. In other words, the singular set S is the graph of this function σ; in particular, it is a codimension-1 embedded submanifold in D [see Fig. 5.9]. If (t_0, p_0) is a t-regular corank-1 singular point, then for sufficiently small δ, all points in the singular set will be of corank 1. (There exists a $(k-1)$-dimensional minor of the matrix $\frac{\partial x}{\partial p}(t_0, p_0)$ that is nonzero, and by continuity this holds in a neighborhood.) We henceforth *assume* that this is the case:

(A)   The singular set S is a codimension-1 embedded submanifold of D that entirely consists of corank-1 singular points and can be described as the graph of a continuously differentiable function $\sigma : P \to (t_0 - \varepsilon, t_0 + \varepsilon), p \mapsto \sigma(p)$,

$$S = \{(t,p) \in \mathrm{int}(D) : t = \sigma(p)\} = \mathfrak{gr}(\sigma).$$

Lemma 5.4.1. *There exists a C^1 vector field $v : P \to \mathbb{S}^{k-1} = \{z \in \mathbb{R}^k : z^T z = 1\}$, $p \mapsto v(p)$, such that*

$$\left(\frac{\partial x}{\partial p}(\sigma(p),p) \right) v(p) = 0.$$

Proof. Let $v_0 \in \mathbb{S}^{k-1}$ be a basis vector for the one-dimensional kernel of $\frac{\partial x}{\partial p}$ $(\sigma(p_0),p_0)$ and define the mapping

$$E : P \times \mathbb{S}^{k-1}, \quad (p,v) \mapsto E(p,v) = \frac{\partial x}{\partial p}(\sigma(p),p)v.$$

The equation $E(p,v) = 0$ has the solution (p_0,v_0), and the partial derivative with respect to v is of rank $k-1$ at (p_0,v_0), the dimension of $\mathbb{S}^{k-1}$. Hence, by the implicit function theorem, the equation $E(p,v) = 0$ can locally be solved for v as a differentiable function of p near p_0. $\square$

The corresponding vector field that gives the eigenvectors in the kernel of the differential for the full flow F is obtained by adding zero as first coordinate,

$$\hat{v} : P \to \mathbb{S}^k, \quad p \mapsto \hat{v}(p) = \begin{pmatrix} 0 \\ v(p) \end{pmatrix}.$$

Corank-1 singularities are broadly classified as fold or cusp points depending on whether the vector field $\hat{v}$ is transversal to the tangent space of the singular set S at (t_0,p_0) or not [108].

Definition 5.4.3 (Fold and cusp points). A corank-1 singular point is called a fold point if

$$T_{(t_0,p_0)}S \oplus \mathrm{lin\ span}\ \{\hat{v}(p_0)\} = \mathbb{R}^{k+1};$$

it is called a cusp point if

$$\hat{v}(p_0) \in T_{(t_0,p_0)}S.$$

For our setup, we have the following simple criterion:

Lemma 5.4.2. *The point (t_0,p_0), $t_0 = \sigma(p_0)$, is a fold point if and only if the Lie derivative of the function σ along the vector field v does not vanish at p_0,*

$$(L_v\sigma)(p_0) = \nabla\sigma(p_0)v(p_0) \neq 0.$$

Proof. Since S is the graph of the function σ, $S = \{(t,p) : t - \sigma(p) = 0\}$, the tangent space $T_{(t_0,p_0)}S$ consists of all vectors that are orthogonal to $(1,-\nabla\sigma(p_0))$. $\square$

By reversing the orientation of the vector field v, without loss of generality we may always *assume* that $\nabla\sigma(p_0)v(p_0)$ *is positive* for a fold point.

5.4.1 Classical Envelopes

We show that the local geometry of a flow of extremals near a fold singularity is identical to that for the flow of catenaries near the envelope in the problem of minimum surfaces of revolution in the classical calculus of variations. In fact, *if we define the manifold M_f of fold points as the image of the singular set S under the flow F, $M_f = F(S)$, then trajectories in the parameterized family $\mathscr{E}$ touch M_f in exactly one point.*

Definition 5.4.4 (Classical envelope). Let $\mathscr{E}$ be a C^1-parameterized family of normal extremals with domain D. A classical envelope for the parameterized family $\mathscr{E}$ is a (possibly small) portion of an admissible controlled trajectory (ξ, η) of the control system defined over some interval $[a, b]$ with the property that there exists a differentiable curve $p : [a, b] \to P$, $t \mapsto p(t)$, such that $\xi(t) = x(t, p(t))$, and for $p = p(t)$ we have that

$$H(t, \lambda(t, p), \xi(t), \eta(t)) = H(t, \lambda(t, p), x(t, p), u(t, p)). \tag{5.47}$$

One obvious way of satisfying condition (5.47) is with $\eta(t) = u(t, p(t))$, and this corresponds to the classical definition from the calculus of variations. For this reason, and also to distinguish it from a more general concept introduced below, we call envelopes of this type *classical*. The main property of envelopes is the agreement of the cost along concatenations of portions of the envelope with the trajectories of the parameterized family.

Theorem 5.4.1 (Envelope theorem). *Let $(\xi, \eta) : [a, b] \to M \times U$ be a classical envelope for a C^1-parameterized family $\mathscr{E}$ of normal extremals with domain*

$$D = \{(t, p) : p \in P, \ t_-(p) \le t \le t_+(p)\}.$$

For any interval $[t_1, t_2] \subset (a, b)$, setting $p_1 = p(t_1)$ and $p_2 = p(t_2)$, we then have that

$$C(t_1, p_1) = \int_{t_1}^{t_2} L(t, \xi(t), \eta(t)) dt + C(t_2, p_2). \tag{5.48}$$

Proof. For $s \in [t_1, t_2]$, define a 1-parameter family of admissible controlled trajectories (ξ_s, η_s) as

$$\eta_s(t) = \begin{cases} \eta(t) & \text{for } t_1 \le t \le s, \\ u(t, p(s)) & \text{for } s < t \le t_+(s), \end{cases}$$

and

$$\xi_s(t) = \begin{cases} \xi(t) & \text{for } t_1 \le t \le s, \\ x(t, p(s)) & \text{for } s < t \le t_+(s). \end{cases}$$

Thus the controlled trajectories (ξ_s, η_s) follow (ξ, η) over the interval $[t_1, s]$ until the point $\xi(s)$ is reached and then switch to the controlled trajectory of the parameterized family $\mathscr{E}$ determined by the parameter $p(s)$ that passes through $\xi(s)$ at time s, $\xi(s) = x(s, p(s))$. In particular, these concatenations satisfy the terminal constraints, since the controlled trajectories in the family $\mathscr{E}$ do so. The corresponding cost $\Gamma(s)$ is given by

$$\Gamma(s) = \int_{t_1}^{s} L(t, \xi(t), \eta(t))dt + C(s, p(s)),$$

and it satisfies $\Gamma(t_1) = C(t_1, p_1)$ and $\Gamma(t_2) = \int_{t_1}^{t_2} L(t, \xi(t), \eta(t))dt + C(t_2, p_2)$. It follows from our general regularity assumptions that the integrand is bounded over a compact interval, and thus Γ is an absolutely continuous function. Hence, Γ is differentiable almost everywhere on $[t_1, t_2]$, and it suffices to show that

$$\Gamma'(s) = \frac{d\Gamma}{ds}(s) \equiv 0,$$

so that Γ is constant on $[t_1, t_2]$.

Differentiating $\Gamma(\cdot)$ gives

$$\Gamma'(s) = L(s, \xi(s), \eta(s)) + \frac{\partial C}{\partial t}(s, p(s)) + \frac{\partial C}{\partial p}(s, p(s))\frac{dp}{ds}(s)$$

$$= L(s, \xi(s), \eta(s)) - L(s, x(s, p(s)), u(s, p(s))) + \lambda(s, p(s))\frac{\partial x}{\partial p}(s, p(s))\frac{dp}{ds}(s),$$

where in the last equation, we use the formulas for the derivatives of C and the shadow-price lemma, Lemma 5.2.2. Since $x(s, p(s)) \equiv \xi(s)$ and ξ is a controlled trajectory, it follows that

$$\frac{\partial x}{\partial t}(s, p(s)) + \frac{\partial x}{\partial p}(s, p(s))\frac{dp}{ds}(s) = \frac{d\xi}{ds}(s) = f(s, \xi(s), \eta(s)),$$

so that

$$\frac{\partial x}{\partial p}(s, p(s))\frac{dp}{ds}(s) = f(s, \xi(s), \eta(s)) - f(s, x(s, p(s)), u(s, p(s))).$$

Hence

$$\Gamma'(s) = H(s, \lambda(s, p(s)), \xi(s), \eta(s)) - H(s, \lambda(s, p(s)), x(s, p(s)), u(s, p(s))) \equiv 0$$

and the result follows. $\square$

Recall that $C(t,p)$ denotes the cost of the trajectories in the parameterized family $\mathscr{E}$, i.e., $C(t,p)$ is the cost for the control $u(\cdot,p)$ with trajectory $x(\cdot,p)$ and initial condition $x(t,p)$ at initial time t. In terms of the boundary points $(t_1,x_1) = (t_1,x(t_1,p_1))$ and $(t_2,x_2) = (t_2,x(t_2,p_2))$, the envelope condition (5.48) can thus equivalently be expressed as

$$V^{\mathscr{E}}(t_1,x_1) = \int_{t_1}^{t_2} L(t,\xi(t),\eta(t))dt + V^{\mathscr{E}}(t_2,x_2).$$

As in the calculus of variations, Eq. (5.48) relates the cost of the controlled trajectories in the family $\mathscr{E}$ to concatenations along the envelope.

We now show that *integral curves of the vector field v defined in Lemma 5.4.1 generate in a canonical way classical envelopes at fold singularities*: Let

$$\pi : (-\varepsilon,\varepsilon) \to P, \quad s \mapsto \pi(s),$$

be the integral curve of the vector field v that passes through the point p_0 at time $t=0$, i.e.,

$$\frac{d\pi}{ds}(s) = v(\pi(s)), \qquad \pi(0) = p_0, \tag{5.49}$$

and let

$$\chi : P \to S, \quad p \mapsto (\sigma(p),p),$$

be the C^1-diffeomorphism that injects the parameter space into the singular set. With $x = x(t,p)$ denoting the parameterized trajectories in the family $\mathscr{E}$, let Γ denote the curve

$$\Gamma : (-\varepsilon,\varepsilon) \to \mathbb{R}^n, \qquad s \mapsto \Gamma(s) = (x \circ \chi \circ \pi)(s) = x(\sigma(\pi(s)),\pi(s)). \tag{5.50}$$

We claim that a reparameterization of this curve is a controlled trajectory of the system. Let $\theta = \sigma \circ \pi$, so that

$$\frac{d\theta}{ds}(0) = \nabla\sigma(p_0) \cdot v(p_0) > 0.$$

Hence, for ε small enough, $t = \theta(s)$ is a strictly increasing function that maps $(-\varepsilon,\varepsilon)$ onto some interval (a,b) with inverse $s = \theta^{-1}(t)$. Let $p = p(t)$ be the curve

$$p : (a,b) \to P, \qquad t \mapsto p(t) = (\pi \circ \theta^{-1})(t),$$

and define

$$\xi : (a,b) \to \mathbb{R}^n, \qquad t \mapsto \xi(t) = x(t,p(t)),$$

and

$$\eta : (a,b) \to \mathbb{R}^n, \qquad t \mapsto \eta(t) = u(t,p(t)).$$

Proposition 5.4.1. *The pair* (ξ, η) *is a portion of an admissible controlled trajectory.*

Proof. By construction, η is a continuous function that takes values in the control set U. (The latter is automatic, since $u = u(t,p)$ is the control of the parameterized families of extremals.) Hence η is an admissible control. Differentiating ξ, it follows that

$$\dot{\xi}(t) = \dot{x}(t,p(t)) + \frac{\partial x}{\partial p}(t,p(t))\dot{p}(t) = f(t,\xi(t),\eta(t)) + \frac{\partial x}{\partial p}(t,p(t))\dot{p}(t).$$

But by construction, the second term vanishes, since $\dot{p}(t)$ is a multiple of the vector $v(p(t))$ in the null space of $\frac{\partial x}{\partial p}(t,p(t))$: writing $t = \theta(s)$, we have that

$$\dot{p}(t) = \pi'(s)\frac{ds}{dt}(t) = v(\pi(s))\frac{ds}{dt}(t) = v(p(t))\frac{ds}{dt}(t)$$

and thus

$$\frac{\partial x}{\partial p}(t,p(t))\dot{p}(t) = \frac{\partial x}{\partial p}(t,p(t))v(p(t))\frac{ds}{dt}(t) = 0. \tag{5.51}$$

As in the proof of Theorem 5.4.1, this segment can then be concatenated at the point $\xi(t) = x(t,p(t))$ with the controlled trajectory in the parameterized family $\mathscr{E}$ for the parameter $p(t)$, and this defines an admissible controlled trajectory.   □

Thus, the controlled trajectory (ξ, η) is a classical envelope. Note that it makes no difference for this argument whether the curve Γ can be reparameterized as an increasing or a decreasing function, since we can simply reverse the orientation of the vector field v. But the result is no longer valid if a reversal of orientation occurs at p_0. This happens at a simple cusp singularity, and thus this result does not extend to cusp points. We shall see below that it is this property that is responsible for the fact that controlled trajectories can still be optimal at a cusp point, while this is never true for fold points.

Thus the classical results of the calculus of variations directly carry over to parameterized families of extremals. They are equally valid for the time-dependent and the time-independent formulations, nor do they require that the flow cover an open set in the state space. We illustrate the usage of this result to show that local optimality of controlled trajectories ceases at fold singularities for the canonical flow of nonsingular extremals constructed in Sect. 5.3 (Theorem 5.3.1).

Theorem 5.4.2. *Let $\mathscr{E}$ be the n-dimensional nicely C^1-parameterized canonical family of nonsingular extremals with domain $D = \{(t,p) : p \in P, t_-(p) \leq t \leq T\}$ for the optimal control problem [OC_T] constructed in Theorem 5.3.1. (Recall that we assume that the controls are the unique minimizers of the Hamiltonian and that they take values in the interior of the control set.) Suppose there exists a function $\sigma : P \to (t_-(p), T)$, $p \mapsto \sigma(p)$, such that the associated flow*

$$F : D \to \mathbb{R}^{n+1}, \qquad (t,p) \mapsto F(t,p) = (t, x(t,p)),$$

is a diffeomorphism when restricted to $D_{\text{opt}} = \{(t,p) : p \in P, \sigma(p) < t < T\}$, *and that it has fold singularities at the points in* $S = \{(t,p) : p \in P, t = \sigma(p)\}$. *Then every controlled trajectory* $(x(\cdot,p), u(\cdot,p))$ *defined over an interval* $[\tau, T]$ *with* $\tau > \sigma(p)$ *provides a relative minimum over the domain* $G = F(D_{\text{opt}})$, *but is no longer optimal over the interval* $[\sigma(p), T]$.

Proof. The statements about local optimality of the controlled trajectories on D_{opt} follow from Corollary 5.2.2. It remains to show that the controlled trajectory $(x(\cdot,p), u(\cdot,p))$ is no longer optimal over the full interval $[\sigma(p), T]$.

Let $p_1 = p$, $t_1 = \sigma(p)$ and $x_1 = x(\sigma(p), p)$. Since (t_1, p_1) is a fold point, on some small interval $[t_1, t_2]$ there exists a classical envelope (ξ, η) through the point $(t_1, x_1) \in S$. Let $x_2 = \xi(t_2)$; since $(t_2, x_2) \in S$, there exists a parameter p_2 such that $t_2 = \sigma(p_2)$ and $x_2 = x(t_2, p_2)$. Define a second controlled trajectory $(\hat{\xi}, \hat{\eta})$ by

$$\hat{\xi}(t) = \begin{cases} \xi(t) & \text{if} \quad t_1 \leq t < t_2, \\ x(t, p_2) & \text{if} \quad t_2 \leq t \leq T, \end{cases} \quad \text{and} \quad \hat{\eta}(t) = \begin{cases} \eta(t) & \text{if} \quad t_1 \leq t < t_2, \\ u(t, p_2) & \text{if} \quad t_2 \leq t \leq T. \end{cases}$$

By Theorem 5.4.1, we have that

$$C(t_1, p_1) = \int_{t_1}^{t_2} L(t, \xi(t), \eta(t)) dt + C(t_2, p_2),$$

and thus the controlled trajectories $(x(\cdot, p_1), u(\cdot, p_1))$ and $(\hat{\xi}, \hat{\eta})$ have the same cost. Hence, if $(x(\cdot, p_1), u(\cdot, p_1))$ is optimal over the interval $[t_1, T]$, then so is $(\hat{\xi}, \hat{\eta})$.

In order to show that this is not the case, we construct a curve of extremals in $\mathscr{E}$ that all project onto the *one* controlled trajectory $(x(\cdot, p_2), u(\cdot, p_2))$ on the interval $[t_2, T]$. For t fixed, the combined (x, λ) flow in the cotangent bundle always has an n-dimensional image, since this is the dimension of the manifold describing the terminal conditions at time T. But by our earlier results, for times t, $t_2 < t < T$, we already know that this flow has an n-dimensional projection into the state space. Contradiction.

If $(\hat{\xi}, \hat{\eta})$ is optimal, then on the interval $[t_2, T]$, there exists a nontrivial solution $\hat{\lambda}$ to the homogeneous linear equation

$$\frac{d\hat{\lambda}}{dt}(t) = -\hat{\lambda}(t) f_x(t, x(t, p_2), u(t, p_2))$$

with terminal condition $\hat{\lambda}(T) = \hat{v} \Psi_x(\xi(p_2))$ for some $\hat{v} \in (\mathbb{R}^{n-k})^*$ that satisfies

$$0 = \hat{\lambda}(t) f_u(t, x(t, p_2), u(t, p_2)).$$

This directly follows from the maximum principle if $(\hat{\gamma},\hat{\eta})$ is an abnormal extremal. If $(\hat{\gamma},\hat{\eta})$ is a normal extremal, then these relations are satisfied by the difference between the adjoint vector for $(\hat{\gamma},\hat{\eta})$ and the multiplier $\lambda(\cdot,p_2)$ from the parameterized family $\mathscr{E}$. Furthermore, because of the different structures of the controls over the interval $[t_1,t_2)$, the two multipliers cannot have the same values for $t = t_2$. Thus, in either case there exists a nontrivial abnormal extremal lift of $(\hat{\xi},\hat{\eta})$ to the cotangent bundle. For $|\varepsilon|$ sufficiently small, the covector

$$\lambda(t;\varepsilon) = \lambda(t,p_2) + \varepsilon\hat{\lambda}(t)$$

is thus a solution of the adjoint equation

$$\dot{\lambda}(t;\varepsilon) = -\lambda(t;\varepsilon)f_x(t,x(t,p_2),u(t,p_2)) - L_x(t,x(t,p_2),u(t,p_2))$$

on the interval $[t_2,T]$ that satisfies the terminal condition

$$\lambda(T;\varepsilon) = \varphi_x(\xi(p_2)) + (v(p_2) + \varepsilon\hat{v})\,\Psi_x(\xi(p_2)).$$

Recall, from the construction of the parameterized family $\mathscr{E}$ in Theorem 5.3.1, that the function $u(t,x,\lambda)$ that defines the parameterized controls is the unique local solution of the equation $H_u(t,\lambda,x,u) = 0$ for u in a neighborhood of the reference extremal. For t_2 and ε small enough, the pair $(x(t,p_2),\lambda(t;\varepsilon))$ will lie in this neighborhood. Since $u(\cdot,p_2)$ is a minimizing control that lies in the interior of the control set, we have that

$$0 = L_u(t,x(t,p_2),u(t,p_2)) + \lambda(t;p_2)f_u(t,x(t,p_2),u(t,p_2)),$$

and since $0 = \hat{\lambda}(t)f_u(t,x(t,p_2),u(t,p_2))$, this gives that

$$0 = L_u(t,x(t,p_2),u(t,p_2)) + \lambda(t;\varepsilon)f_u(t,x(t,p_2),u(t,p_2)).$$

Thus $u(t,p_2)$ is a local solution of the equation

$$0 = H_u(t,\lambda(t;\varepsilon),x(t,p_2),u(t,p_2))$$

as well, and so it follows that

$$u(t,p_2) = u(t,x(t,p_2),\lambda(t;\varepsilon)).$$

By our assumption that controls lie in the interior of the control set and are the unique minimizers of the Hamiltonian around the reference extremal, it furthermore follows from the proof of Theorem 5.3.1 that for given terminal values

$$(x(T,p),\lambda(T,p)) = (\xi(p_2),\varphi(\xi(p_2)) + (v(p_2) + \varepsilon\hat{v})^T\,\Psi(\xi(p_2))),$$

extremals (λ, x, u) are the unique solutions to the following combined system of differential equations and minimality condition:

$$\dot{x} = f(t, x, u), \qquad \dot{\lambda} = -\lambda f_x(t, x, u) - L_x(t, x, u), \qquad 0 = H_u(t, \lambda, x, u).$$

The triple consisting of the multiplier $\lambda(\cdot; \varepsilon)$, state $x(\cdot, p_2)$, and control $u(\cdot, p_2)$ satisfies these equations, and thus for ε near 0, the curves $t \mapsto (\lambda(t; \varepsilon), x(t, p_2), u(t, p_2))$ are extremals for the optimal control problem [OC$_T$], which for $\varepsilon = 0$, reduce to the reference extremal. But extremals around the reference are unique, and hence these extremals are members of the parameterized family $\mathscr{E}$ that all project onto the controlled trajectory $(x(\cdot, p_2), u(\cdot, p_2))$ on the interval $[t_2, T]$. This establishes our contradiction and proves the theorem. □

This result extends to the case that the terminal time T is defined only implicitly as well, and it shows that for the optimal control problem [OC], fold points of the flow F are *conjugate points* in the sense of the calculus of variations: given a parameterized family of extremals, it is a necessary condition for optimality of the reference controlled extremal $(x(\cdot, p_0), u(\cdot, p_0))$ that the flow $F(\cdot, p_0)$ not contain a fold singularity over the interval $[\tau, T)$. We shall show by means of an example in Sect. 5.5 that trajectories can remain optimal on the interval $[\tau, T)$ if $\tau = \sigma(p_0)$ is a singular point that is not a fold point. In fact, this always holds for simple cusp points. But it can be shown in general that there must not be a singular point in the open interval (τ, T). This is the generalization of the *Jacobi condition* to this setup. However, this result is best proven by different means and will not be pursued in this text. In this chapter, we are interested in establishing sufficient conditions for optimality, and we have already seen in the last section that if the flow $F(\cdot, p_0)$ is regular on the half-open interval $[\tau, T)$, then the reference extremal is a relative minimum when compared with other controlled trajectories that lie in some open neighborhood of the reference trajectory. This regularity condition therefore is the *strengthened Jacobi condition*.

5.4.2 The Hilbert Invariant Integral and Control Envelopes

The concept of a classical envelope is closely related to the differential 1-form $\omega = \lambda dx - H dt$, the *Hilbert invariant integral* (Sect. 1.6), and we briefly digress to present this connection here. The following technical computation also leads to a generalization of the concept of a classical envelope that will be useful later on. For simplicity, we consider a parameterized family of normal extremals, but the statement can easily be modified to apply with a general multiplier $\lambda_0 = \lambda_0(p)$ (see [148]).

Lemma 5.4.3. *Let $\mathscr{E}$ be a C^1-parametrized family of normal extremals with domain $D = \{(t, p) : p \in P, t_-(p) \leq t \leq t_+(p)\}$ and suppose $(\xi, \eta) : [t_0, T] \to M \times U$*

is a controlled trajectory. Suppose there exist an interval $[a,b] \subset [t_0, T]$ and differentiable curves $\tau : [\alpha, \beta] \to [a,b]$, $s \mapsto \tau(s)$, and $\Theta : [\alpha, \beta] \to D$, $s \mapsto \Theta(s) = (t(s), p(s))$ such that for all $s \in [\alpha, \beta]$, we have

$$\xi(\tau(s)) = x(t(s), p(s)). \tag{5.52}$$

Let

$$\Gamma(\tau) = \int_\tau^T L(s, \xi(s), \eta(s))ds + \varphi(T, \xi(T))$$

denote the cost-to-go for the controlled trajectory from the point $\xi(\tau)$ at time τ, and for $s \in [\alpha, \beta]$, let

$$\Delta(s) = \Gamma(\tau(s)) - C(t(s), p(s))$$

denote the difference in the cost-to-go functions of (ξ, η) and the controlled trajectory in the parameterized family $\mathcal{E}$ from the initial point $\xi(\tau(s)) = x(t(s), p(s))$. Then we have that

$$\Delta(\beta) - \Delta(\alpha) = \int_{\tau \times \Theta} H(t, \lambda(t,p), x(t,p), u(t,p))dt - H(\tau, \lambda(t,p), \xi(\tau), \eta(\tau))d\tau.$$
$$\tag{5.53}$$

As was the case with envelopes, the parametrizations τ and Θ may represent only a small piece of the controlled trajectory (ξ, η). Also, take note of the fact that the multiplier λ is given by the same expression, namely $\lambda(t, p)$, in both Hamiltonians. This is not a typo.

Proof. This is a direct computation. All curves are evaluated at s, but we suppress the argument:

$$\Delta(\beta) - \Delta(\alpha) = \int_\alpha^\beta \frac{d}{ds}[\Gamma(\tau(s)) - C(t(s), p(s))]\, ds$$

$$= \int_\alpha^\beta \left[\dot{\Gamma}(\tau)\frac{d\tau}{ds} - \frac{\partial C}{\partial t}(t,p)\frac{dt}{ds} - \frac{\partial C}{\partial p}(t,p)\frac{dp}{ds}\right] ds.$$

Given our differentiability assumptions, we always have that

$$\dot{\Gamma}(\tau)\frac{d\tau}{ds} - \frac{\partial C}{\partial t}(t,p)\frac{dt}{ds} = -L(\tau, \xi(\tau), \eta(\tau))\frac{d\tau}{ds} + L(t, x(t,p), u(t,p))\frac{dt}{ds},$$

and it follows from $\xi(\tau(s)) \equiv x(t(s), p(s))$ that

$$f(\tau, \xi(\tau), \eta(\tau))\frac{d\tau}{ds} = \dot{\xi}(\tau)\frac{d\tau}{ds} = \frac{\partial x}{\partial t}(t,p)\frac{dt}{ds} + \frac{\partial x}{\partial p}(t,p)\frac{dp}{ds}$$

$$= f(t, x(t,p), u(t,p))\frac{dt}{ds} + \frac{\partial x}{\partial p}(t,p)\frac{dp}{ds}.$$

Hence,

$$\frac{\partial x}{\partial p}(t,p)\frac{dp}{ds} = f(\tau,\xi(\tau),\eta(\tau))\frac{d\tau}{ds} - f(t,x(t,p),u(t,p))\frac{dt}{ds},$$

and thus by the shadow-price lemma,

$$\frac{\partial C}{\partial p}(t,p)\frac{dp}{ds} = \lambda(t,p)\frac{\partial x}{\partial p}(t,p)\frac{dp}{ds}$$

$$= \lambda(t,p)\left(f(\tau,\xi(\tau),\eta(\tau))\frac{d\tau}{ds} - f(t,x(t,p),u(t,p))\frac{dt}{ds}\right).$$

Combining terms gives the desired result:

$$\Delta(\beta) - \Delta(\alpha)$$

$$= \int_\alpha^\beta \left[H(t,\lambda(t,p),x(t,p),u(t,p))\frac{dt}{ds} - H(\tau,\lambda(t,p),\xi(\tau),\eta(\tau))\frac{d\tau}{ds}\right] ds.$$

$$= \int_{\tau\times\Theta} H(t,\lambda(t,p),x(t,p),u(t,p))dt - H(\tau,\lambda(t,p),\xi(\tau),\eta(\tau))d\tau.$$

$$\square$$

This proof can be rephrased in terms of the differential form ω, but in our calculation the λdx term is being replaced with the shadow-price lemma. This lemma serves the same purpose as the Hilbert invariant integral in the calculus of variations as a tool for eliminating controlled trajectories from optimality. If the parameterizations τ and t are the same, we immediately get the following corollary:

Corollary 5.4.1. *Let $\mathscr{E}$ be a C^1-parametrized family of normal extremals and let $(\xi,\eta) : [t_0,T] \to M \times U$ be a controlled trajectory. If there exists a continuously differentiable curve $\Theta : [\alpha,\beta] \to D$, $s \mapsto (t(s),p(s))$, such that for all $s \in [\alpha,\beta]$ we have that*

$$\xi(t(s)) = x(t(s),p(s)),$$

then

$$C(t(\alpha),p(\alpha)) \le \int_{t(\alpha)}^{t(\beta)} L(t,\xi(t),\eta(t))dt + C(t(\beta),p(\beta)). \qquad (5.54)$$

Proof. In this case, $d\tau = dt$, and the minimum condition of the maximum principle gives that

$$H(t,\lambda(t,p),x(t,p),u(t,p)) \le H(t,\lambda(t,p),\xi(t),\eta(t)).$$

Hence

$$0 \ge \Delta(\beta) - \Delta(\alpha) = \Gamma(t(\beta)) - C(t(\beta),p(\beta)) - \Gamma(t(\alpha)) + C(t(\alpha),p(\alpha)),$$

and thus

$$C(t(\alpha),p(\alpha)) - C(t(\beta),p(\beta)) \leq \Gamma(t(\alpha)) - \Gamma(t(\beta)) = \int_{t(\alpha)}^{t(\beta)} L(t,\xi(t),\eta(t))dt.$$

$\square$

In particular, if the entire trajectory lies in the region covered by the family, i.e., if $[t(\alpha),t(\beta)] = [t_0,T]$, then we have that

$$C(t_0,p_0) \leq \int_{t_0}^{T} L(t,\xi(t),\eta(t))dt + C(T,p(T))$$

$$= \int_{t_0}^{T} L(t,\xi(t),\eta(t))dt + \varphi(T,\xi(T)) = J(\eta), \qquad (5.55)$$

and thus the cost along the controlled trajectory (ξ,η) is no better than the cost along the extremal in the parametrized family of extremals defined by the parameter $p_0 = p(t_0)$ that starts at time t_0. Note that this trajectory itself need not be optimal, but the relation (5.55) is sufficient to exclude (ξ,η) when we are building an optimal synthesis from the extremals in our family.

Our reason for allowing different time parameterizations in Eq. (5.52) lies in another application of Lemma 5.4.3 that gives the more general, and surprisingly useful, concept of control envelopes. These notions go back to the work of H. Sussmann [240, 242], but here we follow our own setup and formalism.

Definition 5.4.5 (Control envelope). Let $\mathscr{E}$ be a C^1-parameterized family of normal extremals with domain D. Let (ξ,η) be a portion of a controlled trajectory defined over an interval $[a,b]$ with the property that there exist differentiable curves $\tau : [\alpha,\beta] \rightarrow [a,b], s \mapsto \tau(s)$, and $\Theta : [\alpha,\beta] \rightarrow D, s \mapsto \Theta(s) = (t(s),p(s))$ such that for all $s \in [\alpha,\beta]$ we have

$$\xi(\tau(s)) = x(t(s),p(s)).$$

We call (ξ,η) a control envelope for the parameterized family $\mathscr{E}$ if along the curves τ and Θ, we have that

$$H(\tau,\lambda(t,p),\xi(\tau),\eta(\tau))d\tau = H(t,\lambda(t,p),x(t,p),u(t,p))dt. \qquad (5.56)$$

The difference between this and the definition of a classical envelope thus is that it is no longer required that the trajectory ξ be generated by means of the controlled trajectories in the parameterized families as $\xi(s) = x(s,p(s))$. In fact, it can be rather arbitrary, as long as the values along the Hamiltonian function relate in the correct way. In the case of a classical envelope, we simply have that $t \equiv \tau$, and thus Eq. (5.56) reduces to Eq. (5.47). While the new definition may appear somewhat artificial, this condition will naturally be satisfied for bang-bang trajectories if the controlled trajectory (ξ, η) lies in a switching surface, and we shall return to it in Sect. 6.1.3. It immediately follows from Lemma 5.4.3 that the envelope theorem remains valid for control envelopes.

Corollary 5.4.2 (Control envelope theorem). *Let* $(\xi, \eta) : [a, b] \to M \times U$ *be a control envelope for a* C^1-*parameterized family of normal extremals with domain* $D = \{(t, p) : p \in P, t_-(p) \leq t \leq t_+(p)\}$. *For any interval* $[s_1, s_2] \subset (a, b)$, *let* $t_1 = t(s_1)$, $p_1 = p(s_1)$, *and* $t_2 = t(s_2)$, $p_2 = p(s_2)$. *Then we have that*

$$C(t_1, p_1) = \int_{t_1}^{t_2} L(s, \xi(s), \eta(s)) ds + C(t_2, p_2).$$

Proof. In the notation of Lemma 5.4.3, we have that $\Delta(s_2) = \Delta(s_1)$; equivalently,

$$C(t_1, p_1) - C(t_2, p_2) = \Gamma(t_1) - \Gamma(t_2) = \int_{t_1}^{t_2} L(s, \xi(s), \eta(s)) ds. \qquad \square$$

5.4.3 Lyapunov–Schmidt Reduction and the Geometry of Fold Singularities

We close this section with developing the geometric properties of an n-dimensional flow of extremal controlled trajectories and its associated value functions $C = C(t, p)$, respectively $V^{\mathscr{E}} = C \circ F^{-1}$, near a fold singularity. As before, we consider the *time-dependent* formulation. In this case, the singular points can be characterized in terms of an eigenvector for the eigenvalue 0 of the matrix $\frac{\partial x}{\partial p}(t, p)$, and this allows us to give a more convenient description of the singular set S than the determinant of $\Delta(t, p)$, the so-called *Lyapunov–Schmidt reduction* [109]. Although geometrically not intrinsic, since it generally depends on a choice of basis for the kernel of $\frac{\partial x}{\partial p}(t, p)$, in the corank-1 case this is not an issue. It is a useful construction that describes the singular set as the zero set of a scalar function defined in terms of a left and right eigenvector of the matrix $\frac{\partial x}{\partial p}(t_0, p_0)$ for the eigenvalue 0. Recall that a right eigenvector, or just eigenvector, of a matrix $A \in \mathbb{R}^{n \times n}$ for the eigenvalue μ is a nonzero column vector $x \in \mathbb{R}^n$ such that $Ax = \mu x$, while a left eigenvector is a nonzero row vector $y \in (\mathbb{R}^n)^*$ such that $yA = \mu y$. In this case, there exist left and right eigenvector fields that can be used to characterize the singular set, and as shown above, we have the following lemma:

Lemma 5.4.4. *There exist a nonzero* C^1 *right-eigenvector field* $v : P \to \mathbb{R}^n$, $v = v(p)$, *and a nonzero* C^1 *left-eigenvector field* $w : P \to (\mathbb{R}^n)^*$, $w = w(p)$, *such that*

$$\left(\frac{\partial x}{\partial p}(\sigma(p), p) \right) v(p) = 0 \quad \text{and} \quad w(p) \left(\frac{\partial x}{\partial p}(\sigma(p), p) \right) = 0.$$

Note that $w(p)$ spans the orthogonal complement of the image of $\frac{\partial x}{\partial p}(\sigma(p), p)$,

$$\text{lin span}\{w(p)\} = \text{Im}\left(\frac{\partial x}{\partial p}(\sigma(p), p)\right)^{\perp}.$$

For the optimal control problem, there exists a remarkable and useful relation between the left and right eigenvectors that follows from the somewhat surprising result below, a corollary of the shadow-price lemma, Lemma 5.2.2.

Proposition 5.4.2. *For any C^2-parametrized family of normal extremals, the matrix*

$$\Xi(t, p) = \frac{\partial \lambda}{\partial p}(t, p)\frac{\partial x}{\partial p}(t, p) \tag{5.57}$$

is symmetric.

Proof. By the shadow price lemma, the partial derivative $\frac{\partial C}{\partial p_j}(t, p)$ of the parameterized cost with respect to the parameter p_j is given by

$$\frac{\partial C}{\partial p_j}(t, p) = \sum_{k=1}^{n} \lambda_k(t, p)\frac{\partial x_k}{\partial p_j}(t, p) = \lambda(t, p)\frac{\partial x}{\partial p_j}(t, p), \qquad j = 1, \dots, n,$$

where $\frac{\partial x}{\partial p_j}(t, p)$ is the column vector of the partial derivatives of the components of x with respect to p_j. Differentiating this equation with respect to p_i gives

$$\frac{\partial^2 C}{\partial p_i \partial p_j}(t, p) = \sum_{k=1}^{n}\left(\frac{\partial \lambda_k}{\partial p_i}(t, p)\frac{\partial x_k}{\partial p_j}(t, p) + \lambda_k(t, p)\frac{\partial^2 x_k}{\partial p_i \partial p_j}(t, p)\right).$$

For a C^2-parameterized family of extremals, the second partial derivatives, $\frac{\partial^2 C}{\partial p_i \partial p_j}$ (t, p) and $\frac{\partial^2 x_k}{\partial p_i \partial p_j}(t, p)$, are equal, and thus we have that

$$\sum_{k=1}^{n}\left(\frac{\partial \lambda_k}{\partial p_i}(t, p)\frac{\partial x_k}{\partial p_j}(t, p)\right) = \sum_{k=1}^{n}\left(\frac{\partial \lambda_k}{\partial p_j}(t, p)\frac{\partial x_k}{\partial p_i}(t, p)\right).$$

But these terms, respectively, are the (i, j) and (j, i) entries of the matrix Ξ. $\square$

Corollary 5.4.3. *If nonzero, then a left-eigenvector field w for the matrix function $p \mapsto \frac{\partial x}{\partial p}(\sigma(p), p)$ on P is given by*

$$w(p) = v^T(p)\frac{\partial \lambda}{\partial p}(\sigma(p), p). \tag{5.58}$$

Proof. We have that

$$w(p)\frac{\partial x}{\partial p}(\sigma(p),p) = v^T(p)\frac{\partial \lambda}{\partial p}(\sigma(p),p)\frac{\partial x}{\partial p}(\sigma(p),p) = v^T(p)\Xi(t,p)$$

$$= (\Xi(t,p)v(p))^T = \left(\frac{\partial \lambda}{\partial p}(\sigma(p),p)\frac{\partial x}{\partial p}(\sigma(p),p)v(p)\right)^T = 0.$$

$\square$

This holds, for example, for the nicely C^1-parameterized canonical family of nonsingular extremals constructed earlier.

Proposition 5.4.3. *Let $\mathscr{E}$ be the n-dimensional, nicely C^1-parameterized, canonical family of nonsingular extremals with domain $D = \{(t,p) : p \in P, t_-(p) \leq t \leq T\}$ for the optimal control problem [OC$_T$] constructed in Theorem 5.3.1. Then $w(p) = v^T(p)\frac{\partial \lambda}{\partial p}(\sigma(p),p)$ is nonzero.*

Proof. If $w(p) = 0$, then the vector functions $t \mapsto \frac{\partial x}{\partial p}(t,p)v(p)$ and $t \mapsto \frac{\partial \lambda^T}{\partial p}(t,p)v(p)$ are solutions to the homogeneous linear system (5.33) that vanish for $t = \sigma(p)$. Hence these functions vanish identically. At the terminal time T, we thus have that $0 = \frac{\partial \xi}{\partial p}(p)v(p)$, and by Eq. (5.36),

$$0 = \frac{\partial \lambda^T}{\partial p}(T,p)v(p)$$

$$= (\varphi_{xx}(\xi(p)) + v(p)\Psi_{xx}(\xi(p)))\frac{\partial x}{\partial p}(T,p)v(p) + \Psi_x^T(\xi(p))\frac{\partial v^T}{\partial p}(p)v(p)$$

$$= \Psi_x^T(\xi(p))\frac{\partial v^T}{\partial p}(p)v(p).$$

The columns of the matrix $\Psi_x^T(\xi(p))$ are linearly independent, and thus overall, we have that

$$\frac{\partial \omega}{\partial p}(p)v(p) = \left(\frac{\partial \xi}{\partial p}(p), \frac{\partial v^T}{\partial p}(p)\right)v(p) = 0.$$

But for the canonical parameterization, the matrix $\frac{\partial \omega}{\partial p}(p)$ is nonsingular, and thus $v(p) = 0$ as well, contradicting the fact that $v(p)$ is an eigenvector. $\square$

We return to the general setup and define the *Lyapunov–Schmidt reduction* in terms of the left- and right-eigenvector fields v and w. Let

$$\zeta : D \to \mathbb{R}, \quad (t,p) \mapsto \zeta(t,p) = w(p)\frac{\partial x}{\partial p}(t,p)v(p). \tag{5.59}$$

Note that for

$$w(p) = v(t)^T \frac{\partial \lambda}{\partial p}(t,p),$$

this function can be expressed in the form

$$\zeta(t,p) = w(p)\frac{\partial x}{\partial p}(t,p)v(p) = v(p)^T \Xi(t,p)v(p) = \langle v(p), \Xi(t,p)v(p)\rangle,$$

so that ζ is a symmetric quadratic form. Denote the zero set of ζ in D by Z. Clearly, Z contains the singular set S. If the gradient of ζ, $\nabla \zeta$, does not vanish at the singular point (t_0, p_0), $t_0 = \sigma(p_0)$, then Z is an embedded n-dimensional manifold near (t_0, p_0) that contains S, under our assumptions itself an embedded n-dimensional manifold. Hence S and Z are equal. The gradient of ζ at a point $(\sigma(p),p) \in S$ is easily calculated: dropping the arguments, we have that

$$\zeta = w\frac{\partial x}{\partial p}v = \sum_{i=1}^{n}\sum_{j=1}^{n} w_i \frac{\partial x_i}{\partial p_j}v_j,$$

and thus the partial derivative with respect to t is simply given by the quadratic form

$$\frac{\partial \zeta}{\partial t} = \sum_{i=1}^{n}\sum_{j=1}^{n} w_i \frac{\partial^2 x_i}{\partial t \partial p_j}v_j = w\frac{\partial^2 x}{\partial t \partial p}v.$$

On the singular set S, and this is where our interest lies, the partial derivatives with respect to p also become simple, since w and v are the left and right eigenvectors of $\frac{\partial x}{\partial p}(\sigma(p),p)$. In general, we have that

$$\frac{\partial}{\partial p_k}\left(w\frac{\partial x}{\partial p}v\right)$$

$$= \sum_{i=1}^{n}\sum_{j=1}^{n}\left(\frac{\partial w_i}{\partial p_k}\frac{\partial x_i}{\partial p_j}v_j + w_i\frac{\partial^2 x_i}{\partial p_k \partial p_j}v_j + w_i\frac{\partial x_i}{\partial p_j}\frac{\partial v_j}{\partial p_k}\right)$$

$$= \sum_{i=1}^{n}\frac{\partial w_i}{\partial p_k}\left(\sum_{j=1}^{n}\frac{\partial x_i}{\partial p_j}v_j\right) + \sum_{i=1}^{n}\sum_{j=1}^{n}w_i\frac{\partial^2 x_i}{\partial p_k \partial p_j}v_j + \sum_{j=1}^{n}\left(\sum_{i=1}^{n}w_i\frac{\partial x_i}{\partial p_j}\right)\frac{\partial v_j}{\partial p_k},$$

and on the singular set S, this reduces to

$$\frac{\partial}{\partial p_k}\left(w\frac{\partial x}{\partial p}v\right) = \sum_{i=1}^{n}\sum_{j=1}^{n} w_i\frac{\partial^2 x_i}{\partial p_k \partial p_j}v_j,$$

since both $\frac{\partial x}{\partial p}v$ and $w\frac{\partial x}{\partial p}$ vanish. These partial derivatives can be expressed in a more convenient and compact form if instead we consider the directional derivative in the direction of a tangent vector z in p-space. For then we have that

$$\langle \nabla_p \zeta(\sigma(p),p),z\rangle = \sum_{k=1}^{n} \frac{\partial}{\partial p_k}\left(w\frac{\partial x}{\partial p}v\right)z_k = \sum_{k=1}^{n}\sum_{i=1}^{n} w_i\left(\sum_{j=1}^{n}\frac{\partial^2 x_i}{\partial p_k \partial p_j}v_j\right)z_k$$

$$= \sum_{i=1}^{n} w_i\left(\sum_{j=1}^{n}\sum_{k=1}^{n}\frac{\partial^2 x_i}{\partial p_k \partial p_j}v_j z_k\right)$$

$$= \left\langle w,\frac{\partial^2 x}{\partial p^2}(v,z)\right\rangle = w\frac{\partial^2 x}{\partial p^2}(v,z).$$

Thus, the directional derivative in the direction of z is given on S by the inner product of the left eigenvector w with the column vector $\frac{\partial^2 x}{\partial p^2}(v,z)$ whose entries are the quadratic forms of the second derivatives of the components x_i of the state with respect to p acting on the vectors v and z.

Summarizing this calculation, at a singular point $(\sigma(p),p)$, $p \in P$, the directional derivative of ζ in the direction of the vector $\mathfrak{z} = (\tau,z)$ can be expressed as

$$\langle \nabla\zeta(\sigma(p),p),\mathfrak{z}\rangle = \left\langle w(p),\frac{\partial^2 x}{\partial t\partial p}(\sigma(p),p)v(p)\tau + \frac{\partial^2 x}{\partial p^2}(\sigma(p),p)\Big(v(p),z\Big)\right\rangle.$$

$$(5.60)$$

We thus have the following statement:

Proposition 5.4.4. *Suppose* (t_0,p_0) *is a t-regular, corank-1 singular point for which the gradient* $\nabla\zeta(\sigma(p_0),p_0)$ *does not vanish. Then there exist an open neighborhood* $W = (t_0 - \varepsilon, t_0 + \varepsilon) \times P$ *of* (t_0,p_0) *and a* C^1-*function* σ *defined on* P, $\sigma: P \to (t_0 - \varepsilon, t_0 + \varepsilon)$ *such that the singular set* S *is given by* $S = (t,p) \in W : t = \sigma(p)\}$. *In terms of a* C^1 *right-eigenvector field* $v: P \to \mathbb{R}^n$,

$$\frac{\partial x}{\partial p}(\sigma(p),p)v(p) \equiv 0,$$

and a C^1 *left-eigenvector field* $w: P \to (\mathbb{R}^n)^*$,

$$w(p)\frac{\partial x}{\partial p}(\sigma(p),p) \equiv 0,$$

the singular set S *can be described as*

$$S = \left\{(t,p) \in W : \zeta(t,p) = w(p)\frac{\partial x}{\partial p}(t,p)v(p) = 0\right\},$$

$$(5.61)$$

and the tangent space to S at $(\sigma(p), p)$, $T_{(\sigma(p),p)}S$, is given by

$$\left\{ \mathfrak{z} = \begin{pmatrix} \tau \\ z \end{pmatrix} \in \mathbb{R}^{n+1} : w(p) \left(\frac{\partial^2 x}{\partial t \partial p}(\sigma(p), p) v(p) \tau + \frac{\partial^2 x}{\partial p^2}(\sigma(p), p) (v(p), z) \right) = 0 \right\}.$$

(5.62)

Corollary 5.4.4. *Under the same assumptions, and setting $v_0 = v(p_0)$ and $w_0 = w(p_0)$, (t_0, p_0) is a fold singularity if and only if*

$$w_0 \frac{\partial^2 x}{\partial p^2}(t_0, p_0)(v_0, v_0) \neq 0. \tag{5.63}$$

Proof. The point (t_0, p_0) is a fold if and only if the vector $\mathfrak{z}_0 = (0, v_0)^T$ is not tangent to S at (t_0, p_0). Assuming that the gradient $\nabla \zeta(t_0, p_0)$ does not vanish, this is equivalent to $w_0 \frac{\partial^2 x}{\partial p^2}(t_0, p_0)(v_0, v_0) \neq 0$. Note that this condition by itself guarantees that $\nabla \zeta(t_0, p_0)$ does not vanish. □

5.4.4 The Geometry of the Flow F and the Graph of the Value Function $V^{\mathscr{E}}$ near a Fold Singular Point

The name for the fold singularity has its origin in the geometric properties of the mapping at such a point that resemble those of a quadratic function: Let $M_f = F(S)$ be the image of the singular manifold S under the flow $F(t, p) = (t, x(t, p))$. For a corank-1 singular point, the differential $DF(t, p)$ is of rank n with null space spanned by $(0, v(p))^T$, and at a fold point this vector does not lie in the tangent space to S at $(\sigma(p), p)$. Hence the restriction of the flow F to its singular set S is a diffeomorphism, and thus M_f is an n-dimensional manifold as well. It follows from the chain rule that the tangent space to M_f at the point $F(\sigma(p), p) = (\sigma(p), x(\sigma(p), p))$ is given by the image of the tangent space to S at the point $(\sigma(p), p)$, $T_{(\sigma(p),p)}S$, under the differential DF. Furthermore,

$$\left(-w(p) \frac{\partial x}{\partial t}(\sigma(p), p), \; w(p) \right) \begin{pmatrix} 1 & 0 \\ \frac{\partial x}{\partial t}(\sigma(p), p) & \frac{\partial x}{\partial p}(\sigma(p), p) \end{pmatrix} = (0, 0),$$

and thus $\mathfrak{n} = \left(-w(p) \frac{\partial x}{\partial t}(\sigma(p), p), \; w(p) \right)$ is a normal vector to $T_{F(\sigma(p),p)}M_f$.

In order to establish the geometric properties of the flow F, let

$$\gamma_p : [-\varepsilon, \varepsilon] \to D, \quad s \mapsto (\sigma(p), p + s v(p)),$$

be small line segments in the parameter space that pass through the point $(\sigma(p), p)$ in S for $s = 0$. Since the vector $(0, v(p))^T$ is transversal to S, without loss of generality we may assume that $\gamma_p(s) \notin S$ for all $0 < |s| \leq \varepsilon$ and p in a sufficiently small neighborhood of p_0. Let $\phi_p = F \circ \gamma_p$ be the images of the line segments γ_p under the flow F. By Taylor's theorem, since $\frac{\partial x}{\partial p}(\sigma(p), p)v(p) = 0$, we have that

$$x(\sigma(p), p + sv(p)) = x(\sigma(p), p) + \frac{1}{2}s^2 \frac{\partial^2 x}{\partial p^2}(\sigma(p), p)(v(p), v(p)) + o(s^2),$$

and thus

$$\phi_p(s) = F(\gamma_p(s)) = \begin{pmatrix} \sigma(p) \\ x(\sigma(p), p) \end{pmatrix} + \frac{1}{2}s^2 \begin{pmatrix} 0 \\ \frac{\partial^2 x}{\partial p^2}(\sigma(p), p)(v(p), v(p)) \end{pmatrix} + o(s^2).$$

Since the singular points are fold singularities, $\frac{\partial^2 x}{\partial p^2}(\sigma(p), p)(v(p), v(p)) \neq 0$, and taking the inner product of the second-order tangent vector

$$\ddot{\phi}_p(0) = \begin{pmatrix} 0 \\ \frac{\partial^2 x}{\partial p^2}(\sigma(p), p)(v(p), v(p)) \end{pmatrix}$$

with the normal vector $\mathfrak{n}$ to $T_{F(\sigma(p), p)} M_f$, we get that

$$\langle \mathfrak{n}, \ddot{\phi} \rangle = w(p) \frac{\partial^2 x}{\partial p^2}(\sigma(p), p)(v(p), v(p)) \neq 0.$$

This expression has constant sign for p near p_0, and thus, except for $s = 0$, where $\phi_p(0) = F(\sigma(p), p) = (\sigma(p), x(\sigma(p), p)) \in M_f$, the curves ϕ_p have first-order contact with the manifold M_f and thus lie to the same side of M_f. (Note that we could also express the curves as two separate directed curves in terms of the parameter $\sqrt{|s|}$ for $s \geq 0$ and $s \leq 0$, and then the one-sided tangent vectors are transversal to and point to the same side of the tangent plane. Hence the curves lie to the same side of the manifold M_f.) For the parameterized family of extremals, this implies that the flow F is $2:1$ near the singular set S. We summarize the mapping properties in the following proposition:

Proposition 5.4.5. *If (t_0, p_0) is a fold singular point, then there exist open neighborhoods W of (t_0, p_0) and G of $(t_0, x_0) = (t_0, x(t_0, p_0))$ such that S divides W into two connected components W_+ and W_-, $W = W_- \cup S \cup W_+$, with the following properties: the flow F restricted to the open subsets W_+ and W_- is a $C^{1,2}$ diffeomorphism, and F maps each of the regions W_+ and W_- of the (t, p)-space onto a region $G_+ \subset G$, $F(W_+) = G_+ = F(W_-)$. Away from the singular set S, the flow F is $2:1$ on W.* ∎

Similar calculations also establish the geometric properties of the parameterized cost near a fold singularity. We augment the flow map F with the parameterized cost and consider the graph of the associated value function $V^{\mathcal{E}} = C \circ F^-$, where

F^- now denotes the set of all inverse images, i.e., $F^-(t,x) = \{(t,p) : x = x(t,p)\}$. Since the flow F is $2 : 1$ near the fold points, this becomes a multivalued function. The mapping

$$\mathfrak{gr}\left(V^{\mathscr{E}}\right) : D \to \mathbb{R}^{n+2}, \qquad (t,p) \mapsto \mathfrak{gr}\left(V^{\mathscr{E}}\right)(t,p) = \begin{pmatrix} t \\ x(t,p) \\ C(t,p) \end{pmatrix}, \qquad (5.64)$$

gives a parameterization of the value $V^{\mathscr{E}} = C \circ F^-$, and when the image is plotted in (t,x,C)-space, it is the graph of this multivalued function $V^{\mathscr{E}} = C \circ F^-$ (also, see Sect. 1.6) By the shadow-price lemma, the Jacobian matrix of $\mathfrak{gr}\left(V^{\mathscr{E}}\right)$ is given by

$$D\mathfrak{gr}\left(V^{\mathscr{E}}\right)(t,p) = \begin{pmatrix} 1 & 0 \\ f(t,x(t,p),u(t,p)) & \frac{\partial x}{\partial p}(t,p) \\ -L(t,x(t,p),u(t,p)) & \lambda(t,p)\frac{\partial x}{\partial p}(t,p) \end{pmatrix},$$

and the nonzero covector

$$\varpi(t,p) = \left(h(t,p), \; -\lambda(t,p), \; 1 \right), \qquad h(t,p) = H(t,\lambda(t,p),x(t,p),u(t,p)),$$

is normal to the image $\mathrm{Im}\, D\mathfrak{gr}\left(V^{\mathscr{E}}\right)(t,p)$ everywhere,

$$\varpi(t,p) \, D\mathfrak{gr}\left(V^{\mathscr{E}}\right)(t,p) \equiv 0 \qquad \text{for all } (t,p) \in D. \qquad (5.65)$$

In particular, the last row of $D\mathfrak{gr}\left(V^{\mathscr{E}}\right)(t,p)$ is a linear combination of the first $n+1$ rows, and therefore the mapping $\mathfrak{gr}\left(V^{\mathscr{E}}\right)$ has the same singular set S as F. Indeed, both maps share the eigenvector field $\hat{v}(p) = (0,v(p))^T$. Away from the singular set, $D\mathfrak{gr}\left(V^{\mathscr{E}}\right)$ is of full rank, and near those regular points, the image of $\mathfrak{gr}\left(V^{\mathscr{E}}\right)$ is locally a codimension-1 embedded submanifold, a hypersurface, with the tangent plane given by the hyperplane $\mathfrak{H}(t,p)$ through the point $\mathfrak{gr}\left(V^{\mathscr{E}}\right)(t,p)$ with normal vector $\varpi(t,p)$. On the singular set S, the dimension drops by 1, and the extra normal vector that annihilates the image is given by

$$\varpi_f(p) = \left(-w(p)f(\sigma(p),x(\sigma(p),p),u(\sigma(p),p)), \; w(p), \; 0 \right),$$

where, as before, $w(p)$ is the left eigenvector for $\frac{\partial x}{\partial p}(\sigma(p),p)$. However, even at the singular points, the hyperplane $\mathfrak{H}(\sigma(p),p)$ is well-defined, and the tangent space to M_f is a codimension-1 subspace of this hyperplane.

Lemma 5.4.5. *Suppose* $w(p) = v(p)^T \frac{\partial \lambda}{\partial p}(\sigma(p),p)$ *is nonzero; then we have the following formulas for the derivatives of* $\mathfrak{gr}\left(V^{\mathscr{E}}\right)$ *on the singular set S, i.e., for* $(t,p) = (\sigma(p),p)$,

$$\varpi \left(\frac{\partial^2 \mathfrak{gr}\left(V^{\mathscr{E}}\right)}{\partial p^2} \right) (v, v) \equiv 0, \tag{5.66}$$

$$\varpi \left(\frac{\partial^3 \mathfrak{gr}\left(V^{\mathscr{E}}\right)}{\partial p^3} \right) (v, v, v) \equiv 2w \frac{\partial^2 x}{\partial p^2}(v, v) \neq 0. \tag{5.67}$$

Proof. All functions are evaluated at (t, p) and the eigenvectors v and w at p, but we drop the argument. For all $(t, p) \in D$ we have that

$$\left(\frac{\partial^2 \mathfrak{gr}\left(V^{\mathscr{E}}\right)}{\partial p^2} \right) (v, v) = \begin{pmatrix} 0 \\ \frac{\partial^2 x}{\partial p^2}(v, v) \\ \frac{\partial^2 C}{\partial p^2}(v, v) \end{pmatrix}.$$

By the shadow-price lemma, we have $\frac{\partial C}{\partial p} = \lambda \frac{\partial x}{\partial p}$ for all (t, p) and thus

$$\frac{\partial^2 C}{\partial p^2} = \frac{\partial \lambda}{\partial p} \frac{\partial x}{\partial p} + \lambda \frac{\partial^2 x}{\partial p^2}.$$

With $w = v^T \frac{\partial \lambda}{\partial p}$ the left eigenvector of $\frac{\partial x}{\partial p}$, this gives

$$\frac{\partial^2 C}{\partial p^2}(v, v) = w^T \frac{\partial x}{\partial p} v + \lambda \frac{\partial^2 x}{\partial p^2}(v, v) = \zeta + \lambda \frac{\partial^2 x}{\partial p^2}(v, v), \tag{5.68}$$

where ζ is the function defining the Lyapunov–Schmidt reduction. On the singular set, $\frac{\partial x}{\partial p} v = 0$, and thus $\frac{\partial^2 C}{\partial p^2}(v, v) = \lambda \frac{\partial^2 x}{\partial p^2}(v, v)$, so that

$$\left(\frac{\partial^2 \mathfrak{gr}\left(V^{\mathscr{E}}\right)}{\partial p^2} \right) (v, v) = \begin{pmatrix} 0 \\ \frac{\partial^2 x}{\partial p^2}(v, v) \\ \lambda \frac{\partial^2 x}{\partial p^2}(v, v) \end{pmatrix}.$$

This vector is orthogonal to $\varpi(\sigma(p), p)$, and Eq. (5.66) follows.

Differentiating Eq. (5.68) once more with respect to p, letting the derivative act on v, and evaluating on the singular set, we obtain that

$$\frac{\partial^3 C}{\partial p^3}(v, v, v) = 2w^T \left(\frac{\partial^2 x}{\partial p^2}(v, v) \right) + \lambda \frac{\partial^3 x}{\partial p^3}(v, v, v).$$

Thus, on the singular set, we have that

$$
\left(\frac{\partial^3 \mathfrak{gr}\left(V^{\mathscr{E}}\right)}{\partial p^3}\right)(v,v,v) = \begin{pmatrix} 0 \\ \frac{\partial^3 x}{\partial p^3}(v,v,v) \\ \frac{\partial^3 C}{\partial p^3}(v,v,v) \end{pmatrix} = \begin{pmatrix} 0 \\ \frac{\partial^3 x}{\partial p^3}(v,v,v) \\ 2w\frac{\partial^2 x}{\partial p^2}(v,v) + \lambda\frac{\partial^3 x}{\partial p^3}(v,v,v) \end{pmatrix},
$$

which gives Eq. (5.67).   □

A third-order Taylor expansion of the mapping $\mathfrak{gr}\left(V^{\mathscr{E}}\right)$ along the curves γ_p takes the general form

$$
\mathfrak{gr}\left(V^{\mathscr{E}}\right)(\gamma_p(s)) = \mathfrak{gr}\left(V^{\mathscr{E}}\right)(\gamma_p(0)) + sD\mathfrak{gr}\left(V^{\mathscr{E}}\right)(\gamma_p(0)) \cdot \dot{\gamma}_p(0)
$$

$$
+ \frac{1}{2}s^2 D^2 \mathfrak{gr}\left(V^{\mathscr{E}}\right)(\gamma_p(0)) \cdot (\dot{\gamma}_p(0), \dot{\gamma}_p(0))
$$

$$
+ \frac{1}{6}s^3 D^3 \mathfrak{gr}\left(V^{\mathscr{E}}\right)(\gamma_p(0)) \cdot (\dot{\gamma}_p(0), \dot{\gamma}_p(0), \dot{\gamma}_p(0)) + o(s^3).
$$

We have $\gamma_p(0) = (\sigma(p), p)$, $\dot{\gamma}_p(0) = (0, v(p))^T$, and thus with v and σ evaluated at p and all functions evaluated at $(\sigma(p), p)$, we get that

$$
\mathfrak{gr}\left(V^{\mathscr{E}}\right)(\gamma_p(s)) = \begin{pmatrix} \sigma \\ x \\ C \end{pmatrix} + \frac{1}{2}s^2 \begin{pmatrix} 0 \\ \frac{\partial^2 x}{\partial p^2}(v,v) \\ \lambda\frac{\partial^2 x}{\partial p^2}(v,v) \end{pmatrix}
$$

$$
+ \frac{1}{6}s^3 \begin{pmatrix} 0 \\ \frac{\partial^3 x}{\partial p^3}(v,v,v) \\ 2w\frac{\partial^2 x}{\partial p^2}(v,v) + \lambda\frac{\partial^3 x}{\partial p^3}(v,v,v) \end{pmatrix} + o(s^3).
$$

Now take the inner product of the normal vector $\varpi(\sigma(p), p)$ to the hyperplane $\mathfrak{H}(\sigma(p), p)$ with the increment in s. This normal vector annihilates the quadratic term and also the $\frac{\partial^3 x}{\partial p^3}(v,v,v)$-terms at s^3, so that

$$
\left\langle \varpi(\sigma(p), p), \mathfrak{gr}\left(V^{\mathscr{E}}\right)(\gamma_p(s)) - \mathfrak{gr}\left(V^{\mathscr{E}}\right)(\gamma_p(0)) \right\rangle = \frac{1}{3}s^3 w\frac{\partial^2 x}{\partial p^2}(v,v) + o(s^3).
$$

Assuming that $w(p) = v(p)^T \frac{\partial \lambda}{\partial p}(\sigma(p), p)$ is nonzero (cf. Proposition 5.4.3), the leading term is a nonzero multiple of the term defining the fold singularity. This time, however, it is of odd order, and thus it follows that the hypersurfaces defined by $\mathfrak{gr}\left(V^{\mathscr{E}}\right)(\gamma_p(s))$ for $s < 0$ and for $s > 0$ lie to opposite sides of the hyperplane $\mathfrak{H}(\sigma(p), p)$. Since the normal vector $\varpi(\sigma(p), p)$ has its last coordinate given by 1, this implies that near

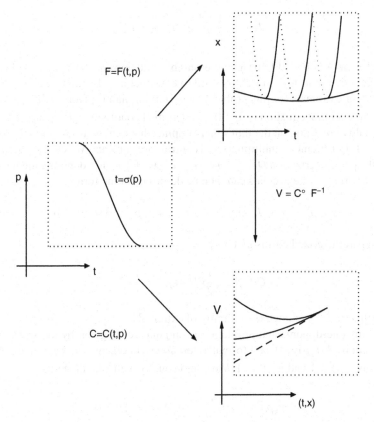

Fig. 5.10 The square on the *left* is the neighborhood W in (t, p)-space with the singular set S given by the inscribed curve. On the *right*, the *top* diagram represents the image of the flow F showing the "fold" in the controlled trajectories; the *bottom* diagram shows a slice of the corresponding value in the state space

$$\mathfrak{gr}\left(V^{\mathscr{E}}\right)(\gamma_p(0)) = \begin{pmatrix} \sigma(p) \\ x(\sigma(p), p) \\ C(\sigma(p), p) \end{pmatrix}$$

one of these surfaces lies below $\mathfrak{H}(\sigma(p), p)$ in the direction of the cost C and the other one lies above $\mathfrak{H}(\sigma(p), p)$. Thus, one of the branches is minimizing, and optimality ceases as the fold singularities are crossed. The overall structure is illustrated in Fig. 5.10

We give a simple, yet canonical, one-dimensional example in which all computations can easily be done.

Example 5.4.1 (Fold singularity). For a fixed terminal time T, consider the problem to minimize the objective

$$J(u) = \frac{1}{2} \int_{t_0}^{T} u^2 dt + \frac{1}{3} x(T)^3$$

over all locally bounded Lebesgue measurable functions $u : [t_0, T] \to \mathbb{R}$ subject to the dynamics $\dot{x} = u$. It is straightforward to construct a real-analytic parameterized family $\mathcal{E}$ of extremals for this problem: The Hamiltonian is given by $H = \frac{1}{2}\lambda_0 u^2 + \lambda u$, and the adjoint equation is $\dot{\lambda} \equiv 0$ with terminal condition $\lambda(T) = \lambda_0 x(T)^2$. The nontriviality condition on the multipliers implies that extremals are normal, and we set $\lambda_0 \equiv 1$. Furthermore, multipliers and controls are constant. We choose as domain D for the parameterization $D = \{(t, p) : t \leq T, \ p \in \mathbb{R}\}$ with p denoting the terminal point, $p = x(T)$. All extremals can then be described in the form

$$x(t, p) = p + (T - t)p^2, \quad u(t, p) = -p^2, \quad \lambda(t, p) = p^2,$$

and the parameterized cost is given by

$$C(t, p) = \frac{1}{2}(T - t)p^4 + \frac{1}{3}p^3.$$

This defines a C^ω-parameterized family of extremals.

For this one-dimensional problem, the singular set is given by the solutions to the equation $\frac{\partial x}{\partial p}(t, p) = 0$, and formally, left and right eigenvectors are given by the constants $v(p) \equiv 1$ and $w(p) \equiv 1$. We could equally well take the term

$$w(p) = v(p)^T \frac{\partial \lambda}{\partial p}(\sigma(p), p) = 2p \neq 0$$

that came up in the theoretical computations. Thus,

$$S = \left\{ (t, p) \in D : t = \sigma(p) = T + \frac{1}{2p}, \ p < 0 \right\}.$$

Since $\frac{\partial^2 x}{\partial p^2}(t, p) = 2(T - t) > 0$, all singular points are fold points. Figure 5.11 shows the flow of extremals for $p < 0$ on the left and the graph of the corresponding multivalued value function $V^{\mathcal{E}} = C \circ F^-$ on the right.

Figure 5.12 shows four slices of this value function in the state space for $t = 0.6$, $t = 0.8$, $t = 1$, and $t = 1.6$. For $t = 1.6$, the fold point does not lie in the range shown, and thus the corresponding section of the value function $V^{\mathcal{E}} = C \circ F^{-1}$ appears singlevalued. For the other three time slices, the value function has two values and shows the characteristic behavior near a fold singularity.

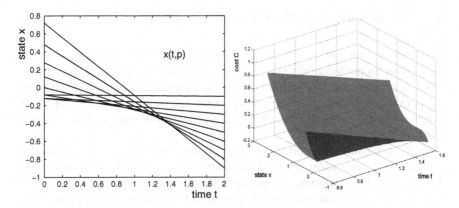

Fig. 5.11 The flow of the parameterized family of extremals $\mathscr{E}$ over the interval $[0, T]$ for $T = 2$ and $p < 0$ (*left*) and the graph of the associated value function $V^{\mathscr{E}} = C \circ F^-$ (*right*)

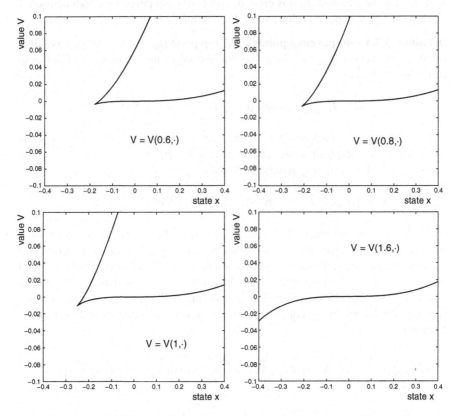

Fig. 5.12 Slices of the value corresponding to the parameterized family $\mathscr{E}$ of extremals for $t =$ const

5.5   Simple Cusp Singularities and Cut-Loci

The geometry of the flow near cusp singularities is more intricate and so are its implications on the local optimality of controlled trajectories. We do not develop these here in full generality, but limit our presentation to illustrate them with another simple example, analogous to the one just given for the case of a fold, where all computations can easily be done. We shall pick up this geometric structure again and in more detail in Chap. 7.

Recall that a singular point (t_0, p_0) of a C^2-parameterized family $\mathscr{E}$ of normal extremals is a *cusp point* if the eigenvector $\hat{v}(p_0) = (0, v(p_0))^T = (0, v_0)^T$ of $DF(t_0, p_0)$ for the eigenvalue 0 is tangent to the singular set S at (t_0, p_0), $\hat{v}(p_0) \in T_{(t_0,p_0)}S$. This requires an extra equality constraint to be satisfied, and thus typically the set of all cusp points is a lower-dimensional subset of S. It is called a simple cusp singularity if this happens in the least-degenerate case, i.e., if no additional equality relations will be satisfied. In this case, the set C of cusp points is a codimension-1 submanifold of S, hence a codimension-2 submanifold D.

Definition 5.5.1 (Simple cusp points). A cusp point (t_0, p_0), $t_0 = \sigma(p_0)$, is called simple if the vector $\hat{v}(p_0) = (0, v_0)^T$ is transversal to the submanifold C of cusp points in $T_{(t_0,p_0)}S$, i.e.,

$$T_{(t_0,p_0)}C \oplus \text{lin span } \{\hat{v}(p_0)\} = T_{(t_0,p_0)}S.$$

We again make *assumption (A)* of the last section. Hence, the singular set is described as the graph of a function σ as $S = \{(t, p) \in D : t = \sigma(p)\}$, and by Lemma 5.4.2, the set of cusp points is given by

$$C = \{(t, p) \in D : t = \sigma(p), \quad L_v\sigma(p) = \nabla\sigma(p)v(p) = 0\},$$

i.e., is the set of points where the Lie derivative of the function σ in the direction of the zero-eigenvalue eigenvector field v vanishes. If the gradient of this function in p does not vanish at p_0, then C is a codimension-1 embedded submanifold of S and $(t_0, p_0) \in C$, $t_0 = \sigma(p_0)$, is simple if and only if the second Lie derivative $L_v^2\sigma$ of σ in the direction of v does not vanish at p_0. (In this case, $\hat{v}(p_0)$ is not tangent to the submanifold defined by the equation $L_v\sigma(p) = 0$, while it is tangent to S at (t_0, p_0).) Note that

$$L_v^2\sigma(p_0) = \left(\frac{\partial}{\partial p}_{|p_0} \nabla\sigma(p)v(p)\right)v_0 = \frac{\partial^2\sigma}{\partial p^2}(p_0)(v_0, v_0) + \nabla\sigma(p_0)\frac{\partial v}{\partial p}(p_0)v_0,$$

with $\frac{\partial^2\sigma}{\partial p^2}(p_0)(v_0, v_0)$ denoting the quadratic form defined by the Hessian matrix of σ acting on the eigenvector v_0, and $\frac{\partial v}{\partial p}(p)$ is the Jacobian matrix of the vector field v.

In terms of the Lyapunov–Schmidt reduction, $\zeta(t,p) = w(p)\frac{\partial x}{\partial p}(t,p)v(p)$, if the gradient $\nabla\zeta(\sigma(p_0),p_0)$ does not vanish, then by Proposition 5.4.4, the set C of cusp points is given as

$$C = \left\{(t,p) \in W : \; w(p)\frac{\partial x}{\partial p}(t,p)v(p) = 0, \quad w(p)\frac{\partial^2 x}{\partial p^2}(t,p)(v(p),v(p)) = 0\right\}.$$

For example, the condition that $\nabla\zeta(\sigma(p_0),p_0) \neq 0$ can simply be guaranteed by

$$\frac{\partial\zeta}{\partial t}(t_0,p_0) = w_0\frac{\partial^2 x}{\partial t\partial p}(t_0,p_0)v_0 \neq 0.$$

Making this assumption, simple cusp points can be described as follows:

Proposition 5.5.1. *Suppose (t_0,p_0) is a t-regular, corank-1 singular point for which $\frac{\partial\zeta}{\partial t}(t_0,p_0) = w_0\frac{\partial^2 x}{\partial t\partial p}(t_0,p_0)v_0 \neq 0$. Then (t_0,p_0) is a simple cusp if and only if*

$$w_0\frac{\partial^2 x}{\partial p^2}(t_0,p_0)(v_0,v_0) = 0$$

and

$$w_0\left(\frac{\partial^3 x}{\partial p^3}(t_0,p_0)(v_0,v_0,v_0) + 3\frac{\partial^2 x}{\partial p^2}(t_0,p_0)\left(\frac{\partial v}{\partial p}(p_0)v_0,v_0\right)\right) \neq 0. \qquad (5.69)$$

Proof. For all $p \in P$ we have that

$$\frac{\partial x}{\partial p}(\sigma(p),p)v(p) = 0.$$

Differentiating this equation, and letting the derivative act upon $v(p)$, gives

$$\frac{\partial^2 x}{\partial t\partial p}(\sigma(p),p)v(p) \cdot L_v\sigma(p) + \frac{\partial^2 x}{\partial p^2}(\sigma(p),p)(v(p),v(p))$$

$$+ \frac{\partial x}{\partial p}(\sigma(p),p)\frac{\partial v}{\partial p}(p)v(p) = 0.$$

Multiplying on the left by the left eigenvector field $w(p)$ annihilates the last term. Furthermore, for a cusp point, $L_v\sigma(p) = 0$, and thus $w_0\frac{\partial^2 x}{\partial p^2}(t_0,p_0)(v_0,v_0) = 0$ follows. Now differentiate this relation once more, evaluate at the cusp point, and multiply on the right with v and on the left with w. In doing so, all terms that include $L_v\sigma(p_0)$, $w_0\frac{\partial x}{\partial p}(t_0,p_0)$ or $w_0\frac{\partial^2 x}{\partial p^2}(t_0,p_0)(v_0,v_0)$ vanish. We are left with

$$\left(w_0 \frac{\partial^2 x}{\partial t \partial p}(t_0, p_0)v_0 \right) L_v^2 \sigma(p_0) + w_0 \frac{\partial^3 x}{\partial p^3}(t_0, p_0)(v_0, v_0, v_0)$$

$$+ 3w_0 \frac{\partial^2 x}{\partial p^2}(t_0, p_0)\left(\frac{\partial v}{\partial p}(p_0)v_0, v_0 \right) = 0.$$

By assumption, $w_0 \frac{\partial^2 x}{\partial t \partial p}(t_0, p_0)v_0$ is nonzero, and thus $L_v^2 \sigma(p_0) \neq 0$ if and only if Eq. (5.69) holds. $\square$

We now illustrate the mapping properties of the flow F of a parameterized family $\mathscr{E}$ of extremals near a simple cusp point. These and their implications on the associated value function are quite intricate, and therefore we use a simple yet canonical example that is based on the so-called normal form of the simple cusp to develop these properties. The geometric properties seen in this example indeed are generally valid features near simple cusp points. We once more consider a one-dimensional optimal control problem similar to the one above, but with a different penalty term that preprograms the cusp singularity.

Example 5.5.1 (Simple cusp singularity). [148] For a fixed terminal time T, consider the problem to minimize the objective

$$J(u) = \frac{1}{2} \int_{t_0}^{T} u^2 dt + \frac{1}{2} \left(x(T)^4 - x(T)^2 \right) \tag{5.70}$$

over all locally bounded Lebesgue measurable functions $u : [t_0, T] \to \mathbb{R}$ subject to the dynamics $\dot{x} = u$.

This simple regulator problem over a finite interval is related to the inviscid Burgers's equation (e.g., see [65, 66]), a fundamental partial differential equation in fluid mechanics that is a prototype for equations that develop shock waves. In the optimal control formulation, this becomes the Hamilton–Jacobi–Bellman equation, and we shall see that it is the simple cusp point that generates a cut-locus of optimal controls where solutions are no longer unique, the shockwave in the language of PDEs.

As before, it is elementary to construct a real-analytic parameterized family $\mathscr{E}$ of extremals. Once again, extremals are normal, and we set $\lambda_0 \equiv 1$. The Hamiltonian is thus given by $H = \frac{1}{2}u^2 + \lambda u$, and the minimizing control satisfies $u = -\lambda$. The adjoint equation is $\dot{\lambda} \equiv 0$ with terminal condition $\lambda(T) = 2x(T)^3 - x(T)$. Again choosing as parameter the terminal point $p = x(T)$, we can parameterize all extremals over the domain $D = \{(t, p) : t \leq T, \, p \in \mathbb{R}\}$ in the form

$$x(t, p) = p + (t - T)(p - 2p^3), \quad u(t, p) = p - 2p^3, \quad \lambda(t, p) = 2p^3 - p.$$

The terminal cost is given by $\gamma(p) = \frac{1}{2}\left(p^4 - p^2 \right)$, and since we start integrating from the terminal manifold $N = \{(t, x) : t = T\}$, the transversality condition (5.13) is automatically satisfied, and the parameterized cost is given by

$$C(t,p) = \frac{1}{2}(T-t)(p-2p^3)^2 + \frac{1}{2}(p^4 - p^2). \tag{5.71}$$

This defines a C^ω-parameterized family of extremals.

All functions, including the parameterized cost, are polynomial, and this allows us to make explicit computations in the analysis of the flow $F(t,p) = (t,x(t,p))$. The singular set S is given by the solutions to the equation $\frac{\partial x}{\partial p}(t,p) = 0$ with the left and right eigenvectors simply given by $v \equiv 1$ and $w \equiv 1$. Solving for t gives

$$t = \sigma(p) = T - \frac{1}{1-6p^2}, \tag{5.72}$$

which is finite and less than T for $|p| < \frac{1}{\sqrt{6}}$. Since

$$\frac{\partial^2 x}{\partial p^2}(t,p) = -12p(t-T),$$

the map has fold singularities at $(\sigma(p),p)$ for $p \in (-\frac{1}{\sqrt{6}}, \frac{1}{\sqrt{6}})$, $p \neq 0$, and $(t,p) = (T-1,0)$ is a simple cusp point. For $\frac{\partial^2 x}{\partial t \partial p}(t,p) = 1 - 6p^2$ is positive on the singular set and $\frac{\partial^3 x}{\partial p^3}(T-1,0) = 12$. Since $\frac{\partial v}{\partial p}(p) \equiv 0$, this is equivalent to the transversality condition (5.69).

Let

$$t_{cp}(p) = \begin{cases} T - \frac{1}{1-6p^2} & \text{for } |p| < \frac{1}{\sqrt{6}}, \\ -\infty & \text{for } |p| \geq \frac{1}{\sqrt{6}}. \end{cases} \tag{5.73}$$

It then follows from Proposition 5.4.2 that controlled trajectories $(x(\cdot,p), u(\cdot,p))$ for $p \neq 0$ defined over an interval $[t_0, T]$ give a strong local minimum if $t_0 > \sigma(p)$, but they are no longer optimal if $t_0 \leq \sigma(p)$. The points $(t_{cp}(p), p)$ for $0 < |p| < \frac{1}{\sqrt{6}}$ are conjugate points of the corresponding trajectories that lose local optimality there. For $|p| \geq \frac{1}{\sqrt{6}}$, no conjugate points exist, and the trajectories can be integrated backward for all times. It therefore might seem reasonable to restrict the domain of the parametrization to

$$D_{cp} = \{(t,p) : t_{cp}(p) \leq t \leq T, \ p \in \mathbb{R}\},$$

but there are advantages to keeping the full family of extremals. And in fact, as will be shown below, the presence of the simple cusp point causes that not all trajectories are globally optimal until $t_{cp}(p)$. This follows from the mapping properties of the flow F that we develop now.

For $\alpha \geq 0$, consider the curves

$$\Phi_\alpha : \left(-\frac{1}{\sqrt{\alpha}}, \frac{1}{\sqrt{\alpha}}\right) \to \mathbb{R}^2, \qquad p \mapsto (t_\alpha(p), p), \quad t_\alpha(p) = T - \frac{1}{1-\alpha p^2},$$

and denote the half-curves for positive or negative values of p by Φ_α^+ and Φ_α^-, respectively. Except for the trivial intersection in the point $(T-1,0)$, which all curves have in common, the curves Φ_α define a foliation of the subset

$$\widehat{D} = \{(t,p) \in D : t \leq T - 1\},$$

and the image of $\widehat{D}$ under F is the subset $\widehat{G} = (-\infty, T-1] \times \mathbb{R}$ of G. But there exist nontrivial overlaps in the map. A direct calculation verifies that

$$x(t_\alpha(p), p) = \frac{(2-\alpha)p^3}{1-\alpha p^2}, \tag{5.74}$$

and using $p^2 = \frac{1}{\alpha}\left(1 - \frac{1}{T-t_\alpha}\right)$, we can represent the image of the curve Φ_α as

$$x = \pm \frac{2-\alpha}{\alpha}\sqrt{\frac{1}{\alpha}(T-t-1)}\sqrt{1 - \frac{1}{T-t}}, \qquad t \leq T-1.$$

In particular, the singular set S of the parametrization is given by the curve Φ_6,

$$S = \left\{(t,p) \in D : t = t_{cp}(p) = t_6(p), \ |p| < \frac{1}{\sqrt{6}}\right\},$$

and its image under the flow F is given by the curve

$$G_c = F(S) = \left\{\left(t, \pm\sqrt{\frac{2}{27}\frac{(T-t-1)^3}{(T-t)}}\right) : t \leq T-1\right\},$$

a cusp, which gives the singularity its name. It follows from Eq. (5.74) that

$$F(t_6(p), p) = F(t_{1.5}(-2p), -2p), \quad |p| < \frac{1}{\sqrt{6}}, \tag{5.75}$$

and thus the curve $\Phi_{1.5}$ also is mapped onto G_c. But note that the curve

$$G_+ = \left\{\left(t, \sqrt{\frac{2}{27}\frac{(T-t-1)^3}{(T-t)}}\right) : t < T-1\right\} = F(\Phi_6^+) = F(\Phi_{1.5}^-)$$

is the image of the half-curves Φ_6^+ and $\Phi_{1.5}^-$, while

$$G_- = \left\{\left(t, -\sqrt{\frac{2}{27}\frac{(T-t-1)^3}{(T-t)}}\right) : t < T-1\right\} = F(\Phi_6^-) = F(\Phi_{1.5}^+)$$

is the image of Φ_6^- and $\Phi_{1.5}^+$.

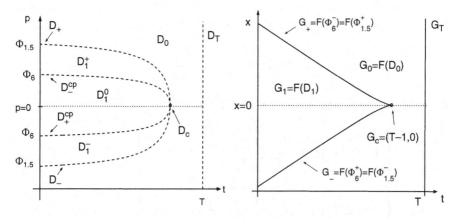

Fig. 5.13 Stratifications of the domain D and the range G that are compatible with the map F

We summarize the mapping properties of the flow F using the same name t_α for the extended function defined as

$$
t_\alpha(p) = \begin{cases} T - \frac{1}{1-\alpha p^2} & \text{for} \quad |p| < \frac{1}{\sqrt{\alpha}}, \\ -\infty & \text{for} \quad |p| \geq \frac{1}{\sqrt{\alpha}}. \end{cases} \tag{5.76}
$$

The regions and submanifolds defined below form a partition of the domain of the parameterized family of extremals into embedded analytic submanifolds,

$$
\mathscr{D} = \{D_T, D_0, D_1^-, D_1^0, D_1^+, D_-^{cp}, D_+^{cp}, D_-, D_+, D_c\}
$$

(see Fig. 5.13), where $D_T = \{T\} \times \mathbb{R}$, $D_c = \{(T-1,0)\}$,

$$
D_0 = \{(t,p) \in D : t_{1.5}(p) < t < T\},
$$
$$
D_1^+ = \{(t,p) \in D : t_6(p) < t < t_{1.5}(p),\ p > 0\},
$$
$$
D_1^- = \{(t,p) \in D : t_6(p) < t < t_{1.5}(p),\ p < 0\},
$$
$$
D_1^0 = \{(t,p) \in D : t < t_6(p),\ p \in \mathbb{R}\},
$$

and

$$
D_+ = \left\{ (t,p) \in D : t = t_{1.5}(p),\ 0 < p < \sqrt{\frac{2}{3}} \right\},
$$

$$
D_- = \left\{ (t,p) \in D : t = t_{1.5}(p),\ -\sqrt{\frac{2}{3}} < p < 0 \right\},
$$

$$D_-^{cp} = \left\{ (t,p) \in D : t = t_6(p), \; 0 < p < \frac{1}{\sqrt{6}} \right\},$$

$$D_+^{cp} = \left\{ (t,p) \in D : t = t_6(p), \; -\frac{1}{\sqrt{6}} < p < 0 \right\}.$$

Accordingly, partition the range of F, $G = F(D)$, into the open sets

$$G_0 = \left\{ (t,x) \in G : |x| > \sqrt{\frac{2}{27} \frac{(T-t-1)^3}{(T-t)}}, \; t \le T - 1 \right\} \cup (T-1,T) \times \mathbb{R},$$

$$G_1 = \left\{ (t,x) \in G : |x| < \sqrt{\frac{2}{27} \frac{(T-t-1)^3}{(T-t)}}, \; t < T - 1 \right\},$$

the curves G_- and G_+ defined above, the cusp point $G_c = \{(T-1,0)\}$, and the terminal manifold $G_T = \{T\} \times \mathbb{R}$ (see Fig. 5.13),

$$\mathscr{G} = \{G_T, G_0, G_-, G_+, G_c, G_1\}.$$

The embedded submanifolds in the collections $\mathscr{D}$ and $\mathscr{G}$ give a decomposition of the domain and the range of the flow of extremals into embedded submanifolds such that F maps each submanifold in $\mathscr{D}$ diffeomorphically onto exactly one of the submanifolds in $\mathscr{G}$. Specifically,

$$G_T = F(D_T), \quad G_0 = F(D_0), \quad G_c = F(D_c),$$
$$G_- = F(D_-) = F(D_-^{cp}), \quad G_+ = F(D_+) = F(D_+^{cp}),$$
$$G_1 = F(D_1^-) = F(D_1^0) = F(D_1^+).$$

Thus the flow F is $3:1$ onto G_1, $2:1$ onto the branches G_- and G_+ of fold points, and $1:1$ otherwise.

These mapping properties can succinctly be expressed using the concepts of stratifications, and we briefly digress to introduce this notion, which will also be used later on. Let M be a C^r-manifold and let A be a subset of M. Also let $\mathscr{A} = \{A_i : i \in I\}$ be a family of connected C^r-embedded submanifolds of M. A set A is the *locally finite union* of the A_i, $i \in I$, if $A = \cup_{i \in I} A_i$ and if every compact subset K of M intersects only finitely many of the A_i. For a subset A of M, its frontier, $\text{Fron} A$, is the set of all boundary points of A in M that do not lie in A. The collection $\mathscr{A}$ is said to satisfy the *frontier axiom* if whenever A_i and A_j are elements of $\mathscr{A}$, $A_i \ne A_j$, such that $A_i \cap \text{Clos} A_j \ne \emptyset$, then $A_i \subset \text{Fron} A_j$ and $\dim A_i < \dim A_j$, i.e., for every $A \in \mathscr{A}$ the frontier of A is a union of members of $\mathscr{A}$ of smaller dimension.

Definition 5.5.2 (Stratification). Let M be a C^r-manifold. A C^r-stratification $\mathscr{A} = \{A_i : i \in I\}$ of M is a locally finite family of connected C^r-embedded submanifolds A_i, $i \in I$, of M that satisfies the frontier axiom. An element of $\mathscr{A}$ is called a stratum.

Definition 5.5.3 (Refinement). A C^r-stratification $\mathscr{A}' = \{A_i' : i \in I\}$ of M is a refinement of the C^r-stratification $\mathscr{A} = \{A_i : i \in I\}$ of M if every stratum $A_i \in \mathscr{A}$ is a union of strata $A_i' \in \mathscr{A}'$.

Definition 5.5.4 (Compatible stratification). A C^r-stratification $\mathscr{A} = \{A_i : i \in I\}$ of M is compatible with a subset $A \subset M$ if A is a union of strata.

Definition 5.5.5 (1:1 compatible stratifications). Let $F : M \to N$ be a C^r-map and let $\mathscr{A} = \{A_i : i \in I\}$ and $\mathscr{B} = \{B_j : j \in J\}$ be C^r-stratifications of M and N that are compatible with subsets A of M and B of N, respectively. We say that the stratifications $\mathscr{A}$ and $\mathscr{B}$ are $1:1$ compatible with the map F over (A, B) if whenever $a \in A_i \subset A$, $b \in B_j \subset B$, and $F(a) = b$, then F maps A_i C^r-diffeomorphically onto B_j.

For this example, the collections $\mathscr{D}$ and $\mathscr{G}$ are stratifications of the domain and range that are $1:1$ compatible with the flow map F.

For each initial point in G_0, there exists only one extremal, and this one is optimal. But we need to see which of the three values is optimal over G_1. The parameterized cost $C(t, p)$ for the family is easily evaluated. The value C along the curves Φ_α is given by

$$C(t_\alpha(p), p) = \frac{(4 - \alpha)p^2 + (\alpha - 3)}{1 - \alpha p^2} p^4.$$

For $\alpha = 2$, it follows that $C(t_2(p), p) = -p^4$, and thus the values for the two trajectories corresponding to $\pm p$ are equal. Furthermore, by Eq. (5.74) the curves

$$D_-^{cl} = \left\{ (t, p) \in D : t = t_2(p), \quad -\frac{1}{\sqrt{2}} < p < 0 \right\}$$

and

$$D_+^{cl} = \left\{ (t, p) \in D : t = t_2(p), \quad 0 < p < \frac{1}{\sqrt{2}} \right\}$$

are both mapped diffeomorphically onto the half-line

$$\overline{G} = \{ (t, 0) \in G_1 : t < T - 1 \}.$$

Thus, given an initial condition $(t, 0) \in \overline{G}$, the two trajectories that are determined by the preimages $(t, \pm p)$ in D_-^{cl} and D_+^{cl} have the same value for the objective. Define the value $V^{\mathscr{E}} = V$ of the parameterized family $\mathscr{E}$ of extremals as $V^{\mathscr{E}} = C \circ F^-$, where F^- again denotes the multivalued map that assigns to a point (t, x) all possible parameter values (t, p) that satisfy $F(t, p) = (t, x)$. Thus the map F^- is single-valued on $G_T \cup G_0 \cup G_c$, has two values on G_- and G_+, and three values on G_1.

If we define π_i, $i = -, 0, +$, as the inverse parameter maps for the restrictions of F to D_1^i, then the associated value $V^{\mathscr{E}}$ has three sections over the set G_1, which we denote by

$$V_i : G_1 \to \mathbb{R}, \qquad V_i(t,x) = C(t, \pi_i(t,x)).$$

We have shown that

$$V_+(t,0) \equiv V_-(t,0) \qquad \text{for } (t,0) \in \overline{G}, \tag{5.77}$$

and V_0 will never be minimal. In fact, $V_0(t,x) > V_{\pm}(t,x)$ for all $(t,x) \in G_1$. Analyzing the map F further, it can be seen that F maps each of the regions

$$D_{\text{opt}}^+ = \{(t,p) \in D_1^+ : t_2(p) < t < t_{1.5}(p)\}$$

and

$$D_{\text{not}}^- = \{(t,p) \in D_2^- : t_6(p) < t < t_2(p)\}$$

diffeomorphically onto $\widehat{G}_+ = G_1 \cap \{(t,x) \in G : x > 0\}$. Analogously, the regions

$$D_{\text{opt}}^- = \{(t,p) \in D_1^- : t_2(p) < t < t_{1.5}(p)\}$$

and

$$D_{\text{not}}^+ = \{(t,p) \in D_1^+ : t_6(p) < t < t_2(p)\}$$

are mapped diffeomorphically onto $\widehat{G}_- = G_1 \cap \{(t,x) \in G : x < 0\}$. Computing the values, one obtains that

$$V_+(t,x) < V_-(t,x) \qquad \text{for all } (t,x) \in \widehat{G}_+$$

and

$$V_+(t,x) > V_-(t,x) \qquad \text{for all } (t,x) \in \widehat{G}_-.$$

Hence, a parametrization of the trajectories that are globally optimal is given if the domain is restricted to

$$D_{\text{opt}} = \{(t,p) \in D : t_2(p) \le t \le T\}.$$

The curve Φ_2 gives a parametrization of the cut-locus Γ of the two branches corresponding to the parameterizations over D_1^+ and D_1^-, respectively, and for initial points on Γ there exist two optimal trajectories in the family. In the interior of D_{opt} the parametrization is an analytic diffeomorphism.

Figure 5.14 illustrates the flow of the parameterized family of extremals: in the top row, it shows the two fields of locally optimal trajectories defined for $p > 0$ and $p < 0$, respectively. While the controlled trajectories $(x(\cdot, p), u(\cdot, p))$ for $p \ne 0$ are strong local minima over the interval $[t_0, T]$ if $t_0 > \sigma(p) = t_6(p)$, they are globally optimal only if $t_0 \ge t_2(p) > t_6(p)$. The simple cusp point at $(T-1, 0)$ generates a

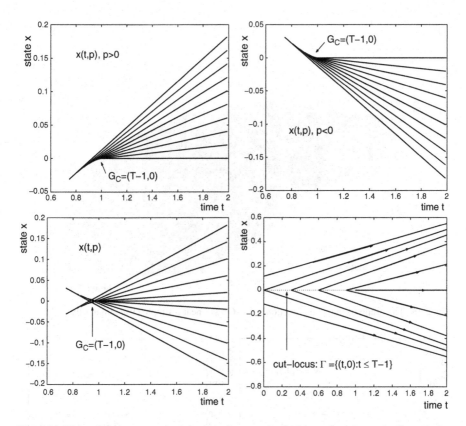

Fig. 5.14 Flow of the parameterized family of extremals $\mathscr{E}$: the *top row* shows the flow on the intervals $[t_{cp}, T]$ for $p > 0$ (*left*) and $p < 0$ (*right*); the *bottom row* gives a combination of these two flows (*left*) and the resulting optimal synthesis of controlled trajectories (*right*)

cut-locus (intersection) between the two branches of the parameterized family for $p > 0$ and $p < 0$ that limits the global optimality of trajectories at time $t_2(p)$, and thus the controlled trajectories for $p \neq 0$ already lose global optimality prior to the conjugate point. The optimal synthesis is shown in the bottom right of Fig. 5.14.

   Figure 5.15 shows the graph of the multivalued cost function $V^{\mathscr{E}} = C \circ F^{-1}$ in the (t,x)-space in a neighborhood of the simple cusp point, and Fig. 5.16 shows various time slices that illustrate the changes in the cost as the simple cusp point is passed. Interestingly, taken together, the graph of this multivalued function is qualitatively identical to the singular set of the swallowtail singularity, the next type of singularity in the order of cusps. The same observation holds for the fold singularities in which case the graph of the corresponding multivalued function $V^{\mathscr{E}}$ takes the shape of the singular set of a simple cusp. These relations also are a consequence of the fundamental relation for optimality expressed in the shadow-price lemma.

   Note that we have $\sigma(0) = t_2(0) = T - 1$ for the extremal $(x(\cdot, 0), u(\cdot, 0))$ that passes through the simple cusp point and that this controlled trajectory remains

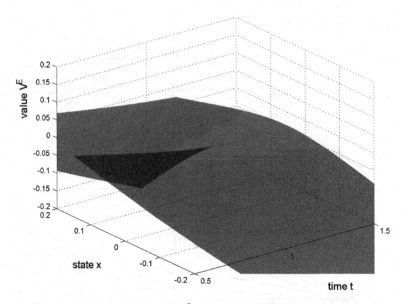

Fig. 5.15 The graph of the value function $V^{\mathscr{E}} = C \circ F^{-}$ near a simple cusp singularity

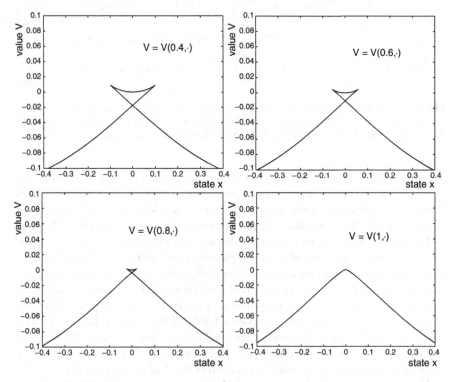

Fig. 5.16 Slices of the multivalued function $V^{\mathscr{E}}$ corresponding to the parameterized family $\mathscr{E}$ of extremals for $t = \text{const}$

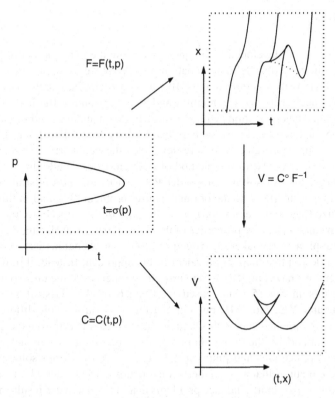

Fig. 5.17 The square on the left is the neighborhood W in (t,p)-space with the singular set S given by the inscribed curve. The *top* diagram on the right represents the image of the flow F near a "simple cusp," and the *bottom* diagram shows a slice of the corresponding value in the state space

globally optimal over the compact interval $[T-1,T]$. Thus, this is an example where the trajectory loses optimality only after the conjugate point has been crossed (we are thinking of integrating the field backward). In contrast, we have seen that trajectories lose optimality at the conjugate point for fold singularities, i.e., they are no longer optimal on the intervals $[t_{cp}(p),T]$ if $t_{cp}(p)$ denotes the corresponding conjugate time.

These geometric properties are valid in general near simple cusp points. This can be shown by transforming the system into normal form and then analyzing the resulting system, which has all the features of the example considered here. However, the computations involving the required changes of coordinates are quite technical and lengthy [134, 135]. The geometry is illustrated in Fig. 5.17. This structure is seen in many practical applications (e.g., optimal paths in a velocity field [64], chemical control of a batch reactor [47], and many more), and in these cases, the cut-locus typically is the determining feature for the overall synthesis of optimal controlled trajectories. We shall take up this topic in detail in Chap. 7.

5.6 Notes

In this chapter, we developed techniques that allow the construction of solutions to the Hamilton–Jacobi–Bellman equation locally. There exists a vast amount of literature on viscosity solutions to the HJB equation, including excellent expositions in book form such as the texts by Fleming and Soner [96] and by Bardi and Capuzzo-Dolcetta [30]. The connections with the value function are explored in many other texts on optimal control theory such as the book by Vinter [254]. But this is not the topic of our text. Rather, we here developed the geometric underpinnings of this theory in the context of the method of characteristics. While this method is a classical procedure for solving first-order PDEs, it is not entirely straightforward to adjust the constructions to the optimal control problem [33], and in this chapter we formalized our own version of how to do this. In this procedure, the shadow-price lemma and injectivity properties of the flow of extremals play the major roles. A similar approach was taken by Young in [260], but we deliberately avoided the notion of a descriptive map so prevalent in his approach. Instead, throughout our development, we have emphasized the mapping properties of the underlying flow.

The results in Sect. 5.3 are based on Chap. 6 of the classical textbook by A. Bryson and Y.C. Ho [64], but our reasoning is quite a bit different in its mathematically oriented approach. Our constructions clearly bring out the geometric properties that underly the important procedure of perturbation feedback control in engineering applications and pinpoint the significance of various solutions to the differential equations that arise in these constructions. The reader who is interested in the classical spacecraft guidance problems that motivated these results may wish to consult [53]. Also, extensions of the theory to problems with control constraints are considered in [93].

Our emphasis on the mapping properties has various other advantages. To begin with, it allows us to relate singularities in the flow of extremals to singularities in the value function. The geometry of the value function near fold and simple cusp points was initially investigated in joint research with M. Kiefer in [135], and some of these results were incorporated into Sects. 5.4 and 5.5. The example considered in Sect. 5.5 is from the papers by C.I. Byrnes et al. [65,66] and was fully analyzed by us and A. Nowakowski in [148] in connection with the Hilbert invariant integral. The connections with shocks in the solutions of the Hamilton–Jacobi–Bellman equation are also the topic of the paper [68] by Caroff and Frankowska. But generally, in our opinion, only the surface has been scratched in exploring these relations, and we believe that this could be an interesting area of research for years to come. Much of the construction given here utilizing the Lyapunov–Schmidt reduction in this context is new. Similarly, the presentation of the material analyzing the value function near a fold singularity based on envelopes in connection with parameterized families of extremals also is new. We have chosen this particular path for the exposition in order to emphasize the direct connections with the methods of the classical calculus of variations. On this topic, we also have elected to forgo the generality that the results of H. Sussmann [240, 242] provide for the benefit of naturally fitting these constructions within our framework. While much of the material presented in this chapter is based on classical ideas, the methods and procedures that we employed to derive them represent our own approach to this topic.

The results presented here and also in the next chapter naturally relate to necessary and sufficient conditions for local minima. Our approach to constructing local solutions to the Hamilton–Jacobi–Bellman equation is equivalent to the construction of local fields of extremals around a reference trajectory and thus gives conditions for strong minima. There exists a wealth of literature on necessary and sufficient conditions for optimality for both weak and strong minima that, analogous to ideas from the calculus of variations, are based on the second variation, e.g., [19, 84, 86, 168] and many more. While the results by Agrachev, Stefani, and Zezza [19] are for strong minima, generally these techniques are more effective in the case of weak local minima. Furthermore, without any practical procedures (such as those presented in Sect. 5.3 or the ones given by Maurer and Oberle in [181]) to verify abstract conditions on coercivity of quadratic forms, these results have a more theoretical character.

Chapter 6
Synthesis of Optimal Controlled Trajectories: From Local to Global Solutions

Our overall objective is to analyze the mapping properties of a flow of extremal controlled trajectories and to show that the cost-to-go function satisfies the Hamilton–Jacobi–Bellman equation in regions where this flow covers an open set of the state injectively (x-space in the time-invariant case, respectively (t,x)-space in the time-dependent case). In the previous chapter, we have assumed that the control $u = u(t,p)$ in the parameterized family of extremals is continuous and differentiable with respect to the parameter. Some regularity properties of the embedding of a reference extremal into a family of extremals are needed, and in our framework, these are guaranteed by the differentiability properties with respect to the parameter p. More general definitions are given in the work of H. Sussmann (e.g., see [196]), but these will not be pursued here. However, continuity of the control in t is neither a natural nor a realistic assumption. As the examples analyzed in this text show, typically various patches defined in terms of parameterized families of extremals need to be combined to obtain the full solutions. It thus remains to show how these patches can be concatenated. This includes both concatenations of parameterized families defined over the same parameter set as they arise naturally for bang-bang trajectories with the same type of switchings, but also situations in which parameterized families for different parameter sets need to be glued together, as was the case in the Fuller problem. The main result in this chapter is a powerful verification theorem for the global optimality of a synthesis of extremal trajectories that has been constructed in this way. This result, Theorem 6.3.3, has grown out of Boltyansky's geometric approach to sufficiency in optimal control problems of constructing a *regular synthesis* of optimal controlled trajectories [41]. The version that we present below gives a significant improvement over Boltyansky's original constructions and was developed over the years in a sequence of publications by H. Sussmann [196,229,243]. While we do not develop this result in its full generality as far as the technical assumptions on the underlying system are concerned—we retain the simpler technical framework of smooth vector fields employed in this text—our argument does provide the essential aspects of the construction, the most important

H. Schättler and U. Ledzewicz, *Geometric Optimal Control: Theory, Methods and Examples*, Interdisciplinary Applied Mathematics 38, DOI 10.1007/978-1-4614-3834-2_6, © Springer Science+Business Media, LLC 2012

one being that the differentiability assumption on the solution to the Hamilton–Jacobi–Bellman equation can be relaxed to be valid only on the complement of a locally finite union of embedded submanifolds of positive codimension.

This chapter is organized as follows: In Sect. 6.1, we consider broken extremals whose controls are discontinuous in time. We emphasize that these constructions are for families that have a finite number of patches. The typical scenario for which these families of broken extremals arise is given by piecewise continuous controls, and the theory applies equally to bang-bang and singular controls. It follows from the shadow-price lemma that the cost-to-go-function is a continuously differentiable solution to the HJB equation on those patches where the associated flow is a diffeomorphism. This property is the main objective of our constructions and becomes the essential assumption in the verification theorem. As an illustration of these arguments, in Sect. 6.2 we carefully work out the construction of a parameterized family of extremals arising in a mathematical model for tumor antiangiogenic treatment that contains singular arcs. The extremals constructed in this example indeed are globally optimal, and the results that were developed in Chap. 5 allow us to prove that the associated value is a continuously differentiable solution of the Hamilton–Jacobi–Bellman equation on open regions that are covered injectively by the flow. It is of no bearing for this argument that the flow actually collapses to a lower-dimensional submanifold along the singular arcs. Section 6.3 then develops a verification result that allows us to prove the global optimality of a synthesis of controlled trajectories that has been constructed by means of the procedures presented here and in Chap. 5. We once more stress that this result, in contrast to the versions of the theory developed in Chap. 5, allows that the corresponding value function $V^{\mathscr{E}}$ need not be differentiable on a locally finite union of embedded submanifolds of positive codimension, the natural and realistic scenario. This proof relies on some perturbation and lower semicontinuity properties to overcome the issues connected with nondifferentiability of the value function. These will be fully developed here. Also, the final result no longer assumes that the verification function is continuous. Optimal control problems for which this is the case are by no means an oddity, but often arise when there is a lack of local controllability properties in the system.

6.1 Parameterized Families of Broken Extremals

We start by dispensing with the condition that the controls $u = u(t, p)$ in a parameterized family of extremals need to be continuous and allow the requirements imposed in the definition to be satisfied piecewise in time.

6.1.1 Concatenations of Parameterized Families of Extremals

Without loss of generality, we take two parameterized families of extremals, and in order to keep the notation unambiguous, *we consider the time-dependent case*, i.e., the flow F is defined in terms of the graphs of the controlled trajectories, $F(t,p) = (t,x(t,p))$. Adjustments to the time-independent situation are immediate and will be considered in examples. Let P be an open subset of $\mathbb{R}^d$ with $1 \leq d \leq n$ and let $\mathscr{E}_1$ be a C^r-parameterized family of extremals with domain

$$D_1 = \{(t,p) : p \in P, t_{1,-}(p) \leq t \leq t_{1,+}(p)\},$$

source

$$N_{1,-} = \{(t,x) : t = t_{1,-}(p), x = \xi_{1,-}(p), \quad p \in P\},$$

target

$$N_{1,+} = \{(t,x) : t = t_{1,+}(p), x = \xi_{1,+}(p), \quad p \in P\},$$

and cost $\gamma_{1,\pm} : P \to \mathbb{R}$, $p \mapsto \gamma_{1,\pm}(p)$, at the source, respectively target. For the same parameter set P, let $\mathscr{E}_2$ be a C^r-parameterized family of normal extremals with domain

$$D_2 = \{(t,p) : p \in P, t_{2,-}(p) \leq t \leq t_{2,+}(p)\},$$

source

$$N_{2,-} = \{(t,x) : t = t_{2,-}(p), x = \xi_{2,-}(p), \quad p \in P\},$$

target

$$N_{2,+} = \{(t,x) : t = t_{2,+}(p), x = \xi_{2,+}(p), \quad p \in P\},$$

and cost $\gamma_{2,\pm} : P \to \mathbb{R}$, $p \mapsto \gamma_{2,\pm}(p)$, at the source, respectively target. We denote the associated controls, trajectories, and multipliers by the corresponding subscript. For example, λ_2 denotes the adjoint variable for the family $\mathscr{E}_2$, and we denote the constant multiplier by $\lambda_{0,2}(p)$.

Two C^r-parameterized families $\mathscr{E}_1$ and $\mathscr{E}_2$ of extremals can be concatenated if for all $p \in P$ we have that (i) $t_{1,+}(p) = t_{2,-}(p)$, $\xi_{1,+}(p) = \xi_{2,-}(p)$, (ii) $\lambda_{0,1}(p) = \lambda_{0,2}(p)$, $\lambda_1(t_{1,+}(p),p) = \lambda_2(t_{2,-}(p),p)$, and (iii) $\gamma_+^1(p) = \gamma_-^2(p)$. Conditions (i) and (ii) enforce that the controlled trajectories of the two flows and their adjoint variables match at the junction $N_{1,+} = N_{2,-}$, while condition (iii) guarantees the agreement of the associated cost functions.

Requiring the agreement of the parameter sets in the two families simplifies our presentation, but does not constitute an essential restriction. Clearly, it would be enough to have a C^r-diffeomorphism that connects the two parameter sets, but we do not concern ourselves here with this more general formulation. Our treatment of the Fuller problem in Sect. 5.2.3 shows how this can be done in a specific case.

To simplify the notation, we henceforth denote the functions defining the concatenation by

$$\tau(p) = t_{1,+}(p) = t_{2,-}(p), \quad \xi(p) = \xi_{1,+}(p) = \xi_{2,-}(p), \quad \gamma(p) = \gamma_+^1(p) = \gamma_-^2(p),$$

$$\lambda_0(p) = \lambda_{0,1}(p) = \lambda_{0,2}(p), \qquad \text{and} \qquad \lambda(p) = \lambda_1(t_{1,+}(p),p) = \lambda_2(t_{2,-}(p),p).$$

Furthermore, it follows from the fact that $\mathscr{E}_1$ and $\mathscr{E}_2$ are C^r-parameterized families of extremals that the controls $u_i = u_i(t,p)$ solve the minimization problems

$$\min_{v \in U} H(t,\lambda_i(t,p),x_i(t,p),v) = H(t,\lambda_i(t,p),x_i(t,p),u_i(t,p)).$$

Hence, also the functions $h_i(t,p) = H(t,\lambda_i(t,p),x_i(t,p),u_i(t,p))$ remain continuous at the junction, and we let

$$h(p) = h_1(t_{1,+}(p),p) = h_2(t_{2,-}(p),p).$$

The minimization condition generally determines when and where the control u_1 must be changed to u_2. The value where the minimum of the Hamiltonian is achieved changes, and typically discontinuities arise in the optimal controls. A common scenario is that of bang-bang junctions, but also concatenations with singular controls can be analyzed within this framework, and we shall give a rather typical example in the next section. Naturally, situations in which regularity properties of the parametrization break down also match this framework. The latter happens, for example, at saturation points when a singular control with values in the interior of the control set reaches the control limits. In this case, although the control may remain continuous, generally differentiability in p will be lost, and this also will give rise to different patches in the parameterizations.

The concatenated family $\mathscr{E} = \mathscr{E}_1 * \mathscr{E}_2$ then is defined as the family of extremals with domain

$$D = \{(t,p) : p \in P, t_{1,-}(p) \le t \le t_{2,+}(p)\},$$

source

$$N_{1,-} = \{(t,x) : t = t_{1,-}(p), \ x = \xi_{1,-}(p), \quad p \in P\},$$

target

$$N_{2,+} = \{(t,x) : t = t_{2,+}(p), \ x = \xi_{2,+}(p), \quad p \in P\},$$

and the controls u, trajectories x, and adjoint variable λ are defined piecewise as

$$u(t,p) = \begin{cases} u_1(t,p) & \text{for} \quad (t,p) \in \text{int}(D_1), \\ u_2(t,p) & \text{for} \quad (t,p) \in D_2, \end{cases}$$

$$x(t,p) = \begin{cases} x_1(t,p) & \text{for} \quad (t,p) \in D_1, \\ x_2(t,p) & \text{for} \quad (t,p) \in D_2, \end{cases}$$

and

$$\lambda(t,p) = \begin{cases} \lambda_1(t,p) & \text{for} \quad (t,p) \in D_1, \\ \lambda_2(t,p) & \text{for} \quad (t,p) \in D_2. \end{cases}$$

(see Fig. 6.1).

Fig. 6.1 Flow F of a family of broken extremals

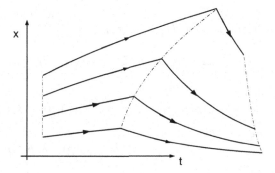

Definition 6.1.1 (Family of broken extremals). A C^r-parameterized family of broken extremals is a finite concatenation $\mathscr{E} = \mathscr{E}_1 * \cdots * \mathscr{E}_k$ of C^r-parameterized families of extremals.

The important feature of families of normal broken extremals $(\lambda_0(p) \equiv 1)$ is that the property that the corresponding value in the state space is a differentiable solution to the Hamilton–Jacobi–Bellman equation propagates along the individual segments and holds whenever the flow covers an open set in the state space injectively. This holds even if later on, this flow collapses onto lower-dimensional manifolds, a common scenario both with bang-bang controls (see Sect. 2.6) and when the controlled trajectories follow singular arcs (see Sect. 6.2). The key observation is that since the trajectories, multipliers, and the cost agree at the junctions, the transversality condition (5.13),

$$\lambda(p)\frac{\partial \xi}{\partial p}(p) = \lambda_0(p)\frac{\partial \gamma}{\partial p}(p) + h(p)\frac{\partial \tau}{\partial p}(p),$$

propagates from one family to the other. This directly follows from Corollary 5.2.1. Recall that it was this transversality condition that guaranteed the validity of the shadow-price lemma (cf. step 1 in the proof of Lemma 5.2.2),

$$\lambda_0(p)\frac{\partial C}{\partial p}(t,p) = \lambda(t,p)\frac{\partial x}{\partial p}(t,p). \tag{6.1}$$

Hence, the shadow-price lemma remains valid as one crosses from one parameterized family of extremals to the next, and Eq. (6.1) holds on the domain D of the concatenated family $\mathscr{E}$ of extremals away from the junction $\mathscr{J} = \{(t,p) : t = \tau(p), \, p \in P\}$. On $\mathscr{J}$, the partial derivatives of C and x with respect to the parameter p generally are discontinuous, but their jumps cancel in the expression (6.1), and this allows us to construct solutions to the Hamilton–Jacobi–Bellman equation for families of broken extremals. We give these formulas for later reference.

Lemma 6.1.1. *The shadow-price identity (6.1) is valid for the concatenated family* $\mathscr{E} = \mathscr{E}_1 * \mathscr{E}_2$ *of broken extremals for all* $(t,p) \in D \setminus \mathscr{J}$. *Along the switching surface*

$\mathcal{T} = \{(t,p) : t = \tau(p),\ p \in P\}$, *it is valid in the limits as $t \to \tau(p)$ from the right and the left. Furthermore, setting*

$$k_0(p) = L(\tau(p), \xi(p), u_2(\tau(p), p)) - L(\tau(p), \xi(p), u_1(\tau(p), p))$$

and

$$k(p) = f(\tau(p), \xi(p), u_2(\tau(p), p)) - f(\tau(p), \xi(p), u_1(\tau(p), p)),$$

we have that

$$\lambda_0(p)k_0(p) + \lambda(p)k(p) = 0$$

and

$$\nabla C_2(\tau(p), p) = \nabla C_1(\tau(p), p) - k_0(p)\left(1, -\frac{\partial \tau}{\partial p}(p)\right), \qquad (6.2)$$

$$Dx_2(\tau(p), p) = Dx_1(\tau(p), p) + k(p)\left(1, -\frac{\partial \tau}{\partial p}(p)\right). \qquad (6.3)$$

Recall that $\frac{\partial \tau}{\partial p}(p)$ denotes the gradient of τ written as a row vector, so that $\frac{\partial x_2}{\partial p}(\tau(p), p)$ is a rank-1 correction of the matrix $\frac{\partial x_1}{\partial p}(\tau(p), p)$.

Proof. It remains to compute the formulas at the switching surface $\mathcal{T}$. Recall that the controls u_i extend as $C^{0,r}$-functions onto an open neighborhood of D_i, and thus the states x_i extend to $C^{1,r}$-functions onto open neighborhoods $\widetilde{D}_i$ of the junction manifold $\mathcal{T}$. Defining, for $i = 1$ and 2,

$$C_i : \widetilde{D}_i \to \mathbb{R}, \quad C_i(t,p) = \int_t^{\tau(p)} L(s, x_i(s,p), u_i(s,p))\,ds + \gamma(p),$$

this formula extends the definition of the cost functions for the two parameterizations onto the neighborhoods $\widetilde{D}_i$. Note that the extensions x_i of the states beyond D_i no longer are extremals, and nor do the extensions of the functions C_i represent the cost of the extremals in the concatenated family. But simply having these extensions allows us to relate their partial derivatives on the switching surface by means of the following elementary lemma.

Lemma 6.1.2. *Let $z_0 \in \mathbb{R}^n$ and let Z be an open neighborhood of z_0. Suppose $g : Z \to \mathbb{R}$ and $h : Z \to \mathbb{R}$ are continuously differentiable functions such that $h(z) = 0$ on $\{z \in Z : g(z) = 0\}$. If $g(z_0) = 0$ and $\nabla g(z_0) \neq 0$, then there exist a neighborhood W of z_0 contained in Z and a continuous function $k : W \to \mathbb{R}$ such that $h(z) = k(z)g(z)$ for $z \in W$ and $\nabla h(z) = k(z)\nabla g(z)$ whenever $g(z) = 0$.*

Proof of the Lemma. Without loss of generality, assume that $\frac{\partial g}{\partial z_1}(z_0) \neq 0$ and replace the first basis vector e_1 in the standard basis defining a change of coordinates near z_0 by

$$\Theta : Z \to \mathbb{R}^n \quad z \mapsto \zeta = \Theta(z) = \begin{pmatrix} g(z) \\ z_2 - z_2^0 \\ \vdots \\ z_n - z_n^0 \end{pmatrix}.$$

The Jacobian matrix of Θ at z_0 is nonsingular, and thus by the inverse function theorem, there exists an open neighborhood W of z_0 so that $\Theta \upharpoonright W \to V = \Theta(W)$ is a diffeomorphism. Defining functions $G : V \to \mathbb{R}$, $G = g \circ \Theta^{-1}$ and $H : V \to \mathbb{R}$, $H = h \circ \Theta^{-1}$, we then have for all $\zeta \in V$ that

$$G(\zeta) = g(\Theta^{-1}(\zeta)) = g(z) = \zeta_1$$

and $H(0, \zeta_2, \ldots, \zeta_n) = 0$. Hence for $\zeta \in V$,

$$H(\zeta) = \zeta_1 \int_0^1 \frac{\partial H}{\partial \zeta_1}(t\zeta_1, \zeta_2, \ldots, \zeta_n) dt.$$

Setting

$$K : V \to \mathbb{R}, \qquad K(\zeta) = \int_0^1 \frac{\partial H}{\partial \zeta_1}(t\zeta_1, \zeta_2, \ldots, \zeta_n) dt,$$

it follows that $H(\zeta) = K(\zeta)G(\zeta)$, and with $k = K \circ \Theta$, we have that

$$h(z) = k(z)g(z) \quad \text{for all } z \in W.$$

The integrand in the definition of K is continuous on $[0,1] \times V$, and thus K and k are continuous. Furthermore, whenever $g(z) = 0$, the product $h = kg$ is differentiable at z with partial derivatives given by

$$\frac{\partial h}{\partial z_i}(z) = \lim_{t \to 0} \frac{k(z + te_i)g(z + te_i) - 0}{t}$$

$$= \left(\lim_{t \to 0} k(z + te_i) \right) \left(\lim_{t \to 0} \frac{g(z + te_i) - 0}{t} \right) = k(z) \frac{\partial g}{\partial z_i}(z).$$

This proves the lemma. $\square$

In our case here, in the time-dependent case, $z = (t, p)$, $g(z) = t - \tau(p)$, and h can be taken as the difference in the cost, $C_2(t, p) - C_1(t, p)$, or the difference in each of the coordinates of the state, $x_2^{(i)}(t, p) - x_1^{(i)}(t, p)$, $i = 1, \ldots, n$. Hence there exist continuous functions k_0 and $k = (k_1, \ldots, k_n)$ defined on a neighborhood W of (t_0, p_0) such that

$$\nabla C_2(\tau(p), p) - \nabla C_1(\tau(p), p) = -k_0(p) \left(1, -\frac{\partial \tau}{\partial p}(p) \right)$$

and

$$Dx_2(\tau(p),p) - Dx_1(\tau(p),p) = k(p)\left(1, -\frac{\partial\tau}{\partial p}(p)\right).$$

From the definition of the parameterized costs, for $i = 1,2$ we have that

$$\frac{\partial C_i}{\partial t}(t,p) = -L(t,x_i(t,p),u_i(t,p)),$$

and thus

$$k_0(p) = L(\tau(p),\xi(p),u_2(\tau(p),p)) - L(\tau(p),\xi(p),u_1(\tau(p),p)).$$

Similarly,

$$\begin{aligned}k(p) &= \dot{x}_2(\tau(p),p) - \dot{x}_1(\tau(p),p)\\ &= f(\tau(p),\xi(p),u_2(\tau(p),p)) - f(\tau(p),\xi(p),u_1(\tau(p),p)).\end{aligned}$$

In particular,

$$\lambda_0(p)k_0(p) + \lambda(p)k(p) = h_2(\tau(p),p) - h_1(\tau(p),p) = 0$$

from the minimum condition of the maximum principle. $\square$

We give a simple, but in many ways canonical, example of a parameterized family of broken extremals when the flow of controlled trajectories collapses and show that this does not invalidate the construction of solutions to the Hamilton–Jacobi–Bellman equation along those segments where the flow is injective.

Example 6.1.1 (Time-optimal control to the origin for the double integrator). Recall (see Sect. 2.6.1) that this is the problem to steer points $x \in \mathbb{R}^2$ into the origin time-optimally by means of the dynamics $\dot{x}_1 = x_2$ and $\dot{x}_2 = u$ with $|u| \le 1$. Optimal trajectories enter the origin through one of the two semiparabolas Γ_+, $x_1 = \frac{1}{2}x_2^2$ for $x_2 < 0$, and Γ_-, $x_1 = -\frac{1}{2}x_2^2$ for $x_2 > 0$, along the controls $u \equiv +1$ and $u \equiv -1$, respectively. There are two parameterized families of normal extremals that combine to form the optimal synthesis. We illustrate the construction by giving all the functions involved in the definition of concatenations of the form XY, i.e., controlled trajectories that start with $u = -1$ and enter the origin along Γ_+.

The problem is time-independent, and thus the flow is given by $F(t,p) = x(t,p)$ with a one-dimensional parameter set P. In Example 5.2.1, Sect. 5.2.1, we have formulated a C^ω-parameterized family of normal extremals $\mathscr{E}_-$ that describes the X-trajectories integrated backward from Γ_+ and uses this switching curve Γ_+ to parameterize the extremals. In this family, Γ_+ is parameterized using $P = (0,\infty)$ as $\xi_+(p) = \left(\frac{1}{2}p^2, -p\right)^T$, and the domain of the parameterization is $D = \{(t,p) : t \le -p < 0\}$ with $t_-(p) = -\infty$, i.e., trajectories are integrated backward in time for all times, and $t_+(p) = -p$. This choice of parameterization implies that for points on Γ_+, the cost γ_+, i.e., the time along the Y-trajectory from $\xi_+(p)$ into the origin, is given by $\gamma_+(p) = p$. Together with the specifications $u(t,p) \equiv -1$, $x(t,p)$ the solution of the dynamics with terminal condition $\xi(p)$, and the adjoint variable $\lambda(t,p)$ the backward solution to the equations $\dot{\lambda}_1 = 0$ and $\dot{\lambda}_2 = -\lambda_1$ with terminal conditions $\lambda(p) = \lambda(t_+(p),p) = \left(\frac{1}{p},0\right)$ at the switching points, it was verified that the transversality condition is satisfied and that this defines a C^ω-parameterized family of extremals $\mathscr{E}_-$. Now simply take this as the first parameterized family $\mathscr{E}_1$ with

$$\tau(p) = -p, \quad \xi(p) = \left(\frac{1}{2}p^2, -p\right)^T, \quad \gamma(p) = p, \quad \text{and} \quad \lambda(p) = \left(\frac{1}{p}, 0\right)$$

defining the source data for a second C^ω-parameterized family of normal extremals $\mathscr{E}_2$ that is obtained by integrating $u = +1$ along Γ_+ into the origin. Thus, with $P = (0,\infty)$, we have $D_2 = \{(t,p): -p \leq t \leq 0 = t_+(p)\}$, the initial and terminal values of the cost are given by $\gamma_{2,-}(p) = p$ and $\gamma_{2,+}(p) = 0$, and the multiplier $\lambda(t,p)$ is defined on D_2 as the forward solution to the adjoint equation with initial condition $\lambda(p)$ at the switching point $\xi(p)$. Thus, the parameterized family of extremals constructed in Sect. 5.2.1 immediately extends to a family of broken extremals that describes the full switching structure. The same holds for the family $\mathscr{E}_+$ that describes the concatenations of the form YX, i.e., controlled trajectories that start with $u = +1$ and enter the origin along Γ_-.

Note that by Corollary 5.2.1, the transversality condition propagates along the subdomain D_2 to the terminal state $\xi_{2,+}(p) = (0,0)$ with cost $\gamma_{2,+}(p) = 0$ at the origin. Naturally, here the transversality condition

$$\lambda(p)\frac{\partial \xi_+}{\partial p}(p) = \lambda_0(p)\frac{\partial \gamma_+}{\partial p}(p) + h(p)\frac{\partial T}{\partial p}(p)$$

is satisfied trivially, since also $h(p) \equiv 0$. Thus, starting the construction at the terminal point, this condition is obviously satisfied and no longer would need to be verified as we did in Sect. 5.2.1. However, in that section, we still insisted on the control being continuous and thus had to verify this condition at the switching points.

This example, albeit simple, illustrates how the framework applies to situations in which the flow of trajectories collapses to a lower-dimensional subset on the subsequent domain D_2. Importantly, this does not impede the construction of solutions to the Hamilton–Jacobi–Bellman equation. For this example, the flow restricted to the domain D_1 of the first parameterized family, $F_1 = F \upharpoonright D_1$, $F(t,p) = x(t,p)$, is a diffeomorphism, and on $G_- = \{(x_1,x_2): x_1 > -\mathrm{sgn}(x_2)\frac{1}{2}x_2^2\}$ the value is simply defined by $V_- = C \circ F_1^{-1}$. Since the transversality condition propagates along a parameterized family of extremals, the shadow-price lemma remains valid, and thus it follows from Theorem 5.2.1 that V is a continuously differentiable function on G_- that solves the Hamilton–Jacobi–Bellman equation.

In fact, $V_-(x) = T(x) = x_2 + 2\sqrt{x_1 + \frac{1}{2}x_2^2}$ on G_-, as was shown earlier, but we do not need to know this explicit form.

Similarly, YX-trajectories form a parameterized family of normal broken extremals that defines a solution of the HJB equation on G_+. Since trajectories collapse onto the curves Γ_+ and Γ_-, the combined value function no longer is differentiable on these two curves. This, however, as will be shown in Sect. 6.3, does not invalidate the optimality of the synthesis. However, it is essential for the argument that the value that arises from the synthesis is differentiable and a solution of the HJB equation on sufficiently rich open subsets. It is this step that is accomplished with parameterized families of broken extremals.

6.1.2 Transversal Crossings

In the example just considered, the flow F collapsed onto a lower-dimensional submanifold along the second parameterized family. While this is a common scenario, equally important are situations in which the individual flows are diffeomorphisms and in which the switching surfaces are crossed transversally. We now consider these cases and assume that the parameter set P is n-dimensional. In this case

$$\mathscr{T} = \{(t,p) : t = \tau(p), \; p \in P\}$$

is an embedded hypersurface (codimension-1 submanifold) of D, and its image in the combined state space is given by

$$\mathscr{S} = \{(t,x) : t = \tau(p), \; x = \xi(p), \; p \in P\}.$$

We want the flow F restricted to $\mathscr{T}$ to be a diffeomorphism so that $\mathscr{S} = F(\mathscr{T})$ also is an n-dimensional embedded submanifold of the extended state space. This property is guaranteed locally if the flow of one of the two parameterized families of extremals is regular, i.e., if its Jacobian matrix is nonsingular.

Lemma 6.1.3. *If the differential $DF_i(t,p)$ of the flow $F_i(t,p) = (t, x_i(t,p))$, $i = 1$ or $i = 2$, is nonsingular for $t = \tau(p)$, then near $(t,x) = (\tau(p), \xi(p))$, the switching surface $\mathscr{S}$ is an embedded n-dimensional submanifold and the flow F_i is transversal to $\mathscr{S}$, i.e., the tangent vectors to the graphs of the trajectories $(1, \dot{x}(t,p))^T$ do not lie in the tangent space to $\mathscr{S}$ at $F_i(t,p)$.*

Proof. By the inverse function theorem, the mapping F_i is a local diffeomorphism that maps a neighborhood of $(t,p) = (\tau(p), p)$ diffeomorphically onto a neighborhood of $(t,x) = (\tau(p), \xi(p))$. Hence it maps the n-dimensional embedded submanifold $\mathscr{T}$ into an n-dimensional embedded submanifold $\mathscr{S}$, and the tangent space to $\mathscr{S}$ at (t,x) is given by the image of the tangent space to $\mathscr{T}$ at $(\tau(p), p)$ under the differential DF_i. Since the direction $(1,0)$ is transversal to $\mathscr{T}$ in the

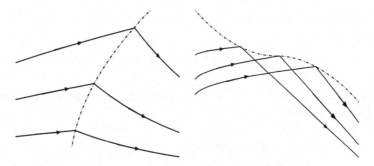

Fig. 6.2 The flow of a parameterized family of broken extremals near a transversal crossing (*left*) and a transversal fold (*right*)

parameter space, the graphs of the trajectories $t \mapsto x(t,p)$ are transversal to $\mathscr{S}$ at $(\tau(p), \xi(p))$ as well. $\square$

Definition 6.1.2 (Transversal crossings and folds). We say that the C^r-parameterized family $\mathscr{E} = \mathscr{E}_1 * \mathscr{E}_2$ of broken extremals has a regular switching point at $(t_0, p_0) = (\tau(p_0), p_0)$ if both flow maps F_1 and F_2 are regular at (t_0, p_0). We call such a switching point a transversal crossing if the graphs of the trajectories $t \mapsto x_i(t,p)$, $i = 1, 2$, cross the switching surface

$$\mathscr{S} = \{(t,x) : t = \tau(p),\ x = \xi(p),\ p \in P\}$$

in the same direction and a transversal fold if they cross it in opposite directions.

Recall that our formulations are for the time-dependent case, but analogous definitions apply to the time-independent formulation.

The main results of this section show that local optimality properties of the flow of extremals are preserved at transversal crossings—in fact, in this case the value function $V^{\mathscr{E}}$ remains continuously differentiable at the corresponding switching surface $\mathscr{S}$—while optimality typically ceases at a switching surface consisting of transversal folds (see Fig. 6.2). As will be seen below, the reason is that if such a surface supports trajectories of the system, these are control envelopes (see Sect. 5.4.1) for the control problem. In this sense, transversal folds correspond to "conjugate points" for parameterized families of broken extremals.

We start with a characterization of transversal crossings and folds in the parameter space. Since we typically integrate the family of extremals backward from the terminal manifold, here we take the second parameterized family $\mathscr{E}_2$ as our starting point.

Proposition 6.1.1. *Suppose* $\frac{\partial x_2}{\partial p}(t,p)$ *is nonsingular for* $t = \tau(p)$ *and let*

$$k(p) = f(\tau(p), \xi(p), u_2(\tau(p), p)) - f(\tau(p), \xi(p), u_1(\tau(p), p)).$$

The point $(\tau(p), p)$ is a regular switching point if and only if

$$1 + \frac{\partial \tau}{\partial p}(p) \left(\frac{\partial x_2}{\partial p}(\tau(p), p) \right)^{-1} k(p) \neq 0.$$

The concatenated family $\mathscr{E} = \mathscr{E}_1 * \mathscr{E}_2$ of normal broken extremals has a transversal crossing at $(\tau(p), p)$ if

$$1 + \frac{\partial \tau}{\partial p}(p) \left(\frac{\partial x_2}{\partial p}(\tau(p), p) \right)^{-1} k(p) > 0 \qquad (6.4)$$

and a transversal fold if

$$1 + \frac{\partial \tau}{\partial p}(p) \left(\frac{\partial x_2}{\partial p}(\tau(p), p) \right)^{-1} k(p) < 0. \qquad (6.5)$$

Proof. The trajectories x_1 and x_2 extend as $C^{1,r}$-functions onto a neighborhood of $\mathscr{T}$, and by (6.3), their partial derivatives are related by the rank-1 correction formula

$$\frac{\partial x_1}{\partial p}(\tau(p), p) = \frac{\partial x_2}{\partial p}(\tau(p), p) + k(p) \frac{\partial \tau}{\partial p}(p).$$

Lemma 6.1.4. *Suppose $A \in \mathbb{R}^{n \times n}$ is nonsingular and let u and v be vectors in $\mathbb{R}^n$. Then $B = A + uv^T$ is nonsingular if and only if $1 + v^T A^{-1} u \neq 0$. In this case,*

$$(A + uv^T)^{-1} = A^{-1} - \frac{A^{-1} u v^T A^{-1}}{1 + v^T A^{-1} u}.$$

Proof of the Lemma. The statement is trivially valid if either one of u and v is the zero vector, and thus we assume that both u and v are nonzero. Writing $B = A \left(\mathrm{Id} + A^{-1} uv^T \right)$, the matrix B is nonsingular if and only if $C = \mathrm{Id} + A^{-1} uv^T$ is. But C has eigenvalue $\lambda = 1$ with geometric multiplicity $n - 1$ (the corresponding eigenspace is the orthogonal complement of the one-dimensional space spanned by v), and the nth eigenvalue is $1 + v^T A^{-1} u$ with eigenvector $w = A^{-1} u$:

$$Cw = w + wv^T w = (1 + v^T w)w = (1 + v^T A^{-1} u)w.$$

The formula for the inverse is readily verified.   □

Hence, the point $(\tau(p), p)$ is a regular switching point if and only if

$$1 + \frac{\partial \tau}{\partial p}(p) \left(\frac{\partial x_2}{\partial p}(\tau(p), p) \right)^{-1} k(p) \neq 0.$$

The sign of this quantity distinguishes between transversal folds and crossings.

Since $\frac{\partial x_2}{\partial p}(t,p)$ is nonsingular at $t = \tau(p)$, the map $F_2 : V \to W$, $(t,p) \mapsto$ $(t, x_2(t,p))$, is a $C^{1,r}$ diffeomorphism and hence invertible between some neighborhoods V of $(\tau(p),p)$ and W of $(\tau(p), \xi(p))$. Denote the inverse by $F_2^{-1} : W \to V$, $(t,x) \mapsto (t, \pi(t,x))$, and define $\Psi : W \to \mathbb{R}$ by

$$\Psi(t,x) = t - \tau(\pi(t,x)). \tag{6.6}$$

Then $\Psi \in C^{1,r}$, $\mathscr{S} = \{(t,x) \in W : \Psi(t,x) = 0\}$, and on $\mathscr{S}$ the gradient $\nabla\Psi(t,x)$ is not zero, since for $(t,x) = (\tau(p), \xi(p))$, and writing $u_2(p) = u_2(\tau(p),p)$, we have that

$$\nabla\Psi(t,x) = \left(1, -\frac{\partial\tau}{\partial p}(p)\right)\begin{pmatrix} 1 & 0 \\ \frac{\partial x_2}{\partial t}(\tau(p),p) & \frac{\partial x_2}{\partial p}(\tau(p),p) \end{pmatrix}^{-1}$$

$$= \left(1, -\frac{\partial\tau}{\partial p}(p)\right)\begin{pmatrix} 1 & 0 \\ -\left(\frac{\partial x_2}{\partial p}(\tau(p),p)\right)^{-1}\frac{\partial x_2}{\partial t}(\tau(p),p) & \left(\frac{\partial x_2}{\partial p}(\tau(p),p)\right)^{-1} \end{pmatrix}$$

$$= \left(1 + \frac{\partial\tau}{\partial p}(p)\left(\frac{\partial x_2}{\partial p}(\tau(p),p)\right)^{-1} f(\tau(p),\xi(p),u_2(p)),\right.$$

$$\left. -\frac{\partial\tau}{\partial p}(p)\left(\frac{\partial x_2}{\partial p}(\tau(p),p)\right)^{-1}\right).$$

In particular, if $\frac{\partial\Psi}{\partial x}(t,x) = 0$, then $\frac{\partial\Psi}{\partial t}(t,x) = 1$ and therefore $\nabla\Psi(t,x) \neq 0$. Furthermore, along the flow F_2 we have that

$$\nabla\Psi(t,x) \cdot \begin{pmatrix} 1 \\ f(t,x,u_2(p)) \end{pmatrix} \equiv 1,$$

and along F_1, setting $u_1(p) = u_1(\tau(p),p)$,

$$\nabla\Psi(t,x) \cdot \begin{pmatrix} 1 \\ f(t,x,u_1(p)) \end{pmatrix} = 1 + \frac{\partial\tau}{\partial p}(p)\left(\frac{\partial x_2}{\partial p}(\tau(p),p)\right)^{-1}k(p).$$

The tangent plane to $\mathscr{S}$ at (t,x) can be described as

$$T_{(t,x)}\mathscr{S} = \left\{(\alpha,z) \in \mathbb{R}^{n+1} : \frac{\partial\Psi}{\partial t}(t,x)\alpha + \frac{\partial\Psi}{\partial x}(t,x)z = 0\right\},$$

and two vectors $(1,v)^T$ and $(1,w)^T$ point to the same side of $T_{(t,x)}\mathscr{S}$ at (t,x) if and only if $\nabla\Psi(t,x) \cdot \left(\begin{smallmatrix}1\\v\end{smallmatrix}\right)$ and $\nabla\Psi(t,x) \cdot \left(\begin{smallmatrix}1\\w\end{smallmatrix}\right)$ have the same sign. Hence, it follows that the switching point at $(\tau(p), p)$ is a transversal crossing if and only if $1 + \nabla \tau(p) \left(\frac{\partial x_2}{\partial p}(\tau(p), p) \right)^{-1} k(p)$ is positive, while it is a transversal fold if and only this quantity is negative. $\square$

We shall return to the question of how to compute $\nabla \tau(p) \left(\frac{\partial x_2}{\partial p}(\tau(p), p) \right)^{-1}$ efficiently below once we have established the main results.

Definition 6.1.3 (Field of broken extremals). We say that the C^r-parameterized family $\mathscr{E} = \mathscr{E}_1 * \mathscr{E}_2$ of broken extremals defines a field of broken extremals over the domain D if (i) each of the two flows $F_i : D_i \to G_i = F_i(D_i)$, is a $C^{1,r}$-diffeomorphism on the domains D_i and (ii) the combined flow map

$$F : D \to G = F(D), \quad (t, p) \mapsto (t, x(t, p)) = \begin{cases} (t, x_1(t, p)) & \text{for} \quad (t, p) \in D_1, \\ (t, x_2(t, p)) & \text{for} \quad (t, p) \in D_2, \end{cases}$$

is injective. The sets G and G_i are defined as the images under these flows, and we have that $G = G_1 \cup \mathscr{S} \cup G_2$.

Theorem 6.1.1. *Let* $\mathscr{E} = \mathscr{E}_1 * \mathscr{E}_2$ *be a* C^r-*parameterized field of normal broken extremals. If* $\mathscr{E}$ *has transversal crossings at all switching points in* $\mathscr{T} = \{(t, p) : t = \tau(p), \ p \in P\}$, *then the associated value function* $V^{\mathscr{E}} : G \to \mathbb{R}$, $V^{\mathscr{E}} = C \circ F^{-1}$, *is a continuously differentiable solution to the Hamilton–Jacobi–Bellman equation on* G.

Proof. It needs to be shown that the value $V^{\mathscr{E}}$ remains continuously differentiable on the switching surface $\mathscr{S}$. The fact that the Hamilton–Jacobi–Bellman equation remains valid on $\mathscr{S}$ then is immediate.

Let $(t_0, x_0) = (t_0, x(t_0, p_0)) \in \mathscr{S}$. There exist $C^{0,1}$ extensions u_0 and u_1 of the controls onto a small open neighborhood W of (t_0, p_0), and since $\mathscr{E}$ has a regular switching at (t_0, p_0), without loss of generality, we may assume that the flow maps F_i, $i = 1, 2$, are diffeomorphisms on W with images $\widetilde{G}_i \supset G_i$. Let C_1 and C_2 be the corresponding extensions of the cost functions defined in the proof of Lemma 6.1.1, i.e., for $i = 1, 2$ and all $(t, p) \in W$ we have that

$$C_i(t, p) = \int_t^{\tau(p)} L(s, x_i(s, p), u_i(s, p)) \, ds + C(\tau(p), p).$$

Then the functions $\widetilde{V}_i : \widetilde{G}_i \to \mathbb{R}$, $\widetilde{V}_i = C_i \circ F_i^{-1}$, are well-defined and continuously differentiable on $\widetilde{G}_i$. However, since the maximum condition on the controls is satisfied only on D_i, the functions $\widetilde{V}_i$ are solutions to the Hamilton–Jacobi–Bellman equation only on $G_i = F_i(D_i)$, and there they agree with the values $V_i^{\mathscr{E}}$ of the individual parameterized families. This, and the fact that the identities below are valid on the sets D_i, follows from Theorem 5.2.1:

$$\frac{\partial V_i^{\mathscr{E}}}{\partial t}(t, x_i(t, p)) = -H(t, \lambda(t, p), x_i(t, p), u_i(t, p)),$$

$$\frac{\partial V_i^{\mathscr{E}}}{\partial x}(t, x_i(t, p)) = \lambda(t, p).$$

But these formulas, coupled with the continuity of the trajectories, the multipliers, and the Hamiltonian on the switching surface, imply that the gradients of $V_1^{\mathscr{E}}$ and $V_2^{\mathscr{E}}$ are equal on $\mathscr{S}$. Thus, $V_1^{\mathscr{E}}$ and $V_2^{\mathscr{E}}$ are continuously differentiable functions defined in a neighborhood of $\mathscr{S}$ that along with their gradients agree for points on $\mathscr{S}$. Hence, the composite function $V^{\mathscr{E}} : G \to \mathbb{R}$ defined by

$$V(t, x) = \begin{cases} V_1^{\mathscr{E}}(t, x) & \text{for} \quad (t, x) \in G_1 \cup \mathscr{S}, \\ V_2^{\mathscr{E}}(t, x) & \text{for} \quad (t, x) \in G_2, \end{cases}$$

is continuously differentiable on $\mathscr{S}$ with $\nabla V^{\mathscr{E}} = \nabla V_1^{\mathscr{E}} = \nabla V_2^{\mathscr{E}}$. □

Corollary 6.1.1. *Let $\mathscr{E}$ be a C^1-parameterized field of normal broken extremals with regular transversal crossings over its domain D. Let $G = F(D)$, $\overline{G} = F(D) \cup N$, and let $V^{\mathscr{E}} : \overline{G} \to \mathbb{R}$, $V^{\mathscr{E}} = C \circ F^{-1}$, be the corresponding value function and $u^* : G \to U$, $u^* = u \circ F^{-1}$, the corresponding feedback control. Then $V^{\mathscr{E}}$ is a continuously differentiable solution to the Hamilton–Jacobi–Bellman equation on G that has a continuous extension to the terminal manifold N. The feedback control is optimal on $\overline{G}$ (i.e., in comparison to any other control for which the corresponding trajectory lies in $\overline{G}$), and the corresponding value function is given by $V^{\mathscr{E}}$.* ∎

Example 6.1.2 (Time-optimal control to the origin for the harmonic oscillator). The canonical example that illustrates these features is given by the optimal synthesis for the time-optimal control problem to the origin for the harmonic oscillator. Again, this is a time-independent control problem, and the flow is given by $F(t, p) = x(t, p)$ with p a one-dimensional parameter.

Figure 6.3 once more shows the optimal synthesis for this problem (see Sect. 2.6.4): optimal trajectories enter the origin along the semicircles

$$\Gamma_+ = \{x \in \mathbb{R}^2 : (x_1 - 1)^2 + x_2^2 = 1, \ x_2 \leq 0\}$$

(with control $u = +1$) and

$$\Gamma_- = \{x \in \mathbb{R}^2 : (x_1 + 1)^2 + x_2^2 = 1, \ x_2 \geq 0\}$$

(with control $u = -1$). Let $\widetilde{\Gamma}_+$ and $\widetilde{\Gamma}_-$ denote the relatively open curves

$$\widetilde{\Gamma}_+ = \{x \in \mathbb{R}^2 : (x_1 - 1)^2 + x_2^2 = 1, \ x_2 < 0\}$$

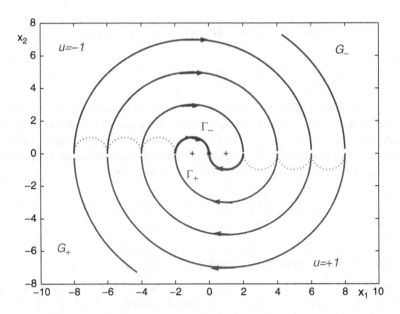

Fig. 6.3 Synthesis of optimal controlled trajectories to the origin for the harmonic oscillator

and

$$\widetilde{\Gamma}_- = \{x \in \mathbb{R}^2 : (x_1 + 1)^2 + x_2^2 = 1, \ x_2 > 0\}.$$

These are one-dimensional, real-analytic embedded submanifolds from which trajectories are integrated backward for controls that alternate between $u = +1$ and $u = -1$ with switchings occurring every π units of time. The corresponding switching surfaces are easily described in the state space. For example, integrating the flow for $u = +1$ backward from $\widetilde{\Gamma}_-$ for π units of time generates the semicircle

$$\mathscr{S}_1 = \{x \in \mathbb{R}^2 : (x_1 - 3)^2 + x_2^2 = 1, \ x_2 < 0\}.$$

Geometrically, this simply is the semicircle $\widetilde{\Gamma}_+$ shifted to the right by 2 units. Since the endpoints, for which $x_2 = 0$, are excluded, both flows for $u = +1$ and $u = -1$ are transversal to $\mathscr{S}_1$ everywhere, and thus the combined flow has a transversal crossing. This is repeated for every other switching. And the same holds for trajectories integrated backward from $\widetilde{\Gamma}_+$. Thus the associated value function $V^{\mathscr{E}}$ is continuously differentiable everywhere in $\mathbb{R}^2$ with the exception of the two special trajectories shown as solid curves in Fig. 6.3 that are obtained by integrating the full curves Γ_+ and Γ_- backward. These controlled trajectories have switching times exactly for $t = -n\pi$ for $n = 0, 1, 2, \ldots$. It was shown in Sect. 2.6.4 that these are strictly abnormal extremals, and indeed it is easily verified that the value function no longer is differentiable in $\mathbb{R}^2$ along these trajectories. Also, note that the flow would be tangential to the switching curves at these points if we were to include the endpoints into the parameterized family of extremals. In principle, this could be done (and this then is an example for such a family with both normal and abnormal extremals), but we would need to allow for the parameter set P to be a manifold with boundary. Also, as for the time-optimal control problem to the origin, we can easily include the last pieces along $\tilde{\Gamma}_+$ and $\tilde{\Gamma}_-$ in the parameterized family, and then once again, trajectories collapse onto these lower-dimensional submanifolds.

6.1.3 Transversal Folds

Transversal folds, on the other hand, typically correspond to surfaces where optimality of the combined flow ceases. Essentially, under some regularity assumptions, in this case the switching surface $\mathscr{S}$ is made up of controlled trajectories (ξ, η) that are control envelopes in the sense of Definition 5.4.5. If it then can be argued that controlled trajectories that lie in $\mathscr{S}$ cannot be optimal, using the envelope theorem (Corollary 5.4.2), it can be shown that $\mathscr{S}$ consists of "conjugate points" where optimality ceases. An argument along these lines often invokes that controls corresponding to controlled trajectories that lie in $\mathscr{S}$ must be singular and that singular controls are not optimal. However, to make this reasoning precise, specific models need to be considered. Quite simply, for systems with an arbitrary control set U, there is no reason to believe that the switching surface $\mathscr{S}$ supports controlled trajectories (ξ, η) to begin with. We therefore here restrict our presentation to the time-optimal control problem to a point for a time-invariant, single-input system

$$\Sigma: \qquad \dot{x} = f(x) + ug(x), \qquad |u| \le 1, \qquad x \in \mathbb{R}^n.$$

Theorem 6.1.2. *Let $\mathscr{E} = \mathscr{E}_1 * \mathscr{E}_2$ be a C^1-parameterized family of normal broken extremals for the time-optimal control problem to a point for the system Σ. Suppose that the controls u_1 and u_2 are given by the bang controls $u = \pm 1$ with a transversal fold at $\mathscr{T} = \{(t, p) : t = \tau(p), \ p \in P\}$ and switching surface*

$$\mathscr{S} = \{x : \ \xi(p) = x(\tau(p), p), \ p \in P\} = F(\mathscr{T}).$$

In this case, the switching surface $\mathscr{S}$ supports controlled trajectories of the system Σ, and these are singular arcs. If singular controlled trajectories that lie in $\mathscr{S}$ are slow (i.e., violate the Legendre–Clebsch condition), then optimality of the flow ceases at $\mathscr{S}$.

Proof. Let $X = f - g$ and $Y = f + g$. By assumption, the vector fields X and Y point to opposite sides of the hypersurface $\mathscr{S}$ at every point. We claim that there exists a continuous function $\eta : \mathscr{S} \to (-1, 1), x \mapsto \eta(x)$, defined on $\mathscr{S}$ with values in the open interval $(-1, 1)$, such that the vector field $S(x) = f(x) + \eta(x)g(x)$ is everywhere on $\mathscr{S}$ tangent to $\mathscr{S}$. Since the family of broken extremals has regular switching points, $\mathscr{S} = F(\mathscr{T})$ is a codimension-1 embedded submanifold that locally can be represented as the zero set of some C^1-function Ψ, $\mathscr{S} = \{x \in W : \Psi(x) = 0\}$, for which the gradient $\nabla\Psi(x)$ is not zero on $\mathscr{S}$ (see Eq. (6.6)). The vector field S is tangent to the switching surface $\mathscr{S}$ if and only if the Lie derivative of Ψ in the direction of S vanishes on $\mathscr{S}$. Hence the function u simply is the solution to the equation $L_S\Psi(x) = \nabla\Psi(x) \cdot (f(x) + ug(x)) = 0$, i.e.,

$$\eta(x) = -\frac{L_f\Psi(x)}{L_g\Psi(x)} = -\frac{L_{X+Y}\Psi(x)}{L_{Y-X}\Psi(x)} = \frac{L_X\Psi(x) + L_Y\Psi(x)}{L_X\Psi(x) - L_Y\Psi(x)}.$$

This defines a continuous function that since $L_X\Psi(x)$ and $L_Y\Psi(x)$ have opposite signs, takes values in the open interval $(-1,1)$. Hence η is an admissible feedback control on $\mathscr{S}$.

The integral curves of this vector field S thus are controlled trajectories (ξ, η) of Σ that lie on the switching surface $\mathscr{S}$. These integral curves are control envelopes for the system: since the switching points are regular, the restriction of the flow to D_1, $F_1 = F \restriction D_1$, extends as a local diffeomorphism into a neighborhood of $\mathscr{T}$. The trajectory $\xi = \xi(s)$ lies in $\mathscr{S}$, and therefore

$$F_1^{-1}(\xi(s)) = (t(s), p(s)) \in D_1 \cap \mathscr{T}$$

with $t(s) = \tau(p(s))$. Thus, if we take $\tilde{\tau}(s) \equiv s$, then we have

$$\xi(\tilde{\tau}(s)) = x(t(s), p(s)),$$

and we need to show that Eq. (5.56) in Definition 5.4.5 holds, i.e., that

$$H(\lambda(t,p), \xi(\tilde{\tau}), \eta(\tilde{\tau}))d\tilde{\tau} = H(\lambda(t,p), x(t,p), u_1(t,p))dt. \qquad (6.7)$$

(We changed the notation for the reparameterization in this definition to $\tilde{\tau}$, since here τ already has been used to denote the switching time.)

Since the problem is time-invariant and the terminal time is free, we have on all of D_1 that

$$H = H(\lambda(t,p), x(t,p), u_1(t,p)) \equiv 0,$$

so that the right-hand side of Eq. (6.7) vanishes. But so does the left-hand side: for normal extremals, the Hamiltonian is given by $H = 1 + \langle \lambda, f(x) + ug(x) \rangle$, and along the switching points, the parameterized switching function $\Phi(t,p) = \langle \lambda(t,p), g(x(t,p)) \rangle$ vanishes identically, $\Phi(\tau(p), p) \equiv 0$. Hence for all $p \in P$,

$$1 + \langle \lambda(\tau(p), p), f(x(\tau(p), p)) \rangle \equiv 0,$$

and thus since $\xi(\tilde{\tau}) = x(t,p) = x(\tau(p), p)$, we in fact have for any control η that

$$H(\lambda(t,p), \xi(\tilde{\tau}), \eta(\tilde{\tau})) = 1 + \langle \lambda(\tau(p), p), f(x(\tau(p), p)) \rangle + 0 \cdot \eta(\tilde{\tau}) \equiv 0.$$

Fig. 6.4 Construction of an envelope for a transversal fold

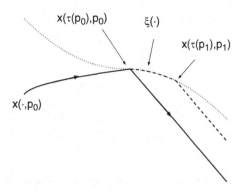

Thus condition (6.7) is trivially satisfied along controlled trajectories that lie in the switching surface.

Corollary 5.4.2 can now be used to exclude the optimality of the parameterized family $\mathscr{E}_1$: we show that for any $p \in P$, the controlled trajectory $x(\cdot, p)$ is not time-optimal for $t \leq \tau(p)$. Fix a parameter p_0 and let $t_0 \leq \tau(p_0)$. The value along the controlled trajectory $\Gamma_0 = (x(\cdot, p_0), u(\cdot, p_0))$ with initial point $x(t_0, p_0)$ is given by $C(t_0, p_0)$. We construct a second controlled trajectory Γ_1 as follows: for $t_0 \leq t \leq \tau(p_0)$, follow the controlled reference trajectory; then, for some small positive time $\varepsilon > 0$, follow the controlled trajectory (ξ, η) that is the integral curve of the vector field $S(x) = f(x) + \eta(x)g(x)$ and starts at the point $\xi(0) = x(\tau(p_0), p_0) \in \mathscr{S}$. The point $\xi(\varepsilon)$ still lies on $\mathscr{S}$, and thus there exists a unique parameter value p_1 such that $\xi(\varepsilon) = x(\tau(p_1), p_1)$. We then follow the controlled trajectory $(x(\cdot, p_1), u(\cdot, p_1))$ for the full interval $[\tau(p_1), t_{2,+}(p_1)]$ of definition in $\mathscr{E}_2$ (see Fig. 6.4).

The cost for the comparison trajectory Γ_1 is given by

$$C(t_0, p_0) - C(\tau(p_0), p_0) + \varepsilon + C(\tau(p_1), p_1).$$

But by Corollary 5.4.2 we have that

$$C(\tau(p_0), p_0) = \varepsilon + C(\tau(p_1), p_1),$$

and thus Γ_0 and Γ_1 take the same time $C(t_0, p_0)$. However, since the control values η for the controlled trajectory ξ lie in the interior of the control interval $[-1, 1]$, this piece of the comparison trajectory is a singular arc. By assumption, this arc is slow, and thus Γ_1 is not time-optimal. Since Γ_0 takes the same time, Γ_0 is not time-optimal either. □

Examples of this type arise in the context of slow singular arcs like the one analyzed for time-optimal control in the plane in Sect. 2.9.3. We shall return to this topic also in Chap. 7 when we consider time-optimal control problems in $\mathbb{R}^3$, where once more, transversal fold points limit the optimality of bang-bang trajectories near slow singular arcs.

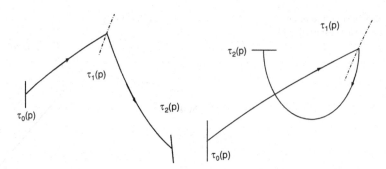

Fig. 6.5 Self-intersection properties of trajectories in the time-dependent (*left*) and time-independent cases (*right*)

We note that several of the arguments made in the proof of Theorem 6.1.2 are applicable if we consider only the first parameterized family $\mathscr{E}_1$. If the switching point for $t = \tau(p)$ is regular, and if there exists a controlled trajectory (ξ, η) that lies in the switching surface $\mathscr{S}$, then it follows that (ξ, η) is a control envelope. For example, this applies to Example 6.1.1, the time-optimal control problem to the origin for the double integrator. In fact, the switching curves Γ_+ and Γ_- are control envelopes in the sense of Definition 5.4.5, and the envelope condition of Theorem 5.4.2 is satisfied. However, in this case, the switching is not a transversal fold, but the second flow follows the switching surface and the control η lies in the boundary of the control set. Thus the step in the argument that implies nonoptimality is not applicable. Similar features arise if the switching surface consists of fast (locally minimizing) singular arcs, and then the flow follows these controlled trajectories. We shall give a detailed example in Sect. 6.2.

6.1.4   Local Analysis of a Flow of Broken Extremals

We now localize the results above. For these formulations, there is one significant difference between the time-dependent and time-independent formulations. The curves in the flow for a time-dependent problem are the graphs of the controlled trajectories, and these cannot self-intersect. This property is not guaranteed for a time-independent system and needs to be assumed (see Fig. 6.5).

Proposition 6.1.2. *Let $\mathscr{E}$ be an n-dimensional C^r-parameterized family of normal broken extremals for a time-dependent optimal control problem with domain $D = \{(t,p) : p \in P, t_-(p) \le t \le t_+(p)\}$ and switching times*

$$t_-(p) = \tau_0(p) < \tau_1(p) < \cdots < \tau_k(p) < \tau_{k+1}(p) = t_+(p).$$

Let $p_0 \in P$ determine a reference controlled trajectory $(x(\cdot,p_0),u(\cdot,p_0))$ and suppose that (i) *the matrix $\frac{\partial x_i}{\partial p}(t,p_0)$ is nonsingular on the intervals $\tau_i(p_0) \leq t \leq \tau_{i+1}(p_0)$ for $i = 0,\ldots,k$, and* (ii) *the trajectory $x(\cdot,p_0)$ has regular and transversal crossings at all switchings. Then there exists a neighborhood $W \subset P$ of p_0 such that the restriction of $\mathscr{E}$ to $\{(t,p): t_-(p) \leq t \leq t_+(p),\ p \in W\}$ defines a C^r-parameterized field of normal broken extremals that has regular and transversal crossings.*

Proof. Without loss of generality, we consider the case $k = 1$ and denote the switching times by $\tau(p)$. By assumption, the flows F_1 and F_2, $F_i(t,p) = (t,x(t,p))$, are regular along the reference controlled trajectory. Thus, by the inverse function theorem, for every time $t \in [t_-(p_0),\tau(p_0)]$ and $t \in [\tau(p_0),t_+(p_0)]$ there exists a neighborhood O_t of (t,p_0) of the form $O_t = (t - \varepsilon_t, t + \varepsilon_t) \times W_t \subset D$, so that the restriction of the flow map F_i to O_t is a C^r-diffeomorphism. Since we have regular switchings, the endpoints of the intervals can be included. By compactness, we can select finite subcovers of the curve $t \mapsto x(t,p_0)$. Overall, there thus exist neighborhoods W_1 and W_2 of p_0 such that the flows F_1 and F_2, restricted to the domains $\tilde{D}_1 = \{(t,p): p \in W_1, t_-(p) \leq t \leq \tau(p)\}$ and $\tilde{D}_2 = \{(t,p): p \in W_2, \tau(p) \leq t \leq t_+(p)\}$, are C^r-diffeomorphisms. Furthermore, since we have a transversal crossing at $(\tau(p_0),p_0)$, the combined flow is injective in a neighborhood of $(\tau(p_0),p_0)$ as well, and thus the restriction of $\mathscr{E}$ to D defines a C^r field of broken extremals. $\square$

For a time-invariant problem, we need to add the condition that the reference controlled trajectory not self-intersect. Then, locally, the same argument can be made.

Proposition 6.1.3. *Let $\mathscr{E}$ be an $(n-1)$-dimensional C^r-parameterized family of normal broken extremals for a time-independent optimal control problem with domain $D = \{(t,p): p \in P, t_-(p) \leq t \leq t_+(p)\}$ and switching times*

$$t_-(p) = \tau_0(p) < \tau_1(p) < \cdots < \tau_k(p) < \tau_{k+1}(p) = t_+(p).$$

Let $p_0 \in P$ determine a reference controlled trajectory $(x(\cdot,p_0),u(\cdot,p_0))$ and suppose that (i) *the matrix $\frac{\partial x_i}{\partial p}(t,p_0)$ is nonsingular on the intervals $\tau_i(p_0) \leq t \leq \tau_{i+1}(p_0)$ for $i = 0,\ldots,k$,* (ii) *the trajectory $x(\cdot,p_0)$ has regular and transversal crossings at all switchings, and* (iii) *there are no self-intersections along the reference trajectory, i.e., the mapping $x(\cdot,p_0): [t_-(p_0),t_+(p_0)] \to M, t \mapsto x(t,p_0)$, is injective. Then there exists a neighborhood $W \subset P$ of p_0 such that the restriction of $\mathscr{E}$ to $\{(t,p): t_-(p) \leq t \leq t_+(p),\ p \in W\}$ defines a C^r-parameterized field of normal broken extremals that has regular and transversal crossings.* ∎

Under the assumptions of either Proposition 6.1.2 or 6.1.3, the combined flow F piecewise is a diffeomorphism on the domain D, and the parameterized value function C gives rise to a continuously differentiable solution to the Hamilton–Jacobi–Bellman equation on $G = F(D)$ with the associated local optimality results. Thus, in order to obtain results about local optimality, we need conditions to verify that the flow is regular and has transversal crossings. For a time-dependent problem, we already gave characterizations of the regularity of the flow in Sect. 5.3, and these results apply. In this case, if $\mathscr{E}$ is a nicely C^1-parametrized family of normal extremal lifts, then as long as the flow is regular, the function

$$Z(t,p) = \frac{\partial \lambda^T}{\partial p}(t,p) \left(\frac{\partial x}{\partial p}(t,p) \right)^{-1}$$

satisfies the differential equation

$$\dot{Z} + Zf_x + f_x^T Z + H_{xx} + (Zf_u + H_{xu}) \frac{\partial u}{\partial p} \left(\frac{\partial x}{\partial p} \right)^{-1} \equiv 0 \qquad (6.8)$$

with the partial derivatives of f and H evaluated along the extremal corresponding to the parameter p. If the control takes values in the interior of the control set, then, as shown in Sect. 5.3, one can solve the equation $\frac{\partial H}{\partial u} = 0$ for $\frac{\partial u}{\partial p}$, and this becomes a Riccati differential equation whose explosion times correspond to the times when regularity of the flow ceases (e.g., Theorems 5.3.2 and 5.3.3). The situation becomes considerably simpler for the case of *bang-bang controls*, a common situation with broken extremals. In this case, the control $u = u(t,p)$ is constant along the individual domains and thus $\frac{\partial u}{\partial p} = 0$. Hence Eq. (6.8) simplifies to the linear Lyapunov differential equation

$$\dot{Z} + Zf_x + f_x^T Z + H_{xx} \equiv 0, \qquad (6.9)$$

and solutions to this linear equation always exist. It therefore follows from Proposition 2.4.1 that $\frac{\partial x}{\partial p}(t,p)$ is nonsingular if and only if it is nonsingular at the initial (or terminal) time. This gives the following result:

Corollary 6.1.2. *Let $\mathscr{E}$ be a nicely C^1-parameterized family of normal extremal lifts for a time-dependent control problem and let $\Gamma_p : [t_-(p),t_+(p)] \to M \times U$ be an extremal for which the corresponding control does not depend on p, i.e., $\frac{\partial u}{\partial p}(t,p) \equiv 0$. Then $\frac{\partial x}{\partial p}(t,p)$ is nonsingular over $[t_-(p),t_+(p)]$ if and only if the matrix $\frac{\partial x}{\partial p}(t,p)$ is nonsingular at the initial time or terminal time.*   ∎

It is clear from the uniqueness of solutions to a differential equation that the flow F corresponding to a constant control cannot have singularities. (In the time-dependent case, this holds as well, since the flow consists of the graphs of the controlled trajectories.) Conjugate points can thus occur only at the switching points, and typically these are transversal folds. While it may be easy to check directly for some problems whether the switchings are transversal crossings or folds (e.g., see the time-optimal control problem to the origin for the harmonic oscillator), for a general nonlinear system this brings us back to the criteria established in Proposition 6.1.1. Indeed, there exist effective ways to evaluate these formulas whenever the switchings occur on surfaces that are defined as the zero sets of smooth functions in (t,p)-space. For example, for bang-bang trajectories, typically the parameterizations $t = \tau(p)$ of the switching surfaces are obtained by solving an equation of the type $\Phi(t, p) = 0$ given by some switching function for t. This indeed allows us to calculate the quantity

$$\frac{\partial \tau}{\partial p}(p) \left(\frac{\partial x}{\partial p}(\tau(p), p) \right)^{-1}$$

without having to take partial derivatives with respect to the parameter p or matrix inversions. However, the specifics depend on the form of the dynamics. We illustrate the procedure for a nonlinear regulator problem for a single-input, control-affine system. Systems of this type are common and arise, for example, in models for cancer chemotherapy over a prescribed therapy horizon (e.g., see [155, 156]).

[OC-BB] For a fixed terminal time T, consider the optimal control problem to minimize an objective of the form

$$J(u) = \int_{t_0}^{T} (L(x) + u) \, dt + \varphi(x(T))$$

subject to the dynamics

$$\dot{x} = f(x) + u g(x), \qquad 0 \le u \le 1, \qquad x \in \mathbb{R}^n.$$

Extremals for this problem are normal: The Hamiltonian is given by

$$H = \lambda_0 (L(x) + u) + \lambda (f(x) + u g(x))$$

and the adjoint equation and terminal condition are

$$\dot{\lambda} = -\lambda (Df(x) + u Dg(x)) - \lambda_0 \nabla L(x), \qquad \lambda(T) = \lambda_0 \nabla \varphi(x(T)).$$

Thus, if $\lambda_0 = 0$, then λ vanishes identically, contradicting the nontriviality of the multipliers. We thus set $\lambda_0 = 1$. A parameterized family of extremals can then be constructed by integrating the dynamics and adjoint equation backward from the terminal time, choosing $p = x(T)$ as parameter and enforcing the minimum condition. The parameterized switching function is given by

$$\Phi(t, p) = 1 + \lambda(t, p) g(x(t, p)),$$

and the control in the parameterized family of extremals satisfies

$$u(t, p) = \begin{cases} 1 & \text{if } \Phi(t, p) < 0, \\ 0 & \text{if } \Phi(t, p) > 0. \end{cases}$$

Suppose controls are bang-bang and let $t = \tau(p)$ define a switching surface $\mathscr{S}$. This function is the solution of the equation $\Phi(t, p) = 0$ near a switching point (t_0, p_0) of the reference trajectory, and by the implicit function theorem such a solution exists locally if the time derivative $\dot{\Phi}(t_0, p_0)$ does not vanish. It then follows from implicit differentiation that

$$\dot{\Phi}(\tau(p), p)\frac{\partial \tau}{\partial p}(p) + \frac{\partial \Phi}{\partial p}(\tau(p), p) \equiv 0.$$

Differentiating $\Phi(t, p)$ with respect to p, and using $Z(t, p) = \frac{\partial \lambda^T}{\partial p}(t, p)\left(\frac{\partial x}{\partial p}(t, p)\right)^{-1}$, we have that

$$\frac{\partial \Phi}{\partial p}(t, p) = \left(\frac{\partial \lambda}{\partial p}(t, p)g(x(t, p))\right)^T + \lambda(t, p)Dg(x(t, p))\frac{\partial x}{\partial p}(t, p)$$

$$= \left(\lambda(t, p)Dg(x(t, p)) + g^T(x(t, p))Z(t, p)\right)\frac{\partial x}{\partial p}(t, p).$$

Setting $t_0 = \tau(p_0)$ and $x_0 = x(t_0, p_0)$, we thus get that

$$\frac{\partial \tau}{\partial p}(p_0)\left(\frac{\partial x}{\partial p}(t_0, p_0)\right)^{-1} = -\frac{\lambda(t_0, p_0)Dg(x_0) + g^T(x_0)Z(t_0, p_0)}{\dot{\Phi}(t_0, p_0)}.$$

Further simplifications in the transversality condition (6.4) can be made because of the special form of the dynamics. Let us denote the left- and right-hand limits of functions at the switching surface by a $-$ and $+$ sign, respectively, and denote the jump in the control at the switching surface by $\Delta u = u_+ - u_-$. It then follows from the minimization property of the controls that $\Delta u = -\text{sgn } \dot{\Phi}(t_0, p_0)$, and thus

$$1 + \frac{\partial \tau}{\partial p}(p_0)\left(\frac{\partial x_2}{\partial p}(t_0, p_0)\right)^{-1}\Delta u g(x_0)$$

$$= 1 + \frac{1}{|\dot{\Phi}(t_0, p_0)|}\left(\lambda(t_0, p_0)Dg(x_0) + g^T(x_0)Z(t_0, p_0)\right)g(x_0) > 0.$$

Summarizing, we have the following result:

Theorem 6.1.3. *Suppose a reference control $u(\cdot, p_0)$ for the optimal control problem [OC-BB] has a bang-bang switch at t_0 with $\dot{\Phi}(t_0, p_0) \neq 0$. Then the switching surface $\mathscr{S}$ is an n-dimensional embedded submanifold near (t_0, x_0), $x_0 = x(t_0, p_0)$, and there exists a continuously differentiable function τ defined in some neighborhood W of p_0 such that $\mathscr{S} = \{(t, x) : t = \tau(p), x = x(\tau(p), p), p \in W\}$. Assuming that $\frac{\partial x}{\partial p}(t_0+, p_0)$ is nonsingular, the combined flow $F : (t, p) \mapsto (t, x(t, p))$ has a regular and transversal crossing at (t_0, x_0) if and only if*

$$|\dot{\Phi}(t_0, p_0)| + \{\lambda(t_0, p_0)Dg(x_0) + g^T(x_0)Z_+(t_0, p_0)\}g(x_0) > 0. \qquad (6.10)$$

Except for the value of the matrix $Z_+(t_0, p_0)$, all the quantities in (6.10) are continuous at the switching time and are readily available or easily computed. But the matrix Z is discontinuous at the switching times and needs to be propagated backward recursively as solution to the Lyapunov equation (6.9) from the terminal time T. Note that we have $x(T, p) \equiv p$ and $\lambda(T, p) = \nabla\varphi(p)$, so that

$$Z_-(T, p) = \frac{\partial \lambda^T}{\partial p}(T, p) \left(\frac{\partial x}{\partial p}(T, p)\right)^{-1} = \frac{\partial^2 \varphi}{\partial x^2}(p).$$

Then the value of $Z_+(t_0, p_0)$ for the last switching simply is the solution to Eq. (6.9) with this terminal condition. However, the partial derivatives $\frac{\partial x}{\partial p}$ and $\frac{\partial \lambda^T}{\partial p}$ are discontinuous at the switching surface, and it becomes necessary to update these formulas. In order to simplify the appearance of these formulas, we drop the arguments. The function τ is evaluated at p, states x and multipliers λ are evaluated at $(\tau(p), p)$, and the left-and right-hand limits of Z, Z_- and Z_+ are also evaluated at $(\tau(p), p)$. It follows from Eq. (6.3) that

$$\frac{\partial x_-}{\partial p} = \frac{\partial x_+}{\partial p} + \Delta u g(x) \frac{\partial \tau}{\partial p},$$

and analogously, we have for the multipliers that

$$\frac{\partial \lambda_-^T}{\partial p} = \frac{\partial \lambda_+^T}{\partial p} - \Delta u Dg(x)^T \lambda^T \frac{\partial \tau}{\partial p}.$$

Hence

$$Z_- = \frac{\partial \lambda_-^T}{\partial p}\left(\frac{\partial x_-}{\partial p}\right)^{-1} = \left(\frac{\partial \lambda_+^T}{\partial p} - \Delta u Dg(x)^T \lambda^T \frac{\partial \tau}{\partial p}\right)\left(\frac{\partial x_+}{\partial p} + \Delta u g(x) \frac{\partial \tau}{\partial p}\right)^{-1}$$

Let

$$R = R(p) = -\frac{\partial \tau}{\partial p}(p)\left(\frac{\partial x_+}{\partial p}(\tau(p), p)\right)^{-1} \qquad \Delta u = \frac{\lambda Dg(x) + g^T(x)Z_+}{|\dot{\Phi}|},$$

so that

$$Z_- = \left(\frac{\partial \lambda_+^T}{\partial p}\left(\frac{\partial x_+}{\partial p}\right)^{-1} - \Delta u Dg(x)^T \lambda^T \frac{\partial \tau}{\partial p}\left(\frac{\partial x_+}{\partial p}\right)^{-1}\right)\left(\frac{\partial x_+}{\partial p}\right)$$

$$\times \left(\frac{\partial x_+}{\partial p} + \Delta u g(x) \frac{\partial \tau}{\partial p}\right)^{-1}$$

$$= (Z_+ + Dg(x)^T \lambda^T R)\left(\frac{\partial x_+}{\partial p}\right)\left(\frac{\partial x_+}{\partial p} + \Delta u g(x) \frac{\partial \tau}{\partial p}\right)^{-1}.$$

Furthermore,

$$1 - Rg(x) = 1 + \frac{\partial \tau}{\partial p} \left(\frac{\partial x_+}{\partial p} \right)^{-1} \Delta u g(x),$$

and this term is positive for a transversal crossing. It then follows from Lemma 6.1.4 that

$$\left(\frac{\partial x_+}{\partial p} \right) \left(\frac{\partial x_+}{\partial p} + \Delta u g(x) \frac{\partial \tau}{\partial p} \right)^{-1} = \mathrm{Id} - \frac{\Delta u g(x) \frac{\partial \tau}{\partial p} \left(\frac{\partial x_+}{\partial p} \right)^{-1}}{1 - Rg(x)} = \mathrm{Id} + \frac{g(x)R}{1 - Rg(x)},$$

and thus we get that

$$Z_- = \left(Z_+ + Dg(x)^T \lambda^T R \right) \left(\mathrm{Id} + \frac{g(x)R}{1 - Rg(x)} \right),$$

giving us the following update formula for $Z(t,p)$ at the switching $t = \tau(p)$:

Lemma 6.1.5. *Suppose an extremal controlled trajectory has a bang-bang switching at the point* (t_0, x_0), $x_0 = x(t_0, p_0)$, *and* $|\dot{\Phi}(t_0, p_0)| \neq 0$. *If this switching is a transversal crossing, then with*

$$R(t_0, p_0) = \frac{\lambda(t_0, p_0)Dg(x_0) + g^T(x_0)Z_+(t_0, p_0)}{|\dot{\Phi}(t_0, p_0)|}, \tag{6.11}$$

the left-sided limit for $Z(t_0, p_0)$ *is given by*

$$Z_-(t_0, p_0) = \left(Z_+(t_0, p_0) + Dg(x_0)^T \lambda(t_0, p_0)^T R(t_0, p_0) \right) \left(\mathrm{Id} + \frac{g(x_0)R(t_0, p_0)}{1 - R(t_0, p_0)g(x_0)} \right). \tag{6.12}$$

This formula provides the required terminal value on the left-hand side of the switching surface, and then the solution Z can be propagated further backward by integrating the Lyapunov equation (6.9). Altogether, a simple algorithmic procedure results to check whether successive switching surfaces are regular ($R(t_0, p_0)$ $g(x_0) \neq 1$) and consist of transversal crossings ($R(t_0, p_0)g(x_0) < 1$) or folds ($R(t_0, p_0)g(x_0) > 1$).

For time-independent problems, similar characterizations of the regularity of the flow F can be given, but in this case the matrix

$$Z(t, p) = D\lambda^T(t, p) \cdot Dx(t, p)^{-1}$$

needs to be considered with D denoting the matrix of partial derivatives with respect to all variables (t, p).

6.2 A Mathematical Model for Tumor Antiangiogenic Treatment

We give an example of a parameterized family of broken extremals that includes singular arcs. Similar to the flow of bang-bang trajectories for the double integrator, the flow F collapses onto a lower-dimensional manifold as the controls follow singular arcs and no longer covers a full set in the state space. However, as we shall see, this does not invalidate the constructions, and the value of the parameterized family still is a solution of the Hamilton–Jacobi–Bellman equation on those sections of the parameterized family along which the flow is a diffeomorphism. The scenario considered here is typical for syntheses involving optimal singular arcs in low dimensions. We give it as an illustration and to show the flexibility of the constructions. We consider the following optimal control problem formulated in the customary notation for this application:

[A] For a free terminal time T, minimize the value $p(T)$ subject to the dynamics

$$\dot{p} = -\xi p \ln\left(\frac{p}{q}\right), \qquad\qquad p(0) = p_0, \qquad (6.13)$$

$$\dot{q} = bp - dp^{\frac{2}{3}}q - \mu q - \gamma u q, \qquad q(0) = q_0, \qquad (6.14)$$

$$\dot{y} = u, \qquad\qquad y(0) = 0, \qquad (6.15)$$

over all Lebesgue measurable functions $u : [0, T] \to [0, a]$ for which the corresponding trajectory satisfies $y(T) \leq A$.

It is not difficult to see that the set $M = \{(p, q, y) : p > 0,\ q > 0,\ 0 \leq y \leq A\}$ has the following invariance property [160]: given any admissible control $u : [0, T] \to [0, a]$, the solution to the dynamics (6.13)–(6.15) exists over the interval $[0, T]$ and its trajectory lies in M. Hence it is not necessary to impose any state-space constraints in the model. The limits in the variable y are easily handled, and the essential part of the state space is the positive quadrant in p and q, an open subset of $\mathbb{R}^2$.

This is a mathematical model for antiangiogenic treatment of a solid tumor formulated by Hahnfeldt, Panigrahy, Folkman, and Hlatky in [114]. The model is based on cell populations and aggregates the main biological features into minimally parameterized equations with two principal variables, the primary tumor volume, p, and the carrying capacity of its vasculature, q. In the forthcoming text [166] on applications of methods of geometric optimal control to biomedical models, a detailed derivation of these equations (based on an asymptotic expansion of an underlying diffusion–consumption equation) will be given and the biological background will be explained in more detail. Here, we are interested in its mathematical aspects and thus only briefly indicate the underlying medical problem.

A growing solid tumor, in its initial stage of so-called avascular growth, is able to obtain enough nutrients and oxygen from its surrounding environment. However, as it reaches about 2 mm in diameter, this is no longer the case, and most tumor cells enter a dormant stage of the cell cycle. This causes a release of stimulators, such as VEGF (vascular endothelial growth factor), that cause the migration of existing blood vessels as well as the creation of new ones that supply the tumor with nutrients and thus enable its further growth. This process is called *tumor angiogenesis* and is essential in the further growth of the tumor. The resulting network of blood vessels and capillaries is called the tumor vasculature. Tumor antiangiogenesis is a treatment approach that aims at depriving the tumor of this needed vasculature. Ideally, without an adequate support network, the tumor shrinks, and its further development is halted. This approach was proposed already in the early seventies by J. Folkman, but became a medical reality only in the nineties with the discovery of inhibitory mechanisms.

In the model above, tumor growth is described by a so-called Gompertzian growth function, Eq. (6.13), and its carrying capacity, the variable q, can be interpreted as the ideal size to which the tumor can grow. If $q > p$, the tumor has ample nutrients available and will proliferate, while it will shrink if $p > q$. The dynamics for the carrying capacity q, Eq. (6.14), consists of the difference of a stimulatory effect, $S = bp$, that is taken proportional to the tumor size, and an inhibitory effect, $I = dp^{\frac{2}{3}}q$, that arises through the interaction of antiangiogenic inhibitors that are released through the tumor surface with the vasculature. The extra term μq represents natural death, and γuq models additional loss to the vasculature achieved through the administration of outside inhibitors. This term represents the control in the model. In this simplified version, the drug's dosage is identified with its concentration, and the maximum dosage is denoted by a. Angiogenic inhibitors are expensive biological agents in limited supply. Thus, the optimal control problem arises as to how a given amount of antiangiogenic agents can be applied in the best possible way. In problem [A], this is formulated by asking for the maximum tumor reduction that can be achieved with a given amount A of antiangiogenic agents. This leads to an optimal control problem with free terminal time T that represents the time when this minimum is achieved. Equation (6.15) simply accounts for the amount of inhibitors that already have been used, and the addition of such an extra state variable is the standard way of including isoperimetric constraints.

Assuming $b > \mu$, the uncontrolled system ($u = 0$) has a unique equilibrium point at $(\bar{p}, \bar{q})$ given by $\bar{p} = \bar{q} = \left(\frac{b-\mu}{d}\right)^{3/2}$, which can be shown to be globally asymptotically stable [87]. Its value generally is too high to be acceptable, and this equilibrium state is malignant. The medically relevant region is contained in the domain

$$\mathscr{D} = \{(p,q) : 0 < p < \bar{p},\ 0 < q < \bar{q}\}.$$

In order to exclude irrelevant discussions about the structure of optimal controls in regions where the model does not represent the underlying medical problem to begin with, we henceforth restrict our discussions to this square domain $\mathscr{D}$.

6.2.1 Preliminary Analysis of Extremals

In [160], we have given a complete solution to this optimal control problem, and we refer the interested reader to our text [166] for a full discussion and numerous further extensions. Here we focus on one specific aspect and make some simplifying assumptions. For example, for initial data (p_0, q_0) with $p_0 < q_0$, i.e., when the tumor is proliferating, then, since $\dot{p}$ is always positive in the set $\{p < q\}$, it is possible that the amount of inhibitors is insufficient to reach a tumor volume that is smaller than p_0. For such a case, the optimal solution to problem [A] is simply given by $T = 0$. However, if A is large enough, this degenerate situation does not arise, and then the optimal terminal time T will be positive and all inhibitors will be used up, $y(T) = A$. We consider only these *well-posed data*.

We start with an overview of basic properties of optimal controls for this problem. The Hamiltonian H is given by

$$H = -\lambda_1 \xi p \ln\left(\frac{p}{q}\right) + \lambda_2\left(bp - \left(\mu + dp^{\frac{2}{3}}\right)q - \gamma uq\right) + \lambda_3 u. \qquad (6.16)$$

If u is an optimal control defined over the interval $[0, T]$ with corresponding trajectory (p, q, y), then there exist a constant $\lambda_0 \geq 0$ and an absolutely continuous covector $\lambda : [0, T] \to (\mathbb{R}^3)^*$ such that the following conditions hold:

1. (Nontriviality condition) $(\lambda_0, \lambda(t)) \neq (0, 0)$ for all $t \in [0, T]$,
2. (Adjoint equations) λ_3 is constant, nonnegative, and λ_1 and λ_2 satisfy the equations

$$\dot{\lambda}_1 = -\frac{\partial H}{\partial p} = \xi \lambda_1 \left(\ln\left(\frac{p}{q}\right) + 1\right) + \lambda_2\left(\frac{2}{3}d\frac{q}{p^{\frac{1}{3}}} - b\right), \qquad (6.17)$$

$$\dot{\lambda}_2 = -\frac{\partial H}{\partial q} = -\xi \lambda_1 \frac{p}{q} + \lambda_2\left(\mu + dp^{\frac{2}{3}} + \gamma u\right), \qquad (6.18)$$

with terminal conditions

$$\lambda_1(T) = \lambda_0 \quad \text{and} \quad \lambda_2(T) = 0,$$

3. (Minimum condition) for almost every time $t \in [0, T]$, the optimal control $u(t)$ minimizes the Hamiltonian along $(\lambda(t), p(t), q(t))$ over the control set $[0, a]$ with minimum value given by 0.

The nonnegativity condition on the multiplier λ_3 follows from the terminal constraint $y(T) \leq A$. A first observation is that if $\lambda_0 = 0$, then λ_1 and λ_2 vanish identically, and in this case λ_3 must be positive. Hence, the control is $u \equiv 0$, which corresponds to ill-posed data that we have excluded. Thus, *extremals for well-posed initial data are normal*, and we henceforth take $\lambda_0 = 1$. In particular, λ_1 and λ_2 cannot vanish simultaneously, and λ_2 has only simple zeroes.

Lemma 6.2.1. *Along an optimal trajectory* (p,q,y), *all available inhibitors are exhausted,* $y(T) = A$, *and at the final time* $p(T) = q(T)$ *holds.*

Proof. Since the cancer volume p is growing for $p < q$ and is shrinking for $p > q$, optimal trajectories can terminate only at times for which $p(T) = q(T)$. For, if $p(T) < q(T)$, then it would simply have been better to stop earlier, since p was increasing over some interval $(T - \varepsilon, T]$. (Since the initial condition is well-posed, the optimal final time T is positive.) On the other hand, if $p(T) > q(T)$, then we can always add another small interval $(T, T + \varepsilon]$ with the control $u = 0$ without violating any of the constraints, and p will decrease along this interval if ε is small enough. Thus, at the final time, necessarily $p(T) = q(T)$. If now $y(T) < A$, then we can still add a small piece of a trajectory for $u = a$ over some interval $[0, \varepsilon]$. Since $p(T) = q(T)$, we have that $\dot{p}(T) = 0$, and in $\mathscr{D}$, on the diagonal it also holds that $\dot{q}(T) < 0$. This implies that the trajectory enters the region $p > q$ where the tumor volume p is still decreasing further. Hence T was not the optimal time.   □

The Hamiltonian $H(\lambda(t), p(t), q(t), u)$ is minimized over the interval $[0, a]$ as a function of u by the optimal control $u(t)$. The *switching function* Φ is given by

$$\Phi(t) = \lambda_3 - \lambda_2(t)\gamma q(t), \tag{6.19}$$

and we have that

$$u(t) = \begin{cases} 0 & \text{if } \Phi(t) > 0, \\ a & \text{if } \Phi(t) < 0, \end{cases} \tag{6.20}$$

while the control can be singular on an interval I where the switching function vanishes identically. For this problem, singular controls indeed are optimal, and a surface of singular trajectories in the state space M determines the structure of optimal controls.

6.2.2 Singular Control and Singular Arcs

We first compute explicit formulas for the singular control and singular arc. Note that if an optimal control is singular over an interval I, then $\lambda_3 > 0$, and also $\lambda_2(t)$ is positive for $t \in I$. Write the state as $z = (p, q, y)^T$; then the drift f and control vector field g are given by

$$f(z) = \begin{pmatrix} -\xi p \ln\left(\dfrac{p}{q}\right) \\ bp - \left(\mu + dp^{\frac{2}{3}}\right) q \\ 0 \end{pmatrix}, \qquad g(z) = \begin{pmatrix} 0 \\ -\gamma q \\ 1 \end{pmatrix}, \tag{6.21}$$

and their first- and second-order Lie brackets are

$$[f,g](z) = \gamma p \begin{pmatrix} \xi \\ -b \\ 0 \end{pmatrix}, \qquad [g,[f,g]](z) = -\gamma^2 b p \begin{pmatrix} 0 \\ 1 \\ 0 \end{pmatrix}, \qquad (6.22)$$

and

$$[f,[f,g]](z) = \gamma p \begin{pmatrix} \xi^2 + \xi b \dfrac{p}{q} \\ \xi b \ln\left(\dfrac{p}{q}\right) + \xi \left(\dfrac{2}{3} d \dfrac{q}{\sqrt[3]{p}} - b\right) - \left(\mu + dp^{\frac{2}{3}}\right) b \\ 0 \end{pmatrix}. \qquad (6.23)$$

If u is singular on an interval I, then all the derivatives of the switching function Φ vanish on I (cf. Sect. 2.8), and we thus have that

$$\dot{\Phi}(t) = \langle \lambda(t), [f,g](z(t)) \rangle \equiv 0,$$
$$\ddot{\Phi}(t) = \langle \lambda(t), [f + ug, [f,g]](z(t)) \rangle \equiv 0.$$

Here, since λ_2 is positive along a singular arc,

$$\langle \lambda(t), [g,[f,g]](z(t)) \rangle = -\lambda_2(t)\gamma^2 b p(t) < 0,$$

and thus singular controls are of order 1, the strengthened Legendre–Clebsch condition is satisfied, and the singular control can be expressed in the form

$$u_{\text{sing}}(t) = -\frac{\langle \lambda(t), [f,[f,g]](z(t)) \rangle}{\langle \lambda(t), [g,[f,g]](z(t)) \rangle}.$$

The vector fields g, $[f,g]$, and $[g,[f,g]]$ are linearly independent everywhere, and a direct calculation verifies that we can express $[f,[f,g]]$ in the form

$$[f,[f,g]] = \left(\xi + b\frac{p}{q}\right)[f,g] + \psi[g,[f,g]]$$

with

$$\psi = \psi(p,q) = -\frac{1}{\gamma}\left(\xi \ln\left(\frac{p}{q}\right) + b\frac{p}{q} + \frac{2}{3}\xi\frac{d}{b}\frac{q}{p^{\frac{1}{3}}} - \left(\mu + dp^{\frac{2}{3}}\right)\right).$$

Thus we have the following result:

Proposition 6.2.1. *If the control u is singular on an open interval (α, β) with corresponding trajectory (p, q), then the singular control is given in feedback form as*

$$u_{sing}(t) = \frac{1}{\gamma}\left(\xi \ln\left(\frac{p(t)}{q(t)}\right) + b\frac{p(t)}{q(t)} + \frac{2}{3}\xi\frac{d}{b}\frac{q(t)}{p^{\frac{1}{3}}(t)} - \left(\mu + dp^{\frac{2}{3}}(t)\right)\right). \qquad (6.24)$$

However, the singular control can be optimal only on a thin set. Since the terminal time T is free, we have $H \equiv 0$, and thus, along a singular arc, in addition to $\langle \lambda, g \rangle \equiv 0$ and $\langle \lambda, [f, g] \rangle \equiv 0$, we also must have that $\langle \lambda, f \rangle \equiv 0$. Since the multiplier λ is nonzero (recall that λ_2 must be positive along a singular arc), these three vector fields must be linearly dependent. For this example, g is always linearly independent of f and $[f, g]$, and thus this simply becomes the set where f and $[f, g]$ are linearly dependent. This relation determines a curve in (p, q)-space given as

$$bp - dp^{\frac{2}{3}}q - bp\ln\left(\frac{p}{q}\right) - \mu q = 0.$$

It is possible to desingularize this equation with a blowup in the variables of the form $p = xq$, $x > 0$, and in terms of the projective coordinate x we have that

$$\mu + dp^{\frac{2}{3}} = bx(1 - \ln x), \qquad (6.25)$$

which clearly brings out the geometry of the singular curve. The quotient $\frac{q}{p}$ is proportional to the *endothelial density*, which is used to replace the carrying capacity of the vasculature as a variable. As it turns out, the singular curve and its corresponding singular control can be expressed solely in terms of the variable x. In these variables, Eq. (6.25) can be rewritten in the form

$$p^2 + \varphi(x)^3 = 0 \qquad \text{with} \qquad \varphi(x) = \frac{bx(\ln x - 1) + \mu}{d}.$$

The function φ is strictly convex with a minimum at $x = 1$ and minimum value $\frac{\mu - b}{d}$. In particular, if $\mu \geq b$, then this equation has no positive solutions, and thus no admissible singular arc exists. The case $\mu < b$ that we consider here is the only medically relevant case. For $\mu = 0$, the zeros of φ are given by $x_1^* = 0$ and $x_2^* = e$, and φ is negative on the interval $(0, e)$. In general, for $\mu > 0$, we have $\varphi(0) = \frac{\mu}{d} = \varphi(e)$ and the zeros x_1^* and x_2^* satisfy $0 < x_1^* < 1 < x_2^* < e$. We thus have the following result:

Proposition 6.2.2. *The singular curve $\mathscr{S}$ lies entirely in the sector $\{(p, q) : x_1^* q < p < x_2^* q\}$, where x_1^* and x_2^* are the unique zeros of the equation $\varphi(x) = 0$ and satisfy*

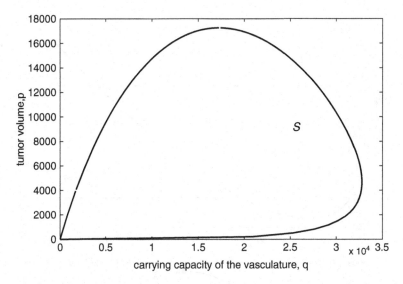

Fig. 6.6 The singular curve $\mathscr{S}$ for $\xi = 0.084, b = 5.85, d = 0.00873$, and $\mu = 0.02$

$0 \leq x_1^* < 1 < x_2^* \leq e$. *In the variables* (p,x), *with* $x = \frac{p}{q}$, *the singular curve can be parameterized in the form*

$$p^2 = \left(\frac{bx(1 - \ln x) - \mu}{d} \right)^3 \qquad \text{for} \quad x_1^* < x < x_2^*. \tag{6.26}$$

Figure 6.6 shows the singular curve for the following parameter values: $\xi = 0.084$ per day, $b = 5.85$ per day, $d = 0.00873$ per mm^2 per day, $\mu = 0.02$ per day. These values are based on data from [114], but we use them only for numerical illustrations; our results are generally valid.

The next result gives an equivalent expression for the singular control along the singular arc in terms of the scalar variable x alone. This relation is valid only on the singular arc $\mathscr{S}$, but it allows us to determine the admissible part of the singular arc, that is, the portion of $\mathscr{S}$ where the singular control takes values in the interval $[0, a]$.

Proposition 6.2.3. *Along the singular curve* $\mathscr{S}$, *the singular control can be expressed as a function of the scalar variable* $x = \frac{p}{q}$ *in the form*

$$\Psi(x) = \frac{1}{\gamma} \left[\left(\frac{1}{3}\xi + bx \right) \ln x + \frac{2}{3}\xi \left(1 - \frac{\mu}{bx} \right) \right]. \tag{6.27}$$

There exists exactly one connected arc on the singular curve $\mathscr{S}$ *along which the control is admissible, i.e., satisfies the bounds* $0 \leq \Psi \leq a$. *This arc is defined over an interval* $[x_\ell^*, x_u^*]$, *where* x_ℓ^* *and* x_u^* *are the unique solutions to the equations* $\Psi(x_\ell^*) = 0$ *and* $\Psi(x_u^*) = a$. *These values satisfy* $x_1^* < x_\ell^* < x_u^* < x_2^*$.

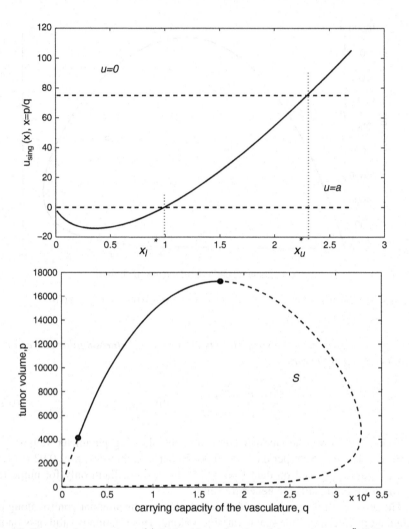

Fig. 6.7 The singular control $u_{\text{sing}}(x)$ is plotted as a function of the quotient $x = \frac{p}{q}$ (*left*) and the singular curve $\mathscr{S}$ is plotted in the (p,q)-plane with the admissible part (where the singular control takes values in the interval $[0, a]$) marked by the solid portion of the curve (*right*). Away from this solid segment, the singular control is either negative or exceeds the maximum allowable limit $a = 75$. In order to better visualize tumor reductions, p is plotted vertically and q horizontally

Figure 6.7 gives a plot of the singular control and the admissible portion of the petal-like singular curve $\mathscr{S}$ for $a = 75$ marked as a solid curve for the same numerical values as before. For the control, we use $\gamma = 0.15$ kg per mg of dose per day with concentration in mg of dose per kg.

Proof. In the variables p and x, the singular control is given by

$$u_{\text{sing}}(t) = \frac{1}{\gamma}\left(\xi \ln x(t) + bx(t) + \frac{2}{3}\xi \frac{dp(t)^{2/3}}{bx(t)} - \left(\mu + dp(t)^{2/3}\right)\right).$$

Along the singular arc, we have that $p^{2/3} = -\varphi(x)$, and thus we obtain the singular control as a feedback function of x alone, $u_{\text{sing}}(t) = \Psi(x(t))$, in the form

$$\Psi(x) = \frac{1}{\gamma}\left(\xi \ln x + bx + \frac{2}{3}\xi \frac{bx(1 - \ln x) - \mu}{bx} - bx(1 - \ln x)\right)$$

$$= \frac{1}{\gamma}\left[\left(\frac{1}{3}\xi + bx\right)\ln x + \frac{2}{3}\xi\left(1 - \frac{\mu}{bx}\right)\right].$$

Note that $\lim_{x \searrow 0} \Psi(x) = -\infty$ and $\lim_{x \to \infty} \Psi(x) = +\infty$. Now

$$\Psi'(x) = \frac{1}{\gamma}\left[b(\ln x + 1) + \frac{1}{3}\xi\left(\frac{1}{x} + 2\frac{\mu}{bx^2}\right)\right],$$

$$\Psi''(x) = \frac{1}{\gamma x^3}\left(bx^2 - \frac{1}{3}\xi x - \frac{4}{3}\xi\frac{\mu}{b}\right),$$

and the second derivative has a unique positive zero at

$$x_* = \frac{1}{6}\frac{\xi}{b}\left(1 + \sqrt{1 + 48\frac{\mu}{\xi}}\right).$$

It follows that Ψ is strictly concave for $0 < x < x_*$ and strictly convex for $x > x_*$. If the function Ψ has no stationary points, then Ψ is strictly increasing, and thus, as claimed, there exists a unique interval $[x_\ell^*, x_u^*]$ on which Ψ takes values in $[0, a]$ and the limits are the unique solutions of the equations $\Psi(x) = 0$ and $\Psi(x) = a$, respectively. The same holds if Ψ has a unique stationary point at x_*. In the remaining case, it follows from the convexity properties that Ψ has a unique local maximum at $\tilde{x}_1 < x_*$ and a unique local minimum at $\tilde{x}_2 > x_*$. It suffices to show that Ψ is negative at its local maximum. For this, as before, implies that Ψ is strictly increasing when it is positive. Suppose now that $\Psi'(\tilde{x}) = 0$. Then

$$-b\ln\tilde{x} = b + \frac{1}{3}\xi\left(\frac{1}{\tilde{x}} + 2\frac{\mu}{b\tilde{x}^2}\right) > 0$$

and thus

$$\Psi(\tilde{x}) = \frac{1}{\gamma}\left[\left(\frac{1}{3}\xi + b\tilde{x}\right)\left(-1 - \frac{1}{3}\frac{\xi}{b}\left(\frac{1}{\tilde{x}} + 2\frac{\mu}{b\tilde{x}^2}\right)\right) + \frac{2}{3}\xi\left(1 - \frac{\mu}{b\tilde{x}}\right)\right]$$

$$= \frac{1}{\gamma}\left[-b\tilde{x} - \frac{1}{9}\frac{\xi^2}{b}\left(\frac{1}{\tilde{x}} + 2\frac{\mu}{b\tilde{x}^2}\right) - \frac{4}{3}\frac{\xi\mu}{b\tilde{x}}\right] < 0.$$

Hence Ψ is negative at every stationary point.   □

6.2.3 A Family of Broken Extremals with Singular Arcs

Optimal controlled trajectories then need to be synthesized from the constant controls $u = 0$ and $u = a$ and the singular control. This has been done for the problem [A] in [160]. We briefly describe some features of this synthesis, but focus on the case that arises for initial conditions (p_0, q_0) with $p_0 < q_0$ when enough inhibitors are available to lower the tumor volume. This case is the most important and typical situation. Optimal controls are initially given at full dose, $u \equiv a$, until the singular arc is reached. For small tumor volumes, this point may lie where the singular control no longer is admissible, and in this case, optimal controls simply give all available inhibitors in one full dose segment. If the singular arc is reached in the section where the singular control is admissible, then at this point, optimal controls switch to the singular regime and follow the singular arc. In the case that inhibitors are plentiful, so that it would be possible to reach the lower saturation point where $u_{\mathrm{sing}} = a$, optimal controlled trajectories actually leave the singular arc prior to saturation and inhibitors are exhausted along another full dose segment for $u \equiv a$. However, this is rather the exception, and typically inhibitors are given along the singular arc until they are exhausted, $y = A$. At that time, the state of the system lies on the singular arc in the region $p < q$, and thus the tumor volume will still be decreasing for a while even as no more inhibitors are administered, $u = 0$, and the minimum tumor volume is realized as the corresponding trajectory crosses the diagonal in (p, q)-space, $p = q$. Consequently, for these initial conditions, optimal controlled trajectories are concatenations of the form **as0** consisting of a full dose bang arc followed by a singular segment and another bang arc when no inhibitors are given. The switching times are the unique times when the singular arc is reached and when all inhibitors have been exhausted, respectively, and are easily computed for given initial conditions.

We construct a C^ω-parameterized family of broken extremals that describes this situation and show that the associated value function satisfies the Hamilton–Jacobi–Bellman equation along the initial bang segment where the flow F is a diffeomorphism. Along the middle segment, when the flow follows the singular arc, the image collapses onto the singular surface $\mathscr{S}$ in the three-dimensional state-space, but this does not invalidate the reasoning, and solutions to the Hamilton–Jacobi–Bellman equation can be propagated along such a section. These considerations are local, but then, together with other parameterized families, they can be used to piece together a global synthesis. These considerations will be taken up in the next section. Arguments like the one given here generally *establish the crucial property that the value function associated with a memoryless synthesis of extremals is a solution of the Hamilton–Jacobi–Bellman equation on open regions that are covered diffeomorphically by the flow F*.

Let $(\bar{z}, \bar{u})$ be a controlled trajectory of the type **as0** with initial condition $z_0 = (p_0, q_0, 0)$ defined over an interval $[0, \overline{T}]$. More precisely, suppose the control is given by

$$\bar{u}(t) = \begin{cases} a & \text{for } 0 \le t \le \bar{t}_1, \\ u_{\text{sing}}(t) & \text{for } \bar{t}_1 < t \le \bar{t}_2, \\ 0 & \text{for } \bar{t}_2 < t \le \overline{T}, \end{cases}$$

where $\bar{t}_1$ is the time when the trajectory corresponding to the control $u \equiv a$ intersects the admissible segment of the singular arc, $\bar{t}_2$ is the time along the admissible singular arc when all inhibitors become exhausted, $y(\bar{t}_2) = A$, and $\overline{T}$ is the time when the trajectory corresponding to the control $u \equiv 0$ reaches the diagonal, $p(\overline{T}) = q(\overline{T})$. We denote the switching points by $\bar{z}_1 = \bar{z}(\bar{t}_1)$ and $\bar{z}_2 = \bar{z}(\bar{t}_2)$.

We start the construction by embedding this reference controlled trajectory into the obvious parameterized family of controlled trajectories of the same type and will only later verify that they are extremal. Since p denotes one of the state variables in this problem, here we use w to denote the parameters. Since this is a time-independent problem, the flow to be constructed is $F(t,w) = z(t,w)$ and the parameter set is two-dimensional. We simply vary the initial conditions in a neighborhood of the initial condition $z_0 = (p_0, q_0)$ and let $W \subset \mathbb{R}^2$ be a small neighborhood of $w_0 = z_0$. In principle, we shall shrink this neighborhood without mention whenever necessary for the argument. But in fact, for this construction to be valid, the parameter set W can be quite a large subset of $\{(p_0, q_0) : p_0 < q_0\}$ that needs to be restricted only due to ill-posed data (with an insufficient amount of inhibitors A) or when the concatenation structure changes due to saturation phenomena (which we do not describe here).

Denote the solutions of the dynamics for the constant controls $u \equiv a$, $u \equiv 0$ and the singular control $u = u_{\text{sing}}$ with initial condition z_0 at initial time t_0 by $\Phi^a(t; t_0, z_0)$, $\Phi^0(t; t_0, z_0)$, and $\Phi^{\text{sing}}(t; t_0, z_0)$, respectively. For example, we have that

$$\bar{z}_1 = \Phi^a(\bar{t}_1; 0, z_0) \qquad \text{and} \qquad \bar{z}_2 = \Phi^{\text{sing}}(\bar{t}_2; \bar{t}_1, \bar{z}_1).$$

Essentially, the argument in constructing a C^r-parameterized family of broken controlled trajectories simply is that if the flow of a C^r vector field intersects a codimension-1 embedded C^r submanifold transversally, then the times of intersection are given by a C^r function. This is simply a consequence of the implicit function theorem.

Lemma 6.2.2. *There exists a real-analytic function t_1 defined on W, $t_1 : W \to (0, \overline{T})$, $w \mapsto t_1(w)$, that satisfies $t_1(w_0) = \bar{t}_1$ and is such that for all $w \in W$, the points $\Phi^a(t_1(w); 0, w)$ describe the intersections of the trajectories corresponding to the constant controls $u \equiv a$ with the singular curve $\mathscr{S}$.*

Proof. While the flow F is in $\mathbb{R}^3$, the variable y accounts only for the amount of inhibitors that have been used and does not affect the two-dimensional geometry described in Sect. 6.2.2. The singular surface $\mathscr{S}$ simply becomes a vertical hypersurface in y that lies over the singular arc in (p,q)-space, and we use the same letter for both. Except for the upper and lower saturation points along the singular surface $\mathscr{S}$, it is easily verified that the full dose trajectories meet the admissible portion of the singular surface $\mathscr{S}$ transversally. In fact, this trajectory is tangential to the singular surface $\mathscr{S}$ precisely if the singular control is given by $u_{\text{sing}} = a$. By Proposition 6.2.2, $\mathscr{S}$ is an embedded C^ω-submanifold that can locally be described as the zero set of some real analytic function. That is, for every point $z \in \mathscr{S}$ there exist some neighborhood Z of z and a function $G : Z \to \mathbb{R}$ such that $\mathscr{S} \cap Z = \{z \in Z : G(z) = 0\}$. The composite mapping

$$\eta = G \circ \Phi^a : W \to \mathbb{R}, \quad \eta(t,w) = G(\Phi^a(t;0,w)),$$

is differentiable and satisfies $\eta(\bar{t}_1, w_0) = 0$. Furthermore, the transversal intersection is equivalent to

$$\frac{\partial \eta}{\partial t}(\bar{t}_1, w_0) = \nabla G(\bar{z}_1) \cdot (f(\bar{z}_1) + a g(\bar{z}_1)) \neq 0$$

and thus by the implicit function theorem, the equation $\eta(t,w) = 0$ can be solved in a neighborhood of $(\bar{t}_1, w_0)$ for t as a function of w by a real-analytic function. □

Thus the first switching can be described by a real-analytic function. The very same argument applies to the second and third switchings as well.

Lemma 6.2.3. *There exists a real-analytic function t_2 defined on W, $t_2 : W \to (0, \bar{T})$, $w \mapsto t_2(w)$, that satisfies $t_2(w) > t_1(w)$ for all $w \in W$, $t_2(w_0) = \bar{t}_2$, and is such that for $w \in W$ the points*

$$\Phi^{\text{sing}}(t_2(w); t_1(w), \Phi^a(t_1(w); 0, w))$$

describe the points on the singular surface $\mathscr{S}$ where the inhibitors are exhausted, i.e.,

$$y\left(\Phi^{\text{sing}}(t_2(w); t_1(w), \Phi^a(t_1(w); 0, w))\right) = A.$$

Proof. At the first switching point, $\bar{z}_1$, inhibitors are left. We are assuming—this is the particular scenario for which we construct this particular parameterized family of extremals—that these are used up fully and without saturation along the controlled reference trajectory at time $\bar{t}_2$. Thus, for a sufficiently small neighborhood W of w_0, the same holds for all initial conditions in W. The codimension-1 submanifold that the flow intersects now is simply the hyperplane determined by $y = A$. Since $\dot{y} = u_{\text{sing}}(t) > 0$, again trajectories meet this hyperplane transversally, and by the implicit function theorem, we once more can solve the equation $y\left(\Phi^{\text{sing}}(t; t_1(w), \Phi^a(t_1(w); 0, w))\right) = A$ by a real-analytic function of the parameter near the point $(\bar{t}_2, w_0)$. □

Lemma 6.2.4. *There exists a real-analytic function T defined on W, $T : W \to (0, \infty)$, $w \mapsto T(w)$, that satisfies $T(w) > t_2(w)$ for all $w \in W$, $T(w_0) = \bar{T}$, and is such that for $w \in W$, the endpoints*

$$\Xi(w) = \Phi^0(T(w); t_2(w), \Phi^{\text{sing}}(t_2(w); t_1(w), \Phi^a(t_1(w); 0, w)))$$

lie on the diagonal $p = q$.

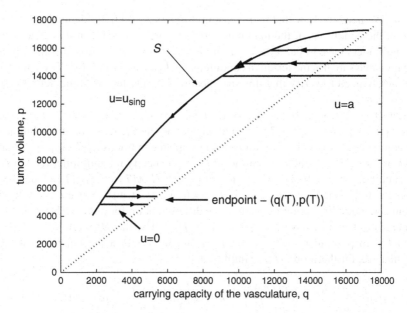

Fig. 6.8 A projection of the flow F for the parameterized family $\mathscr{E}$ of type **as0** into (q,p)-space

Proof. Along the third segment the variable y is constant, $y(t) \equiv A$, and thus we consider only p and q and the corresponding two-dimensional flow. On the diagonal, $\mathscr{D}_0 = \{(p,q) \in \mathscr{D} : p = q\}$, we have that $\dot{p} = 0$ and $\dot{q}$ is positive for $u = 0$ in the biologically relevant domain $\mathscr{D}$. Thus again the flow is transversal and the same argument applies.   □

Overall, we therefore have a C^{ω}-parameterized family $\mathscr{T}$ of broken controlled trajectories of the type **as0** with parameter set W and domain

$$D = \{(t,w) : 0 \equiv t_-(w) \leq t \leq T(w) = t_+(w)\}$$

(see Fig. 6.8). The parameterized cost for this family is given by the value of the coordinate p at the endpoint, $C(t,w) = \langle e_1, \Xi(w) \rangle$, and thus also is a real-analytic function of w. Since w represents the initial condition for the family of controlled trajectories, it also follows that the associated value $V^{\mathscr{E}} = V^{\mathscr{E}}(z)$ in the state space is a real-analytic function of the initial condition z_0.

It should be clear from the construction that this parameterized family $\mathscr{T}$ is extremal, and we verify that this is the case.

Theorem 6.2.1. *The parameterized family $\mathscr{T}$ gives rise to a C^{ω}-parameterized family $\mathscr{E}$ of broken extremals.*

Proof. We need to show that there exist multipliers $\lambda = \lambda(t, w)$ defined on D such that all the conditions of the maximum principle are satisfied. But it is clear, what the multipliers have to be: $\lambda_0(w) \equiv 1$ and $\lambda = \lambda(t, w)$ is the solution of the adjoint equation that satisfies the terminal conditions $\lambda_1(T(w), w) \equiv 1$ and $\lambda_2(T(w), w) \equiv 0$. We merely need to verify that this specification is consistent with the definition of the controls.

Let $D_1 = \{(t, w) : 0 \leq t \leq t_1(w)\}$, $D_2 = \{(t, w) : t_1(w) \leq t \leq t_2(w)\}$, and $D_3 = \{(t, w) : t_2(w) \leq t \leq T(w)\}$ and let $\lambda_1(t, w)$ and $\lambda_2(t, w)$ be the solution to the adjoint equation with $u \equiv 0$ on D_3 and terminal conditions $\lambda_1(T(w), w) \equiv 1$ and $\lambda_2(T(w), w) \equiv 0$. For the moment, we do not yet specify the multiplier λ_3. On D_3, the Hamiltonian $H = H(\lambda(t, w), z(t, w), 0)$ satisfies $H(\lambda(T(w), w), z(T(w), w), 0) = 0$ at the terminal time T and is given by $H = \langle \lambda(t, w), f(z(t, w)) \rangle$. Since λ is a solution to the corresponding adjoint equation, the differentiation formula in Theorem 2.8.2 remains valid, and we have that $\frac{d}{dt} H = \langle \lambda, [f, f] \rangle \equiv 0$, so that $H(\lambda(t, w), z(t, w), 0) \equiv 0$ on D_3. In particular, $\langle \lambda(t_2(w), w), f(z(t_2(w), w)) \rangle = 0$ at the junction with the singular arc. Furthermore, $H \equiv 0$ implies that

$$\lambda_1(t, w) \xi \frac{p(t, w)}{q(t, w)} \ln \left(\frac{p(t, w)}{q(t, w)} \right) = \lambda_2(t, w) \left(b \frac{p(t, w)}{q(t, w)} - \left(\mu + d p(t, w)^{\frac{2}{3}} \right) \right).$$

The trajectory $z(t, w)$ lies in the region $p > q$, and therefore the coefficient at $\lambda_1(t, w)$ is positive for $t_2(w) \leq t < T(w)$. Furthermore, in the region $\mathcal{D} = \{(p, q) : 0 < p < \bar{p}, \ 0 < q < \bar{q}\}$ we also have that

$$b \frac{p}{q} - \left(\mu + d p^{2/3} \right) > b - \mu - d p^{2/3} > b - \mu - d \bar{p}^{2/3} = 0,$$

and thus the coefficient at $\lambda_2(t, w)$ is positive as well. Since λ_1 and λ_2 cannot vanish simultaneously, it follows that $\lambda_1(t, w)$ and $\lambda_2(t, w)$ have the same sign for $t \in [t_2(w), T(w))$. Since $\lambda_1(T(w), w) \equiv 1$, both λ_1 and λ_2 are positive over $[t_2(w), T(w))$.

Along the singular arc $\mathscr{S}$, the vector fields f and $[f, g]$ are linearly dependent and thus we also have that $\langle \lambda(t_2(w), w), [f, g](z(t_2(w), w)) \rangle = 0$. The parameterized switching function is given by,

$$\Phi(t, w) = \langle \lambda(t, w), g(z(t, w)) \rangle = \lambda_3(w) - \lambda_2(t, w) \gamma q(t, w)$$

and we now define the constant multiplier λ_3 as

$$\lambda_3(w) = \lambda_2(t_2(w), w) \gamma q(t_2(w), w) > 0.$$

With this specification, we ensure that $\langle \lambda(t_2(w), w), g(z(t_2(w), w)) \rangle = 0$, and thus all the conditions required for the multiplier to have a singular junction at $t_2(w)$ are satisfied. We then define $\lambda(t, w)$ on D_2 as the solution to the adjoint equation for the singular control. The controlled trajectories $(z(t, w), u(t, w))$ and the multiplier

$\lambda(t,w)$ thus satisfy all the conditions of the maximum principle along singular controls on D_2, and in addition, since $\lambda_2(t,w) > 0$, the singular control is of order 1 and the strengthened Legendre–Clebsch condition for minimality is satisfied. Once more, $H(\lambda(t,w), z(t,w), u_{\text{sing}}(t,w)) \equiv 0$ is a direct corollary of the fact that the multiplier λ satisfies the adjoint equation. Finally, on D_1 we let $\lambda(t,w)$ be the solution to the adjoint equation for $u \equiv a$ with terminal conditions $\lambda(t_1(w), w)$.

These specifications define a multiplier $\lambda = \lambda(t,w)$ that satisfies the adjoint equation and the transversality conditions at the terminal time; furthermore,

$$H(\lambda(t,w), z(t,w), u(t,w)) \equiv 0,$$

and the switching function $\Phi(t,w)$ vanishes identically on D_2. It remains to show that the switching function is positive on the interior of D_3 and negative on the interior of D_1. This then implies that the minimization condition on the Hamiltonian is satisfied.

We first consider the domain D_3 at the terminal time and show that the switching function $\Phi(t,w)$ is positive on $\tilde{D}_3 = \{(t,w) : t_2(w) < t \leq T(w)\}$. In the definition of this parameterized family of extremals we are assuming that inhibitors are exhausted along the singular arc prior to saturation and thus the value of the singular control is strictly smaller than a. Since singular controls are of order 1, it follows from Proposition 2.8.4 that the switching function indeed is positive for $t > t_2(w)$ close enough to $t_2(w)$ so that a concatenation with $u = 0$ at $t_2(w)$ satisfies the conditions of the maximum principle locally. Furthermore, $\Phi(T(w), w) = \lambda_3(w) > 0$.

Lemma 6.2.5. *If $\Phi(\tau, w) = 0$ for some $\tau \in (t_2(w), T(w))$, then $\dot{\Phi}(\tau, w) > 0$.*

Thus the switching function can have at most one zero on the interval $(t_2(w), T(w))$. But this is not possible, since the values near both ends of the interval are positive. It remains to prove the lemma: on the interval $(t_2(w), T(w))$, trajectories lie in the region $p > q$, and there the vector fields f, g and the constant third coordinate vector field $h = (0,0,1)^T$ are linearly independent. Hence the Lie bracket $[f,g]$ can be written as a linear combination of these vector fields, say

$$[f,g](z) = \rho(z)f(z) + \sigma(z)g(z) + \zeta(z)h.$$

On $(t_2(w), T(w))$, the multiplier λ vanishes against f, and thus if the switching function has a zero at time τ, then we have that

$$\dot{\Phi}(\tau, w) = \lambda_3(w)\zeta(z(\tau, w)).$$

An elementary computation verifies that

$$\zeta(z) = \frac{b\left(\frac{p}{q}\right)\left(1 - \ln\left(\frac{p}{q}\right)\right) - \left(\mu + dp^{\frac{2}{3}}\right)}{\ln\left(\frac{p}{q}\right)}.$$

The denominator of σ is positive, since $p > q$ and the zero set of the numerator is exactly the locus where the vector fields f and $[f, g]$ are linearly dependent, i.e., the singular curve $\mathscr{S}$ (see Eq. (6.26)). Under our assumption that $b > \mu$, the numerator is positive inside the singular loop and thus $\dot{\Phi}(\tau, w) > 0$. This proves the lemma and thus verifies that $\Phi(t, w)$ is positive on $\tilde{D}_3 = \{(t, w) : t_2(w) < t \le T(w)\}$. Hence the minimizing control is $u \equiv 0$.

The situation is different on the domain D_1. As above, it follows from Proposition 2.8.4 that the switching function indeed is negative for $t < t_1(w)$ close enough to $t_1(w)$, so that a concatenation with $u = a$ at $t_1(w)$ satisfies the conditions of the maximum principle locally. But now, in principle, it is possible that there exists another switch to the control $u = 0$ as trajectories are integrated backward. This happens if the junction points are too close to the upper saturation point on the singular arc. Indeed, near this saturation point, a local synthesis of optimal controls takes the form **0as0** (also, see Sect. 7.5) and thus one additional switching is possible if we integrate the system and adjoint equations backward from the singular junctions at time $t_1(w)$. However, this is not the case if these singular junctions lie at a sufficient distance to the saturation point. Thus, here we terminate this particular parameterized family as such a switching is encountered. In this construction, however, we simply assume that this is not the case along the reference controlled trajectory $(\bar{z}, \bar{u})$, and thus this will also be the case in a sufficiently small neighborhood W of the initial condition. Consequently, all the controlled trajectories in the family $\mathscr{T}$ are extremals.

It remains to verify the transversality condition (5.13) in Definition 5.2.4: Let

$$\bar{\xi}(w) = z(T(w), w) = (p(T(w), w), p(T(w), w), A)^T$$

and

$$\bar{\lambda}(w) = \lambda(T(w), w) = (1, 0, \lambda_3(w))$$

denote the state and multipliers at the terminal time $T(w)$ and denote the cost at the terminal points by $\bar{\gamma}(w) = C(T(w), w) = p(T(w), w)$. Since extremals are normal, $\lambda_0(w) \equiv 1$, and since the system has a free terminal time, $\bar{h}(w) = H(\bar{\lambda}(w), \bar{\xi}(w), 0) \equiv 0$, the transversality condition (5.13) required for the shadow-price lemma to hold takes the form

$$\bar{\lambda}(w) \frac{\partial \bar{\xi}}{\partial w}(w) = \lambda_0(w) \frac{\partial \bar{\gamma}}{\partial w}(w) + \bar{h}(w) \frac{\partial T}{\partial w}(w) = \frac{\partial \bar{\gamma}}{\partial w}(w).$$

This simply is the transversality condition of the maximum principle, and it is satisfied by construction:

$$\bar{\lambda}(w) \frac{\partial \bar{\xi}}{\partial w}(w) = (1, 0, \lambda_3(w)) \cdot \frac{\partial}{\partial w} \begin{pmatrix} p(T(w), w) \\ p(T(w), w) \\ A \end{pmatrix} = \frac{\partial}{\partial w} p(T(w), w) = \frac{\partial \bar{\gamma}}{\partial w}(w).$$

By Corollary 5.2.1, this transversality condition propagates along the subdomain D_3 to the states $\tilde{\xi}_2(w) = z(t_2(w), w)$, costates $\tilde{\lambda}_2(w) = \lambda(t_2(w), w)$, and cost $\tilde{\gamma}_2(w) = C(t_2(w), w)$. Since these define the terminal values for the second subdomain D_2, the transversality condition is also satisfied on D_2 and it analogously propagates onto D_1 and thus is valid on all of D. This concludes the construction. $\square$

6.2.4   Analysis of the Corresponding Flow and Value Function

The flow of trajectories corresponding to this parameterized family of trajectories is given by $F : D \to M$, $(t, w) \mapsto F(t, w) = z(t, w)$. Let $F_i : D_i \to G_i = F(D_i)$, $i = 1, 2, 3$, denote the restrictions of the flow to the three closed subdomains D_1, D_2, and D_3 and let $\tilde{D}_1 = \{(t, w) : 0 \le t < t_1(w)\}$, $\tilde{D}_2 = \{(t, w) : t_1(w) < t < t_2(w)\}$, and $\tilde{D}_3 = \{(t, w) : t_2(w) < t \le T(w)\}$ denote the relatively open subsets in the domain D. Also let $M_- = \{(0, w) : w \in W\}$, $M_1 = \{(t_1(w), w) : w \in W\}$, $M_2 = \{(t_2(w), w) : w \in W\}$, and $M_+ = \{(T(w), w) : w \in W\}$ denote the parametrizations of the initial conditions, the switching surfaces, and the terminal set in the parameter space, respectively, and denote their images under the flow F by N_-, N_1, N_2, and N_+, respectively.

Since the initial control $u \equiv a$ is constant, it follows from the uniqueness of solutions to an ODE that the flow F_1 is a diffeomorphism (with an appropriate extension at the transversal boundary segments defined by the initial time $t_-(w) \equiv 0$ and the first switching time $t_1(w)$). Thus $\tilde{G}_1 = F(\tilde{D}_1)$ is an open set with boundary segments given by N_- and $N_1 \subset \mathcal{S}$. We can therefore define the value on the set G_1 in the state space as $V^{\mathcal{E}} : G_1 \to \mathbb{R}$, $z \mapsto V^{\mathcal{E}}(z) = C \circ F_1^{-1}$, and because the manifolds N_- and N_1 are crossed transversally by the flow, $V^{\mathcal{E}}$ extends as a C^1-diffeomorphism into a neighborhood of $\tilde{G}_1$. By construction, the shadow price lemma is valid on $\tilde{D}_1$, and it thus follows from the version of Theorem 5.2.1 for a time-independent problem that $V^{\mathcal{E}}$ is a real-analytic solution of the Hamilton–Jacobi–Bellman equation that extends to the boundary segments N_- and N_1.

Corollary 6.2.1. *The value function $V^{\mathcal{E}}$ is a real-analytic solution to the Hamilton–Jacobi–Bellman equation on G_1.*   ∎

Over the second domain, D_2, the flow collapses onto the singular surface $\mathcal{S}$ (see Fig. 6.8) and then remains on lower-dimensional sets as $y \equiv A$ on D_3. However, this does not invalidate that the HJB equation is valid on the first part where the flow is a diffeomorphism. Since $\mathcal{S}$ is a thin set, it does not make sense to consider the Hamilton–Jacobi–Bellman equation for these values, and, as we shall see next, this is not needed.

Generally, given a parameterized family of broken extremals, the combined flow is easily analyzed, even if it collapses onto lower-dimensional sets as here. The important feature is that it is still possible to show that the associated value function is a solution to the Hamilton–Jacobi–Bellman equation on initial segments where the flow is a diffeomorphism. This is the indispensable property needed if one wants to solve optimal control problems globally, and it generally also gives rise to local optimality results. For example, for the problem considered here, as already used a couple of times before, it follows from Proposition 2.8.4 that we can integrate the constant controls $u = 0$ and $u = a$ backward from every point of the singular arc for small times and satisfy the conditions of the maximum principle. In this way one can construct a local field of extremals that covers a neighborhood of the controlled reference trajectory $(\bar{z}, \bar{u})$, and indeed this proves the local optimality of the reference controlled trajectory. The technical difficulty lies in the fact that the associated value function no longer is differentiable along the portion of $\mathscr{S}$ covered by the flow. Using approximation procedures, the differentiability assumption on the value function can indeed be relaxed on a thin set (essentially, a locally finite union of embedded submanifolds of positive codimension), and optimality of the synthesis can be proven under these relaxed conditions. This was the key idea of Boltyansky that led him to the formulation of a *regular synthesis*, the construction of a globally optimal field of controlled trajectories. This will be our next topic. In this theory, one indispensable ingredient is that the associated value function is a differentiable solution to the Hamilton–Jacobi–Bellman equation on the complementary open subsets. This, of course, is a local property, and the techniques developed here and in Chap. 5 provide effective tools to verify this condition.

6.3 Sufficient Conditions for a Global Minimum: Syntheses of Optimal Controlled Trajectories

We now develop a verification theorem that allows us to prove the optimality of a family of extremal controlled trajectories that has been obtained by combining various local parameterized fields of extremals. The key feature is that the globally defined value function $V^{\mathscr{E}} : G \to \mathbb{R}$ needs to be differentiable only on a sufficiently rich open subset of G. We once more restrict our presentation to the problem [OC] defined in Chap. 5, but here we consider the time-independent formulation with free terminal time. Analogous constructions are valid for a time-varying problem. For the reader's convenience, we recall the problem formulation and the precise assumptions that will be made:

[OC] minimize the functional

$$\mathscr{I}(u;z) = \int_0^T L(x(s), u(s))ds + \varphi(x(T))$$

over all locally bounded Lebesgue measurable functions u that take values in the control set $U \subset \mathbb{R}^m$ a.e., $u : [0, T] \to U, t \mapsto u(t)$, for which the corresponding trajectory x,

$$\dot{x}(t) = f(x(t), u(t)), \quad x(0) = z,$$

satisfies the terminal constraint $x(T) \in N = \{x \in \mathbb{R}^n : \Psi(x) = 0\}$.

We use the variable z to denote the initial condition for the problem. It is assumed that

1. for each $u \in U$ fixed, the vector field $f(\cdot, u)$ and the Lagrangian $L(\cdot, u)$ are continuously differentiable on the state space M and the functions $(x, u) \mapsto f(x, u)$ and $(x, u) \mapsto \frac{\partial f}{\partial x}(x, u)$, respectively $(x, u) \mapsto L(x, u)$ and $(x, u) \mapsto \frac{\partial L}{\partial x}(x, u)$, are jointly continuous;
2. the terminal set N is an embedded k-dimensional submanifold, and $\varphi : N \to \mathbb{R}$ is continuously differentiable. In particular, the function φ thus extends to a continuously differentiable function into a full neighborhood of N in the ambient state space.

These assumptions will be in effect throughout this section.

Definition 6.3.1 (Synthesis). A synthesis for the optimal control problem [OC] over a domain $G \subset M$ consists of a family of controlled trajectories $\mathscr{S} = \{(x_z, u_z) : z \in G\}$ that start at the point $z \in G$, $x_z(0) = z$. A synthesis is called optimal (respectively, extremal) if each controlled trajectory in the family $\mathscr{S}$ is optimal (respectively, extremal).

Clearly, an optimal synthesis is extremal, and the aim of all of our developments is to give conditions that guarantee that an extremal synthesis that has been constructed through an analysis of necessary conditions for optimality is optimal. The following result, in a simplified formulation, provides such a statement and gives the main result of this section. We recall that a union of sets $S_i \subset M$, $i \in I$, is said to be *locally finite* if every compact subset K of M intersects only a finite number of the sets S_i.

Theorem 6.3.1 (Simple verification theorem). *Let $G \subset M$ be a domain with N in its boundary and suppose $V : G \cup N \to \mathbb{R}$ is a continuous function defined on G that satisfies $V(z) \leq \varphi(z)$ for $z \in N$. Suppose there exists a locally finite union of embedded submanifolds M_i, $i \in \mathbb{N}$, of positive codimensions such that the function V is continuously differentiable on the complement of these submanifolds in M, $M_g = M \setminus \cup_{i \in \mathbb{N}} M_i$, and satisfies the Hamilton–Jacobi–Bellman inequality*

$$\frac{\partial V}{\partial z}(z)f(z, u) + L(z, u) \geq 0 \qquad \text{for all} \quad z \in M_g \quad \text{and} \quad u \in U. \qquad (6.28)$$

Then, for every controlled trajectory (x, u) that starts at a point $z \in M$ whose trajectory lies in the region G over the interval $[0, T)$ and ends in N at time T, we have that

$$J(u) \geq V(z). \qquad (6.29)$$

Several comments about the theorem are in order. First of all, requiring that V be continuous makes for a smoother formulation, but is not necessary. In many practical situations, continuity will be satisfied, and thus it seemed worthwhile to give this simpler formulation separately. However, the proof of this result naturally leads to a more general setting for functions V that satisfy *Sussmann's weak continuity requirement* (see Definition 6.3.3 below), and we shall prove the result in this form. This will also include a weaker form of the inequality $V(z) \leq \varphi(z)$ at the terminal set. Note that the formulation above allows for the possibility that an optimal trajectory x might pass through a point $y = x(\tau) \in N$ in the terminal manifold without actually terminating at this point. This occurs if because of a high value $\varphi(y)$ of the penalty function, the objective can still be improved upon by continuing and steering the system over an interval $[\tau, T]$ into another terminal point $x(T)$ for which

$$\int_\tau^T L(x(s), u(s))ds + \varphi(x(T)) < \varphi(y).$$

In such a case, $V(y) < \varphi(y)$. Naturally, often these situations can easily be remedied by redefining the manifold N and deleting those portions. From a practical point of view, the problem formulation might have been a bad one to begin with, but it is no problem to allow for these scenarios in the verification theorem.

While these are technical aspects, the fundamental question is how the verification theorem relates to the optimality of controlled trajectories. Generally, the function V is defined as the value of the objective for some synthesis of controlled trajectories for initial conditions $z \in G$ that are constructed through an analysis of extremals, as has been the topic throughout most of this text, especially Chaps. 5 and 6. In fact, parameterized families of extremals, $\mathscr{E}$, and their associated value functions $V^\mathscr{E}$ naturally give rise to piecewise defined functions V as are needed in this verification theorem. Constructions like those carried out above in Sect. 6.2 can be used to establish the continuous differentiability of $V^\mathscr{E}$ on open subsets, and this, at the same time, guarantees that the Hamilton–Jacobi–Bellman equation is valid. Not only implies this that the inequality (6.28) is automatically satisfied, but it also gives a control $u = u_z$ in the parameterized family for which equality holds, $V(z) = J(u)$. In principle, the function V could come from an arbitrary selection of extremal controlled trajectories (x_z, u_z) that steer the points $z \in G$ into N, and it can even be allowed that there is more than one member of this family for some values $z \in G$ as long as they give the same value of the objective. This, for example, is of interest in cases like those considered in Sect. 5.5 when there exists a cut-locus of optimal trajectories and when this allows us to keep all the optimal controlled trajectories in the synthesis. Often, however, the function V comes from a unique specification in terms of what is called a *memoryless* synthesis.

Definition 6.3.2 (Memoryless synthesis). A synthesis $\mathscr{S} = \{(x_z, u_z) : z \in G\}$ for the optimal control problem [OC] over a domain $G \subset M$ is called memoryless if whenever (x_z, u_z) is a controlled trajectory defined on $[0, T]$ and $\tilde{z} = x(\tau)$ is a point on the trajectory for $\tau > 0$, then the controlled trajectory $(x_{\tilde{z}}, u_{\tilde{z}})$ in the family starting at the point $\tilde{z}$ is given by the restriction of the controlled trajectory (x_z, u_z) to the interval $[\tau, T]$.

While pursuing the construction of parameterized families of extremals, lower-dimensional submanifolds M_i where differentiability fails arise naturally. These include sections where the flow of the parameterized families collapse to follow lower-dimensional submanifolds (as is often the case for time-optimal bang-bang trajectories or along singular trajectories in small dimensions), but also may be composed of submanifolds where the value function is no longer continuous. This phenomenon typically is related to a lack of local controllability properties in the system, and we shall give an example for a time-optimal control problem in Sect. 7.2. Being able to include these lower-dimensional subsets gives the results a global nature. Clearly, if the set G is small, we have similar results for a relative minimum as before. But various parameterized families of extremals can simply be glued together as long as the resulting value function is continuous. For example, in a full solution for the problem considered in Sect. 6.2, the concatenation structure needs to be changed from **as0** to **asa0** near the point where the singular control saturates [160] and there is a loss of differentiability for the value function on a codimension-1 submanifold near the saturation point because of this phenomenon; but continuity of the combined value is automatic. Altogether, the construction of parameterized families of extremals canonically leads to the definition of functions V that not only satisfy these requirements, but at the same time realize the values $V(z)$. Thus these are *optimal collections of controlled trajectories over the domain G*.

Finally, our focus in this text is on establishing applicable criteria that allow us to prove the optimality of syntheses of controlled trajectories, but we do not address the related issue of when such syntheses exist. If one wants to establish results of this type in great generality, a considerably more technical framework needs to be employed as far as the assumptions on the data are concerned, and for this we refer the interested reader to the paper by Piccoli and Sussmann [196].

We now proceed to prove Theorem 6.3.1 and start with the observation that the result is trivial if V is differentiable everywhere on G. This simply is the argument considered earlier in Theorem 5.1.1: for any controlled trajectory that is defined over an interval $[0,T]$ we then have that

$$\frac{d}{dt}V(x(t)) = \frac{\partial V}{\partial z}(x(t))f(x(t),u(t)) \geq -L(x(t),u(t))$$

and thus

$$V(x(T)) - V(x(0)) \geq -\int_0^T L(x(s),u(s))ds.$$

Hence

$$V(x(0)) \leq \int_0^T L(x(s),u(s))ds + V(x(T))$$

$$\leq \int_0^T L(x(s),u(s))ds + \varphi(x(T)) = J(u).$$

This simple-minded reasoning breaks down if there exist lower-dimensional submanifolds along which V is not differentiable. In principle, the set of times when a given controlled trajectory x lies in such a submanifold can be an arbitrary closed subset of the interval $[0,T]$ (cf. Proposition 2.8.1), and it simply is no longer possible to differentiate the function V along such a trajectory. Dealing with this problem is a nontrivial technical matter. The idea, which goes back to Boltyansky's original approach of a so-called *regular synthesis* [41], is to perturb the given nominal trajectory in such a way that the resulting trajectory has a value that is close to that of the original trajectory but meets the manifolds where V is not differentiable only in a finite set of times. Then the argument above can be carried out piecewise, and the result follows in the limit as the approximations approach the given controlled trajectory. In Boltyansky's original definition, several, at times stringent, assumptions were made that guarantee these properties. Not all of these conditions are necessary, and here we carry out this approximation procedure following Sussmann's arguments that lead to his weaker continuity requirements on the value function V as well. We first dispose of some of the technical aspects of this construction starting with the following fundamental approximation result, which sets the stage for the argument. In the proofs below, we need to use somewhat more advanced, but still standard, results from measurable functions (the Vitali covering lemma) and differential geometry (Sard's theorem). For a proof of these statements, we refer the reader to the literature on the respective subjects.

Proposition 6.3.1. [239] *Given any controlled trajectory* (x,u) *defined over an interval* $[0,T]$ *that starts at* z, $x(0) = z$, *there exists a sequence of controlled trajectories* $\{(x_n, u_n)\}_{n \in \mathbb{N}}$, *also defined over the interval* $[0,T]$ *and starting at* z, $x_n(0) = z$ *for all* $n \in \mathbb{N}$, *with the following properties:*

1. *The controls* u_n *are piecewise constant, take values in a compact subset* K *of* U, *and* u_n *converges to* u *almost everywhere on* $[0,T]$;
2. *The trajectories* x_n *converge to* x *uniformly over the interval* $[0,T]$ *and*

$$\int_0^T L(x_n(s), u_n(s))ds \to \int_0^T L(x(s), u(s))ds.$$

The important property here is that the controls are *piecewise constant*, i.e., have only a finite number of switchings, and are not just so-called simple measurable controls with a finite number of values. In fact (e.g., see [112] and the argument given below), using the Vitali covering lemma, it can be shown that a function $u : \mathbb{R} \to \mathbb{R}$ is measurable if and only if there exists a sequence of piecewise constant functions u_n that converges to u almost everywhere on $\mathbb{R}$.

Proof. We first construct the approximating controls. Since (x,u) is an admissible controlled trajectory, the control u takes values in a compact subset K of U almost everywhere. Without loss of generality, we assume that this is true for all $t \in [0,T]$. Fix $n \in \mathbb{N}$ and select a finite subcover of this set K with balls B_i with center $u_i \in U$ and radius $\frac{1}{n}$, $i = 1, \ldots, r$, with the ordering arbitrary, but fixed. Then, for each time t, define a function v by choosing as value the point u_i with i the smallest index such that $u(t) \in B_i$,

$$v(t) = u_{i(t)}, \qquad \text{where} \quad i(t) = \min\{i : u(t) \in B_i\}.$$

The inverse image of the value u_i under v is given by the inverse image of the set where the control u takes values in B_i, but not in the sets B_j for $j = 1, \ldots, i-1$,

$$E_i = \{t \in [a,b] : v(t) = u_i\} = \{t \in [a,b] : u(t) \in B_i \setminus \cup_{j=1}^{i-1} B_j\},$$

and this set is measurable, since the control u is a measurable function. Hence v is measurable as well. However, v is only a simple function (i.e., its range is finite, but the sets on which the values are taken are only Lebesgue measurable), and we still need to approximate it with piecewise constant controls (i.e., the sets on which the values are taken are *intervals*). Such a function can be constructed by means of the Vitali covering lemma [257, Corollary (7.18)]. This result implies that if E is a set of finite Lebesgue measure in $\mathbb{R}$, then for every $\varepsilon > 0$, there exists a finite collection of disjoint intervals I_k, $k = 1, \ldots, \ell$, such that we have

$$\mu\left(E \setminus \cup_{k=1}^{\ell} I_k\right) < \varepsilon \qquad \text{and} \qquad \mu\left(\cup_{k=1}^{\ell} I_k\right) < (1+\varepsilon)\mu(E).$$

In particular, the Lebesgue measure of the symmetric difference of E and $\cup_{k=1}^{\ell} I_k$, $E \Delta \cup_{k=1}^{\ell} I_k$, the set of points that lie in one set but not the other, is of order ε:

$$(1+\varepsilon)\mu(E) > \mu\left(\cup_{k=1}^{\ell} I_k\right) = \mu\left(\cup_{k=1}^{\ell} I_k \cap E^c\right) + \mu\left(\cup_{k=1}^{\ell} I_k \cap E\right)$$

and

$$\mu(E) = \mu\left(E \cap \cup_{k=1}^{\ell} I_k\right) + \mu\left(E \setminus \cup_{k=1}^{\ell} I_k\right)$$

implies that $\mu\left(\cup_{k=1}^{\ell} I_k \cap E\right) \geq \mu(E) - \varepsilon$, which gives

$$\mu\left(\cup_{k=1}^{\ell} I_k \setminus E\right) = \mu\left(\cup_{k=1}^{\ell} I_k \cap E^c\right) < (1+\varepsilon)\mu(E) - \mu(E) + \varepsilon = \varepsilon(1+\mu(E)),$$

so that

$$\mu\left(E \Delta \cup_{k=1}^{\ell} I_k\right) = \mu\left(E \setminus \cup_{k=1}^{\ell} I_k\right) + \mu\left(\cup_{k=1}^{\ell} I_k \setminus E\right) < \varepsilon(2 + \mu(E)).$$

Applying this result to E_1, there exist disjoint intervals $I_{1,k}$, $k = 1, \ldots, \ell_1$, that are such that the Lebesgue measure of the symmetric difference between E_1 and the union of the intervals $I_{1,k}$ is less than $\frac{1}{2r(2+T)n}$. On these intervals, define the control $\tilde{u}_n$ to be constant given by the first value u_1. It then follows that $\|u(t) - \tilde{u}_n(t)\| < \frac{1}{n}$ for all $t \in E_1 \cap \left(\cup_{k=1}^{\ell_1} I_{1,k}\right)$, and the Lebesgue measure of the subset of $E_1 \cup \left(\cup_{k=1}^{\ell_1} I_{1,k}\right)$ where this is not valid is at most $\frac{1}{2rn}$. Then repeat this procedure for the set $E_2 \setminus \cup_{k=1}^{\ell_1} I_{1,k}$ and choose the new intervals $I_{2,k}$, $k = 1, \ldots, \ell_2$, to be disjoint from all previously constructed intervals $I_{1,k}$. Since there are only finitely many intervals, we can always drop possible overlaps when we define the control $\tilde{u}_n$ by the second value u_2. The procedure is finite and results in a finite collection of disjoint intervals on which the control is given by one of the values $u_1, \ldots, u_r$ with the property that $\|u(t) - \tilde{u}_n(t)\| < \frac{1}{n}$ with the possible exception of a set of measure $\frac{1}{2n}$. By construction, the complement of these disjoint intervals, itself a disjoint union of intervals, also has Lebesgue measure less than $\frac{1}{2n}$, and on these intervals we arbitrarily define the controls to be given by the first value u_1. Thus $\tilde{u}_n$ is a piecewise constant control that satisfies

$$\mu\left(\left\{t \in [a,b] : \|u(t) - \tilde{u}_n(t)\| \geq \frac{1}{n}\right\}\right) < \frac{1}{n}.$$

Thus the sequence $\{\tilde{u}_n\}_{n \in \mathbb{N}}$ converges in measure, and it is a well-known fact about measurable functions that there exists a subsequence, which we label $\{u_n\}_{n \in \mathbb{N}}$, that converges almost everywhere to u (see Appendix D). This verifies the first condition.

The second one follows from a standard argument about solutions to ordinary differential equations. The corresponding trajectory x_n is the solution to the differential equation $\dot{x} = f(x, u_n(t))$ with initial condition $x_n(0) = z$. The nominal controlled trajectory x defines a compact curve $C = \{x(t) \in M : 0 \leq t \leq T\}$ in the state space, and for $R > 0$, let M_R denote the compact tubular neighborhood of this curve consisting of all points that have distance less than or equal to R,

$$M_R = \{z \in M : \text{dist}(z, C) \leq R\}.$$

As before, let K be the compact set in the control set that contains the values of u. The function f is continuous on $M_R \times K$, and thus the infinity norm is bounded, say $N = \max_{M_R \times K} \|f(x, u)\|_\infty$. It then follows that whenever y is a point in $M_{\frac{1}{2}R}$, then for any admissible control u, the solution to the initial value problem $\dot{x} = f(x, u(t))$, $x_n(\tau) = y$, exists at least for an interval of length $\kappa = \frac{R}{2N}$ and lies in the compact set M_R. This simply is a consequence of the a priori bound

$$\|x(t) - y\|_\infty \leq \int_\tau^t \|f(x(s), u(s))\|_\infty ds \leq N |t - \tau|.$$

On the compact set $M_R \times K$, the function f also satisfies a Lipschitz condition in x (see Appendix B); that is, there exists a constant L such that for all points x and y in M_R and all control values $u \in K$ we have that

$$\|f(x, u) - f(y, u)\|_\infty \leq L \|x - y\|_\infty.$$

As long as the trajectory x_n lies in the compact set M_R, we can therefore estimate

$$\|x_n(t) - x(t)\|_\infty = \left\| \int_0^t f(x_n(s), u_n(s)) - f(x(s), u(s)) ds \right\|_\infty$$

$$\leq \int_0^t \|f(x_n(s), u_n(s)) - f(x(s), u(s))\|_\infty ds$$

$$\leq \int_0^t \|f(x_n(s), u_n(s)) - f(x(s), u_n(s))\|_\infty \, ds$$

$$+ \int_0^t \|f(x(s), u_n(s)) - f(x(s), u(s))\|_\infty \, ds$$

$$\leq \int_0^t [L\|x_n(s) - x(s)\|_\infty + b_n(s)] \, ds,$$

where b_n is a bounded function that converges to zero a.e. Defining

$$\Delta_n(t) = \max_{0 \leq s \leq t} \|x_n(s) - x(s)\|_\infty,$$

it follows that

$$\Delta_n(t) \leq \int_0^t [L\Delta_n(s) + b_n(s)] \, ds, \qquad \Delta_n(0) = 0,$$

and the Gronwall–Bellman inequality (Proposition B.1.1 in Appendix B) gives that

$$\Delta_n(t) \leq \int_0^t \left[L \left(\int_0^s b_n(r) dr \right) \exp\left(L(t-s) + b_n(s)\right) \right] ds.$$

Since the sequence $\{b_n\}_{n \in \mathbb{N}}$ is bounded and converges to zero a.e., the dominated convergence theorem implies that $\lim_{n \to \infty} \Delta_n(t) = 0$ as long as the sequence $\{x_n\}_{n \in \mathbb{N}}$ lies in M_R. But by the a priori bound, we know that this will hold at least on the interval $[0, \kappa]$. By choosing n large enough, we can then ensure that $\Delta_n(\kappa) \leq \frac{R}{2}$ for all $n \geq N_1$. But then, the a priori bound once more allows us to conclude that the trajectories will still remain in the larger compact set M_R up to time 2κ. Choosing $n \geq N_2$ will guarantee that $\Delta_n(2\kappa) \leq \frac{R}{2}$. Since we can always extend the domain by at least a constant length $\kappa > 0$, for n sufficiently large, all trajectories will exist on the full interval $[0, T]$ and $\lim_{n \to \infty} \Delta_n(T) = 0$, i.e., the trajectories converge uniformly to x on $[0, T]$.

The last statement is an immediate consequence of the dominated convergence theorem. Since the trajectories remain in the compact set $M_R \times K$, the integrand L is bounded and $L(x_n(s), u_n(s)) \to L(x(s), u(s))$ a.e. This proves the proposition. $\square$

In the proof of the verification theorem, we approximate a given controlled trajectory (x, u) by another controlled trajectory that meets the locally finite union of embedded submanifolds where the function V is not continuously differentiable in a finite set. The above proposition allows us to reduce this perturbation problem to the case of a constant control or smooth vector fields. But then this becomes a simple transversality argument: if the flow $\Phi_t^X(\cdot)$ of a vector field X at a point p is transversal to an embedded submanifold M_b of M that has positive codimension, then there exists an $\varepsilon > 0$ such that the flow does not lie in M_b for $0 < |t| \leq \varepsilon$ and thus t is an isolated point of the set of times when the flow meets M_b. One just needs to make sure that this will be the case for enough trajectories, and Sard's theorem (see below) allows us to formalize the argument. We recall that we assume that all manifolds are connected and second countable (cf. Definition 4.3.1).

Let X be a C^1-vector field on M and denote its flow by $\Phi_t^X(\cdot)$, i.e., $\Phi_t^X(p)$ is the point on the integral curve of X that starts at p at time 0. It follows from standard results on uniqueness of solutions to an ODE that this integral curve is defined on a maximal open interval $I_p = (t_-(p), t_+(p))$ and the domain Ω of the flow is given by $\Omega = \{(t,p) : t \in I_p\}$. Let M_b be an embedded submanifold of M with positive codimension and denote the set of all times for which the integral curve of X that passes through p at time 0 meets M_b by $\mathscr{T}_p$, i.e.,

$$\mathscr{T}_p = \{t \in I_p : \Phi_t^X(p) \in M_p\}.$$

The set $\mathscr{T}_p$ is *discrete* if for every time $t \in \mathscr{T}_p$ there exists an open interval that contains t and does not contain any other points from $\mathscr{T}_p$. Let G ("good") denote the set of all points $p \in M$ for which the set $\mathscr{T}_p$ is discrete, i.e.,

$$G = \{p \in M : \text{the set } \mathscr{T}_p \text{ is discrete}\}$$

and denote its complement by B ("bad"). Then we have the following result:

Proposition 6.3.2. *For any embedded submanifold M_b of positive codimension, G is a set of full measure in M; equivalently, B has Lebesgue measure zero. If the codimension of M_b is greater than 1, then the set $\mathscr{T}_p$ is empty for a set of full measure in M. For these points the trajectory starting at p does not intersect M_b at all.*

Proof. Let $\Omega_b = \Omega \cap (\mathbb{R} \times M_b)$ and let $F : \Omega_b \to M$, $(t,p) \mapsto q = F(t,p) = \Phi_t^X(p)$, denote the restriction of the flow to points in Ω_b. The mapping F is differentiable on Ω with the t-derivative of F given by the vector field X at the corresponding point,

$$\frac{\partial F}{\partial t}(t,p) = X\left(\Phi_t^X(p)\right) = X(q),$$

and the partial derivative with respect to p is given by the differential of the flow map for fixed time t, $\Phi_t^X(\cdot)$. Since

$$\Phi_t^X(p + \varepsilon v) = \Phi_t^X(p) + \varepsilon\left(\Phi_t^X(p)\right)_* v + o(\varepsilon),$$

we simply have that

$$\frac{\partial F}{\partial p}(t,p) \cdot v = \left(\Phi_t^X(p)\right)_* v$$

for any tangent vector $v \in T_p(M)$. The differential $\left(\Phi_t^X(p)\right)_*$ is a bijective linear mapping from the tangent space $T_p M$ to $T_q M$. In the mapping F, its domain is restricted to the subspace $T_p(M_b)$, and its image thus is the subspace

$$W_q = \left(\Phi_t^X(p)\right)_*(T_p(M_b))$$

of $T_q M$ of the same dimension as $T_p(M_b)$. Therefore, the rank of the differential of F at the point (t, p), $DF(t, p)$, is given by

$$\operatorname{rk} DF(t, p) = \begin{cases} \dim M_b + 1 & \text{if } X(q) \notin W_q, \\ \dim M_b & \text{if } X(q) \in W_q, \end{cases}$$

i.e., it increases by 1 whenever the vector $X(q) \in T_q(M)$ is transversal to the subspace W_q. Clearly, this is a geometrically obvious fact. We evaluate this condition at the point p by moving the vector $X(q)$ and the subspace W_q from q back to p along the flow of the vector field X. This simply gives

$$\left(\Phi^X_{-t}(q)\right)_* X(q) = \left(\Phi^X_{-t}(q)\right)_* X(\Phi^X_t(p)) = X(p)$$

(cf. Proposition 4.5.1 for $Y = X$), and since $\left(\Phi^X_{-t}(q)\right)_*$ is the inverse to $\left(\Phi^X_t(p)\right)_*$, the image of the subspace W_q under this transformation simply is the tangent space $T_p(M_b)$. Moving vectors along the flow of a vector field is a diffeomorphism, and thus the vector $X(p)$ is transversal to the manifold M_b at p if and only if $X(q)$ is transversal to W_q. We thus have that

$$\operatorname{rk} DF(t, p) = \begin{cases} \dim M_b + 1 & \text{if } X(p) \notin T_p(M_b), \\ \dim M_b & \text{if } X(p) \in T_p(M_b), \end{cases}$$

independently of t.

The singular set of a smooth mapping between manifolds is defined analogously to the definitions given in Sect. 5.4. If $g : M \to N$ is a smooth mapping between manifolds, then a point $q \in M$ is said to be a *regular point* if the differential at q maps the tangent space $T_q M$ onto the tangent space $T_{g(q)} N$ of N at $g(q)$, and it is called a *singular point* if the image of $T_q M$ under the differential is of dimension smaller than $\dim N$. Thus, it follows that all points of the mapping F are singular if the codimension of M_b is greater than 1, and if it is equal to 1, then still all points p where $X(p)$ is tangent to M_b are singular as well. Sard's theorem (for a proof, see [184]) establishes that the image of the singular set in N is small.

Theorem 6.3.2 (Sard). *If $g : M \to N$ is a smooth mapping between manifolds, then the image of the singular set S,*

$$S = \{q \in M : \operatorname{rk} Dg(q) < \dim N\},$$

has Lebesgue measure 0 in N, $\mu(g(S)) = 0$. ∎

In particular, if the codimension of M_b is greater than 1, then the set of all points $q \in M$ for which the trajectory of X through q intersects M_b has Lebesgue measure 0. For if $p = \Phi^X_t(q) \in M_b$, then also $q = \Phi^X_{-t}(p) = F(-t, p)$, and so q lies in the image $F(\Omega_b)$, which has measure zero.

So now assume that M_b is of codimension 1 and suppose the point $q \in F(\Omega_b) \subset M$ does not lie in the image of the singular set. We claim that the associated set $\mathscr{T}_q$ is discrete. Let $t \in \mathscr{T}_q$ and set $p = \Phi_t^X(q) \in M_b$. As above, we then also have that $q = \Phi_{-t}^X(p) = F(-t,p)$, and since q does not lie in the image of the singular set, it follows that $X(p) \notin T_p(M_b)$. Hence there exists an $\varepsilon > 0$ such that $\Phi_t^X(p) \notin M_b$ for $0 < |t| \leq \varepsilon$. This proves that the set $\mathscr{T}_q$ is discrete. Thus the set B is contained in the image of the singular set under F, and by Sard's theorem, this is a set of measure 0. $\square$

Note that since the maximal intervals of definition for integral curves are open, in any compact interval $[a,b] \subset I_p$ there can be only a finite number of times $t \in \mathscr{T}_p$. Hence, away from $p \in B$, integral curves have only a finite number of intersections with the manifold M_b over compact intervals and the same then holds for a locally finite union of such manifolds.

We now are ready to give the proof of Theorem 6.3.1. This proof does not use the continuity of the function V, but only the following three properties identified by H. Sussmann.

Definition 6.3.3 (Sussmann's weak continuity requirement). [196,229] Let $G \subset M$ be a domain with N in its boundary and suppose $V : G \cup N \to \mathbb{R}$ is a function defined on G. We say that the function V satisfies Sussmann's weak continuity requirement if the following three conditions are satisfied:

1. The function V is *lower semicontinuous* on G, i.e., whenever $\{z_n\}_{n \in \mathbb{N}}$ is a sequence of points from G that converges to some limit $z \in G$, then

$$V(z) \leq \liminf_{n \to \infty} V(z_n).$$

2. For every constant control $u \in U$, the function V has the *no-downward-jumps* property along the vector field $x \mapsto f(x,u)$, i.e., if γ is an integral curve of such a vector field defined on a compact interval $[a,b]$, $\gamma : [a,b] \to G, t \mapsto \gamma(t)$, then for all $s \in (a,b]$ we have that

$$\liminf_{h \searrow 0} V(\gamma(s-h)) \leq V(\gamma(s)).$$

3. For every point $n \in N$ and every $\varepsilon > 0$, there exists a nonempty open set $\Omega \subset G \cap B_\varepsilon(n)$ such that for all $z \in \Omega$ we have that

$$V(z) \leq \varphi(n) + \varepsilon.$$

Note that all these requirements are satisfied if the function V is continuous on $G \cup N$. It can be shown that the value function of the optimal control problem [OC] satisfies the first two of these conditions, and thus these are necessary conditions that need to be satisfied by any verification function V. While the second condition may look somewhat odd, this is a natural property that will be needed when approximating trajectories cross lower-dimensional submanifolds where the function V is not differentiable. The third condition allows us to handle discontinuities in the value function at the terminal manifold N, and it simply requires that for any potential target point $n \in N$, there exist sufficiently rich sets that are close such that the function V still has some upper continuity property along sequences converging to n along these sets. We shall give an example of a value function exhibiting this property in Sect. 7.2.

We then have the following generalization of the simple verification theorem stated earlier.

Theorem 6.3.3 (Main verification theorem). *[196] Let $G \subset M$ be a domain with N in its boundary and suppose $V : G \cup N \to \mathbb{R}$ is a function defined on G that satisfies Sussmann's weak continuity condition. Suppose there exists a locally finite union of embedded submanifolds M_i, $i \in \mathbb{N}$, of positive codimensions such that the function V is continuously differentiable on the complement of these submanifolds in M, $M_g = M \setminus \cup_{i \in \mathbb{N}} M_i$, and satisfies the Hamilton–Jacobi–Bellman inequality*

$$\frac{\partial V}{\partial z}(z) f(z, u) + L(z, u) \geq 0 \qquad \text{for all} \quad z \in M_g \quad \text{and} \quad u \in U. \tag{6.30}$$

Then, for every controlled trajectory (x, u) that starts at a point $z \in M$ whose trajectory lies in the region G over the interval $[0, T)$ and ends in N at time T, we have that

$$J(u) \geq V(z). \tag{6.31}$$

Proof. Let (x, u) be an arbitrary nominal controlled trajectory defined over an interval $[0, T]$ that starts at $z \in M$, ends in N at time T, and is such that the trajectory x lies in the region G over the interval $[0, T)$. For any $\varepsilon > 0$, by the third property in the definition of the weak continuity requirement, there exists a nonempty open set $\Omega \subset G \cap B_\varepsilon(x(T))$ such that for all $q \in \Omega$ we have that

$$V(q) \leq \varphi(x(T)) + \varepsilon.$$

In a first step, we simply perturb the terminal condition to lie in Ω (see Fig. 6.9). Using the nominal control u, it follows from the continuous dependence of a solution of a differential equation on initial data (which remains valid for solutions defined in the Carathéodory sense) that for ε small enough, the solution $x_q = x(\cdot; q)$ to the terminal value problem

$$\dot{x} = f(x, u(t)), \qquad x(T) = q,$$

exists on the full interval $[0, T]$. It suffices to show that there exists a point $q \in \Omega$ such that

$$V(x_q(0)) \leq \varphi(x(T)) + \int_0^T L(x_q(s), u(s)) ds + \varepsilon. \tag{6.32}$$

Fig. 6.9 Perturbation
argument near the terminal
manifold N

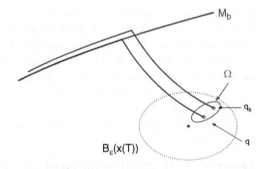

For since $\varepsilon > 0$ is arbitrary, there exists a sequence of points q_k, $k \in \mathbb{N}$, that satisfy this inequality with $\varepsilon = \frac{1}{k}$ and thus, in particular, $q_k \to x(T)$. But as $q_k \to x(T)$, the solutions $x_{q_k}(t)$ uniformly converge to the nominal trajectory $x(t)$ on the interval $[0,T]$. These trajectories lie in a compact set in the state space, and the admissible control u also takes values in a compact subset K of U. Since the Lagrangian L is continuous, it follows that $L(x_{q_k}(t),u(t)) \to L(x(t),u(t))$ as $q_k \to x(T)$, and thus, by the dominated convergence theorem, we have that

$$\lim_{k \to \infty} \int_0^T L(x_{q_k}(t),u(t))dt = \int_0^T L(x(t),u(t))dt.$$

Since V is lower semicontinuous, taking the limit as $k \to \infty$ in Eq. (6.32), we obtain the desired conclusion:

$$V(z) = V(x(0)) \le \lim_{k \to \infty} \inf V(x_{q_k}(0)) \le \lim_{k \to \infty} \left(\varphi(x(T)) + \int_0^T L(x_{q_k}(s),u(s))ds + \frac{1}{k} \right)$$

$$= \varphi(x(T)) + \int_0^T L(x(s),u(s))ds = J(u).$$

It thus remains only to verify Eq. (6.32).

Using Proposition 6.3.1, approximate the nominal control u by a sequence of piecewise constant controls u_n such that u_n converges to u a.e. on $[0,T]$. For n large enough, the corresponding trajectories $x_{n,q}$ that satisfy the terminal condition $x_{n,q}(T) = q$ exist on $[0,T]$ and also converge uniformly on this interval to the solution x_q (see Fig. 6.9). But these trajectories may intersect the bad set B, $B = \cup_{i \in \mathbb{N}} M_i$, for uncountably many times t, and thus the point q needs to be chosen to make sure that these trajectories meet B only for at most a finite number of times. Fix the index n, and dropping the dependence on this index in our notation, suppose the final interval is given by $[\tau, T]$ and denote the constant control on this interval by v. It then follows from Proposition 6.3.2 that the set of terminal conditions $q \in \Omega$ that have the property that the trajectory corresponding to the constant control v and ending at q meets a particular manifold M_i for a discrete set of times is a set of full measure in Ω. In particular, for such a point q, the set of times that lie in $[\tau, T]$ is finite. Since we are assuming that the union of the manifolds M_i is locally finite, it follows that there exists a set of full measure in Ω for which the trajectories ending at one of those points meet B only a finite number of times t_i, $\tau < t_1 < \cdots < t_r < T$. Without loss of generality we can also assume that τ and T are not such times, but write $\tau = t_0$ and $t_{r+1} = T$. On each of these intervals, the trajectory $x_{q,n}$ lies in the open set M_g, and thus the standard Hamilton–Jacobi–Bellman argument applies: for $\delta > 0$, the function V is differentiable over the interval $[t_i + \delta, t_{i+1} - \delta]$ along the trajectory $x_{n,q}$, and it follows from Eq. (6.30) that

$$\frac{d}{dt} V(x_{n,q}(t)) = \frac{\partial V}{\partial z}(x_{n,q}(t))f(x_q(t),v) \geq -L(x_{n,q}(t),v),$$

which gives

$$V(x_{n,q}(t_i + \delta)) \leq V(x_{n,q}(t_{i+1} - \delta)) + \int_{t_i+\delta}^{t_{i+1}-\delta} L(x_{n,q}(s),v)ds.$$

Taking the limit $\delta \searrow 0$, we have that

$$x_{n,q}(t_i + \delta) \to x_{n,q}(t_i) \qquad \text{and} \qquad x_{n,q}(t_{i+1} - \delta) \to x_{n,q}(t_{i+1}),$$

and the integral converges to $\int_{t_i}^{t_{i+1}} L(x_{n,q}(s),v)ds$. Let

$$V(x_{n,q}^+(t_i)) = \lim_{\delta \searrow 0} \inf V(x_{n,q}(t_i + \delta)) \quad \text{and} \quad V(x_{n,q}^-(t_{i+1})) = \lim_{\delta \searrow 0} \inf V(x_{n,q}(t_{i+1} - \delta))$$

denote the lower limits, respectively, of the function V along this trajectory as δ decreases to zero. We then have that

$$V(x_{n,q}^+(t_i)) \leq V(x_{n,q}^-(t_{i+1})) + \int_{t_i}^{t_{i+1}} L(x_{n,q}(s),v)ds,$$

or equivalently,

$$V(x_{n,q}^+(t_i)) - V(x_{n,q}^-(t_{i+1})) \leq \int_{t_i}^{t_{i+1}} L(x_{n,q}(s),v)ds.$$

Adding these inequalities over all segments gives

$$V(x_{n,q}^+(\tau)) + \left(\sum_{i=1}^{r} V(x_{n,q}^+(t_i)) - V(x_{n,q}^-(t_i)) \right) - V(x_{n,q}^-(T)) \leq \int_{\tau}^{T} L(x_{n,q}(s),v)ds.$$

It follows from the no-downward-jump property of the verification function V that the value of the objective cannot improve as the trajectory for the constant control v

crosses a manifold in the bad set B at time t_i, and therefore, for all $i = 1, \ldots, r$, we have that

$$V(x_q^-(t_i)) = \lim_{\delta \searrow 0} \inf V(x_{n,q}(t_i - \delta)) \leq V(x_q(t_i)).$$

But by the lower semi-continuity of V we have in general (i.e., for arbitrary sequences, not just those along trajectories corresponding to constant controls) that

$$V(x_q^\pm(t_i)) = \lim_{\delta \searrow 0} \inf V(x_{n,q}(t_i \pm \delta)) \geq V(x_q(t_i)),$$

and thus it follows that

$$V(x_{n,q}^+(t_i)) - V(x_{n,q}^-(t_i)) \geq 0$$

for all $i = 1, \ldots, r$. Hence

$$V(x_{n,q}^+(\tau+)) \leq V(x_{n,q}^-(T)) + \int_\tau^T L(x_{n,q}(s), v) ds.$$

Furthermore, since $x_{n,q}(T - \delta) \to q$ as $\delta \searrow 0$, for δ small enough, the points $x_{n,q}(T - \delta)$ lie in Ω, and thus $V(x_{n,q}(T - \delta)) \leq \varphi(x(T)) + \varepsilon$. But then also

$$V(x_{n,q}^-(T)) = \lim_{\delta \searrow 0} \inf V(x_{n,q}(T - \delta)) \leq \varphi(x(T)) + \varepsilon.$$

Altogether, we thus have that

$$V(x_{n,q}^+(\tau+)) \leq \varphi(x(T)) + \int_\tau^T L(x_{n,q}(s), u_n(s)) ds + \varepsilon.$$

Note that at the expense of adding ε, the penalty term φ is evaluated at the terminal point of the nominal trajectory x.

This argument can be iterated. The image of the set Ω under the flow of the solutions for the constant control v at time τ is another open set $\tilde{\Omega}$, and as before, it follows that there exists a subset of full measure with the property that solutions that end in this set meet the bad set B only finitely many times. If necessary, we can perturb the terminal point $q \in \Omega$ so that $x_q(\tau) \in \tilde{\Omega}$ and then repeat the argument for the next interval. Since there is only a finite number of such intervals, it is possible to find a terminal point $q \in \Omega$ that works for all these segments. Once more invoking the no-downward-jump property of the verification function V at the switching times, it then follows that

$$V(x_{n,q}^+(0)) \leq \varphi(x(T)) + \int_0^T L(x_{n,q}(s), u_n(s)) ds + \varepsilon.$$

Finally, the lower semicontinuity of V still gives that $V(x_q(0)) \leq V(x_q^+(0))$, and thus it follows that

$$V(x_{n,q}(0)) \leq \varphi(x(T)) + \int_0^T L(x_{n,q}(s), u_n(s)) ds + \varepsilon.$$

Taking the limit as $n \to \infty$, the trajectories $x_{n,q}$ converge uniformly on $[0,T]$ to the solution x_q, and thus it follows, once more using the lower semicontinuity of V and Proposition 6.3.1, that

$$V(x_q(0)) \le \liminf_{n \to \infty} V(x_{n,q}(0)) \le \varphi(x(T)) + \int_0^T L(x_q(s), u(s))ds + \varepsilon.$$

This establishes the inequality (6.32) and completes the proof of the theorem. □

The result in [196] gives a version of this theorem under significantly weaker differentiability assumptions on the data of the problem. It also allows for the union of the exceptional manifolds M_i not to be locally finite by using a technical lemma that allows for a countable number of switchings along a subarc. But this appears to be less of a practical concern, and therefore we did not use this formulation, which could easily still be included in the argument.

Once more, we do not concern ourselves here with the general question of establishing the existence of syntheses that would give rise to value functions V that would have the required properties. The definitions given by Boltyansky [41] and Brunovsky [62] provide some examples for this, with the most general version currently known in the paper by Piccoli and Sussmann [196], and we refer the interested reader to this paper for this topic. Our experience is that indeed, the construction of parameterized families of extremals naturally leads to verification functions that have all the required properties, and we shall illustrate this once more in the last chapter for the time-optimal control problem to a point in low dimensions.

6.4 Notes

The results presented here about families of broken extremals have their origin in joint work with J. Noble, and a first version was published in [189]. Our presentation here gives a much improved exposition of these results and has added a detailed analysis of the mapping properties near transversal crossings and folds. This research has been part of a very strong renewed interest in recent years in local optimality results for bang-bang trajectories, e.g., the papers by A. Sarychev [207, 208], A. Agrachev, G. Stefani, and P.L. Zezza [20], U. Felgenhauer [89, 90], H. Maurer, N. Osmolovskii, and coworkers [180, 182, 183], and L. Poggiolini and coworkers [197, 198]. While there exist various versions of these results—some numerical, others highly abstract in a Hamiltonian framework—essentially, for single-input problems, the common underlying framework is that the derivative of the switching function for the controlled reference trajectory needs to be nonzero at bang-bang junctions (which generates well-defined switching surfaces), and some kind of coercivity properties need to be satisfied that ensure transversal crossings. For multi-input systems, these results are directly applicable if the switching times do not coincide, but the situation becomes much more difficult in case of simultaneous switchings, and we refer the interested reader to the recent paper by Poggiolini and Spadini [197].

Much of our own research on this topic has been motivated by mathematical models that describe cell-cycle-specific cancer chemotherapy [155, 156, 253]—control problems over a prescribed therapy horizon $[0, T]$—where optimal controls are bang-bang and the results developed in this chapter are used to construct fields of locally optimal bang-bang extremals. We refer the interested reader to [155, 156] and our forthcoming text [166] for an exposition of these results. Also, the example presented in Sect. 6.2 is a result of our increasing interest in mathematical models that describe medical models in connection with cancer treatment approaches. The particular concatenation sequence arising there—bang-singular-bang—is the most typical one, not only in low dimensions, and quite recently there has been strong activity in the area of necessary and sufficient optimality conditions for concatenations of bang with singular arcs in general [22, 199]. In our approach, results about strong local optimality become a consequence of embedding such a controlled reference trajectory into a local field. This directly connects with the verification results derived in Sect. 6.3 that apply both locally and globally. For example, an almost complete analysis of the problem considered in Sect. 6.2 has been given (to the extent that this is possible in a page-limited research publication) in [160], and a local synthesis can easily be constructed based on Proposition 2.8.4. The strong local optimality (in the sense of the most restrictive classical definitions) of the particular controlled trajectories considered there is immediate from the verification theorem proven in Sect. 6.3. This section gives a streamlined exposition of Sussmann's arguments devoid of the technical issues that arise if one considers the nonsmooth setting [196]. The construction presented here contains the core of the argument, and to the best of our knowledge, has not been presented in a coherent and simplified exposition before.

Chapter 7
Control-Affine Systems in Low Dimensions: From Small-Time Reachable Sets to Time-Optimal Syntheses

We have seen in Chap. 4 that necessary conditions for optimality follow from separation results using convex approximations for the reachable set from a point. If the reachable sets are known exactly, not only necessary conditions, but complete solutions can be obtained for related optimal control problems (e.g., the time-optimal control problem). In general, determining these sets is as difficult a problem as solving an optimal control problem. However, in low dimensions, the Lie algebraic formalism introduced in Sect. 4.5 provides effective tools to accomplish this for control affine-systems of the form Σ,

$$\Sigma : \qquad \dot{x} = f(x) + ug(x), \qquad |u| \leq 1, \qquad x \in M \subseteq \mathbb{R}^n, \qquad (7.1)$$

where M is a small open neighborhood of some reference point p in $\mathbb{R}^n$, f and g are smooth (respectively, sufficiently often continuously differentiable) vector fields on M, and as always, admissible controls are Lebesgue measurable functions u that take values in the closed interval $[-1, 1]$ almost everywhere. Obviously, the precise structure of these sets will depend on the vector fields f and g, and even for a system of the form (7.1), it can be very complicated. Thus, once more, we make it our guiding principle to proceed from the "general" to the "special." In the same way as the local behavior of a function near a point is determined by its Taylor coefficients, the local behavior of an analytic system Σ is determined by the values of the drift vector field f, the control vector field g, and all their Lie brackets at a reference point p [231]. Recall that the most typical, or *codimension-0* situation, arises if all the vector fields f, g and relevant Lie brackets are in general position, that is, satisfy no linear dependencies beyond those required by antisymmetry and the Jacobi identity. The concept of *codimension* is used to organize the Lie-bracket conditions into groups of increasing degrees of degeneracy. Depending on how many nontrivial independent equality relations exist, as was done in Sects. 2.9 and 2.10, cases of positive codimension are distinguished. We loosely call the dependencies of the vector fields f and g and their Lie brackets at p, the *Lie bracket configuration* of the system at p. We shall see that as both the dimension n of the state space

H. Schättler and U. Ledzewicz, *Geometric Optimal Control: Theory, Methods and Examples*, Interdisciplinary Applied Mathematics 38, DOI 10.1007/978-1-4614-3834-2_7, © Springer Science+Business Media, LLC 2012

and the codimension k of the Lie bracket configuration of Σ at the initial point p increase, the structure of the small-time reachable sets becomes increasingly more complex, and while it can be determined completely from the conditions of the maximum principle for small n and k, as these numbers increase, these conditions increasingly need to be supplemented by more and more sophisticated global considerations that go well beyond a direct application of necessary conditions for optimality. Since the full small-time reachable set is constructed, not only do we consider trajectories that lie close to some reference trajectory, as is the case in local optimality considerations, but indeed, on the level of controlled trajectories, global aspects are taken into account automatically. We shall see that this is already the case for $n = 3$ and $k = 0$, where cut-loci and transversal folds of bang-bang extremals determine the overall structure. Our strategy is to establish the geometric properties of these reachable sets by inductively building on the results for lower dimensional systems, and then to add to these by analyzing the geometric properties of hypersurfaces that are defined by increasingly more complex concatenation structures until the full small-time reachable sets have been obtained. Obviously, the heuristic reasoning is that we expect a rather regular structure that inductively builds on the simpler lower-dimensional features. This indeed will be the case for all the cases considered here.

The main results in this last chapter give a complete description of the small-time reachable sets and the corresponding time-optimal local syntheses to a point in dimension 3 under codimension-0 and -1 assumptions. For the codimension-0 case, this synthesis combines bang-bang trajectories with two switchings, BBB, whose optimality is limited by a cut-locus that is generated by a slow singular arc, with the typical BSB synthesis near a fast singular arc and thus, in a uniform picture, presents the typical local syntheses near singular arcs. In the codimension-1 case, saturation phenomena on the singular arc need to be taken into account, and then one more segment needs to be added to the concatenation sequences, which now become $BBSB$, respectively $BSBB$. These results also provide a full description of the boundary trajectories in the codimension-0 four-dimensional case. For the cases analyzed here, we shall see that the structures of the lower-codimension cases in higher dimensions succinctly unify all the possible cases for the higher-codimension cases in lower dimensions. For example, the trajectories in the boundary of the small-time reachable set in the codimension-0 four-dimensional case follow the optimal concatenation sequences for the codimension-1 three-dimensional case, which itself combines all codimension-2 two-dimensional cases.

The results we give are not merely descriptions of these local structures, but the local results can then be combined to form global solutions. For example, for real analytic systems, there exists a wealth of literature about when these procedures can be carried out mechanically in the context of stratification theorems about subanalytic sets (e.g., [60–62, 233, 243]), but these results lie beyond the scope of our text. Yet the important point we want to make is that these solutions are not just mere examples, but are general and point the way to the structure of globally optimal controlled trajectories for optimal control problems. For example, for the three-dimensional mathematical model for tumor antiangiogenesis considered in Sect. 6.2, in the paper [160], we have given a complete solution that is completely determined by the features of the codimension-0 and -1 situations described in this text that involve an optimal singular arc and two saturation points. This particular synthesis will be fully developed in [166].

This chapter is organized as follows: After establishing a framework for our geometric constructions in Sect. 7.1, we briefly summarize the implications that the results on time-optimal control presented in Sects. 2.9 and 2.10 have on the structure of the small-time reachable sets. In Sect. 7.3 we present a landmark result by C. Lobry on boundary trajectories in dimension three that formed one of the corner-stones in the development of nonlinear control theory as a whole [170] and then determine the small-time reachable set under codimension-0 assumptions. These results naturally connect with the boundary trajectories for the four-dimensional system and give us the local time-optimal synthesis to an equilibrium point of f in $\mathbb{R}^3$. These results will be developed in Sect. 7.4 and we shall describe the analogous results for the codimension-1 case in Sect. 7.5.

7.1 Basic Topological Properties of Small-Time Reachable Sets

The small-time reachable set is the set $\text{Reach}_{\Sigma, \leq T}(p)$,

$$\text{Reach}_{\Sigma, \leq T}(p) = \{x(t) \in \mathbb{R}^n : \text{there exists a control } u(\cdot) \in \mathscr{U} \text{ defined on an interval}$$
$$[0,t] \text{ with } t \leq T \text{ such that } (x(\cdot), u(\cdot)) \text{ is a controlled}$$
$$\text{trajectory defined over } [0,t] \text{ satisfying } x(0) = p\},$$

when the overall time T is kept small. If this set is known, then solutions for the corresponding time-optimal control problems can generally be given rather easily using an inversion procedure on the boundary trajectories. We first establish some basic compactness properties of these sets for control-affine systems.

Theorem 7.1.1. *For T small, the reachable sets $\text{Reach}_{\Sigma, \leq T}(p)$ are compact.*

Proof. Let O be an open neighborhood of p with compact closure K in M. The norm of the right-hand side of the differential equation (7.1), $F(x,u) = f(x) + ug(x)$, is bounded for $x \in K$ and $|u| \leq 1$, say $\|F(x,u)\|_\infty \leq C < \infty$. Thus there exists a $\overline{T} > 0$ such that for any admissible control u defined on $[0,\overline{T}]$, the solution to the differential equation $\dot{x} = f(x) + u(t)g(x)$, $x(0) = p$, exists on $[0,\overline{T}]$ and lies in O. We henceforth always take $T \leq \overline{T}$.

Let $\mathscr{T}$ be the collection of all trajectories defined over $[0,T]$ that start at p. This family $\mathscr{T}$ is globally Lipschitz (see Appendix B): for all $s,t \in [0,T]$ and any controlled trajectory (x,u), we have that

$$\|x(t) - x(s)\|_\infty = \left\| \int_s^t f(x(r)) + u(r)g(x(r)) dr \right\|_\infty \leq C|t - s|. \tag{7.2}$$

In particular, taking $s = 0$, it follows that $\mathscr{T}$ is bounded, and thus $\text{Reach}_{\Sigma, \leq T}(p)$ is bounded.

It remains to show that $\text{Reach}_{\Sigma, \leq T}(p)$ is closed. Let $\{q_n\}_{n \in \mathbb{N}}$ be a sequence of points in $\text{Reach}_{\Sigma, \leq T}(p)$ that converges to q as $n \to \infty$. Let $\{(x_n(\cdot), u_n(\cdot))\}_{n \in \mathbb{N}}$ be a sequence of controlled trajectories defined over intervals $[0, t_n]$ such that $x_n(t_n) = q_n$. Extending the controls to the interval $[t_n, T]$, without loss of generality we may assume that all controls and trajectories are defined on $[0, T]$. Furthermore, by taking a subsequence if necessary, we may assume that $t_n \to \bar{t}$. It suffices to show that there exists a controlled trajectory $(x(\cdot), u(\cdot))$ defined on $[0, T]$ such that $x(\bar{t}) = q$.

The key observation is that the family $\mathscr{T}$ is *equicontinuous* (see Appendix A): a family $\mathscr{T}$ of continuous curves defined over the compact interval $[0, T]$ is equicontinuous if for every time $t \in [0, T]$ and every $\varepsilon > 0$ there exists a $\delta > 0$ such that for all the curves $x \in \mathscr{T}$ we have that $\|x(t) - x(s)\| < \varepsilon$ whenever $|t - s| < \delta$. This property is an immediate consequence of the global Lipschitz condition (7.2). The Arzelá–Ascoli theorem, a classical result in analysis (see Corollary A.2.1), implies that there exists a subsequence $\{x_{n_k}(\cdot)\}_{k \in \mathbb{N}}$ that converges uniformly on $[0, T]$ to a continuous limit $\bar{x}$. By Theorem 3.2.1, the sequence $\{u_{n_k}(\cdot)\}_{k \in \mathbb{N}}$ of controls is weakly sequentially compact in $L^1([0, T])$, and thus, taking another subsequence if necessary, we may also assume that u_{n_k} converges weakly to some admissible control $\bar{u}$. Without loss of generality, simply assume that $x_n \to \bar{x}$ uniformly on $[0, T]$ and that $u_n \rightharpoonup \bar{u}$ weakly in $L^1([0, T])$. We claim that $(\bar{x}, \bar{u})$ is a controlled trajectory, i.e., that $\bar{x}$ is the trajectory corresponding to $\bar{u}$. For all $t \in [0, T]$ we have that

$$\bar{x}(t) = \lim_{n \to \infty} x_n(t) = \lim_{n \to \infty} \left(p + \int_0^t f(x_n(s)) + u_n(s) g(x_n(s)) ds \right)$$

$$= p + \int_0^t f(\bar{x}(s)) ds + \lim_{n \to \infty} \int_0^t u_n(s) g(x_n(s)) ds$$

and

$$\lim_{n \to \infty} \int_0^t u_n(s) g(x_n(s)) - \bar{u}(s) g(\bar{x}(s)) ds$$

$$= \lim_{n \to \infty} \int_0^t u_n(s) \left[g(x_n(s)) - g(\bar{x}(s)) \right] ds + \lim_{n \to \infty} \int_0^t [u_n(s) - \bar{u}(s)] g(\bar{x}(s)) ds.$$

Using Lebesgue's dominated convergence theorem (see Appendix D), the first integral converges to zero, since $x_n(s) \to \bar{x}(s)$ pointwise and g and the controls are bounded; the second integral converges to zero by the weak convergence of u_n to $\bar{u}$. Hence

$$\bar{x}(t) = p + \int_0^t f(\bar{x}(s)) + \bar{u}(s) g(\bar{x}(s)) ds,$$

and thus $(\bar{x}(\cdot), \bar{u}(\cdot))$ is a controlled trajectory.

Finally, given $\varepsilon > 0$, choose $\delta = \delta(\varepsilon) > 0$ such that $\|\bar{x}(\bar{t}) - \bar{x}(s)\| < \frac{\varepsilon}{2}$ whenever $|\bar{t} - s| < \delta$ and choose $N = N(\varepsilon)$ such that $\|x_n - \bar{x}\|_\infty < \frac{\varepsilon}{2}$ and $|\bar{t} - t_n| < \delta$ for all $n \geq N$. Then for all $n \geq N$,

$$\|\bar{x}(\bar{t}) - q_n\| \leq \|\bar{x}(\bar{t}) - \bar{x}(t_n)\| + \|\bar{x}(t_n) - x_n(t_n)\| < \varepsilon$$

and thus $q = \lim_{n\to\infty} q_n = \bar{x}(\bar{t}) \in \text{Reach}_{\Sigma, \leq T}(p)$.   □

Corollary 7.1.1. *For small t, the time-t-reachable sets* $\text{Reach}_{\Sigma, t}(p)$ *are compact.*

Proof. This argument is identical to the one just given, except that all trajectories are defined on $[0, t]$ and all points are reached in time t exactly.   □

The assumption that Σ is control-affine matters for this result, and we give a simple example that shows that small-time reachable sets need not be closed for a general nonlinear system.

Example. Let R denote the reachable set at time T from the origin for the two-dimensional system

$$\dot{x}_1 = (1 - x_2^2)u^2, \qquad \dot{x}_2 = u, \qquad |u| \leq 1.$$

Partition the interval $[0, T]$ into $2n$ equidistant intervals of the form $I_j = \left(\frac{jT}{2n}, \frac{(j+1)T}{2n} \right]$, $j = 1, \ldots, 2n - 1$, and define a control u_n that alternates between $+1$ and -1 on these intervals starting with $u = +1$. If we denote the corresponding trajectory by $x^{(n)}(\cdot)$, then we have $\|x_2^{(n)}\|_\infty = \frac{T}{2n}$ and $x_2^{(n)}(T) = 0$. Furthermore, since $u^2 \equiv 1$,

$$\left| x_1^{(n)}(T) - T \right| = \left| \int_0^T x_2^{(n)}(s)^2 ds \right| \leq \frac{T^3}{4n^2},$$

so that

$$\lim_{n\to\infty} (x_1^{(n)}(T), x_2^{(n)}(T)) = (T, 0) \in \text{clos } R.$$

But this point is not reachable in small time T: for any admissible control defined over an interval $[0, T]$ with $T \leq 1$, we have that $x_2^2(t) \leq 1$ and thus $\dot{x}_1 = (1 - x_2^2)u^2 \leq 1 - x_2^2$. Hence the only way to realize $x_1(T) = T$ is for $x_2(t) \equiv 0$. But this requires $u(t) \equiv 0$ and thus $x_1(T) = 0$ as well. Hence $(T, 0) \notin \text{clos } R$.   □

The construction of small-time reachable sets gives us direct information about time-optimal controls. It provides a geometric approach that complements analytic techniques of the type that were already described and used for planar systems in Sects. 2.9 and 2.10. We have seen in Sect. 3.5 that there are close connections between trajectories that lie in the boundary of the reachable set and time-optimal controls for linear systems. For nonlinear systems, the same properties are sometimes still valid for specific models, but they are no longer true in general, since reachable sets need not be convex. However, the following simple observation remains true and will be used frequently.

Lemma 7.1.1. *If a controlled trajectory* $(x(\cdot), u(\cdot))$ *steers a point* $p = x(0)$ *into* $q = x(T)$ *in time* T *time-optimally, then* $x(t) \in \partial \operatorname{Reach}_{\Sigma, t}(p)$ *for all times* $t \in [0, T)$.

Proof. Clearly, if $x(\tau) \in \operatorname{int}(\operatorname{Reach}_{\Sigma, \tau}(p))$, for some $\tau \in [0, T)$, then there exists an $\varepsilon > 0$ such that $x(\tau + \varepsilon)$ is also reachable in time τ. Concatenating the control that steers p into $x(\tau + \varepsilon)$ in time τ with the restriction of the control $u(\cdot)$ to the interval $[\tau + \varepsilon, T]$, this control steers p into q in time $T - \varepsilon$. Contradiction. $\square$

More generally, trajectories that lie in the boundary of the small-time reachable set for all times play a major role in establishing the "borders" of the reachable sets $\operatorname{Reach}_{\Sigma, \leq T}(p)$; we call them boundary trajectories.

Definition 7.1.1 (Boundary trajectory). A controlled trajectory (x, u) defined over the interval $[0, T]$ is a boundary trajectory for the small-time reachable set if $x(t) \in \partial \operatorname{Reach}_{\Sigma, \leq T}(p)$ for all times $t \in [0, T]$.

We close this section with establishing our notation and introducing the type of coordinates that will be used throughout our explicit constructions. Especially, a "good" set of coordinates is indispensable. Depending on this choice, for a given system, the problem may become simple or extremely complicated. It is here that the Lie-bracket configuration at p matters, and quite frankly, in nondegenerate situations it dictates a choice of so-called *canonical coordinates*. These are defined in terms of the times spent along the vector fields $X = f - g$ and $Y = f = g$ and their low-order Lie brackets that are linearly independent.

Definition 7.1.2 (Canonical coordinates). Suppose M is an open neighborhood of some point $p \in \mathbb{R}^n$ and let $Z_1, \ldots, Z_n$ be n smooth vector fields defined on M such that the vectors $Z_1(p), \ldots, Z_n(p)$, are linearly independent. Then there exists an open ball $B_\varepsilon(0)$ such that the mappings $\Psi_i : B_\varepsilon(0) \to M$, $i = 1, 2$,

$$\Psi_1 : (\xi_1, \ldots, \xi_n) \mapsto p \exp(\xi_n Z_n + \xi_{n-1} Z_{n-1} + \cdots + \xi_2 Z_2 + \xi_1 Z_1) \qquad (7.3)$$

and

$$\Psi_2 : (\xi_1, \ldots, \xi_n) \mapsto p \exp(\xi_n Z_n) \exp(\xi_{n-1} Z_{n-1}) \cdots \exp(\xi_2 Z_2) \exp(\xi_1 Z_1) \quad (7.4)$$

define smooth coordinates in a neighborhood of p in M. The $(\xi_1, \ldots, \xi_n)$ are said to be canonical coordinated of the first and second kind, respectively.

Note that for both $i = 1$ and 2 we have that $\Psi_i(0, \ldots, 0) = p$ and that the jth column of the Jacobian matrix $D\Psi_i(0, \ldots, 0)$ is given by the vector field $Z_j(p)$, or using the exponential notation introduced in Sect. 4.5,

$$\frac{\partial \Psi_1}{\partial \xi_j}(0, \ldots, 0) = \frac{\partial \Psi_2}{\partial \xi_j}(0, \ldots, 0) = \frac{\partial}{\partial \xi_j} p \exp(\xi_j Z_j) = p Z_j.$$

Since these vectors are linearly independent, the Jacobian matrix is nonsingular, and it follows from the inverse function theorem that there exists a neighborhood of 0 on which the mappings Ψ_1 and Ψ_2 are C^1-diffeomorphisms. Hence Ψ_i is a well-defined change of coordinates near p. In our calculations below, we mainly use canonical coordinates of the second kind with basis vectors Z_i that are chosen depending on the Lie-bracket configuration of the system at p. Without loss of generality, by shrinking the neighborhood M if necessary, we always assume that any linear independence relations that hold at p are valid on all of M.

We close these introductory comments with establishing our notation for various concatenations of trajectories. As in Chap. 2, we always write $X = f - g$, $Y = f + g$ and denote the vector field corresponding to a singular control by S. Concatenations of bang and singular arcs are simply denoted by the corresponding sequence of letters, i.e., $XYSX$ denotes a controlled trajectory that initially is given by an X-arc (corresponding to the constant control $u \equiv -1$) on some interval $[0, t_1]$, then by a Y-arc (corresponding to the constant control $u \equiv 1$) over some interval $[t_1, t_2]$, followed by a singular arc over an interval $[t_2, t_3]$ and one more Y-arc over the final interval $[t_3, T]$. We shall always assume that T is a fixed, sufficiently small time so that all trajectories are defined on $[0, T]$. Furthermore, it becomes convenient to allow for the possibility that these structures collapse. Thus, for example, we allow that $t_1 = t_2$, and in this case we actually have a controlled trajectory of type XSX.

We still set up a notation that will allow us to give concise descriptions of the small-time reachable sets $\text{Reach}_{\Sigma, \leq T}(p)$ and their boundaries. In all examples considered, these indeed will be cell complexes in the sense of algebraic topology and can best be described in terms of the concatenation structures that define these *cells*. We denote cells by $\mathscr{C}$ (with subscripts that indicate the concatenation sequence); for example, $\mathscr{C}_0 = \{p\}$ denotes the trivial zero-dimensional cell consisting of only the initial point. More interestingly,

$$\mathscr{C}_X = \{p\exp(rX) : 0 \leq r \leq T\},$$

$$\mathscr{C}_{XY} = \{p\exp(rX)\exp(sY) : 0 \leq r, \, 0 \leq s, \, r+s \leq T\},$$

$$\mathscr{C}_{XYX} = \{p\exp(rX)\exp(sY)\exp(tX) : 0 \leq r, \, 0 \leq s, \, 0 \leq t, \, r+s+t \leq T\},$$

and so on, with analogous notation for different concatenation sequences. Each of these cells is a compact set that contains the cells with shorter concatenation sequences, i.e., for example, $\mathscr{C}_0 \subset \mathscr{C}_X \subset \mathscr{C}_{XY} \subset \mathscr{C}_{XYX} \subset \cdots$, and under general position assumptions, these sets form the boundary of these cell complexes. In fact, typically we have a stratification (see Sect. 5.5) and it becomes useful to have this concept available here. For this reason, we still prefer to have a separate notation for the submanifolds that make up these stratifications of the cell complexes. We denote the submanifolds that consist of trajectories of the system with the letter $\mathscr{T}$, i.e.,

$$\mathscr{T}_X = \{p\exp(rX) : 0 < r < T\},$$

$$\mathscr{T}_{XY} = \{p\exp(rX)\exp(sY) : 0 < r, \, 0 < s, \, r+s < T\},$$

$$\mathscr{T}_{XYX} = \{p\exp(rX)\exp(sY)\exp(tX) : 0 < r, \, 0 < s, \, 0 < t, \, r+s+t < T\},$$

and so on. We also need labels for the submanifolds that define the endpoints of the controlled trajectories at time T and denote these by the letter $\mathcal{E}$, i.e.,

$$\mathcal{E}_X = \{p\exp(TX)\},$$

$$\mathcal{E}_{XY} = \{p\exp(rX)\exp(sY): 0 \leq r,\ 0 \leq s,\ r+s = T\},$$

$$\mathcal{E}_{XYX} = \{p\exp(rX)\exp(sY)\exp(tX): 0 \leq r,\ 0 \leq r,\ 0 \leq t,\ r+s+t = T\}.$$

Finally, the wedge product $\wedge$ provides a convenient notation for linear independence relations of vectors, and we use this notation both to formulate our assumptions and also extensively in the computations. For our purposes, it suffices to state that the wedge product of n vectors $v_1, \ldots, v_n$ in $\mathbb{R}^n$ is the multilinear form given by the determinant of the matrix whose columns are the ordered vectors $v_1, \ldots, v_n$,

$$v_1 \wedge \cdots \wedge v_n = \det(v_1, \ldots, v_n).$$

In particular, $v_1 \wedge \cdots \wedge v_n = 0$ if and only if the vectors v_i are linearly dependent.

7.2   Small-Time Reachable Sets in Dimension 2

In this section, we summarize the implications of the results from Sects. 2.9 and 2.10 on the structure of small-time reachable sets in dimension two. We always assume that the neighborhood M of the initial point p and the terminal time T are chosen small, and throughout this section we assume that

(A) the vector fields f and g are linearly independent on M,

$$f(p) \wedge g(p) \neq 0 \qquad \Longleftrightarrow \qquad X(p) \wedge Y(p) \neq 0.$$

Under this assumption, the integral curves of the vector fields X and Y emanating from p are the only boundary trajectories.

Proposition 7.2.1. *Under assumption (A), if (x,u) is a boundary trajectory, then u is constant given by $+1$ or -1. The boundary portion of the small-time reachable set* Reach$_{\Sigma, \leq T}(p)$ *that is not due to the restriction of the time to T is therefore composed on an X- and a Y-arc starting at p.*

Proof. By Theorem 4.2.2, there exists a nontrivial covector field λ such that a.e. on $[0,T]$,

$$0 = \langle \lambda(t), f(x(t)) + u(t)g(x(t)) \rangle = \min_{u \in U} \langle \lambda(t), f(x(t)) + u(t)g(x(t)) \rangle.$$

If the control u has a switching at some time τ in the interior of the domain, then this implies that we must have both

$$0 = \langle \lambda(\tau), g(x(\tau)) \rangle \qquad \text{and} \qquad 0 = \langle \lambda(\tau), f(x(\tau)) \rangle.$$

Actually, since we know only that the conditions of the maximum principle are valid almost everywhere, in principle it could be the case that these conditions are not valid for the specific time τ. However, it is easily seen that these conditions always hold at switching times. For suppose the control switches from $u = +1$ to $u = -1$ at τ. Then for some small $\delta > 0$, there exist measurable subsets E_+ of $(\tau - \delta, \tau)$ and E_- of $(\tau, \tau + \delta)$ of full measure δ such that $\langle \lambda(t), g(x(t)) \rangle < 0$ for $t \in E_+$ and $\langle \lambda(t), g(x(t)) \rangle > 0$ for $t \in E_-$. Taking sequences $\{t_n^-\}_{n \in \mathbb{N}} \subset E_+$ and $\{t_n^+\}_{n \in \mathbb{N}} \subset E_-$ that converge to τ, it follows from the continuity of the switching function $\Phi(t) = \langle \lambda(t), g(x(t)) \rangle$ that we must have $\Phi(\tau) = 0$. The analogous argument for the function $\langle \lambda(t), f(x(\tau)) \rangle$ then also gives that $\langle \lambda(\tau), f(x(\tau)) \rangle = 0$. This, however, contradicts the nontriviality of λ. Hence controls that steer p into boundary points of the (full) reachable set must be constant and thus are given by $u \equiv +1$ or $u \equiv -1$. $\square$

Since admissible controls are convex combinations of $u = -1$ and $u = +1$, the linear independence of X and Y implies that the small-time reachable set $\text{Reach}_{\Sigma, \leq T}(p)$ must somehow lie "between" these X- and Y-trajectories emanating from p. Thus, apart from the boundary points of $\text{Reach}_{\Sigma, \leq T}(p)$ that lie in the exact time-T-reachable set, those portions that lie in the boundary for all small times T are given by the union of the sets $\mathscr{C}_0 = \{p\}$, $\mathscr{C}_X$, and $\mathscr{C}_Y$. The remaining portions in $\partial \text{Reach}_{\Sigma, \leq T}(p)$ lie in $\text{Reach}_{\Sigma, T}(p)$, but these sections are no longer determined by assumption (A) alone. Clearly, $\mathscr{C}_{XY} \subset \text{Reach}_{\Sigma, \leq T}(p)$ and $\mathscr{C}_{YX} \subset \text{Reach}_{\Sigma, \leq T}(p)$, but $\text{Reach}_{\Sigma, \leq T}(p)$ may be strictly larger than either of these sets. The problem of describing $\text{Reach}_{\Sigma, \leq T}(p)$ exactly, even for small T, is equivalent to solving the two-dimensional optimal control problem for Σ locally assuming only that f and g are linearly independent at p. As we have already seen to some extent in Sects. 2.9 and 2.10, the structure of optimal controls that lie in a sufficiently small neighborhood of p depends on the Lie-bracket configuration at p, and depending on its codimension, infinitely many inequivalent cases exist. But the results of these sections allow us to determine the structure of the small-time reachable sets $\text{Reach}_{\Sigma, \leq T}(p)$ under the same conditions. In fact, since we always start at the point p, rather than considering arbitrary trajectories that lie near p, our controls here will generally be even simpler. We briefly summarize the resulting structures for the small-time reachable sets and then illustrate them with some examples. As in Chap. 2, we write

$$[f, g](x) = \alpha(x) f(x) + \beta(x) g(x) \tag{7.5}$$

with smooth functions α and β on M. Then the results of Sect. 2.9 immediately imply the following structures for the small-time reachable sets:

Theorem 7.2.1. *Under assumption (A), the small-time reachable sets from p for Σ, $\text{Reach}_{\Sigma, \leq T}(p)$, are given by the following sets:*

1. in codimension 0, we have that

$$\text{Reach}_{\Sigma, \leq T}(p) = \begin{cases} \mathscr{C}_{XY} & \text{if } \alpha(p) > 0, \\ \mathscr{C}_{YX} & \text{if } \alpha(p) < 0. \end{cases}$$

2. *in codimension* 1, *for* $\alpha(p) = 0$, *we have that*

$$\text{Reach}_{\Sigma, \leq T}(p) = \begin{cases} \mathscr{C}_{XY} & \text{if} \quad L_X \alpha(p) > 0 \quad \text{and} \quad L_Y \alpha(p) > 0, \\ \mathscr{C}_{YX} & \text{if} \quad L_X \alpha(p) < 0 \quad \text{and} \quad L_Y \alpha(p) < 0, \end{cases}$$

if $L_X \alpha(p)$ *and* $L_Y \alpha(p)$ *have the same sign, and by*

$$\text{Reach}_{\Sigma, \leq T}(p) = \begin{cases} \mathscr{C}_{BB} = \mathscr{C}_{XY} \cup \mathscr{C}_{YX} & \text{if} \quad L_X \alpha(p) > 0 \quad \text{and} \quad L_Y \alpha(p) < 0, \\ \mathscr{C}_{SB} = \mathscr{C}_{SY} \cup \mathscr{C}_{SX} & \text{if} \quad L_X \alpha(p) < 0 \quad \text{and} \quad L_Y \alpha(p) > 0, \end{cases}$$

if $L_X \alpha(p)$ *and* $L_Y \alpha(p)$ *have opposite signs.*
3. *in codimension* 2, *for* $\alpha(p) = 0$ *and* $L_Y \alpha(p) = 0$, *no new geometric structures arise, but depending on the Lie-bracket configuration, the small-time reachable set is given by one of the sets* $\mathscr{C}_{XY}$, $\mathscr{C}_{YX}$, *or* $\mathscr{C}_{SB}$.

Proof. These geometric structures are an immediate consequence of the results proven in Sects. 2.9 and 2.10, and we only briefly recall these constructions.

It is the coefficient α that determines the structure of time-optimal controls in codimension-0 conditions: if α is positive on M, then time-optimal trajectories are of type XY, and if α is negative on M, time-optimal trajectories are of type YX (Proposition 2.9.1). This gives the first statement.

Under codimension-1 conditions, the function α vanishes at p, but the vector fields X and Y are transversal to the curve $\mathscr{S} = \{x \in M : \alpha(x) = 0\}$. If X and Y point to the same side of $\mathscr{S}$ on M, then the structure of the reachable set from p is the same as in the corresponding codimension-0 case: if X and Y point into the region where α is positive, then only switchings from $u = -1$ to $u = +1$ are time optimal there, and thus $\text{Reach}_{\Sigma, \leq T}(p) = \mathscr{C}_{XY}$. Analogously, $\text{Reach}_{\Sigma, \leq T}(p) = \mathscr{C}_{YX}$ if X and Y point into the region where α is negative. In comparison with the structure of time-optimal trajectories near p, here the situation simplifies somewhat since all trajectories start at the point $p \in \mathscr{S}$. Indeed, the same structure of trajectories remains optimal for cases of higher codimension as long as all the trajectories lie in $M_+ = \{x \in M : \alpha(x) > 0\}$ or $M_- = \{x \in M : \alpha(x) < 0\}$. For this reason, for example, the small-time reachable sets are also given by $\mathscr{C}_{XY}$ in the codimension-2 case described in Proposition 2.10.1.

But different structures arise if X and Y point to opposite sides of $\mathscr{S}$. In this case, $\mathscr{S}$ is a singular arc that is fast if the Lie derivative $L_X \alpha$ is negative and is slow if this Lie derivative is positive (Proposition 2.9.3). For a slow singular arc, time-optimal controls are still bang-bang with at most one switching (Proposition 2.9.5), but now no longer is one of the cells $\mathscr{C}_{XY}$ or $\mathscr{C}_{YX}$ contained in the other, but the small-time reachable set is given by their union,

$$\text{Reach}_{\Sigma, \leq T}(p) = \mathscr{C}_{XY} \cup \mathscr{C}_{YX},$$

with a cut-locus arising between these two cells that determines the time-optimal trajectories. In the case that the singular arc is fast, the structure of the small-time reachable set is simpler. Once more, since all trajectories start at a point on the singular arc, compared with the structure of time-optimal trajectories that lie in a neighborhood of p that is of the form BSB (Proposition 2.9.4), only SB-trajectories matter for the small-time reachable set. Indeed, since X and Y point to opposite sides of the singular arc, we now have that

$$\text{Reach}_{\Sigma, \leq T}(p) = \mathscr{C}_{SY} \cup \mathscr{C}_{SX},$$

and the two cells $\mathscr{C}_{SY}$ and $\mathscr{C}_{SX}$ only intersect along their trivial common boundary stratum $\mathscr{C}_S$. Note that X and Y point to sides of $\mathscr{S}$ in which it is not optimal to switch away from X respectively Y.

No new structures for small-time reachable sets arise for the codimension-2 cases analyzed in Sect. 2.10. In these cases, while one of the vector fields X and Y still is transversal to the singular arc, the other is tangent, but has order of contact equal to 1 (see Figs. 2.16–2.18). As under the codimension-1 assumptions, it is still true that all trajectories starting from p move into regions of the state space where no further switchings are possible, and thus again the small-time reachable set is given by one of the sets $\mathscr{C}_{XY}$, $\mathscr{C}_{YX}$, $\mathscr{C}_{SB}$. □

We give examples of the small-time reachable sets for each of these four cases. If we introduce canonical coordinates of the second type of the form

$$(\xi_1, \xi_2) \mapsto p \exp(\xi_1 X) \exp(\xi_2 Y),$$

then concatenations of the form $p e^{sX} e^{tY}$ that consist of an X-trajectory starting from p for time s followed by a Y-trajectory for time t simply have coordinates $(\xi_1, \xi_2) = (s, t)$, and thus $\mathscr{C}_{XY}$ becomes the triangle

$$\mathscr{C}_{XY} = \{(\xi_1, \xi_2) : 0 \leq \xi_1, 0 \leq \xi_2, \xi_1 + \xi_2 \leq T\}.$$

Figure 7.1 illustrates this structure, and the trajectories show how points in this set are reached time-optimally if $\alpha(p) > 0$.

As a comparison, Fig. 7.2 shows the cells $\mathscr{C}_{YX}$ for the system given by the vector fields

$$X = \begin{pmatrix} 1 \\ -ax_2 \end{pmatrix} \quad \text{and} \quad Y = \begin{pmatrix} 0 \\ 1 \end{pmatrix},$$

or equivalently,

$$f = \frac{1}{2}(X + Y) = \frac{1}{2}\begin{pmatrix} 1 \\ 1 - ax_2 \end{pmatrix} \quad \text{and} \quad g = \frac{1}{2}(Y - X) = \frac{1}{2}\begin{pmatrix} -1 \\ 1 + ax_2 \end{pmatrix}.$$

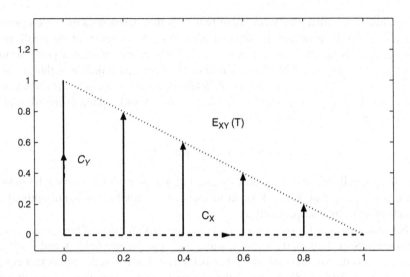

Fig. 7.1 The cell $\mathscr{C}_{XY}$ in canonical coordinates $(\xi_1, \xi_2) \mapsto p \exp(\xi_1 X) \exp(\xi_2 Y)$

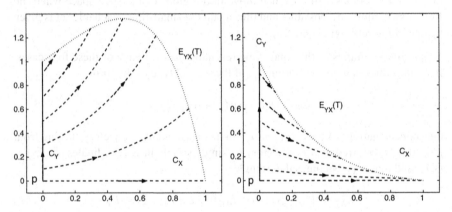

Fig. 7.2 The cell $\mathscr{C}_{YX}$ for the vector field $X = (1, -ax_2)^T$ for $a = -1$ (on the *left*) and $a = +1$ (on the *right*)

In this case, we have that

$$[f,g](x) = \frac{1}{2}[X,Y](x) = -\frac{1}{2}DX(x)Y = \frac{1}{2}\begin{pmatrix} 0 \\ a \end{pmatrix} = \frac{a}{2}f + \frac{a}{2}g,$$

and thus $\alpha(x) \equiv \frac{a}{2}$. As the figure clearly illustrates, we have that $\mathscr{C}_{XY} \subset \mathscr{C}_{YX}$ for $\alpha < 0$ and $\mathscr{C}_{XY} \supset \mathscr{C}_{YX}$ for $\alpha > 0$.

Figure 7.3 shows the synthesis of controlled trajectories if the initial point p lies on a slow singular arc. The figure was created using the vector fields

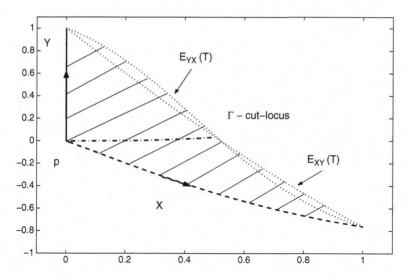

Fig. 7.3 The small-time reachable set from a point p on a slow singular arc

$$X = \begin{pmatrix} 1 \\ -1 + x_2^2 \end{pmatrix} \quad \text{and} \quad Y = \begin{pmatrix} 0 \\ 1 \end{pmatrix},$$

but the geometric structure shown is generally valid under the corresponding codimension-1 assumptions. For this example,

$$f = \frac{1}{2}(X + Y) = \frac{1}{2}\begin{pmatrix} 1 \\ x_2^2 \end{pmatrix} \quad \text{and} \quad g = \frac{1}{2}(Y - X) = \frac{1}{2}\begin{pmatrix} -1 \\ 2 - x_2^2 \end{pmatrix},$$

and we have that

$$[f, g](x) = -\frac{1}{2}DX(x)Y = \begin{pmatrix} 0 \\ -x_2 \end{pmatrix} = (-x_2)(f + g).$$

Hence $\alpha(x) = -x_2$, the singular arc is given by $\mathscr{S} = \{x \in M : x_2 = 0\}$, and

$$L_X \alpha(x) = (0, -1)X(x) = 1 - x_2^2 \quad \text{and} \quad L_Y \alpha(x) = (0, -1)Y(x) = -1.$$

Thus X and Y point to opposite sides of the singular arc, but the Legendre–Clebsch condition is violated and the singular arc is slow. For this example, all equations can be integrated explicitly: the X-trajectory starting at the origin at time 0 is given by $x(t) = (t, -\tanh(t))$, and concatenating this solution with the vertical lines that are the integral curves of Y, the endpoints corresponding to an XY-trajectory of the form $0e^{sX}e^{(T-s)Y}$ are given by

$$\mathscr{E}_{XY} = \{(x_1, x_2) : x_1 = s, \; x_2 = T - s - \tanh(s), \; 0 \le r \le T\}.$$

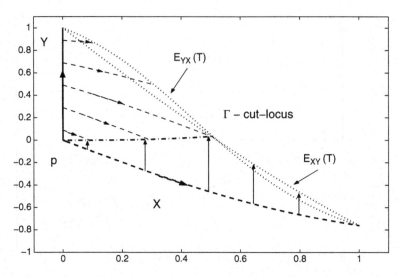

Fig. 7.4 Time-optimal trajectories from a point p on a slow singular arc

If we reverse the order of integrations, $0e^{(T-s)Y}e^{sX}$, we similarly obtain

$$\mathscr{E}_{YX} = \left\{ (x_1, x_2) : x_1 = s, \; x_2 = \frac{(1+s)e^{-(T-s)} - (1-s)e^{T-s}}{(1+s)e^{-(T-s)} + (1-s)e^{T-s}}, \; 0 \le r \le T \right\}.$$

These two curves intersect in a nontrivial cut-locus determined by the solution of the transcendental equation

$$T - s - \tanh(s) = \frac{(1+s)e^{-(T-s)} - (1-s)e^{T-s}}{(1+s)e^{-(T-s)} + (1-s)e^{T-s}}.$$

It is not difficult to compute the solutions numerically, and Fig. 7.4 identifies the cut-locus Γ for this example and also shows the structure of time-optimal trajectories. Time-optimal XY- and YX-trajectories reach Γ in the same time.

By simply changing the sign of the quadratic term in the vector field X above,

$$X = \begin{pmatrix} 1 \\ -1 - x_2^2 \end{pmatrix},$$

and retaining Y, we obtain an example for a fast singular arc. For this model,

$$f = \frac{1}{2}(X + Y) = \frac{1}{2}\begin{pmatrix} 1 \\ -x_2^2 \end{pmatrix}, \qquad g = \frac{1}{2}(Y - X) = \frac{1}{2}\begin{pmatrix} -1 \\ 2 + x_2^2 \end{pmatrix},$$

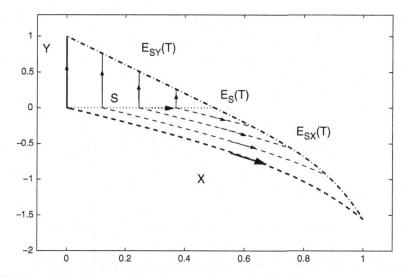

Fig. 7.5 The small-time reachable set from a point p on a fast singular arc

and we now have that

$$[f,g](x) = \begin{pmatrix} 0 \\ x_2 \end{pmatrix} = x_2 (f + g).$$

Thus $\alpha(x) = x_2$, the singular arc is still given by $\mathscr{S} = \{x \in M : x_2 = 0\}$, and now

$$L_X \alpha(x) = (0,1) X(x) = -1 - x_2^2 \quad\text{and}\quad L_Y \alpha(x) = (0,1) Y(x) = 1.$$

Again X and Y point to opposite sides of the singular arc, but now the Legendre–Clebsch condition is satisfied and the singular arc is fast. Figure 7.5 shows the small-time reachable set $\text{Reach}_{\Sigma,\leq T}(p)$ for this system.

As all these examples show, the X- and Y-trajectories provide a *barrier that cannot be crossed in small time* by controlled trajectories. It is not possible to circumvent this border in a sufficiently small neighborhood of p by "going around." For longer times, however, clearly this is an option, and there may exist points in the reachable sets $\text{Reach}_{\Sigma,\leq T}(p)$ that converge to p, but lie outside of this sector. We close this section with an example that not only illustrates this behavior as reachable sets evolve over large times T, but also is an example of an optimal synthesis for which the value function is not continuous.

Example 7.2.1. We consider the reachable sets $\text{Reach}_{\Sigma,\leq T}(0)$ from the origin as the total time T evolves for the system Σ given by

$$X = \begin{pmatrix} 1 - x_2 \\ x_1 \end{pmatrix} \quad\text{and}\quad Y = \begin{pmatrix} 0 \\ 1 \end{pmatrix},$$

or equivalently, the control and drift vector fields

$$f = \frac{1}{2}\begin{pmatrix} 1-x_2 \\ 1+x_1 \end{pmatrix} \quad \text{and} \quad g = \frac{1}{2}\begin{pmatrix} -1+x_2 \\ 1-x_1 \end{pmatrix}.$$

We have that

$$[f,g](x) = -\frac{1}{2}DX(x)Y = -\frac{1}{2}\begin{pmatrix} 0 & -1 \\ 1 & 0 \end{pmatrix}\begin{pmatrix} 0 \\ 1 \end{pmatrix} = \frac{1}{2}\begin{pmatrix} 1 \\ 0 \end{pmatrix},$$

and thus the functions α and β are given by

$$\alpha(x) = \frac{1}{2}\frac{1-x_1}{1-x_2} \quad \text{and} \quad \beta(x) = -\frac{1}{2}\frac{1+x_1}{1-x_2}.$$

It thus follows from Proposition 2.9.1 that optimal XY-junctions must lie in the regions $\{(x_1,x_2): x_1 < 1, x_2 < 1\}$ and $\{(x_1,x_2): x_1 > 1, x_2 > 1\}$ and optimal YX-junctions must lie in the regions $\{(x_1,x_2): x_1 > 1, x_2 < 1\}$ and $\{(x_1,x_2): x_1 < 1, x_2 > 1\}$. Along a singular arc S, the vector fields g and $[f,g]$ must be linearly dependent, and this implies that singular arcs can lie only on the vertical line $x_1 \equiv 0$. But this corresponds to the vector field Y (and the control $u \equiv +1$), and thus for this system, although a singular arc exists, it plays no special role, and we need to analyze only concatenations of X- and Y-trajectories.

Figure 7.6 shows the evolution of the reachable sets $\text{Reach}_{\Sigma, \leq T}(0)$ as $T \to \infty$. Observe that X-trajectories are the circles $x_1^2 + (x_2 - 1)^2 = \text{const}$, all of which have period 2π, and Y-trajectories are vertical lines oriented upward. At the origin, $f(0)$ and $g(0)$ are linearly independent and $\alpha(0) > 0$. Hence the small-time reachable set from the origin initially is given by the cell $\mathscr{C}_{XY}$, and this is valid for all times T that satisfy $0 < T \leq 1$. At time $T = 1$, the Y-trajectory reaches the equilibrium point $(0,1)$ of the vector field X, and for times $T > 1$, along this Y-trajectory, X now points to the opposite side of the x_2-axis and switchings from Y to X become optimal. This generates a second portion of the reachable sets that lies in $\{x_1 < 0\}$. If these YX-junctions occur at times τ between 1 and 2, these corresponding trajectories add to the reachable set until, after time π along the X-trajectory, they again reach the positive x_2-axis, at which point they simply terminate. For switching times $\tau > 2$, these trajectories all will be optimal for $\frac{3\pi}{2}$ units of time until they reach the half-line $\{x_1 \geq 1, x_2 = 1\}$, at which point optimal controls switch to $u = +1$. In some sense, the anchor point that determines the structure of time-optimal trajectories and of the reachable sets $\text{Reach}_{\Sigma, \leq T}(0)$ is the point $(1,1)$ that is reached along the X-trajectory that starts at the origin at time $T = \frac{\pi}{2}$. This is a conjugate point to the origin, and it is no longer optimal to follow X. As Fig. 7.6 shows, this trajectory now would enter the interior of the reachable sets. At this point, the control switches to $u = +1$ and afterward follows a Y-trajectory without any further switchings. In fact, for all points $0 \leq x_1 \leq 1$ with x_2 above the X-trajectory emanating from the origin, optimal

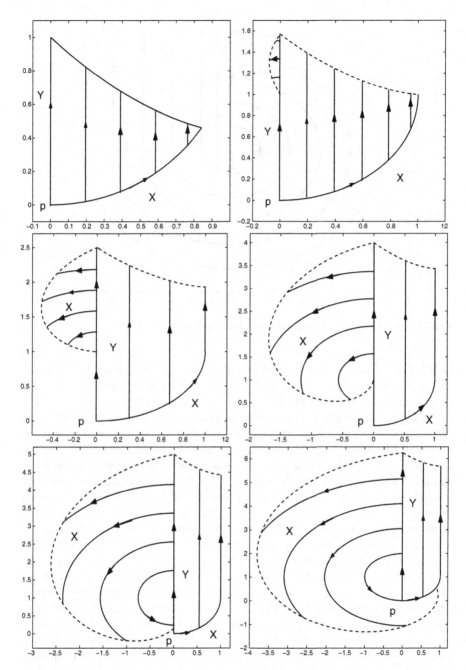

Fig. 7.6 The small-time reachable sets Reach$_{\Sigma, \leq T}(0)$ for $T = 1$ (*top row, left*), $T = \frac{\pi}{2}$ (*top row right*), $T = 2.5$ (*middle row, left*), $T = 4$ (*middle row, right*), $T = 5$ (*bottom row, left*), and $T = 6.25$ (*bottom row, right*)

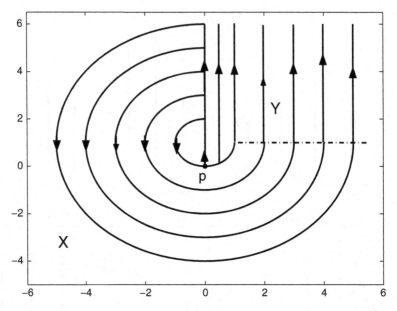

Fig. 7.7 Synthesis of time-optimal trajectories from the origin

trajectories are XY-arcs, while optimal trajectories that steer the origin into points in $\{x_1 > 1,\ x_2 > 1\}$ are of the type YXY, the longest time-optimal concatenation sequence to reach any point in $\mathbb{R}^2$.

Figure 7.7 shows the synthesis of time-optimal controls from the origin to an arbitrary point $q \in \mathbb{R}^2$. Note in particular that there exists a cycle consisting of an initial Y-trajectory over the interval $[0,2]$ followed by an X-trajectory over the interval $(2, 2 + \frac{\pi}{2})$ that at time $T = 2 + \frac{\pi}{2}$ reaches the origin again. This trajectory indeed is time-optimal over the interval $[0, 2 + \frac{\pi}{2})$ excluding the final point. As a result, the optimal value function for this problem, which here, for a given point $q \in \mathbb{R}^2$, denotes the minimum time when q is reachable from the origin, is discontinuous along the quarter-circle that connects the origin with $(1,1)$, the vertical half-line $\{x_1 = 1, x_2 \geq 1\}$, and the segment $\{x_1 = 0, 0 \leq x_2 \leq 1\}$; it also is not differentiable along the vertical half-line $\{x_1 = 0, x_2 \geq 1\}$. But clearly, the value function is lower semicontinuous at the discontinuities, and the synthesis shown in Fig. 7.7 satisfies all the requirements of the main verification theorem, Theorem 6.3.3. Overall, the set where $V^{\mathscr{E}}$ is not differentiable is a finite union of embedded submanifolds that define a stratification of the bad set B. It is not difficult to set up parameterized families of extremals that show that the value function is continuously differentiable away from these sets. Hence this synthesis is optimal. Naturally, one could also consider the time-optimal control problem to the origin with reversed directions of the vector fields, and the same results remain valid.

The fact that the value function is not continuous at the origin is caused by a lack of local controllability in the system. The system is completely controllable from the origin, i.e., $\text{Reach}_\Sigma(0) = \mathbb{R}^2$, but points (q_1, q_2) close to the origin that satisfy $q_1 < 0$ cannot be reached in small time, and this is the property that matters. As this example shows, a constructive approach of building fields of extremals is not hampered by such issues, and the theory easily applies to cover situations in which the value functions are discontinuous.

7.3 Small-Time Reachable Sets in Dimension 3

As the dimension n of the state space increases, features that are seen only in cases of positive codimension in lower dimensions emerge to become the nondegenerate scenarios in higher dimensions. In a certain sense, the higher-dimensional codimension-0 models provide an *unfolding* of the geometric structures of the small-time reachable sets $\text{Reach}_{\Sigma, \leq T}(p)$ from positive-codimension scenarios in lower dimensions. We illustrate this general feature for dimensions $n = 3$ and $n = 4$.

7.3.1 Boundary Trajectories in Dimension 3: Lobry's Example

We assume that

(B1) the vector fields f, g and their Lie bracket $[f, g]$ are linearly independent on a neighborhood M of a reference point $p \in \mathbb{R}^3$,

$$f(p) \wedge g(p) \wedge [f, g](p) \neq 0 \qquad \Longleftrightarrow \qquad X(p) \wedge Y(p) \wedge [X, Y](p) \neq 0.$$

The structure of boundary trajectories under these conditions was first analyzed by Lobry in his landmark paper [170], one of earliest papers on nonlinear control theory, that became a foundation for the development of the theory as a whole. The result is an immediate generalization from the two- to the three-dimensional case.

Proposition 7.3.1. *Under assumption (B1), if (x, u) is a boundary trajectory, then u is bang-bang with at most one switching. The boundary portion of the small-time reachable set $\text{Reach}_{\Sigma, \leq T}(p)$ that is not due to the restriction of the time to T is therefore composed of the cells $\mathscr{C}_{XY}$ and $\mathscr{C}_{YX}$.*

Proof. Let λ be a nontrivial covector field such that the conditions of Theorem 4.2.2 are satisfied. If the switching function $\Phi(t) = \langle \lambda(t), g(x(t)) \rangle$ has a zero at time t_1, then it follows that $\langle \lambda(t_1), f(x(t_1)) \rangle = 0$ as well, and so $\lambda(t_1)$ vanishes against both $X(x(t_1))$ and $Y(x(t_1))$. Since f, g, and $[f, g]$ are linearly independent on M, and since $\lambda(t_1)$ is nonzero, it cannot vanish against $[f, g]$ at $x(t_1)$. Hence

$$\dot{\Phi}(t_1) = \langle \lambda(t), [f, g](x(t)) \rangle \neq 0,$$

and thus t_1 is a bang-bang switch. Thus boundary trajectories are bang-bang with isolated switchings.

We show that there can be at most one. Suppose there exists a second switching at time $t_2 > t_1$, and for the sake of argument, assume that $u(t) \equiv -1$ on (t_1, t_2). Then $\lambda(t_2)$ also vanishes against both X and Y at $x(t_2)$. Let $q_1 = x(t_1)$, $q_2 = x(t_2)$ and move the vector $Y(q_2)$ back to q_1 along the flow of X. Setting $\tau = t_2 - t_1$, in exponential notation, this simply becomes

$$q_2 Y \exp(-\tau X) = q_1 \exp(\tau X) Y \exp(-\tau X)$$
$$\sim \exp(\tau \operatorname{ad} X) Y(q_1) = Y(q_1) + \tau[X, Y](q_1) + o(\tau),$$

where we made use of the asymptotic expansion (4.31) from Sect. 4.5. The covector λ is moved backward along the flow of X by integrating the covariational equation, which by Proposition 4.4.2, is the adjoint equation. Thus we simply get $\lambda(t_1)$. Furthermore, by Proposition 4.4.1, we have that

$$\langle \lambda(t_1), q_2 Y \exp(-\tau X) \rangle = \langle \lambda(t_1) (\exp(-\tau X))^*, q_2 Y \rangle = \langle \lambda(t_2), Y(q_2) \rangle = 0.$$

Hence $\lambda(t_1)$ vanishes against $X(q_1)$, $Y(q_1)$ and $\exp(\tau \operatorname{ad} X) Y(q_2)$. By the nontriviality of $\lambda(t_1)$, these vectors must be linearly dependent, and thus we have that

$$0 = X \wedge Y \wedge \exp(\tau \operatorname{ad} X) Y = X \wedge Y \wedge Y + \tau[X, Y] + o(\tau)$$
$$= \tau(1 + o(\tau)) (X \wedge Y \wedge [X, Y])$$

which contradicts the linear independence of X, Y and $[X, Y]$. Thus boundary trajectories are bang-bang with at most one switching. □

The underlying geometry becomes clear if we choose a good set of coordinates. Both canonical coordinates of the first and second kind will do here, and we use this simple structure to illustrate the computations for both cases. Coordinates of the *first kind* are defined by

$$(\xi_1, \xi_2, \xi_3) \mapsto p \exp(\xi_1 X + \xi_2 Y + \xi_3 [X, Y]).$$

In particular, by the uniqueness of the coordinate representation, the canonical coordinates for the X-trajectory $p \exp(sX)$, $0 \le s \le T$, are simply given by

$$\mathscr{C}_X = \{\xi \in \mathbb{R}^3 : 0 \le \xi_1 \le T, \ \xi_2 = 0, \ \xi_3 = 0\},$$

and the canonical coordinates for the Y-trajectory $p \exp(tY)$, $0 \le s \le T$, are

$$\mathscr{C}_Y = \{\xi \in \mathbb{R}^3 : \xi_1 = 0, \ 0 \le \xi_2 \le T, \ \xi_3 = 0\}.$$

The canonical coordinates for the surfaces $\mathscr{T}_{XY}$ and $\mathscr{T}_{YX}$ of XY- and YX-trajectories can easily be computed using the Baker–Campbell–Hausdorff formula. It follows from Corollary 4.5.5 that

$$p\exp(sX)\exp(tY) = p\exp\left(sX + tY + \frac{1}{2}st[X,Y] + st\rho\right),$$

with the remainder term ρ determined by higher-order brackets that are at least of order 3. Each such commutator has the factor st, and at least one more term s or t as an additional factor. Thus ρ is of order $O(T)$ in time, where once more, we use the Landau notation $O(T^k)$ to denote terms that can be bounded by CT^k for some constant $C < \infty$. If we express the remainder as a linear combination of the basis vector fields X, Y, and $[X,Y]$, then these terms only add higher-order perturbations of the form $stO(T)$ to the existing coefficients at X, Y and $[X,Y]$ and we get that

$$p\exp(sX)\exp(tY)$$
$$= p\exp\left(s(1+O(T^2))X + t(1+O(T^2))Y + \frac{1}{2}st(1+O(T))[X,Y]\right),$$

i.e., the ξ-coordinates of a point $q = p\exp(sX)\exp(tY) \in \mathscr{T}_{XY}$ are given by

$$\xi_1(q) = s(1+O(T^2)), \qquad \xi_2(q) = t(1+O(T^2)), \qquad \xi_3(q) = \frac{1}{2}st(1+O(T)).$$

The equations for ξ_1 and ξ_2 can be solved for s and t near $(0,0)$, and hence for ε sufficiently small, we can express the ξ_3-coordinate on the surface $\mathscr{T}_{XY}$ as a function of ξ_1 and ξ_2 in the form

$$\xi_3^{XY} = \frac{1}{2}\xi_1\xi_2(1+O(\|\xi\|)).$$

This expression is positive for small $\|\xi\|$. Hence we have the following statement:

Lemma 7.3.1. *In canonical coordinates of the first kind, the surface $\mathscr{T}_{XY}$ of XY-trajectories can be described as the graph of a smooth function $\xi_3^{XY}(\xi_1,\xi_2)$ that is positive on the first quadrant in the (ξ_1,ξ_2)-plane and has the cells $\mathscr{C}_X$ and $\mathscr{C}_Y$ along the positive ξ_1- and ξ_2-axes in its boundary.* ∎

Canonical coordinates for the surface $\mathscr{T}_{YX}$ follow in the same way, except that the orders of X and Y are reversed and this introduces a minus sign. Now we have that

$$p\exp(tY)\exp(sX)$$
$$= p\exp\left(s(1+O(T^2))X + t(1+O(T^2))Y - \frac{1}{2}st(1+O(T))[X,Y]\right),$$

and thus the ξ-coordinates of a point $q = p\exp(tY)\exp(sX) \in \mathscr{T}_{YX}$ are given by

$$\xi_1(q) = s(1+O(T^2)), \qquad \xi_2(q) = t(1+O(T^2)), \qquad \xi_3(q) = -\frac{1}{2}st(1+O(T)).$$

Hence, the ξ_3-coordinate on the surface $\mathscr{T}_{YX}$ also is given as a function of ξ_1 and ξ_2 in the form

$$\xi_3^{YX} = -\frac{1}{2}\xi_1\xi_2(1+O(\|\xi\|)).$$

We thus also have the following analogous statement:

Lemma 7.3.2. *In canonical coordinates of the first kind, the surface $\mathscr{T}_{YX}$ of YX-trajectories can be described as the graph of a smooth function $\xi_3^{XY}(\xi_1,\xi_2)$ that is negative on the first quadrant in the (ξ_1,ξ_2)-plane and also has the cells $\mathscr{C}_X$ and $\mathscr{C}_Y$ along the positive ξ_1- and ξ_2-axes in its boundary.* ∎

Indeed, the small-time reachable sets $\text{Reach}_{\Sigma,\leq T}(p)$ will lie between these two surfaces, but as before, their precise structures depend on the Lie-bracket configuration at p.

Canonical coordinates of the first kind are somewhat easier to use in simple low-dimensional examples, but as the dimension increases, and concatenations of longer sequences of trajectories need to be considered, the asymptotic product expansion in Proposition 4.5.2 becomes far superior in its ease of use. We give the computations of canonical coordinates of the *second kind* as well in order to explain the procedure at this simpler example. These coordinates are defined by

$$(\xi_1,\xi_2,\xi_3) \mapsto p\exp(\xi_3[X,Y])\exp(\xi_2Y)\exp(\xi_1X),$$

and now the coordinates of a point $q = p\exp(sY)\exp(tX)$ corresponding to a YX-trajectory are immediate, $\xi_1(q) = t$, $\xi_2(q) = s$, and $\xi_3(q) = 0$. Thus, as before, the cells $\mathscr{C}_X$ and $\mathscr{C}_Y$ lie on the positive ξ_1- and ξ_2-axes,

$$\mathscr{C}_X = \{\xi \in \mathbb{R}^3 : 0 \leq \xi_1 \leq T, \, \xi_2 = 0, \, \xi_3 = 0\},$$

$$\mathscr{C}_Y = \{\xi \in \mathbb{R}^3 : \xi_1 = 0, \, 0 \leq \xi_2 \leq T, \, \xi_2 = 0\},$$

but now also $\mathscr{C}_{YX}$ has a trivial coordinate representation and is simply a triangle in the positive (ξ_1,ξ_2)-quadrant,

$$\mathscr{C}_{YX} = \{\xi \in \mathbb{R}^3 : 0 \leq \xi_1, \, 0 \leq \xi_2, \, \xi_1+\xi_2 \leq T, \, \xi_3 = 0\}.$$

Canonical coordinates of the second kind for the surface $\mathscr{T}_{XY}$ are computed using the commutator formula in Corollary 4.5.2. We have that

$$q_0\exp(sX)\exp(tY) = q_0\exp(st\rho)\exp(st[X,Y])\exp(tY)\exp(sX),$$

with the remainder term ρ again determined by higher order brackets that are at least of order 3 and of order $O(T)$ in time. The remainder ρ again will be expressed as a linear combination of the basis vector fields X, Y, and $[X,Y]$, and when these terms are combined with the exponentials of X, Y, and $[X,Y]$, they only add higher-order terms of order $stO(T)$ to the existing coefficients at X, Y, and $[X,Y]$. We therefore get that

$$q_0 \exp(sX)\exp(tY) = q_0 \exp(st(1+O(T))[X,Y])$$
$$\times \exp\left(t(1+O(T^2))Y\right)\exp\left(s(1+O(T^2))X\right),$$

i.e., the ξ-coordinates of $q = q_0 \exp(sX)\exp(tY)$ are given by

$$\xi_1(q) = s(1+O(T^2)), \qquad \xi_2(q) = t(1+O(T^2)), \qquad \xi_3(q) = st(1+O(T)).$$

Hence, for ε sufficiently small, we again can express the ξ_3-coordinate on the cell $\mathscr{C}_{XY}$ as a function of ξ_1 and ξ_2 in the form $\xi_3 = \xi_1\xi_2(1+O(\|\xi\|))$, and thus, once more, the surface $\mathscr{T}_{XY}$ of XY-trajectories can be described as the graph of a smooth function $\xi_3^{XY}(\xi_1,\xi_2)$ that is positive on the first quadrant in the (ξ_1,ξ_2)-plane. The domain $D = \operatorname{dom}\left(\xi_3^{XY}\right)$ of this function lies in the first quadrant and has the coordinate axes in its boundary. Since the surface $\mathscr{T}_{YX}$ of YX-trajectories is described in the same way with the trivial function $\xi_3^{YX}(\xi_1,\xi_2) \equiv 0$, it follows that $\mathscr{T}_{XY}$ lies above $\mathscr{T}_{YX}$ in the direction of ξ_3 with the cells $\mathscr{C}_X$ and $\mathscr{C}_Y$ as their common boundary.

This argument illustrates the power of using a good set of coordinates, a pervasive feature in all our constructions. It also already points to the inductive nature of these constructions. The boundary trajectories for the small-time reachable set in Lobry's case in dimension 3 are given by the two cells $\mathscr{C}_{XY}$ and $\mathscr{C}_{YX}$ that describe the full small-time reachable sets $\operatorname{Reach}_{\Sigma,\leq T}(p)$ in dimension 2 in the codimension-0 case. The full three-dimensional small-time reachable set $\operatorname{Reach}_{\Sigma,\leq T}(p)$ is then constructed relative to this frame.

7.3.2 Small-Time Reachable Sets under Codimension-0 Assumptions

Throughout this section, assumption (B1) remains in effect, and we write the second-order brackets as linear combinations of the basis X, Y, and $[X,Y]$,

$$[X,[X,Y]] = a_1X + a_2Y + a_3[X,Y] = \alpha f + \cdots,$$
$$[Y,[X,Y]] = b_1X + b_2Y + b_3[X,Y] = \beta f + \cdots.$$

We determine the small-time reachable set $\text{Reach}_{\Sigma,\leq T}(p)$, and more generally, as in Sects. 2.9 and 2.10 for the two-dimensional case, we give the structure of time-optimal controlled trajectories that lie in a sufficiently small neighborhood of p, under the following codimension-0 assumptions:

(B2) The functions α and β do not vanish at p; equivalently, the vector fields g, $[f,g]$, and $[f \pm g, [f,g]]$ are linearly independent at p:

$$g(p) \wedge [f,g](p) \wedge [X,[f,g]](p) \neq 0 \quad \text{and} \quad g(p) \wedge [f,g](p) \wedge [Y,[f,g]](p) \neq 0.$$

Geometrically, the functions α and β indicate to which side of the plane generated by g and $[f,g]$ the vector fields $[X,[X,Y]]$ and $[Y,[X,Y]]$ point. The wedge product of three vectors vanishes if and only if these vectors are linearly dependent. Thus under assumption (B1), the wedge product of g and $[f,g]$ with a third vector Z vanishes if and only if Z lies in the plane generated by the linearly independent vectors g and $[f,g]$. In particular, $[X,[f,g]]$ and $[Y,[f,g]]$ point to the same side of the plane spanned by g and $[f,g]$ if and only if the wedge products $g \wedge [f,g] \wedge [X,[f,g]]$ and $g \wedge [f,g] \wedge [Y,[f,g]]$ have the same sign, and to opposite sides if these signs are different. As in the two-dimensional case, these geometric properties are connected with the existence and optimality properties of singular arcs for the time-optimal control problem.

In this section, we prove the following result, and, as a corollary, determine the full structure of the small-time reachable sets $\text{Reach}_{\Sigma,\leq T}(p)$.

Theorem 7.3.1. *Let M be a sufficiently small neighborhood of a point p for which conditions (B1) and (B2) are satisfied. Then time-optimal controlled trajectories that lie in M are concatenations of at most three pieces of the following type:*

$$
\begin{array}{llll}
YXY & \text{if} & \alpha(p) > 0 & \text{and} \quad \beta(p) > 0, \\
XYX \text{ or } YXY & \text{if} & \alpha(p) > 0 & \text{and} \quad \beta(p) < 0, \\
BSB & \text{if} & \alpha(p) < 0 & \text{and} \quad \beta(p) > 0, \\
XYX & \text{if} & \alpha(p) < 0 & \text{and} \quad \beta(p) < 0.
\end{array}
$$

This is a substantial result with a long proof that we develop in several steps. Suppose (x,u) is a time-optimal controlled trajectory that lies in M. By Theorem 4.2.3, there exist a constant $\lambda_0 \geq 0$ and a nontrivial solution $\lambda : [0,T] \to (\mathbb{R}^3)^*$ of the adjoint equation along (x,u) such that we have

$$u(t) = -\text{sgn } \Phi(t) = -\text{sgn } \langle \lambda(t), g(x(t)) \rangle$$

almost everywhere on $[0,T]$ and

$$H = \lambda_0 + \langle \lambda(t), f(x(t)) + u(t)g(x(t)) \rangle = 0.$$

It was already shown above that these relations are *always valid at switching times*, and we freely use this fact henceforth.

Lemma 7.3.3. *On a sufficiently small neighborhood M of p, the only abnormal extremals are XY- and YX-trajectories.*

Proof. For an abnormal extremal, the multiplier λ vanishes against both vector fields f and g at every switching time. Thus, if (x,u) is a time-optimal controlled trajectory that has two switchings at times $t_1 < t_2$, then setting $q_i = x(t_i)$, $\lambda(t_1)$ vanishes against $f(q_1)$, $g(q_1)$, but also against the vector $g(q_2)$ moved back along the flow to the point q_1. Regardless of the value of the control between t_1 and t_2, this generates a third vector of the form $\pm[f,g](q_1) + O(T)$, which by assumption (B1), is linearly independent of $f(q_1)$ and $g(q_1)$. Contradiction. $\qquad\square$

We henceforth consider only normal extremals and set $\lambda_0 = 1$. We begin the proof of Theorem 7.3.1 with the analysis of possible singular arcs.

Proposition 7.3.2. *For a point $q \in M$, if $\alpha(q)$ and $\beta(q)$ have the same sign, then there does not exist an admissible singular arc through q. If $\alpha(q)$ and $\beta(q)$ have opposite signs, then there exists a unique singular arc passing through q, and it is given by the integral curve of the vector field*

$$S = \frac{\alpha Y - \beta X}{\alpha - \beta}.$$

The singular arc is fast if $\alpha(q) < 0$ and it is slow if $\alpha(q) > 0$.

Proof. For notational clarity, we often drop the argument t in our computations. The second derivative of the switching function $\Phi(t) = \langle \lambda(t), g(x(t)) \rangle$ can be expressed in the form

$$\ddot{\Phi}(t) = \langle \lambda, [f,[f,g]](x) \rangle + u \langle \lambda, [g,[f,g]](x) \rangle$$
$$= \frac{1}{2}(1-u)\langle \lambda, [X,[X,Y]](x) \rangle + \frac{1}{2}(1+u)\langle \lambda(t), [Y,[X,Y]](x) \rangle.$$

If the control u is singular over an open interval, then λ vanishes against $g(x)$ and $[f,g](x)$, and $\langle \lambda, f(x) \rangle = -1$. Expressing the second-order brackets in terms of f, g, and $[f,g]$, the condition $\ddot{\Phi}_\Gamma(t) \equiv 0$ is equivalent to

$$0 = (1-u)\alpha(x) + (1+u)\beta(x),$$

which determines the singular control as

$$u_{\text{sing}}(t) = \frac{\alpha(x(t)) + \beta(x(t))}{\alpha(x(t)) - \beta(x(t))}.$$

Note that $\left|u_{\text{sing}}(t)\right| < 1$ if α and β have opposite signs, while $\left|u_{\text{sing}}(t)\right| > 1$ if α and β have the same sign. Hence, the singular control is not admissible in the second case and no singular arc exists through q. In the first case,

$$S = f + u_{\text{sing}}g = \frac{1}{2}\left(X + Y + \frac{\alpha+\beta}{\alpha-\beta}(Y-X)\right) = \frac{\alpha Y - \beta X}{\alpha-\beta},$$

defines a smooth vector field whose integral curves are admissible singular arcs. The Legendre–Clebsch condition for optimality requires that

$$0 \geq \langle \lambda, [g,[f,g]](x)\rangle = \frac{1}{2}\langle \lambda, [Y-X,[X,Y]](x)\rangle = \frac{1}{2}(\alpha(x) - \beta(x)),$$

where we use that $\langle \lambda, f(x)\rangle = -1$. Thus the Legendre–Clebsch condition is satisfied for $\alpha < 0$ and violated for $\alpha > 0$.   □

In contrast to the two-dimensional case, we now have a well-defined singular vector field that can be used everywhere, not just on one singular arc.

An important concept in our analysis of the various cases (which has already been used in the proof of Proposition 7.3.1 and, implicitly, in the proof of the lemma about abnormal extremals above) is what we called g-dependent points earlier, and what, in the context here, generally are called conjugate triples (e.g., [210–212, 240, 242]). We shall show below that indeed these points are conjugate, and we therefore retain both terminologies.

Definition 7.3.1 (g-Dependent points/conjugate triples.). Given an extremal controlled trajectory (x,u), three consecutive points $q_1 = x(t_1)$, $q_2 = x(t_2)$, and $q_3 = x(t_3)$, $t_1 \leq t_2 \leq t_3$, along the trajectory x where the switching function vanishes, $\Phi(t_i) = 0$ for $i = 1, 2, 3$, are called g-dependent, and the points (q_1, q_2, q_3) form a conjugate triple. We allow that two of the points agree if the switching function has a double zero at the corresponding time, and in this case we call (q_1, q_2, q_3) a singular conjugate triple.

Conjugate triples impose nontrivial relations on the times along the controlled trajectory between the consecutive switchings that can easily be computed.

Lemma 7.3.4. Let (q_1, q_2, q_3) be a conjugate triple along an $\cdot XY \cdot$-trajectory with switching points q_1, $q_2 = q_1 \exp(s_1 X)$, $s_1 > 0$, and $q_3 = q_2 \exp(s_2 Y)$, $s_2 > 0$. Then $g(q_1)$ and the vectors obtained by moving $g(q_2)$ and $g(q_3)$ back to q_1 along the flow of the extremal are linearly dependent. These two vectors are given by

$$q_2 g \exp(-s_1 X) = q_1 \exp(s_1 X) g \exp(-s_1 X) \sim \exp(s_1 \operatorname{ad} X) g(q_2)$$

and

$$q_3 g Y \exp(-s_2 Y) \exp(-s_1 X) = q_1 \exp(s_1 X) \exp(s_2 Y) g \exp(-s_2 Y) \exp(-s_1 X)$$

$$\sim \exp(s_1 \operatorname{ad} X) \exp(s_2 \operatorname{ad} Y) g(q_3).$$

Recall that the exponential notation provides a common framework for the flows of the vector fields as well as for moving tangent vectors along these flows. Thus, $q_2 g \exp(-s_1 X)$ denotes the vector $g(q_2)$ moved back along the flow of the vector field X to the point q_1, and so on. Also, we consistently use the notation $\cdot XY\cdot$ with the dot $\cdot$ indicating that there is a switching point at the initial point q_1 and at the endpoint $q_3 = q_1 \exp(s_1 X) \exp(s_2 Y)$.

Proof. The lemma is a direct consequence of the conditions of the maximum principle, and a similar argument has already been made several times: if we take the time at the junction q_1 to be zero, then there exists a nontrivial solution λ to the adjoint equation such that $\langle \lambda(s_1), g(q_2) \rangle = 0$ and $\langle \lambda(s_1 + s_2), g(q_3) \rangle = 0$. By Proposition 4.4.1, the value of this inner product does not change if both λ and the vectors g are moved back along the flow to the point q_1, i.e.,

$$0 = \langle \lambda(0), \exp(s_1 \, \mathrm{ad}X) g(q_2) \rangle$$

and

$$0 = \langle \lambda(0), \exp(s_1 \, \mathrm{ad}X) \exp(s_2 \mathrm{ad}Y) g(q_3) \rangle .$$

Since $\lambda(0)$ is nonzero, it follows that the three vectors $g(q_1)$, $\exp(s_1 \, \mathrm{ad}X) g(q_2)$, and $\exp(s_1 \, \mathrm{ad}X) \exp(s_2 \, \mathrm{ad}Y) g(q_3)$ are linearly dependent.   □

This linear dependency relation can easily be evaluated. Recall that the wedge product is a multi-linear form that obeys the same rules as the determinant. Thus, when taking the wedge product of

$$\exp(s_1 \mathrm{ad}X) g(q_2) = g(q_1) + s_1 [X,g](q_1) + o(s_1)$$

with $g(q_1)$, the constant term in this expansion gets canceled, and we therefore can factor s_1. Using the operator notation

$$\left(\frac{\exp(s_1 \, \mathrm{ad}X) - \mathrm{id}}{s_1} \right) g = [X,g] + o(1),$$

and also dropping the first switching point q_1 for notational convenience, it follows that

$$0 = g \wedge \exp(s_1 \, \mathrm{ad}X) g \wedge \exp(s_1 \, \mathrm{ad}X) \exp(s_2 \mathrm{ad}Y) g$$

$$= s_1 s_2 \left(g \wedge \left(\frac{\exp(s_1 \, \mathrm{ad}X) - \mathrm{id}}{s_1} \right) g \wedge \exp(s_1 \, \mathrm{ad}X) \left(\frac{\exp(s_2 \, \mathrm{ad}Y) - \mathrm{id}}{s_2} \right) g \right)$$

$$= s_1 s_2 \left(g \wedge [X,g] + \frac{1}{2} s_1 [X,[X,g]] + \cdots \right.$$

$$\left. \wedge \exp(s_1 \, \mathrm{ad}X) \left([Y,g] + \frac{1}{2} s_2 [Y,[Y,g]] + \cdots \right) \right)$$

$$= s_1 s_2 \left(g \wedge [X,g] + \frac{1}{2} s_1 [X,[X,g]] + \cdots \right.$$

$$\left. \wedge [Y,g] + s_1 [X,[Y,g]] + \frac{1}{2} s_2 [Y,[Y,g]] + \cdots \right).$$

Dividing by s_1 and s_2, and using the coordinate expressions for $[X,[f,g]]$ and $[Y,[f,g]]$, we obtain that

$$0 = g \wedge [f,g] + \frac{1}{4} \alpha s_1 f + \cdots \wedge [f,g] + \left(\frac{1}{2} \alpha s_1 + \frac{1}{4} \beta s_2 + \cdots \right) f$$

$$= \begin{vmatrix} 0 & 1 & 0 \\ \frac{1}{4} \alpha s_1 + \cdots & 0 & 1 + \cdots \\ \frac{1}{2} \alpha s_1 + \frac{1}{4} \beta s_2 + \cdots & 0 & 1 + \cdots \end{vmatrix} (f \wedge g \wedge [f,g])$$

$$= -\frac{1}{4} (\alpha s_1 + \beta s_2 + \cdots)(f \wedge g \wedge [f,g])$$

which implies that

$$\alpha(q_1) s_1 + \beta(q_1) s_2 + \cdots = 0. \tag{7.6}$$

We call this equation a *conjugate point relation*. Similar computations, or just the use of input symmetries (see Sect. 2.10), lead to analogous formulas for $\cdot YX \cdot$-trajectories, where as above, the dots indicate that there are switching points at the initial and final points of this segment. We want to stress the ease of these computations that give *necessary conditions for optimality when bang-bang trajectories have a larger number of switchings*. This will become more and more important as the dimension of the state space increases. Note that Eq. (7.6) cannot be satisfied with small times s_1 and s_2 near p if $\alpha(p)$ and $\beta(p)$ have the same sign. Hence in such a case there cannot be optimal bang-bang trajectories with more than two switchings.

Proposition 7.3.3. *If $\alpha(p)$ and $\beta(p)$ have the same sign, then time-optimal controlled trajectories that lie in a sufficiently small neighborhood M of p are at most of the type YXY if $\alpha(p) > 0$ and of the type XYX if $\alpha(p) < 0$. In particular, the small-time reachable set from p is given by*

$$\mathrm{Reach}_{\Sigma, \leq T}(p) = \begin{cases} \mathscr{C}_{YXY} & \text{if} \quad \alpha(p) > 0, \\ \mathscr{C}_{XYX} & \text{if} \quad \alpha(p) < 0. \end{cases}$$

Proof. On a sufficiently small neighborhood M of p, the linear terms in Eq. (7.6) dominate the higher-order remainders, and thus there cannot exist optimal $\cdot XY \cdot$-concatenations. Applying the input symmetry ρ of Sect. 2.10 that interchanges X with Y, the same follows for $\cdot YX \cdot$-concatenations.

The specific structures follow from some convexity-type properties of the switching function in these cases. Without loss of generality, we consider the case in which $\alpha(p)$ and $\beta(p)$ are positive on M. We *claim* that with the exception of some cases in which optimal controls are bang-bang with at most one switching, on a sufficiently small neighborhood M of p, the second derivative $\ddot{\Phi}(\tau)$ of the switching function is always negative along an optimal Y-trajectory when $\dot{\Phi}(\tau) = 0$. Once this fact has been established, if $I = (t_1, t_2)$ is a maximal open subinterval of $[0, T]$ where $u = +1$, then it follows from the fact that the switching function is negative along Y-trajectories that we must have either $t_1 = 0$ or $t_2 = T$. For if $0 < t_1 < t_2 < T$, then the switching function has a negative minimum at some time $\tau \in (t_1, t_2)$, contradicting the fact that $\ddot{\Phi}(\tau) < 0$. (Note that the control $u \equiv +1$ cannot be singular if α and β have the same sign, and thus the switching function cannot vanish identically on I; in principle, this a possible degenerate case.) Thus Y-trajectories necessarily need to lie at the beginning or at the end of $[0, T]$, and optimal controlled trajectories are at most of the form YXY.

It remains to establish our claim above. The second derivative of Φ along Y for a time τ where $\dot{\Phi}(\tau) = \langle \lambda(\tau), [X, Y](x) \rangle = 0$ is given by

$$\ddot{\Phi}(\tau) = \langle \lambda, [Y, [f, g]](x) \rangle = \beta(x) \langle \lambda, f(x) \rangle + (b_2(x) - b_1(x)) \langle \lambda, g(x) \rangle.$$

It follows from $H \equiv 0$ that $\langle \lambda, Y(x) \rangle = -1$ along an extremal Y-trajectory, and thus we have that

$$\ddot{\Phi}(\tau) = \langle \lambda, [Y, [f, g]](x) \rangle = -\beta(x) + \omega(x) \langle \lambda, g(x) \rangle$$

where $\omega = b_2 - b_1 - \beta$ is a smooth function. It is not difficult, but somewhat tedious, to show that after normalizing the multiplier $\lambda(0)$ so that $\|\lambda(0)\|_1 = 1$, for every $\kappa > 0$ there exists a small enough neighborhood M of p such that optimal trajectories that lie in M are bang-bang with at most one switching whenever $|\langle \lambda(0), g(p) \rangle| \geq 1 - \kappa$. (We refer the interested reader to [209, 210] for this argument.) As a consequence, it is possible to choose the neighborhood small enough that the term $-\beta$ dominates $\omega(x) \langle \lambda, g(x) \rangle$ for all the remaining choices of the multipliers. Consequently, the sign of the second derivative $\ddot{\Phi}(\tau)$ is given by the negative sign of $\beta(p)$, which proves the result. $\qquad \square$

In fact, this argument could be used to establish the following stronger convexity properties of the switching function.

Lemma 7.3.5. *[209] There exists a neighborhood M of p such that for all extremals that are not bang-bang with at most one switching, the switching function Φ has the following convexity properties: Φ is strictly convex along any X-trajectory if $\alpha(p) < 0$ and strictly concave if $\alpha(p) > 0$ and Φ is strictly convex along any Y-trajectory if $\beta(p) < 0$ and strictly concave if $\beta(p) > 0$.* $\qquad \blacksquare$

Proposition 7.3.4. *If $\alpha(p) < 0$ and $\beta(p) > 0$, then time-optimal controlled trajectories that lie in a sufficiently small neighborhood M of p are concatenations of*

Fig. 7.8 A qualitative sketch
of the structure of the
endpoints in the boundary
Reach$_{\Sigma,T}(p)$ of the
small-time reachable set for
$\alpha(p) < 0$ and $\beta(p) > 0$

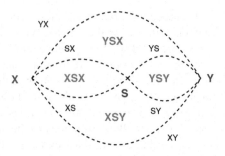

at most the type BSB, where B is either X or Y and S denotes a singular arc. The
small-time reachable set from p is given by the union of the corresponding cells
$\mathscr{C}_{XSX}$, $\mathscr{C}_{XSY}$, $\mathscr{C}_{YSX}$, and $\mathscr{C}_{YSY}$,

$$\text{Reach}_{\Sigma,\leq T}(p) = \mathscr{C}_{BSB}.$$

Figure 7.8 illustrates the qualitative structure of the boundary portion Reach$_{\Sigma,T}(p)$
of the small-time reachable set that is defined by the points that are reachable time-
optimally in time T.

Proof. In this case, and ignoring the cases in which optimal controls a priori
are bang-bang with at most one switching, the switching function Φ is strictly
convex along X-trajectories and strictly concave along Y-trajectories. Both of these
convexity properties prevent further switchings, and it follows that both X- and
Y-trajectories need to lie at the end of the interval $[0,T]$. Singular controls are
admissible and satisfy the Legendre–Clebsch condition and thus overall, only BSB
can be time-optimal.

It is illustrative to employ a singular conjugate point relation in this case as well.
Let $*X\cdot$ denote a concatenation of the form $q_2 = q_1 \exp(sX)$ with the $*$ indicating
that there is a singular junction at q_1 and the dot representing another junction at q_2.
Setting the time along the trajectory at q_1 to 0, it then follows that the multiplier
$\lambda(0)$ vanishes against the vectors $g(q_1)$, $[f,g](q_1)$, and against the vector $g(q_2)$
transported back to q_1 along the flow of X. As before, this vector is given by

$$\exp(sadX)g(q_2) = g(q_1) + s[f,g](q_1) + \frac{1}{2}s^2[X,f,g]](q_1) + o(s^2).$$

But in the wedge product with $g(q_1)$ and $[f,g](q_1)$, the constant and linear terms get
canceled and thus, writing the second-order bracket $[X,f,g]]$ as a linear combination
of f, g, and $[f,g]$, the linear dependence of these three vectors is equivalent to
$\alpha(q_1) + o(1) = 0$. This is not possible on a small enough neighborhood of p. $\quad\square$

Proposition 7.3.5. *If $\alpha(p) > 0$ and $\beta(p) < 0$, then time-optimal controlled trajectories that lie in a sufficiently small neighborhood M of p are trajectories of at most
type XYX or YXY.*

Fig. 7.9 A variation along a
$YXYX$-trajectory

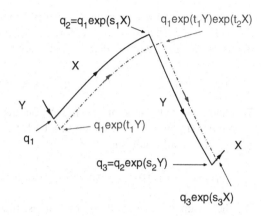

This result is the three-dimensional analogue of Proposition 2.9.5 from Sect. 2.10 that describes time-optimal trajectories near a slow singular arc. As in the planar case, this result lies considerably deeper than the previous propositions, and we develop its proof in several steps.

In this case, the convexity properties of the switching function are conducive to a large number of switchings: Φ is strictly concave along X-trajectories where it is positive and strictly convex along Y-trajectories where it is negative. In the work by Kupka and Bonnard (e.g., [43, 49, 142]), this situation is called *elliptic*, reflecting the characteristic convexity properties of the switching function, while the case in which α and β have the same sign is called *hyperbolic* in view of the fact that for at least one of the constant controls $u = +1$ and $u = -1$, the curvature works against zeros for the switching function. The case $\alpha < 0$ and $\beta > 0$ would be *parabolic* in this classification. We prefer to classify these cases according to the existence and local optimality of singular arcs.

Ignoring situations in which optimal controls are bang-bang with at most one switching a priori, the convexity properties of the switching function imply that switchings occur whenever the switching function vanishes. Thus time-optimal extremals that lie in a sufficiently small neighborhood of p are bang-bang. But this time, *the number of switchings can be large, and indeed there exist bang-bang extremals with an arbitrarily large number of switchings.* But they are not optimal. In order to exclude the optimality of bang-bang controls with three switchings, we set up a specific variation that is not a point variation and indeed is quite different from these variations considered earlier (see Fig. 7.9). Essentially, *our construction below sets up a parameterized family of bang-bang extremals* as defined in Sect. 6.1 for which *the third switching points are transversal folds.* It is in this construction that the Lie-algebraic computations show their full force, and we use the proof of this result to showcase these geometric techniques. They center on the computation of canonical coordinates using the exponential formalism and associated with it the determination of tangent spaces for surfaces defined by various concatenations of trajectories of the system.

Consider a $YXYX$ concatenation that lies in M and denote the switching points by $q_1, q_2,$ and q_3. Thus (q_1, q_2, q_3) form a conjugate triple. Let $s_1 > 0$ and $s_2 > 0$ be the times along the intermediate X- and Y-arcs, i.e.,

$$q_2 = q_1 \exp(s_1 X) \qquad \text{and} \qquad q_3 = q_2 \exp(s_2 Y) = q_1 \exp(s_1 X) \exp(s_2 Y).$$

We shall show that the point $q_3 \exp(s_3 X)$ on this $YXYX$-trajectory is also reachable from the first junction q_1 by means of a YXY-trajectory, i.e., that there exist positive times t_1, t_2, t_3, all small, such that

$$q_1 \exp(s_1 X) \exp(s_2 Y) \exp(s_3 X) = q_1 \exp(t_1 Y) \exp(t_2 X) \exp(t_3 Y). \tag{7.7}$$

We then want to compare these total times. If we consider Eq. (7.7) simply as an equation in the variables s_i and t_j, then this equation has the trivial solutions $0 = t_1$, $s_1 = t_2$, $s_2 = t_3$, $s_3 = 0$ and $s_1 = 0$, $s_2 = t_1$, $s_3 = t_2$, $0 = t_3$. We are interested in solutions that lie near the first of these two trivial solutions and therefore desingularize this behavior through a blowup of s_3 in the direction of s_1. That is, rather than using s_3 as a variable, we make a transformation of the form $s_3 = \tau s_1$ and instead use τ as the third variable. Thus, we consider the following form of Eq. (7.7):

$$q_1 \exp(s_1 X) \exp(s_2 Y) \exp(s_1 \tau X) = q_1 \exp(t_1 Y) \exp(t_2 X) \exp(t_3 Y) \tag{7.8}$$

and denote the difference in time by

$$\Delta = \Delta(\tau) = t_1 + t_2 + t_3 - s_1 - s_2 - s_1 \tau.$$

We first show that given (s_1, s_2) and $\tau > 0$ small, there exist unique times t_j that are defined by smooth functions of (s_1, s_2) and τ that solve this equation and then compute a second-order Taylor expansion for Δ in terms of τ for (s_1, s_2) fixed. We start with expressing both sides of Eq. (7.8) in terms of canonical coordinates of the second kind of the form

$$(\xi_1, \xi_2, \xi_3) \mapsto q_1 \exp(\xi_3[X, Y]) \exp(\xi_2 Y) \exp(\xi_1 X).$$

Using Corollary 4.5.2 to calculate the coordinates in terms of the s_i and t_j, for XYX-trajectories, we obtain

$$q_1 \exp(s_1 X) \exp(s_2 Y) \exp(s_1 \tau X)$$

$$= q_1 \cdots \exp\left(\frac{1}{2} s_1^2 s_2 [X, [X, Y]]\right) \exp\left(\frac{1}{2} s_1 s_2^2 [Y, [X, Y]]\right)$$

$$\times \exp(s_1 s_2 [X, Y]) \exp(s_2 Y) \exp(s_1 (1 + \tau) X). \tag{7.9}$$

The higher-order brackets will be expressed as linear combinations of the basis, and we write

$$[X,[X,Y]] = a_1(q_1)X + a_2(q_1)Y + a_3(q_1)[X,Y] + R_a,$$

where the coefficients a_i are evaluated at the initial point q_1 and R_a is the remainder defined by this equation. Thus $R_a(q_1) = 0$, or equivalently, $q_1 = q_1 \exp(R_a)$. Substituting this expression into Eq. (7.9) and expanding into a product of exponentials gives

$$q_1 \cdots \exp\left(\frac{1}{2} s_1^2 s_2 [X,[X,Y]]\right) \cdots$$

$$= q_1 \cdots \exp\left(\frac{1}{2} s_1^2 s_2 \left(a_1(q_1)X + a_2(q_1)Y + a_3(q_1)[X,Y] + R_a\right)\right) \cdots$$

$$= q_1 \cdots \exp\left(\frac{1}{2} s_1^2 s_2 R_a\right) \exp\left(\frac{1}{2} s_1^2 s_2 a_3(q_1)[X,Y]\right)$$

$$\times \exp\left(\frac{1}{2} s_1^2 s_2 a_2(q_1)Y\right) \exp\left(\frac{1}{2} s_1^2 s_2 a_1(q_1)X\right) \cdots,$$

where all commutator terms have at least the factor $s_1^4 s_2^2$ and become part of higher-order terms. The X-, Y-, and $[X,Y]$-exponentials now need to be rearranged so that they can be combined with other like terms. For example, $\exp\left(\frac{1}{2} s_1^2 s_2 a_1(q_1)X\right)$ needs to be combined with the term $\exp(s_1(1 + \tau)X)$. In order to do so, it needs to be commuted with other terms to become adjacent to this term. In doing so, additional commutator terms arise. But each of them has at least the factor $s_1^2 s_2^2$ and thus is of higher order when compared with the existing terms. Hence, modulo commutators that only contribute to higher-order terms, we can simply replace the third-order brackets by their linear combinations and then combine like terms. This gives

$$q_1 \exp(s_1 X) \exp(s_2 Y) \exp(s_1 \tau X)$$

$$= q_1 \exp\left(s_1 s_2 \left(1 + \frac{1}{2} a_3(q_1)s_1 + \frac{1}{2} b_3(q_1)s_2 + o(s)\right)[X,Y]\right)$$

$$\times \exp\left(\left(s_2 + \frac{1}{2} s_1 s_2 [a_2(q_1)s_1 + b_2(q_1)s_2 + o(s)]\right)Y\right)$$

$$\times \exp\left(\left(s_1(1 + \tau) + \frac{1}{2} s_1 s_2 [a_1(q_1)s_1 + b_1(q_1)s_2 + o(s)]\right)X\right),$$

with the expression $o(s)$ once more denoting higher-order terms $\rho(s)$ that have the property that $\lim_{\|s\| \to 0} \frac{\rho(s)}{\|s\|} = 0$, where $\|s\| = s_1 + s_2$ and the times s_1 and s_2 are positive. In fact, if the vector fields f and g are three time continuously differentiable, all these higher-order terms are at least quadratic in s_1 and s_2. Notice, however, that τ never partakes in the commutation procedures and thus does not come up in these higher-order terms. The coordinates therefore are exact in τ. This gives us the following formulas:

Lemma 7.3.6. *The canonical coordinates for $q_1 \exp(s_1 X)\exp(s_2 Y)\exp(s_1 \tau X)$ are given by*

$$\xi_1^{XYX}(s) = s_1(1+\tau) + \frac{1}{2}s_1 s_2 \left(a_1(q_1)s_1 + b_1(q_1)s_2 + o(s) \right),$$

$$\xi_2^{XYX}(s) = s_2 + \frac{1}{2}s_1 s_2 \left(a_2(q_1)s_1 + b_2(q_1)s_2 + o(s) \right),$$

$$\xi_3^{XYX}(s) = s_1 s_2 \left(1 + \frac{1}{2}a_3(q_1)s_1 + \frac{1}{2}b_3(q_1)s_2 + o(s) \right).$$

Similarly, for YXY-trajectories, we obtain that

$q_1 \exp(t_1 Y)\exp(t_2 X)\exp(t_3 Y)$

$$= q_1 \exp(t_1 Y)\cdots\exp\left(\frac{1}{2}t_2^2 t_3[X,[X,Y]] \right)\exp\left(\frac{1}{2}t_2 t_3^2[Y,[X,Y]] \right)$$

$$\times \exp(t_2 t_3[X,Y])\exp(t_3 Y)\exp(t_2 X)$$

$$= q_1 \cdots \exp\left(\frac{1}{2}t_2^2 t_3[X,[X,Y]] \right)\cdots\exp\left(\frac{1}{2}t_2 t_3^2[Y,[X,Y]] \right)\exp(t_1 Y)$$

$$\times \exp(t_2 t_3[X,Y])\exp(t_3 Y)\exp(t_2 X)$$

$$= q_1 \cdots \exp\left(\frac{1}{2}t_2^2 t_3[X,[X,Y]] \right)\exp\left(\frac{1}{2}t_2 t_3^2[Y,[X,Y]] \right)\cdots\exp(t_1 t_2 t_3[Y,[X,Y]])$$

$$\times \exp(t_2 t_3[X,Y])\exp((t_1 + t_3)Y)\exp(t_2 X).$$

Note that an extra commutator term that is cubic in the times t_j arises when $\exp(t_1 Y)$ and $\exp(t_1 t_2[X,Y])$ are commuted. As above, we express the third-order brackets as linear combinations of the basis and overall obtain

$q_1 \exp(t_1 Y)\exp(t_2 X)\exp(t_3 Y)$

$$= q_1 \exp\left(t_2 t_3 \left(1 + \frac{1}{2}t_2 a_3(q_0) + \left(\frac{1}{2}t_3 + t_1 \right) b_3(q_0) + o(t) \right)[X,Y] \right)$$

$$\times \exp\left(\left((t_1 + t_3 + t_2 t_3 \left[\frac{1}{2}t_2 a_2(q_0) + \left(\frac{1}{2}t_3 + t_1 \right) b_2(q_0) + o(t) \right] \right)Y \right)$$

$$\times \exp\left(\left((t_2 + t_2 t_3 \left[\frac{1}{2}t_2 a_1(q_0) + \left(\frac{1}{2}t_3 + t_1 \right) b_1(q_0) + o(t) \right] \right)X \right),$$

which gives the following formulas:

Lemma 7.3.7. *The* canonical *coordinates for* $q_1 \exp(t_1 Y) \exp(t_2 X) \exp(t_3 Y)$ *are given by*

$$\xi_1^{YXY}(t) = t_2 + t_2 t_3 \left[\frac{1}{2} a_1(q_1) t_2 + b_1(q_1) \left(\frac{1}{2} t_3 + t_1 \right) + o(t) \right],$$

$$\xi_2^{YXY}(t) = t_1 + t_3 + t_2 t_3 \left[\frac{1}{2} a_2(q_1) t_2 + b_2(q_1) \left(\frac{1}{2} t_3 + t_1 \right) + o(t) \right],$$

$$\xi_3^{YXY}(t) = t_2 t_3 \left(1 + \frac{1}{2} a_3(q_1) t_2 + b_3(q_1) \left(\frac{1}{2} t_3 + t_1 \right) + o(t) \right).$$

Lemma 7.3.8. *There exists a neighborhood M of p with the property that for arbitrary points $q_1 \in M$, sufficiently small positive times s_1 and s_2, and a positive parameter τ, the equation*

$$q_1 \exp(s_1 X) \exp(s_2 Y) \exp(s_1 \tau X) = q_1 \exp(t_1 Y) \exp(t_2 X) \exp(t_3 Y)$$

has a unique solution in t that is given by smooth functions $t_i(\tau; s)$, $i = 1, 2, 3$, of s_1, s_2, and τ. Furthermore, s_2 divides t_1 and t_3, and s_1 divides t_2. Low-order expansions for the solutions have the form

$$t_1 = s_2 \left(\frac{\tau}{1+\tau} + o(s_1) \right), \quad t_2 = s_1 \left[(1+\tau) + o(s_2) \right], \quad t_3 = s_2 \left(\frac{1}{1+\tau} + o(s_1) \right),$$

$$(7.10)$$

where $o(s_1)$ and $o(s_2)$ denote higher-order terms that after division by s_1 and s_2, respectively, vanish at the origin.

Proof. The function ζ defined by

$$\zeta(t) = \frac{\xi_3^{YXY}(t)}{\xi_1^{YXY}(t)} = \frac{t_2 t_3 (1 + o(1))}{t_2 + t_2 t_3 o(1)} = \frac{t_3 (1 + o(1))}{1 + o(t_3)} = t_3 (1 + o(1))$$

is differentiable near the origin and satisfies $\zeta(t_1, t_2, 0) = 0$. Thus, the mapping

$$\Xi : (t_1, t_2, t_3) \mapsto \Xi(t) = (\xi_1^{YXY}(t), \xi_2^{YXY}(t), \zeta(t))$$

is differentiable, satisfies $\Xi(0) = 0$, and has a non-singular Jacobian at the origin,

$$D\Xi(0) = \begin{pmatrix} 0 & 1 & 0 \\ 1 & 0 & 1 \\ 0 & 0 & 1 \end{pmatrix}.$$

Thus this transformation is invertible near 0 with a continuously differentiable inverse Ξ^{-1} (see Appendix A, Theorem A.3.1). Furthermore, the corresponding expression for the XYX-coordinates is given by

$$\zeta(s) = \frac{\xi_3^{XYX}(s)}{\xi_1^{XYX}(s)} = \frac{s_1 s_2 (1+o(1))}{s_1 (1+\tau) + s_1 s_2 o(1)} = \frac{s_2 (1+o(1))}{1+\tau+o(s_2)} = s_2 (1+o(1)),$$

and thus also the transformation $\Theta : (s_1, s_2, \tau) \mapsto \Theta(s) = (\xi_1(s), \xi_2(s), \zeta(s))$ is differentiable near the origin. Hence the composition

$$(s_1, s_2, \tau) \mapsto (t_1, t_2, t_3) = \left(\Xi^{-1} \circ \Theta \right)(s_1, s_2, \tau)$$

is a differentiable mapping defined in a neighborhood of $(0,0,0)$. Thus, given any sufficiently small times (s_1, s_2, τ), there exist unique times (t_1, t_2, t_3) such that Eq. (7.8) is satisfied, and these times are differentiable functions of (s_1, s_2, τ). The function ζ represents the blowup of s_3 in the direction of s_1 that desingularizes Eq. (7.8) and leads to unique and smooth solutions. Note that $t_1 = t_3 = 0$ if $s_2 = 0$ and $t_2 = 0$ if $s_1 = 0$; thus s_2 divides t_1 and t_3, while s_1 divides t_2. Low-order expansions for the functions t_j can be computed by equating the canonical coordinates,

$$s_1 (1 + \tau) + o(s_1 s_2) = t_2 + o(t_2 t_3),$$

$$s_2 + o(s_1 s_2) = t_1 + t_3 + o(t_2 t_3),$$

$$s_1 s_2 (1 + O(s)) = t_2 t_3 (1 + O(t)),$$

which gives Eq. (7.10).   □

Lemma 7.3.9. *For small times s_1 and s_2, the function Δ has the following second-order Taylor expansion in τ:*

$$\Delta(\tau; s_1, s_2) = \frac{1}{2} s_1 s_2 \left[-\tau \left(\alpha(q_1) s_1 + \beta(q_1) s_2 + o(s) \right) + \tau^2 \left(\beta(q_1) s_2 + o(s) \right) + o(\tau^2) \right].$$

In particular,

$$\Delta'(0) = \frac{d\Delta}{d\tau}(0) = -\frac{1}{2} s_1 s_2 \left[\alpha(q_1) s_1 + \beta(q_1) s_2 + o(s) \right],$$

$$\Delta''(0) = \frac{d^2\Delta}{d\tau^2}(0) = s_1 s_2 \left[\beta(q_1) s_2 + o(s) \right].$$

Proof. Adding the ξ_1- and ξ_2-coordinates, and using the low-order expansions in Eq. (7.10) for the t_j, gives

$$\Delta(\tau) = t_1 + t_2 + t_3 - s_1 - s_2 - s_1 \tau$$

$$= \frac{1}{2}\alpha(q_1)\left[s_1^2 s_2 - t_2^2 t_3 \right] + \beta(q_1) \left[\frac{1}{2} s_1 s_2^2 - t_2 t_3 \left(t_1 + \frac{1}{2} t_3 \right) \right] + s_1 s_2 o(s)$$

$$= \frac{1}{2}\alpha(q_1)s_1^2 s_2 \left[1 - (1+\tau) + o(s)\right]$$

$$+ \beta(q_1)s_1 s_2^2 \left[\frac{1}{2} - \frac{\tau}{1+\tau} - \frac{1}{2}\frac{1}{1+\tau} + o(s)\right] + s_1 s_2 o(s)$$

$$= \frac{1}{2}s_1 s_2 \left[-\alpha(q_1)s_1\tau - \beta(q_1)\frac{s_2\tau}{1+\tau} + o(s)\right]$$

$$= \frac{1}{2}s_1 s_2 \left[-\alpha(q_1)s_1\tau - \beta(q_1)s_2\tau + \beta(q_1)s_2\tau^2 + o(\tau^2;s)\right].$$

The formulas for the first and second derivatives in τ follow from this expansion. $\square$

This construction is valid regardless of the signs of α and β. If α and β have the same sign, then the linear terms are dominant and the first-order term $\Delta'(0)$ implies that the XYX-trajectory

$$q_1 \exp(s_1 X)\exp(s_2 Y)\exp(s_1\tau X)$$

is faster than the corresponding YXY-trajectory

$$q_1 \exp(t_1 Y)\exp(t_2 X)\exp(t_3 Y)$$

if these terms are negative, and it is slower if they are positive. This is consistent with the results already derived in Proposition 7.3.3. However, in the case we are considering now, $\alpha > 0$ and $\beta < 0$, the first-order term $\Delta'(0)$ is inconclusive. Indeed, this term is the conjugate point relation along the extremal defined earlier.

Proposition 7.3.6. *For the conjugate triple* (q_1, q_2, q_3), *we have that* $\Delta'(0) = 0$.

Proof. Differentiating both sides of Eq. (7.8) with respect to τ at $\tau = 0$, it follows that

$$s_1 X(q_3) = \dot{t}_1(0)\mathfrak{v}_1 + \dot{t}_2(0)\mathfrak{v}_2 + \dot{t}_1(0)\mathfrak{v}_3, \tag{7.11}$$

where the vectors $\mathfrak{v}_i$, $i = 1, 2, 3$, are given by

$$\mathfrak{v}_1 = q_1 \exp(t_1(0)Y)Y\exp(t_2(0)X)\exp(t_3(0)Y) = q_1 Y\exp(s_1 X)\exp(s_2 Y),$$

$$\mathfrak{v}_2 = q_1 \exp(t_1(0)Y)\exp(t_2(0)X)X\exp(t_3(0)Y) = q_2 X\exp(s_2 Y),$$

$$\mathfrak{v}_3 = q_1 \exp(t_1(0)Y)Y\exp(t_2(0)X)\exp(t_3(0)Y)Y = q_3 Y.$$

Thus, $\mathfrak{v}_1$ is the vector $Y(q_1)$ moved forward along the trajectory to its endpoint q_3, and so on. We transform this vector equation into a scalar equation by factoring out the linear subspace W spanned by the vectors $g(q_3)$,

$$\exp(-s_2 \mathrm{ad}Y)g(q_2) = g(q_3) - s_2[f,g](q_3) + o(s_2),$$

and

$$\exp\left(-s_2\operatorname{ad}Y\right)\exp\left(-s_1\operatorname{ad}X\right)g(q_1) = g(q_3) - (s_1+s_2)[f,g](q_3) + o(s_1,s_2)$$

that define the conjugate point relation. Because of assumption (B1), the vectors $X(q_3)$ and $Y(q_3)$ do not lie in W, and thus $X(q_3)$ mod $W \neq 0$. Note that $\mathfrak{v}_3 = Y(q_3) = X(q_3) + 2g(q_3) = X(q_3)$ mod W. Furthermore,

$$\begin{aligned}
\mathfrak{v}_1 &= q_1 Y \exp(s_1 X)\exp(s_2 Y) = q_1\left(X+2g\right)\exp(s_1 X)\exp(s_2 Y)\\
&= q_2 X \exp(s_2 Y) + 2q_1 g \exp(s_1 X)\exp(s_2 Y)\\
&= q_2\left(Y-2g\right)\exp(s_2 Y) + 2q_3 \exp(-s_2 Y)\exp(-s_1 X)g\exp(s_1 X)\exp(s_2 Y)\\
&= q_3 Y - 2q_3 \exp(-s_2 Y)g\exp(s_2 Y)\\
&\quad + 2q_3 \exp(-s_2 Y)\exp(-s_1 X)g\exp(s_1 X)\exp(s_2 Y)\\
&\sim X(q_3) + 2g(q_3) - 2\exp(-s_2\operatorname{ad}Y)g(q_2) + 2\exp(-s_2\operatorname{ad}Y)\exp(-s_1\operatorname{ad}X)g(q_1)\\
&= X(q_3) \text{ mod } W
\end{aligned}$$

and

$$\begin{aligned}
\mathfrak{v}_2 &= q_2 X \exp(s_2 Y) = q_2\left(Y+2g\right)\exp(s_2 Y)\\
&= q_3 Y + 2q_3 \exp(-s_2 Y)g\exp(s_2 Y)\\
&\sim X(q_3) + 2g(q_3) - 2\exp(-s_2\operatorname{ad}Y)g(q_2) = X(q_3) \text{ mod } W.
\end{aligned}$$

Thus all vectors $\mathfrak{v}_i$ are congruent to $X(q_2)$ mod W, which is nonzero. Hence Eq. (7.11) becomes $s_1 = \dot{t}_1(0) + \dot{t}_2(0) + \dot{t}_1(0)$, i.e., $\Delta'(0) = 0$.   □

Thus, we have that $\alpha(q_1)s_1 + \beta(q_1)s_2 + o(s) = 0$. In these computations, the higher-order terms all have at least one of s_1 and s_2 as a factor, and we can thus write

$$\alpha(q_1)s_1(1+o(1)) = -\beta(q_1)s_2(1+o(1)).$$

Since $\alpha(q_1)$ and $\beta(q_1)$ are nonzero and have opposite signs, it follows that there exist a neighborhood M of p and positive constants c and C that depend only on this neighborhood such that whenever $q_1 \exp(s_1 X)\exp(s_2 Y)$ is a time-optimal $\cdot XY\cdot$-trajectory, then $cs_1 \leq s_2 \leq Cs_1$. We say the times s_1 and s_2 along time-optimal $\cdot XY\cdot$-trajectories in M are *comparable*. But then the second derivative $\Delta''(0)$ is negative: since the times are comparable, we get that $\|s\| = s_1 + s_2 \geq \left(\frac{1}{C}+1\right)s_2$, and thus for small enough $\|s\|$ we have that

$$\beta(q_1)s_2 + o(s) \leq \beta(q_1)s_2\left(1+\frac{o(s)}{\|s\|}\right) < \frac{1}{2}\beta(q_1)s_2 < 0.$$

Recall that since $f(p)$ is nonzero, we can always enforce this by making the neighborhood M small enough. But this means that for small positive τ, the YXY-trajectory

$$q_1 \exp(t_1 Y) \exp(t_2 X) \exp(t_3 Y)$$

is faster than the XYX-trajectory

$$q_1 \exp(s_1 X) \exp(s_2 Y) \exp(s_1 \tau X).$$

Overall, see Fig. 7.9, this implies that $YXYX$-trajectories are not time-optimal near p if $\alpha(p) > 0$ and $\beta(p) < 0$.

We use an input symmetry to exclude the optimality of $XYXY$-trajectories as well. Let ρ denote the input symmetry that interchanges X and Y, or equivalently, $\rho(f) = f$ and $\rho(g) = -g$ (see Sect. 2.10). Under this transformation, the identity

$$[X, [X, Y]] = \alpha f + \cdots$$

becomes

$$[Y, [Y, X]] = \rho(\alpha) f + \cdots$$

and since

$$[Y, [X, Y]] = \beta f + \cdots$$

it follows that $\rho(\alpha) = -\beta$; similarly, $\rho(\beta) = -\alpha$. Hence, the assumption that α is positive and that β is negative is invariant under this transformation, while it interchanges the roles of X and Y. Our calculation thus immediately implies that $XYXY$-trajectories also are not time-optimal near p. Since we already know that time-optimal controls for trajectories that lie in a sufficiently small neighborhood M of p are bang-bang, it follows that the corresponding *time-optimal controls are bang-bang with at most two switchings*. This concludes the proof of Proposition 7.3.5. □

We point out that this construction provides an alternative geometric approach for the analysis of bang-bang extremals near a slow singular arc in $\mathbb{R}^2$ that was given in Sect. 2.9. Also, this argument verifies that the local optimality of bang-bang trajectories ceases at the third switching points, and thus it is an appropriate terminology to call such a triple of switching points a *conjugate triple* and the relation that exists between the switching times a *conjugate point relation*. However, our analysis for the case $\alpha(p) > 0$ and $\beta(p) < 0$ is not complete yet. This is the three-dimensional generalization of Fig. 7.4 from dimension 2, and as in that case not every extremal bang-bang control with two switchings is optimal. Once more there exists a nontrivial cut-locus between XYX- and YXY-trajectories that limits the optimality of these trajectories.

Theorem 7.3.2. *Suppose $\alpha(p) > 0$ and $\beta(p) < 0$ and let M be a sufficiently small neighborhood of p. Then, for every initial point $q \in M$, the set of points that are reachable time-optimally in time t, $0 < t < T$, by means of both an XYX- and a*

Fig. 7.10 A qualitative
sketch of the structure of the
endpoints in the boundary
Reach$_{\Sigma,T}(p)$ of the
small-time reachable set for
$\alpha(p) > 0$ and $\beta(p) < 0$

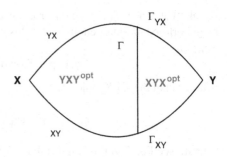

YXY-trajectory is a surface Γ. *This surface contains the X- and Y-trajectories,* $\mathscr{T}_X$
and $\mathscr{T}_Y$, *in its relative boundary, and for t fixed, the curve* $\Gamma_t = \Gamma \cap \mathrm{Reach}_{\Sigma,t}(p)$ *is
the transversal intersection (cut-locus) of the surfaces*

$$\mathscr{E}^t_{XYX} = \{q\exp(s_1X)\exp(s_2Y)\exp(s_3X) : \ s_1 + s_2 + s_3 = t\}$$

and

$$\mathscr{E}^t_{YXY} = \{q\exp(t_1Y)\exp(t_2X)\exp(t_3Y) : \ t_1 + t_2 + t_3 = t\}$$

*of endpoints of all XYX- respectively YXY-trajectories that are reachable in time t.
This cut-locus* Γ *separates optimal from nonoptimal concatenations.*

Figure 7.10 illustrates the qualitative structure of the portion Reach$_{\Sigma,T}(p)$ in the
boundary of the small-time reachable set Reach$_{\Sigma,\leq T}(p)$ that is defined by the points
that are reachable time-optimally in time T for this case.

Proof. We again carry out the proof in various steps and start with determining
the *cut-locus* $\Gamma_t = \mathscr{E}^t_{XYX} \cap \mathscr{E}^t_{YXY}$, i.e., the set where arbitrary XYX- and YXY-
trajectories defined over the same interval $[0,t]$ intersect. Both surfaces include
trivial parameterizations of the X- and Y-trajectories, and thus these trajectories
lie in the relative boundary of Γ. For example, the X-trajectory can be described
by setting $s_2 = 0$ in $\mathscr{E}^t_{XYX}$ or $t_2 = t$ in $\mathscr{E}^t_{YXY}$. The nontrivial cut-locus is computed
by equating the canonical coordinates. Without loss of generality, we consider the
initial point p, but it will be clear that these results hold as this point varies over a
sufficiently small neighborhood M of p. It has already been shown that there exists
a neighborhood M of p with the property that for sufficiently small positive times s_1
and s_2 and positive parameter τ, the equation

$$p\exp(s_1X)\exp(s_2Y)\exp(s_1\tau X) = p\exp(t_1Y)\exp(t_2X)\exp(t_3Y)$$

has a unique solution that is given by smooth functions $t_i(\tau;s_1,s_2)$, $i = 1,2,3$, that
satisfy Eq. (7.10). A nontrivial cut-locus Γ, if it exists, is determined by imposing
the additional equation $\Delta(\tau) = 0$, or

$$t_1 + t_2 + t_3 = s_1 + s_2 + s_1\tau.$$

Using this relation in the sum of $\xi_1 + \xi_2$ and dividing by ξ_3, we obtain that

$$\frac{1}{2}\alpha(p)t_2 + \beta(p)\left(t_1 + \frac{1}{2}t_3\right) + o(t) = \frac{1}{2}\alpha(p)s_1 + \frac{1}{2}\beta(p)s_2 + o(s).$$

We now substitute the low-order approximations (7.10) into this relation to get

$$\frac{1}{2}\alpha(p)s_1(1+\tau) + \beta(p)s_2\left(\frac{\tau}{1+\tau} + \frac{1}{2}\frac{1}{1+\tau}\right) - \frac{1}{2}\alpha(p)s_1 - \frac{1}{2}\beta(p)s_2 + o(s) = 0,$$

and upon simplification, we obtain the following equation defining the cut-locus:

$$\alpha(p)s_1(1+\tau) + \beta(p)s_2 + o(s) = 0. \tag{7.12}$$

This equation can be solved for s_2 as a function of s_1 and τ,

$$s_2 = -\frac{\alpha(p)}{\beta(p)}s_1(1+\tau) + o(s_1), \tag{7.13}$$

and s_2 is positive. Thus, if we let the total time t vary between 0 and T, then the full cut-locus $\Gamma = \cup_{0 < t < T} \Gamma_t$ is a surface that can be described as the image of the mapping

$$\Omega_{XYX}: \quad (s_1, \tau) \mapsto p\exp(s_1 X)\exp(s_2 Y)\exp(s_1 \tau X)$$

where s_2 is given by Eq. (7.13). Equivalently, the cut-locus is also given as the image

$$\Omega_{YXY}: \quad (s_1, \tau) \mapsto (t_1, t_2, t_3) \mapsto p\exp(t_1 Y)\exp(t_2 X)\exp(t_3 Y)$$

with the times t_j the unique solutions of Eq. (7.8).

It is more useful to parameterize the cut-locus in terms of the time after the second switching. If instead of $s_1\tau$, we denote this time by s_{cl}, then Eq. (7.12) becomes

$$s_{cl} = s_{cl}(s_1, s_2) = -s_1 - \frac{\beta(p)}{\alpha(p)}s_2 + o(s). \tag{7.14}$$

An important observation then is the following:

Lemma 7.3.10. *For an XYX-trajectory, denote the switching points by*

$$q_1 = p\exp(s_1 X) \quad \text{and} \quad q_2 = q_1\exp(s_2 Y)$$

and let

$$q_{cl} = p\exp(s_1 X)\exp(s_2 Y)\exp(s_{cl}X)$$

be the point on the cut-locus. Let s_{cp} denote the time along this $\cdot YX\cdot$-trajectory when the third switching point $q_3 = q_2\exp(s_{cp}X)$ occurs. Then $s_{cl} < s_{cp}$, i.e., the conjugate point q_3 lies after *the point* q_{cl} *on the cut-locus. In particular, the* XYX-*trajectory still is an extremal at time* s_{cl}. *The analogous statement holds for* YXY-*trajectories.*

Proof. We compare Eq. (7.12) that defines the cut-locus with the conjugate point equation along the $\cdot YX\cdot$-trajectory with first switching point q_1. The easiest way to obtain the conjugate point relation along the $\cdot YX\cdot$-trajectory is to use the input symmetry ρ that interchanges X with Y, and we get that

$$\beta(q_1)s_2 + \alpha(q_1)s_{cp} + o(s) = 0.$$

We have seen above that the conjugate point relation along an $\cdot XY\cdot$-trajectory with times s_2 and s_3 along the X- and Y-trajectories is given by

$$\alpha(q_1)s_2 + \beta(q_1)s_3 + o(s) = 0.$$

Since $\rho(\alpha) = -\beta$ and $\rho(\beta) = -\alpha$, this becomes

$$-\beta(q_1)s_2 - \alpha(q_1)s_3 + o(s) = 0.$$

The conjugate point relation thus defines the time s_{cp} along the second $\cdot X\cdot$-leg as

$$s_{cp} = -\frac{\beta(q_1)}{\alpha(q_1)}s_2\left(1 + o(s_2)\right).$$

On the other hand, it follows from Eq. (7.12) which defines the cut-locus that

$$s_1 + s_{cl} = s_1(1 + \tau) = -\frac{\beta(p)}{\alpha(p)}s_2 + o(s).$$

Since $q_1 = p\exp(s_1 X)$, we have that $\alpha(q_1) = \alpha(p) + o(s_1)$ and $\beta(q_1) = \beta(p) + o(s_1)$ and thus also

$$s_1 + s_{cl} = -\frac{\beta(q_1)}{\alpha(q_1)}s_2(1 + o(s_1)) + o(s).$$

Hence

$$s_{cp} - s_{cl} = s_1 + o(s).$$

Since the times s_1 and s_2 are comparable by Eq. (7.13), this expression is positive. Thus the point on the cut-locus happens before the third conjugate point q_{cp} along the $\cdot YX\cdot$-trajectory (see Fig. 7.11). Once more, using the input symmetry ρ gives that the same result is valid for YXY-trajectories.   □

Therefore, all XYX-trajectories that are defined for times that do not violate the cut-locus equation, i.e., for $s_3 \leq s_{cl}$, are extremals and similarly for YXY-trajectories. But so are the ones that correspond to times s_3 that still satisfy $s_{cl} < s_3 \leq s_{cp}$. As we show next, these trajectories are no longer time-optimal near p.

Fig. 7.11 The times s_{cl} and s_{cp} along an extremal XYX-trajectory

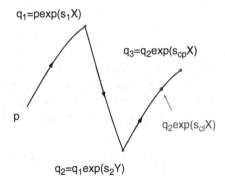

Lemma 7.3.11. *The surfaces $\mathscr{E}^t_{XYX}$ and $\mathscr{E}^t_{YXY}$ intersect transversally in the cut-locus $\Gamma_t = \mathscr{E}^t_{XYX} \cap \mathscr{E}^t_{YXY}$, i.e., together their tangent spaces at a point $q \in \Gamma_t$ span the full space,*

$$T_q \mathscr{E}^t_{XYX} + T_q \mathscr{E}^t_{YXY} = \mathbb{R}^3.$$

Proof. We first compute the tangent space to the surface $\mathscr{E}^t_{XYX}$. Let $q_1 = p \exp(s_1 X)$, $q_2 = q_1 \exp(s_2 Y)$, and $q = q_2 \exp(s_3 X)$. Since the total time t is fixed, we have $s_3 = t - s_1 - s_2$ and s_3 becomes this function of s_1 and s_2. Two tangent vectors to $\mathscr{E}^t_{XYX}$ at q are then given by the partial derivatives of the expression defining the surface with respect to s_1 and s_2. Using the exponential notation, these vectors can be computed as follows:

$$\mathfrak{v}_1 = \frac{\partial}{\partial s_1} p \exp(s_1 X) \exp(s_2 Y) \exp(s_3 X)$$

$$= p \exp(s_1 X) X \exp(s_2 Y) \exp(s_1 \tau X) + qX \frac{\partial s_3}{\partial s_1}$$

$$= q \exp(-s_3 X) \exp(-s_2 Y) X \exp(s_2 Y) \exp(s_3 X) - qX$$

$$\sim \exp(-s_3 \operatorname{ad} X) \exp(-s_2 \operatorname{ad} Y) X(q_1) - X(q)$$

$$= \exp(-s_3 \operatorname{ad} X) [\exp(-s_2 \operatorname{ad} Y) X(q_1) - X(q_2)]$$

$$= \exp(-s_3 \operatorname{ad} X) [s_2 [X, Y](q) + o(s_2)]$$

$$= s_2 ([X, Y](q) + o(1)).$$

Similarly,

$$\mathfrak{v}_2 = \frac{\partial}{\partial s_2} p \exp(s_1 X) \exp(s_2 Y) \exp(s_3 X)$$

$$= p \exp(s_1 X) \exp(s_2 Y) Y \exp(s_1 \tau X) + qX \frac{\partial s_3}{\partial s_2}$$

$$= q \exp(-s_3 X) Y \exp(s_3 X) - qX$$

$$\sim \exp(-s_3 \operatorname{ad} X) Y(q_2) - X(q)$$

$$= Y(q) - s_3 \left([X,Y](q) + o(1) \right) - X(q)$$
$$= 2g(q) - s_3 \left([X,Y](q) + o(1) \right).$$

Since the vector fields g and $[X,Y]$ are linearly independent, these two vectors are linearly independent and therefore span the tangent space to $\mathscr{E}^t_{XYX}$ at q. Similarly, with $\tilde{q}_1 = p\exp(t_1 Y)$ and $\tilde{q}_2 = \tilde{q}_1 \exp(t_2 X)$, the tangent space to $\mathscr{E}^t_{YXY}$ at $q = p\exp(t_1 Y)\exp(t_2 X)\exp(t_3 Y)$ is spanned by

$$\mathfrak{w}_1 = \frac{\partial}{\partial t_1} p\exp(t_1 Y)\exp(t_2 X)\exp(t_3 Y)$$

$$= p\exp(t_1 Y) Y \exp(t_2 X)\exp(t_3 Y) + p\exp(t_1 Y)\exp(t_2 X)\exp(t_3 Y)\frac{\partial t_3}{\partial t_1}Y$$

$$= q\exp(-t_3 Y)\exp(-t_2 X) Y \exp(t_2 X)\exp(t_3 Y) - qY$$

$$\sim \exp(-t_3 \operatorname{ad} Y)\left[\exp(-t_2 \operatorname{ad} X) Y(\tilde{q}_1) - Y(q)\right]$$

$$= \exp(-t_3 \operatorname{ad} Y)\left[t_2 \left(-[X,Y](\tilde{q}_2) + o(1)\right)\right]$$

$$= -t_2 \left([X,Y](q) + o(1)\right)$$

and

$$\mathfrak{w}_2 = \frac{\partial}{\partial t_2} p\exp(t_1 Y)\exp(t_2 X)\exp(t_3 Y)$$

$$= p\exp(-t_3 Y) X \exp(t_3 Y)$$

$$= X(q) + t_3[X,Y](q) + o(t_3).$$

The three vectors $\mathfrak{v}_1$, $\mathfrak{v}_2$, and $\mathfrak{w}_2$ are linearly independent,

$$\mathfrak{v}_1 \wedge \mathfrak{v}_2 \wedge \mathfrak{w}_2 = s_2 \begin{vmatrix} 0 & 0 & 1 \\ 0 & 2 & -s_3 \\ 1 & -1 & t_3 \end{vmatrix} (1+o(t))(f \wedge g \wedge [X,Y]) \neq 0,$$

and thus $\mathscr{E}^t_{XYX}$ and $\mathscr{E}^t_{YXY}$ intersect transversally.   □

The cut-locus Γ divides the surfaces $\mathscr{E}^t_{XYX}$ and $\mathscr{E}^t_{YXY}$ into two connected components each. Let $\mathscr{E}^{t,opt}_{XYX}$ be the one were the time s_3 along the third leg does not exceed the time s_{cl} until the cut-locus and let $\mathscr{E}^{t,nopt}_{XYX}$ be the section where $s_3 > s_{cl}$. Note that the component $\mathscr{E}^{t,opt}_{XYX}$ is attached to the Y-trajectory $\mathscr{C}_Y$, while the section $\mathscr{E}^{t,nopt}_{XYX}$ is attached to the X-trajectory $\mathscr{C}_X$. Reversing the order, let $\mathscr{E}^{t,opt}_{YXY}$ denote the section of $\mathscr{E}^t_{YXY}$ that is attached to the X-trajectory $\mathscr{C}_X$ and $\mathscr{E}^{t,nopt}_{YXY}$ the one attached to the Y-trajectory $\mathscr{C}_Y$. Since the surfaces $\mathscr{E}^t_{XYX}$ and $\mathscr{E}^t_{YXY}$ intersect transversally in Γ, it follows that for each of these surfaces one of these two components lies in the interior of the small-time reachable set $\operatorname{Reach}_{\Sigma,\leq t}(p)$. It is geometrically clear, and this can be verified using the canonical coordinates, that $\mathscr{E}_{XYX}^{t,opt}$ and $\mathscr{E}_{YXY}^{t,opt}$ (which contain the trajectories with s_3, respectively t_3, values near 0) are the sections that lie in the boundary of the small-time reachable set, while the sections $\mathscr{E}_{XYX}^{t,nopt}$ and $\mathscr{E}_{YXY}^{t,nopt}$ lie in the interior and thus are no longer time-optimal from p. This concludes the proof of Theorem 7.3.2. $\square$

Corollary 7.3.1. *The small-time reachable set from a point p where $\alpha(p) > 0$ and $\beta(p) < 0$ is given by the union of the two cells*

$$\mathscr{C}_{XYX}^{opt} = \{p\exp(s_1 X)\exp(s_2 Y)\exp(s_3 X) : s_3 \leq s_{cl}(s_1, s_2)\}$$

and

$$\mathscr{C}_{YXY}^{opt} = \{p\exp(t_1 Y)\exp(t_2 X)\exp(t_3 Y) : t_3 \leq t_{cl}(t_1, t_2)\}$$

corresponding to the optimal concatenations,

$$\text{Reach}_{\Sigma, \leq T}(p) = \mathscr{C}_{XYX}^{opt} \cup \mathscr{C}_{YXY}^{opt}.$$

Summarizing, under *codimension-0 assumptions in dimension* 3, there exist four qualitatively different scenarios that can be grouped according to the existence and local optimality properties of singular controlled trajectories:

(a) in the **totally bang-bang case** (α and β have the same sign on M), singular controls are inadmissible and optimal controls near p are bang-bang with at most two switchings. In this case, the lengths between the switchings are unrestricted and optimal controlled trajectories near p are of type YXY if α is positive and of type XYX if α is negative.
(b) in the **singular case** (α and β have opposite signs on M), singular controls are admissible, but they can be fast or slow.

   (b.1) If singular trajectories are **fast**, time-optimal controlled trajectories near p are of type BSB; that is, they are concatenations of at most a bang arc B (that is, either X or Y), followed by a singular arc S and another bang arc B. In this case, the times along the individual segments are also unrestricted and all possible combinations are optimal. Note that we allow in our notation that pieces in the concatenation sequence are absent; equivalently, the time along a specific arc may be 0. Thus, for example, BSB-trajectories include bang-bang trajectories with one or no switching.
   (b.2) In the case of a **slow** singular arc, even in a small neighborhood of p, there exist bang-bang extremals (that satisfy the necessary conditions of the maximum principle) that have an arbitrarily large number of switchings. However, these are not optimal. Optimal controls have at most two switchings, and the times along the third leg are limited by the time when a nontrivial cut-locus $\Gamma = \mathscr{T}_{XYX} \cap \mathscr{T}_{YXY}$ is reached. This time occurs prior to the time when a conjugate point relation forces the next switching.

While the cases (a) and (b.1) follow from the conditions of the maximum principle and higher-order necessary conditions for optimality, this no longer is true for case (b.2). The explicit geometric constructions that were presented and the resulting small-time reachable sets go well beyond the analysis of necessary conditions for local optimality and indeed analyze the full class of admissible controlled trajectories. For as we have seen, *XYX- and YXY-trajectories are no longer time-optimal near p once the cut-locus has been passed, but* as we shall show next, *they are local minima in the sense of providing a strong local minimum over a neighborhood of the trajectory until the third switching time is reached.* In fact, we show below that the third switching point is the conjugate point along this trajectory, and it is only here that local optimality is lost. Thus, the optimality of these trajectories near p cannot be excluded based on necessary conditions for optimality alone. We summarize the optimality properties of bang-bang trajectories in the theorem below, which, without loss of generality, is formulated for an XYX-trajectory. The analogous result holds for a YXY-trajectory.

Theorem 7.3.3. *Suppose* $\alpha(p) > 0$ *and* $\beta(p) < 0$ *and let M be a sufficiently small neighborhood of p. Consider an XYX-trajectory with switching points* $q_1 = p\exp(s_1 X)$ *and* $q_2 = q_1\exp(s_2 Y)$ *and let* s_3 *denote the time along the last X-trajectory. Let* s_{cl} *be the time when the point* $q_{cl} = p\exp(s_1 X)\exp(s_2 Y)\exp(s_{cl} X)$ *on the cut-locus is reached and let* s_{cp} *denote the time when the third switching point* $q_3 = q_2\exp(s_{cp} X)$ *is reached. Then this XYX-trajectory (i) is time-optimal near p if* $s_3 \leq s_{cl}$, *(ii) is still locally time-optimal in the sense of providing a strong local minimum over a neighborhood of the trajectory if* $s_{cl} < s_3 < s_{cp}$, *(iii) is an extremal, but no longer locally optimal, if* $s_3 = s_{cp}$ *and (iv) is no longer an extremal if* $s_3 > s_{cp}$. *The third switching point* q_3 *is a transversal fold, and at this point local optimality ceases. Continuations of the XYX-trajectory through additional switchings that satisfy the conjugate point relations are still extremals, but they are no longer locally optimal.*

Proof. Statement (i) was already proven above in Theorem 7.3.2.

The second claim (ii) follows from the results in Sect. 6.1 about the method of characteristics for broken extremals: choose small times $\sigma_1 > s_1$ and $\sigma_2 > s_2$ and define a C^1-parameterized family $\mathscr{F}_{XYX}$ of bang-bang trajectories of type XYX by letting the times t_1 and t_2 along the first two legs vary freely over $(0, \sigma_1)$ and $(0, \sigma_2)$, respectively, and then restricting the third time by $0 < t_3 < s_{cp}$, where s_{cp} is the time of the third switching time, i.e.,

$$\mathscr{F}_{XYX} = \{p\exp(t_1 X)\exp(t_2 Y)\exp(t_3 X) : 0 < t_1 < \sigma_1, \ 0 < t_2 < \sigma_2, \ 0 < t_3 < s_{cp}\}.$$

It follows from the proof of Lemma 7.3.10 that for each pair (t_1, t_2), the time s_{cp} is uniquely defined as a differentiable function φ of t_1 and t_2 by the fact that the points $\tilde{q}_1 = p\exp(t_1 X)$, $\tilde{q}_2 = \tilde{q}_1\exp(t_2 Y)$ and the endpoint $\tilde{q}_3 = \tilde{q}_2\exp(t_3 X)$ form a conjugate triple, and it is given by

$$s_{cp} = \varphi(t_1, t_2) = -\frac{\beta(p)}{\alpha(p)} t_2 (1 + o(t_1, t_2)). \tag{7.15}$$

Fig. 7.12 The parameterized
family $\mathscr{F}_{XYX}$

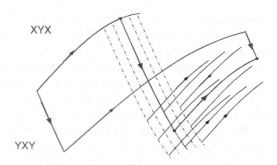

XYX

YXY

Since $\tilde{q}_1$ and $\tilde{q}_2$ are switching points, it follows that the corresponding adjoint vector $\tilde{\lambda}(t_1)$ satisfies the relations

$$\left\langle \tilde{\lambda}(t_1), g(\tilde{q}_1) \right\rangle = 0 \quad \text{and} \quad \left\langle \tilde{\lambda}(t_1), \exp(t_2 \operatorname{ad} Y) g(\tilde{q}_1) \right\rangle = 0$$

and the vectors $g(\tilde{q}_1)$ and

$$\exp(t_2 \operatorname{ad} Y) g(\tilde{q}_1) = g(\tilde{q}_1) + t_2 [Y, g](\tilde{q}_1) + o(t_2)$$

are linearly independent. Normalizing $\left\| \tilde{\lambda}(t_1) \right\| = 1$, and using the fact that the switching function $\Phi(t) = \left\langle \tilde{\lambda}(t), g(\tilde{x}(t)) \right\rangle$ needs to be negative along the Y-arc, this then determines $\tilde{\lambda}(t_1)$ uniquely, and the corresponding covector $\tilde{\lambda}(t)$ simply is the unique solution of the corresponding adjoint equation. Since there are no conjugate triples over the interval $(0, s_{cp})$, the minimum condition in the maximum principle is satisfied, and this defines a C^1-parameterized family of extremals. Since the controls are constant, the flow of the X-trajectories starting from the points $p \exp(t_1 X) \exp(t_2 Y)$ is a diffeomorphism (by the uniqueness of solutions of the corresponding ODE) and covers a neighborhood of the third X-leg of the reference trajectory (see Fig. 7.12). Hence this portion of $\mathscr{F}_{XYX}$ defines *a local field of extremals*.

Denote the controlled reference trajectory in this field corresponding to the times s_i, $i = 1, 2, 3$, by $(\bar{x}, \bar{u})$ and let $q = \bar{x}(T)$, $T = s_1 + s_2 + s_3$. It follows from the general results of Chap. 5 that the reference trajectory is faster than any other controlled trajectory that lies in the set covered by the field. (Note that this field does not include the Y-trajectory starting from p, and thus the YXY-trajectories that steer p into q in time less than T are not covered by this field.) For $\kappa > 0$, let

$$V_\kappa = \{ x \in M : \| x - \bar{x}(t) \|_\infty < \kappa \}$$

be a tubular neighborhood of the reference trajectory. The family $\mathscr{F}_{XYX}$ does not cover the full neighborhood V_κ, but it follows from our results above that for small κ, it does cover the portion of V_κ that lies in the reachable set $\text{Reach}_{\Sigma, \leq \bar{T}}(p)$ for some suitable $\tilde{T} > T$. Thus, any trajectory that lies in V_κ, but at one time passes outside of the set covered by the field $\mathscr{F}_{XYX}$, must already have left this reachable set and thus overall takes a time longer than T, and thus is no better than the reference trajectory. In particular, by choosing κ small enough, we can ensure that the Y-trajectory starting from p lies in V_κ for only a very short time, and thus the optimal YXY-trajectory does not lie in V_κ. It follows that $(\bar{x}, \bar{u})$ is a local minimum over all controlled trajectories that lie in V_κ.

We next show that the third switching point q_3 is a transversal fold point in the sense of Definition 6.1.2 in Sect. 6.1. This implies that there exists a control envelope through q_3, and thus XYX-trajectories are no longer optimal for $s_3 = s_{cp}$, proving (iii). Let

$$\mathscr{S}_{cp} = \{ p \exp(t_1 X) \exp(t_2 Y) \exp(s_{cp}(t_1, t_2) X) : |t_i - s_i| < \varepsilon, \, i = 1, 2 \}$$

be the surface consisting of the third switching points as the times t_1 and t_2 vary near the times s_1 and s_2 along the reference trajectory. We need to show that X and Y point to opposite sides of the tangent space to $\mathscr{S}_{cp}$ at q_3. This tangent space is spanned by the vectors

$$\begin{aligned}
\mathfrak{v}_1 &= p \exp(s_1 X) X \exp(s_2 Y) \exp(s_{cp} X) + p \exp(s_1 X) \exp(s_2 Y) \exp(s_{cp} X) X \frac{\partial s_{cp}}{\partial t_1} \\
&= \exp(-s_{cp} \operatorname{ad} X) \exp(-s_2 \operatorname{ad} Y) X(q_1) + \frac{\partial s_{cp}}{\partial t_1} X(q_3)
\end{aligned}$$

and

$$\begin{aligned}
\mathfrak{v}_2 &= p \exp(s_1 X) \exp(s_2 Y) Y \exp(s_{cp} X) + p \exp(s_1 X) \exp(s_2 Y) \exp(s_{cp} X) Y \frac{\partial s_{cp}}{\partial t_2} \\
&= \exp(-s_{cp} \operatorname{ad} X) Y(q_1) + \frac{\partial s_{cp}}{\partial t_2} X(q_3).
\end{aligned}$$

It follows from Eq. (7.15) that

$$\frac{\partial s_{cp}}{\partial t_1} = o(s) \qquad \text{and} \qquad \frac{\partial s_{cp}}{\partial t_2} = -\frac{\beta(p)}{\alpha(p)} + o(s),$$

and thus we have that

$$\mathfrak{v}_1 = X(q_3) + s_2 [X, Y](q_3) + o(s)$$

and

$$\mathfrak{v}_2 = Y(q_3) - \frac{\beta(p)}{\alpha(p)} X(q_3) - s_3 [X, Y](q_3) + o(s).$$

These two vectors are linearly independent and thus span the tangent space to $\mathscr{S}_{cp}$ at q_3. Taking the wedge product of these two tangent vectors with $X(q_3)$ and $Y(q_3)$ gives

$$\mathfrak{v}_1 \wedge \mathfrak{v}_2 \wedge X(q_3) = \begin{vmatrix} 1 & 0 & s_2 \\ -\dfrac{\beta(p)}{\alpha(p)} & 1 & -s_3 \\ 1 & 0 & 0 \end{vmatrix} (1+o(s))(X \wedge Y \wedge [X,Y])(q_3)$$

$$= -s_2(1+o(s))(X \wedge Y \wedge [X,Y])(q_3)$$

and

$$\mathfrak{v}_1 \wedge \mathfrak{v}_2 \wedge Y(q_3) = \begin{vmatrix} 1 & 0 & s_2 \\ -\dfrac{\beta(p)}{\alpha(p)} & 1 & -s_3 \\ 0 & 1 & 0 \end{vmatrix} (1+o(s))(X \wedge Y \wedge [X,Y])(q_3)$$

$$= \left(-\dfrac{\beta(p)}{\alpha(p)}s_2 + s_3\right)(1+o(s))(X \wedge Y \wedge [X,Y])(q_3).$$

Since α and β have opposite signs, it follows that the coefficients at the wedge product $(X \wedge Y \wedge [X,Y])(q_3)$ have opposite signs, and thus X and Y point to opposite sides of $\mathscr{S}_{cp}$ at q_3. Hence the parameterized family $\mathscr{F}_{XYX}$ has a transversal fold at the third switching point. Since we know that singular arcs are slow in this case, the result now follows from Theorem 6.1.2.

Finally, condition (iv) simply is a consequence of the fact that XYX-trajectories are no longer extremal if $s > s_{cp}$. If switchings are made according to the conjugate point relations, then indeed extremals with an arbitrarily large number of switching can be constructed that lie in an arbitrarily small neighborhood of p, but these trajectories are no longer optimal in any sense after their third switching. This clearly makes the point by how much the class of extremals is larger than that of optimal trajectories.   □

Time-optimal control in $\mathbb{R}^3$ has been and still is the focus of several papers, e.g., [18, 43, 54, 210, 211, 224]. Our results presented here give the geometric structures that small-time reachable sets and time-optimal controls have locally under codimension-0 conditions. The resulting structures are simple—just concatenations of three bang and singular pieces—yet the proofs of these results become highly nontrivial. Cases of higher codimension have been analyzed in [209], and some of these results are summarized in [210] and [211]. While the concatenation sequences naturally get more complicated, it still holds in many of the cases considered there that an upper bound on the length of the number of concatenations can be given in the form $3 + k$, where k denotes the codimension of the Lie-bracket configuration at p. This bound is also supported in the research of Agrachev and Sigalotti [18, 224]. However, the structure of time-optimal trajectories near a point $p \in \mathbb{R}^3$ under generic conditions still is not fully resolved.

7.4   From Boundary Trajectories in Dimension 4 to Time-Optimal Control in $\mathbb{R}^3$

In the transition from the two- to the three-dimensional time-optimal control problem, the following relation is evident: time-optimal controlled trajectories for the two codimension-0 cases for the two-dimensional problem, XY and YX, form the boundary trajectories for the three-dimensional setup, $\mathscr{C}_{XY}$ and $\mathscr{C}_{YX}$. The same holds for the transition from the three- to the four-dimensional problem, but this is less obvious and requires a proof. In this section, based on computations analogous to those just carried out, we give a complete description of the boundary trajectories for the small-time reachable set in dimension four and show that they comprise all four classes of time-optimal trajectories for the codimension-0 three-dimensional problem. Many of the arguments are analogous, and thus we only highlight some of the differences and more geometric aspects. We also use this setup as a nontrivial example to construct a time-optimal synthesis to an equilibrium point of f for a corresponding three-dimensional system. Such a problem, since $f(p) = 0$, actually is an example of a codimension-3 Lie-bracket configuration for the underlying system Σ, since the requirement that f vanish corresponds to three independent equality constraints imposed on the system. As such, in the hierarchy of Lie-bracket configurations, it lies well beyond the cases analyzed so far. Nevertheless, this geometric approach rather directly leads to the full solution.

7.4.1   Boundary Trajectories in Dimension 4 under Codimension-0 Assumptions

We consider the system $\Sigma : \dot{x} = f(x) + ug(x)$, $|u| \leq 1$, near a reference point $p \in \mathbb{R}^4$ and make the following assumption:

(C1)   the vector fields $X, Y, [X,Y]$, and $[Y,[X,Y]]$ are linearly independent at p,

$$X(p) \wedge Y(p) \wedge [X,Y](p) \wedge [Y,[X,Y]](p) \neq 0.$$

We use these vector fields as a basis near p, and on some neighborhood M of p, write

$$[X,[X,Y]] = aX + bY + c[X,Y] + d[Y,[X,Y]] \qquad (7.16)$$

with smooth functions $a, \dots, d$ defined on M. We also assume that

(C2) the function d does not vanish at p, $d(p) \neq 0$; equivalently, the vector fields $X, Y, [X,Y]$, and $[X,[X,Y]]$ are linearly independent at p,

$$X(p) \wedge Y(p) \wedge [X,Y](p) \wedge [X,[X,Y]](p) \neq 0.$$

It will be seen that the sign of d distinguishes between the totally bang-bang case in which no admissible singular controls exist and boundary trajectories are bang-bang with two switchings ($d > 0$) and the case that a singular vector field is well-defined ($d < 0$). In this case, boundary trajectories consist of both the four BSB cells $\mathscr{C}_{BSB} = \mathscr{C}_{XSX} \cup \mathscr{C}_{XSY} \cup \mathscr{C}_{YSX} \cup \mathscr{C}_{YSY}$ and the two BBB cells $\mathscr{C}_{XYX}^{opt}$ and $\mathscr{C}_{XYX}^{opt}$ that arise through a cut-locus.

We again choose canonical coordinates of the second kind with the vector fields from (C1) as the basis, but for reasons that have to do with the construction of a three-dimensional time-optimal synthesis later on, we label the first coordinate as ξ_0,

$$(\xi_0, \xi_1, \xi_2, \xi_3) \mapsto p \exp(\xi_3[Y,[X,Y]]) \exp(\xi_2[X,Y]) \exp(\xi_1 Y) \exp(\xi_0 X).$$

Note that the projection onto the plane $\{\xi_3 = 0\}$ reduces to Lobry's setup analyzed above, and it follows that the restriction of the reachable set to the three-dimensional subspace spanned by (ξ_0, ξ_1, ξ_2) has a similar geometric structure. The cells $\mathscr{C}_{XY}$ and $\mathscr{C}_{YX}$ will lie in the boundary of the small-time reachable set and anchor the cells $\mathscr{T}_{XYX}$ and $\mathscr{T}_{YXY}$ of all bang-bang trajectories with exactly two switchings. In the four-dimensional problem, these manifolds become hypersurfaces that can be described as graphs of functions of (ξ_0, ξ_1, ξ_2).

Lemma 7.4.1. *The cells $\mathscr{T}_{XYX}$ and $\mathscr{T}_{YXY}$ are graphs of smooth functions $\phi_{\mp}$ of the variables (ξ_0, ξ_1, ξ_2), $\xi_3 = \phi_{\mp}(\xi_0, \xi_1, \xi_2)$. These functions have smooth extensions to the boundary strata $\mathscr{T}_{XY}$ and $\mathscr{T}_{YX}$; the function ϕ_- defining $\mathscr{T}_{XYX}$ has a smooth extension to $\mathscr{T}_Y$ and a continuous extension to $\mathscr{T}_X$, while the function ϕ_+ defining $\mathscr{T}_{YXY}$ has a smooth extension to $\mathscr{T}_X$ and a continuous extension to $\mathscr{T}_Y$.*

Proof. Without loss of generality, we consider the hypersurface $\mathscr{T}_{XYX}$,

$$\mathscr{T}_{XYX} = \{q = p \exp(s_1 X) \exp(s_2 Y) \exp(s_3 X) : s_1, s_2, s_3 > 0, \ s_1 + s_2 + s_3 < T\}.$$

Denote the junctions by $q_1 = p \exp(s_1 X)$ and $q_2 = q_1 \exp(s_2 Y)$. We show that $\mathscr{T}_{XYX}$ is a three-dimensional manifold and that the last coordinate vector field, $\frac{\partial}{\partial \xi_3}$, always points to one side of $\mathscr{T}_{XYX}$. Hence $\mathscr{T}_{XYX}$ satisfies the vertical line test in canonical coordinates, and this implies that it can be represented as the graph of some function ϕ_-. As above, the partial derivatives with respect to s_1, s_2, and s_3 at the end-point q are given by

$$\frac{\partial q}{\partial s_1} = p \exp(s_1 X) X \exp(s_2 Y) \exp(s_3 X),$$

$$\frac{\partial q}{\partial s_2} = p \exp(s_1 X) \exp(s_2 Y) Y \exp(s_3 X),$$

$$\frac{\partial q}{\partial s_3} = p \exp(s_1 X) \exp(s_2 Y) \exp(s_3 X) X \sim X(q),$$

and are tangent vectors to $\mathscr{T}_{XYX}$ at q. In order to prove that $\frac{\partial}{\partial \xi_3}$ always points to one side of $\mathscr{T}_{XYX}$, we verify that the wedge product of these three vectors with $\frac{\partial}{\partial \xi_3}$ has constant nonzero sign. Hence, these four vectors are linearly independent, and in particular, the three vectors above span the tangent space to $\mathscr{T}_{XYX}$ at q. Rather than evaluating the wedge product at q, it is computationally more advantageous to move all vectors back to the first switching point q_1 along the trajectory. As we have already argued several times, this is a diffeomorphism, and thus linear independence relations are preserved. In particular, the wedge product cannot change sign during this operation. The transport of the tangent vectors generates the following three vectors:

$$\frac{\partial q}{\partial s_1} \exp(-s_3 X) \exp(-s_2 Y) = p \exp(s_1 X) X \sim X(q_1),$$

$$\frac{\partial q}{\partial s_2} \exp(-s_3 X) \exp(-s_2 Y) = q_1 \exp(s_2 Y) Y \exp(-s_2 Y) \sim Y(q_1),$$

$$\frac{\partial q}{\partial s_3} \exp(-s_3 X) \exp(-s_2 Y) = q_1 \exp(s_2 Y) X \exp(-s_2 Y) \sim \exp(s_2 \mathrm{ad} Y) X(q_2).$$

In the wedge product of these vectors, the vector $X(q_1)$ will cancel the $X(q_1)$-term coming from $\exp(s_2 \mathrm{ad} Y) X(q_2)$, and we factor s_2. Also note that

$$\frac{\partial}{\partial \xi_3}(0,0,0,0) = [Y,[X,Y]](p),$$

and the vector $\frac{\partial}{\partial \xi_3}$ transported back to q_1 will simply be of the form

$$\exp(s_2 \mathrm{ad} Y) \exp(s_3 \mathrm{ad} X) \frac{\partial}{\partial \xi_3} = [Y,[X,Y]](p) + O(T).$$

Hence, with all the vectors evaluated at the point q_1, we have that

$$X \wedge Y \wedge \exp(s_2 \mathrm{ad} Y) X \wedge \exp(s_2 \mathrm{ad} Y) \exp(s_3 \mathrm{ad} X) \frac{\partial}{\partial \xi_3} =$$

$$= s_2 \left(X \wedge Y \wedge \left(\frac{\exp(s_2 \mathrm{ad} Y) - \mathrm{id}}{s_2} \right) X \wedge \exp(s_2 \mathrm{ad} Y) \exp(s_3 \mathrm{ad} X) \frac{\partial q}{\partial \xi_3} \right)$$

$$= s_2 \left(X \wedge Y \wedge [Y,X] + O(s_2) \wedge [Y,[X,Y]] + O(T) \right)$$

$$= s_2 \begin{vmatrix} 1 & 0 & 0 & 0 \\ 0 & 1 & 0 & 0 \\ * & * & -1 & O(T) \\ * & * & O(T) & 1+O(T) \end{vmatrix} (X \wedge Y \wedge [X,Y] \wedge [Y,[X,Y]])$$

$$= s_2 \left(-1 + O(T) \right) (X \wedge Y \wedge [X,Y] \wedge [Y,[X,Y]]), \qquad (7.17)$$

which is nonzero. Thus the tangent vectors $\frac{\partial q}{\partial s_i}$, $i = 1,2,3$ and $\frac{\partial}{\partial \xi_3}$ are linearly independent. Writing

$$\frac{\partial q}{\partial s_1} = \frac{\partial q}{\partial \xi_0} \frac{\partial \xi_0}{\partial s_1} + \frac{\partial q}{\partial \xi_1} \frac{\partial \xi_1}{\partial s_1} + \frac{\partial q}{\partial \xi_2} \frac{\partial \xi_2}{\partial s_1} + \frac{\partial q}{\partial \xi_3} \frac{\partial \xi_3}{\partial s_1},$$

$$\frac{\partial q}{\partial s_2} = \frac{\partial q}{\partial \xi_0} \frac{\partial \xi_0}{\partial s_2} + \frac{\partial q}{\partial \xi_1} \frac{\partial \xi_1}{\partial s_2} + \frac{\partial q}{\partial \xi_2} \frac{\partial \xi_2}{\partial s_2} + \frac{\partial q}{\partial \xi_3} \frac{\partial \xi_3}{\partial s_2},$$

$$\frac{\partial q}{\partial s_3} = \frac{\partial q}{\partial \xi_0} \frac{\partial \xi_0}{\partial s_3} + \frac{\partial q}{\partial \xi_1} \frac{\partial \xi_1}{\partial s_3} + \frac{\partial q}{\partial \xi_2} \frac{\partial \xi_2}{\partial s_3} + \frac{\partial q}{\partial \xi_3} \frac{\partial \xi_3}{\partial s_3},$$

it follows that the matrix

$$\begin{pmatrix} \frac{\partial \xi_0}{\partial s_1} & \frac{\partial \xi_1}{\partial s_1} & \frac{\partial \xi_2}{\partial s_1} & \frac{\partial \xi_3}{\partial s_1} \\[2mm] \frac{\partial \xi_0}{\partial s_2} & \frac{\partial \xi_1}{\partial s_2} & \frac{\partial \xi_2}{\partial s_2} & \frac{\partial \xi_3}{\partial s_2} \\[2mm] \frac{\partial \xi_0}{\partial s_3} & \frac{\partial \xi_1}{\partial s_3} & \frac{\partial \xi_2}{\partial s_3} & \frac{\partial \xi_3}{\partial s_3} \\[2mm] 0 & 0 & 0 & 1 \end{pmatrix}$$

is nonsingular and so is then its principal 3×3 minor. By the inverse function theorem, it is possible to solve for the times (s_1, s_2, s_3) as smooth functions of (ξ_0, ξ_1, ξ_2) near q provided $s_2 \neq 0$. Since $\frac{\partial}{\partial \xi_3}$ is transversal to $\mathscr{T}_{XYX}$ everywhere, the entire 3-manifold is the graph of a smooth function ϕ_-. Furthermore, these calculations show that ϕ_- has a smooth extension to a neighborhood of $s_1 = 0$ and $s_3 = 0$ provided $s_2 > 0$. It is clear that these extensions agree with $\mathscr{T}_Y$, $\mathscr{T}_{XY}$, and $\mathscr{T}_{YX}$. As should be clear, the vertical line test breaks down for $\{s_2 = 0\}$, i.e., along the X-trajectory. In this case, the hypersurface $\mathscr{T}_{XYX}$ reduces to $\mathscr{T}_X$, and therefore ϕ_- still extends continuously to $\mathscr{T}_X = \{(\xi_0, 0, 0, 0) : \xi_0 > 0\}$. This proves the lemma for $\mathscr{T}_{XYX}$. $\qquad\square$

In this geometric argument, we considered all bang-bang trajectories with at most two switchings, regardless of whether they are extremals. Analyzing whether this is the case, once again is easily pursued within the framework of conjugate triples. Although the setting now is in $\mathbb{R}^4$, since we are considering boundary trajectories, at every switching the multiplier λ vanishes not only against g, but also against f, or equivalently, against X and Y. Hence, three consecutive switching points q_1, q_2, and q_3 still form a conjugate triple (q_1, q_2, q_3). But now, for example, for an $\cdot XY\cdot$ boundary trajectory with switching points q_1, $q_2 = q_1 \exp(s_1 X)$ and $q_3 = q_2 \exp(s_2 Y)$, the vectors $X(q_1)$, $Y(q_1)$,

$$q_2 Y \exp(-s_1 X) = q_1 \exp(s_1 X) Y \exp(-sX) \sim \exp(s_1 \mathrm{ad} X) Y(q_2)$$

and

$$q_3 X \exp(-s_2 Y) \exp(-s_1 X) \sim \exp(s_1 \mathrm{ad} X) \exp(s_2 \mathrm{ad} Y) X(q_1)$$

are linearly independent, and this generates a corresponding conjugate point relation. As above, we factor out s_1 and s_2 and use Eq. (7.16) to write $[X, [X, Y]]$ as a linear combination of the basis, to obtain that

$$0 = X \wedge Y \wedge \exp(s_1 \mathrm{ad} X) Y \wedge \exp(s_1 \mathrm{ad} X) \exp(s_2 \mathrm{ad} Y) X$$

$$= s_1 s_2 \left(X \wedge Y \wedge \left(\frac{\exp(s_1 \mathrm{ad} X) - \mathrm{id}}{s_1} \right) Y \wedge \exp(s_1 \mathrm{ad} X) \left(\frac{\exp(s_2 \mathrm{ad} Y) - \mathrm{id}}{s_2} \right) X \right)$$

$$= s_1 s_2 \left(X \wedge Y \wedge [X, Y] + \frac{1}{2} s_1 [X, [X, Y]] + \cdots \right.$$

$$\left. \wedge - [X, Y] - \frac{1}{2} s_2 [Y, [X, Y]] + \cdots - s_1 [X, [X, Y]] + \cdots \right) \qquad (7.18)$$

$$= s_1 s_2 \begin{vmatrix} 1 & 0 & 0 & 0 \\ 0 & 1 & 0 & 0 \\ * & * & 1 + O(s_1) & \frac{1}{2} s_1 d + O(s_1^2) \\ * & * & -1 + O(T) & -\frac{1}{2} s_2 - s_1 d + O(T^2) \end{vmatrix} (X \wedge Y \wedge [X, Y] \wedge [Y, [X, Y]])$$

$$= -\frac{1}{2} s_1 s_2 \left(s_2 + s_1 d + O(T^2) \right) (X \wedge Y \wedge [X, Y] \wedge [Y, [X, Y]]). \qquad (7.19)$$

More generally, let us write this relation in the form

$$X \wedge Y \wedge \left(\frac{\exp(s_1 \mathrm{ad} X) - \mathrm{id}}{s_1} \right) Y \wedge \exp(s_1 \mathrm{ad} X) \left(\frac{\exp(s_2 \mathrm{ad} Y) - \mathrm{id}}{s_2} \right) X$$

$$= \Psi_{XY}(q_1; s_1, s_2) \cdot (X \wedge Y \wedge [X, Y] \wedge [Y, [X, Y]]),$$

where the function $\Psi(q_1; s, t)$ is defined as the quotient of these two wedge products of 4-vectors. The equation

$$0 = \Psi_{XY}(q_1; s_1, s_2) \Leftrightarrow 0 = ds_1 + s_2 + O(T^2) \tag{7.20}$$

then defines the *conjugate point relation* as a function of the first switching point q_1 and the times s_1 and s_2 along the consecutive bang segments X and Y. The relation for a $\cdot YX \cdot$-trajectory follows analogously or by using the input symmetry ρ of Sect. 2.10.

By assumption (C2), we have $d(p) \neq 0$, and without loss of generality we may assume that M is chosen so small that all values of d on M dominate T. Then the linear terms in this expansion dominate the remainders, and therefore the expression $ds_1 + s_2 + o(T^2)$ does not vanish if $d > 0$. Hence, in this case no conjugate triples exist, and bang-bang trajectories that lie in the boundary of the small-time reachable set have at most two switchings. The following result is the immediate analogue of Proposition 7.3.3.

Proposition 7.4.1. *If d is positive on M, then the boundary trajectories of the small-time reachable set* $\text{Reach}_{\Sigma, \leq T}(p)$ *are given by all bang-bang trajectories in the cells* $\mathscr{C}_{XYX}$ *and* $\mathscr{C}_{YXY}$. *The surface* $\mathscr{T}_{XYX}$ *lies below* $\mathscr{T}_{YXY}$ *in the direction of the coordinate* ξ_3 *everywhere.* ∎

The last statement follows from the cut-locus computation given below, and we shall comment on this later. Thus, in this case, the structure of the boundary of the small-time reachable set is the direct extension of the two- and three-dimensional cases.

Henceforth we assume that $d < 0$. This case corresponds to the situation that singular controls exist and now the boundary trajectories of the small-time reachable set in dimension four are made up of both the stratified hypersurface $\mathscr{C}_{BSB}$ that combines concatenations of the X- and Y-trajectories with a singular arc and the hypersurface $\mathscr{C}_{BBB}$ that arises form a cut-locus in the bang-bang trajectories. We first discuss the cut-locus

$$p \exp(s_1 X) \exp(s_2 Y) \exp(s_3 X) = p \exp(t_1 Y) \exp(t_2 X) \exp(t_3 Y). \tag{7.21}$$

A cut-locus is a globally defined object that is rather difficult to find, and it generally needs to be analyzed by direct computational arguments. The following result thus is remarkable in the sense that it locates a cut-locus by means of easily verifiable conditions. In fact, *conjugate triples imply the existence of a cut-locus.* Let Γ_{XY} denote the set of all nontrivial points in $\mathscr{T}_{XY}$ that satisfy the conjugate point relation for $\cdot XY \cdot$-trajectories,

$$\Gamma_{XY} = \{q = p \exp(s_1 X) \exp(s_2 Y) : \Psi_{XY}(p; s_1, s_2) = 0, \, s_1 > 0, s_2 > 0\},$$

and similarly

$$\Gamma_{YX} = \{q = p \exp(t_1 Y) \exp(t_2 X) : \Psi_{YX}(p; t_1, t_2) = 0, \, t_1 > 0, t_2 > 0\}.$$

Proposition 7.4.2. *[212] In the case $d(p) < 0$, a nontrivial cut-locus*

$$\Gamma = \mathscr{T}_{XYX} \cap \mathscr{T}_{YXY}$$

bifurcates from the curves Γ_{XY} and Γ_{YX} of conjugate triples.

Proof. By the implicit function theorem, Eq. (7.21) can be solved uniquely for (s_1,s_2) as a function of $(s_3;t_1,t_2,t_3)$ near a point q on the cut-locus if the Jacobian matrix with respect to $(s_3;t_1,t_2,t_3)$ is nonsingular. Let $q_1 = p\exp(t_1 Y)$, $q_2 = q_1\exp(t_2 Y)$, and $q = q_3 = q_2\exp(t_3 X)$. Then these partial derivatives are given by

$$\frac{\partial q}{\partial t_1} = q_1 Y \exp(t_2 X)\exp(t_3 Y), \qquad \frac{\partial q}{\partial t_2} = q_2 X \exp(t_3 Y),$$

$$\frac{\partial q}{\partial t_3} = q_3 Y, \qquad \text{and} \qquad \frac{\partial q}{\partial s_3} = q_3 X.$$

But these are the four vectors that arise in the conjugate point relation along the last XY-portion of the YXY-trajectory. Hence, these vectors are linearly dependent if and only if the points (q_1,q_2,q_3) form a conjugate triple. By the implicit function theorem, the cut-locus equation (7.21) therefore has a unique, and thus only the trivial, solution $s_1 = t_2$, $s_2 = t_3$, $s_3 = t_1 = 0$ everywhere on $\mathscr{T}_{XY}$ except possibly along the curve Γ_{XY} defined by the solutions to the equation $\Psi_{XY}(p;t_2,t_3) = 0$.

Along this curve, indeed a nontrivial solution to Eq. (7.21) bifurcates. This can be seen geometrically by considering the tangent space to $\mathscr{T}_{YXY}$ at q. If we transport the tangent vectors $\frac{\partial q}{\partial t_1}$, $\frac{\partial q}{\partial t_2}$, and $\frac{\partial q}{\partial t_3}$ back to the first junction q_1, we get the vectors $X(q_1)$, $Y(q_1)$, and $\exp(t_2 \mathrm{ad} X)Y(q_2)$. This gives at q_1 that

$$X \wedge Y \wedge \exp(t_2 \mathrm{ad} X)Y \wedge \exp(t_2 \mathrm{ad} X)\exp(t_3 \mathrm{ad} Y)\frac{\partial}{\partial \xi_3}$$

$$\equiv t_2(1+O(T))(X \wedge Y \wedge [X,Y] \wedge [Y,[X,Y]]) \tag{7.22}$$

where $\frac{\partial}{\partial \xi_3}$ again denotes the last coordinate vector field in canonical coordinates. (Recall that $\frac{\partial}{\partial \xi_3}(0) = [Y,[X,Y]](p)$.) But from the conjugate point relation for (q_1,q_2,q_3) we also get at q_1 that

$$X \wedge Y \wedge \exp(t_2 \mathrm{ad} X)Y \wedge \exp(t_2 \mathrm{ad} X)\exp(t_3 \mathrm{ad} Y)X$$

$$\equiv t_2 t_3 \Psi_{XY}(p;t_2,t_3)(X \wedge Y \wedge [X,Y] \wedge [Y,[X,Y]]). \tag{7.23}$$

Hence the vectors $\frac{\partial}{\partial \xi_3}$ and $X(q)$ point to the same side of the tangent space $T_q(\mathscr{T}_{YXY})$ if $\Psi_{XY}(p;t_2,t_3) > 0$ and to opposite sides if $\Psi_{XY}(p;t_2,t_3) < 0$. This implies that the surfaces $\mathscr{T}_{YXY}$ and $\mathscr{T}_{XYX}$ cross each other near $\Gamma_{XY} \subset \mathscr{T}_{XY}$. Elementary geometric considerations show that trajectories that maximize the coordinate ξ_3 near the surface $\mathscr{T}_{XY}$ are of the form YXY if $\Psi_{XY}(p;t_2,t_3) > 0$ and of the form XYX if $\Psi_{XY}(p;t_2,t_3) < 0$. $\qquad\square$

While this result guarantees the existence of a nontrivial cut-locus, the set itself needs to be determined through explicit calculations. As before, this is done by equating the canonical coordinates of the left- and right-hand sides of Eq. (7.21). We give only the resulting equations:

$$(\xi_0) \qquad s_1 + s_3 + O(S^3) = t_2 + O(T^3), \tag{7.24}$$

$$(\xi_1) \qquad s_2 + O(S^3) = t_1 + t_3 + O(T^3), \tag{7.25}$$

$$(\xi_2) \qquad s_1 s_2 (1 + O(S)) = t_2 t_3 (1 + O(T)), \tag{7.26}$$

$$(\xi_3) \qquad \frac{1}{2} s_1 s_2 (s_1 d + s_2 + O(S)) = \frac{1}{2} t_2 t_3 (2 t_1 + t_2 d + t_3 + O(T)), \tag{7.27}$$

where $S = s_1 + s_2 + s_3$ and $T = t_1 + t_2 + t_3$. Dividing ξ_3 by ξ_2, we furthermore obtain that

$$s_1 d + s_2 + O(S) = 2 t_1 + t_2 d + t_3 + O(T). \tag{7.28}$$

Equations (7.24), (7.25), and (7.28) have dominant linear terms and can be solved uniquely for s as a function of t,

$$s_1 = \frac{1}{d} t_1 + t_2 + O(T^2), \qquad s_2 = t_1 + t_3 + O(T^3), \qquad s_3 = -\frac{1}{d} t_1 + O(T^2).$$

The conjugate point relations imply that all these times (and also t_3 calculated below) are nonnegative for extremals. Now substitute these functions for s into Eq. (7.26) to obtain

$$t_1 \left(\frac{1}{d} t_1 + t_2 + \frac{1}{d} t_3 \right) + O(T^3) = 0.$$

In general, for several variables, specific quadratic terms need not dominate arbitrary cubic remainders. However, in our situation it follows from the conjugate point relations that the times t_i satisfy a relation of the type

$$t_1 \geq \varepsilon T = \varepsilon (t_1 + t_2 + t_3) \quad \Leftrightarrow \quad t_1 \geq \frac{\varepsilon}{1 - \varepsilon} (t_2 + t_3),$$

and thus this equation can be solved for t_3 as

$$t_3 = -t_1 - d t_2 + O(T^2). \tag{7.29}$$

This solution is well-defined near $\{t_3 = 0\}$ (i.e., the curve Γ_{YX}), and thus a nontrivial cut-locus Γ, $\Gamma = \mathscr{T}_{YXY} \cap \mathscr{T}_{XYX}$, extends beyond $\mathscr{T}_{YX}$ along Γ_{YX}. Similarly, by solving Eqs. (7.24), (7.25), and (7.28) for t as a function of s, it follows that Γ also extends beyond $\mathscr{T}_{XY}$ at Γ_{XY}. It is not difficult to verify that the hypersurfaces $\mathscr{T}_{YXY}$ and $\mathscr{T}_{XYX}$ intersect transversally in the cut-locus Γ, and thus we have the following result:

Proposition 7.4.3. *If $d < 0$, then the hypersurfaces $\mathcal{T}_{YXY}$ and $\mathcal{T}_{XYX}$ intersect transversally along a two-dimensional surface Γ. This surface extends smoothly across $\mathcal{T}_{YX}$ and $\mathcal{T}_{XY}$, and intersects these surfaces transversally in the curves Γ_{YX} and Γ_{XY} of conjugate triples.* ∎

We next establish the geometric location of the surfaces $\mathcal{T}_{YXY}$ and $\mathcal{T}_{XYX}$, but revert to the case $d(p) > 0$ for a moment. Using Eqs. (7.24)–(7.26), the difference in the ξ_3 coordinates of YXY- and XYX-trajectories over a common base point (ξ_0, ξ_1, ξ_2) can be expressed in the form

$$
\begin{aligned}
\xi_3^{YXY} - \xi_3^{XYX} &= \frac{1}{2} t_2 t_3 \left[2t_1 + t_2 d + t_3 - s_1 d - s_2 + \cdots \right] \\
&= \frac{1}{2} t_2 t_3 \left[t_1 + s_3 d + \cdots \right],
\end{aligned}
$$

and for trajectories along which all times are comparable, this quantity is positive. Since there do no exist nontrivial intersections between the surfaces $\mathcal{T}_{YXY}$ and $\mathcal{T}_{XYX}$ in the case $d > 0$, it follows that $\mathcal{T}_{XYX}$ always lies below $\mathcal{T}_{YXY}$, as claimed in Proposition 7.4.1.

In the case that d is negative, these locations change along the cut-locus. As a subset of $\mathcal{T}_{XYX}$ (or $\mathcal{T}_{YXY}$), Γ can be described as the graph of the function $\xi_3 = \xi_3^{XYX}(\xi_0, \xi_1, \xi_2)$, and it separates the portion of the hypersurface $\mathcal{T}_{XYX}$ that lies above $\mathcal{T}_{YXY}$ in the direction of ξ_3 from the one that lies below $\mathcal{T}_{YXY}$. If we denote the corresponding substrata by a superscript $\pm$, then only the trajectories in the "southern hemisphere" $\mathcal{S} = \mathcal{T}_{XYX}^{-} \cup \Gamma \cup \mathcal{T}_{YXY}^{+}$ that correspond to the two sections that minimize the ξ_3-coordinate are boundary trajectories. The remaining portions of the hypersurfaces, $\mathcal{T}_{XYX}^{+}$ and $\mathcal{T}_{YXY}^{-}$, lie in the interior of the small-time reachable set, and therefore these trajectories cannot be time-optimal. Within our construction, the natural way to see this is to complement the construction with a stratified hypersurface $\mathcal{N}$ that closes the small-time reachable set with a "northern hemisphere" $\mathcal{N}$ in the coordinates $(\xi_0, \xi_1, \xi_2, \xi_3)$. This is indeed possible [141], and $\mathcal{N}$ is made of concatenations with singular arcs. The analysis of the northern hemisphere is the same as in the three-dimensional case, and we derive only the formulas for the singular control and arc, and leave the rest to the interested reader (also, see [141]).

Proposition 7.4.4. *If $\Gamma = (x, u)$ is an extremal controlled trajectory pair with the property that the control u is singular on an open subinterval I, then for all $t \in I$, the singular control is given as a feedback function of x in the form*

$$
u_{sing}(t) = \frac{d(x) + 1}{d(x) - 1}. \tag{7.30}
$$

Proof. The multiplier λ vanishes on I against X, Y, and $[X,Y]$ along x. Since X, Y, $[X,Y]$, and $[Y,[X,Y]]$ are linearly independent, it cannot vanish against $[Y,[X,Y]]$, and thus

$$2\langle \lambda,[g,[f,g]](x)\rangle = \langle \lambda,[Y-X,[X,Y]](x)\rangle = (1-d(x))\langle \lambda,[Y,[X,Y]](x)\rangle \neq 0. \tag{7.31}$$

Hence $\ddot{\Phi} = \langle \lambda,[f+ug,[f,g]](x)\rangle \equiv 0$ gives that

$$u_{\text{sing}}(t) = -\frac{\langle \lambda,[Y+X,[X,Y]](x)\rangle}{\langle \lambda,[Y-X,[X,Y]](x)\rangle} = -\frac{1+d(x)}{1-d(x)}$$

as desired. □

The singular control thus is a smooth feedback control defined on M. Note that it is admissible if and only if $d < 0$. Also, in a sufficiently small neighborhood of p, the quotient $(d+1)/(d-1)$ is bounded away from ± 1, and so no saturation is possible in M. This phenomenon will be investigated in the last section. As above, singular conjugate point relations can be invoked to show that boundary trajectories are at most of the form BSB, and it can be shown that the surfaces $\mathscr{T}_{XSX}$, $\mathscr{T}_{XSY}$, $\mathscr{T}_{YSX}$, and $\mathscr{T}_{YSY}$ form a stratified hypersurface $\mathscr{S}_{BSB}$ that can be described as the graph of a piecewise smooth function $\phi_S = \phi_S(\xi_0,\xi_1,\xi_2)$. In principle, this verification is the same as for bang-bang trajectories, but the fact that the singular control is a smooth feedback function must be taken into account in the calculations. We leave it to the attentive reader to supply the details or to consult [215], where a similar calculation is carried out. This concludes the construction of the small-time reachable set for the nondegenerate four-dimensional case, and we get the following analogue to Propositions 7.3.4 and 7.3.5.

Proposition 7.4.5. *If $d(p) < 0$, then there exists a sufficiently small neighborhood M of p such that the boundary trajectories of the small-time reachable set* $\text{Reach}_{\Sigma,\leq T}(p)$ *are given by a stratified hypersurface*

$$\mathscr{S} = \mathscr{T}_{XYX}^- \cup \mathscr{T}_{YXY}^+$$

of bang-bang trajectories that minimize the coordinate ξ_3 (over the small-time reachable set for fixed (ξ_0,ξ_1,ξ_2)-coordinates) and a stratified hypersurface

$$\mathscr{N} = \mathscr{S}_{BSB} = \mathscr{T}_{XSX} \cup \mathscr{T}_{XSY} \cup \mathscr{T}_{YSX} \cup \mathscr{T}_{YSY}$$

of trajectories that contain a singular arc and maximize the coordinate ξ_3. ∎

Figure 7.13 illustrates the structure of the boundary trajectories for the case that $d(p)$ is negative.

The statements about which trajectories minimize, respectively maximize the coordinate ξ_3 can be verified using formulas for the canonical coordinates. They can also be deduced rather directly from the proof of the maximum principle given

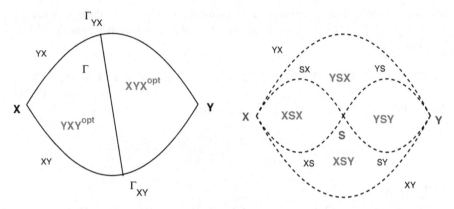

Fig. 7.13 Schematic illustration of the structure of boundary trajectories in the "southern hemisphere" $\mathscr{S}$ (*left*) and "northern hemisphere" (*right*) for $d(p) < 0$

Fig. 7.14 Entry and exit points for integral curves of the vector field $[Y, [X, Y]]$

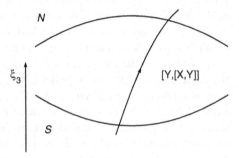

in Sect. 4.2.3. Recall that the multiplier λ in Theorem 4.2.2 has been chosen to be non-negative on the vectors from the approximating cone $\mathscr{K}$ to the reachable set at the endpoint $q \in \text{Reach}_{\Sigma, T}(p)$. By the Legendre–Clebsch condition, Theorem 4.6.1, the vector $-[g, [f, g]](q) + O(T)$ lies in this approximating cone. For trajectories that contain singular arcs, it follows from Eq. (7.31) that $\langle \lambda, [g, [f, g]](x) \rangle$ and $\langle \lambda, [Y, [X, Y]](q) \rangle$ are nonzero and have the same sign. In particular, for small T we have that both $\langle \lambda, [g, [f, g]](q) \rangle$ and $\langle \lambda, [Y, [X, Y]](q) \rangle$ are negative and thus the vector $[Y, [X, Y]](q)$ is separated from $\mathscr{K}$ by $\lambda(T)$. Since it is the vector field $[Y, [X, Y]]$ that defines the third coordinate ξ_3, it follows that trajectories that contain a singular arc maximize ξ_3. Similarly, for trajectories that minimize ξ_3, the vector $[Y, [X, Y]](q)$ points into the reachable set and thus we have $\langle \lambda, [Y, [X, Y]](q) \rangle > 0$ and these are the bang-bang trajectories. Another way to look at the underlying geometry is to say that points in the lower hemisphere $\mathscr{S}$ form entry points into the reachable sets along integral curves of the vector field $[Y, [X, Y]]$ and points in the northern hemisphere are exit points (see Fig. 7.14).

7.4.2 Construction of a Local Time-Optimal Synthesis to an Equilibrium Point in Dimension 3

For a system $\Sigma : \dot{x} = f(x) + ug(x)$, $|u| \leq 1$, in $\mathbb{R}^3$, we consider the local time-optimal control problem to steer points q in a neighborhood M of an equilibrium point p of the vector field f, $f(p) = 0$, into this equilibrium point p in minimum time under the following assumptions:

(1) The vectors $g(p)$, $[f,g](p)$, and $[g,[f,g]](p)$ are linearly independent;
(2) If we express $[f,[f,g]]$ as a linear combination of this basis near p,

$$[f,[f,g]] = \alpha g + \beta [f,g] + \gamma [g,[f,g]], \qquad (7.32)$$

then we have $-1 < \gamma(p) < 0$.

Our aim is to illustrate that an optimal synthesis of controlled trajectories for this problem can be constructed in a relatively straightforward way from the structure of the small-time reachable set and the accompanying analysis of boundary trajectories that has been carried out for an associated augmented four-dimensional system $\tilde{\Sigma}$ where time has been added as an extra coordinate, $\dot{x}_0 \equiv 1$. If the underlying system Σ is small-time locally controllable (in the sense that p is an interior point of the small-time reachable set $\text{Reach}_{\Sigma, \leq T}(p)$ for any $T > 0$), a local time-optimal synthesis can generally be obtained by projecting the time slices $\text{Reach}_{\tilde{\Sigma}, t}(p)$ of the reachable set of the augmented system $\tilde{\Sigma}$ into the original state space. Under the assumptions made above, the augmented system $\tilde{\Sigma}$ is given by the system Σ considered in Sect. 7.4.1. For if $\tilde{x} = (x_0, x)$, $x \in \mathbb{R}^3$, and the vector fields $\tilde{f}$ and $\tilde{g}$ are defined by

$$\tilde{f}(\tilde{x}) = \begin{pmatrix} 1 \\ f(x) \end{pmatrix} \quad \text{and} \quad \tilde{g}(\tilde{x}) = \begin{pmatrix} 0 \\ g(x) \end{pmatrix},$$

then it follows that

$$[\tilde{X}, \tilde{Y}](\tilde{x}) = \begin{pmatrix} 0 \\ [X,Y](x) \end{pmatrix}, \qquad [\tilde{X}, [\tilde{X}, \tilde{Y}]](\tilde{x}) = \begin{pmatrix} 0 \\ [X,[X,Y]](x) \end{pmatrix},$$

and so on. Hence the assumptions (C1) and (C2) of Sect. 7.4.1 that the vector fields $\tilde{X} = \tilde{f} - \tilde{g}$, $\tilde{Y} = \tilde{f} + \tilde{g}$, $[\tilde{X}, \tilde{Y}]$, and $[\tilde{X}, [\tilde{X}, \tilde{Y}]](\tilde{x})$, respectively $[\tilde{Y}, [\tilde{X}, \tilde{Y}]](\tilde{x})$, are linearly independent at p are equivalent to the condition that $|\gamma(p)| \neq 1$. Note that it follows from Eq. (7.32) that the singular control for the problem is given by the feedback control $u_{\text{sing}}(x) = -\gamma(x)$. The assumption $|\gamma(p)| > 1$ thus corresponds to the totally bang-bang case of Proposition 7.4.1 when optimal controls are bang-bang with at most two switchings, and in this case the corresponding local optimal synthesis is a straightforward extension of the time-optimal synthesis for the double integrator from dimension two to three. The case $|\gamma(p)| < 1$ corresponds to Proposition 7.4.5 and has a nontrivial synthesis that we develop here. It is clear that we can normalize the sign of γ by simply replacing g with $-g$, and thus, without loss of generality, we may assume that $-1 < \gamma(p) \leq 0$. Some comments about the extra assumption $\gamma(p) \neq 0$ are in order. It follows from results about the local controllability of nonlinear systems (for example, see [235]) that the system Σ is small-time locally controllable from p if $\gamma(p) \neq 0$, but this need not hold if $\gamma(p) = 0$. Thus the values $\gamma(p) = 0$ and $|\gamma(p)| = 1$ are bifurcation values for the structure of the optimal synthesis, i.e., correspond to more degenerate scenarios in which additional assumptions need to be made to determine the optimal solutions. For this reason, we restrict $\gamma(p)$ to lie in the interval $(-1,0)$ with the negative sign chosen arbitrarily. For γ close to -1, the singular vector field $S(x) = f(x) - \gamma(x)g(x)$ is therefore close to Y.

Our aim here is merely to describe the optimal synthesis—we shall give a stratification of a neighborhood of p and define the optimal control on each of the strata—but we shall skip some of the computational details. Most of the arguments needed to verify our geometric claims are immediate extensions of the reasoning given above. But we also give explicit formulas for the relevant structures of the synthesis that have been computed using canonical coordinates of the second kind of the form

$$(x_1,x_2,x_3) \mapsto p\exp(x_3[g,[f,g]])\exp(x_2[f,g])\exp(x_1g)$$

and these computations, especially when the singular feedback control is involved, become somewhat lengthy and technical. In these formulas, we assign a weight i to x_i and use the symbol $\doteq$ to indicate equality modulo terms of higher weight. We refer the reader to the paper [215] for the details of these computations.

For the **example**

$$\Sigma_3: \quad \dot{x}_1 = u, \qquad \dot{x}_2 = -x_1, \qquad \dot{x}_3 = -\frac{1}{2}x_1^2 - \gamma x_2, \qquad (7.33)$$

with γ a constant, we have that

$$f(x) = \begin{pmatrix} 0 \\ -x_1 \\ -\frac{1}{2}x_1^2 - \gamma x_2 \end{pmatrix}, \qquad g = \begin{pmatrix} 1 \\ 0 \\ 0 \end{pmatrix}, \qquad [f,g](x) = \begin{pmatrix} 0 \\ 1 \\ x_1 \end{pmatrix},$$

$$[f,[f,g]] = \gamma \begin{pmatrix} 0 \\ 0 \\ 1 \end{pmatrix}, \qquad [g,[f,g]] = \begin{pmatrix} 0 \\ 0 \\ 1 \end{pmatrix},$$

and all brackets of orders three and higher vanish, i.e., the system Σ_3 is nilpotent of order 3. For this example the formulas we give are exact.

Since the objective is to steer points into p time-optimally, it makes sense to build the synthesis inductively by integrating trajectories backward from p. Compared with the construction of the small-time reachable set for the augmented system $\tilde{\Sigma}$, which is done forward in time, this introduces a time-reversal input symmetry that changes f into $-f$ and g into $-g$. In particular, second-order Lie brackets like $[g,[f,g]]$ and $[Y,[X,Y]]$ reverse sign, and as a consequence the roles of the northern and southern hemispheres in the construction of the small-time reachable set become reversed. Thus, in the synthesis, bang-bang will now "maximize" in the direction of x_3, while those trajectories that contain a singular arc will be "minimizing." Otherwise, no qualitative changes arise. As before, we use notations like $\mathscr{T}_{XY}$ to denote actual XY trajectories when the system is integrated forward in time, i.e.,

$$\mathscr{T}_{XY} = \{p\exp(-tY)\exp(-sX) : s > 0, t > 0\}.$$

and with $q = p\exp(-tY)\exp(-sX)$, this becomes the forward trajectory

$$p = q\exp(sX)\exp(tY).$$

Lemma 7.4.2. *[215] The set $\mathscr{M} = \mathscr{C}_{XY} \cup \mathscr{C}_{YX}$ is a stratified surface that near p can be represented as the graph of a piecewise smooth function $x_3 = F_{\mathscr{M}}(x_1, x_2)$. It divides a small neighborhood M of p in the direction of the coordinate vector field $\frac{\partial}{\partial x_3}$ into an upper region $\mathscr{M}_+$ and a lower region $\mathscr{M}_-$. The submanifolds $\mathscr{C}_0 = \{p\}$, $\mathscr{T}_X$, $\mathscr{T}_Y$, $\mathscr{T}_{XY}$, and $\mathscr{T}_{YX}$ form the strata of $\mathscr{M}$, and on $\mathscr{M}$ the optimal feedback control is given by*

$$u_*(x) = \begin{cases} -1 & \text{if } x \in \mathscr{T}_X \cup \mathscr{T}_{XY}, \\ +1 & \text{if } x \in \mathscr{T}_Y \cup \mathscr{T}_{YX}. \end{cases} \qquad \blacksquare$$

This lemma can easily be verified by computing the tangent spaces to the surfaces $\mathscr{T}_{XY}$ and $\mathscr{T}_{YX}$ and by verifying that the third coordinate vector field $\frac{\partial}{\partial x_3}$, which at the equilibrium point p is given by $[g,[f,g]](p)$, always points to the same side. These computations are identical to those done above. The local optimal feedback flow on $\mathscr{M}$ is the one for the double integrator and is indicated in Fig. 7.15, but naturally $\mathscr{M}$ still warps in 3-space.

We show that optimal trajectories are bang-bang in the region $\mathscr{M}_+$ that lies above $\mathscr{M}$ in the direction of the coordinate vector field $\frac{\partial}{\partial x_3}$, while they contain a singular arc in the region $\mathscr{M}_-$ that lies below $\mathscr{M}$. In the construction of the optimal synthesis, X-trajectories are now integrated backward from $\mathscr{T}_{YX}$ and Y-trajectories are integrated backward from $\mathscr{T}_{XY}$. However, in contrast to the totally bang-bang case, this can no longer be done optimally from every point. The reason lies in the conjugate point relations. If we consider a YXY-trajectory of the form

$$p\exp(-tY)\exp(-sX)\exp(-rY) = q$$

Fig. 7.15 The stratified
surface $\mathcal{M} = \mathscr{C}_{XY} \cup \mathscr{C}_{YX}$

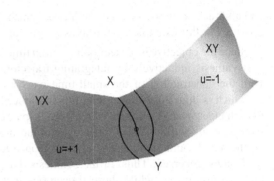

with positive times r, s, and t, then the corresponding trajectory has switchings at
$q_1 = q\exp(rY)$ and at $q_2 = q_1\exp(sX)$. The time s along the intermediate X-arc
then determines the times $\tilde{t}$ and $\tilde{r}$ until the next switchings through a conjugate
point relation. For $\tilde{r}$ this need not concern us, since as follows from the structure
of the small-time reachable set, the synthesis will be terminated at a cut-locus prior
to this conjugate point anyhow. However, if the initial time t is longer than $\tilde{t}$, then
this trajectory no longer is an extremal, and it thus cannot be part of the optimal
synthesis. It is therefore necessary to restrict the length of t to obey the conjugate
point relation. These are precisely the values for which (q_1, q_2, p) form a conjugate
triple. Evaluating the corresponding conjugate point relation at the equilibrium point
p, for YXY-trajectories, we define the function $\Psi_{XY}(s,t)$ through the equation

$$g(p) \wedge \left(\frac{\exp(-t\operatorname{ad}Y) - \operatorname{id}}{t} \right) g(q_1) \wedge \exp(-t\operatorname{ad}Y) \left(\frac{\exp(-s\operatorname{ad}X) - \operatorname{id}}{s} \right) g(q_2)$$

$$= \Psi_{XY}(s,t)\left(g(p) \wedge [f,g](p) \wedge [g,[f,g]](p) \right).$$

It then follows that the first Y-leg of an optimal YXY-trajectory needs to *arrive* in
the surface $\mathcal{T}_{XY}$ at a point $p\exp(-tY)\exp(-sX)$ that lies in the set

$$\mathscr{A}_{XY} = \{ p\exp(-tY)\exp(-sX) : s > 0,\ t > 0 \quad \text{and} \quad \Psi_{XY}(s,t) > 0 \}.$$

The sign of the function Ψ_{XY} is chosen from the fact that

$$\Psi_{XY}(s,0) = \frac{1}{2}(1 - \gamma)s + o(s) > 0,$$

and for $t = 0$, the conjugate point relation is always satisfied. A by now standard
computation shows that with $\bar{\gamma} = \gamma(p)$, this relation imposes the restriction that

$$t \le t_{cp}(s) = \frac{1 - \bar{\gamma}}{1 + \bar{\gamma}}s + o(s).$$

Fig. 7.16 The arrival regions $\mathscr{A}_{XY} \subset \mathscr{T}_{XY}$ and $\mathscr{A}_{YX} \subset \mathscr{T}_{YX}$ in the stratified surface $\mathscr{M}$

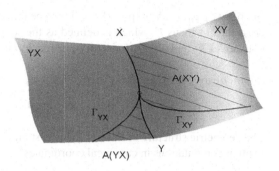

The boundary sections for this arrival region $\mathscr{A}_{XY}$ are given by the X-trajectory $\mathscr{T}_X$ and the curve Γ_{XY} of conjugate points,

$$\Gamma_{XY} = \{p\exp(-tY)\exp(-sX) : s > 0, \, t > 0 \quad \text{and} \quad \Psi_{XY}(s,t) = 0\}.$$

Analogously, for an XYX-trajectory of the form

$$p\exp(-tX)\exp(-sY)\exp(-rX) = q$$

with $q_1 = q\exp(rX)$ and $q_2 = q_1\exp(sY)$, define a function $\Psi_{YX}(s,t)$ through the equation

$$g(p) \wedge \left(\frac{\exp(-t\,\mathrm{ad}X) - \mathrm{id}}{t}\right) g(q_1) \wedge \exp(-t\,\mathrm{ad}X) \left(\frac{\exp(-s\,\mathrm{ad}Y) - \mathrm{id}}{s}\right) g(q_2)$$

$$= \Psi_{YX}(s,t)\,(g(p) \wedge [f,g](p) \wedge [g,[f,g]](p)),$$

and for this case, it follows that the first X-leg of an optimal XYX-trajectory needs to *arrive* in the surface $\mathscr{T}_{YX}$ at a point $p\exp(-tX)\exp(-sY)$ that lies in the arrival region

$$\mathscr{A}_{YX} = \{p\exp(-tX)\exp(-sY) : s > 0, \, t > 0 \quad \text{and} \quad \Psi_{YX}(s,t) < 0\}.$$

Similarly, the boundary sections for $\mathscr{A}_{YX}$ are given by the Y-trajectory $\mathscr{T}_Y$ and the curve Γ_{YX} of conjugate points,

$$\Gamma_{YX} = \{p\exp(-tX)\exp(-sY) : s > 0, \, t > 0 \quad \text{and} \quad \Psi_{YX}(s,t) = 0\}.$$

These arrival regions and their boundary sections are illustrated in Fig. 7.16.

Integral curves of the vector field Y are now integrated backward from the arrival region $\mathscr{A}_{XY} \subset \mathscr{T}_{XY}$. It is not difficult to verify that forward in time, these curves cross $\mathscr{T}_{XY}$ going from $\mathscr{M}_+$ into $\mathscr{M}_-$. Actually, these directions reverse along the curve Γ_{YX} of conjugate points. Thus, when integrated backward from $\mathscr{A}_{XY}$, the region $\mathscr{T}_{YXY}$ lies in $\mathscr{M}_+$. The time r along these trajectories of the form

$p \exp(-tY) \exp(-sX) \exp(-rY)$ will then be limited so that this trajectory does not cross the cut-locus Γ, which is defined as the set of points q for which there exist both an XYX- and a YXY-trajectory that reach p in the same time,

$$\Gamma = \{ p\exp(-t_3Y)\exp(-t_2X)\exp(-t_1Y) = p\exp(-s_3X)\exp(-s_2Y)\exp(-s_1X)$$

$$t_1 + t_2 + t_3 = s_1 + s_2 + s_3, \quad s_i, t_j \text{ positive and small} \}.$$

The geometric structure of this cut-locus is summarized below. It can be verified by explicit computations in canonical coordinates.

Lemma 7.4.3. *[215] The cut-locus Γ is a surface that lies in the region $\mathcal{M}_+$ and has the curves Γ_{XY} and Γ_{YX} of conjugate points and the terminal point p in its relative boundary. In canonical coordinates, it can be represented as the graph of a smooth function $x_3 = F_\Gamma(x_1, x_2)$ whose domain is defined by the curves Γ_{XY} and Γ_{YX} in $\mathcal{M}$. Modulo terms of higher weights and with $\bar{\gamma} = \gamma(p)$, the projections of the curves Γ_{XY} and Γ_{YX} of conjugate points into (x_1, x_2)-space are given by*

$$\Gamma_{XY}: \qquad 4\bar{\gamma}^2 x_2 \stackrel{\circ}{=} (\bar{\gamma}^2 + 2\bar{\gamma} - 1) x_1^2, \qquad \text{and} \qquad x_1 < 0$$

and

$$\Gamma_{YX}: \qquad 4\bar{\gamma}^2 x_2 \stackrel{\circ}{=} (-\bar{\gamma}^2 + 2\bar{\gamma} + 1) x_1^2, \qquad \text{and} \qquad x_1 < 0$$

with the domain $\mathrm{dom}(\Gamma)$ for the cut-locus given by the set

$$\mathrm{dom}(\Gamma) = \{ (x_1, x_2): (\bar{\gamma}^2 + 2\bar{\gamma} - 1) x_1^2 \le 4\bar{\gamma}^2 x_2 \le (-\bar{\gamma}^2 + 2\bar{\gamma} + 1) x_1^2, x_1 < 0 \}.$$

Over this domain, the cut-locus Γ is given modulo terms of weight 5 by the solutions to the equation

$$x_3 x_1 \stackrel{\circ}{=} \frac{\bar{\gamma}^4 - 6\bar{\gamma}^2 + 1}{96\bar{\gamma}^4} x_1^4 + \frac{1}{2} x_1^2 x_2 + \frac{1}{2} \bar{\gamma} x_2^2.$$

∎

The cut-locus Γ thus divides the region $\mathcal{M}_+$ into two connected components, a (small) set N_- that lies between the stratified surface $\mathcal{M}$ and the cut-locus Γ, and the (large) set N_+ that lies above the cut-locus Γ, respectively $\mathcal{M}$, in regions that are not covered by the cut-locus. The geometry is illustrated in Fig. 7.17, which shows the cut-locus for the nilpotent example Σ_3 from Eq. (7.33). The cut-locus Γ is the solid surface that extends from Γ_{XY} to Γ_{YX} and lies above $\mathcal{M}$, and the region N_+ is everything that lies above Γ, respectively $\mathcal{M}$. The figure also shows an example of a time-optimal YXY-trajectory that starts at a point on Γ and as it intersects the surface $\mathcal{T}_{XY}$, follows the XY-structure on this surface into the equilibrium point.

Lemma 7.4.4. *[215] For every point $q = p\exp(-tY)\exp(-sX) \in \mathscr{A}_{XY}$, there exists a unique first positive time $r = r_\Gamma(q)$ described by a differentiable function such that $q\exp(-rY) \in \Gamma$. The mapping $G_Y: \mathscr{A}_{XY} \to \Gamma$, $q \mapsto q\exp(-r_\Gamma Y)$, is a diffeomorphism between the arrival region and the cut-locus, and the flow of the*

Fig. 7.17 The cut-locus Γ in the region $\mathcal{M}_+$ and a YXY-trajectory starting from Γ

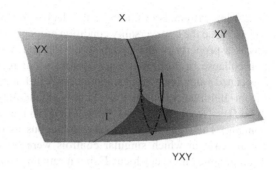

vector field Y, $\Phi_Y : \{(r,q) : q \in \mathcal{A}_{XY}, 0 < r < r_\Gamma(q)\} \to N_+$ is a diffeomorphism onto the region N_+. ∎

An analogous result holds for X-trajectories and the region N_-. In particular, starting at points on the cut-locus Γ, integral curves of the vector field Y enter the region N_+ and integral curves of the vector field X enter N_-. It thus follows that time-optimal controls on $\mathcal{M}_+$ are given by

$$u_*(x) = \begin{cases} -1 & \text{if} \quad x \in N_- \cup \Gamma, \\ +1 & \text{if} \quad x \in N_+ \cup \Gamma. \end{cases} \qquad (7.34)$$

For points x in either of these two regions, away from the cut-locus, the corresponding XYX- or YXY-trajectories define a parameterized family $\mathcal{E}$ of broken extremals. The associated value function $V = V^{\mathcal{E}}$ is a continuously differentiable solution to the Hamilton–Jacobi–Bellman equation on the sets N_+ and N_-. For example, suppose $x_0 \in N_-$. As parameter set, pick a small surface Q that is transversal to the flow of the vector field X at x_0 and let $x(t,q) = q \exp(tX)$ denote the flow with $q_0 = x_0$. By means of a coordinate chart, we could think of Q as an open subset of $\mathbb{R}^2$ (the definition of a parameterized family was formulated in this way), but this clearly is not necessary. However, in order to keep the notation unambiguous, here we use q to denote the parameter. Choose Q so small that all these trajectories meet the arrival region $\mathcal{A}_{YX}$ transversally. Then the time $t_1 = t_1(q)$ defined as the solution to the equation $x(t_1(q), q) \in \mathcal{A}_{YX}$ is a continuously differentiable function. At this point, the control switches, and the trajectories in the parameterized family are now given by $x(t,q) = x(t_1(q), q) \exp(tY)$. The flow collapses onto the surface $\mathcal{T}_{YX}$, but it still meets the stratum $\mathcal{T}_X$ transversally, and thus there also exists a differentiable function $t_2 = t_2(q)$ such that $x(t_2(q), q) \in \mathcal{T}_X$. Once more the control switches, and on the last segment the trajectory is given by $x(t,q) = x(t_2(q), q) \exp(tX)$ until the origin is reached at time $T = T(q)$. Again, this function is differentiable in q. It follows from our constructions above that these trajectories are all extremals, and the corresponding multiplier $\lambda = \lambda(t,q)$ is uniquely determined (modulo a positive scalar multiple) by the two switching points. At the terminal point, the transversality conditions are satisfied trivially, and thus this defines a C^r-parameterized family of bang-bang extremals (with $r \geq 2$ the degree of smoothness of the vector fields f and g). Since the transversality condition propagates across the switching surfaces— this is a property of the parameterization and holds regardless of whether there is a collapse in the flow—it follows from our results in Chaps. 5 and 6 that the associated value function is a continuously differentiable solution to the Hamilton–Jacobi–Bellman equation at x_0. (Simply extend the domain of the parameterization to be $D = \{(t,q) : q \in Q, -\varepsilon \leq t \leq T(q)\}$, so that $(0,x_0)$ lies in the interior of this domain.) This reasoning is totally analogous to that as illustrated in Sect. 6.2.3 for the case in which singular controls were present. The value function is not differentiable on the cut-locus Γ, but it remains continuous on $\mathcal{M}_+ \cup \mathcal{M}$.

In the region $\mathcal{M}_-$, optimal controls contain a segment along a singular arc. As for $\mathcal{M}_+$, concatenations of the singular trajectories with X- and Y-trajectories define two surfaces $\mathcal{T}_{SX}$ and $\mathcal{T}_{SY}$ that split the region $\mathcal{M}_-$ into two connected components S_+ and S_- where the optimal controls are $+1$ and -1, respectively. The singular control is given by the feedback function $u_{\text{sing}}(x) = -\gamma(x)$, and the singular vector field is $S(x) = f(x) - \gamma(x)g(x)$. Let $\mathcal{T}_S = \{p\exp(-tS) : t > 0\}$ denote the backward orbit of the singular trajectory through p and let

$$\mathcal{T}_{SX} = \{p\exp(-sX)\exp(-tS) : s > 0, t > 0\}$$

and

$$\mathcal{T}_{SY} = \{p\exp(-sY)\exp(-tS) : s > 0, t > 0\}.$$

The following geometric properties are once more verified through explicit calculations using canonical coordinates. But there is a difference from the cases considered so far in that the singular control itself is a feedback function. This necessitates the use of Lemma 2.8.1 in computing canonical coordinates. However, since we carry out these computations only modulo terms of higher order, it is possible to replace the function γ near p by its value $\bar{\gamma} = \gamma(p)$. Intuitively, the justification is that, for example, we have that

$$[f, \gamma g] = \gamma[f,g] + L_f(\gamma)g.$$

In all expansions, the Lie bracket will be carrying quadratic terms in the times, while the principal terms that arise at the vector field g come with linear coefficients. Thus, the extra terms that arise from the Lie derivatives of γ will be of higher order when compared with other terms that already exist. Naturally, this needs to be traced carefully in the computations, but it does work out under the assumptions made here. We have the following results about the surfaces $\mathcal{T}_{SX}$ and $\mathcal{T}_{SY}$:

Lemma 7.4.5. *[215] Both $\mathcal{T}_{SX}$ and $\mathcal{T}_{SY}$ are smooth surfaces that lie in $\mathcal{M}_-$ and can be represented as graphs of functions $x_3 = F_{S,\pm}(x_1, x_2)$ in canonical coordinates. Modulo terms of higher weights, the domain $\mathrm{dom}(\mathcal{T}_{SX})$ for $\mathcal{T}_{SX}$ is given by*

$$\mathrm{dom}(\mathcal{T}_{SX}) = \left\{(x_1, x_2) : x_2 > \frac{1}{2\bar{\gamma}}x_1^2 \text{ for } x_1 < 0 \quad \text{and} \quad x_2 > \frac{1}{2}x_1^2 \text{ for } x_1 > 0\right\},$$

Fig. 7.18 The stratified two-dimensional set $\Xi = \mathscr{C}_{SX} \cup \mathscr{C}_{SY}$ in the region $\mathscr{M}_-$

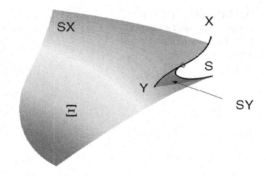

and over this domain, the points on $\mathcal{T}_{SX}$ satisfy

$$x_3 \stackrel{\circ}{=} -\frac{1}{6\bar{\gamma}}x_1^3 + x_1 x_2 - \frac{1}{3}\sqrt{2\bar{\gamma}(\bar{\gamma}-1)}\left|x_2 - \frac{1}{2\bar{\gamma}}x_1^2\right|^{3/2}.$$

Similarly, the domain $\mathrm{dom}(\mathcal{T}_{SY})$ for $\mathcal{T}_{SY}$ is given by

$$\mathrm{dom}(\mathcal{T}_{SY}) = \left\{(x_1, x_2) : \frac{1}{2\bar{\gamma}}x_1^2 < x_2 < -\frac{1}{2}x_1^2 \text{ for } x_1 < 0\right\},$$

and over this domain, the points on $\mathcal{T}_{SY}$ satisfy

$$x_3 \stackrel{\circ}{=} -\frac{1}{6\bar{\gamma}}x_1^3 + x_1 x_2 + \frac{1}{3}\sqrt{-2\bar{\gamma}(\bar{\gamma}+1)}\left|x_2 - \frac{1}{2\bar{\gamma}}x_1^2\right|^{3/2}.$$

The domain $\mathrm{dom}(\mathcal{T}_{SY})$ is contained in $\mathrm{dom}(\mathcal{T}_{SX})$, $\mathrm{dom}(\mathcal{T}_{SY}) \subset \mathrm{dom}(\mathcal{T}_{SX})$, and over $\mathrm{dom}(\mathcal{T}_{SY})$ the surface $\mathcal{T}_{SY}$ lies above the surface $\mathcal{T}_{SX}$. The two surfaces meet along the singular curve $\mathcal{T}_S$ that is given by

$$x_2 \stackrel{\circ}{=} \frac{1}{2\bar{\gamma}}x_1^2 \quad \text{and} \quad x_3 \stackrel{\circ}{=} \frac{1}{3\bar{\gamma}}x_1^3 \quad \text{for} \quad x_1 \leq 0. \qquad \blacksquare$$

Figure 7.18 illustrates the cusp-like behavior of the stratified set $\Xi = \mathscr{C}_{SX} \cup \mathscr{C}_{SY}$. This figure once more has been made using the exact formulas for the nilpotent system Σ_3 and $\gamma = -\frac{1}{4}$. The entire set Ξ lies below the stratified surface $\mathscr{M}$ and it divides $\mathscr{M}_-$ into two disjoint connected components S_- and S_+. We label the components so that S_- is the set where the optimal control is given by $u \equiv -1$ and on S_+ we have that $u \equiv +1$. The set S_+ is the connected component of $\mathscr{M}_-$ that lies below the union of the cells $\mathscr{C}_{YX} \cup \mathscr{C}_{SY}$ in the direction of the coordinate x_3, but above the singular surface $\mathcal{T}_{SX}$. An equivalent way of describing this set is that it is the component that besides the set Ξ, contains the surface $\mathcal{T}_{YX}$ of YX-trajectories in its boundary. Correspondingly, the complement S_- is the other connected component that contains the surface $\mathcal{T}_{XY}$ of XY-trajectories in

Fig. 7.19 The connected component S_+ and a sample YSX-trajectory in the region $\mathcal{M}_-$

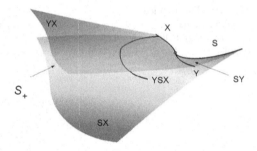

its boundary. This geometry, along with a sample YSX-trajectory, are illustrated in Fig. 7.19. The sample trajectory starts at a point in S_+ with the control $u = +1$ until the corresponding trajectory meets the surface $\mathscr{T}_{SX}$. At this point, the control switches to the singular control $u = -\gamma(x)$ and the corresponding trajectory remains on the surface $\mathscr{T}_{SX}$ until it meets the boundary stratum $\mathscr{T}_X$. At that point, the control switches to $u = -1$ and follows the X-trajectory into the equilibrium point p.

The control corresponding to this synthesis is then given by

$$u_*(x) = \begin{cases} -1 & \text{if } x \in S_-, \\ -\gamma(x) & \text{if } x \in \Xi = \mathscr{T}_{SX} \cup \mathscr{T}_S \cup \mathscr{T}_{SY}, \\ +1 & \text{if } x \in S_+. \end{cases} \qquad (7.35)$$

As with the synthesis in $\mathcal{M}_+$, it follows that the associated value function $V = V^{\mathscr{E}}$ is a continuously differentiable solution to the Hamilton–Jacobi–Bellman equation on the sets S_+ and S_-. In fact, this situation is identical to the one that was analyzed in Sect. 6.2.3. The value function is not differentiable on the stratified surface Ξ, but it remains continuous on $\mathcal{M}_- \cup \mathcal{M}$ and thus is continuous everywhere. Overall, we have the following result:

Theorem 7.4.1. *The controlled trajectories associated with the feedback control $u_*(x)$ defined in Eqs. (7.34) and (7.35) are a local time-optimal synthesis.*

Proof. The controlled trajectories and the stratification that were constructed define a memoryless synthesis for which the assumptions of the simple verification theorem, Theorem 6.3.1, are satisfied: All controlled trajectories included in the synthesis are extremal, and except for the cut-locus Γ, cover a neighborhood of p injectively. The associated value-function $V = V^{\mathscr{E}}$ is continuous and is a continuously differentiable solution to the Hamilton–Jacobi–Bellman equation on the open sets N_+, N_-, S_+, and S_-. The complement of these open sets is a finite union of embedded submanifolds of positive codimension consisting of the strata in $\mathcal{M}$, Γ, and Ξ. Thus, by Theorem 6.3.1, the synthesis is optimal. $\qquad \square$

7.5   The Codimension-1 Case in Dimension 4: Saturating Singular Arcs

We close with a description of the structure of the boundary trajectories in the small-time reachable sets in dimension four and the corresponding three-dimensional optimal synthesis when singular controls saturate. This corresponds to the next degenerate or codimension-1 case. In the context of the four-dimensional system $\Sigma : \dot{x} = f(x) + ug(x)$, $|u| \le 1$, $x \in \mathbb{R}^4$, we still assume that

(**C1**)   the vector fields X, Y, $[X,Y]$, and $[Y,[X,Y]]$ are linearly independent at p,

$$X(p) \wedge Y(p) \wedge [X,Y](p) \wedge [Y,[X,Y]](p) \ne 0,$$

and use the same canonical coordinates as in Sect. 7.4.1,

$$(\xi_0, \xi_1, \xi_2, \xi_3) \mapsto p\exp(\xi_3[Y,[X,Y]])\exp(\xi_2[X,Y])\exp(\xi_1 Y)\exp(\xi_0 X).$$

We again write

$$[X,[X,Y]] = aX + bY + c[X,Y] + d[Y,[X,Y]], \tag{7.36}$$

but now assume that $d(p) = 0$ and that the Lie derivatives of d along X and Y, $L_X d$ and $L_Y d$, do not vanish at p. Recall that for $d(p) > 0$, boundary trajectories of the small-time reachable set $\text{Reach}_{\Sigma, \le T}(p)$ are given by all bang-bang trajectories in the cells $\mathscr{C}_{XYX}$ and $\mathscr{C}_{YXY}$ with the surface $\mathscr{T}_{YXY}$ lying above $\mathscr{T}_{XYX}$ in the direction of the coordinate ξ_3 (Proposition 7.4.1). For $d(p) < 0$, boundary trajectories consist of the bang-bang surface $\mathscr{N} = \mathscr{T}_{XYX}^- \cup \mathscr{T}_{YXY}^+$ determined by the cut-locus of bang-bang trajectories with two switchings that minimize the coordinate ξ_3 and a stratified hypersurface $\mathscr{S} = \mathscr{S}_{BSB}$ of trajectories that contain a singular arc and maximize ξ_3 (Proposition 7.4.5). If the Lie derivatives $L_X d(p)$ and $L_Y d(p)$ have the same sign, then trajectories lie in the regions $d(p) > 0$, respectively $d(p) < 0$. It easily follows that boundary trajectories have the same structure as in the totally bang-bang case if $L_X d(p) > 0$ and as in the mixed bang-singular case for $L_X d(p) < 0$. A new situation arises if $L_X d(p)$ and $L_Y d(p)$ have opposite signs. Using an input symmetry, we can normalize these signs and make the following assumption:

(**C3**) $d(p) = 0$, $L_X d(p) > 0$,   and $L_Y d(p) < 0$.

We again assume that all relevant inequalities are satisfied in a sufficiently small neighborhood M of p. As should be expected in this case, which represents the simplest step in a complicated bifurcation sequence of the reachable sets from the totally bang-bang case $d(p) > 0$ to the bang-singular case $d(p) < 0$, boundary trajectories become an amalgam of these two cases. It is the conjugate point relation along $\cdot XY \cdot$ and $\cdot YX \cdot$-trajectories that becomes the key to unlocking these structures, and we first compute these equations.

Consider an $\cdot XY \cdot$-trajectory with switching points q_1, $q_2 = q_1 \exp(sX)$ and $q_3 = q_2 \exp(tY)$. Recall from Sect. 7.4.1 that the points q_1, q_2, and q_3 form a conjugate triple if and only if the vectors $X(q_1)$, $Y(q_1)$, $\exp(s\,\mathrm{ad}X)Y(q_2)$, and $\exp(s\,\mathrm{ad}X)\exp(t\,\mathrm{ad}Y)X(q_3)$ are linearly dependent and that this is equivalent to a conjugate point relation of the form $\Psi_{XY}(q_1;s,t)=0$, where the function Ψ_{XY} is defined as a quotient of two 4-vectors:

$$X \wedge Y \wedge \left(\frac{\exp(s\,\mathrm{ad}X)-\mathrm{id}}{s}\right) Y \wedge \exp(s\,\mathrm{ad}X) \left(\frac{\exp(t\,\mathrm{ad}Y)-\mathrm{id}}{t}\right) X$$

$$= \Psi_{XY}(q_1;s,t) \cdot (X \wedge Y \wedge [X,Y] \wedge [Y,[X,Y]]).$$

Under assumption (C3), we now need a more accurate expansion for this function. In the same way as before, and only keeping leading terms, we have that

$$X \wedge Y \wedge \left(\frac{\exp(s\,\mathrm{ad}X)-\mathrm{id}}{s}\right) Y \wedge \exp(s\,\mathrm{ad}X) \left(\frac{\exp(t\,\mathrm{ad}Y)-\mathrm{id}}{t}\right) X$$

$$= X \wedge Y \wedge [X,Y] + \frac{1}{2}s[X,[X,Y]] + \frac{1}{6}s^2[X,[X,[X,Y]]] + \cdots$$

$$\wedge -[X,Y] - \frac{1}{2}t[Y,[X,Y]] + \cdots - s[X,[X,Y]] - \frac{1}{2}s^2[X,[X,[X,Y]]] + \cdots.$$
$$\tag{7.37}$$

We again write vectors in this expansion as linear combinations of the basis, but we also need this representation for the third-order bracket $[X,[X,[X,Y]]]$. It follows from Eq. (7.36) that

$$[X,[X,[X,Y]]] = [X, aX + bY + c[X,Y] + d[Y,[X,Y]]]$$

$$= L_X(a)X + L_X(b)Y + (L_X(c)+b)[X,Y]$$

$$+ c[X,[X,Y]] + L_X(d)[Y,[X,Y]] + d[X,[Y,[X,Y]]].$$

Using the coordinate representation Eq. (7.36) for $[X,[X,Y]]$, and writing $[X,[Y,[X,Y]]]$ as a linear combination of the basis, we see that the coefficient φ at the vector field $[Y,[X,Y]]$,

$$[X,[X,[X,Y]]] = \cdots + \varphi[Y,[X,Y]], \tag{7.38}$$

is a smooth function on M that satisfies $\varphi(p) = L_X(d)(p)$. Substituting into Eq. (7.37) then gives that

$$\Psi_{XY}(q_1;s,t) = \begin{vmatrix} 1+o(1) & \frac{1}{2}sd + \frac{1}{6}s^2\varphi + o(s^2) \\[2ex] -1+o(1) & -\frac{1}{2}t - sd - \frac{1}{2}s^2\varphi + o(t) + o(s^2) \end{vmatrix}$$

$$= -\frac{1}{2}\left(t + sd + \frac{2}{3}s^2\varphi + o(t) + o(s^2)\right). \tag{7.39}$$

Similarly, the conjugate point relation for a conjugate triple (q_1, q_2, q_3) along a $\cdot YX\cdot$-trajectory of the form $q_3 = q_1 \exp(tY)\exp(sX)$ is given by

$$\Psi_{YX}(q_1; s, t) = -\frac{1}{2}\left(t + sd + \frac{1}{3}s^2\varphi + o(t) + o(s^2)\right) \tag{7.40}$$

with, formally, the only difference in the coefficient at the s^2-term.

These two formulas allow us to determine the structure of optimal bang-bang trajectories in this case.

Proposition 7.5.1. *The hypersurface $\mathcal{T}_{XYX}$ is the graph of a smooth function ϕ_S of the variables (ξ_0, ξ_1, ξ_2), $\xi_3 = \phi_S(\xi_0, \xi_1, \xi_2)$. The vector field Y points to the same side of $\mathcal{T}_{XYX}$ as the coordinate vector field $\frac{\partial}{\partial \xi_3}$ everywhere, XYX-trajectories minimize the coordinate ξ_3 (over the small-time reachable set for fixed (ξ_0, ξ_1, ξ_2)-coordinates) and all XYX-trajectories are extremal.*

Proof. These geometric properties are verified by a direct computation of the tangent space to $\mathcal{T}_{XYX}$ and then taking the wedge product with Y and $\frac{\partial}{\partial \xi_3}$, respectively. These are the same computations as done earlier, and we only verify the statement about extremals. Consider an XYX-trajectory starting from p of the form

$$q = p\exp(s_1 X)\exp(s_2 Y)\exp(s_3 X)$$

and denote the switching points by $q_1 = p\exp(s_1 X)$ and $q_2 = q_1\exp(s_2 Y)$. The points (p, q_1, q_2) form a conjugate triple if and only if $\Psi_{XY}(p; s_1, s_2) = 0$ (cf. Eq. (7.39)). Since $d(p) = 0$ and $\varphi(p) = L_X(d)(p)$, this is equivalent to

$$0 = s_2(1 + o(1)) + \frac{2}{3}L_X(d)(p)s_1^2(1 + o(1)).$$

But $L_X(d)(p) > 0$, and thus this equation has no solution. Hence the first two times s_1 and s_2 are not limited by a conjugate point relation. The same also holds for s_3, but for a different reason. The points (q_1, q_2, q) form a conjugate triple if and only if $\Psi_{YX}(q_1; s_2, s_3) = 0$ (cf. Eq. (7.40)). This relation now takes the form

$$0 = s_2(1 + o(1)) + d(q_1)s_3 + \frac{1}{3}\varphi(q_1)s_3^2(1 + o(1)).$$

But $q_1 = p\exp(s_1 X)$, and in case (C3), the Lie derivative of d along X is positive in a sufficiently small neighborhood M of p. Hence $d(q_1) > 0$. However, this term generally is small and not able to dominate higher-order terms. For this reason, we need the quadratic term. Using a Taylor expansion, we simply have that

$$\varphi(q_1) = \varphi(p\exp(s_1 X)) = \varphi(p) + O(s_1) = L_X(d)(p) + O(s_1) > 0,$$

and thus the quadratic term is also positive near p. Hence, there also exists no positive solution s_3 to the conjugate point relation $\Psi_{YX}(q_1; s_2, s_3) = 0$. Thus the times along XYX-trajectories can vary freely and all these trajectories are extremals. $\quad\square$

This is no longer true for YXY-trajectories. Consider a trajectory of the form

$$q = p\exp(t_1 Y)\exp(t_2 X)\exp(t_3 Y)$$

and again denote the switching points by $q_1 = p\exp(t_1 Y)$ and $q_2 = q_1 \exp(t_2 X)$. Since trajectories start from the point p, it still holds that the first two times t_1 and t_2 are free. For as above, in this case the conjugate point relation $\Psi_{YX}(p; t_1, t_2) = 0$ simplifies to

$$0 = t_1\left(1 + o(1)\right) + \frac{1}{3}L_X(d)(p)t_2^2\left(1 + o(1)\right),$$

which again has no positive solutions. But this is no longer the case for the second conjugate point relation $\Psi_{XY}(q_1; t_2, t_3) = 0$. We now have that

$$0 = t_3\left(1 + o(1)\right) + d(q_1)t_2 + \frac{2}{3}\varphi(q_1)t_2^2\left(1 + o(1)\right),$$

and this equation has a unique solution for t_3 as a function of t_1 and t_2, $\bar{t}_3 = \tau(t_1, t_2)$ of the form

$$\tau = -d(q_1)t_2\left(1 + o(t_2)\right) - \frac{2}{3}\varphi(q_1)t_2^2\left(1 + o(t_2)\right)$$

with the dependence on the first time t_1 coming in through evaluation of the functions d and φ at the first junction $q_1 = p\exp(t_1 Y)$. Since $d(p) = 0$, it follows that expressions of the type $d(q_1)t_2^2$ that arise in solving for t_3 are of size $O(t_1 t_2^2)$ and can be incorporated into the higher-order remainders at the quadratic term. We thus get that

$$\bar{t}_3 = \tau(t_1, t_2) = -d(q_1)t_2 - \frac{2}{3}\varphi(q_1)t_2^2\left(1 + o(1)\right). \tag{7.41}$$

Factoring out t_2, and using Taylor's theorem to evaluate the functions at the reference point p, we obtain that

$$\bar{t}_3 = \tau(t_1, t_2) = -t_2\left(L_Y(p)t_1 + \frac{2}{3}L_X(p)t_2 + \cdots\right).$$

Let $\bar{t}_2 = \theta(t_1)$ be the time when the quantity in parentheses vanishes, i.e.,

$$\bar{t}_2 = \theta(t_1) = -\frac{3}{2}\frac{L_Y(p)}{L_X(p)}t_1 + \cdots. \tag{7.42}$$

Since $L_X(p)$ and $L_Y(p)$ have opposite signs, this quantity is always positive. For $t_2 \geq \bar{t}_2 = \theta(t_1)$, it follows that $\bar{t}_3 = \tau(q_1; t_2) \leq 0$, and thus the conjugate point relation $\Psi_{XY}(q_1; t_2, t_3) = 0$ puts no restrictions on the time t_3, i.e., if t_2 is large enough, then the corresponding YXY-trajectory is extremal. However, if $t_2 < \bar{t}_2 = \theta(t_1)$, then the time $\bar{t}_3 = \tau(t_1, t_2)$ is positive, and the third time t_3 needs to satisfy $t_3 \leq \bar{t}_3 = \tau(t_1, t_2)$. If we define the sets

$$U = \left\{ (t_1, t_2, t_3) \in \mathbb{R}^3_+ : t_2 < \bar{t}_2 = \theta(t_1), \ t_3 > \bar{t}_3 = \tau(t_1, t_2) \right\},$$

$$P = \left\{ (t_1, t_2, t_3) \in \mathbb{R}^3_+ : t_2 < \bar{t}_2 = \theta(t_1), \ t_3 = \bar{t}_3 = \tau(t_1, t_2) \right\},$$

$$D = \left\{ (t_1, t_2, t_3) \in \mathbb{R}^3_+ : t_2 < \bar{t}_2 = \theta(t_1), \ t_3 < \bar{t}_3 = \tau(t_1, t_2) \right\},$$

then the following geometric properties can be verified with analogous computations to those that were made earlier.

Proposition 7.5.2. *The hypersurface $\mathcal{T}_{YXY}$ is the graph of a smooth function ϕ_N of the variables (ξ_0, ξ_1, ξ_2), $\xi_3 = \phi_N(\xi_0, \xi_1, \xi_2)$. The vector field X points to the same side of $\mathcal{T}_{YXY}$ as the coordinate vector field $\frac{\partial}{\partial \xi_3}$ at points defined by times in U, is tangent to $\mathcal{T}_{YXY}$ for points defined by times in P, and points to the opposite side of $\mathcal{T}_{YXY}$ than $\frac{\partial}{\partial \xi_3}$ for points defined by times in D. The surface $\mathcal{T}_{YXY}$ everywhere lies above the surface $\mathcal{T}_{XYX}$ in direction of the coordinate vector field $\frac{\partial}{\partial \xi_3}$.* ■

The portions of the hypersurface $\mathcal{T}_{YXY}$ that are close to the X-trajectory (which is obtained in the limit $t_2 \to T$) indeed will provide the upper closure for the boundary trajectories (these are the boundary trajectories that maximize the coordinate ξ_3). But trajectories that satisfy $t_3 > \bar{t}_3 = \tau(t_1, t_2)$ are not boundary trajectories, also reiterated by the fact that X points in the same direction as the coordinate vector field $\frac{\partial}{\partial \xi_3}$, i.e., upward, and thus these points clearly cannot lie in the northern hemisphere. We need to restrict the YXY-trajectories and properly complement the boundary portions of $\mathcal{T}_{YXY}$ to find the northern hemisphere $\mathcal{N}$; the southern hemisphere and the corresponding "equator" are simply given by $\mathcal{S} = \mathcal{C}_{XYX}$. The missing pieces are provided through trajectories that contain singular arcs.

Proposition 7.5.3. *Optimal boundary trajectories in* $\mathrm{Reach}_{\Sigma, \leq T}(p)$ *that contain a singular arc are at most of the form $YSXY$.*

Proof. This proposition is an application of singular conjugate point relations. We denote a junction of a singular arc S with a bang trajectory by an asterisk, $*$, and, as before, denote ordinary junctions by a dot, $\cdot$. We first show that there do not exist any concatenations of the form $\cdot Y*$ and $*Y\cdot$. Suppose $q_2 = q_1 \exp(tY)$ and the trajectory has a singular junction at q_1 and a switching point at q_2. Then there exists a nontrivial multiplier λ that vanishes against the vector fields X, Y, and $[X,Y]$ at q_1 and also vanishes against the vector that is obtained by transporting the vector $X(q_2)$ back to q_1 along the flow of Y, i.e.,

$$0 = X(q_1) \wedge Y(q_1) \wedge [X,Y](q_1) \wedge \exp(t \operatorname{ad} Y) X(q_2)$$

$$= \frac{1}{2} t^2 \left(X(q_1) \wedge Y(q_1) \wedge [X,Y](q_1) \wedge \left(\frac{\exp(t \operatorname{ad} Y) - (\operatorname{id} + t \operatorname{ad} Y)}{\frac{1}{2} t^2} \right) X(q_2) \right)$$

$$= \frac{1}{2}t^2\left(X(q_1) \wedge Y(q_1) \wedge [X,Y](q_1) \wedge -[Y,[X,Y]](q_1)(1+o(t))\right).$$

This contradicts assumption (C1), the linear independence of these vector fields. Similarly, $\cdot Y*$-concatenations are not possible, and thus whenever there is a singular junction SY or YS, the time t along the vector field Y is unrestricted.

This does not hold for singular concatenations with the vector field X. Suppose first that $q_2 = q_1 \exp(sX)$ and the trajectory is of the type $\cdot X*$, i.e., has a switching point at q_1 and a singular junction at q_2. Then, as above, but evaluating all the vector fields at the singular junction q_2, it follows that the vectors X, Y, and $[X,Y]$ at q_2 and the vector $Y(q_1)$ moved forward to q_2 along the flow of X are linearly dependent. Thus

$$0 = X(q_2) \wedge Y(q_2) \wedge [X,Y](q_2) \wedge \left(\frac{\exp(-s\operatorname{ad}X) - (\operatorname{id}-s\operatorname{ad}X)}{\frac{1}{2}s^2} \right) Y(q_2)$$

$$= X(q_2) \wedge Y(q_2) \wedge [X,Y](q_2) \wedge \left([X,[X,Y]](q_2) - \frac{1}{3}s[X,[X,[X,Y]]](q_2) + o(s) \right).$$

The coefficient of the last term at $[Y,[X,Y]](q_2)$ therefore must vanish. This coefficient is given by

$$d(q_2) - \frac{1}{3}\varphi(q_2)s + o(s).$$

Since there is a singular junction at q_2, we have $d(q_2) \le 0$. (Otherwise, the singular control is not admissible.) Furthermore, in small time, we always have $\varphi(q_2) = \varphi(p) + O(T) = L_X(d)(p) + O(T) > 0$. Hence $\cdot X*$-concatenations are not extremal, and this excludes concatenations of the form YXS. Since trajectories start at p, trajectories of the type $p\exp(sX)\exp(rS)$ are not possible either. For since $L_X(d) > 0$, it follows that $d(p\exp(sX)) > 0$, and thus the singular control is inadmissible at $p\exp(sX)$. Thus, trajectories that contain a singular arc are initially of the form YS with the time along the Y-arc free.

Now consider a junction of type $*X\cdot$ as can arise at the end of a singular arc and suppose $q_2 = q_1 \exp(sX)$ is another junction point. Define a function $\Psi_{*X}(q_1;s)$ by moving the vector $Y(q_2)$ back to q_1 and writing

$$X(q_1) \wedge Y(q_1) \wedge [X,Y](q_1) \wedge \left(\frac{\exp(s\operatorname{ad}X) - (\operatorname{id}+s\operatorname{ad}X)}{\frac{1}{2}s^2} \right) Y(q_2)$$

$$= \Psi_{*X}(q_1;s) \left(X(q_1) \wedge Y(q_1) \wedge [X,Y](q_1) \wedge [Y,[X,Y]](q_1) \right).$$

A similar computation as above gives that

$$\Psi_{*X}(q_1;s) = d(q_1) + \frac{1}{3}\varphi(q_1)s(1+o(1)). \tag{7.43}$$

Assuming that the singular control does not saturate at the point q_1, we have that $d(q_1) < 0$, and therefore, by the implicit function theorem, the equation $\Psi_{*X}(q_1; s) = 0$ has a unique solution $\bar{s} = \sigma(q_1)$, and by Eq. (7.43), this solution is positive. Thus the time s along an X-arc following a singular junction is limited, and at the time $\bar{s}$ a switch in the control to $u = +1$ occurs. But then no further switchings are possible: if $q_3 = q_2 \exp(tY)$ were another junction, then the triple (q_1, q_2, q_3) would be conjugate; but by (7.41) the time $\bar{t} = \tau(q_1, s)$ until the third junction is given by

$$\bar{t}_3 = \tau(q_1, t_2) = -d(q_1)s - \frac{2}{3}\varphi(q_1)s^2(1 + o(1)),$$

and thus

$$\bar{t}_3 = \tau(q_1, t_2) = -\frac{1}{3}\varphi(q_1)s^2(1 + o(1)) = -\frac{1}{3}L_X(d)(p)s^2(1 + O(T)) < 0.$$

Contradiction. Thus boundary trajectories that contain a singular arc have at most the structure $YSXY$. ☐

Not all of these segments need to be present in every such boundary trajectory, however, and there are limitations on the times along these trajectories. Consider a trajectory of the form

$$p \exp(tY) \exp(rS) \exp(sX)$$

with switchings at the points $q_1 = p \exp(tY)$ and $q_2 = q_1 \exp(rS)$. Then, the time s along the X-arc is restricted by the equation

$$s \leq \bar{s}(t, r) = \sigma(q_2) = \sigma(p \exp(tY) \exp(rS))$$

with σ the solution to Eq. (7.43). Since

$$d(q_2) = L_Y(d)(p)t + L_X(d)(p)r + \cdots$$

and

$$\varphi(q_2) = L_X(d)(p) + \cdots,$$

it follows that

$$\bar{s}(t, r) = -3\left(\frac{L_Y(d)(p)}{L_X(d)(p)}t + r + \cdots\right).$$

In particular, for $r = 0$, denote by $\mathcal{T}_{Y\bar{X}} \subset \mathcal{T}_{YX}$ and $\mathcal{T}_{Y\bar{X}Y} \subset \mathcal{T}_{YXY}$, respectively, the curve and surfaces of endpoints obtained when the time along the X-arc is given by $s = \bar{s}(t, 0)$,

$$\mathcal{T}_{Y\bar{X}} = \{p \exp(tY) \exp(sX) : s = \bar{s}(t, 0)\},$$
$$\mathcal{T}_{Y\bar{X}Y} = \{p \exp(tY) \exp(sX) \exp(uY) : s = \bar{s}(t, 0)\}.$$

Comparing the formula for $\bar{s}(t,0)$ with the formula Eq. (7.42), we see that $\bar{s}(t,0) > \theta(t)$, and hence the conjugate point relation for a $\cdot XY\cdot$ junction will never become active. Thus the time u along the second Y-arc is unrestricted and these are all extremal YXY-trajectories. In fact, it is this surface where the portion of $\mathscr{T}_{YXY}$ that lies in the boundary of the small-time reachable set meets those trajectories that contain a singular arc.

We also need to make sure that the singular controls remain admissible. Let

$$\Delta = \{q \in M : d(q) = 0\};$$

since the Lie derivative of d along X does not vanish, Δ is an embedded three-dimensional manifold near p. The singular vector field at the point p is given by $S(p) = X(p)$ (see Proposition 7.4.4), and therefore the Lie derivative of d along S is positive in a sufficiently small neighborhood M of p. Hence, singular arcs starting at a point $p\exp(tY)$ will eventually reach Δ, where they saturate. It is easy to see that there exists a well-defined smooth function $\bar{r} = \bar{r}(t)$ such that

$$p\exp(tY)\exp(\bar{r}(t)S) \in \Delta.$$

An expansion for $\bar{r}$ is easily computed from

$$d(p\exp(tY)\exp(\bar{r}(t)S)) = d(p) + L_Y(d)(p)t + L_S(d)(p)\bar{r} + \cdots$$

to be

$$\bar{r}(t) = -\frac{L_Y(d)(p)}{L_X(d)(p)}t + o(t).$$

If the singular control saturates, then the control actually needs to switch.

Lemma 7.5.1. *It is not optimal to continue with the control $u = -1$ at a saturation point $p\exp(tY)\exp(\bar{r}(t)S)$.*

Proof. This junction condition follows from a calculation like those in Proposition 2.8.4. Suppose the control $u \equiv -1$ is being used after saturation and let λ be the value of the corresponding multiplier at the saturation time such that the conditions of the maximum principle are satisfied. Since q is a singular junction, λ vanishes against the vectors $X(q)$, $Y(q)$, and $[X,Y](q)$. By assumption (C1), λ therefore does not vanish against $[Y,[X,Y]](q)$. Furthermore, the second and third derivatives of the switching function from the right are given by

$$\ddot{\Phi} = \frac{1}{2}\langle\lambda,[X,[X,Y]](q)\rangle \quad \text{and} \quad \dddot{\Phi} = \frac{1}{2}\langle\lambda,[X,[X,[X,Y]]](q)\rangle.$$

But, since $q \in \Delta$,

$$\langle\lambda,[X,[X,Y]](q)\rangle = d(q)\langle\lambda,[Y,[X,Y]](q)\rangle = 0,$$

and from Eq. (7.38) we get that

$$\langle \lambda, [X, [X, [X, Y]]](q) \rangle = \varphi(q) \langle \lambda, [Y, [X, Y]](q) \rangle,$$

with a positive function φ. But by Eq. (7.31),

$$2\langle \lambda, [g, [f, g]](q) \rangle = \langle \lambda, [Y - X, [X, Y]](q) \rangle = \langle \lambda, [Y, [X, Y]](q) \rangle$$

and thus $\langle \lambda, [Y, [X, Y]](q) \rangle$ is negative by the Legendre–Clebsch condition. But then the switching function is negative to the right of the saturation point, contradicting the minimum property of the maximum principle. $\square$

Hence, at a saturation point, the control switches to $u = +1$. Overall, we therefore get the following concatenations for boundary trajectories that contain singular arcs:

$$\mathscr{T}_{YS} = \{p \exp(tY) \exp(rS) : t > 0,\ 0 < r < \bar{r}(t)\},$$

$$\mathscr{T}_{Y\bar{S}} = \{p \exp(tY) \exp(\bar{r}(t)S) : t > 0\},$$

$$\mathscr{T}_{YSY} = \{p \exp(tY) \exp(rS) \exp(uY) : t > 0,\ 0 < r < \bar{r}(t),\ u > 0\},$$

$$\mathscr{T}_{Y\bar{S}Y} = \{p \exp(tY) \exp(\bar{r}(t)S) \exp(uY) : t > 0,\ u > 0\},$$

$$\mathscr{T}_{YSX} = \{p \exp(tY) \exp(rS) \exp(sX) : t > 0,\ 0 < r < \bar{r}(t),\ 0 < s < \bar{s}(t,r)\},$$

$$\mathscr{T}_{YS\bar{X}} = \{p \exp(tY) \exp(rS) \exp(\bar{s}(t,r)X) : t > 0,\ 0 < r < \bar{r}(t)\},$$

$$\mathscr{T}_{YS\bar{X}Y} = \{p \exp(tY) \exp(rS) \exp(\bar{s}(t,r)X) \exp(uY) : t > 0,\ 0 < r < \bar{r}(t) u > 0\}.$$

Denoting the corresponding sets when also equalities are allowed by $\mathscr{C}$, this defines a singular cell

$$\mathscr{C}_{\text{sing}} = \mathscr{C}_{YSY} \cup \mathscr{C}_{YSX} \cup \mathscr{C}_{YS\bar{X}Y}$$

that combines with $\mathscr{T}_{YXY}$ to form the northern hemisphere of the boundary of the small-time reachable set. Modulo some standard geometric considerations, which can be found in [221], we have shown the following result:

Proposition 7.5.4. *The stratified hypersurface $\mathscr{C}_{\text{sing}}$ is the graph of a piecewise defined, smooth function ϕ_N of the variables (ξ_0, ξ_1, ξ_2), $\xi_3 = \phi_N(\xi_0, \xi_1, \xi_2)$. The restrictions of ϕ_N to the domains for the surfaces $\mathscr{T}_{YSY}$, $\mathscr{T}_{YSX}$, and $\mathscr{T}_{YS\bar{X}Y}$ have smooth extensions to their relative boundaries. The surface $\mathscr{C}_{\text{sing}}$ lies above the surface $\mathscr{T}_{YXY}$ (in direction of the coordinate vector field $\frac{\partial}{\partial \xi_3}$) everywhere in its domain, and these two surfaces meet along the stratum $\mathscr{T}_{Y\bar{X}Y}$ in the relative boundary of $\mathscr{C}_{\text{sing}}$.* $\blacksquare$

The geometric structure of the trajectories that make up this northern hemisphere is illustrated in Fig. 7.20.

Altogether, our results give a complete description of the boundary trajectories in the small-time reachable set $\text{Reach}_{\Sigma, \leq T}(p)$ under the codimension-1 assumptions

Fig. 7.20 A qualitative sketch of the structure of the endpoints in the northern hemisphere $\mathcal{N}$ of the boundary $\text{Reach}_{\Sigma,T}(p)$ of the small-time reachable set under assumptions (C1) and (C3)

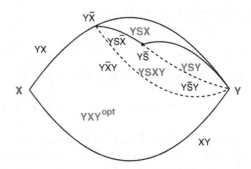

(C1) and (C3). It is now possible, exactly as in Sect. 7.4.2, to construct a time-optimal synthesis to an equilibrium point for the corresponding system in $\mathbb{R}^3$, and we briefly describe this result. The underlying system again is of the form

$$\Sigma : \dot{x} = f(x) + ug(x), \qquad x \in \mathbb{R}^3, \qquad |u| \le 1,$$

and we are assuming that

(1) $f(p) = 0$,
(2) the vectors $g(p)$, $[f,g](p)$, and $[g,[f,g]](p)$ are linearly independent.

As before, we write

$$[f,[f,g]] = \alpha g + \beta [f,g] + \gamma [g,[f,g]].$$

The coefficient $\tilde{d}$ for the augmented system $\tilde{\Sigma}$ with state $\tilde{x} = (x_0, x)^T$ where time has been added as extra coordinate, $\dot{x}_0 \equiv 1$, agrees with the function d defined through the relation

$$[f - g, [f,g]] = bg + c[f,g] + d[f+g, [f,g]]$$

in $\mathbb{R}^3$, and we have that

$$d(x) = \frac{\gamma(x) - 1}{\gamma(x) + 1}.$$

Thus, the assumption $d(p) = 0$ corresponds to $\gamma(p) = 1$. The singular control again is $u_{\text{sing}}(x) = -\gamma(x)$, and it saturates on the set $\Delta = \{x \in M : \gamma(x) = 1\}$ with the singular vector field at saturation given by X. (By changing g into $-g$, the situation that was considered in Sect. 7.4.2 arises.) Assuming that the Lie derivative of γ along X does not vanish at p, $L_X(\gamma)(p) \ne 0$, this becomes a two-dimensional embedded submanifold. A direct calculation gives the relations between the Lie derivatives of d and γ, and we have that

$$L_X(d) = \frac{2L_X(\gamma)}{(1+\gamma)^2} \qquad \text{and} \qquad L_X(\gamma) = \frac{2L_X(d)}{(d-1)^2}$$

and analogously for Y. Hence the conditions that correspond to (C3) can be expressed in one of the following two equivalent forms:

(3) the function d vanishes at p, $d(p) = 0$, but has nonzero Lie derivatives along X and Y. Without loss of generality, we normalize the Lie derivatives so that $L_X d(p) > 0$ and $L_Y d(p) < 0$. Equivalently,

$$\gamma(p) = 1, \qquad L_X(\gamma)(p) > 0, \qquad \text{and} \qquad L_Y(\gamma)(p) < 0.$$

Under these conditions, the system once more is small-time locally controllable (i.e., p is an interior point of the small-time reachable set $\text{Reach}_{\Sigma, \leq T}(p)$ for any $T > 0$), and as before, a time-optimal synthesis for steering points in a sufficiently small neighborhood M of p into the equilibrium point p can be constructed from the structure of the small-time reachable set for the augmented system by projecting the boundary trajectories into the original state space.

As above, this synthesis is built inductively by integrating trajectories backward form p, and thus the roles of trajectories in the northern and southern hemispheres interchange. Again, we use notation like $\mathcal{T}_{XY}$ to denote actual XY trajectories when the system is integrated forward in time, i.e.,

$$\mathcal{T}_{XY} = \{p \exp(-tY)\exp(-sX) : s > 0, \, t > 0\}.$$

As in Sect. 7.4.2, the stratified hypersurface $\mathcal{M} = \mathcal{C}_{XY} \cup \mathcal{C}_{YX}$ is the graph of a piecewise smooth function $x_3 = F_{\mathcal{M}}(x_1, x_2)$ near p, and it divides a small neighborhood M of p into a region $\mathcal{M}_+$ that lies above $\mathcal{M}$ in the direction of the coordinate vector field $\frac{\partial}{\partial x_3}$ and a region $\mathcal{M}_-$ that lies below $\mathcal{M}$. As before, time-optimal trajectories are bang-bang in $\mathcal{M}_+$ and trajectories in $\mathcal{M}_-$ contain a singular arc.

For this problem, because of the absence of a cut-locus, the bang-bang synthesis simplifies; in fact, it becomes trivial in the region $\mathcal{M}_+$. Since boundary trajectories in the northern hemisphere are of type XYX, we simply have that

$$u_*(x) \equiv -1 \qquad \text{for all} \quad x \in \mathcal{M}_+.$$

This reflects the fact that the structure of the small-time reachable set in this bifurcation scenario consists of an amalgam of the simple totally bang-bang case ($d > 0$) and the complicated bang-singular scenario ($d < 0$). The lower part $\mathcal{M}_-$ is divided into two regions S_+ and S_- where the controls, respectively, are given by $u = +1$ and $u = -1$. The boundary between these two regions comprises two surfaces,

$$\Xi = \Xi_1 \cup \Xi_2,$$

where Ξ_1 is an integral manifold of the singular vector field and Ξ_2 is the surface in the boundary of S_- where the controls switch from $u = -1$ to $u = +1$. Contrary to the other strata (and also in contrast to the strata constructed in Sect. 7.4.2), the surface Ξ_2 does not support trajectories, but only describes the transit of trajectories. These surfaces are given explicitly in the form

$$\Xi_1 = \mathcal{T}_{SY} = \{p\exp(-tY)\exp(-rS) : t > 0, \, 0 < r < \bar{r}(t)\}$$

where $\bar{r}$ denotes the time of saturation for the singular control, and

$$\Xi_2 = \mathcal{T}_{\bar{X}SY} = \{p\exp(-tY)\exp(-rS)\exp(-\bar{s}(t,r)X) : t > 0, \, 0 < r < \bar{r}(t)\}$$

with $\bar{s}(t,r)$ the time determined by the singular conjugate point relation

$$\Psi_{*X}(p\exp(-tY)\exp(-rS);-s) = 0.$$

The region S_- is the set

$$\mathcal{T}_{XSY} = \{p\exp(-tY)\exp(-rS)\exp(-sX) : t > 0, \, 0 < r < \bar{r}(t), \, 0 < s < \bar{s}(t,r)\},$$

and it is the subset of $\mathcal{M}_-$ that has Ξ and a part of the surface $\mathcal{T}_{XY}$ in its boundary, but does not border the surface $\mathcal{T}_{YX}$. The region S_+ is the complement of this set in $\mathcal{M}_-$. It consists of the cell $\mathcal{T}_{YXY}^{opt}$ that corresponds to optimal YXY-trajectories that lie in the boundary of the small-time reachable set,

$$\mathcal{T}_{YXY} = \{p\exp(-tY)\exp(-sX)\exp(-uY) : t > 0, \, s > \bar{s}(t,0), \, u > 0\},$$

its main frontier stratum $\mathcal{T}_{Y\bar{X}Y}$,

$$\mathcal{T}_{Y\bar{X}Y} = \{p\exp(-tY)\exp(-sX)\exp(-uY) : t > 0, \, s = \bar{s}(t,0), \, u > 0\},$$

and the cell

$$\begin{aligned}
\mathscr{C}_{YSXY} &= \mathcal{T}_{YS\bar{X}Y} \cup \mathcal{T}_{Y\bar{S}Y} \cup \mathcal{T}_{YSY} \\
&= \{p\exp(-tY)\exp(-rS)\exp(-sX)\exp(-uY) : \\
&\qquad t > 0, \, 0 < r \le \bar{r}(t), \, 0 \le s \le \bar{s}(t,r), \, u > 0\}.
\end{aligned}$$

Summarizing, we have the following optimal synthesis:

Theorem 7.5.1. *The controlled trajectories corresponding to the feedback control* $u_*(x)$ *defined by*

$$u_*(x) = \begin{cases} -1 & \text{if } x \in N_- \cup \mathcal{T}_{XY} \cup \mathcal{T}_X \cup \mathcal{M}_-, \\ -\gamma(x) & \text{if } x \in \mathcal{T}_{SY}, \\ +1 & \text{if } x \in N_+ \cup \mathcal{T}_{YX} \cup \mathcal{T}_Y, \end{cases}$$

together with the associated stratification of the state-space define a local time-optimal synthesis of controlled trajectories.

Proof. As in the codimension-0 case, the controlled trajectories and the stratification that were constructed satisfy the assumptions of the simple verification theorem, Theorem 6.3.1: Setting up a parameterized family of extremals, it is easily seen that the value-function $V = V^{\mathscr{E}}$ defined by this synthesis is continuous and is a continuously differentiable solution to the Hamilton–Jacobi–Bellman equation on all the open strata. The complement of these open sets is a finite union of embedded submanifolds of positive codimension, and thus, by Theorem 6.3.1, the synthesis is optimal. $\qquad\qquad\square$

7.6 Notes

The results presented in this chapter were developed in the work of H. Schättler et al. (e.g., [141, 209–211, 215, 221]). They illustrate how constructive geometric arguments can be used in connection with explicit Lie-algebraic computations in canonical coordinates to give precise results about small-time reachable sets and how these, in turn, in certain cases, directly give rise to a time-optimal synthesis of controlled trajectories for related lower-dimensional optimal control problems. Clearly, the procedures are tied to nondegenerate-case or general-position-type assumptions, but they can be carried out under assumptions made only on the Lie-bracket configuration of the system. And the fact that cases of higher codimension or problems in higher dimensions become harder does not invalidate the procedure. The results obtained make a clear connection between the structure of time-optimal controlled trajectories for higher-codimension situations in lower dimensions and the structure of these trajectories for lower-codimension cases in higher dimensions. The structure of optimal controls in the codimension-2 two-dimensional scenario considered in Sect. 2.10 is the same as the structure of the optimal trajectories in the codimension-1 three-dimensional case just considered, and these concatenations form the boundary trajectories for the codimension-0 four-dimensional small-time reachable set. These results are not mere special cases; indeed they comprise general structures. Even in the general optimal control problem, it is these structures that determine optimal syntheses. For instance, the most degenerate situation that arises in the optimal synthesis for the mathematical model for tumor antiangiogenesis considered in Sect. 6.2 is that of two saturation points along an optimal singular arc, the situation just considered. One of them gives rise to optimal controls at most of the form *XYSB*, and the second generates concatenations of the form *BSYX*, leading overall to a synthesis that consists of *XYSYX* trajectories that generate the global solution to this problem.

The cases considered in this last chapter of our text give the structure of an optimal synthesis near general phenomena (*BSB* near fast singular arcs, *BBB* with a cut-locus near slow singular arcs, *BSBB* or *BBSB* near points where singular arcs saturate) that often are the optimal solutions in general when these phenomena are involved and more degenerate situations do not arise. In this sense, these structures typically directly apply to particular situations in which, based on the structures that are optimal here, optimal solutions can be conjectured and often verified easily. For example, this is how we solved the problem from Sect. 6.2. But this will have to be left for some other time.

Appendix A
A Review of Some Basic Results from Advanced Calculus

We briefly review some fundamental concepts and results from advanced calculus that are used frequently throughout the text. Most of these (with possibly the exception of Sard's theorem) are standard, and we generally refer the reader to textbooks on this subject for a more detailed exposition and proofs of these results. We assume that the reader is familiar with basic topological notions (such as open and closed sets, convergence, etc.) and matrix algebra (vector space, linear operators, matrix and vector computations, etc.).

A.1 Topology and Convergence in Normed Vector Spaces

Definition A.1.1 (Normed space). Let V be a vector space over $\mathbb{R}$. A function $\|\cdot\| : \mathbb{R} \to [0, \infty)$ is called a norm if for all vectors $v, w \in V$ and all real numbers λ the following three conditions are satisfied: (i) (positive definite) $\|v\| = 0 \iff v = 0$, (ii) (positive homogeneity) $\|\lambda v\| = |\lambda| \, \|v\|$, and (iii) (triangle inequality) $\|v + w\| \leq \|v\| + \|w\|$. A vector space V endowed with a norm is called a *normed space*.

Examples of norms that we use throughout the text are the $\|\cdot\|_p$ norms for $1 \leq p \leq \infty$. For $V = \mathbb{R}^n$, these are defined by

$$\|x\|_1 = \sum_{i=1}^{n} |x_i|, \qquad \|x\|_p = \left(\sum_{i=1}^{n} |x_i|^p \right)^{1/p}, \qquad \text{for } 1 < p < \infty,$$

and

$$\|x\|_\infty = \max_{i=1,\dots,n} |x_i|.$$

It is obvious that each function is positive definite and positively homogeneous. The triangle inequality for $\|\cdot\|_1$ and $\|\cdot\|_\infty$ is an immediate consequence of the triangle inequality in $\mathbb{R}$. For $1 < p < \infty$, the triangle inequality is a consequence

H. Schättler and U. Ledzewicz, *Geometric Optimal Control: Theory, Methods and Examples*, Interdisciplinary Applied Mathematics 38,
DOI 10.1007/978-1-4614-3834-2, © Springer Science+Business Media, LLC 2012

of Hölder's inequality, which we discuss more generally in the context of L_p-spaces in Appendix D. For $p = 2$, the so-called *Euclidean norm*, it also follows from the Cauchy–Schwarz inequality. The Euclidean norm is unique among the $\|\cdot\|_p$ norms in the sense that it is induced by an inner product.

Definition A.1.2 (Inner product space). Let V be a vector space over $\mathbb{R}$. An *inner product* on V is a positive definite symmetric bilinear form $(\cdot, \cdot)$, i.e., a mapping $(\cdot, \cdot) : V \times V \to \mathbb{R}$, $(v, w) \mapsto (v, w)$, such that for all vectors v and w of V and all real numbers λ we have that (i) (bilinear form) $(v_1 + \lambda v_2, w) = (v_1, w) + \lambda (v_2, w)$, (ii) (symmetric) $(v, w) = (w, v)$, and (iii) (positive definite) $(v, v) \geq 0$ and $(v, v) = 0$ if and only if $v = 0$.

The Euclidean inner product on $\mathbb{R}^n$ is given by $(v, w) = v^T w$. If $(\cdot, \cdot)$ is an inner product on V, then $\|v\| = \sqrt{(v, v)}$ defines a norm. It is clear that this function is positive definite and positively homogeneous. The triangle inequality follows from the *Cauchy–Schwarz inequality*:

$$|(v, w)| \leq \|v\| \, \|w\| \qquad \text{for all } v, w \in V.$$

This inequality allows us to define the angle φ between two vectors v and w in an inner product space by

$$\cos \varphi = \frac{(v, w)}{\|v\| \, \|w\|}.$$

In particular, two vectors v and w are orthogonal if and only if their inner product is zero. It is not difficult to show that a norm $\|\cdot\|$ is induced by an inner product if and only if the parallelogram identity,

$$\|v + w\|^2 + \|v - w\|^2 = 2 \left(\|v\|^2 + \|w\|^2 \right),$$

is valid for all vectors v and w from V.

Norms allow us to define the basic topological concepts of neighborhoods and convergence in the same natural way as in $\mathbb{R}^n$. We briefly summarize these fundamental definitions.

Open and closed sets. Given a normed space $(V, \|\cdot\|)$, for $x \in V$ and $\varepsilon > 0$, define the ball with radius ε around x as $B_\varepsilon(x) = \{y \in V : \|y - x\| < \varepsilon\}$. A set $E \subset V$ is *open* if for every $x \in E$ there exists an $\varepsilon > 0$ such that the ball $B_\varepsilon(x)$ lies entirely in E, $B_\varepsilon(x) \subset E$. It easily follows from the triangle inequality that the balls $B_\varepsilon(x)$ are open. A set F is *closed* if its complement, $F^c = \{x \in V : x \notin F\}$, is open.

Convergence. A sequence $\{x_n\}_{n \in \mathbb{N}} \subset V$ converges to a limit $x \in V$ if

$$\lim_{n \to \infty} \|x_n - x\| = 0,$$

and we simply write $x_n \to x$. Closed sets are more importantly characterized through convergence properties. It is easily seen that a set $F \subset V$ is closed if and only if whenever a sequence $\{x_n\}_{n\in\mathbb{N}} \subset F$ of points in the set F converges to a point $x \in V$, then it follows that $x \in F$. A sequence $\{x_n\}_{n\in\mathbb{N}} \subset V$ is said to be a Cauchy sequence if for every $\varepsilon > 0$, there exists an integer N such that $\|x_n - x_m\| < \varepsilon$ for all indices $n, m \geq N$.

Definition A.1.3 (Banach space). A normed space $(V, \|\cdot\|)$ in which every Cauchy sequence converges is said to be *complete*. If the underlying vector space is infinite-dimensional, a complete normed space is called a *Banach space*.

Continuity. A mapping $F : V \to W$ between normed spaces is *continuous* at a point $x \in V$ if for every $\varepsilon > 0$ there exists a $\delta = \delta(\varepsilon) > 0$ such that $F(B_\delta(x)) \subset B_\varepsilon(F(x))$. It is easily seen that F is continuous at x if and only if whenever $\{x_n\}_{n\in\mathbb{N}}$ is a sequence that converges to x, then $F(x_n) \to F(x)$. The mapping $F : V \to W$ is continuous if it is continuous at every point. Equivalently, from a topological point of view, $F : V \to W$ is continuous if and only if inverse images of open sets in W are open in V. Note that the norm itself, $\|\cdot\| : V \to \mathbb{R}$, is a continuous function. It follows from the triangle inequality that $|\|x\| - \|y\|| \leq \|x - y\|$ for all x and y. Hence, if $x_n \to x$, then also $\|x_n\| \to \|x\|$. In particular, if $x_n \to x$, then $\{\|x_n\|\}_{n\in\mathbb{N}}$ is bounded. Similarly, an inner product is continuous in both variables: if $v_n \to v$ and $w_n \to w$, then $(v_n, w_n) \to (v, w)$.

Definition A.1.4 (Compact). Let V be a normed vector space. A family $\{U_i\}_{i\in I}$ of open sets U_i for i in some arbitrary index set I is an open cover for the set $E \subset V$ if $E \subset \cup_{i\in I} U_i$. A set $K \subset V$ is compact, if every open cover of K contains a finite subcover.

It is easily seen from the definition that compact sets are closed and bounded. These conditions, however, are not sufficient in an infinite-dimensional vector space, but they characterize compact sets in finite-dimensional spaces.

Theorem A.1.1 (Heine–Borel). *A subset $K \subset \mathbb{R}^n$ is compact if and only if it is closed and bounded.* ∎

Corollary A.1.1 (Weierstrass). *Every bounded sequence $\{x_n\}_{n\in\mathbb{N}} \subset \mathbb{R}^n$ contains a convergent subsequence.* ∎

The next result is one of the most important properties of continuous functions on compact sets, and we include its simple proof to give a typical compactness argument.

Theorem A.1.2. *A continuous function $f : K \to \mathbb{R}$ defined on a compact set K in $\mathbb{R}^n$ attains its minimum and maximum on K.*

Proof. Let $m = \inf_{x\in K} f(x)$ and $M = \sup_{x\in K} f(x)$. A priori, we have only that $-\infty \leq m \leq M \leq \infty$, and it is claimed that there exist points $x_* \in K$ and $x^* \in K$ such that $-\infty < m = f(x_*) \leq f(x^*) = M < \infty$. Without loss of generality, we only consider the case of the minimum.

It follows from the definition of the infimum as the greatest lower bound that there exists a sequence $\{x_n\}_{n \in \mathbb{N}} \subset K$ such that $f(x_n) \to m$, a so-called minimizing sequence. Since K is compact, there exists a convergent subsequence, and we may simply assume that $x_n \to x_*$. Since K is closed, the limit x_* lies in K. Since f is continuous on K, it follows that $f(x_n) \to f(x_*)$, which must be finite. Thus $-\infty < m = f(x_*)$. $\qquad\square$

The Heine–Borel theorem implies that the unit sphere $S = \{x \in V : \|x\| = 1\}$ in a finite-dimensional normed space is compact. Indeed, it can be shown that this is correct if and only if V is finite-dimensional. On $\mathbb{R}^n$, this implies that all norms are equivalent in the following sense:

Definition A.1.5 (Equivalent norms). Two norms $|\cdot|$ and $\|\cdot\|$ on a vector space V are equivalent if there exist positive constants m and M such that for all vectors $v \in V$ we have that $m|v| \leq \|v\| \leq M|v|$.

Equivalent norms have the same convergent sequences, and as far as the definition of open sets or convergence is concerned, there is no need to distinguish between equivalent norms.

Proposition A.1.1. *All norms on $\mathbb{R}^n$ are equivalent.*

Proof. The unit ball $S = \{v \in \mathbb{R}^n : \|v\|_2 = v^T v = 1\}$ is a compact set in $\mathbb{R}^n$, and as a positive definite continuous function, the norm $\|\cdot\|$ attains a positive minimum m and finite maximum M on S. Thus we have for all $v \in S$ that $0 < m \leq \|v\| \leq M < \infty$. Given any nonzero vector $w \in \mathbb{R}^n$, there exists a $v \in S$ and a $\lambda > 0$ such that $w = \lambda v$ and thus, since norms are positively homogeneous, the inequality

$$m \|w\|_2 = m\lambda \leq \lambda \|v\| = \|w\| \leq \lambda M = M \|w\|_2$$

holds for all vectors $w \in \mathbb{R}^n$. The general result easily follows by combining the bounds obtained in this way for two arbitrary norms. $\qquad\square$

Thus all norms on $\mathbb{R}^n$ induce the same topology as the standard Euclidean norm $\|v\|_2 = \sqrt{\sum_{i=1}^n v_i^2}$, and we do not need to specify which norm is being used on $\mathbb{R}^n$ when we talk of convergence. The compactness of the unit spheres also allows us to define "good" matrix norms for $A \in \mathbb{R}^{m \times n}$ that are compatible with a given vector norm, the so-called *operator* or *lub-norm*. Let $\|\cdot\|_a$ be a norm in $\mathbb{R}^n$ and let $\|\cdot\|_b$ be a norm in $\mathbb{R}^m$. Then the subordinate matrix norm $\|A\|_{a,b}$ is defined as

$$\|A\|_{a,b} = \max_{x \neq 0} \frac{\|Ax\|_b}{\|x\|_a} = \max_{\|x\|_a = 1} \|Ax\|_b .$$

The function $A \mapsto \|A\|_{a,b}$ is well-defined, and it is easily verified that it satisfies the defining properties of a norm. Furthermore, it is immediate from the definition that the matrix norm is compatible with the vector norm in the sense that

$$\|Ax\|_b \leq \|A\|_{b,a} \cdot \|x\|_a .$$

If $A \in \mathbb{R}^{n \times n}$ and the same norm is used, we delete the subscripts. It is not difficult to see that the operator norm corresponding to the Euclidean norm is given by the largest singular value of A, i.e.,

$$\|A\|_2 = \max_{x^T x = 1} \sqrt{x^T A^T A x} = \sqrt{\lambda_{\max}(A^T A)},$$

with $\lambda_{\max}$ denoting the largest eigenvalue of the positive semidefinite matrix $A^T A$. (Since $A^T A$ is symmetric, there exists an orthogonal matrix $Q \in \mathbb{R}^{n \times n}$, $Q^T Q = \mathrm{Id}$, such that $Q^T A^T A Q = \Lambda$ with $\Lambda = \mathrm{diag}(\lambda_1, \dots, \lambda_n)$ a diagonal matrix. The claim immediately follows from this by making a linear change of coordinates $x = Qy$ and observing that $\|Qy\|_2 = \|x\|_2 = 1$.) Similarly,

$$\|A\|_\infty = \max_{\|x\|_\infty = 1} \left\{ \max_{i=1,\dots,n} \left| \sum_{k=1}^{n} a_{ik} x_k \right| \right\} = \max_{i=1,\dots,n} \sum_{k=1}^{n} |a_{ik}|.$$

A.2 Uniform Convergence and the Banach Space $C(K)$

Definition A.2.1 (Uniform convergence). Let V be a normed vector space. A sequence $\{f_n\}_{n \in \mathbb{N}}$ of functions $f_n : E \to \mathbb{R}^m$ defined on some subset E of V converges uniformly on E to some limit f if for every $\varepsilon > 0$, there exists an integer $N = N(\varepsilon)$ such that for all $n \geq N$ and all points $x \in E$ we have that $\|f_n(x) - f(x)\| < \varepsilon$.

If the sequence f_n converges pointwise in E, then for each $x \in E$ and every $\varepsilon > 0$, there exists an integer $N = N(\varepsilon; x)$ such that $\|f_n(x) - f(x)\| < \varepsilon$ for all $n \geq N$, but this number N depends on the point x as well. Uniform convergence means that an integer N can be chosen that works for all points in the set E. The significance of the concept lies with the following simple fact:

Proposition A.2.1. *If* $\{f_n\}_{n \in \mathbb{N}}$ *is a sequence of continuous functions* $f_n : E \to \mathbb{R}^m$ *that converges uniformly on* E, *then the limit* $f : E \to \mathbb{R}$ *is continuous.*

Proof. Given $x \in E$, let $\varepsilon > 0$. Since $\{f_n\}_{n \in \mathbb{N}}$ converges uniformly to f on E, there exists an integer N such that $\|f_N(y) - f(y)\| < \frac{\varepsilon}{3}$ for all $y \in E$. Since f_N is continuous, there exists a $\delta = \delta(\varepsilon) > 0$ such that $\|f_N(y) - f_N(x)\| < \frac{\varepsilon}{3}$ whenever $\|y - x\| < \delta$. Hence, we have for all $y \in B_\delta(x)$ that

$$\|f(y) - f(x)\| \leq \|f(y) - f_N(y)\| + \|f_N(y) - f_N(x)\| + \|f_N(x) - f(x)\| < \varepsilon,$$

and thus f is continuous at x. $\square$

Often such a sequence of functions is given as the partial sums of a series, i.e., $f_n = \sum_{i=0}^{n} g_i$, where the g_i are continuous functions defined on E. In this case we have the following simple criterion for uniform convergence due to Weierstrass.

Lemma A.2.1. *If there exists a convergent series $\sum_{i=0}^{\infty} \varepsilon_i$ of positive numbers ε_i such that $\|g_i(x)\| \leq \varepsilon_i$ for all $x \in E$ and all $i \in \mathbb{N}$, then the sequence $\{f_n\}_{n \in \mathbb{N}}$, $f_n(x) = \sum_{i=0}^{n} g_i(x)$, converges uniformly on E to the continuous limit $f(x) = \sum_{i=0}^{\infty} g_i(x)$.* ∎

Definition A.2.2 (C(K)). Let K be a compact set of $\mathbb{R}^n$; $C(K)$ is the normed vector space of all continuous functions $f : K \to \mathbb{R}^m$ equipped with the supremum norm

$$\|f\|_\infty = \max_{x \in K} \|f(x)\|.$$

Since K is compact, $\|f\|_\infty$ is well-defined. It is clear that it defines a norm on the space of continuous functions defined on K, and convergence in the supremum norm is the same as uniform convergence on K. Proposition A.2.1 implies that this space is complete.

Proposition A.2.2. *The space $C(K)$ is a Banach space.*

Proof. Suppose $\{f_n\}_{n \in \mathbb{N}}$ is a sequence of continuous functions $f_n : K \to \mathbb{R}^m$ that is Cauchy in the supremum norm, i.e., for every $\varepsilon > 0$ there exists an integer $N = N(\varepsilon)$ such that for all integers $k, n \geq N(\varepsilon)$, we have that $\|f_n - f_k\|_\infty < \varepsilon$. But then, for every $x \in K$, the sequence $\{f_n(x)\}_{n \in \mathbb{N}}$ is Cauchy in $\mathbb{R}$ and thus has a limit, call it $f(x)$. Taking the limit as $m \to \infty$, we obtain for $n \geq N(\varepsilon)$ that

$$\|f_n - f\|_\infty = \lim_{k \to \infty} \left(\max_{x \in K} |f_n(x) - f_k(x)| \right) \leq \varepsilon,$$

i.e., $\{f_n\}_{n \in \mathbb{N}}$ converges to f in the supremum norm, or equivalently, uniformly on the set K. But then the limit f is continuous by Proposition A.2.1 and thus $\{f_n\}_{n \in \mathbb{N}}$ converges in the supremum norm to an element in the space. □

In the context of control systems, it is often important to be able to select subsequences of controlled trajectories $\{(x_n, u_n)\}_{n \in \mathbb{N}}$ for which the sequence of trajectories converges uniformly. The Arzelà–Ascoli theorem provides a criterion for this to be possible. Equivalently, it gives conditions for a subset $\mathcal{K}$ in the Banach space $C(K)$ to be compact. We need the following terminology:

Definition A.2.3 (Equicontinuous). A family $\mathcal{F}$ of continuous functions $f : E \to \mathbb{R}^m$ is said to be equicontinuous if for every $\varepsilon > 0$ there exists a $\delta > 0$, depending only on ε, such that for all $f \in \mathcal{F}$ and all points x and y from K that satisfy $\|x - y\| < \delta$, we have that $\|f(x) - f(y)\| < \varepsilon$.

Note that any finite family $\mathcal{F}$ of continuous functions $f : \overset{'}{K} \to \mathbb{R}^m$ defined on a compact set K is equicontinuous. For this definition is equivalent to a single function being uniformly continuous, and it follows from a standard compactness argument that continuous functions are uniformly continuous on compact sets. Thus an equicontinuous family is a family of continuous functions that is uniformly continuous with the same ε and δ working for every function in the family. For example, all functions $f : K \to \mathbb{R}^n$ that satisfy a Lipschitz condition of the form $\|f(x) - f(y)\| \leq L\|x - y\|$ with a fixed Lipschitz constant L are equicontinuous.

Definition A.2.4 (Pointwise compact). Let K be a compact subset of $\mathbb{R}^n$ and let $\mathscr{F}$ be a family of continuous functions f defined on K, $f : K \to \mathbb{R}^m$. The family $\mathscr{F}$ is said to be point wise compact if for every point $x \in K$, the set $\{f(x) : f \in \mathscr{F}\} \subset \mathbb{R}^m$ is compact.

Then the following result holds [177, Sect. 5.6]:

Theorem A.2.1 (Arzelà–Ascoli). *Let K be a compact subset of $\mathbb{R}^n$ and let $\mathscr{F}$ be a family of continuous functions f defined on K, $f : K \to \mathbb{R}^m$. Then the family $\mathscr{F}$ is compact in $C(K)$ if and only if $\mathscr{F}$ is closed, equicontinuous, and pointwise compact.* ■

Corollary A.2.1. *Let $\{f_n\}_{n \in \mathbb{N}}$ be a sequence of continuous functions $f_n : K \to \mathbb{R}^m$ defined on a compact set that is equicontinuous and pointwise bounded. Then there exists a subsequence $\{f_{n_k}\}_{k \in \mathbb{N}}$ that converges uniformly.*

A.3   Differentiable Mappings and the Implicit Function Theorem

One of the most important results of advanced calculus is the implicit function theorem which gives information about the local solvability of an equation of the form $F(x, y) = 0$ near a solution point (x_0, y_0). It is equivalent to the inverse function theorem, which establishes the local invertibility of a mapping $F : \mathbb{R}^n \to \mathbb{R}^n$ near a point x_0 where the matrix of the partial derivatives of F, $DF(x_0)$, is nonsingular. We briefly review these important results that will be used many times in this text. We recall the fundamental definitions:

Definition A.3.1 (Differentiable mapping). Let U be an open subset of $\mathbb{R}^n$. A mapping $F : M \to \mathbb{R}^m$ is differentiable at a point $x_0 \in U$ if there exist a linear mapping, denoted by $DF(x_0) : \mathbb{R}^n \to \mathbb{R}^m$, and a continuous function $r : M \to \mathbb{R}^m$ that vanishes at x_0, $r(x_0) = 0$, such that for all $x \in U$ we have that

$$F(x) = F(x_0) + DF(x_0)(x - x_0) + r(x)(x - x_0).$$

The matrix $DF(x_0)$ is called the derivative or the Jacobian matrix of F at x_0. The mapping F is differentiable on U if it is differentiable at every point $x_0 \in M$.

Suppose the mapping F has components $f_i : U \to \mathbb{R}$, $i = 1, \ldots, m$. If F is differentiable at x_0, then the limits

$$\frac{\partial f_i}{\partial x_j}(x_0) = \lim_{h \to 0} \frac{f_i(x_1, \ldots, x_{j-1}, x_j + h, x_j, \ldots, x_n) - f_i(x_1, \ldots, x_{j-1}, x_j, x_j, \ldots, x_n)}{h}$$

exist and are called the *partial derivatives* of the components f_i at x_0; $\frac{\partial f_i}{\partial x_j}(x_0)$ is the (i,j)th entry of the Jacobian matrix. The row vector

$$\nabla f_i(x_0) = \left(\frac{\partial f_i}{\partial x_1}(x_0), \ldots, \frac{\partial f_i}{\partial x_n}(x_0) \right)$$

is the *gradient* of the function f_i. Conversely, it can be shown that if all partial derivatives exist and are continuous on U, then F is differentiable on U with the Jacobian matrix given by the matrix whose entries are the partial derivatives. In this case, we say that F is continuously differentiable on M.

We refer the reader to the textbook [177] for these and other fundamental results about differentiable functions such as the chain rule and the mean value theorem. We just note that if a function $f : U \rightarrow \mathbb{R}$ is differentiable at a point $x_0 \in U$, and $v \in \mathbb{R}^n$ is any vector, then by the chain rule, the composition $\varphi : (-\varepsilon, \varepsilon) \rightarrow \mathbb{R}$ defined by $\varphi(t) = f(x_0 + tv)$ is differentiable at $t = 0$ with derivative given by $\varphi'(0) = \nabla f(x_0)v$. This is the *directional derivative* of the function f in the direction v at x_0, also called the *Lie derivative* of f along v.

Higher-order derivatives are defined inductively as the derivatives of the mappings given by the lower-order derivatives (for example, see [177, Chap. 6]). Our computations in Chap. 5 make extensive use of these multilinear forms and their associated multidimensional Taylor expansions. For a scalar function $f : M \subset \mathbb{R}^n \rightarrow \mathbb{R}$, these derivatives are easily expressed in coordinates. For example, the quadratic Taylor expansion reads

$$f(x_0 + h) = f(x_0) + \sum_{i=1}^{n} \frac{\partial f}{\partial x_i}(x_0)h_i + \frac{1}{2} \sum_{i=1}^{n} \sum_{j=1}^{n} \frac{\partial^2 f}{\partial x_i \partial x_j}(x_0)h_i h_j + \cdots,$$

and it is clear what the Taylor expansion up to an arbitrary order r will look like. In vector notation, this becomes

$$f(x_0 + h) = f(x_0) + \nabla f(x_0)h + \frac{1}{2}\frac{\partial^2 f}{\partial x^2}(x_0)(h, h) + o\left(\|h\|^2 \right),$$

where $\frac{\partial^2 f}{\partial x^2}(x_0)(h, h)$ denotes the bilinear form defined by the Hessian matrix of the second-order partial derivatives of f evaluated at x_0 and acting on h. For a vector-valued function $F : M \rightarrow \mathbb{R}^m$, we then write this equation componentwise in the form

$$F(x_0 + h) = F(x_0) + DF(x_0)h + \frac{1}{2}D^2 F(x_0)(h, h) + o\left(\|h\|^2 \right),$$

with the understanding that $D^2 F(x_0)(h, h)$ denotes the vector whose ith entry is given by the quadratic forms $\frac{\partial^2 f_i}{\partial x^2}(x_0)(h, h)$ defined by the ith component f_i. Similar notation can be used for arbitrary orders.

Definition A.3.2 (C^r). A function $F : M \to \mathbb{R}^m$ is of class C^r, $r \geq 1$, if the first r derivatives exist and are continuous.

Equivalently, the partial derivatives of all components f_i of F up to order r exist and are continuous functions.

Definition A.3.3 (C^r-Diffeomorphism). Let U and V be open subsets of $\mathbb{R}^n$. The mapping $F : U \to V$ is a C^r-diffeomorphism if F is one-to-one and onto, r-times continuously differentiable on U, and has an r-times continuously differentiable inverse $F^{-1} : V \to U$.

Theorem A.3.1 (Inverse function theorem). *Let M be an open subset of $\mathbb{R}^n$ and suppose the mapping $F : M \to \mathbb{R}^n$ is r-times continuously differentiable (of class C^r) at a point $x_0 \in M$ with a nonsingular Jacobian matrix $DF(x_0)$. Then there exist a neighborhood U of x_0, $U \subset M$, and a neighborhood V of the image point $y_0 = F(x_0)$ such that the mapping $F : U \to V$ is a C^r-diffeomorphism.* ∎

The implicit function theorem is an equivalent version of the inverse function theorem in the sense that one follows from the other.

Theorem A.3.2 (Implicit function theorem). *Let $M \subset \mathbb{R}^n \times \mathbb{R}^m$ be an open subset and suppose the mapping $F : M \to \mathbb{R}^m$ is r-times continuously differentiable (of class C^r) at a point $(x_0, y_0) \in M$ where $F(x_0, y_0) = 0$. If the matrix of the partial derivatives $\frac{\partial F}{\partial y}(x_0, y_0)$ is nonsingular, then there exist a neighborhood U of x_0, a neighborhood V of y_0, and a unique function $f : U \to V$ that is r-times continuously differentiable (of class C^r) on U such that (i) $f(x_0) = y_0$ and (ii) $F(x, y) = 0$ and $(x, y) \in U \times V$ if and only if $y = f(x)$. In particular, the points on the graph of f are the only solutions to the equation $F(x, y) = 0$ that lie in the neighborhood $U \times V$.*

The implicit function theorem simply asserts that the equation $F(x, y) = 0$ can locally be solved uniquely for y as a function of x, $y = f(x)$, near a point (x_0, y_0) where the partial derivatives with respect to y are "nonzero".

Proof. The proof is an immediate corollary of the inverse function theorem. Define a mapping

$$G : \mathbb{R}^n \times \mathbb{R}^m \to \mathbb{R}^n \times \mathbb{R}^m, \qquad (x, y) \mapsto G(x, y) = (x, F(x, y)).$$

This mapping has a nonsingular Jacobian at the point (x_0, y_0), and thus, by the inverse function theorem, G is a C^r-diffeomorphism near (x_0, y_0). The inverse G^{-1} is of the form

$$G^{-1} : \mathbb{R}^n \times \mathbb{R}^m \to \mathbb{R}^n \times \mathbb{R}^m, \qquad (x, y) \mapsto G^{-1}(x, y) = (x, H(x, y)),$$

with some function $H \in C^r$. In particular, $F(x, H(x, y)) \equiv y$, and thus $f(x) = H(x, 0)$ provides a solution of class C^r. It follows from the inverse function theorem that this solution is unique. □

A.4   Regular and Singular Values: Sard's Theorem

Definition A.4.1 (Regular and singular points and values). Let U and V be open subsets of $\mathbb{R}^n$ and let $F : U \to V$ be a C^r mapping. A point $x_0 \in U$ is called regular if the Jacobian matrix $DF(x_0)$ is nonsingular; it is called singular (or critical) if $DF(x_0)$ is singular. A point $y \in V$ is called a regular value if every point x in the inverse image of y,

$$F^{-1}(\{y\}) = \{x \in U : F(x) = y\},$$

is a regular point; y is called singular (or critical) if there exists an $x \in F^{-1}(\{y\})$ that is singular.

Definition A.4.2 (Rank of a mapping at a point). For a mapping $F : \mathbb{R}^n \to \mathbb{R}^m$, the rank of F at the point x is defined as the rank of the Jacobian matrix $DF(x)$.

Definition A.4.3 (Singular set of a mapping). Let $U \subset \mathbb{R}^n$ be an open set and let $F : U \to \mathbb{R}^m$ be a C^r-mapping. Then the singular set S is the set of all points $x \in U$ for which the rank is less than m, the dimension of the image:

$$S = \{x \in U : \operatorname{rk}(DF(x)) < m\}.$$

If $m > n$, then this is always true, and every point is singular. While Sard's theorem below holds in either case, the interesting case is $m \leq n$. Essentially, Sard's theorem states that the image of the singular set is "small." For a proof of this result, we highly recommend Milnor's notes [184], which give a proof based on an argument of Pontryagin.

Theorem A.4.1 (Sard's theorem). *Let $U \subset \mathbb{R}^n$ be an open set and let $F : U \to \mathbb{R}^m$ be a C^r-mapping. Then the image of the singular set S under F, $F(S)$, has Lebesgue measure 0, $\mu(F(S)) = 0$.* ∎

The basic notions of Lebesgue measure are reviewed in Appendix D. This result states that for every $\varepsilon > 0$, there exists a sequence of cubes $\{Q_k\}_{k \in \mathbb{N}}$ that covers $F(S)$ such that the volumes of the cubes add up to a number less than ε. Since $\mathbb{R}^n$ can be covered by a countable union of open balls, it follows that the regular values, the complement of $F(S)$, are everywhere dense in $\mathbb{R}^m$ (Brown's theorem).

Appendix B
Ordinary Differential Equations

The general theory of ordinary differential equations—that is, results about existence and uniqueness of solutions to ordinary differential equations, $\dot{x} = f(t,x)$, as
well as the fact that these solutions depend continuously, respectively differentiably,
on initial conditions and parameters—is of fundamental importance in the text and is
used pervasively everywhere. Furthermore, several times in the text (e.g., Sects. 5.3
or 6.3, to mention just a couple of instances), we need to make use of the actual
constructions that establish these theorems. For this reason, we include a proof of
these results under the assumption that the time-varying vector fields $f = f(t,x)$ are
continuous in the time variable t. The equally classical theory for vector fields that
are only Lebesgue measurable in t, to the extent that it is needed in the later chapters
of the book, will be described in Sect. D.4.

B.1 Existence and Uniqueness of Solutions of Ordinary Differential Equations

Let G be a domain in $\mathbb{R} \times \mathbb{R}^n$, i.e., an open and connected subset. A mapping $f :
G \to \mathbb{R}^n$, $(t,x) \mapsto f(t,x)$, will be called a *time-varying vector field* on G. Vector
fields f that have discontinuities arise in syntheses of solutions of optimal control
problems, and there are plenty of examples in the text that illustrate the structure of
solutions for some of these cases. But no general existence results can be valid for
this case, since geometric properties of the vector field at the discontinuities matter.
Here, we always assume that f is continuous on G, and we write $f \in C^{0,r}(G)$ if for
t fixed, the mapping $x \mapsto f(t,x)$ is r-times continuously differentiable in x with all
partial derivatives continuous in both variables t and x on G. We call the system
time-invariant or autonomous if the vector field f does not depend on t, and in this
case we simply consider G as a region in the state space $\mathbb{R}^n$ and f as a vector field
on G.

H. Schättler and U. Ledzewicz, *Geometric Optimal Control: Theory, Methods
and Examples*, Interdisciplinary Applied Mathematics 38,
DOI 10.1007/978-1-4614-3834-2, © Springer Science+Business Media, LLC 2012

Definition B.1.1 (Solution to the initial value problem). Given a point $(t_0, x_0) \in G$, a continuously differentiable curve $x : I \to \mathbb{R}^n$, $t \mapsto x(t)$, defined on some open interval I containing t_0 is a solution to the initial value problem

$$\dot{x} = f(t,x), \qquad x(t_0) = x_0, \tag{B.1}$$

in G if it satisfies the following three conditions: (i) $\dot{x}(t) = f(t, x(t))$ for all $t \in I$, (ii) $x(t_0) = x_0$, and (iii) the graph of x lies in the region G, i.e., $(t, x(t)) \in G$ for all $t \in I$.

Theorem B.1.1 (Peano). *If f is continuous on G, then every initial value problem (B.1) has a solution.* ∎

This result gives the most general existence result in the theory of ordinary differential equations. Essentially, it is proven by an application of the Arzelà–Ascoli theorem, Corollary A.2.1, to a family of approximating curves that are generated using a simple Euler scheme,

$$x(t+h) = x(t) + hf(t, x(t)).$$

However, continuity of f does not guarantee uniqueness of solutions. For example, the autonomous differential equation $\dot{x} = \sqrt[3]{x}$, $x(0) = 0$, has the trivial solution $x(t) \equiv 0$, but also $x(t) = \sqrt{\left(\frac{2}{3}t\right)^3}$ is a solution. Thus

$$x_\tau(t) = \begin{cases} 0 & \text{for } 0 \le t \le \tau, \\ \sqrt{\left(\frac{2}{3}(t-\tau)\right)^3} & \text{for } t > \tau, \end{cases}$$

defines a 1-parameter family of solutions. From a practical point of view, uniqueness of solutions is as important as existence (if not more so, in engineering problems) and here we therefore develop the classical theory that guarantees both existence and uniqueness of solutions. The extra requirement that gives uniqueness of solutions is the Lipschitz condition.

Definition B.1.2 (Lipschitz condition). Let $f : G \to \mathbb{R}^n$, $(t,x) \mapsto f(t,x)$, be a continuous time-varying vector field on a domain G. We say that f satisfies a Lipschitz condition in x on a subset U of G if there exists a constant L, the Lipschitz constant, such that for all points (t,x) and (t,y) that lie in G, we have that

$$\|f(t,x) - f(t,y)\| \le L\|x - y\|.$$

We say that f is Lipschitz continuous in x on G if every point $(t,x) \in G$ has a neighborhood U such that f satisfies a Lipschitz condition in x on U.

Since the state space is finite-dimensional, all norms are equivalent, and there is no need to specify the norm in the definition. But note that the numerical value of the Lipschitz constant L very much depends on the norm that is being used. As an example, any norm $\|\cdot\| : V \to \mathbb{R}$ on a normed vector space V satisfies a global Lipschitz condition with Lipschitz constant $L = 1$. This simply is the statement of the identity $|\|x\| - \|y\|| \leq \|x - y\|$.

Lemma B.1.1. *Suppose $f : G \to \mathbb{R}^n$, $(t,x) \mapsto f(t,x)$, is bounded and Lipschitz continuous in x. Then for every compact subset $K \subset G$ there exists a Lipschitz constant $L = L(K)$ such that*

$$\|f(t,x) - f(t,y)\| \leq L\|x - y\| \qquad \text{for all} \quad (t,x) \in G \text{ and } (t,y) \in K.$$

In particular, the function f satisfies a Lipschitz condition in x on K.

Proof. Let K be a compact subset of G. Since f is Lipschitz continuous in x, for every point $(\tilde{t}, \tilde{x}) \in K$ there exist a neighborhood $\tilde{N}$, which without loss of generality we may take in the form $\tilde{N} = (\tilde{t} - \varepsilon, \tilde{t} + \varepsilon) \times B_\varepsilon(\tilde{x})$ for some $\varepsilon > 0$, and a constant $\tilde{L}$ such that for all (t,x) and (t,y) in $\tilde{N}$ we have that

$$\|f(t,x) - f(t,y)\| \leq \tilde{L}\|x - y\|. \tag{B.2}$$

Let $N = (\tilde{t} - \varepsilon, \tilde{t} + \varepsilon) \times B_{\frac{\varepsilon}{2}}(\tilde{x})$. Then, for $(t,x) \notin \tilde{N}$ and $(t,y) \in N$, the norm $\|x - y\|$ is bounded away from zero, $\|x - y\| \geq \frac{\varepsilon}{2}$, and since the function f is bounded, it follows that for some finite constant C we have that

$$\frac{\|f(t,x) - f(t,y)\|}{\|x - y\|} \leq C < \infty \qquad \text{for all} \quad (t,x) \notin \tilde{N} \text{ and } (t,y) \in N.$$

For $(t,x) \in \tilde{N}$, the bound (B.2) is valid by the Lipschitz condition. Thus by increasing the Lipschitz constant $\tilde{L}$ if necessary, we therefore have that

$$\|f(t,x) - f(t,y)\| \leq \tilde{L}\|x - y\| \qquad \text{for all} \quad (t,x) \in G \text{ and } (t,y) \in N.$$

Since K is compact, we can cover K with a finite number of neighborhoods of type N, and choosing L as the largest of the corresponding Lipschitz constants, we get that

$$\|f(t,x) - f(t,y)\| \leq L\|x - y\| \qquad \text{for all} \quad (t,x) \in G \text{ and } (t,y) \in K.$$

$\square$

Lemma B.1.2. *Suppose $f \in C^{0,1}(G)$ (i.e., f is continuous and continuously differentiable in x with the partial derivatives continuous in both variables (t,x)). Then f is Lipschitz continuous in x on G.*

Proof. It suffices to prove the result for the Euclidean norm $\|\cdot\|_2$. Given $(\tilde{t}, \tilde{x}) \in G$, choose $\varepsilon > 0$ small enough that $\tilde{N} = (\tilde{t} - \varepsilon, \tilde{t} + \varepsilon) \times B_\varepsilon(\tilde{x}) \subset G$ has compact closure in G. Suppose (t,x) and (t,y) are points in $\tilde{N}$ and, without loss of generality, assume that $f(t,x) \neq f(t,y)$. Otherwise, any kind of Lipschitz condition is trivially valid for these two points. Define the function

$$\varphi : [0,1] \to \mathbb{R}, \quad s \mapsto \varphi(s) = \left\langle \frac{f(t,x) - f(t,y)}{\|f(t,x) - f(t,y)\|_2}, f(t, sx + (1-s)y) \right\rangle,$$

where $\langle \cdot, \cdot \rangle$ denotes the standard inner product on $\mathbb{R}^n$. This function is continuously differentiable, and by the mean value theorem, there exists a point $s_* \in (0,1)$ such that

$$\|f(t,x) - f(t,y)\|_2 = \varphi(1) - \varphi(0) = \varphi'(s_*)$$

$$= \left\langle \frac{f(t,x) - f(t,y)}{\|f(t,x) - f(t,y)\|_2}, \frac{\partial f}{\partial x}(t, s_* x + (1-s_*)y)(x-y) \right\rangle.$$

By the Cauchy–Schwarz inequality, we therefore have that

$$\|f(t,x) - f(t,y)\|_2 \leq \left\| \frac{\partial f}{\partial x}(t, s_* x + (1-s_*)y)(x-y) \right\|_2$$

$$\leq \mathrm{lub}_2 \left(\frac{\partial f}{\partial x}(t, s_* x + (1-s_*)y) \right) \|x-y\|_2$$

$$\leq \left[\max_{(t,z) \in \bar{N}} \mathrm{lub}_2 \left(\frac{\partial f}{\partial x}(t,z) \right) \right] \|x-y\|_2.$$

Here $\mathrm{lub}_2(A)$ denotes the least upper bound, or operator matrix norm associated with the Euclidean vector norm, i.e.,

$$\mathrm{lub}_2(A) = \max_{\|x\|_2 = 1} \|Ax\|_2 = \sqrt{\lambda_{\max}(A^T A)},$$

the square root of the largest eigenvalue of the positive semidefinite matrix $A^T A$. Since the partial derivatives $\frac{\partial f}{\partial x}$ are continuous in G, it follows that the largest eigenvalue of the matrix $\left(\frac{\partial f}{\partial x}(t,z) \right)^T \left(\frac{\partial f}{\partial x}(t,z) \right)$ for (t,z) in the compact closure of $\bar{N}$ is finite, and thus f satisfies a Lipschitz condition in x on the neighborhood $\bar{N}$. $\square$

Theorem B.1.2 (Existence and uniqueness of solutions). *Let $G \subset \mathbb{R} \times \mathbb{R}^n$ be a domain and suppose $f : G \to \mathbb{R}^n$, $(t,x) \mapsto f(t,x)$, is continuous and Lipschitz continuous in x on G. Then for every $(t_0, x_0) \in G$, there exists a unique solution $x = x(t; t_0, x_0)$ to the initial value problem*

$$\dot{x} = f(t,x), \qquad x(t_0) = x_0, \qquad (t,x) \in G, \tag{B.3}$$

defined on a maximal interval $(t_-(t_0, x_0), t_+(t_0, x_0))$ of definition (i.e., it is not possible to extend the solution beyond any of these times without leaving the domain G).

We develop the proof in a series of lemmas. The existence of local solutions directly follows from the Picard–Lindelöf iteration given below.

Lemma B.1.3 (Picard–Lindelöf iteration). *With $D_\beta(x_0) = \{x \in \mathbb{R}^n : \|x - x_0\|_\infty \leq \beta\}$, let $K = K(\alpha, \beta) = [t_0 - \alpha, t_0 + \alpha] \times D_\beta(x_0) \subset G$. If for $M = \max_{(t,x) \in K} \|f(t,x)\|_\infty$, we have that $\alpha M \leq \beta$, then Eq. (B.3) has a solution x that is defined over the full interval $[t_0 - \alpha, t_0 + \alpha]$ and takes values in $D_\beta(x_0)$, $x : [t_0 - \alpha, t_0 + \alpha] \rightarrow D_\beta(x_0)$.*

Proof. Inductively define a sequence of functions $\{x_n\}_{n \in \mathbb{N}}$ as $x_0(t) \equiv x_0$ and

$$x_{n+1}(t) = x_0 + \int_{t_0}^t f(s, x_n(s))ds.$$

We claim that all functions are defined over the full interval $[t_0 - \alpha, t_0 + \alpha]$ and have values in $D_\beta(x_0)$. Let L be a Lipschitz constant for the function f in x over the compact set K in the supremum norm $\|\cdot\|_\infty$. We show inductively that for all $t \in [t_0 - \alpha, t_0 + \alpha]$,

$$\|x_n(t) - x_0\|_\infty \leq \beta \quad \text{and} \quad \|x_{n+1}(t) - x_n(t)\|_\infty \leq ML^n \frac{|t - t_0|^{n+1}}{(n+1)!}. \tag{B.4}$$

For $n = 0$, the first assertion is trivial, and we have that

$$\|x_1(t) - x_0\|_\infty = \left\| \int_{t_0}^t f(s, x_0)ds \right\|_\infty \leq \left| \int_{t_0}^t \|f(s, x_0)\|_\infty ds \right| = M|t - t_0|.$$

Thus assume inductively that both assertions hold for n. We then have, as above,

$$\|x_{n+1}(t) - x_0\|_\infty = \left\| \int_{t_0}^t f(s, x_n(s))ds \right\|_\infty \leq \left| \int_{t_0}^t \|f(s, x_n(s))\|_\infty ds \right|$$

$$\leq M|t - t_0| \leq \alpha M \leq \beta$$

and thus the function x_{n+1} takes values in $D_\beta(x_0)$ for all $t \in [t_0 - \alpha, t_0 + \alpha]$. Furthermore, we can estimate the difference as

$$\|x_{n+1}(t) - x_n(t)\|_\infty = \left\| \int_{t_0}^t f(s, x_n(s)) - f(s, x_{n-1}(s))ds \right\|_\infty$$

$$\leq \left| \int_{t_0}^t \|f(s, x_n(s)) - f(s, x_{n-1}(s))\|_\infty ds \right|$$

$$\leq \left| \int_{t_0}^t L \|x_n(s) - x_{n-1}(s)\|_\infty \, ds \right|$$

$$\leq \left| \int_{t_0}^t L \cdot ML^{n-1} \frac{|s-t_0|^n}{n!} \, ds \right| = ML^n \frac{|t-t_0|^{n+1}}{(n+1)!},$$

establishing Eq. (B.4) for all $n \in \mathbb{N}$.

It follows from Weierstrass's criterion (see Appendix A) that the sequence $\{x_n\}_{n \in \mathbb{N}}$ converges uniformly on the interval $[t_0 - \alpha, t_0 + \alpha]$ to some limit x and that x itself is a continuous function with values in $D_\beta(x_0)$. Since

$$\|f(s, x_n(s)) - f(s, x(s))\|_\infty \leq L \|x_n(t) - x(t)\|_\infty,$$

also $f(t, x_n(t))$ converges to $f(t, x(t))$ uniformly on $[t_0 - \alpha, t_0 + \alpha]$ and therefore

$$x_0 + \int_{t_0}^t f(s, x_n(s)) ds \to x_0 + \int_{t_0}^t f(s, x(s)) ds$$

uniformly on $[t_0 - \alpha, t_0 + \alpha]$. In the limit $n \to \infty$, we therefore have that

$$x(t) = x_0 + \int_{t_0}^t f(s, x(s)) ds,$$

and thus x is a solution to Eq. (B.3) on the interval $[t_0 - \alpha, t_0 + \alpha]$.   □

This lemma immediately implies the existence of a local solution: given (t_0, x_0), simply take any compact set $K(\varepsilon, \beta) = [t_0 - \varepsilon, t_0 + \varepsilon] \times D_\beta(x_0)$ that lies in G and let M be the maximum of the continuous function $\|f(t, x)\|_\infty$ over the compact set $K(\varepsilon, \beta)$. Then choose $\alpha = \min\{\varepsilon, \frac{\beta}{M}\}$, and the condition $\alpha M \leq \beta$ is automatically satisfied. Note that this lemma actually establishes a generally useful relation between the size of the interval over which a solution is guaranteed to exist and the size of the norm of the dynamics.

We next show that this solution is unique in the set K. Suppose y is another solution to Eq. (B.3) defined on some interval $I \subset [t_0 - \alpha, t_0 + \alpha]$ and with values in $D_\beta(x_0)$ for all $t \in I$. Define $\Delta(t) = \|x(t) - y(t)\|_\infty$, so that $\Delta(t_0) = 0$. Since the graphs of both x and y lie in G, as above we have for all $t \in I$ that

$$\Delta(t) = \left\| \int_{t_0}^t f(s, x(s)) - f(s, y(s)) ds \right\|_\infty$$

$$\leq \left| \int_{t_0}^t \|f(s, x(s)) - f(s, y(s))\|_\infty \, ds \right|$$

$$\leq \left| \int_{t_0}^t L \|x(s) - y(s)\|_\infty \, ds \right| = L \left| \int_{t_0}^t \Delta(s) ds \right|.$$

But by the Gronwall–Bellman inequality below, this relation implies that Δ must be identically zero on the interval I, and this in turn implies that $I = [t_0 - \alpha, t_0 + \alpha]$.

Proposition B.1.1 (Gronwall–Bellman inequality). *Suppose λ and μ are continuous functions on the compact interval $[a,b]$ and μ is nonnegative. If Δ is a continuous function that satisfies*

$$\Delta(t) \le \lambda(t) + \int_a^t \mu(s)\Delta(s)ds \qquad \text{for all} \quad t \in [a,b], \tag{B.5}$$

then

$$\Delta(t) \le \lambda(t) + \int_a^t \lambda(s)\mu(s)\exp\left(\int_s^t \mu(r)dr\right)ds \qquad \text{for all} \quad t \in [a,b]. \tag{B.6}$$

Inequality Eq. (B.6) gives an a priori estimate for Δ in terms of the functions λ and μ. In our application, we have $\lambda \equiv 0$ and thus $\Delta(t) \le 0$ for all $t \ge t_0$, which gives $\Delta(t) \equiv 0$ on this interval. A version of the Gronwall–Bellman inequality that reverses the direction of time and fixes $\Delta(b) = 0$ (which we leave for the reader to formulate), then gives $\Delta(t) \equiv 0$ on I as claimed and establishes the uniqueness of the solution in K.

Proof. Set

$$\varsigma(t) = \int_a^t \mu(s)\Delta(s)ds \qquad \text{and} \qquad v(t) = \lambda(t) + \varsigma(t) - \Delta(t) \ge 0.$$

Then ς satisfies the first-order ordinary differential equation

$$\dot{\varsigma} = \mu\Delta = \mu\varsigma + \mu(\lambda - v)$$

with initial condition $\varsigma(a) = 0$. Hence

$$\varsigma(t) = \int_a^t \exp\left(\int_s^t \mu(r)dr\right)\mu(s)[\lambda(s) - v(s)]ds$$

$$\le \int_a^t \exp\left(\int_s^t \mu(r)dr\right)\mu(s)\lambda(s)ds.$$

Substituting this inequality into Eq. (B.5) gives the desired inequality. $\square$

Corollary B.1.1. *If $\lambda(t)$ is constant, $\lambda(t) \equiv \lambda$, then*

$$\Delta(t) \le \lambda \exp\left(\int_a^t \mu(s)ds\right) \qquad \text{for all} \quad t \in [a,b]. \tag{B.7}$$

Proof. In this case

$$\int_a^t \exp\left(\int_s^t \mu(r)dr\right)\mu(s)ds = -\int_a^t \frac{d}{ds}\left[\exp\left(\int_s^t \mu(r)dr\right)\right]ds$$

$$= -\exp\left(\int_s^t \mu(r)dr\right)\bigg|_a^t = \exp\left(\int_a^t \mu(r)dr\right) - 1,$$

which gives the simplified inequality.   □

Corollary B.1.2. *If both λ and μ are constant, $\lambda(t) \equiv \lambda$ and $\mu(t) \equiv \mu \geq 0$, then $\Delta(t) \leq \lambda \exp(\mu(t-a))$.*   ■

It remains to show that the solution $x = x(t;t_0,x_0)$ exists and lies in the domain G over a maximal open interval $(t_-(t_0,x_0),t_+(t_0,x_0))$. If x is a solution to the initial value problem (B.3) in G defined over some interval I, and if y is another solution to the initial value problem (B.3) in G defined over some interval J, then it follows from the local uniqueness of solutions just established that $x(t) = y(t)$ for all $t \in I \cap J$. This allows us to define a solution z to (B.3) on the interval $I \cup J$ by concatenating the two solutions as

$$z(t) = \begin{cases} x(t) & \text{if } t \in I, \\ y(t) & \text{if } t \in J. \end{cases}$$

In this way, we can construct a solution on a maximal interval $(t_-(t_0,x_0),t_+(t_0,x_0))$. This interval is open, since otherwise, the point $(t_\pm(t_0,x_0),x(t_\pm(t_0,x_0)))$ would lie in G and thus, by Lemma B.1.3, the solution could be extended further. This concludes the proof of Theorem B.1.2.   □

We close this section with a useful criterion for when the solution to a differential equation exists on the full interval.

Definition B.1.3 (Linearly bounded). Let $G = [a,b] \times \mathbb{R}^n$; we say that a function $f : (a,b) \times \mathbb{R}^n \to \mathbb{R}^n$, $(t,x) \mapsto f(t,x)$, is linearly bounded over the interval $[a,b]$ if there exist constants μ and φ such that

$$\|f(t,x)\| \leq \mu \|x\| + \varphi \qquad \text{for all} \quad t \in [a,b].$$

Proposition B.1.2. *Let $G = [a,b] \times \mathbb{R}^n$ and suppose $f : [a,b] \times \mathbb{R}^n \to \mathbb{R}^n$, $(t,x) \mapsto f(t,x)$, is continuous, Lipschitz continuous in x, and linearly bounded over $[a,b]$. Then for every $(t_0,x_0) \in G$, the solution $x = x(t;t_0,x_0)$ exists on the full interval $[a,b]$.*

Proof. Since all norms on $\mathbb{R}^n$ are equivalent, we may use the ∞-norm. Given $(t_0,x_0) \in G$, let $\alpha = \frac{1}{2\mu} > 0$ and $\beta = \|x_0\| + \frac{\varphi}{\mu}$. Without loss of generality, we assume that the full interval $[t_0 - \alpha,t_0 + \alpha]$ lies in $[a,b]$. (Otherwise limit the interval at the left or right endpoint, respectively.) Thus $K = [t_0 - \alpha,t_0 + \alpha] \times D_\beta(x_0) \subset G$ and for all $(t,x) \in K$ we have that

$$M = \max_{(t,x)\in K} \|f(t,x)\| \le \mu \|x\| + \varphi \le \mu (\|x_0\| + \beta) + \varphi.$$

Hence

$$\alpha M \le \frac{1}{2\mu} [\mu (\|x_0\| + \beta) + \varphi] = \frac{1}{2\mu} \left[\mu \left(2\|x_0\| + \frac{\varphi}{\mu} \right) + \varphi \right] = \beta.$$

By Lemma B.1.3, the solution x to the initial value problem with initial condition $x(t_0) = x_0$ therefore exists on the full interval $[t_0 - \alpha, t_0 + \alpha]$. Thus, regardless of where the actual initial condition $(t_0, x_0) \in G$ lies, the solution can always be continued for an interval of length $\frac{1}{\mu} > 0$ to the left and the right until the boundary points a and b are reached. Thus solutions can be continued onto the full interval $[a, b]$. $\qquad\square$

The following important special cases are immediate corollaries:

Corollary B.1.3. *Let* $G = [a, b] \times \mathbb{R}^n$ *and suppose* $f : [a, b] \times \mathbb{R}^n \to \mathbb{R}^n$, $(t, x) \mapsto f(t, x) = A(t)x + b(t)$, *where* A *and* b *are continuous functions on* $[a, b]$. *Then for any initial condition* $(t_0, x_0) \in G$ *the solution* $x = x(t; t_0, x_0)$ *exists on the full interval* $[a, b]$. $\qquad\blacksquare$

Corollary B.1.4. *Let* $G = \mathbb{R}^n$ *and suppose* $f : \mathbb{R}^n \to \mathbb{R}^n$, $x \mapsto f(x)$ *is bounded. Then the solution* $x = x(t; x_0)$ *of the time-invariant system* $\dot{x} = f(x)$ *exists for all times* $t \in \mathbb{R}$. $\qquad\blacksquare$

B.2 Dependence of Solutions on Initial Conditions and Parameters

Let G be a domain in $\mathbb{R} \times \mathbb{R}^n \times \mathbb{R}^k$ and let $f : G \to \mathbb{R}^n$, $(t, x, p) \mapsto f(t, x; p)$, be a time-varying vector field that depends on a k-dimensional parameter p. Throughout this section, we assume that f is continuous and Lipschitz continuous in (x, p). It thus follows from the general theorem on existence of solutions that for each (t_0, x_0, p_0), the initial value problem

$$\dot{x} = f(t, x; p_0), \qquad x(t_0) = x_0, \qquad (t, x, p) \in G, \tag{B.8}$$

has a unique solution $x = x(t; t_0, x_0, p_0)$ defined on a maximal intervalv

$$I(t_0, x_0, p_0) = (t_\pm(t_0, x_0, p_0), x(t_\pm(t_0, x_0, p_0))).$$

(Simply add the trivial differential equation $\dot{p} = 0$ to the system.) The set

$$D = \{(t; t_0, x_0, p_0) : t \in I(t_0, x_0, p_0), (t_0, x_0, p_0) \in G\} \tag{B.9}$$

is called the *domain of the general solution* $x = x(t; t_0, x_0, p_0)$.

Theorem B.2.1. *Let G be a domain in $\mathbb{R} \times \mathbb{R}^n \times \mathbb{R}^k$ and suppose the parameter-dependent, time-varying vector field $f : G \to \mathbb{R}^n$, $(t, x, p) \mapsto f(t, x; p)$, is continuous and Lipschitz continuous in (x, p) on G. Then D is open, and the general solution $x = x(t; t_0, x_0, p_0)$ is a Lipschitz continuous function of all variables on D.*

Proof. Without loss of generality, we again use the supremum norm throughout our calculations. The solution x is a continuously differentiable function in t, and thus, by Lemma B.1.2, it is Lipschitz continuous in t. We therefore consider only the remaining variables $(t_0, x_0, p_0) \in G$. Fix (t_0, x_0, p_0) and denote the corresponding solution by $x(t)$; let (s_0, y_0, q_0) be another point from G and denote the corresponding solution by $y(t)$. Without loss of generality, consider a time $t > t_0$ and let $[a, b]$ be a compact interval contained in $I(t_0, x_0, p_0)$ such that $[t_0, t] \subset (a, b) \subset [a, b]$. Let $D_\varepsilon(p_0) = \{p \in \mathbb{R}^k : \|p - p_0\| \leq \varepsilon\}$ be the closed disk and let U_ε be the tubular neighborhood

$$U_\varepsilon = \{(t, x) \in [a, b] \times \mathbb{R}^n : \|x - x(t)\| \leq \varepsilon\}.$$

For ε sufficiently small, the compact set $U_\varepsilon \times D_\varepsilon(p_0)$ lies in G, and the function f satisfies a global Lipschitz condition in (x, p) on $U_\varepsilon \times D_\varepsilon(p_0)$ with some Lipschitz constant L,

$$\|f(t, x, p) - f(t, y, q)\| \leq L(\|x - y\| + \|p - q\|).$$

Also, let M be an upper bound for $\|f(t, x, p)\|$ on this set $U_\varepsilon \times D_\varepsilon(p_0)$. We show that there exists a positive $\delta = \delta(\varepsilon)$ such that whenever

$$\|(s_0, y_0, q_0) - (t_0, x_0, p_0)\| < \delta,$$

then the solution $y = y(t)$ exists on the interval $[a, b]$, lies in U_ε, and satisfies a Lipschitz condition in all variables. This will prove the theorem.

For δ small enough, we have that

$$\|y_0 - x(s_0)\| = \left\| y_0 - x_0 + \int_{t_0}^{s_0} f(r, x(r), p_0) dr \right\| \leq \|y_0 - x_0\| + M|s_0 - t_0| < \varepsilon,$$

and thus there exists a closed interval $J = [\alpha, \beta]$ that contains s_0 in its interior for which the graph of the solution $y = y(t; s_0, y_0, q_0)$ lies in the neighborhood U_ε of the graph of x. Using the Gronwall–Bellman inequality, it follows that this interval actually is $[a, b]$, provided δ is small enough. This is because for all $t \in [\alpha, \beta]$ we have that

$$\|y(t) - x(t)\| = \left\| y_0 + \int_{s_0}^t f(r,y(r),q_0)dr - x_0 + \int_{t_0}^t f(r,x(r),p_0)dr \right\|$$

$$= \left\| y_0 - x(s_0) + \int_{s_0}^t f(r,y(r),q_0) - f(r,x(r),p_0)dr \right\|$$

$$\leq \|y_0 - x_0\| + M\,|s_0 - t_0| + \left| \int_{s_0}^t \|f(r,y(r),q_0) - f(r,x(r),p_0)\|\,dr \right|$$

$$\leq \|y_0 - x_0\| + M\,|s_0 - t_0| + L\left| \int_{s_0}^t \|y(r) - x(r)\| + \|q_0 - p_0\|\,dr \right|$$

$$= \|y_0 - x_0\| + M\,|s_0 - t_0| + L(b-a)\,\|q_0 - p_0\|$$

$$+ L\left| \int_{s_0}^t \|y(r) - x(r)\|\,dr \right|.$$

The absolute values in this derivation are included, since possibly $t < s_0$. But then it follows from the Gronwall–Bellman inequality (Corollary B.1.2) that

$$\|y(t) - x(t)\| \leq (\|y_0 - x_0\| + M\,|s_0 - t_0| + L(b-a)\,\|q_0 - p_0\|) \exp(L(b-a)).$$

Thus, the difference of the solutions satisfies a Lipschitz condition in the variables (t_0,x_0,p_0) as long as $t \in [\alpha,\beta]$. By choosing δ small enough, we can guarantee that the right-hand side in the last inequality is arbitrarily small as well, and thus we can always get that $\|y(t) - x(t)\| \leq \varepsilon$ on the full interval $[a,b]$. Thus y exists for all $t \in [a,b]$ and its graph lies in U_ε.

Finally, the same computation as was carried out above, but for two arbitrary solutions, verifies that the general solution $x = x(t;t_0,x_0,p_0)$ satisfies a Lipschitz condition in the variables (t_0,x_0,p_0) on the set $[a,b] \times U_\varepsilon \times D_\varepsilon(p_0)$ and thus is Lipschitz continuous in these variables. $\square$

Not only do Lipschitz properties propagate from the dynamics to the solutions, but so do differentiability properties. For the case of a C^1-vector field, this leads to the so-called *variational* or *sensitivity equations* that are of fundamental importance and are an essential tool in the calculus of variations and optimal control theory.

Theorem B.2.2. *Let G be a domain in $\mathbb{R} \times \mathbb{R}^n \times \mathbb{R}^k$ and suppose the parameter-dependent, time-varying vector field $f : G \to \mathbb{R}^n$, $(t,x,p) \mapsto f(t,x;p)$, is continuous and continuously differentiable in (x,p) on G. Then the general solution $x = x(t;t_0,x_0,p_0)$ to the initial value problem (B.8) is continuously differentiable in all variables, and the partial derivatives can be computed as solutions to the linear differential equations that are obtained by formally interchanging differentiation with respect to the variables (t_0,x_0,p_0) and t. Thus,*

1. *the partial derivative with respect to the initial condition x_0,*

$$Y(t;t_0,x_0,p_0) = \frac{\partial x}{\partial x_0}(t;t_0,x_0,p_0) \in \mathbb{R}^{n \times n},$$

is the so-called fundamental solution of the variational equation, i.e., it satisfies

$$\dot{Y} = \frac{\partial f}{\partial x}(t,x(t;t_0,x_0,p_0),p_0)Y, \qquad Y(t_0) = \mathrm{Id}; \qquad (B.10)$$

2. *the partial derivative with respect to the initial time t_0,*

$$y(t;t_0,x_0,p_0) = \frac{\partial x}{\partial t_0}(t;t_0,x_0,p_0) \in \mathbb{R}^n,$$

is the solution of the homogeneous initial value problem

$$\dot{y} = \frac{\partial f}{\partial x}(t,x(t;t_0,x_0,p_0),p_0)y, \qquad y(t_0) = -f(t_0,x_0,p_0); \qquad (B.11)$$

3. *the partial derivative with respect to the parameter p_0,*

$$S(t;t_0,x_0,p_0) = \frac{\partial x}{\partial p_0}(t;t_0,x_0,p_0) \in \mathbb{R}^{n \times k},$$

is the solution of the inhomogeneous initial value problem

$$\dot{S} = \frac{\partial f}{\partial x}(t,x(t;t_0,x_0,p_0),p_0)S + \frac{\partial f}{\partial p}(t,x(t;t_0,x_0,p_0),p_0), \qquad S(t_0) = 0; \quad (B.12)$$

Note that all these differential equations are formally obtained from the identities

$$\dot{x}(t;t_0,x_0,p) = f(t,x(t;t_0,x_0,p),p), \qquad x(t_0;t_0,x_0) \equiv x_0,$$

by differentiating with respect to the variables t_0, x_0, and p, respectively, and interchanging the order of derivatives with t. While this is a simple way of remembering these equations, the proof nevertheless needs to justify this procedure.

Proof. The proofs for these cases are analogous, and we consider only

$$Y(t;t_0,x_0,p_0) = \frac{\partial x}{\partial x_0}(t;t_0,x_0,p_0).$$

Equation (B.10) is a homogeneous linear matrix differential equation, and it thus follows from Corollary B.1.3 that it has a well-defined solution $Y(t;t_0,x_0,p_0)$ over the full maximal interval of definition $I(t_0,x_0,p_0)$ of the solution $x = x(t;t_0,x_0,p_0)$.

It suffices to consider directional derivatives, and we shall show that for any vector $h \in \mathbb{R}^n$, we have that

$$\lim_{\varepsilon \to 0} \frac{1}{\varepsilon} (x(t;t_0,x_0 + \varepsilon h, p_0) - x(t;t_0,x_0,p_0)) = Y(t;t_0,x_0,p_0)h.$$

Let

$$\Delta(t,\varepsilon) = \frac{1}{\varepsilon} (x(t;t_0,x_0 + \varepsilon h, p_0) - x(t;t_0,x_0,p_0)) - Y(t;t_0,x_0,p_0)h,$$

dropping from now on the variables t_0 and p_0, which are held constant in the notation. Let $[a,b]$ be a compact subinterval of $I(x_0)$ and let ε_0 be a sufficiently small positive number. By Theorem B.2.1, the general solution $x = x(t;x_0)$ is Lipschitz continuous, and it thus follows that $\Delta(t,\varepsilon)$ is continuous and bounded on the set

$$\{(t,\varepsilon) : a \leq t \leq b, 0 < \varepsilon \leq \varepsilon_0\}.$$

As a function of t, $\Delta(\cdot,\varepsilon)$ is continuously differentiable, and it follows from Taylor's theorem that it satisfies a linear differential equation of the form

$$\dot{\Delta}(t,\varepsilon) = \frac{1}{\varepsilon} (f(t;x(t,x_0 + \varepsilon h)) - f(t;x(t,x_0))) - \frac{\partial f}{\partial x}(t,x(t;x_0))Y(t;x_0)h$$

$$= \frac{\partial f}{\partial x}(t,x(t;x_0))\Delta(t,\varepsilon) + r(t,\varepsilon),$$

where the remainder term $r(t,\varepsilon)$ satisfies $\lim_{\varepsilon \to 0} r(t,\varepsilon) = 0$, uniformly over the interval $[a,b]$. Furthermore,

$$\Delta(t_0,\varepsilon) = \frac{1}{\varepsilon} (x_0 + \varepsilon h - x_0) - h = 0.$$

Since $Y(\cdot;x_0)$ is the solution of the initial value problem (B.10), it follows from the variation of constants formula for linear differential equations that

$$\Delta(t,\varepsilon) = Y(t;x_0) \left(0 + \int_{t_0}^{t} Y(s;x_0)^{-1} r(s,\varepsilon)ds \right),$$

and thus

$$\lim_{\varepsilon \to 0} \Delta(t,\varepsilon) = Y(t;x_0) \int_{t_0}^{t} Y(s;x_0)^{-1} \left(\lim_{\varepsilon \to 0} r(s,\varepsilon) \right) ds = 0.$$

Hence, for every vector h, the directional derivative of $x_0 \to x(t,x_0)$ in the direction of h is given by $Y(t;x_0)h$. This proves that $x(t,x_0)$ is differentiable in x_0 with derivative $Y(t;x_0)$. $\qquad \square$

Appendix C
An Introduction to Differentiable Manifolds

We give a brief introduction to manifolds and some of the language of differential geometry that provides the framework for nonlinear control theory. After introducing the fundamental concepts and giving some examples, we discuss those concepts and tools that we need in this text, especially in Chap. 4 in the proof of the maximum principle and Chap. 7 in the construction of small-time reachable sets. We forgo generality for a more intuitive approach that makes the connection with advanced calculus. This indeed is the context within which we almost exclusively deal with manifolds in this text. For more general expositions, we refer the reader to any textbook on differential geometry, e.g., [38, 102, 256]. We highly recommend Milnor's notes [184] and Boothby's text [50] to the novice on the subject.

C.1 Embedded Submanifolds of $\mathbb{R}^k$

Historically, the concept of a manifold was abstracted from the notions of curves and surfaces in $\mathbb{R}^3$, and many of the general ideas and concepts of differential geometry are best visualized in this context. We therefore start with embedded submanifolds in $\mathbb{R}^k$.

Definition C.1.1 (d-dimensional C^r-embedded submanifold of $\mathbb{R}^k$). A subset M of $\mathbb{R}^k$ is said to be a d-dimensional C^r-embedded submanifold of $\mathbb{R}^k$ if for every point $p \in M$ there exist a neighborhood V of p in $\mathbb{R}^k$ and an r-times continuously differentiable mapping $F : V \to \mathbb{R}^\ell$ into $\mathbb{R}^\ell$ for some $\ell \in \mathbb{N}$ that is of rank $k - d$ everywhere on V and is such that

$$M \cap V = F^{-1}(0) = \{x \in V \subset \mathbb{R}^k : F(x) = 0\}.$$

Recall that a mapping F is of rank r at a point p if the Jacobian matrix $DF(p)$ has rank r. Intuitively, embedded submanifolds are curves, surfaces, etc. in $\mathbb{R}^k$ that locally have a rather simple structure. In order to see this, reorder the coordinates x_i so that the partial derivatives of F with respect to the variables $x_1, \ldots, x_{k-d}$ are

H. Schättler and U. Ledzewicz, *Geometric Optimal Control: Theory, Methods and Examples*, Interdisciplinary Applied Mathematics 38, DOI 10.1007/978-1-4614-3834-2, © Springer Science+Business Media, LLC 2012

Fig. C.1 Tangent space to an
embedded submanifold

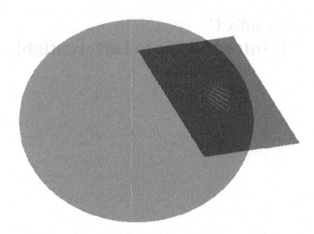

linearly independent at p. If we partition x into vectors $y = (x_1, \ldots, x_{k-d})^T$ and
$m = (x_{k-d+1}, \ldots, x_k)^T$ and write $p = (p_y, p_m)^T$, then it follows from the implicit
function theorem that there exist a neighborhood $W \subset V$ of p of the form $W =
B_\varepsilon^{k-d}(p_y) \times B_\delta^d(p_m)$ and a C^r-function $f : B_\delta^d(p_m) \to B_\varepsilon^{k-d}(p_y)$, $m \mapsto f(m)$, such that
for all $x = (y, m)^T \in W$ we have that $F(x) = 0$ if and only if $y = f(m)$. Hence

$$M \cap W = \{(y, m) \in W \subset \mathbb{R}^k : y = f(m)\}.$$

If we define a new mapping $\tilde{F} : B_\delta^d(p_m) \to W$, $m \mapsto (m, f(m))$, then $\tilde{F}$ is one-to-one,
and it represents the embedded manifold M as the graph of a C^r-function near p.

In control theory, we consider these manifolds, even when they are embedded
in $\mathbb{R}^k$, as stand-alone objects that become models for the state space of a nonlinear
system. This will be appropriate only if the evolution of the system $t \mapsto x(t)$ occurs
on M. This clearly puts restrictions on the tangent vector to this curve. The set of all
directions that are tangent to M at a point p therefore is of intrinsic importance, and
this leads to the notion of the tangent space to M at p (see Fig. C.1).

**Definition C.1.2 (Tangent space to a d-dimensional C^r-embedded submanifold
of $\mathbb{R}^k$ at a point).** Given a d-dimensional C^r-embedded submanifold M of $\mathbb{R}^k$ and a
point $p \in M$, the tangent space to M at p, T_pM, is the set of all vectors $v = \dot{\gamma}(0) \in \mathbb{R}^k$
that are the derivatives at time $t = 0$ of differentiable curves γ, $\gamma : (-\varepsilon, \varepsilon) \to M$,
$t \mapsto \gamma(t)$, that satisfy $\gamma(0) = p$ and lie in M.

Proposition C.1.1. *Given a d-dimensional embedded submanifold M of $\mathbb{R}^k$ and a
point $p \in M$, let $F : V \to \mathbb{R}^\ell$ be a function that represents M near p (i.e., F has
constant rank $k - d$ and $M \cap V = F^{-1}(0)$). Then the tangent space to M at p, T_pM,
is the d-dimensional subspace given by*

$$T_pM = \{v \in \mathbb{R}^k : DF(p)v = 0\} = \ker DF(p).$$

*This space does not depend on the particular function F chosen to represent M
near p.*

Proof. We first note that the definition does not depend on the specific function F that has been chosen to represent M near p. If F and $\tilde{F}$ are two such functions, then there exists an open neighborhood V of p in $\mathbb{R}^k$ such that

$$\{x \in V : F(x) = 0\} = M \cap V = \{x \in V : \tilde{F}(x) = 0\}.$$

Hence, if $\gamma : (-\varepsilon, \varepsilon) \to M, t \mapsto \gamma(t)$, $\gamma(0) = p$, is any curve that lies in $M \cap V$, then both $F(\gamma(t))$ and $\tilde{F}(\gamma(t))$ vanish identically near $t = 0$ and thus

$$\frac{d}{dt}_{|t=0} F(\gamma(t)) = DF(p)\dot{\gamma}(0) = 0 = \frac{d}{dt}_{|t=0} \tilde{F}(\gamma(t)) = D\tilde{F}(p)\dot{\gamma}(0).$$

In particular, the tangent space T_pM is contained in the kernel of any differentiable function F that locally represents M. Since both of these subspaces are d-dimensional, it suffices to verify that T_pM contains a subspace of dimension d. In fact, it is not even clear from our definition that T_pM is a subspace to begin with. But this easily follows if we change from the coordinates x in the ambient space $\mathbb{R}^k$ to the set of coordinates $x = (y, m)^T$ introduced above that give a faithful representation of the set M. The new mapping $\tilde{F} : B_\delta^d(p_m) \to W, m \mapsto (m, f(m))$, represents the embedded manifold M near p as the graph of a C^r-function defined on $\mathbb{R}^d$. For any vector $v \in \mathbb{R}^d$, the image of some small straight line segment $\ell_v : (-\varepsilon, \varepsilon) \to M, t \mapsto \ell_v(t) = p_m + tv$, under $\tilde{F}$ therefore is a curve in M whose tangent vector lies in T_pM. But this image is a d-dimensional subspace, and thus we have $T_pM = \ker DF(p)$. $\qquad\qquad\square$

Examples:

(1) The n-sphere

$$S^n = \{x = (x_0, \ldots, x_n)^T \in \mathbb{R}^{n+1} : F(x) = x_0^2 + \cdots + x_n^2 - 1 = 0\}$$

is an n-dimensional real-analytic embedded submanifold of $\mathbb{R}^{n+1}$. The gradient of the function F is given by $\nabla F(x) = 2x^T$ and is everywhere nonzero on S^n. The tangent space T_pS^n is given by

$$T_pS^n = \{v \in \mathbb{R}^{n+1} : \langle v, p \rangle = 0\},$$

where $\langle \cdot, \cdot \rangle$ denotes the standard inner product on $\mathbb{R}^{n+1}$. Thus these are all vectors perpendicular, or normal, to p, the ordinary tangent plane to S^n at p.

(2) The n-dimensional torus T^n is the direct product of n circles S^1,

$$T^n = S^1 \times \cdots \times S^1,$$

Fig. C.2 The two-dimensional sphere S^2 and torus T^2

and can be embedded into $\mathbb{R}^{2n}$ by means of the equation

$$T^n = \{x \in \mathbb{R}^{2n} : F(x) = (x_1^2 + x_2^2 - 1, \ldots, x_{2n-1}^2 + x_{2n}^2 - 1) = 0\}.$$

The Jacobian $DF(x)$,

$$DF(x) = \begin{pmatrix} 2x_1 & 2x_2 & 0 & 0 & \cdots\cdots & 0 & 0 \\ 0 & 0 & 2x_3 & 2x_4 & \cdots\cdots & 0 & 0 \\ \vdots & \vdots & \vdots & \vdots & \ddots\ddots & \vdots & \vdots \\ 0 & 0 & 0 & 0 & \cdots\cdots & 2x_{2n-1} & 2x_{2n} \end{pmatrix},$$

is of rank n everywhere on the torus, and thus T^n is an n-dimensional real-analytic embedded submanifold of $\mathbb{R}^{2n}$. Figure C.2 shows the more common embedding of the 2-torus into $\mathbb{R}^3$ given by

$$F : S^1 \times S^1 \to \mathbb{R}^3, \quad (\theta, \phi) \mapsto ((2 + \cos\theta)\cos\phi, (2 + \cos\theta)\sin\phi, \sin\theta).$$

The n-dimensional torus T^n also is a *group*. A group G is an algebraic structure consisting of elements g together with an associative operation $\circ : G \times G \to G$, $(a, b) \mapsto a \circ b$, called "multiplication" for which there exists a neutral element e, $e \circ a = a = a \circ e$ for all $a \in G$, and such that every element $a \in G$ is invertible, i.e., there exists a (unique) element in the group, called the inverse of a and denoted by a^{-1}, such that $a \circ a^{-1} = e = a^{-1} \circ a$. The circle S^1 can be identified with the complex numbers of absolute value 1, $S^1 \simeq \{z \in \mathbb{C} : |z| = 1\}$, and ordinary multiplication becomes the operation $\circ$ in the group with the inverse given by $z^{-1} = \frac{1}{z}$. Both multiplication and taking the inverse are defined in terms of analytic operations. More generally, real-analytic manifolds that are groups and for which the operation $G \times G \to G$, $(a, b) \mapsto a \circ b^{-1}$, also is real-analytic, are called *Lie groups*.

(3) Common examples of Lie groups are given by the so-called matrix groups. Consider a matrix $A \in \mathbb{R}^{n \times n}$ as an element of $\mathbb{R}^{n^2}$ with the matrix entries as the coordinates. It is trivial from the definition that open subsets of $\mathbb{R}^k$ are k-dimensional embedded submanifolds of $\mathbb{R}^k$ (simply take F as the identity map). Thus the set of nonsingular matrices,

$$\mathrm{GL}(n) = \{A \in \mathbb{R}^{n \times n} : \det(A) \neq 0\},$$

is an n^2-dimensional analytic submanifold of $\mathbb{R}^{n^2}$, called the *general linear group*. (Here the group operations are matrix multiplication and taking the inverse. The representation of the inverse of a matrix in terms of its classical adjoint shows that these operations are real-analytic functions of the matrix entries.) Since $\det(AB) = \det(A)\det(B)$, the set

$$SL(n) = \{A \in \mathbb{R}^{n \times n} : \det(A) = 1\}$$

is a subgroup, the *special linear group*, and it also is an n^2-dimensional analytic submanifold of $\mathbb{R}^{n^2}$: if we take $F(A) = \det(A) - 1$, then the partial derivative of F with respect to the element a_{ij} is given by its (i,j) minor,

$$\frac{\partial F}{\partial a_{ij}}(A) = \sum_{\pi \in S_n : \pi(i)=j} \mathrm{sgn}\,\pi \cdot a_{1,\pi(1)} \cdots a_{i-1,\pi(i-1)} a_{i+1,\pi(i+1)} \cdots a_{n,\pi(n)},$$

and not all of these minors vanish if A is nonsingular.

(4) A more interesting example of a subgroup is the *orthogonal group* $O(n)$ of all orthogonal matrices,

$$O(n) = \{A \in \mathbb{R}^{n \times n} : AA^T = \mathrm{Id}\}.$$

In this case, $F : \mathbb{R}^{n^2} \to \mathbb{R}^{n^2}$, $F(A) = AA^T - \mathrm{Id}$. For any other matrix $H \in \mathbb{R}^{n \times n}$, we have that

$$F(A + H) = AA^T - \mathrm{Id} + (AH^T + HA^T) + HH^T.$$

Thus H lies in the kernel of $DF(A)$ if and only if $AH^T + HA^T = 0$, i.e., if and only if AH^T is skew-symmetric. Writing $S = HA^T$, we therefore have that

$$\ker DF(A) = \{H \in \mathbb{R}^{n \times n} : H = SA, \quad \text{where } S + S^T = 0\}.$$

The vector space of skew-symmetric $n \times n$ matrices has dimension $\frac{1}{2}n(n-1)$ (all diagonal entries must be 0 and $s_{ji} = -s_{ij}$), and thus the rank of $DF(A)$ is constant, given by $\frac{1}{2}n(n-1)$. Hence the *orthogonal group* $O(n)$ is a $\frac{1}{2}n(n-1)$-dimensional embedded analytic submanifold of $\mathbb{R}^{n^2}$, and its tangent space can be identified with the vector space $\mathfrak{s}(n)$ of all $n \times n$ skew-symmetric matrices. This is the tangent space at the identity, $\mathrm{Id} \in O(n)$, and the tangent space $T_A O(n)$ can be obtained simply by multiplying these matrices by A. Note that the commutator of skew-symmetric matrices again is skew-symmetric:

$$(RS - SR)^T = S^T R^T - R^T S^T = (-S)(-R) - (-R)(-S) = -(RS - SR).$$

Thus, $\mathfrak{s}(n)$ is a Lie algebra of matrices. A *Lie algebra* over $\mathbb{R}$ is a real vector space $\mathfrak{G}$ together with a bilinear operator $[\cdot, \cdot] : \mathfrak{G} \times \mathfrak{G} \to \mathfrak{G}$ such that for all X, Y, and $Z \in \mathfrak{G}$ we have that $[X, Y] = -[Y, X]$ and

$$[X, [Y, Z]] + [Y, [Z, X]] + [Z, [X, Y]] = 0.$$

The last identity is called the Jacobi-identity. It can be rewritten in the form

$$[X, [Y, Z]] = [[X, Y], Z] + [Y, [X, Z]],$$

and if we think of X as "differentiating" the bracket $[Y, Z]$, then this simply demands that the product rule of differentiation be satisfied. It is true in general that the tangent space at the identity for arbitrary Lie groups is a Lie algebra.

An embedded d-dimensional manifold M is connected if it is not possible to write M as the disjoint union of two embedded d-dimensional manifolds. For example, $S^0 = \{x = \mathbb{R} : F(x) = x^2 - 1 = 0\}$ consists of the two points $\{-1\}$ and $\{+1\}$, which are zero-dimensional manifolds, and thus is not connected. It is not too difficult to show that if a manifold M is connected, then it is also path connected in the sense that for any two points p and q in M, there exists a C^r-curve $\gamma : [0, 1] \to M$ such that $\gamma(0) = p$ and $\gamma(1) = q$ (e.g., see [38, Proposition 1.5.2]). The orthogonal group also is not connected: The determinant of orthogonal matrices can be $+1$ or -1. Along any curve γ that lies in $O(n)$ the determinant is continuous, and thus it must be constant. Hence there exist two connected components. The one defined by $\det(A) = +1$ is a subgroup, and called the special orthogonal group $SO(n)$,

$$SO(n) = \{A \in O(n) : \det(A) = 1\}.$$

For example, $SO(2)$ consists of all rotation matrices

$$A_\theta = \begin{pmatrix} \cos\theta & \sin\theta \\ -\sin\theta & \cos\theta \end{pmatrix}, \qquad \theta \in [0, 2\pi),$$

and it is clear that this is the same manifold as S^1, $SO(2) \simeq S^1$, by means of the identification $\theta \mapsto A_\theta$.

We close this section with a proof that the tangent space to an embedded submanifold is a Boltyansky approximating cone. (We refer the reader to Sect. 4.1

for the definition of a Boltyansky approximating cone.) The small difference to the definition of the tangent space is that a differentiable mapping is constructed that uniformly approximates an arbitrary finite collection of tangent vectors. It is intuitively clear that this can be done. Since this result is of somewhat fundamental importance in the development of necessary conditions for optimality in Sect. 4.2.4, we include its proof here. It simply combines the proof of the inverse function theorem (e.g., see [177, Chap. 7]) with the definition of the tangent space.

Proposition C.1.2. *Let $F : \mathbb{R}^n \to \mathbb{R}^m$, $m < n$, be a continuously differentiable mapping and let $M = \{x \in \mathbb{R}^n : F(x) = 0\}$. If the Jacobian matrix $DF(q)$ is of full rank m at a point $q \in A$, then the tangent space to M at q, $T_q M = \ker DF(q)$, is a Boltyansky approximating cone to M at q.*

Proof. Given a fixed finite collection of vectors $v_1, \ldots, v_k$ from $C = \ker DF(q)$, define a linear functional ℓ as $\ell : \mathbb{R}^k \to \mathbb{R}$, $z \mapsto \ell(z) = \sum_{i=1}^k z_i v_i$. For this case, there is no need to introduce the approximating vectors v_i' that are present in Definition 4.1.2, and we simply take $v_i' = v_i$. Furthermore, since C is a subspace, i.e., with $v \in C$ also $-v \in C$, the approximating map naturally needs to be defined not only on a cube Q_δ^k of nonnegative numbers, but on a full neighborhood $B_\delta^k(0)$ of zero,

$$\Xi : B_\delta^k(0) \to M, \qquad z \mapsto \Xi(z).$$

Thus, we need to show that there exist a $\delta > 0$ and a continuous function $r = r(z)$ defined on $B_\delta^k(0)$ that is of order $o(\|z\|)$ as $z \to 0$ such that $\Xi(z) = q + \ell(z) + r(z) \in M$, i.e.,

$$F(q + \ell(z) + r(z)) = 0.$$

This function r is computed by solving the equation $F(z) = 0$ near q using a quasi-Newton algorithm [228], and the superlinear convergence of the procedure implies that r is of the desired order. (If F is twice continuously differentiable, the Newton algorithm can be used, and then its quadratic convergence implies that the remainder $r = r(z)$ is actually of order $O(\|z\|^2)$ as $z \to 0$.) We provide the details.

Since the matrix $DF(q)$ is of full rank, there exists a matrix $B \in \mathbb{R}^{n \times m}$ such that $DF(q)B = \mathrm{Id}$, the identity matrix in $\mathbb{R}^{m \times m}$. For example, we may simply take the pseudoinverse $B = DF(q)^T \left(DF(q)DF(q)^T\right)^{-1}$. Let $K = \|B\|_2$ with the norm the lub-norm induced by the Euclidean norm and choose $\varepsilon > 0$ such that $\|DF(p) - DF(q)\| \leq \frac{1}{2K}$ whenever $p \in B_\varepsilon^n(q)$. The function $\varphi(z) = F(q + \ell(z))$ is continuously differentiable, and it follows from Taylor's theorem that $\varphi(z)$ is of order $o(\|z\|)$ as $z \to 0$. For since $F(q) = 0$ and $DF(q)\ell(z) = 0$, we have that

$$\varphi(z) = F(q + \ell(z)) = F(q) + DF(q)\ell(z) + o(\|z\|) = o(\|z\|).$$

Choose $\delta > 0$ such that both $\|\varphi(z)\| < \frac{\varepsilon}{4K}$ and $\|\ell(z)\| < \frac{\varepsilon}{2}$ hold for all $z \in B_\delta^k(0)$. Then, inductively define a sequence of continuous functions $x_i : B_\delta^k(0) \to B_\varepsilon^n(q)$ as follows: set $x_0(z) = q + \ell(z)$ and define

$$x_{i+1}(z) = x_i(z) - BF(x_i(z)) \qquad \text{and} \qquad y_i(z) = F(x_i(z)). \qquad \text{(C.1)}$$

Claim: the functions x_i all take values in $B_{\varepsilon}^n(q)$, and for $i \in \mathbb{N}$, the following bounds are satisfied in the Euclidean norm:

$$\|y_i(z)\| \le \left(\frac{1}{2}\right)^i \|\varphi(z)\| \quad \text{and} \quad \|x_{i+1}(z) - x_i(z)\| \le K \left(\frac{1}{2}\right)^i \|\varphi(z)\|.$$

The verification is by induction on i. By definition, $y_0(z) = F(q + \ell(z)) = \varphi(z)$, and thus the first inequality holds trivially. Furthermore,

$$\|x_1(z) - x_0(z)\| = \|BF(x_0(z))\| \le \|B\| \, \|F(q + \ell(z))\| = K \|\varphi(z)\|$$

and

$$\|x_1(z) - q\| \le \|x_1(z) - x_0(z)\| + \|\ell(z)\| \le K \|\varphi(z)\| + \|\ell(z)\| < \varepsilon.$$

Thus assume inductively that these inequalities hold for all indices less than or equal to i. In particular, we then have that

$$\|x_{i+1}(z) - q\| \le \|x_{i+1}(z) - x_i(z)\| + \|x_i(z) - x_{i-1}(z)\| + \cdots + \|x_1(z) - x_0(z)\| + \|\ell(z)\|$$

$$\le K \sum_{j=0}^{i} \left(\frac{1}{2}\right)^j \|\varphi(z)\| + \|\ell(z)\| \le \frac{\varepsilon}{4} \left(\sum_{j=0}^{i} \left(\frac{1}{2}\right)^j\right) + \frac{\varepsilon}{2} < \varepsilon,$$

and so $x_{i+1}(z) \in B_{\varepsilon}^n(q)$. By definition of the sequence, we have that

$$DF(q)(x_{i+1}(z) - x_i(z)) = -DF(q)BF(x_i(z)) = -F(x_i(z)) = -y_i(z),$$

and thus also (and dropping the argument z)

$$y_{i+1} = F(x_{i+1}) = F(x_{i+1}) - F(x_i) + y_i = F(x_{i+1}) - F(x_i) - DF(q)(x_{i+1} - x_i)$$

$$= \left(\int_0^1 \frac{d}{dt}(F(tx_{i+1} + (1-t)x_i))\,dt\right) - DF(q)(x_{i+1} - x_i)$$

$$= \left(\int_0^1 DF(tx_{i+1} + (1-t)x_i)\,dt - DF(q)\right)(x_{i+1} - x_i).$$

Since all the points along the line segment from x_i to x_{i+1} lie in $B_{\varepsilon}^n(q)$, it follows that

$$\|y_{i+1}\| \le \left(\int_0^1 \|DF(tx_{i+1} + (1-t)x_i)\,dt - DF(q)\|\,dt\right) \|x_{i+1} - x_i\|$$

$$\le \frac{1}{2K} \cdot K \left(\frac{1}{2}\right)^i \|\varphi(z)\| = \left(\frac{1}{2}\right)^{i+1} \|\varphi(z)\|.$$

This immediately gives

$$\|x_{i+2}(z) - x_{i+1}(z)\| = \|BF(x_{i+1}(z))\| \le K \|y_{i+1}(z)\| \le K \left(\frac{1}{2}\right)^{i+1} \|\varphi(z)\|,$$

concluding the inductive argument.

These inequalities imply that $y_i(z) \to 0$ uniformly on $B_\delta^k(0)$. Furthermore, the sequence $\{x_i(z)\}_{i \in \mathbb{N}}$ also converges uniformly to a continuous limit $\bar{x}(z) = \lim_{i \to \infty} x_i(z)$ on $B_\delta^k(0)$. By continuity of F, it follows that

$$0 = \lim_{i \to \infty} y_i(z) = \lim_{i \to \infty} F(x_i(z)) = F(\bar{x}(z))$$

and thus $\bar{x}(z) \in M$. Defining $r(z)$ as $r(z) = \bar{x}(z) - q - \ell(z)$, we obtain $F(q + \ell(z) + r(z)) = 0$ and

$$\|r(z)\| = \left\| \lim_{i \to \infty} x_i(z) - x_0(z) \right\| = \left\| \lim_{i \to \infty} \sum_{j=0}^{i} (x_j(z) - x_{j-1}(z)) \right\|$$

$$\le \lim_{i \to \infty} \sum_{j=1}^{i} \|x_i(z) - x_{i-1}(z)\| \le \lim_{i \to \infty} \sum_{j=1}^{i} K \left(\frac{1}{2}\right)^{j-1} \|\varphi(z)\| \le 2K \|\varphi(z)\|.$$

Thus the remainder $r = r(z)$ is of order $o(\|z\|)$ as $z \to 0$. This concludes the proof. $\square$

C.2 Manifolds: The General Case

Locally, all d-dimensional embedded submanifolds of $\mathbb{R}^k$ look alike: by means of a change of coordinates they can diffeomorphically be represented as the image of some neighborhood of $\mathbb{R}^d$. This main feature leads to the general definition of a manifold in which it is no longer assumed that the set lies in some Euclidean space.

We always take M as a second-countable topological space that is Hausdorff. We briefly recall these definitions. A *topological space* is a nonempty set Ω together with a collection of distinguished sets, called open, that contains the empty set $\emptyset$ and the full space Ω, and is closed under arbitrary unions and finite intersections. It is said to be *second countable* if there exists a countable basis $\mathscr{B} = \{O_i\}_{i \in \mathbb{N}}$ of open sets such that every open set can be written as a union of some of the sets O_i. For example, in $\mathbb{R}^k$, the set of all balls $B_\varepsilon(p)$ with centers $p \in \mathbb{Q}^n$ and radius $\varepsilon \in \mathbb{Q}_+$ is such a collection, and hence $\mathbb{R}^k$ is second countable. The topology on Ω is *Hausdorff* if whenever p and q are different points in X, then there exist neighborhoods U of p and V of q that are disjoint, $U \cap V = \emptyset$. (A neighborhood of a point p is any set that contains an open set containing p.) A topological space Ω is said to be *locally Euclidean of dimension d* if every point $p \in \Omega$ has a neighborhood U that is homeomorphic to some open subset V of $\mathbb{R}^d$. A homeomorphism is a bijective continuous map that has a continuous inverse. A homeomorphism φ from a neighborhood U in Ω onto a neighborhood V in $\mathbb{R}^d$, $\varphi : U \to V = \varphi(U) \subset \mathbb{R}^d$,

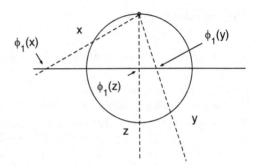

is called a *coordinate chart*, and the components $z_1, \ldots, z_d$ of φ are a local set of coordinates on U. We say that a coordinate chart $\varphi : U \to V$ is centered at $p \in U$ if $\varphi(p) = 0$ and V is a neighborhood of $0 \in \mathbb{R}^d$. If $\varphi_1 : U_1 \to V_1$ and $\varphi_2 : U_2 \to V_2$ are two coordinate charts for which U_1 and U_2 overlap, $U_1 \cap U_2 \neq \emptyset$, then the composition $\psi = \varphi_1 \circ \varphi_2^{-1}$,

$$\psi = \varphi_1 \circ \varphi_2^{-1} : V_1 \cap V_2 \to V_1 \cap V_2,$$

describes the change of coordinates in $\mathbb{R}^d$. By construction, this mapping is a homeomorphism. But ψ is a function between open subsets of $\mathbb{R}^d$, and thus one may ask whether these changes of coordinates have additional smoothness properties, for example, differentiability properties. The two coordinate charts φ_1 and φ_2 are said to be C^r-*compatible* if the change of coordinates $\varphi_1 \circ \varphi_2^{-1}$ is a C^r-diffeomorphism. Differentiable manifolds are topological manifolds together with a maximal family of C^r-compatible coordinate charts.

Definition C.2.1 (C^r-atlas). Let M be a d-dimensional locally Euclidean space. A C^r-atlas consists of an open cover $\{U_\iota\}_{\iota \in I}$ of M, $M = \cup_{\iota \in I} U_\iota$, together with a C^r-compatible family $\{\varphi_\iota\}_{\iota \in I}$ of coordinate charts $\varphi_\iota : U_\iota \to V_\iota \subset \mathbb{R}^d$.

Example: On S^n, let $N = (1, \ldots, 0, 0)$ and $S = (-1, \ldots, 0, 0)$ denote the north and south poles, respectively. The stereographic projections (see Fig. C.3)

$$\varphi_1 : U_1 = S^n \backslash \{N\} \to \mathbb{R}^n, \qquad z \mapsto \varphi_1(z) = \frac{(z_1, \ldots, z_n)}{1 - z_0},$$

and

$$\varphi_2 : U_2 = S^n \backslash \{S\} \to \mathbb{R}^n, \qquad z \mapsto \varphi_2(z) = \frac{(z_1, \ldots, z_n)}{1 + z_0},$$

are homeomorphisms, and their composition $\varphi_1 \circ \varphi_2^{-1}$ is the inversion at the sphere S^n,

$$\varphi_1 \circ \varphi_2^{-1}(z) = \frac{z}{\|z\|^2},$$

i.e., a real-analytic diffeomorphism. Thus these two maps define a C^ω-atlas for S^n.

It is not difficult to see that under the assumption that M is second countable and Hausdorff, given any C^r-atlas $\{U_\iota\}_{\iota \in I}$ of M, there exists a unique maximal atlas that contains all other coordinate charts that are C^r-compatible with the given atlas [50, Theorem 1.3]. Such a maximal atlas is called a *differentiable structure*.

Definition C.2.2 (C^r-differentiable manifold). A d-dimensional C^r-differentiable manifold M is a d-dimensional locally Euclidean space together with a differentiable structure, i.e., a maximal C^r-atlas $\{\varphi_\iota\}_{\iota \in I}$ of coordinate charts $\varphi_\iota : U_\iota \to V_\iota \subset \mathbb{R}^d$ that cover M. Thus, whenever U_ι and U_κ overlap, then the change of coordinates $\varphi_\iota \circ \varphi_\kappa^{-1}$ is a C^r-diffeomorphism.

This definition no longer assumes that M is a subset of some Euclidean space $\mathbb{R}^k$. If $M \subset \mathbb{R}^k$, then M is said to be *embedded* into $\mathbb{R}^k$ if the topology on M is the relative topology from $\mathbb{R}^k$, i.e., open sets (neighborhoods) U in M are intersections of open sets (neighborhoods) V of $\mathbb{R}^k$ with M, $U = V \cap M$. All the manifolds considered in the examples above were embedded in this sense. In general, however, this need not be the case, even if M is a subset of some $\mathbb{R}^k$. To give examples, we first need to consider mappings between manifolds. Since coordinate changes in the manifolds are given by C^r diffeomorphisms, the following definition does not depend on the specific coordinate chart chosen.

Definition C.2.3 (C^r-mapping). Let M and N be C^r-differentiable manifolds of dimensions m and n, respectively. A continuous mapping $F : M \to N$ is called a C^r-mapping if for every point $p \in M$ there exist a coordinate chart (U, φ) centered at p and a coordinate chart (V, ψ) centered at $F(p)$ with $F(U) \subset V$ such that the coordinate representation $\psi \circ F \circ \varphi^{-1} : \varphi(U) \to \psi(V)$ is a C^r mapping from $\mathbb{R}^m$ into $\mathbb{R}^n$.

Definition C.2.4 (C^r-immersion, -submersion, -diffeomorphism). Let M and N be C^r-differentiable manifolds of dimensions m and n, respectively, and let $F : M \to N$ be a C^r-mapping. For a point $p \in M$, let (U, φ) be a coordinate chart centered at p and let (V, ψ) be a coordinate chart centered at $F(p)$ with $F(U) \subset V$. The mapping F is an immersion at p if the Jacobian matrix of the coordinate representation $\psi \circ F \circ \varphi^{-1}$ at p is injective (one-to-one), a submersion if it is surjective (onto), and a diffeomorphism if it is bijective (injective and surjective).

Definition C.2.5 (C^r-immersed and embedded submanifolds). Let M and N be C^r-differentiable manifolds of dimensions m and n, respectively, and let $F : M \to N$ be a (globally) injective C^r-immersion. Then $\tilde{N} = F(N)$, endowed with the topology that makes the mapping F a C^r-diffeomorphism, is an immersed submanifold of N. If $\tilde{N}$ carries the relative topology from N (open sets $\tilde{V}$ of $\tilde{N}$ are intersections of open sets V of N with $\tilde{N}$, $\tilde{V} = V \cap \tilde{N}$), then $\tilde{N}$ is called an embedded submanifold of N and F is called an embedding.

Example (Dense line on a torus). Consider the 2-torus $T^2 \subset \mathbb{R}^4$ defined above, and for $\alpha \in \mathbb{R}$ let F_α be the map $F_\alpha : \mathbb{R} \to T^2, t \mapsto F_\alpha(t) = (e^{2\pi i t}, e^{2\pi \alpha i t})$. The rank

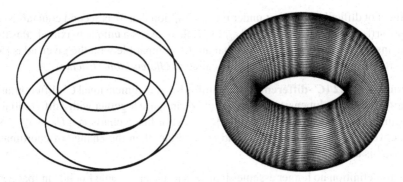

Fig. C.4 Embedded (*left*) versus immersed (*right*) submanifolds of the two-dimensional torus T^2

is 1 everywhere, and so F_α is an analytic immersion. If α is rational, say $\alpha = \frac{m}{n}$, then this generates a periodic curve on T^2, $F_\alpha(n) = F_\alpha(0)$, and in this case the image $N_\alpha = F_\alpha(\mathbb{R})$ is an embedded submanifold of T^2. But if α is irrational, the image is a dense line on T^2, i.e., given a point $q \in T^2$ and any neighborhood V of q in T^2, the intersection $N_\alpha \cap V$ is nonempty. (For a proof, see, for example, [50, p. 86].) In this case, the topology carried over from $\mathbb{R}$ onto N_α does not coincide with the relative topology from T^2, since in any small neighborhood V of T^2, there exist points from N_α that are far from each other in the topology of N_α that is carried over from $\mathbb{R}$ (see Fig. C.4).

It is a famous result due to Whitney, the *Whitney embedding theorem*, that any d-dimensional manifold M can be embedded into $\mathbb{R}^{2d+1}$. In these notes, we shall almost exclusively deal with embedded submanifolds of some $\mathbb{R}^k$.

C.3   Tangent and Cotangent Spaces

We now proceed to the definition of the tangent space at a point p, $T_p M$, for a general d-dimensional C^r-differentiable manifold M. When M was embedded in $\mathbb{R}^k$, we could simply lean on the differentiable structure in $\mathbb{R}^k$ and differentiate curves to generate tangent vectors. In the abstract framework formulated above, this no longer is straightforward, since there is no natural notion of distance on a manifold—when embedded into $\mathbb{R}^k$, this simply was inherited from the ambient space—and thus also speed is not well-defined. But the notion of a tangent vector precisely combines these two concepts—direction and speed [38]. These thus need to be carried over from $\mathbb{R}^d$ onto the manifold by means of coordinate charts. But then changes in the coordinates must be taken into account, and this leads to (somewhat cumbersome) definitions of tangent vectors in terms of equivalence classes that involve both coordinate charts and vectors in $\mathbb{R}^d$. There exist more elegant, abstract ways to define tangent vectors as derivations on C^∞-manifolds (see below), but because of the associated loss of differentiability, these actually do not work equally well on

C^r-manifolds if r is finite [38]. Also, we generally prefer the intuitive geometric over the abstract algebraic approach.

Let M be a d-dimensional C^r-differentiable manifold and let p be a point in M. Given two coordinate charts (U, φ) and (V, ψ) centered at p and two vectors u and v in $\mathbb{R}^d$, we say that the triples (U, φ, u) and (V, ψ, v) are *equivalent*, $(U, \varphi, u) \sim (V, \psi, v)$, if

$$u = D\left(\varphi \circ \psi^{-1}\right)(\psi(p))v. \tag{C.2}$$

That is, if

$$\gamma : (-\varepsilon, \varepsilon) \to \mathbb{R}^k, \qquad t \mapsto \gamma(t),$$

is a curve in $\psi(V)$ that passes through $\psi(p)$ with tangent vector v at time $t = 0$, then the curve

$$\zeta : (-\varepsilon, \varepsilon) \to \mathbb{R}^k, \qquad t \mapsto \zeta(t) = \left(\varphi \circ \psi^{-1} \circ \gamma\right)(t),$$

defined through the corresponding change of coordinates has tangent vector u at time $t = 0$.

Definition C.3.1 (Tangent space of M at p, T_pM). A tangent vector to M at p consists of an equivalence class under the equivalence relation $(U, \varphi, u) \sim (V, \psi, v)$, and the tangent space to M at p, T_pM, is the set of all equivalence classes.

Proposition C.3.1. *T_pM is a d-dimensional vector space.*

Proof. Arbitrarily select a "distinguished" coordinate chart (U, φ) centered at p. The mapping $\tau = \tau_{(U, \varphi)}$

$$\tau : \mathbb{R}^d \to T_pM, \quad u \mapsto \text{class}(U, \varphi, u), \tag{C.3}$$

that assigns to each vector $u \in \mathbb{R}^d$ its corresponding equivalence class is an isomorphism. Clearly,

$$\tau(\alpha u + \beta v) = \text{class}(U, \varphi, \alpha u + \beta v)$$
$$= \alpha \text{class}(U, \varphi, u) + \beta \text{class}(U, \varphi, v) = \alpha \tau(u) + \beta \tau(v),$$

and thus τ is a linear map. Furthermore, given any two distinct vectors, $u_1 \neq u_2$, the associated classes are different, since the equivalence relation (C.2) is not satisfied if we take (V, ψ) as the distinguished chart, $(V, \psi) = (U, \varphi)$. Hence the mapping is one-to-one. On the other hand, each equivalence class has a representative in the coordinate chart (U, φ), and thus this mapping is also surjective. $\qquad \square$

Let

$$\varphi : U \to \mathbb{R}^d, \qquad q \mapsto \varphi(q) = (z_1(q), \ldots, z_d(q)),$$

be a coordinate chart centered at p with coordinates $z = (z_1, \ldots, z_d)$ and denote the corresponding basis vectors in $\mathbb{R}^d$ with the 1 in the ith spot by $e_i = (0, \ldots, 1, \ldots, 0)^T$, $i = 1, \ldots, d$. Then a basis for the tangent space T_pM is given by the tangent vectors

$$\partial_i = \left(\frac{\partial}{\partial z_i}\right)_{|p} = \text{class}(U, \varphi, e_i),$$

and every tangent vector $t \in T_p M$ can be expressed as a linear combination of these vectors,

$$t = \sum_{i=1}^{d} u_i \left(\frac{\partial}{\partial z_i}\right)_{|p},$$

with $u = (u_1, \ldots, u_d)^T$ the coordinates of t relative to this basis. If

$$\psi : V \to \mathbb{R}^d, \qquad q \mapsto \psi(q) = (y_1(q), \ldots, y_d(q)),$$

is another set of coordinates centered at p, then t also has a coordinate representation in terms of this coordinate chart of the form

$$t = \sum_{i=1}^{d} v_i \left(\frac{\partial}{\partial y_i}\right)_{|p},$$

with coordinates $v = (v_1, \ldots, v_d)^T$. The formula for the corresponding change of coordinates for the tangent vector t then is given by the equivalence relation (C.2), $u = D\left(\varphi \circ \psi^{-1}\right) v$.

This formula for the change of coordinates follows if one considers how tangent vectors act on functions by taking directional derivatives. If f is a C^r-function defined near p, then its coordinate representation is given by

$$f_\varphi : \varphi(U) \to \mathbb{R}, \qquad z \mapsto f_\varphi(z) = \left(f \circ \varphi^{-1}\right)(z),$$

and the directional derivative of f_φ in the direction of the vector u is

$$\langle \nabla f_\varphi(0), u \rangle = \sum_{i=1}^{d} \frac{\partial f_\varphi}{\partial z_i}(z) u_i = \sum_{i=1}^{d} u_i \frac{\partial \left(f \circ \varphi^{-1}\right)}{\partial z_i}(\varphi(p)) = t(f).$$

If we think of the coordinate x_i as a smooth function of x, then

$$t(z_i) = \sum_{j=1}^{d} u_j \left(\frac{\partial z_i}{\partial z_j}\right)_{|p} = \sum_{j=1}^{d} u_j \delta_{ij} = u_i.$$

By means of the change of coordinates, we can also consider x_i as a C^r-function of y and then get the desired formula:

$$u_i = t(z_i) = \sum_{j=1}^{d} \left(\frac{\partial z_i}{\partial y_j}\right)_{|p} v_j.$$

Clearly, the matrix $\left(\frac{\partial z_i}{\partial y_j}\right)_{1\leq i,j\leq d}$ is the Jacobian matrix of the change of coordinates, $x = \left(\varphi \circ \psi^{-1}\right)(y)$, and thus this is the relation defining the equivalence of tangent vectors, $u = D\left(\varphi \circ \psi^{-1}\right)v$.

Corollary C.3.1. *If (U,φ) is a coordinate chart centered at p with coordinates $z_1,\ldots,z_d$, then every tangent vector $\mathfrak{t} \in T_pM$ has a coordinate representation in the form*

$$\mathfrak{t} = \sum_{i=1}^{d} \mathfrak{t}(z_i)\left(\frac{\partial}{\partial z_i}\right)_{|p}.$$

In view of these formulas, tangent vectors $\mathfrak{t}$ can also be defined as mappings that take the directional or Lie derivative of functions that are defined in a neighborhood of p in the direction of $\mathfrak{t}$. It follows from the rules of differentiation that for all α, $\beta \in \mathbb{R}$ and arbitrary C^r-functions f and g defined near p, we have that

$$\mathfrak{t}(\alpha f + \beta g) = \alpha\mathfrak{t}(f) + \beta\mathfrak{t}(g) \qquad \text{and} \qquad \mathfrak{t}(fg) = \mathfrak{t}(f)g + f\mathfrak{t}(g). \tag{C.4}$$

Any mapping $\mathfrak{t}$ that satisfies these properties is called a *derivation*, and an equivalent definition of the tangent space for C^∞-manifolds is that it is the vector space of all derivations. However, for C^r-manifolds these two concepts are not equivalent [38].

Definition C.3.2 (Tangent bundle). The tangent bundle TM is the disjoint union of the tangent spaces T_pM for all points p in the manifold,

$$TM = \bigcup_{p\in M} T_pM = \{(p,v): p \in M, \, v \in T_pM\}.$$

Thus the tangent bundle consists of all pairs of a point $p \in M$ and a tangent vector $v \in T_pM$ to M at p.

Proposition C.3.2. *Let M be a C^r-differentiable manifold of dimension d for $r \geq 1$. Then the tangent bundle TM can canonically be made into a $2d$-dimensional C^{r-1}-differentiable manifold.*

Proof (Outline [102]). One needs to define a topology on TM that makes it into a $2d$-dimensional locally Euclidean space and endow it with a differentiable structure. These all naturally carry over from M. Let $\pi : TM \to M$, $(p,v) \mapsto p$, denote the projection onto the base point. For any chart (U,φ) on M centered at p, let $\tau = \tau_{(U,\varphi)}$ be the isomorphism (C.3) between $\mathbb{R}^d$ and T_pM that assigns to a vector $v \in \mathbb{R}^d$ its corresponding equivalence class. The mapping

$$\Phi : \pi^{-1}(U) \to \mathbb{R}^d \times \mathbb{R}^d, \qquad \Phi(q,w) = \left(\varphi(q), \tau_{(U,\varphi)}^{-1}w\right),$$

is bijective, and we can thus make TM into a topological space by taking as neighborhoods of (p,v) the inverse images of neighborhoods of $\Phi(p,v)$ in

$\varphi(U) \times \mathbb{R}^d$. By construction, all the maps Φ are homeomorphisms, and since M is second countable and Hausdorff, these properties carry over. Furthermore, if (U, φ) and (V, ψ) are two coordinate charts centered at p, then the transition functions for the associated coordinate charts $(\pi^{-1}(U), \Phi)$ and $(\pi^{-1}(V), \Psi)$ on TM are given by the change of coordinates on the manifolds and by the equivalence relation (C.2) for the corresponding vectors,

$$\Psi \circ \Phi^{-1} : \varphi(U \cap V) \times \mathbb{R}^d \to \psi(U \cap V) \times \mathbb{R}^d,$$

$$(y, v) \mapsto \left(\psi \circ \varphi^{-1}(y), D \left(\psi \circ \varphi^{-1} \right)(y) v \right).$$

These mappings are C^{r-1} and define a differentiable structure on TM.   □

A C^r-mapping $F : M \to N$ between C^r-manifolds M and N induces a linear mapping between the corresponding tangent spaces, the so-called differential. Let (U, φ) be a coordinate chart centered at p and let (V, ψ) be a coordinate chart centered at $q = F(p)$, such that the coordinate representation of the mapping F is given by

$$F_\varphi : \varphi(U) \to \psi(V), \qquad z \mapsto F_\varphi(z) = \left(\psi \circ F \circ \varphi^{-1} \right)(z).$$

A tangent vector $u \in T_p M$ can be represented by the equivalence class (U, φ, u), and the derivative of the coordinate representation F_φ at 0, $DF_\varphi(0)$, defines a tangent vector $F_*(u) \in T_q N$ represented by the equivalence class $(V, \psi, F_*(u))$, where $F_*(u) = DF_\varphi(0)u$ is simply the image of u induced by the coordinate representation of the mapping F. Clearly, any differentiable curve

$$\gamma : (-\varepsilon, \varepsilon) \to \varphi(U), \qquad s \mapsto \gamma(s), \qquad \gamma(0) = 0,$$

is mapped into a differentiable curve

$$F_\varphi \circ \gamma : (-\varepsilon, \varepsilon) \to \psi(V), \qquad s \mapsto F_\varphi(\gamma(s)), \qquad F_\varphi(\gamma(0)) = 0,$$

with derivative

$$\frac{d}{ds} \left(F_\varphi \circ \gamma \right)(0) = DF_\varphi(0) \dot{\gamma}(0).$$

It is a simple exercise to verify that this definition does not depend on the coordinate charts taken, and thus this is a well-defined and linear mapping from the tangent bundle TM into the tangent bundle TN. This mapping is called the differential of F.

Definition C.3.3 (Differential of a C^r-mapping). Let $F : M \to N$ be a C^r-mapping between C^r-manifolds M and N of dimensions m and n, respectively. The differential of F, F_*, is the mapping $F_* : TM \to TN$ that assigns to a tangent vector $\mathfrak{t}$ to M at p, $\mathfrak{t} \in T_p M$, represented by the class (U, φ, u), the tangent vector $F_* \mathfrak{t}$ to N at $q = F(p)$ represented by the class $(V, \psi, F_*(u))$, where $F_*(u) = DF_\varphi(0)u$.

It is clear that the differential is a linear mapping, and if $F : M \to N$ and $G : N \to P$ are C^r-mappings between C^r-manifolds, then the chain rule in Euclidean space immediately implies that $(G \circ F)_* = G_* \circ F_*$.

For C^∞-manifolds, the differential $F_* : TM \to TN$ can equivalently also be defined in terms of the action of tangent vectors on smooth functions f. As before, if $F : M \to N$ is a C^∞-mapping between C^∞-manifolds M and N, and $\mathfrak{t}$ is a tangent vector to M at p, $\mathfrak{t} \in T_pM$, then $F_*\mathfrak{t}$ is the tangent vector to N at q that differentiates a smooth function g defined near q according to the rule

$$(F_*\mathfrak{t})(g) = \mathfrak{t}(g \circ F),$$

i.e., is given by the Lie derivative of the composite function $g \circ F$ (which is smooth near p) with respect to $\mathfrak{t}$.

The cotangent space and cotangent bundle are the dual notions of the tangent space and the tangent bundle.

Definition C.3.4 (Cotangent space of M at p, T_p^*M). The cotangent space to M at p is the dual space to the tangent space of M at p, i.e., the space of all linear functionals defined on T_pM.

Since M is a finite-dimensional manifold, the space of all linear functionals on the tangent space to M at p is isomorphic to T_pM as a vector space, but it is better not to identify these spaces (as is often done in linear algebra) and instead distinguish between cotangent vectors or covectors $\lambda \in T_p^*M$ and tangent vectors $v \in T_pM$. We use $\langle \lambda, v \rangle$ to denote the action of a covector λ on a tangent vector v; note that the base points for λ and v must agree for this to be defined. In coordinates, we consistently write covectors as row vectors, while we write tangent vectors as column vectors. This significantly simplifies the notation, and it is a much better reflection of the underlying geometric concepts.

Definition C.3.5 (Cotangent bundle). The cotangent bundle T^*M is the disjoint union of the cotangent spaces T_p^*M for all points p in the manifold,

$$T^*M = \bigcup_{p \in M} T_p^*M = \{(p, \lambda) : p \in M, \lambda \in T_p^*M\}.$$

Thus the cotangent bundle consists of all pairs of a point $p \in M$ and a cotangent vector $\lambda \in T_p^*M$ to M at p.

Since a C^r-mapping $F : M \to N$ between C^r-manifolds M and N induces a linear mapping between the tangent spaces, $F_* : TM \to TN$, it also generates a linear mapping in the reverse direction between the cotangent spaces, $F^* : T^*N \to T^*M$, through the adjoint mapping of the differential.

Definition C.3.6 (Pullback of a C^r-mapping). Let $F : M \to N$ be a C^r-mapping between C^r-manifolds M and N of dimensions m and n, respectively. The pullback F^* is the linear map from the cotangent bundle T^*N into the cotangent bundle T^*M,

$F^* : T^*N \to T^*M$, defined as the dual (or adjoint) mapping to the differential $F_* :$ $TM \to TN$, i.e., if $F(p) = q$ and $\lambda \in T_q^*N$, then $F^*(\lambda)$ is the unique covector in T_p^*N that satisfies

$$\langle F^*(\lambda), v \rangle = \langle \lambda, F_*(v) \rangle \qquad \text{for all } v \in T_pM.$$

C.4   Vector Fields and Lie Brackets

Definition C.4.1 (C^r vector and covector fields). Let M be a C^{r+1}-manifold. A C^r vector field X on M is a C^r-section of the tangent bundle, i.e., a C^r mapping $X : M \to TM$ that assigns to every point $p \in M$ a unique tangent vector $X(p) \in T_pM$. Similarly, a C^r covector field λ on M is a C^r-section of the cotangent bundle, i.e., a C^r mapping $\lambda : M \to T^*M$ that assigns to every point $p \in M$ a unique covector $\lambda(p) \in T_p^*M$.

Let (U, φ) be a coordinate chart with coordinates $z_1, \ldots, z_d$. Then, for $q \in U$, the tangent vectors $\left(\frac{\partial}{\partial z_i} \right)_{|q}$ are a basis for T_qM, and a C^r vector field X can be expressed in the form

$$X(q) = \sum_{i=1}^{d} X_i(q) \left(\frac{\partial}{\partial z_i} \right)_{|q},$$

where the X_i are C^r functions defined on U, $X_i : U \to \mathbb{R}$. The coordinate expressions of these functions X_i are given by r-times continuously differentiable real-valued functions

$$f_i = (X_i)_\varphi : \varphi(U) \to \mathbb{R}, \qquad z \mapsto f_i(z) = \left(X_i \circ \varphi^{-1} \right)(z),$$

which, in turn, define the coordinate expression X_φ of the vector field X on $\varphi(U)$,

$$X_\varphi(z) = \sum_{i=1}^{d} f_i(z) \left(\frac{\partial}{\partial z_i} \right)_{|z}.$$

Similarly, the coordinate expression of a covector field is of the form

$$\lambda_\varphi(z) = \sum_{i=1}^{d} g_i(z) \, (dz_i)_z,$$

where the g_i are r-times continuously differentiable real-valued functions and $(dz_i)_{|z}$ denotes the dual basis to the basis $\left(\frac{\partial}{\partial z_i} \right)_{|z}$ of the tangent space, i.e., the linear functionals defined by

$$\left\langle dz_i, \frac{\partial}{\partial z_j} \right\rangle = \begin{cases} 1 & \text{if } i = j, \\ 0 & \text{if } i \neq j. \end{cases}$$

Definition C.4.2 (Integral curve). A C^1 curve $\gamma : I \to M, t \mapsto \gamma(t)$, defined on an open interval $I \subset \mathbb{R}$ is said to be an integral curve of the vector field X if at every point $\gamma(t)$, the tangent vector $\dot{\gamma}(t)$ to the curve γ is given by the value of the vector field X at $\gamma(t)$,

$$\dot{\gamma}(t) = X(\gamma(t)) \qquad \text{for all} \quad t \in I.$$

The curve γ is a C^1 mapping from the one-dimensional manifold $I \subset \mathbb{R}$ into M. The tangent space to $\mathbb{R}$ is one-dimensional with a canonical basis vector given by $\frac{\partial}{\partial t}$ with $t \in I$ the natural coordinate. Thus, the tangent vector $\dot{\gamma}(t)$ to the curve γ at the point $\gamma(t)$ is given by the image of $\frac{\partial}{\partial t}$ under the differential of the mapping γ, i.e., in the notation established above, we have that

$$\dot{\gamma}(t) = \gamma_* \left(\frac{\partial}{\partial t} \right).$$

In the coordinate chart (U, φ), we can express the curve $\gamma(t)$ as a d-tuple $\gamma(t) = (\gamma_1(t), \ldots, \gamma_d(t))^T$, where $\gamma_i = z_i \circ \gamma \in \mathbb{R}$ is the ith coordinate of the curve γ. Hence, the tangent vector $\gamma_* \left(\frac{\partial}{\partial t} \right)$ has the coordinate representation

$$\sum_{i=1}^d \frac{d\gamma_i}{dt} \left(\frac{\partial}{\partial z_i} \right)_{|\gamma(t)}.$$

Comparing this with the coordinate representation of the vector field X along γ, which is given by

$$\sum_{i=1}^d f_i(\gamma(t)) \left(\frac{\partial}{\partial z_i} \right)_{|\gamma(t)},$$

it follows that

$$\frac{d\gamma_i}{dt} = f_i(\gamma(t)),$$

and in local coordinates we thus have the following statement:

Proposition C.4.1. *A curve $\gamma : I \to M, t \mapsto \gamma(t)$, is an integral curve of the vector field X if and only if for every coordinate chart (U, φ) with coordinates $z_1, \ldots, z_d$, the coordinate expressions $\gamma_i = z_i \circ \gamma$ of the curve γ and f_i of the vector field X satisfy the following system of ordinary differential equations:*

$$\frac{d\gamma_i}{dt} = f_i(\gamma_1(t), \ldots, \gamma_d(t)).$$

Thus, vector fields are a way to describe ordinary differential equations on manifolds, and integral curves simply are the corresponding solutions. All the fundamental results about solutions to ordinary differential equations from Appendix B carry over by means of local coordinates, and for example, we get the following result:

Proposition C.4.2. *Let X be a C^r vector field on a C^r manifold M. Then, for every point $p \in M$, there exists a maximal open interval I_p in $\mathbb{R}$ containing the origin such that the integral curve γ of X that passes through p at time $t = 0$ exists on I_p. The corresponding flow $\Phi_t^X(p)$ is defined on the domain $D = \{(t, p : p \in M, t \in I_p\}$ and is a C^r mapping from D to M. Furthermore,*

$$\Phi_{s+t}^X(p) = \Phi_t^X(\Phi_s^X(p)),$$

whenever both sides are defined.   ∎

We close with a formal definition of the Lie derivative and Lie bracket. These are the most fundamental tools used in the text, and we have therefore generally included proofs of the results below in the sections of the text where these concepts came up for the first time, especially Sects. 2.8, 2.9, and 4.5. For the reader's convenience, we include a concise statement of the relevant notions here as well.

For simplicity, let M be a C^∞ manifold and denote the space (module) of all infinitely often continuously differentiable functions on M by $C^\infty(M)$. Also, denote the space of all C^∞ vector fields on M by $V^\infty(M)$. The vector field X can be viewed as defining a first-order differential operator from the space $C^\infty(M)$ into $C^\infty(M)$ by taking, at every point $p \in M$, the derivative of a function $\alpha \in C^\infty(M)$ in the direction of the vector field $X(p)$.

Definition C.4.3 (Lie derivative). The Lie derivative L_X is the first-order differential operator $L_X : C^\infty(M) \to C^\infty(M)$, $\alpha \mapsto L_X(\alpha)$, defined at every point $p \in M$ by taking the derivative of the function α in the direction of the vector $X(p)$.

Given a coordinate chart (U, φ) centered at p, the coordinate representation of α is given by

$$\alpha_\varphi : \varphi(U) \to \mathbb{R}, \qquad z \mapsto \alpha_\varphi(z) = \left(\alpha \circ \varphi^{-1}\right)(z),$$

and if

$$X_\varphi(z) = \sum_{i=1}^d f_i(z) \left(\frac{\partial}{\partial z_i}\right)_{|z}$$

denotes the coordinate representation of the vector field X, then for $q = \varphi^{-1}(z) \in U$, the Lie derivative at q, $L_X(\alpha)(q)$, has the coordinate expression

$$\sum_{i=1}^d \frac{\partial \alpha_\varphi}{\partial z_i}(z) f_i(z) = \left\langle \nabla \alpha_\varphi(z), X_\varphi(z) \right\rangle,$$

with $\nabla \alpha_\varphi$ denoting the gradient of the coordinate representation of α. This gradient $\nabla \alpha_\varphi$ defines a covector that acts on the tangent vector X_φ, and we also write

$$L_X(\alpha)(p) = \left\langle \nabla \alpha(p), X(p) \right\rangle,$$

identifying α and the vector fields X with their coordinate representations.

Given two vector fields, the commutator

$$[L_X, L_Y] = L_X \circ L_Y - L_Y \circ L_X$$

is formally a second-order differential operator. However, in this expression all the terms that are associated with second derivatives cancel (this can be shown through a computation in canonical coordinates analogous to the one given in Sect. 2.8.3), and indeed the commutator is the Lie derivative of another C^∞ vector field Z that is denoted by $[X, Y]$ and called the Lie bracket of the vector fields X and Y.

Definition C.4.4 (Lie bracket). The Lie bracket of two vector fields X and Y defined on M is the vector field $[X, Y]$ such that

$$L_{[X,Y]} = [L_X, L_Y] = L_X \circ L_Y - L_Y \circ L_X.$$

If coordinate expressions for the vector fields X and Y are given by

$$X_\varphi(z) = \sum_{i=1}^{d} f_i(z) \left(\frac{\partial}{\partial z_i} \right)_{|z} \quad \text{and} \quad Y_\varphi(x) = \sum_{i=1}^{d} g_i(z) \left(\frac{\partial}{\partial z_i} \right)_{|z},$$

and if we write $F(z) = (f_1(z), \ldots, f_d(z))^T$ and $G(z) = (g_1(z), \ldots, g_d(z))^T$, then, similarly as was shown in Sect. 2.8.3, the coordinate expression for the Lie bracket $[X, Y]$ is given by

$$[X, Y]_\varphi(z) = DG(z) \cdot F(z) - DF(z) \cdot G(z).$$

A similar computation also verifies that the Jacobi-identity is satisfied, and thus the set $V^\infty(M)$ of all C^∞ vector fields defined on M is a Lie algebra with the Lie bracket defining the bilinear operation $[\cdot, \cdot]$.

The Lie bracket of two vector fields X and Y measures the extent to which these vector fields commute. If we denote the flows of these vector fields by Φ^X and Φ^Y, respectively, then for every neighborhood U of a point $p \in M$ there exists an $\varepsilon > 0$ such that integral curves of X and Y that start at points in U are well-defined for times $t^2 < \varepsilon$. For $t \in [0, \varepsilon)$, define a curve $\gamma \colon [0, \varepsilon) \to M$ by

$$\gamma(t) = \Phi^Y_{-\sqrt{t}} \circ \Phi^X_{-\sqrt{t}} \circ \Phi^Y_{\sqrt{t}} \circ \Phi^X_{\sqrt{t}}(p).$$

It then follows that

$$\dot{\gamma}(0) = [X, Y](p).$$

This relation follows, for example, from the Baker–Campbell–Hausdorff formula, which was proven in Sect. 4.5. More importantly, it is also shown in that section that the curve

$$\gamma \colon (-\varepsilon, \varepsilon) \to T_p M, \qquad \gamma(t) = \left(\Phi^X_{-t} \right)_* Y(\Phi^X_t(p)),$$

i.e., γ is the vector field Y evaluated at the point $\Phi_t^X(p)$ on the integral curve of X through p and moved back to p along the flow of X, has derivative

$$\dot{\gamma}(0) = [X,Y](p).$$

It is this relation from which all the other formulas derived in Sect. 4.5 follow. And the importance of the Lie bracket for nonlinear systems is rooted in this identity, which gives rise to the adjoint representation (see Corollary 4.5.1).

Appendix D
Some Facts from Real Analysis

We briefly summarize some facts about Lebesgue measurable sets and functions. This section only serves the purpose of providing a convenient summary of the results from real analysis that are used in the text, but this area is too vast to be presented comprehensively here. For this, we need to refer the reader to any of the many excellent treatments of this topic available in the literature, e.g., [125, 174, 257].

D.1 Lebesgue Measure and Lebesgue Measurable Functions in $\mathbb{R}^n$

A σ-algebra $\mathscr{F}$ of subsets of $\mathbb{R}^n$ is a nonempty collection of subsets E of $\mathbb{R}^n$ that is closed under taking countable unions and complements. It follows that $\mathscr{F}$ is closed under arbitrary countable set-theoretic operations that combine unions, complements, and intersections. The sets that are contained in $\mathscr{F}$ are called $\mathscr{F}$-measurable. A *measure* is a nonnegative σ-additive set function μ defined on the sets in a σ-algebra $\mathscr{F}$, $\mu : \mathscr{F} \to \mathbb{R}_+$, i.e., if E_i, $i = 1, 2, \ldots$, is a sequence of disjoint sets from $\mathscr{F}$, $E_i \cap E_j = \varnothing$ for $i \neq j$, then

$$ \mu \left(\bigcup_{i=1}^{\infty} E_i \right) = \sum_{i=1}^{\infty} \mu(E_i). $$

The smallest possible σ-algebra on $\mathbb{R}^n$ is given by $\mathscr{F} = \{\varnothing, \mathbb{R}^n\}$, and the largest is the so-called power set, consisting of all subsets of $\mathbb{R}^n$. Given an arbitrary collection $\mathscr{C} = \{C_i : i \in I\}$ of subsets of $\mathbb{R}^n$, there exists a smallest σ-algebra that contains all the sets C_i in the collection $\mathscr{C}$, namely the intersection over all σ-algebras that contain all the sets C_i. This σ-algebra is called the σ-algebra generated by $\mathscr{C}$. The *Borel σ-algebra*, $\mathscr{B}$, is the σ-algebra generated by the open subsets of $\mathbb{R}^n$. It contains all closed sets as well as, for example, countable intersections of open sets (so-called G_δ-sets) and countable unions of closed sets (so-called F_σ-sets) and many more. Generally, Borel measurable sets provide a rich enough collection of measurable subsets of $\mathbb{R}^n$ for many of the results that appear in the text to hold. The *Borel measure* on $\mathscr{B}$ is the unique measure that can be defined on the Borel sets that for compact intervals $I = [a_1, b_1] \times \cdots \times [a_n, b_n]$ agrees with the ordinary volume in $\mathbb{R}^n$,

$$\mu(I) = \Pi_{i=1}^n (b_i - a_i).$$

The Lebesgue σ-algebra, $\mathscr{L}$, is the completion of the Borel σ-algebra under this measure; that is, all Borel measurable sets B are Lebesgue measurable and are given the same measure, and in addition, if $A \subset B$ and B is Borel measurable with measure $\mu(B) = 0$, then the set A is made Lebesgue measurable with Lebesgue measure 0. Lebesgue measurable sets that have measure 0 are called *null sets*. Since points are Borel measurable sets with measure 0, it follows that all countable sets are null sets. The Cantor set C is a well-known example of a null set that is uncountable [257, Chap. 3]. We have the following characterization of Lebesgue measurable sets:

Proposition D.1.1. *The following statements are equivalent:*

1. *The set $E \subset \mathbb{R}^n$ is Lebesgue measurable.*
2. *For any $\varepsilon > 0$, there exist a closed set $F \subset E$ and an open set $G \supset E$ such that $\mu(G \setminus F) < \varepsilon$.*
3. *There exist a Borel measurable set F of type F_σ, $F = \cup_{i=1}^\infty A_i$ with A_i closed, and a null set N such that $E = F \cup N$.*
4. *There exist a Borel measurable set G of type G_δ, $G = \cap_{i=1}^\infty B_i$ with B_i open, and a null set N such that $G = E \cup N$.*

A function $f : \mathbb{R}^n \to \mathbb{R}$ is Lebesgue (respectively, Borel) measurable if for any open set B in $\mathbb{R}$ the inverse image of B,

$$f^{-1}(B) = \{x \in \mathbb{R}^n : f(x) \in B\},$$

is a Lebesgue (respectively, Borel) measurable subset of $\mathbb{R}^n$. We henceforth simply call sets and functions measurable with the understanding that the underlying measure is either Borel or Lebesgue measure. It is easily seen that measurable functions are closed under the standard algebraic operations (sum, product, and quotient). An important property of measurable functions is that they are also closed under taking limits.

Proposition D.1.2. *If $\{f_n\}_{n \in \mathbb{N}}$ is a sequence of measurable functions, then the infimum, $\inf_{n \in \mathbb{N}}(f_n)$, and supremum, $\sup_{n \in \mathbb{N}}(f_n)$, are measurable functions as well. In particular, the smallest and largest accumulation points, $\liminf$ and $\limsup$, of the sequence are measurable. The set of points for which the sequence converges is measurable and the limit is a measurable function.* ∎

The characteristic function χ_E of a set E in $\mathbb{R}^n$ is defined by

$$\chi_E(x) = \begin{cases} 1 & \text{if } x \in E, \\ 0 & \text{if } x \notin E, \end{cases}$$

and is an $\mathscr{F}$-measurable function if and only if $E \in \mathscr{F}$. A function s with finite range of the form

$$s(x) = \sum_{i=1}^{N} c_i \chi_{E_i}(x)$$

is called *simple*, and without loss of generality, we may assume that the sets E_i are disjoint and that the coefficients c_i are distinct, $c_i \neq c_j$ for $i \neq j$. With these normalizations, a simple function s is $\mathscr{F}$-measurable if and only if all sets E_i are $\mathscr{F}$-measurable.

Proposition D.1.3. *Given any nonnegative measurable function $f : E \to [0,\infty)$, there exists a sequence $\{s_n\}_{n\in\mathbb{N}}$ of simple nonnegative $\mathscr{F}$-measurable functions such that $s_n(x)$ converges monotonically to $f(x)$, $s_n(x) \leq s_{n+1}(x) \leq f(x)$ for all $x \in E$.* ∎

Definition D.1.1 (Convergence of measurable functions). Let $\{f_n\}_{n\in\mathbb{N}}$ be a sequence of measurable functions, $f : \mathbb{R}^n \to \mathbb{R}$, and let E be a measurable set in $\mathbb{R}^n$. The sequence $\{f_n\}$ converges to f in measure on E if for any $\varepsilon > 0$ we have that

$$\lim_{n\to\infty} \mu\{x \in E : |f_n(x) - f(x)| > \varepsilon\} = 0;$$

it converges to f almost everywhere (a.e.) in E if there exists a null set $N \subset E$ such that

$$\lim_{n\to\infty} f_n(x) = f(x) \qquad \text{for all} \qquad x \notin E.$$

The fundamental relations between these concepts of convergence are summarized below:

Proposition D.1.4. *Let $\{f_n\}_{n\in\mathbb{N}}$ be a sequence of measurable functions that are defined and finite on a measurable set E in $\mathbb{R}^n$, $f : E \subset \mathbb{R}^n \to \mathbb{R}$. If E has finite measure and f_n converges to f a.e. on E, then f_n also converges to f in measure. Conversely, if f_n converges to f in measure, the sequence $f_n(x)$ need not converge to $f(x)$ for a single point x, but there always exists a subsequence $\{f_{n_k}\}_{k\in\mathbb{N}}$ that converges to f a.e. on E.*

The following two results, Egorov's theorem and Lusin's theorem, essentially state that "modulo sets of arbitrarily small measure," measurable functions are continuous, and convergence a.e. is uniform convergence.

Theorem D.1.1 (Lusin's theorem). *Let $E \subset \mathbb{R}^n$ be a measurable set and $f : E \to \mathbb{R}$ a measurable function that is finite on E. Then there exists a closed subset $F \subset E$ such that $\mu(E \setminus F) < \varepsilon$ and the restriction of f to F is continuous.* ∎

Theorem D.1.2 (Egorov's theorem). *Let $E \subset \mathbb{R}^n$ be a measurable set that has finite measure, $\mu(E) < \infty$, and suppose $\{f_n\}_{n \in \mathbb{N}}$ is a sequence of measurable functions, $f_n : E \to \mathbb{R}$, that are finite on E and converge to a limit f a.e. on E. Then, given any $\varepsilon > 0$, there exists a closed subset F of E such that $\mu(E \setminus F) < \varepsilon$ and f_n converges uniformly to f on F.* ∎

D.2 The Lebesgue Integral in $\mathbb{R}^n$

For a simple nonnegative Lebesgue measurable function $f(x) = \sum_{i=1}^{N} c_i \chi_{E_i}(x)$, the Lebesgue integral of s over a measurable set $E \subset \mathbb{R}^n$ is defined as

$$\int_E s d\mu = \sum_{i=1}^{N} c_i \mu(E_i \cap E).$$

If the simple function is normalized such that the sets E_i are disjoint and that the coefficients c_i are distinct, this gives a unique specification of the Lebesgue integral for s, but more generally, even without these normalizations the definition applies, and it is easy to see that it does not depend on the representation of the simple function. Given an arbitrary nonnegative Lebesgue measurable function $f : \mathbb{R}^n \to [0, \infty)$, the Lebesgue integral then is defined as

$$\int_E f d\mu = \sup \left\{ \int_E s \, d\mu : s \text{ is simple, Lebesgue measurable,} \right.$$

$$\left. \text{and } s(x) \leq f(x) \text{ for all } x \in E \right\}.$$

Definition D.2.1 (Lebesgue integrable). Given a Lebesgue measurable function $f : E \subset \mathbb{R}^n \to \mathbb{R}$, let $f_+ = \max(f, 0)$ and $f_- = -\min(f, 0)$ be its positive and negative parts, respectively. The Lebesgue integral of f over E is defined as

$$\int_E f \, d\mu = \int_E f_+ \, d\mu - \int_E f_- \, d\mu$$

provided not both of the integrals $\int_E f_+ \, d\mu$ and $\int_E f_- \, d\mu$ are infinite. The function f is said to be Lebesgue integrable over E with Lebesgue integral $\int_E f \, d\mu$ if both $\int_E f_+ \, d\mu$ and $\int_E f_- \, d\mu$ are finite.

Note that a function f is Lebesgue integrable if and only if

$$\int_E |f| \, d\mu = \int_E f_+ \, d\mu + \int_E f_- \, d\mu < \infty.$$

Like the Riemann integral, the Lebesgue integral is a linear operator,

$$\int_E (\alpha f + \beta g)\, d\mu = \alpha \int_E f\, d\mu + \beta \int_E g\, d\mu.$$

However, it provides a much more powerful tool for taking limits. Indeed, the results below were the original motivation behind the construction of this integral.

Proposition D.2.1 (Monotone convergence theorem). *Let $\{f_n\}_{n \in \mathbb{N}}$ be a sequence of nonnegative measurable functions that are defined on a measurable set E in $\mathbb{R}^n$, $f_n : E \to \mathbb{R}$, and are monotonically increasing, $f_n(x) \le f_{n+1}(x)$ for all $x \in E$. Then the limit*

$$f(x) = \lim_{n \to \infty} f_n(x) = \sup_{n \in \mathbb{N}} f_n(x) \le \infty$$

exists, is measurable, and satisfies

$$\int_E f\, d\mu = \lim_{n \to \infty} \int_E f_n\, d\mu.$$

Note that the implications are that if one of the two sides in the last equation is finite, respectively infinite, so is the other.

Proposition D.2.2 (Fatou's lemma). *Let $\{f_n\}_{n \in \mathbb{N}}$ be a sequence of nonnegative measurable functions that are defined on a measurable set E in $\mathbb{R}^n$, $f_n : E \to \mathbb{R}$. Then*

$$\int_E \left(\liminf_{n \to \infty} f_n \right) d\mu = \liminf_{n \to \infty} \left(\int_E f_n\, d\mu \right).$$

Proposition D.2.3 (Dominated convergence theorem). *Let $\{f_n\}_{n \in \mathbb{N}}$ be a sequence of measurable functions that are defined on a measurable set E in $\mathbb{R}^n$, $f_n : E \to \mathbb{R}$, and converge a.e. on E to some limit f, $f(x) = \lim_{n \to \infty} f_n(x)$ for a.e. $x \in E$. If there exists a Lebesgue integrable function g such that $|f_n| \le g$ a.e. on E, then*

$$\int_E f\, d\mu = \int_E \left(\lim_{n \to \infty} f_n \right) d\mu = \lim_{n \to \infty} \left(\int_E f_n\, d\mu \right).$$

It is these results allowing us to interchange limits with the integral that for many applications make the Lebesgue integral a superior tool to the Riemann integral for which uniform convergence is required for this operation. On the other hand, the Riemann integral offers a superior formalism for computing integrals, the fundamental theorem of calculus. It is therefore important that these two integrals generally agree if they both exist. More precisely, if $f : E \to \mathbb{R}$ is a bounded function that is Riemann integrable, then f is Lebesgue integrable and the values of the Riemann and the Lebesgue integrals agree. However, the Lebesgue integral exists for a much larger class of functions. For example, the characteristic function of the rationals $\chi_{\mathbb{Q}}$ actually is a simple function in the sense of measurable functions, and for any compact interval $I \subset \mathbb{R}$, by definition, we simply have that $\int_I \chi_{\mathbb{Q}} d\mu = 0$, while this function is not Riemann integrable. There also exist functions that are Riemann integrable, but not Lebesgue integrable. For example, the Riemann integral of $f(x) = \frac{\sin(x)}{x}$ over $[0, \infty)$ exists, but the Lebesgue integral does not. The reason is that the Riemann integral allows for cancellations of positive with negative terms, while these are not tolerated in the measure-oriented definition of the Lebesgue integral, and the Lebesgue integral $\int_0^\infty \frac{\sin(x)}{x} dx$ leads to the indefinite form $\infty - \infty$. But in most cases the Lebesgue integral is more general and provides the preferred mechanism when an interchange of limits with integration is required.

This broader generality in the integration process, however, comes at the expense of a more difficult and much more cumbersome differentiation theory for the resulting integral

$$F(E) = \int_E f \, d\mu,$$

understood as a set-valued function. But the following result, also valid in $\mathbb{R}^n$, holds:

Theorem D.2.1 (Lebesgue's differentiation theorem). *If the function $f : \mathbb{R} \to \mathbb{R}$ is Lebesgue integrable, then the indefinite integral*

$$F(x) = \int_{(-\infty, x]} f \, d\mu$$

is differentiable a.e. on $\mathbb{R}$ with derivative given by $F'(x) = f(x)$. ∎

The proof of this result is much more difficult than for the Riemann integral, and it is related to so-called covering lemmas by Vitali that will also be needed in the text.

Proposition D.2.4 (Vitali covering lemma). *Let E be a Lebesgue measurable set with finite measure and suppose a family $\mathcal{Q}$ of intervals covers E and is such that for any $x \in E$ and any $\varepsilon > 0$ there exists an interval I in the family $\mathcal{Q}$ that contains x and has length less than ε. (For example, this trivially holds if the family $\mathcal{Q}$ consists of all intervals.) Then, for every $\varepsilon > 0$, there exists a finite collection of disjoint intervals I_j, $j = 1, \ldots, N$, such that*

$$\mu\left(E \setminus \bigcup_{j=1}^N I_j\right) < \varepsilon \qquad \text{and} \qquad \sum_{i=1}^N \mu(I_j) < (1 + \varepsilon)\mu(E).$$

The notion of the indefinite integral also is closely related to the Radon–Nikodym theorem and the concept of an absolutely continuous function.

Definition D.2.2 (Absolutely continuous). A set function $F : \mathcal{L} \to \mathbb{R}$, $E \mapsto F(E)$, defined on the Lebesgue measurable sets is said to be absolutely continuous with respect to Lebesgue measure if whenever $\mu(E) = 0$, then $F(E) = 0$.

If f is an integrable Lebesgue measurable function, then it is easily seen from the definition of the Lebesgue integral that $\int_E f d\mu = 0$ whenever $\mu(E) = 0$. Thus $F(E) = \int_E f d\mu$ defines a σ-additive set function that is absolutely continuous with respect to Lebesgue measure. The Radon–Nikodym theorem states that any such set function is of this form.

Theorem D.2.2 (Radon–Nikodym). *Let Φ be a σ-additive set function defined on the σ-algebra of Lebesgue measurable sets in $\mathbb{R}^n$ that is absolutely continuous with respect to Lebesgue measure. Then there exists a Lebesgue integrable function $f : \mathbb{R}^n \to \mathbb{R}$ such that*

$$\Phi(E) = \int_E f \, d\mu.$$

D.3 L_p-Spaces

On function spaces, there exist many ways of defining norms that are not equivalent in the sense of Definition A.1.5. For example, on the space of continuous functions $f : [a,b] \to \mathbb{R}$ defined on a compact interval $[a,b]$, other commonly used norms are the L_p-norms defined by

$$\|f\|_p = \left(\int_a^b f(t)^p dt \right)^{1/p}$$

for $p \geq 1$. For example, the sequence of functions $f_n(t) = t^n$ does not converge to $f = 0$ in the supremum norm in $C([0,1])$ (the limit is not continuous), but for every $p \geq 1$,

$$\|f_n\|_p = \sqrt[p]{\int_0^1 t^{np} dt} = \sqrt[p]{\frac{1}{np+1}} \to 0 \qquad \text{as} \quad n \to \infty,$$

such that f_n converges to $f \equiv 0$ for any $p \geq 1$ in the L_p-norm in the space of continuous functions on $[0,1]$. As this simple example shows, in infinite-dimensional spaces, convergence properties very much depend on the norm that is being used. But clearly, the space of continuous functions on a compact interval $[a,b]$ is not complete in any of the L_p-norms. For example, the sequence of functions $\{f_n\}_{n \in \mathbb{N}}$ given by

$$f_n(t) = \begin{cases} -1 & \text{for } a \leq t \leq \dfrac{a+b}{2} - \dfrac{1}{n}, \\[2mm] n\left(t - \dfrac{a+b}{2}\right) & \text{for } \dfrac{a+b}{2} - \dfrac{1}{n} \leq t \leq \dfrac{a+b}{2} + \dfrac{1}{n}, \\[2mm] +1 & \text{for } \dfrac{a+b}{2} + \dfrac{1}{n} \leq t \leq b, \end{cases}$$

converges to the discontinuous function

$$f(t) = \begin{cases} -1 & \text{for } a \leq t < \frac{a+b}{2}, \\[2mm] 0 & \text{for } t = \frac{a+b}{2}, \\[2mm] +1 & \text{for } \frac{a+b}{2} < t \leq b, \end{cases}$$

in the L_p-norm, and thus the continuous functions on $[a,b]$ are not a Banach space under these norms. Obviously the problem is not one of convergence—the sequence has a well-defined and easily computed limit—but the limit no longer lies in the space under consideration. The L_p-spaces are the completions of the space of continuous functions under these norms.

Definition D.3.1 (L_p). For $1 \leq p < \infty$, the space $L_p = L_p(\mathbb{R}^n)$ is the normed space consisting of all (equivalence classes of) Lebesgue measurable functions $f : \mathbb{R}^n \to \mathbb{R}$ endowed with the norm $\|f\|_p$. The space $L_\infty = L_\infty(\mathbb{R}^n)$ is the normed space consisting of all (equivalence classes of) bounded Lebesgue measurable functions $f : \mathbb{R}^n \to \mathbb{R}$ with the norm $\|f\|_\infty = \inf\{C : |f(x)| \leq C \text{ a.e.}\}$.

Two Lebesgue measurable functions f and g are said to be equivalent if $f(x) \neq g(x)$ only for x in a null set N. This simply makes the elements of the space unique and defines a unique zero vector. It is clear that $\|f\|_p$ is positively homogeneous; the triangle inequality, also called the Minkowski inequality, is a direct consequence of Hölder's inequality.

Proposition D.3.1 (Hölder's inequality). *Let $1 \leq p \leq \infty$ and define the conjugate exponent q by the relation $\frac{1}{p} + \frac{1}{q} = 1$. If $f \in L_p$ and $g \in L_q$, then $fg \in L_1$ and*

$$\|fg\|_1 = \int_{\mathbb{R}^n} |fg| \, d\mu \, leq \left(\int_{\mathbb{R}^n} |f|^p \, d\mu \right)^{1/p} \left(\int_{\mathbb{R}^n} |g|^q \, d\mu \right)^{1/q} = \|f\|_p \|g\|_q$$

This is an important result that will be used several times and we include its proof.

Proof. The result is clear if $p = 1$ or $p = \infty$, and thus assume that $1 < p < \infty$. If $\|f\|_p = 0$, it follows that $f = 0$ a.e., and once more the result is trivial. Without loss of generality, we thus assume that $\|f\|_p$ and $\|g\|_q$ are positive and define $F = \frac{|f|}{\|f\|_p}$ and $G = \frac{|g|}{\|g\|_q}$. Hölder's inequality then is equivalent to the statement that

$$\int_{\mathbb{R}^n} |FG| \, d\mu \leq 1.$$

A simple geometric argument verifies that if $\varphi : [0,\infty) \to [0,\infty)$ is a strictly increasing continuous function that satisfies $\varphi(0) = 0$, and if ψ denotes its inverse function, then for all $a, b \geq 0$ we have that

$$ab \leq \int_0^a \varphi(x) \, dx + \int_0^b \psi(y) \, dy,$$

with equality if and only if $b = \varphi(a)$. Applying this inequality to the function $\varphi(x) = x^{p-1}$, it follows that

$$ab \leq \frac{a^p}{p} + \frac{b^q}{q}.$$

Hence, for all $x \in \mathbb{R}^n$,

$$F(x)G(x) \leq \frac{F(x)^p}{p} + \frac{G(x)^q}{q},$$

and integrating over $\mathbb{R}^n$ gives

$$\int_{\mathbb{R}^n} |FG| \, d\mu \leq \frac{1}{p} \frac{\int_{\mathbb{R}^n} |f|^p \, d\mu}{\|f\|_p^p} + \frac{1}{q} \frac{\int_{\mathbb{R}^n} |g|^q \, d\mu}{\|g\|_q^q} = \frac{1}{p} + \frac{1}{q} = 1.$$

This proves the result. □

Theorem D.3.1. *[257, Thm. 8.14] The spaces L_p, $1 \leq p \leq \infty$, are complete, i.e., are Banach spaces.* ■

In particular, note that for $1 \leq p < \infty$, the dominated convergence theorem also gives sufficient conditions for a sequence $\{f_n\}_{n \in \mathbb{N}} \subset L_p$ to converge in the space L_p, i.e., in the norm $\|\cdot\|_p$.

Another direct consequence of Hölder's inequality is that a function $g \in L_q$ defines a continuous functional on the space L_p through the specification

$$\ell(f) = \int_{\mathbb{R}^n} fg \, d\mu. \tag{D.1}$$

In fact, for $p < \infty$, these are all continuous functionals on L_p.

Theorem D.3.2 (Dual space). *Let $1 \leq p < \infty$ and let q by the conjugate exponent $\frac{1}{p} + \frac{1}{q} = 1$. If ℓ is a continuous linear functional defined on L_p, then there exists a unique $g \in L_q$ such that*

$$\ell(f) = \int_{\mathbb{R}^n} fg \, d\mu.$$

The dual space to L_p thus is isomorphic to L_q, $(L_p)' = L_q$. The dual space to L_∞ contains L_1, but is strictly larger. ■

D.4 Solutions to Ordinary Differential Equations with Lebesgue Measurable Right-Hand Sides

All the fundamental results on existence and uniqueness of solutions to ordinary differential equations as well as continuous and differentiable dependence on initial conditions and parameters are equally valid for right-hand sides that are Lebesgue measurable in time t if the appropriate assumptions are made. These are known as the Carathéodory conditions. We consider a parameter-dependent differential equation

$$\dot{x} = f(t, x, p),$$

where f is defined on some open set $I \times O \times P \subset \mathbb{R} \times \mathbb{R}^n \times \mathbb{R}^k$. Solutions to the differential equation then are given by absolutely continuous curves that satisfy the differential equation a.e.

Definition D.4.1 (Absolutely continuous). A continuous curve $\xi : I \to \mathbb{R}^n$, $t \mapsto \xi(t)$, is said to be absolutely continuous, $\xi \in AC(I;\mathbb{R}^n)$, if there exists a Lebesgue measurable integrable function $f : I \to \mathbb{R}^n$, $t \mapsto f(t)$, $v \in L^1(I;\mathbb{R}^n)$, such that for some $t_0 \in I$,

$$\xi(t) = \int_{[t_0,t]} f(s)ds,$$

with ds denoting integration against Lebesgue measure.

Definition D.4.2 (Solution to the initial value problem). Given a point $(t_0, x_0) \in I \times O$, an absolutely continuous curve $x : J \to \mathbb{R}^n$, $t \mapsto x(t)$, defined on some open interval J containing t_0 is a solution to the initial value problem

$$\dot{x} = f(t,x), \qquad x(t_0) = x_0, \tag{D.2}$$

on J if it satisfies the following three conditions: (i) $\dot{x}(t) = f(t, x(t))$ almost everywhere in J, (ii) $x(t_0) = x_0$, and (iii) the graph of x lies in $I \times O$.

Definition D.4.3 (C^1-Carathéodory conditions). The vector field f satisfies the C^1-Carathéodory conditions if the following assumptions are satisfied:

1. f is measurable, jointly in (t, x),
2. For each $t \in I$ fixed, the function $x \mapsto f(t, x, p)$ is continuously differentiable in (x, p) on $O \times P$,
3. For every compact set $K \subset O$, there exist integrable functions g and h, $g, h \in L_1(I)$, such that for all $x \in K$ and all $t \in I$ we have that

$$\|f(t,x)\| \leq g(t) \qquad \text{and} \qquad \left\|D_{(x,p)}f(t,x)\right\| \leq h(t),$$

4. For every compact set $K \subset O$, there exists an integrable function k, $k \in L_1(I)$, such that for all $x \in K$ and all $t \in I$,

$$\|f(t,x) - f(t,y)\| \leq k(t)\|x - y\|.$$

If the vector field f satisfies the C^1-Carathéodory conditions, then all the results from the standard theory of differential equations formulated in Appendix B remain valid. The classical reference for these results is [175].

References

[1] *Differential Geometric Control Theory*, R. Brockett, R. Millman, H. Sussmann, eds., Progress in Mathematics, Birkhäuser, Boston, 1983.

[2] *Differential Geometry and Control*, Proc. Symposia in Pure Mathematics, Vol. 64, G. Ferreyra, R. Gardner, H. Hermes, H. Sussmann, eds., American Mathematical Society, 1999.

[3] *Fifty Years of Optimal Control*, A. Ioffe. K. Malanowski, and F. Tröltzsch, eds., *Control and Cybernetics*, **38** (2009).

[4] *Geometry of Feedback and Optimal Control*, B. Jakubczyk and W. Respondek, eds., Marcel Dekker, New York, 1998.

[5] *Modern Optimal Control*, E.O. Roxin, ed., Marcel Dekker, New York, 1989.

[6] *Nonlinear Controllability and Optimal Control*, H. Sussmann, ed., Marcel Dekker, New York, 1990.

[7] *Nonlinear Synthesis*, C.I. Byrnes and A. Kurzhansky, eds., Birkhäuser, Boston, 1991.

[8] *Nonsmooth Analysis and Geometric Methods in Deterministic Optimal Control*, B.S. Mordukhovich and H. Sussmann, eds., The IMA Volumes in Mathematics and Its Applications, Springer-Verlag, New York, 1996.

[9] *Optimal Control of Differential Equations*, N. Pavel, ed., Marcel Dekker, New York, 1994.

[10] *Optimal Control: Theory, Algorithms and Applications*, W.W. Hager and P.M. Pardalos, Kluwer Academic Publishers, 1998.

[11] A.A. Agrachev and R.V. Gamkrelidze, Exponential representation of flows and chronological calculus, *Math. USSR Sbornik*, **35** (1979), pp. 727–785.

[12] A.A. Agrachev and R.V. Gamkrelidze, Chronological algebras and nonstationary vector fields, *J. of Soviet Mathematics*, **17** (1979), pp. 1650–1672.

[13] A.A. Agrachev and R.V. Gamkrelidze, Symplectic geometry for optimal control, in: *Nonlinear Controllability and Optimal Control* (H. Sussmann, ed.), Marcel Dekker, (1990), pp. 263–277.

[14] A.A. Agrachev and R.V. Gamkrelidze, Symplectic methods for optimization and control, in: *Geometry of Feedback and Optimal Control*, (B. Jakubczyk and W. Respondek, eds.), Marcel Dekker, (1998), pp. 19–77.

[15] A.A. Agrachev and Y. Sachkov, *Control Theory from the Geometric Viewpoint*, Springer-Verlag, 2004.

[16] A.A. Agrachev and A.V. Sarychev, On abnormal extremals for Lagrange variational problems, *J. of Mathematical Systems, Estimation and Control*, **5** (1995), pp. 127–130.

[17] A.A. Agrachev and A.V. Sarychev, Abnormal sub-Riemannian geodesics: Morse index and rigidity, *Ann. Inst. Henri Poincaré*, **13** (1996), pp. 635–690.

[18] A.A. Agrachev and M. Sigalotti, On the local structure of optimal trajectories in $\mathbb{R}^3$, *SIAM J. on Control and Optimization*, **42** (2003), pp. 513—531.

H. Schättler and U. Ledzewicz, *Geometric Optimal Control: Theory, Methods and Examples*, Interdisciplinary Applied Mathematics 38, DOI 10.1007/978-1-4614-3834-2, © Springer Science+Business Media, LLC 2012

[19] A.A. Agrachev, G. Stefani, and P.L. Zezza, A Hamiltonian approach to strong minima in optimal control, in: *Differential Geometry and Control*, (G. Ferreyra, R. Gardner, H. Hermes, H. Sussmann, eds.), American Mathematical Society, 1999, pp. 11–22.

[20] A.A. Agrachev, G. Stefani, and P.L. Zezza, Strong optimality for a bang-bang trajectory, *SIAM J. Control and Optimization*, **41** (2002), pp. 991–1014.

[21] V.M. Alekseev, V.M. Tikhomirov, and S.V. Fomin, *Optimal Control*, Contemporary Soviet Mathematics, 1987.

[22] M.S. Aronna, J.F. Bonnans, A.V. Dmitruk and P.A. Lotito, Quadratic conditions for bang-singular extremals, *Numerical Algebra, Control and Optimization*, (2012), to appear.

[23] A.V. Arutyunov, Higher-order conditions in anormal extremal problems with constraints of equality type, *Soviet Math. Dokl.*, **42** (1991), pp. 799–804.

[24] A.V. Arutyunov, Second-order conditions in extremal problems. The abnormal points, *Trans. of the American Mathematical Society*, **350** (1998), pp. 4341–4365.

[25] M. Athans and P. Falb, *Optimal Control*, McGraw Hill, 1966.

[26] J.P. Aubin and H. Frankowska, *Set–Valued Analysis*, Birkhäuser, Boston, 1990.

[27] E.R. Avakov, Extremum conditions for smooth problems with equality–type constraints, *USSR Comput. Math. and Math. Phys.*, **25** (1985), pp. 24–32, [translated from *Zh. Vychisl. Mat. Fiz.*, **25** (1985)].

[28] E.R. Avakov, Necessary conditions for a minimum for nonregular problems in Banach spaces. Maximum principle for abnormal problems of optimal control, *Trudy Mat. Inst. AN. SSSR*, **185** (1988), pp. 3–29 [in Russian].

[29] E.R. Avakov, Necessary extremum conditions for smooth abnormal problems with equality and inequality-type constraints, *Math. Zametki*, **45** (1989), pp. 3–11.

[30] M. Bardi and I. Capuzzo-Dolcetta, *Optimal Control and Viscosity Solutions of Hamilton–Jacobi–Bellman Equations*, Modern Birkhäuser Classics, Birkhäuser, 2008.

[31] D.J. Bell and D.H. Jacobson, *Singular Optimal Control Problems*, Academic Press, New York, 1975.

[32] R. Bellman, *Dynamic Programming*, Princeton University Press, 1961.

[33] L.D. Berkovitz, *Optimal Control Theory*, Springer-Verlag, 1974.

[34] L.D. Berkovitz and H. Pollard, A non-classical variational problem arising from an optimal filter problem, *Arch. Rational Mech. Anal.*, **26** (1967), pp. 281–304.

[35] L.D. Berkovitz and H. Pollard, A non-classical variational problem arising from an optimal filter problem II, *Arch. Rational Mech. Anal.*, **38** (1971), pp. 161–172.

[36] R.M. Bianchini, Good needle-like variations, Proceedings of Symposia in Pure Mathematics, Vol. **64**, American Mathematical Society, (1999), pp. 91–101.

[37] R.M. Bianchini and M. Kawski, Needle variations that cannot be summed, *SIAM J. Control and Optimization*, **42** (2003), pp. 218–238.

[38] R.L. Bishop and S.I. Goldberg, *Tensor Analysis on Manifolds*, Dover, (1980).

[39] G. Bliss, *Calculus of Variations*, The Mathematical Association of America, 1925.

[40] G. Bliss, *Lectures on the Calculus of Variations*, The University of Chicago Press, Chicago and London, 1946.

[41] V.G. Boltyansky, Sufficient conditions for optimality and the justification of the dynamic programming method, *SIAM J. on Control*, **4** (1966), pp. 326–361.

[42] V.G. Boltyansky, *Mathematical Methods of Optimal Control*, Holt, Rinehart and Winston, Inc., (1971).

[43] B. Bonnard, On singular extremals in the time minimal control problem in $\mathbb{R}^3$, *SIAM J. Control and Optimization*, **23** (1985), pp. 794–802.

[44] B. Bonnard and M. Chyba, *Singular Trajectories and Their Role in Control Theory*, Mathématiques & Applications, Vol. 40, Springer-Verlag, Paris, 2003.

[45] B. Bonnard, L. Faubourg, G. Launay, and E. Trélat, Optimal control with state space constraints and the space shuttle re–entry problem, *J. of Dynamical and Control Systems*, **9** (2003), pp. 155–199.

[46] B. Bonnard, J.P. Gauthier, and J. de Morant, Geometric time-optimal control for batch reactors, Part I, in: *Analysis of Controlled Dynamical Systems*, (B. Bonnard, B. Bride, J.P. Gauthier and I. Kupka, eds.), Birkhäuser, (1991); Part II, in: Proceedings of the 30th IEEE Conference on Decision and Control, Brighton, United Kingdom, (1991).

[47] B. Bonnard and J. de Morant, Toward a geometric theory in the time-minimal control of chemical batch reactors, *SIAM J. Control and Optimization*, **33** (1995), pp. 1279–1311.

[48] B. Bonnard and I.A.K. Kupka, Théorie des singularités de l'application entrée/sortie et optimalité des trajectoires singulières dans le problème du temps minimal, *Forum Matematicum*, **5** (1993), pp. 111–159.

[49] B. Bonnard and I.A.K. Kupka, Generic properties of singular trajectories, *A. Inst. H. Poincaré, Anal. Non Linéaire*, **14** (1997), pp. 167–186.

[50] W.M. Boothby, *An Introduction to Differentiable Manifolds and Riemannian Geometry*, Academic Press, New York, 1975.

[51] U. Boscain and B. Piccoli, *Optimal Syntheses for Control Systems on 2-D Manifolds*, Mathématiques & Applications, Vol. 43, Springer-Verlag, Paris, 2004.

[52] N. Bourbaki, *Elements of Mathematics: Lie Groups and Lie Algebras*, Chapters 1–3, Springer-Verlag, 1989.

[53] J.V. Breakwell, J.L. Speyer, and A.E. Bryson, Optimization and control of nonlinear systems using the second variation, *SIAM J. on Control*, **1** (1963), pp. 193–223

[54] A. Bressan, The generic local time-optimal stabilizing controls in dimension 3, *SIAM J. Control Optim.*, **24** (1986), pp. 177–190.

[55] A. Bressan and B. Piccoli, A generic classification of time-optimal planar stabilizing feedbacks, *SIAM J. Control Optim.*, **36** (1998), pp. 12–32.

[56] A. Bressan and B. Piccoli, *Introduction to the Mathematical Theory of Control*, American Institute of Mathematical Sciences (AIMS), 2007.

[57] R.W. Brockett, *Finite Dimensional Linear Systems*, John Wiley and Sons, New York, 1970.

[58] R.W. Brockett, System theory on group manifolds and coset spaces, *SIAM J. Control Optim.*, **10** (1972), pp. 265–284.

[59] A.B. Bruckner, J.B. Bruckner and B.S. Thomson, *Real Analysis*, Prentice Hall, 1997.

[60] P. Brunovsky, Every normal linear system has a regular time–optimal synthesis, *Math. Slovaca*, **28** (1979), pp. 81–100.

[61] P. Brunovsky, Regular synthesis for the linear-quadratic optimal control problem with linear control constraints, *J. of Differential Equations*, **38** (1980), pp. 344–360.

[62] P. Brunovsky, Existence of a regular synthesis for general control problems, *J. of Differential Equations*, **38** (1980), pp. 81–100.

[63] J. Burdick, On the inverse kinematics of redundant manipulators: characterization of the self-motion manifolds, in: Proc. of the 1989 IEEE International Conference on Robotics and Automation, Vol. 1, (1989), pp. 264–270.

[64] A.E. Bryson, Jr. and Y.C. Ho, *Applied Optimal Control*, Revised Printing, Hemisphere Publishing Company, New York, 1975.

[65] C.I. Byrnes and H. Frankowska, Unicité des solutions optimales et absence de chocs pour les équations d'Hamilton–Jacobi–Bellman et de Riccati, *C.R. Acad. Sci. Paris*, **315**, Série I (1992), pp. 427–431.

[66] C.I. Byrnes and A. Jhemi, Shock waves for Riccati Partial Differential Equations Arising in Nonlinear Optimal Control, in: *Systems, Models and Feedback: Theory and Applications*, (A. Isidori and T.J. Tarn, eds.), Birkhäuser, (1992), pp. 211–225.

[67] C. Carathéodory, *Variationsrechnung und Partielle Differential Gleichungen erster Ordnung*, Teubner Verlag, Leipzig, 1936; translated as *Calculus of Variations and Partial Differential Equations of First Order*, American Mathematical Society, 3rd edition, 1999.

[68] N. Caroff and H. Frankowska, Conjugate points and shocks in nonlinear optimal control, *Trans. of the American Mathematical Society*, **348** (1996), pp. 3133–3153.

[69] A. Cernea and H. Frankowska, The connection between the maximum principle and the value function for optimal control problems under state constraints, Proc. of the 43rd IEEE Conference on Decision and Control, Nassau, The Bahamas, (2004), pp. 893–898.

[70] L. Cesari, *Optimization - Theory and Applications*, Springer-Verlag, 1983.

[71] Y. Chitour, F. Jean, E. Trélat, Propriétés génériques des trajectoires singulères, *C.R. Math. Acad. Sci. Paris*, **337** (2003), pp. 49–52.

[72] Y. Chitour, F. Jean, E. Trélat, Genericity results for singular curves, *J. Differential Geom.*, **73** (2006), pp. 45–73.

[73] Y. Chitour, F. Jean, E. Trélat, Singular trajectories of control-affine systems, *SIAM J. Control and Optimization*, **47** (2008), pp. 1078–1095.

[74] M. Chyba and T. Haberkorn, Autonomous underwater vehicles: singular extremals and chattering, in: *Systems, Control, Modeling and Optimization*, (F. Cergioli et al., eds.), Springer-Verlag, (2003), pp. 103–113.

[75] M. Chyba, H. Sussmann, H. Maurer, and G. Vossen, Underwater vehicles: The minimum time problem, Proc. of the 43rd IEEE Conference on Decision and Control, Paradise Island, Bahamas, (2004), pp. 1370–1375.

[76] F.H. Clarke, The maximum principle under minimal hypothesis, *SIAM J. Control Optim.*, **14** (1976), pp. 1078–1091.

[77] F.H. Clarke, *Optimization and Nonsmooth Analysis*, Wiley–Interscience, 1983.

[78] F.H. Clarke and M.D.R. de Pinho, The nonsmooth maximum principle, *Contol and Cybernetics*, **38** (2009), pp. 1151–1168.

[79] F. Colonius and W. Kliemann, *The Dynamics of Control*, Birkhäuser, Boston, 2000.

[80] M.G. Crandall and P.L. Lions, Viscosity Solutions of Hamilton–Jacobi Equations, *Trans. of the American Mathematical Society*, **277** (1983), pp. 1–42.

[81] F.H. Croom, *Principles of Topology*, Saunders Publishing, 1983.

[82] M.d.R. de Pinho and M.M.A. Ferreira, *Optimal Control Problems with Constraints*, Seria Matematica Aplicata Si Industriala, Editura Universitii Din Pitesti, Romania, 2002.

[83] M.d.R. de Pinho and R.B. Vinter, Necessary conditions for optimal control problems involving nonlinear differential algebraic equations, *J. Math. Anal. Appl.*, **212** (1997), pp. 493–516.

[84] A. Dmitruk, Quadratic conditions for a weak minimum for singular regimes in optimal control problems, *Soviet Math. Doklady*, **18** (1977).

[85] A. Dmitruk, Quadratic conditions for a Pontryagin minimum in an optimal control problem, linear in the control, *Mathematics of the USSR, Izvestija*, **28** (1987), pp. 275–303.

[86] A. Dmitruk, Jacobi type conditions for singular extremals, *Control and Cybernetics*, **37** (2008), pp. 285–306.

[87] A. d'Onofrio and A. Gandolfi, Tumour eradication by antiangiogenic therapy: analysis and extensions of the model by Hahnfeldt et al. (1999), *Math. Biosciences*, **191** (2004), pp. 159–184.

[88] T. Duncan, B. Pasik–Duncan, and L. Stettner, Parameter continuity of the ergodic cost for a growth optimal portfolio with proportional transaction costs, Proc. of the 47th IEEE Conference on Decision and Control, Cancun, Mexico, (2008), pp. 4275–4279.

[89] U. Felgenhauer, On stability of bang-bang type controls, *SIAM J. Control Optim.*, **41** (2003), pp. 1843–1867.

[90] U. Felgenhauer, Lipschitz stability of broken extremals in bang-bang control problems, in: Large-Scale Scientific Computing (Sozopol 2007), (I. Lirkov et al., eds.), Lecture Notes in Computer Science, vol. 4818, Springer-Verlag (2008), pp. 317–325.

[91] U. Felgenhauer, L. Poggiolini, and G. Stefani, Optimality and stability result for bang-bang optimal controls with simple and double switch behaviour. in: *50 Years of Optimal Control*, (A. Ioffe, K. Malanowski, F. Tröltzsch, eds.), *Control and Cybernetics*, vol. 38(4B) (2009), pp. 1305–1325.

[92] M.M. Ferreira, U. Ledzewicz, M. do Rosario de Pinho, and H. Schättler, A model for cancer chemotherapy with state space constraints, *Nonlinear Analysis*, **63** (2005), pp. 2591–2602.

[93] M.E. Fisher, W.J. Grantham, and K.L. Teo, Neighbouring extremals for nonlinear systems with control constraints, *Dynamics and Control*, **5** (1995), pp. 225–240.

[94] A.T. Fuller, Study of an optimum non–linear system, *J. Electronics Control*, **15** (1963), pp. 63–71.

[95] W.H. Fleming and R.W. Rishel, *Deterministic and Stochastic Optimal Control*, Springer-Verlag, 1975.

[96] W.H. Fleming and M. Soner, *Controlled Markov Processes and Viscosity Solutions*, Springer-Verlag, New York, 1993.

[97] H. Frankowska, An open mapping principle for set-valued maps, *J. Math. Analysis and Applications*, **127** (1987), pp. 172–180.

[98] H. Frankowska, Contingent cones to reachable sets of control systems, *SIAM J. Control Optim.*, **27** (1989), pp. 170–198.

[99] H. Frankowska and R. Vinter, Existence of neighbouring feasible trajectories: applications to dynamic programming for state constrained optimal control problems, *J. Optimization Theory and Applications*, **104** (2000), pp. 21–40.

[100] P. Freeman, Minimum jerk trajectory planning for trajectory constrained redundant robots, *D. Sc. thesis*, Washington University, 2012

[101] R. Gabasov and F.M. Kirillova, High order necessary conditions for optimality, *SIAM J. Control*, **10** (1972), pp. 127–168.

[102] S. Gallot, D. Hulin, and J. Lafontaine, *Riemannian Geometry*, Springer-Verlag, 2nd edition, 1990.

[103] R.V. Gamkrelidze, Hamiltonian form of the Maximum Principle, *Control and Cybernetics*, **38** (2009), pp. 959–972.

[104] H. Gardner-Moyer, Sufficient conditions for a strong minimum in singular control problems, *SIAM J. Control*, **11** (1973), pp. 620–636.

[105] I.M. Gelfand and S.V. Fomin, *Calculus of Variations*, Prentice Hall, Englewood Cliffs, 1963.

[106] I.V. Girsanov, *Lectures on Mathematical Theory of Extremum Problems*, Lecture Notes in Economics and Mathematical Systems, Vol. 67, Springer-Verlag, Heidelberg, 1972.

[107] B.S. Goh, Necessary conditions for singular extremals involving multiple control variables, *SIAM J. Control*, **5** (1966), pp. 716–731.

[108] M. Golubitsky and V. Guillemin, *Stable Mappings and their Singularities*, Springer-Verlag, New York, 1973.

[109] M. Golubitsky and D.G. Schaefer, *Singularities and Groups in Bifurcation Theory, Vol. I*, Applied Mathematical Sciences, vol. 51, Springer-Verlag, New York, 1985.

[110] M. Golubitsky, I.N. Stewart, and D.G. Schaefer, *Singularities and Groups in Bifurcation Theory, Vol. II*, Applied Mathematical Sciences, vol. 69, Springer-Verlag, New York, 1988.

[111] K. Grasse, Reachability of interior states by piecewise constant controls, *Forum Matematicum*, **7** (1995), pp. 607–628.

[112] K. Grasse and H. Sussmann, Global controllability by nice controls, in: *Nonlinear Controllability and Optimal Control*, (H. Sussmann, ed.), Marcel Dekker, New York, (1990), pp. 33–79.

[113] W. Greub, *Linear Algebra*, Graduate Texts in Mathematics, Vol. 23, Springer-Verlag, New York, 1975.

[114] P. Hahnfeldt, D. Panigrahy, J. Folkman, and L. Hlatky, Tumor development under angiogenic signaling: a dynamical theory of tumor growth, treatment response, and postvascular dormancy, *Cancer Research*, **59** (1999), pp. 4770–4775.

[115] H. Hermes, Feedback synthesis and positive local solutions to Hamilton–Jacobi–Bellman equations, Proc. of the 1987 Conference on Mathematical Theory of Networks and Systems, (MTNS), Phoenix, AZ, (1987), pp. 155–164

[116] H. Hermes, Nilpotent approximations of control systems, in: *Modern Optimal Control*, (E.O. Roxin, ed.), Marcel Dekker, New York, 1989, pp. 157–172.

[117] H. Hermes and J.P. Lasalle, *Functional Analysis and Time Optimal Control*, Academic Press, New York, 1969.

[118] M.R. Hestenes, *Calculus of Variations and Optimal Control Theory*, Krieger Publishing Co., Huntington, New York, 1980.

[119] A.D. Ioffe and V.M. Tikhomirov, *Theory of Extremal Problems*, North-Holland, Amsterdam, 1979.

[120] A. Isidori, *Nonlinear Contol Systems*, Springer-Verlag, Berlin, 1989.

[121] D.H. Jacobson, A new necessary condition for optimality of singular control problems, *SIAM J. Control*, **7** (1969), pp. 578–595.

[122] D.H. Jacobson and J.L. Speyer, Necessary and sufficient conditions for optimality for singular control problems: a limit approach, *J. Mathematical Analysis and Applications*, **34** (1971), pp. 239–266.

[123] N. Jacobson, *Lie Algebras*, Dover, (1979).

[124] F. John, *Partial Differential Equations*, Springer-Verlag, 1982.

[125] F. Jones, *Lebesgue Integration on Euclidean Space*, Jones and Bartlett Publishers, Boston, 1993.

[126] V. Jurdjevic, *Geometric Control Theory*, Cambridge Studies in Advanced Mathematics, Vol. 51, Cambridge University Press, 1977.

[127] T. Kailath, *Linear Systems*, Prentice Hall, Englewood Cliffs, NJ, 1980.

[128] M. Kawski, Control variations with an increasing number of switchings, *Bul. of the American Mathematical Society*, **18** (1988), pp. 149–152.

[129] M. Kawski, High-order small-time local controllability, in: *Nonlinear Controllability and Optimal Control*, (H. Sussmann, ed.), Marcel Dekker, New York, (1990), pp. 431–467.

[130] M. Kawski and H. Sussmann, Noncommutative power series and formal Lie-algebraic techniques in nonlinear control theory, in: *Operators, Systems and Linear Algebra: Three Decades of Algebraic Systems Theory*, (U. Helmke, D. Praetzel-Wolters, and E. Zerz, eds.), B.G. Teubner, Stuttgart, (1997), pp. 111–129.

[131] H.J. Kelley, A second variation test for singular extremals, *AIAA (American Institute of Aeronautics and Astronautics) J.*, **2** (1964), pp. 1380–1382.

[132] H.J. Kelley, R. Kopp, and H.G. Moyer, Singular extremals, in: *Topics in Optimization* (G. Leitmann, ed.), Academic Press, 1967.

[133] H.K. Khalil, *Nonlinear Systems*, 3rd. ed., Prentice Hall, 2002.

[134] M. Kiefer, *On Singularities in Solutions to the Hamilton–Jacobi–Bellman Equation and Their Implications for the Optimal Control Problem*, D.Sc. Thesis, Washington University, 1997.

[135] M. Kiefer and H. Schättler, Parametrized families of extremals and singularities in solutions to the Hamilton–Jacobi–Bellman equation, *SIAM J. on Control and Optimization*, **37** (1999), pp. 1346–1371.

[136] A. Kneser, *Lehrbuch der Variationsrechnung*, Braunschweig, 1925.

[137] H.W. Knobloch, *Higher Order Necessary Conditions in Optimal Control Theory*, Lecture Notes in Control and Information Sciences, Vol. 34, Springer-Verlag, Berlin, 1981.

[138] H.W. Knobloch and H. Kwakernaak, *Lineare Kontrolltheorie*, Springer-Verlag, Berlin, 1985.

[139] G. Knowles, *An Introduction to Applied Optimal Control*, Academic Press, 1981.

[140] A.J. Krener, The high order maximum principle and its application to singular extremals, *SIAM J. Control Optim.*, **15** (1977), pp. 256–293.

[141] A.J. Krener and H. Schättler, The structure of small time reachable sets in low dimension, *SIAM J. Control Optim.*, **27** (1989), pp. 120–147.

[142] I.A.K. Kupka, Geometric theory of extremals in optimal control problems I: The fold and Maxwell cases, *Trans. of the American Mathematical Society*, **299** (1987), pp. 225–243.

[143] I.A.K. Kupka, The ubiquity of Fuller's phenomenon, in: *Nonlinear Controllability and Optimal Control* (H. Sussmann, ed.), Marcel Dekker, (1990), pp. 313–350.

[144] H. Kwakernaak and R. Sivan, *Linear Optimal Control Systems*, Wiley–Interscience, (1972).

[145] I. Lasiecka and R. Triggiani, *Control Theory for Partial Differential Equations: Continuous and Approximation Theories*, Vol. I: Abstract Parabolic Systems, Cambridge University Press, 2000.

[146] I. Lasiecka and R. Triggiani, *Control Theory for Partial Differential Equations: Continuous and Approximation Theories*, Vol. II: Abstract Hyperbolic-Like Systems over a Finite Time-Horizon, Cambridge University Press, 2000.

[147] E.B. Lee and L. Marcus, *Foundations of Optimal Control Theory*, Wiley, New York, 1967.

[148] U. Ledzewicz, A. Nowakowski, and H. Schättler, Stratifiable families of extremals and sufficient conditions for optimality in optimal control problems, *J. of Optimization Theory and Applications (JOTA)*, **122** (2004), pp. 105–130.

[149] U. Ledzewicz and H. Schättler, Second order conditions for extremum problems with nonregular equality constraints, *J. of Optimization Theory and Applications (JOTA)*, **86** (1995), pp. 113–144.

[150] U. Ledzewicz and H. Schättler, An extended maximum principle, *Nonlinear Analysis*, **29** (1997), pp. 159–183.

[151] U. Ledzewicz and H. Schättler, High order extended maximum principles for optimal control problems with non-regular constraints, in: *Optimal Control: Theory, Algorithms and Applications*, (W.W. Hager and P.M. Pardalos, eds.), Kluwer Academic Publishers, (1998), pp. 298–325.

[152] U. Ledzewicz and H. Schättler, A high-order generalization of the Lyusternik theorem, *Nonlinear Analysis*, **34** (1998), pp. 793–815.

[153] U. Ledzewicz and H. Schättler, High-order approximations and generalized necessary conditions for optimality, *SIAM J. Contr. Optim.*, **37** (1999), pp. 33–53.

[154] U. Ledzewicz and H. Schättler, A high-order generalized local Maximum Principle, *SIAM J. on Control and Optimization*, **38** (2000), pp. 823–854.

[155] U. Ledzewicz and H. Schättler, Optimal bang-bang controls for a 2-compartment model in cancer chemotherapy, *J. of Optimization Theory and Applications - JOTA*, **114** (2002), pp. 609–637.

[156] U. Ledzewicz and H. Schättler, Analysis of a cell-cycle specific model for cancer chemotherapy, *J. of Biological Systems*, **10** (2002), pp. 183–206.

[157] U. Ledzewicz and H. Schättler, *A synthesis of optimal controls for a model of tumor growth under angiogenic inhibitors*, Proc. of the 44th IEEE Conference on Decision and Control, Sevilla, Spain, (2005), pp. 934–939.

[158] U. Ledzewicz and H. Schättler, Drug resistance in cancer chemotherapy as an optimal control problem, *Discrete and Continuous Dynamical Systems, Series B*, **6** (2006), pp. 129–150.

[159] U. Ledzewicz and H. Schättler, Application of optimal control to a system describing tumor anti–angiogenesis, Proc. of the *17th International Symposium on Mathematical Theory of Networks and Systems* (MTNS), Kyoto, Japan, (2006), pp. 478–484.

[160] U. Ledzewicz and H. Schättler, Anti-angiogenic therapy in cancer treatment as an optimal control problem, *SIAM J. on Control and Optimization*, **46** (2007), pp. 1052–1079.

[161] U. Ledzewicz and H. Schättler, Optimal controls for a model with pharmacokinetics maximizing bone marrow in cancer chemotherapy, *Mathematical Biosciences*, **206** (2007), pp. 320–342.

[162] U. Ledzewicz and H. Schättler, Optimal and suboptimal protocols for a class of mathematical models of tumor anti–angiogenesis, *J. of Theoretical Biology*, **252** (2008), pp. 295–312.

[163] U. Ledzewicz and H. Schättler, Analysis of a mathematical model for tumor anti-angiogenesis, *Optimal Control, Applications and Methods*, **29** (2008), pp. 41–57.

[164] U. Ledzewicz and H. Schättler, On the optimality of singular controls for a class of mathematical models for tumor anti-angiogenesis, *Discrete and Continuous Dynamical Systems, Series B*, **11** (2009), pp. 691–715.

[165] U. Ledzewicz and H. Schaettler, Singular controls and chattering arcs in optimal control problems arising in biomedicine, *Control and Cybernetics*, 38 (2009), pp. 1501–1523.

[166] U. Ledzewicz and H. Schaettler, *Applications of Geometric Optimal Control to Biomedical Problems*, Springer Verlag, to appear.

[167] S. Lenhart and J.T. Workman, *Optimal Control Applied to Biological Models*, Chapman & Hall/CRC, Mathematical & Computational Biology, 2007.

[168] E.S. Levitin, A.A. Milyutin, and N.P. Osmolovskii, Higher order conditions for local minima in problems with constraints, *Uspekhi Mat. Nauk*, translated as: *Russian Mathematical Surveys*, **33** (1978) pp. 97–165.

[169] R.M. Lewis, Definitions of order and junction conditions in singular optimal control problems, *SIAM J. Control and Optimization*, **18** (1980), pp. 21–32.

[170] C. Lobry, Contrôlabilité des Systèmes nonlinéaires, *SIAM J. Control*, **8** (1970), pp. 573–605.

[171] R. Martin and K.L. Teo, *Optimal Control of Drug Administration in Cancer Chemotherapy*, World Scientific, Singapore, 1994.

[172] C. Marchal, Chattering arcs and chattering controls, *J. of Optimization Theory and Applications (JOTA)*, **11**, (1973), pp. 441–468.

[173] J.P. McDanell and W.J. Powers, Necessary conditions for joining optimal singular and nonsingular subarcs, *SIAM J. Control*, **9** (1971), pp. 161–173.

[174] J. McDonald and N. Weiss, *A Course in Real Analysis*, Academic Press, 1999.

[175] E.J. McShane, *Integration*, Princeton University Press, 1944.

[176] H. Nijmeier and A. van der Schaft, *Nonlinear Dynamical Control Systems*, Springer Verlag, 1990.

[177] J.E. Marsden and M.J. Hoffman, *Elementary Classical Analysis*, W.H. Freeman, New York, second edition, 1993.

[178] H. Maurer, An example of a continuous junction for a singular control problem of even order, *SIAM J. Control*, **13** (1975), pp. 899–903.

[179] H. Maurer, On optimal control problems with bounded state variables and control appearing linearly, *SIAM J. on Control and Optimization*, **15** (1977), pp. 345–362.

[180] H. Maurer, C. Büskens, J.H. Kim, and Y. Kaja, Optimization techniques for the verification of second-order sufficient conditions for bang-bang controls, *Optimal Control, Applications and Methods*, **26** (2005), pp. 129–156.

[181] H. Maurer and H.J. Oberle, Second order sufficient conditions for optimal control problems with free final time: the Riccati approach *SIAM J. Control and Optimization*, **41** (2002), pp. 380–403.

[182] H. Maurer and N. Osmolovskii, Second order optimality conditions for bang-bang control problems, *Control and Cybernetics*, **32** (2003), pp. 555–584.

[183] H. Maurer and N. Osmolovskii, Second order sufficient conditions for time-optimal bang-bang control problems, *SIAM J. Control and Optimization*, **42** (2003), pp. 2239–2263.

[184] J.W. Milnor, *Topology from the Differentiable Viewpoint*, The University Press of Virginia, Charlottesville, VA, 1965.

[185] A.A. Milyutin and N.P. Osmolovskii, *Calculus of Variations and Optimal Control*, American Mathematical Society, 1998.

[186] B. Mordukhovich, Variational Analysis and Generalized Differentiation, I: Basic Theory; Grundlehren Series (Fundamental Principles of Mathematical Sciences), Vol. 330, Springer-Verlag, 2006.

[187] B. Mordukhovich, Variational Analysis and Generalized Differentiation, II: Applications, Grundlehren Series (Fundamental Principles of Mathematical Sciences), Vol. 331, Springer-Verlag, 2006.

[188] H. Nijmeijer and A. van der Schaft, Controlled Invariance for nonlinear systems: two worked examples, *IEEE Transactions on Automatic Control*, **29** (1984), pp. 361–364.

[189] J. Noble and H. Schättler, Sufficient conditions for relative minima of broken extremals, *J. of Mathematical Analysis and Applications*, **269** (2002), pp. 98–128.

[190] A. Nowakowski, Field theories in the modern calculus of variations, *Transactions of the Americam Mathematical Society*, **309** (1988), pp. 725–752.

[191] P.J. Olver, *Applications of Lie Groups to Differential Equations*, Graduate Texts in Mathematics, Vol. 107, Springer-Verlag, New York, 1993.

[192] N.P. Osmolovskii, Quadratic extremality conditions for broken extremals in the general problem of the calculus of variations, *J. of Mathematical Sciences*, **123** (2004), pp. 3987–4122.

[193] L.S. Pontryagin, V.G. Boltyanskii, R.V. Gamkrelidze, and E.F. Mishchenko, *The Mathematical Theory of Optimal Processes*, Macmillan, New York, 1964.

[194] H.J. Pesch and M. Plail, The maximum principle of optimal control: a history of ingenious ideas and missed opportunities, *Control and Cybernetics*, **38** (2009), pp. 973–996.

[195] B. Piccoli, Classification of generic singularities for the planar time-optimal synthesis, *SIAM J. Control and Optimization*, **34** (1996), pp. 1914–1946.

[196] B. Piccoli and H. Sussmann, Regular synthesis and sufficient conditions for optimality, *SIAM J. on Control and Optimization*, **39** (2000), pp. 359–410.

[197] L. Poggiolini and M. Spadini, Strong local optimality for a bang-bang trajectory in a Mayer problem, *SIAM J. Control and Optimization*, **49** (2011), pp. 140–161.

[198] L. Poggiolini and G. Stefani, Sufficient optimality conditions for a bang-bang trajectory, in: *Proc. 45th IEEE Conference on Decision and Control*, (2006).

[199] L. Poggiolini and G. Stefani, Sufficient optimality conditions for a bangsingular extremal in the minimum time problem, *Control and Cybernetics*, **37** (2008), pp. 469–490.

[200] V. Ramakrishna and H. Schättler, Controlled invariant distributions and group invariance, *J. of Mathematical Systems and Control*, **1** (1991), pp. 209–240.

[201] D. Rebhuhn, On the stability of the existence of singular controls under perturbation of the control system, *SIAM J. Control and Optimization*, **18** (1978), pp. 463–472.

[202] R.T. Rockafellar, *Convex Analysis*, Princeton University Press, 1970.

[203] E.O. Roxin, Reachable zones in autonomous differential systems, *Bol. Soc. Mat. Mexicana*, (1960), pp. 125–135.

[204] E.O. Roxin, Reachable sets, limit sets and holding sets in control systems, in: *Nonlinear Analysis and Applications*, (V. Lakshmikantham, ed.,) Marcel Dekker, (1987), pp. 401–407.

[205] E.O. Roxin, *Control Theory and Its Applications*, Gordon and Breach Scientific Publishers, 1997.

[206] A. Sarychev, The index of second variation of a control system, *Math. USSR Sbornik*, **41** (1982), pp. 383–401.

[207] A. Sarychev, Morse index and sufficient optimality conditions for bang-bang Pontryagin extremals, in: *System Modeling and Optimization*, Lecture Notes in Control and Information Sciences, Vol. 180, (1992), pp. 440–448.

[208] A. Sarychev, First and second order sufficient optimality conditions for bang-bang controls, *SIAM J. on Control and Optimization*, **35**, (1997), pp. 315–340.

[209] H. Schättler, On the local structure of time-optimal trajectories for a single-input control-linear system in dimension 3, Ph.D. thesis, Rutgers, The State University of New Jersey, 1986.

[210] H. Schättler, On the local structure of time-optimal bang-bang trajectories in $\mathbb{R}^3$, *SIAM J. on Control and Optimization*, **26** 1988, pp. 186–204.

[211] H. Schättler, The local structure of time-optimal trajectories in dimension 3 under generic conditions, *SIAM J. Control Optim.*, **26** (1988), pp. 899–918.

[212] H. Schättler, Conjugate points and intersections of bang-bang trajectories, Proc. of the 28th IEEE Conference on Decision and Control, Tampa, Florida, (1989), pp. 1121–1126.

[213] H. Schättler, Regularity properties of optimal trajectories: recently developed techniques, in: *Nonlinear Controllability and Optimal Control* (H. Sussmann, ed.), Marcel Dekker, (1990), pp. 351–381.

[214] H. Schättler, Extremal trajectories, small-time reachable sets and local feedback synthesis: a synopsis of the three-dimensional case, in: *Nonlinear Synthesis*, (C.I. Byrnes and A. Kurzhansky, eds.), Birkhäuser, Boston, September 1991, pp. 258–269.

[215] H. Schättler, A local feedback-synthesis of time-optimal stabilizing controls in dimension three, *Mathematics of Control, Signals and Systems*, **4** (1991), pp. 293–313.

[216] H. Schättler, A geometric approach to optimal control, in: *WCNA-92, Proceedings of the First World Congress of Nonlinear Analysts, Vol. II*, (V. Lakshmikantham, ed.), Walter de Gruyter Publishers, (1996), pp. 1579–1590.

[217] H. Schättler, Small-time reachable sets and time-optimal feedback control for nonlinear systems, in: *Nonsmooth Analysis and Geometric Methods in Deterministic Optimal Control*, (B. Mordukhovich and H.J. Sussmann, eds.), IMA Volumes in Mathematics and its Applications, Vol. 78, Springer-Verlag, 1996, pp. 203–225.

[218] H. Schättler, Time-optimal feedback control for nonlinear systems, in: *Geometry of Feedback and Optimal Control*, (B. Jakubczyk and W. Respondek, eds.), Marcel Dekker, New York, (1998), pp. 383–421.

[219] H. Schättler, On classical envelopes in optimal control theory, Proc. of the 49th IEEE Conference on Decision and Control, Atlanta, USA, (2010), pp. 1879–1884.

[220] H. Schättler, A local field of extremals for optimal control problems with state constraints of relative degree 1, *J. of Dynamical and Control Systems*, **12** (2006), pp. 563–599.

[221] H. Schättler and M. Jankovic, A synthesis of time-optimal controls in the presence of saturated singular arcs, *Forum Matematicum*, **5** (1993), pp. 203–241.

[222] Ch.E. Shin, On the structure of time-optimal stabilizing controls in $\mathbb{R}^4$, *Bollettino U.M.I.*, **7** (1995), pp. 299–320.

[223] Ch.E. Shin, Time-optimal bang-bang trajectories using bifurcation results, *J. Korean Math.Soc.*, **34** (1997), pp. 553–567.

[224] M. Sigalotti, Regularity properties of optimal trajectories of single-input control systems in dimension three, *J. of Math. Sci.*, **126** (2005), pp. 1561–1573.

[225] E.D. Sontag, *Mathematical Control Theory*, second edition, Springer-Verlag, (1998).

[226] G. Stefani, Higher order variations: how can they be defined in order to have good properties? in: *Nonsmooth Analysis and Geometric Methods in Deterministic Optimal Control*, IMA Volumes in Mathematics and Its Applications, Vol. 78, Springer-Verlag, New York, (1996), pp. 227–237.

[227] G. Stefani and P.L. Zezza, Optimality conditions for a constrained control problem, *SIAM J. Control and Optimization*, **34** (1996), pp. 635–659.

[228] J. Stoer and R. Bulirsch, *Introduction to Numerical Analysis*, Springer-Verlag, New York, 1990.

[229] H.J. Sussmann, Lie brackets, real analyticity and geometric control, in: *Differential Geometric Control Theory*, (R. Brockett, R. Millman, H. Sussmann, eds.), Progress in Mathematics, Birkhäuser, Boston, (1983), pp. 1–116.

[230] H.J. Sussmann, Time-optimal control in the plane, in: *Feedback Control of Linear and Nonlinear Systems*, Lecture Notes in Control and Information Sciences, Vol. 39, Springer-Verlag, Berlin, (1982), pp. 244–260.

[231] H.J. Sussmann, Lie brackets and real analyticity in control theory, in: *Mathematical Control Theory*, Banach Center Publications, Vol. 14, Polish Scientific Publishers, Warsaw, Poland, (1985), pp. 515–542.

[232] H.J. Sussmann, A bang-bang theorem with bounds on the number of switchings, *SIAM J. Control Optim.*, **17** (1979), pp. 629–651.

[233] H.J. Sussmann, Subanalytic sets and feedback control, *J. of Differential Equations*, **31** (1979), pp. 31–52.

[234] H.J. Sussmann, A product expansion for the Chen series, in: *Theory and Applications of Nonlinear Control Systems*, (C. Byrnes and A. Lindquist, eds.), North-Holland, Amsterdam, (1986), pp. 323–335.

[235] H.J. Sussmann, A general theorem on local controllability, *SIAM J. Control Optim.*, **25** (1987), pp. 158–194.

[236] H.J. Sussmann, The structure of time-optimal trajectories for single-input systems in the plane: the C^∞ nonsingular case, *SIAM J. Control Optim.*, **25** (1987), pp. 433–465.

[237] H.J. Sussmann, The structure of time–optimal trajectories for single-input systems in the plane: the general real analytic case, *SIAM J. Control Optim.*, **25** (1987), pp. 868–904.

[238] H.J. Sussmann, Regular synthesis for time-optimal control of single-input real analytic systems in the plane, *SIAM J. Control Optim.*, **25** (1987), pp. 1145–1162.

[239] H.J. Sussmann, Recent developments in the regularity theory of optimal trajectories, *Rend. Sem. Mat. Univ. Politec. Torino*, Fasc. Spec. Control Theory, (1987), pp. 149–182.

[240] H.J. Sussmann, Envelopes, conjugate points, and optimal bang-bang extremals, Proc. of the 1985 Paris Conference on Nonlinear Systems (M. Fliess, M. Hazewinkel, eds.), Reidel Publ., The Netherlands, (1987).

[241] H.J. Sussmann, Thirty years of optimal control: was the path unique? *Modern Optimal Control*, (E.O. Roxin, ed.), Marcel Dekker, New York, 1989, pp. 359–375.

[242] H.J. Sussmann, Envelopes, high-order optimality conditions and Lie brackets, Proc. of the 28th IEEE Conference on Decision and Control, Tampa, Florida, December 1989, pp. 1107–1112.

[243] H.J. Sussmann, Synthesis, presynthesis, sufficient conditions for optimality and subanalytic sets, in: *Nonlinear Controllability and Optimal Control*, (H. Sussmann, ed.), Marcel Dekker, (1990), pp. 1–19.

[244] H.J. Sussmann, A strong version of the Lojasiewicz Maximum Principle, in: *Optimal Control of Differential Equations*, (N. Pavel, ed.), Marcel Dekker, New York, 1994, pp. 293–310.

[245] H.J. Sussmann and J.C. Willems, 300 years of optimal control: from the Brachistochrone to the maximum principle, *IEEE Control Systems*, (1997), pp. 32–44.

[246] H.J. Sussmann, Uniqueness results for the value function via direct trajectory-construction methods, in: Proc. of the 42nd IEEE Conference on Decision and Control, Maui, Hawaii, (2003), pp. 3293–3298.

[247] H.J. Sussmann, Multidifferential calculus: chain rule, open mapping and transversal intersection theorems, in: *Optimal Control: Theory, Algorithms and Applications*, (W.W. Hager and P.M. Pardalos, eds.), Kluwer Academic Publishers, (1998), pp. 436–487.

[248] H.J. Sussmann, Set separation, transversality and the Lipschitz maximum principle, *J. of Differential Equations*, **243** (2007), pp. 446–488.

[249] H.J. Sussmann and G. Tang, Shortest paths for the Reeds-Shepp car: a worked out example of the use of geometric techniques in nonlinear optimal control, Report SYCON 91–10, 1991.

[250] R. Thom, Les singularités des applications différentiables, *Ann. Inst. Fourier*, **6** (1955–56), pp. 43–87.

[251] O.A. Brezhneva and A.A. Tret'yakov, Optimality Conditions for Degenerate Extremum Problems with Equality Constraints, *SIAM J. Control Optim.*, **42** (2003), pp. 729–743.

[252] G.W. Swan, Role of optimal control in cancer chemotherapy, *Math. Biosci.*, **101** (1990), pp. 237–284.

[253] A. Swierniak, A. Polanski, and M. Kimmel, Optimal control problems arising in cell-cycle-specific cancer chemotherapy, *Cell prolif.*, **29** (1996), pp. 117–139.

[254] R.B. Vinter, *Optimal Control Theory*, Birkhäuser, Boston, 2000.

[255] S. Walczak, On some properties of cones in normed spaces and their application to investigating extremal problems, *J. of Optimization Theory and Applications*, **42** (1984), pp. 559–582.

[256] F.W. Warner, *Foundations of Differentiable Manifolds and Lie Groups*, Springer Verlag, 1983.

[257] R.L. Wheeden and A. Zygmund, *Measure and Integral*, Marcel Dekker, New York, 1977.

[258] H. Whitney, Elementary structure of real algebraic varieties, *Ann. Math.*, **66** (1957), pp. 545–556.

[259] W.M. Wonham, Note on a problem in optimal nonlinear control, *J. Electronics Control*, **15** (1963), pp. 59–62.

[260] L.C. Young, *Lectures on the Calculus of Variations and Optimal Control Theory*, W.B. Saunders, Philadelphia, 1969.

[261] M.I. Zelikin and V.F. Borisov, Optimal synthesis containing chattering arcs and singular arcs of the second order, in: *Nonlinear Synthesis*, (C.I. Byrnes and A. Kurzhansky, eds.), Birkhäuser, Boston, (1991), pp. 283–296.

[262] M.I. Zelikin and V.F. Borisov, *Theory of Chattering Control with Applications to Astronautics, Robotics, Economics and Engineering*, Birkhäuser, 1994.

[263] M.I. Zelikin and L.F. Zelikina, The structure of optimal synthesis in a neighborhood of singular manifolds for problems that are affine in control, *Sbornik: Mathematics*, **189** (1998), pp. 1467–1484.

Index

H. Schättler and U. Ledzewicz, *Geometric Optimal Control: Theory, Methods and Examples*, Interdisciplinary Applied Mathematics 38, DOI 10.1007/978-1-4614-3834-2, © Springer Science+Business Media, LLC 2012